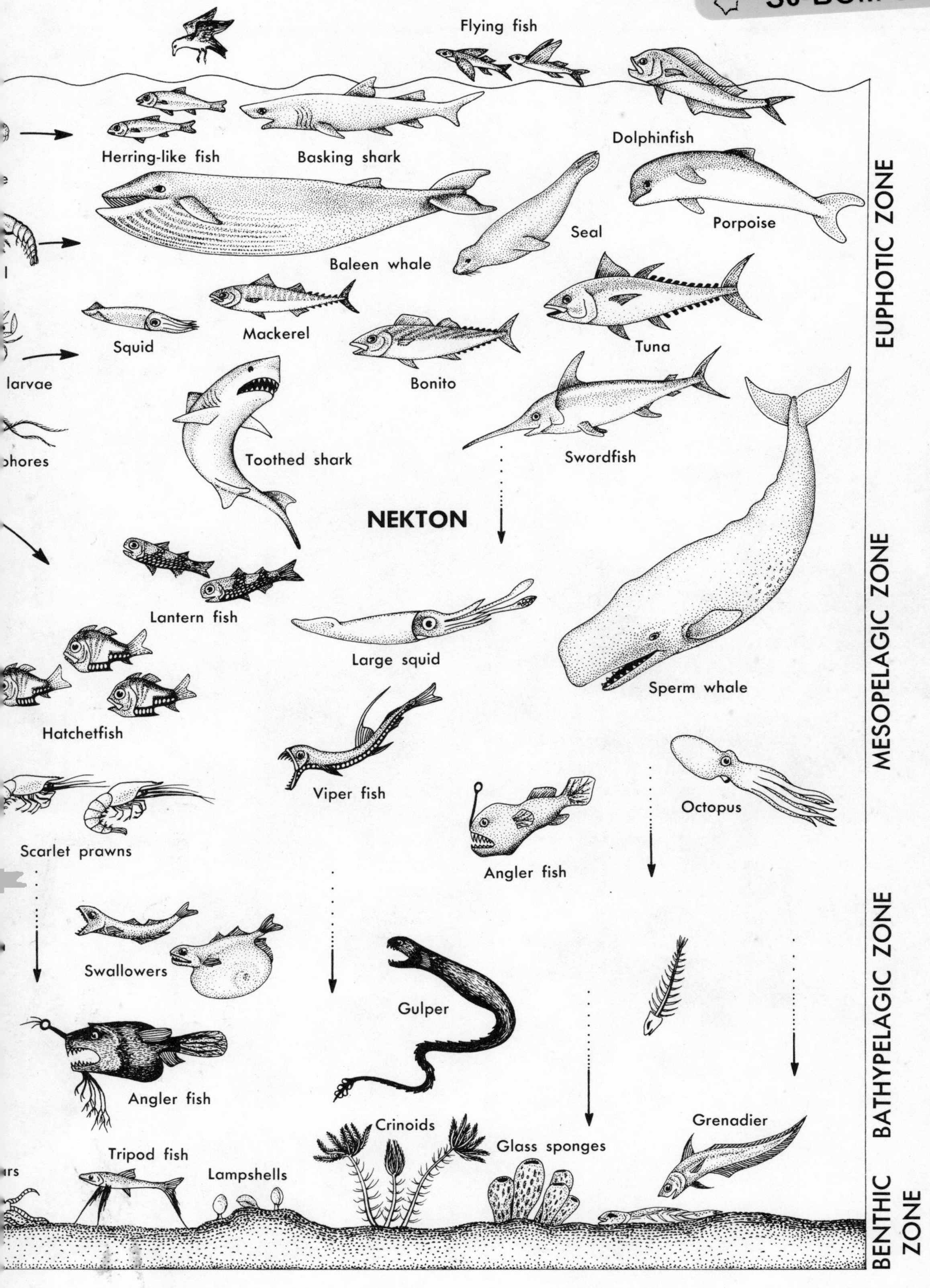

Flying fish
Dolphinfish
Herring-like fish
Basking shark
Porpoise
Seal
Baleen whale
Squid
Mackerel
Tuna
larvae
Bonito
Swordfish
Toothed shark
NEKTON
Lantern fish
Large squid
Sperm whale
Hatchetfish
Viper fish
Octopus
Scarlet prawns
Angler fish
Swallowers
Gulper
Angler fish
Crinoids
Glass sponges
Grenadier
Tripod fish
Lampshells
EUPHOTIC ZONE
MESOPELAGIC ZONE
BATHYPELAGIC ZONE
BENTHIC ZONE

INTRODUCTION TO MARINE BIOLOGY

THIRD EDITION

INTRODUCTION TO MARINE BIOLOGY

BAYARD H. McCONNAUGHEY, Ph.D.

Professor of Biology, Department of Biology,
University of Oregon,
Eugene, Oregon

with 429 illustrations

THE C. V. MOSBY COMPANY

Saint Louis 1978

Cover photograph by Carl Roessler, San Francisco, California

THIRD EDITION

Printed in the United States of America

The C. V. Mosby Company
11830 Westline Industrial Drive, St. Louis, Missouri 63141

Library of Congress Cataloging in Publication Data

McConnaughey, Bayard Harlow, 1916-
Introduction to marine biology.

Bibliography: p.
Includes index.
1. Marine biology. I. Title.
QH91.M23 1978 574.92 77-25826
ISBN 0-8016-3258-7

CB/CB/B 9 8 7 6 5 4 3 2 1

To

WILLIAM A. HILTON

longtime professor of zoology at Pomona College, California.

Professor Hilton for many years single-handedly ran a small summer marine station in the old church at Laguna Beach, in the days when Laguna Beach was still a tiny, beautiful artists' colony. Here and at Pomona College he communicated to generations of students his boundless curiosity about the world of which he was a part, his infectious good humor, and his respect and enthusiasm for life and for all living things.

The search for truth is in one way hard and in another easy, for it is evident that no one can master it fully nor miss it wholly. But each adds a little to our knowledge of nature, and from all the facts assembled there arises a certain grandeur.

Aristotle

Inscribed in Greek on the National Academy of Science Building, Washington, D.C.

. . . no one with an unbiased mind can study any living creature, however humble, without being struck with enthusiasm at its marvellous structure and properties.

Darwin

The Descent of Man

1874

PREFACE

Marine biology embraces many diverse disciplines, some of which are undergoing radical changes in concepts and methods. To avoid presenting merely brief excerpts from such diverse subjects as planktonology, invertebrate and vertebrate zoology, microbiology, submarine geology, ecology, physiology, chemistry, ethology, paleontology, and other disciplines, it has been necessary to select a few broad unifying concepts, or points of view, and build the book around these. Such a selection and limitation entails the elimination of much that is equally valid and important. There are other points of view from which the subject could and should be presented.

This book is organized around the general theme of the production and consumption of food, the special directions and limitations that marine environments impose, and the ways in which various organisms have become adapted for exploiting marine resources and interacting with each other.

A second theme emerges from this. The ecosystem itself, far from being a static equilibrium in which the effects brought about by various organisms exactly counterbalance each other, is in a state of continuous dynamic change and evolution. Living organisms are responsible for many of these changes. Others are brought about by geological and even cosmic agencies.

At the present moment of the earth's history, man is exerting effects on the ecosystem more profound and more rapid than those brought about by any other species throughout the entire history of life. The fate of the ecosystem as we know it today—of man and of all the animals and plants that populate the lands and the seas—may well be contingent on our ability to understand and control these changes, a problem best thought of in terms of *self*-control rather than in terms of controlling nature.

Any text of this kind owes its existence to the labors of countless students of science whose careful research and keen insight have produced the conceptual and factual framework on which it is based. Some of these scientists and their works are listed in the bibliography. They are responsible for any correspondence between the text and the reality it attempts to elucidate. My role as the writer has been largely confined to the production of whatever errors and mistakes still lurk in these pages.

For illustrations I am chiefly indebted to two top-notch photographers, Mr. Milo W. Williams and Dr. Richard H. Chesher, and to four fine artists, Miss Kathryn Torvik, Miss Joanne Salley, Mrs. Lynn Rudy, and Mr. Larry McQueen. Others who kindly contributed photographs include Dr. R. J. M. Riedl, Dr. Kenneth Mesolella, Dr. Donald Wilson, Dr. D. M. Ross, Dr. Ralph Buchsbaum, Dr. Peter V. Fankboner, Dr. John Evans, Mr. James Houk, Mrs. Helen Finlayson, and Robin Anderson. Credits are given in the captions. Photographs credited to the Oregon Institute of Marine Biology are from sets taken from time to time by students and staff of the Institute at Charleston, Oregon, and at the Department of Biology of the University of Oregon in Eugene. These were incorporated into one large set, and so it has not been possible to be sure who took which photographs in order to credit the proper individuals.

I also wish to express my sincere thanks and appreciation to Dr. Larry Oglesby, Dr. Theodore J. Smayda, Dr. Mohammed Saeed Mulkana, Mr. Michael C. Robinson, Mr. David J. Fassler, and others for reading and criticizing previous editions. Their comments and suggestions have been a great help to me in improving the book for this edition.

Bayard H. McConnaughey

CONTENTS

PART ONE

MARINE ENVIRONMENT

general considerations

FIG. 1-1. Gateway to the open ocean. Looking west from the seacliff along the coast of southern Oregon. Beyond the outlying rocks the neritic waters over the continental shelf extend past the horizon, merging with the open ocean, which extends uninterrupted from America to Asia.

CHAPTER

1 THE OPEN OCEAN

The oceans of the world cover approximately 361 million km.^{2}, about 71% of the earth's surface, to an average depth of about 4 km. The deepest parts are over 10 km. deep, plunging farther below sea level than our highest mountains rise above it. The seas are the cradle of life. They are so vast and complex and the assemblages of living organisms in them are so varied that no book can give a complete view of even a few of the many facets of marine science.

STABILITY OF THE OCEANS

Although many characteristics of the oceans seem relatively stable or constant, this stability does not result from inactivity but rather from a dynamic equilibrium among a great number of processes. For example, annual evaporation of many millions of tons of water is balanced by precipitation on the oceans and runoff from the land. Oxygen production by photosynthetic organisms, as well as some dissolving of oxygen from the air at the ocean surface, is balanced by oxygen consumption and loss of oxygen to the atmosphere. The rates of production and consumption of organic matter by organisms balance the mineralization of organic matter by bacteria. Currents are compensated by countercurrents, turbulence, eddies, and other water movements. Everything in the oceans is in a state of flux, but the major processes compensate for each other in such a way as to produce some remarkable uniformities.

All parts of the oceans interact and form one continuous, unified water mass covering most of the earth's surface. All events in the oceans must follow the laws of a continuous medium. Events anywhere in the oceans may produce effects at very great distances. For example, the North Atlantic Gulf Stream, which exerts profound effects on the climate of northern Europe, has its origins and driving force in tropical equatorial trade winds. The oxygenation of deep waters of all latitudes results from the sinking of cold oxygen-rich waters near the poles and their spreading out over the ocean floors. An earthquake in Alaska may produce tsunamis, or long period waves, which exert destructive effects along the coasts of Hawaii and Japan.

The temperatures of ocean waters from pole to pole and from top to bottom vary only a little more than 35° C. The dissolved salts and minerals are everywhere present in remarkably constant proportions, although their total concentration, the salinity, varies somewhat. These ratios are so constant that a measure of any one of them can serve as a basis for reliable estimates of the quantities of most of the others. The exceptions are those which are incorporated as components of living organisms.

REGIONAL DIFFERENCES AND ISOLATION

In spite of these striking uniformities, regional differences are also always apparent. They are maintained by external agencies such as regional weather and amount of runoff.

The extent to which these external agencies affect the characteristics of the ocean waters depends largely on the degree of isolation of these waters. Largely inclosed oceanic basins, such as the Black Sea, Baltic Sea, and Red Sea, have a high degree of isolation and display the most marked idiosyncratic regional characteristics, while regions such as the Atlantic and Pacific coastlines in middle latitudes are much less isolated.

Isolation may be either horizontal or vertical, and is brought about by a variety of different sorts of barriers. Among the most important are physical barriers such as land masses and submarine mountain ranges and ridges; temperature gradients; and gradients in illumination. Of somewhat less importance on a worldwide basis but often of local importance are other gradients such as differences in dissolved oxygen, in plant nutrients, and in salinity, and the fact that existing biotic communities may themselves constitute a barrier to other organisms either settling in or passing through them.

Coastal waters commonly have lower salinity and greater annual and diurnal temperature variation than do offshore oceanic waters in the same latitudes. The boundary of such coastal waters can be placed in those zones where no further salinity increase, nor further decrease in temperature range of surface waters, occurs as one goes farther seaward. Often this boundary corresponds fairly well with the outer limits of the continental shelves. In general, although there is mixing in such a boundary zone, there is little or no net transport of salt across a boundary at sea. Coastal waters can be treated as a separate entity. Open oceanic waters are commonly termed **pelagic,** whereas coastal waters over the continental shelves are described as **neritic.**

VERTICAL STRATIFICATION

In general, vertical gradients change far more rapidly than do horizontal gradients in the oceans, often by a factor of several thousand; thus vertical stratifications are more abrupt and well marked than are horizontal boundaries. On the other hand, horizontal movements of water are far more massive and extensive than are vertical movements. Since dense water will sink beneath less dense water, there is a tendency for the oceans to become vertically stratified with respect to density.

Photic zones

By far the most dramatic and biologically significant change in the oceans correlated with depth is in the amount of light, the wavelengths of light that penetrate to various depths during the day, and the proportion of the time each depth is illuminated. The deeper portions of the photic zone are illuminated much more briefly than are the near surface portions, and the deepest part only when the sun is directly overhead. Sunlight impinging on the ocean surface is partly reflected back into the sky. The portion that enters the water is selectively absorbed and scattered in such a way that the light is rapidly attenuated. Shorter wavelengths penetrate more deeply than do longer ones. Thus the ultraviolet, red, and yellow penetrate relatively short distances into the water, whereas the green and blue wavelengths reach much greater depths (Fig. 1-2, *B*).

The amount of light at various depths and the depths to which light penetrates are greatly influenced by the turbidity of the water (caused by suspended microorganisms and minute organic and inorganic particles), the angle of the incident light, the roughness of the surface, and the cloudiness of the atmosphere.

The overall result is that there is a thin

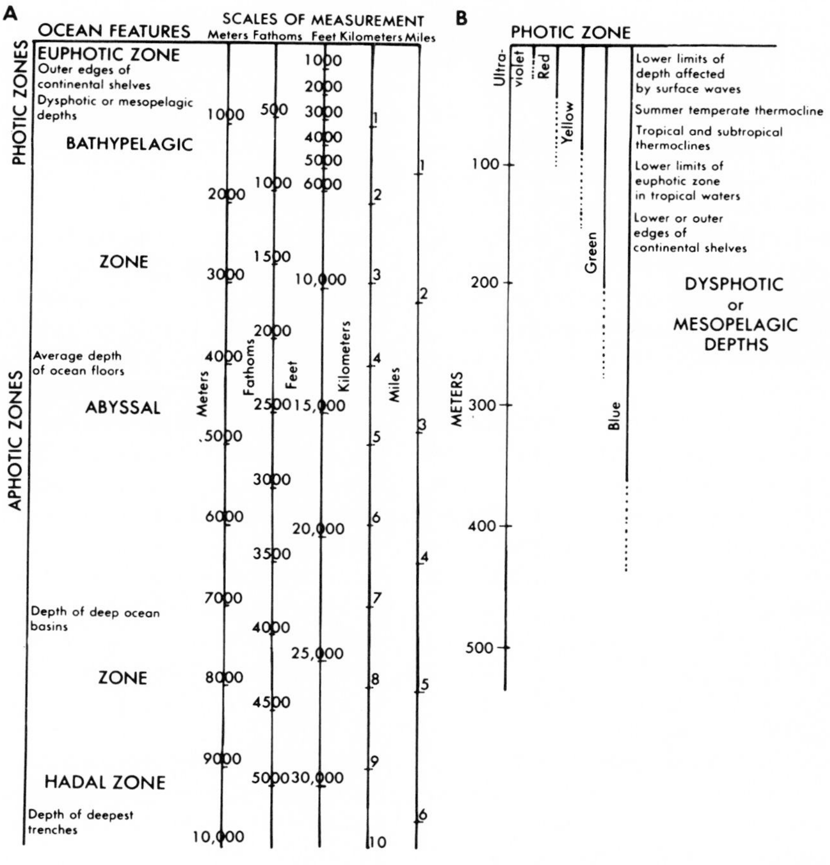

FIG. 1-2. A, Vertical zonation in the sea and the most commonly used units for denoting depth. **B,** Photic zone.

stratum of varying thickness, the **photic zone,** overlying an immense, eternally pitch-black **aphotic zone** comprising by far the greater part of the oceans, to which no light penetrates at any time. The only light in these depths would be transitory gleams from the ghostlike cold light of bioluminescent organisms, and rare flashing of submarine volcanic activity.

Even in clear tropical waters the photic zone is usually only a few hundred meters in depth. It is divisible into an upper portion, the **euphotic zone,** in which there is enough light, at least for part of the year, to support photosynthesis, and a lower portion, the **dysphotic zone,** in which the light is too dim and too briefly present to support photosynthesis. The euphotic zone is seldom if ever more than about 150 meters deep and is usually much less in most of the world.

When thinking about the photic zone it is important to bear in mind the diurnal, seasonal, and regional periodicities and irregularities of the illumination.

Oceanic waters of the euphotic zone are described as **epipelagic** and those of the dysphotic zone as **mesopelagic.** The aphotic zone is divided into an upper **bathyl,** or **bathypelagic,** and a lower **abyssal,** or **abyssopelagic** zone. Unusually deep waters are sometimes termed **hadal.** All these zones, especially the deeper ones, grade insensibly into each other.

The terms are not always used in precisely the same sense by all writers. The term bathypelagic is sometimes used to describe the dysphotic as well as the upper part of the aphotic zone, and the term abyssal is not infrequently used in referring to the entire aphotic zone.

Temperature stratification

Sunlight not only illumines the water into which it penetrates but also warms it. Because of the high heat-holding capacity of water, the temperature does not drop abruptly when illumination ceases at night. The surface waters tend to remain warmer than the underlying strata. Since warm water is less dense than cold water, it tends to remain at the surface.

At the polar regions throughout the year, in subpolar and boreal regions in winter, and even in the higher temperate latitudes, surface water may be sufficiently chilled to sink below the underlying strata, resulting in a seasonal turnover and mixing of waters of approximately the upper thousand meters. In polar regions the cold water may sink all the way to the bottom, making the entire water column very cold and rather uniform in temperature.

Since salinity also increases the density of water, there is a tendency for the waters of the oceans to stratify vertically with respect to both temperature and salinity, the lighter, less saline surface waters overlying increasingly dense, more saline, colder deeper waters.

TEMPERATURE

Temperature is one of the major factors influencing the distribution of marine organisms. Except near the polar regions, surface temperatures are higher and more variable than the temperatures of deeper strata.

Temperature relations are commonly diagrammed by drawing **isotherms**—lines connecting points with the same temperature. Various types of data may be used, depending on the purpose of a diagram. For example, one may wish to determine mean annual temperatures, annual maximal, or annual minimal temperatures. In a given region, isotherms divide the water masses of different depths from each other and can be visualized as warped horizontal planes in the water. Isotherms for surface waters have a general east-west trend, indicating progressively colder waters as one goes north or south from the equator. However, isotherms are very much modified by currents, upwellings, convergences, and local weather. Marine waters show a temperature range of −7° to about 42° C. The warmest temperatures are found during the day in shallow tropical lagoons or tide pools. Open surface waters are rarely warmer than 30° C. The coldest temperatures occur in certain abyssal basins of the Antarctic, where

TABLE 1-1. Surface temperatures of open waters in degrees centigrade

	Minimum	Maximum	Mean	Approximate percent of world oceans with that range
Tropical	20	30	24-27	31
Subtropical	16-18	25-27	20-22	16.5
Temperate	9-16	24-26	16-20	24
Boreal and antiboreal	1-9	10-17	8-14	16
Arctic and antarctic	−3 to +1	8-10	−1 to +5	12.5

TABLE 1-2. Mean annual temperatures at various depths and latitudes in the eastern Atlantic Ocean in degrees centigrade*

Depth (meters)	60° S.	40° S.	20° S.	0°	20° N.	40° N.	60° N.	80° N.
0	Below 0	15	17	27	20	16	9	2
200	Below 1	10	11	15	15	12	6	2
400	Below 1	10	9	9	12	12	9	1
800	Below 1	4	5	5	8	11	8	0
1000	Below 1	3	4	5	6	9	7	Below 0
2000	Below 0	2	3	3.3	4	4	3	−1

*From Ekman, S. 1953. Zoogeography of the seas. Sidgwick & Jackson, Ltd., London.

the water does not freeze because of its salinity and great pressure.

Table 1-1 gives a rough idea of surface temperatures in the various zones that are commonly recognized. It should be emphasized that in the oceans, communities of organisms, especially of pelagic organisms, are associated with water masses displaying certain hydrographic conditions, rather than with particular localities. Movements of water masses may sometimes carry a given community of organisms, or parts of it, for great distances.

Table 1-2 gives some representative temperatures of waters of various depths at different latitudes in the eastern Atlantic. Note that with increasing depth the temperatures tend to be lower and to fluctuate within narrower ranges. Conditions in deep water are commonly more uniform over greater distances than are those in the upper layers. For this reason communities of deep-sea organisms often show much wider north-south distribution than do the communities of organisms above them. A rather uniform deep-water community may underlie several different surface communities.

Thermocline

Extensive areas of relatively stabilized waters, such as occur in tropical regions the year round and in temperate regions during the summer, tend to show less overall productivity than do areas of vertical mixing. The warm surface waters become somewhat impoverished in plant nutrients, and, because of their lesser density, tend to remain at the surface, separated from the colder, richer waters below by a relatively stable thermocline. A thermocline is a stratum where a fairly rapid transition, or change, occurs between the warmer, less dense, less saline surface water mass and the colder, denser, more saline water masses below. In tropical waters the thermocline may occupy a depth of between 100 and 300 meters and be relatively stable throughout the year. In temperate and boreal waters the thermocline is primarily a seasonal phenomenon occurring only in the spring and summer and tends to be at a higher level.

In the oceans, thermoclines are not as abrupt and well marked as they are in freshwater bodies, where thermoclines are usually associated with changes of 1° C. or more per meter. In temperate waters and in high latitudes the surface layers are somewhat warmed during the summer but are kept mixed by winds and by small convection currents set up by evaporation and cooling so that the surface waters are relatively homogeneous. Below them, the increase in density tends to isolate the upper water from the colder layers beneath, but temperature does not change nearly as rapidly with depth as it does in freshwater bodies.

In parts of the eastern North Atlantic, warm, dense, highly saline water flowing out from the Mediterranean creates a mass of warmer water at intermediate depths, so that little fall in temperature occurs between 100 and 1000 meters.

Since seawater continues to increase in density with cooling, to its freezing point at −1.9° C., it sinks as it cools in winter. However, its sinking is limited by the permanently cold deeper layers so that there is not a complete spring and autumn turnover except in shallow waters. The spring and autumn mixing involves just the upper 1000 meters. Only near the polar regions do surface waters become sufficiently cold to sink to the bottom in relatively deep water, causing a complete mixing, with the entire water column becoming very cold.

In temperate waters there is a summer thermocline at about 50 meters, beneath which the temperature continues to decrease more gradually, forming a permanent, much less steep thermoclinal region down to about 1000 meters. As the tropics are approached, the summer thermocline becomes more marked, deeper in the water, and longer lasting, until in tropical and subtropical waters there is a permanent, well-marked thermocline between 50 and 500 meters in depth.

Thermoclines are especially important with respect to vertical circulation and mixing of waters because little or no exchange of water or of dissolved substances occurs through a well-established thermocline. The waters above and below it are effectively isolated from each other. In tropical and subtropical regions, where the thermoclines are permanent, this isolation is a major factor contributing to the exhaustion of plant nutrients above the thermocline and to the formation of oxygen minimum layers just below it.

CIRCULATION

Horizontal movements

Diurnal and seasonal temperature changes, together with the constantly shifting gravitational pull of the moon and sun and the forces set up by the earth's rotation, combine to produce complex movements and circulation of both the atmosphere and the ocean waters.

This circulation is much more active now than it was during the Mesozoic and most previous eras, when the distribution of continental masses was such that the oceans were more continuous than at present, the poles were not thermally isolated, and extensive continental areas were covered by shallow seas.

The movement of Antarctica to the south polar region, and of North America and Eurasia into positions nearly cutting off the north polar sea from communication with the rest of the world ocean, has caused both poles to be thermally isolated, allowing them to freeze over and thus producing a marked cooling and strong zonation of world climates. Formerly broad tropical regions are now restricted to a relatively narrow belt at the equator.

Greater continental emergence—perhaps due in part to withdrawal of water from the oceans and its deposition as ice on Antarctica and in part to the filling of the Atlantic basin as the Americas moved away from the Eurasian-African complex—has contributed to generally harsher, more contrasting climates. Neither air nor land has the heat-holding capacity of water. The present position of the continents blocks effective thermal interchange between the oceans except at the extreme south.

Since air is much lighter, contracts and expands readily with changes in temperature, and is continuous over the entire world, its movements are far more rapid and unconstrained than are those of water. The air also directly circulates immense quantities of water in the form of clouds and water vapor over vast distances. The lakes and rivers of the continents would soon run dry were they not replenished by water condensing from the air as rain and snow.

Wind-driven currents

An air current, or wind, moving over water, imparts to the water some of the kinetic energy of the moving air and sets the water in motion too. Thus there is formed a current of moving water bounded on either side by water of a different character and not moving in the same manner or direction. The frictional drag of water molecules against each other produces a slowing down so that subsurface water, dragged along with the current, moves more slowly than the surface water—the speed of the current rapidly diminishing with depth.

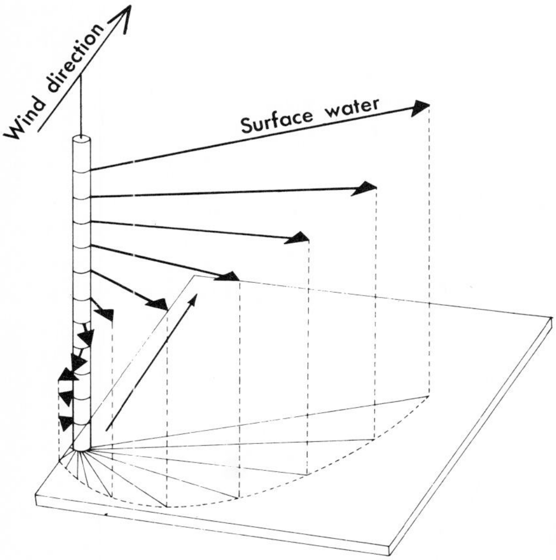

FIG. 1-3. Ekman spiral, showing water movements caused by a steady wind in the Northern Hemisphere. Water is deflected to the right of the wind direction by the Coriolis force, which is set up by the earth's rotation. Frictional drag between water molecules causes similar but slower movement, with increasing deflection at increasing depths.

Ideally a long-continued current passing over a given area of open ocean would eventually set the entire water column in motion. The Coriolis force, which is set up by the earth's rotation, causes this motion to be deflected to the right in the Northern Hemisphere and to the left in the Southern, producing a sort of spiral motion in which the water is deflected more and moves more slowly with increasing depth. Since the presence of horizontal temperature and density gradients interferes with these movements, the effects of wind-driven currents are generally limited to the upper water layers.

It is evident, therefore, that even in the open sea, currents would not flow in precisely the direction of the wind. The Coriolis force increases with distance from the equator and also exerts stronger effects the faster the water is moving.

The overall result is that as a major current moves over the ocean surface, the faster-moving surface waters are also deflected at an angle to the overall direction of the current, to the right in the Northern Hemisphere, and to the left in the Southern. In the Northern Hemisphere this deflection produces compensatory upwelling and convection currents along the left side of the current to replace the surface water, while along the right side there is a piling up of surface water, resulting in downwelling. The cross section of a major current would then be asymmetrical, being somewhat deeper along the right side and with the thermocline sloping from left to right.

The situation is actually more complex than indicated in the previous discussion. The currents do not maintain rigidly fixed courses but meander to considerable extent. The major currents are very broad and break up into sinuous strands that are not all going in exactly the same direction or at the same rate. Turbulence, convection cur-

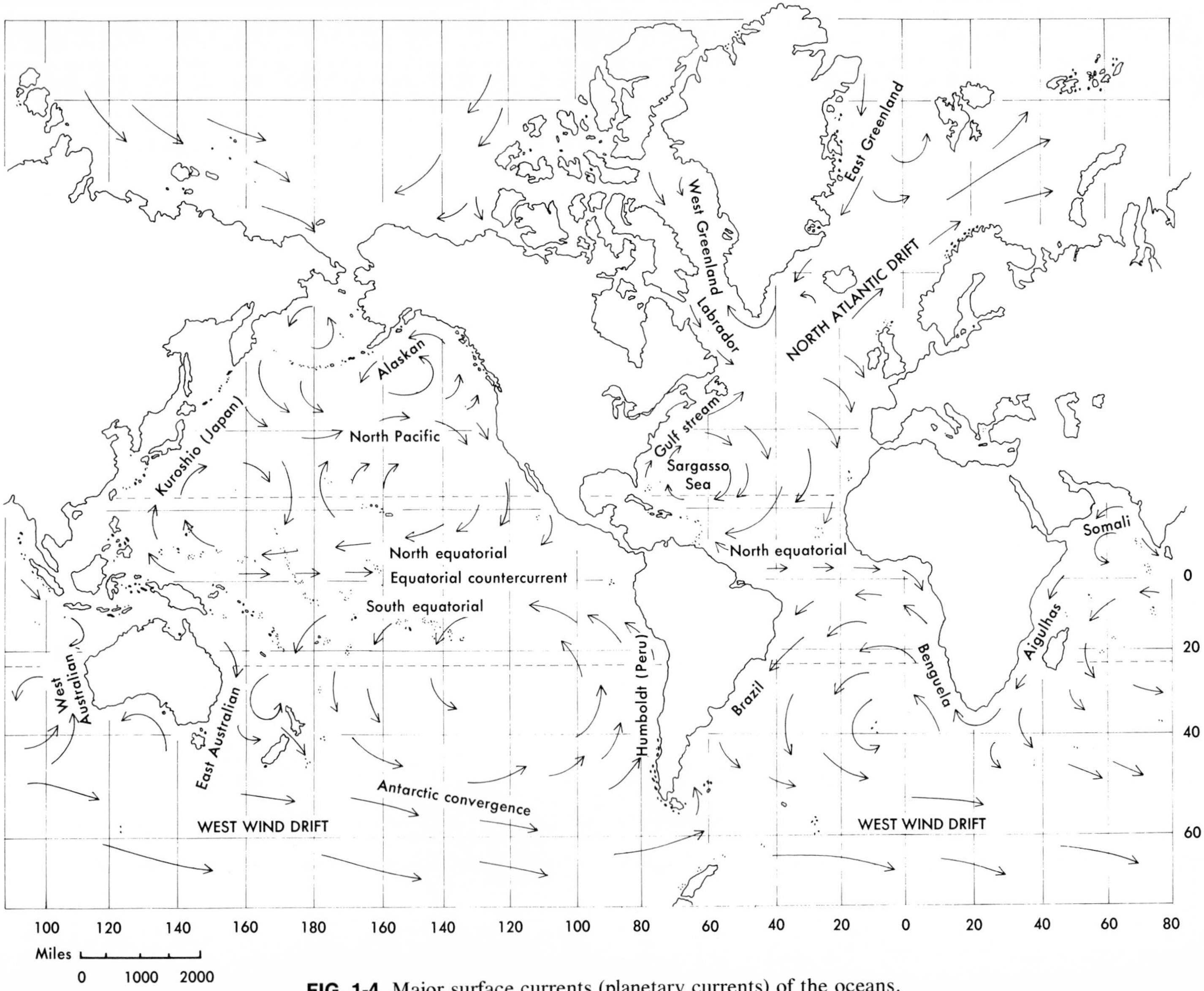

FIG. 1-4. Major surface currents (planetary currents) of the oceans.

rents, eddies, local or regional changes in intensity and direction of winds, and deflection by land masses all complicate the patterns of horizontal surface flow.

Any major current eventually approaches continental masses and is deflected in directions more or less parallel to the shoreline, regardless of wind direction.

Planetary currents

The major ocean currents, because of prevailing wind directions, the Coriolis force, and continental barriers, form great gyres moving in a clockwise direction in the Northern Hemisphere and a counterclockwise direction in the Southern (Fig. 1-4).

In the far south the ocean is continuous around the world, and there is an eastward movement of surface waters around the world between Antarctica and the southern parts of Australia, South America, and Africa.

Other horizontal water movements

Water movements are complicated by many other factors such as storms, changes in weather patterns, convection currents set up because of differences in temperature and salinity between different water masses, changes in barometric pressure, and seasonal changes. Where water masses of different densities meet, the lighter water tends to flow over the denser water.

Offshore winds, by sweeping surface water away from the shoreline, may cause compensatory shoreward movement of subsurface water and its upwelling near the shore. Onshore winds tend to pile up the surface water near shore, causing downwelling with an offshore movement of subsurface water. Strong, continuous winds parallel to the shoreline may have similar effects because of the net transport of surface water at an angle to the wind direction.

Other factors that may cause horizontal water movements include barometric pressure—highs tending to slightly depress the surface beneath them, forcing water toward the periphery, and its movement back when the high is replaced by normal or low pressure. Temperature and salinity differences between adjacent water masses also set up convection currents.

Tidal currents. Another type of current often evident near and over continental shelves, but not in the open ocean, are tidal currents.

The gravitational pull from the moon, and to a lesser extent the sun, is exerted uniformly over very broad open-ocean areas. No special currents are set up by this pull in open waters, but rather an imperceptible shift, or leaning of the water in the direction of the pull, occurs. However, when this shift impinges on the continental slopes and shelves, where it can be strongly influenced by features of local topography, and become either concentrated or dispersed, fairly strong tidal currents are commonly set up. This phenomenon occurs most noticeably along the outer margins of continental shelves, and in the mouths of bays and estuaries—wherever the topography concentrates or funnels the water.

Since these currents are reversed with every change of tide, they do not by themselves result in any great net transport of water or suspended materials. However, by stirring up the finer sediments, which are then subjected to any other general movement of the water, they may create large-scale transport. The outer portions of continental shelves are sometimes marked by coarser, sandier sediments than some of the intermediate shelf areas, apparently as a result of tidal currents. This result is contrary to the general expectation that sediments should become finer with increasing distance from shore because finer particles are carried farther before settling.

Basins. Depressions and basins tend to be somewhat isolated from the general circulation in their region because the densest (coldest, most saline) water tends to collect in them. Other, slightly less dense water usually moves over the basin rather than into it. Thus the water of basins tends to be quieter, less actively circulated, and less well oxygenated than surrounding or overhead waters and to have finer, deeper sediments. In the midocean floors, however, cold, oxygen-rich arctic or antarctic water that has sunk to the floor of the ocean and is spreading out over the sea floor toward the equator may displace the water in basins as well as over the sea floor generally.

Plotting currents

Careful study of water movements as they are related to temperature and salinity enables oceanographers, given the data from appropriately spaced stations in a given region at suitable time intervals, to accurately plot major currents and water movements, both surface and subsurface.

Needless to say, an accurate knowledge of water movements is of great practical and theoretical value. It enables one to understand many otherwise puzzling features of the distribution, movements, and seasonal activities of the biota—of vital importance, for example, to the fishing industry. Such knowledge is also of importance for navigation and many other activities on or in the ocean. Since given water masses tend to carry their own distinctive plankton and microbial flora with them, systematically collected data on plankton are also useful in tracking the movements of water.

Vertical water movements

From a biological standpoint, the vertical movements of water, though generally slower and less evident than the horizontal movements, are of far greater importance. There are two reasons for this fact.

In the first place, all of the dissolved oxy gen in the ocean enters from the top—either by being liberated from plant cells in the euphotic zone during the day, or by dissolving into the ocean water from the atmosphere. Were it not for vertical circulation, the deep waters of the oceans would be stagnant, anaerobic, and unfit for any life except anaerobic bacteria—like the deeper parts of the Black Sea, which has very little vertical circulation.

Second, there is a continual loss of plant nutrients from surface waters as dead organisms, fecal particles, arthropod exoskeletons, and suspended or dissolved organic molecules either drift down or are carried down below the euphotic zone before their mineralization by bacteria is completed. Decomposition of these organic remains, completed in deep water, liberates plant nutrients such as nitrates, phosphates, and silica as dissolved ions and molecules in the water. Since there are not actively growing photosynthesizing plant cells below the euphotic zone to utilize these nutrients, they accumulate there, and the deeper waters become richer in dissolved plant nutrient. Only as the eventual upward movement of deep, nutrient-rich waters brings nutrients into the euphotic zone again is the productive capacity of the oceans maintained. Once brought to the surface, such nutrients may be widely circulated by currents and other horizontal water movements.

Several causes of vertical water movements have already been mentioned. Wind-caused turbulence and wave action produce some mixing of waters near the surface. The rougher the weather, the greater is the mixing; thus surface water to the depth strongly influenced by wave action may be well mixed and have uniform characteristics. Large surface currents produce some vertical mixing because of the movement of the surface layers of the current across the current as well as in the dirction of the current. The result is some upwelling along one side, and some downwelling on the other side. The sinking of one water mass beneath a less dense water mass, when they meet, has also been mentioned. A good example is seen fairly regularly in the Australian region, where cold antarctic water masses moving northeast to the Antarctic Convergence sink beneath warmer waters sweeping across the Tasman Sea. They may remain at intermediate depths for several years and sometimes upwell again on the other side along the coast of Australia, still characterized by the presence of cold-water antarctic species of diatoms, radiolarians, etc. The upwelling and downwelling near shore caused by winds has also been mentioned.

Zones of convergence and divergence

Where currents flowing at different rates or water masses of different temperatures or salinities meet, zones of convergence may form where the surface waters from either side are carried below the surface. Some of these zones are sufficiently constant or recurrent to form permanent or semipermanent regional features. Particularly well known are the Antarctic and Antiboreal (Subtropical) convergences of the Southern Hemisphere. A zone of convergence tends to concentrate surface-living forms along its length and is often marked by great numbers of predators attracted to and taking advantage of this concentration. Areas of divergence, where subsurface water is coming to the surface, tend to show fewer surface-living forms and are often marked by the presence, at or near the surface, of organisms usually found at greater depths. Divergences tend to be broad and less well marked. Convergences are often long, narrow, and sharply delimited.

Ocean currents flowing side by side, especially when the water masses of each are of different densities or the currents are of different speed, may form dramatic boundaries, known as **rips.** William Beebe (1928) described one observed by the *Arcturus* in 1926 in the equatorial Pacific, where such a rip extended from horizon to horizon, as a band marked by foam and whitecaps, about 60 feet wide. The *Arcturus* followed it more than 100 miles. Both currents were warm and flowing in the same direction. The water on the south side of the rip was several degrees cooler than that on the north side and was noticeably darker, rougher, and faster flowing.

The water in the rip contained much more life than that on either side of it, a distance of as much as 10 yards to either side making a great difference. The rip was soupy, with smaller plankton and young fish concentrated from both sides by downwelling. Predators of all kinds—sharks, dolphins, sea turtles, large fish, and great numbers of sea birds—were gathered along the rip in vast numbers. There were also a number of barnacle-covered logs of palm, coconut, bamboo, and trumpet wood that were riddled with shipworm. The differences in the water masses and the boundary between them extended down to a depth of at least 1000 feet.

Smaller zones of convergence and divergence are set up by a variety of causes, such as changes in temperature, rate of evaporation, and winds. Perhaps most widespread and significant is the so-called **Langmuir circulation,** in which light winds blowing steadily over the water surface set up small parallel zones of alternating divergence and convergence. Dissolved and particulate organic matter, together with planktonic organisms, are concentrated in the lines of convergence, facilitating the feeding of small animals. The Langmuir circulation also causes the concentration and polymerization of surface-active dissolved organic substances into larger micelles and facilitates their adsorption onto the surfaces of larger particles, thus making some of the dissolved organic matter indirectly available to detritus and filter feeders. (See Fox and co-workers, 1952, on marine leptopel.)

Evaporation

Excessive evaporation, such as occurs in the Mediterranean and the Red Sea and to some extent in other tropical regions, may, by increasing the salinity of the surface water, cause it to become slightly denser than the immediately underlying subsurface water and, even though warmer, to sink beneath it. In the Mediterranean these warm subsurface waters form the bulk of the countercurrent flowing out from Gibraltar and into the Atlantic beneath the main surface inflow of water from the Atlantic to replace the water lost by evaporation. Since this warm, more saline water does not rise immediately, adjacent parts of the Atlantic containing it show much less decrease in temperature with depth, for a considerable depth, than do most oceanic waters. In the Mediterranean and the Red Sea themselves the deeper waters are warmer than most marine waters of comparable depths, and shallow-water biota limited to relatively warm waters extends into deeper water than usual for it elsewhere.

Chilling and seasonal turnover

In the higher temperate latitudes of both hemispheres and in the boreal and antiboreal zones, winter chilling of surface water increases its density sufficiently to cause it to sink, producing an annual turnover and mixing of surface water and deeper strata. Since cold water holds more dissolved oxygen than does warm water, and since deeper water contains more plant nutrients than does surface water, this exchange serves both to oxygenate the deeper strata and to enrich the surface waters, setting the stage for dramatic outbursts of plant cell growth in the early spring, when the water again stabilizes and daylight is sufficient to promote active photosynthesis. As the sun comes to a more overhead position, the potential depth of the euphotic zone also increases, but where dense clouds of plant cells form, they may produce sufficient turbidity to shade the underlying water and decrease the depth of the euphotic zone.

Polar sinking and bottom circulation

In the far northern and southern polar regions the surface waters become so cold in winter that they are as dense as the water at the bottom. The entire water column is very cold, dense, and oxygen rich. This dense, cold water at the bottom of the water column spreads slowly over the ocean floors (Fig. 1-5) toward the equator, or well beyond the equator in the case of the antarctic water. As it spreads, it very slowly displaces the bottom water already there upward, and is itself very gradually and slightly warmed. Deep oceanic waters are thus oxygenated from the bottom by the gradual displacement upward of the oxygen-rich bottom water as new antarctic or arctic water continues to displace it.

The complete cycling may take hundreds of years, during which time, whatever animal life and bacterial life are present in the water continuously deplete the oxygen.

Oxygen minimum layer

In the middle latitudes, where the warmer, lighter surface waters tend to stay on top throughout the year, separated by a well-marked thermocline and density boundary from the colder waters beneath,

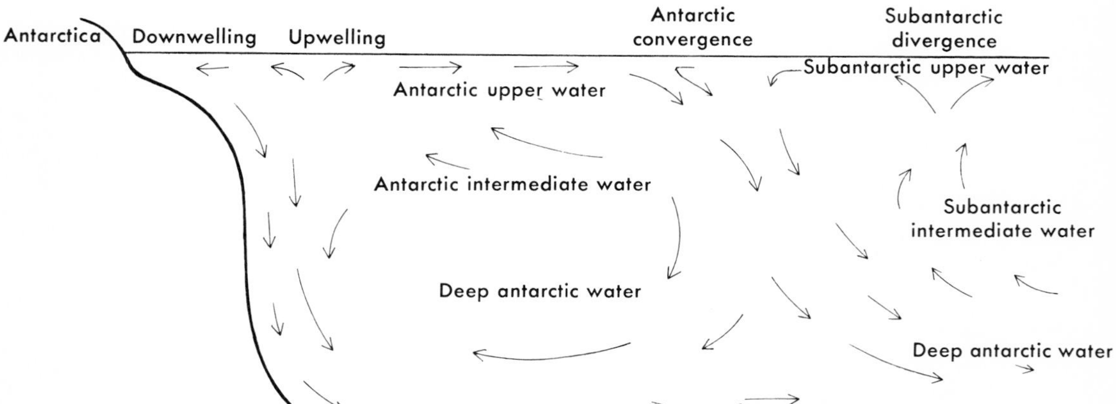

FIG. 1-5. General character of water circulation in and near the Antarctic Ocean.

these rising intermediate depth waters never reach the surface at all. Having no source of oxygen renewal, they continue to be oxygen depleted, forming in some regions an extensive oxygen minimum layer in which dissolved oxygen levels are too low to support most animal life and may become almost undetectable (Fig. 1-6). Yet some plankton animals in apparently good condition are found even in these strata.

Childress (1968) studied the mysid *Gnathophausia ingens* in the eastern Pacific. In this species very young, small individuals are found above the oxygen minimum layer; most intermediate-sized individuals, 35 to 125 mm. long and weighing 0.3 to 15 gm., are found in the oxygen minimum layer; and most of the largest, mature individuals are found below it. Studying the respiration of these mysids in chambers simulating conditions in the oxygen minimum layer, Childress found that they were able to regulate their oxygen consumption remarkably well in nearly anaerobic water, with oxygen pressures as low as 0.14 ml. oxygen per liter (most animals begin to fail at around 1.2 ml./liter). Thus they can live aerobically in waters with oxygen pressures far below those necessary for most animals and can contribute significantly to the biological depletion of oxygen thought to be responsible for the formation and maintenance of the oxygen minimum layers. Their failure to regulate below 0.14 ml./liter would also explain their absence from areas of the Pacific where oxygen values in the oxygen minimum layer fall below this value. Both *Gnathophausia* and *Euphausia mucronata* found in the oxygen minimum layer from the same regions are able to consume the oxygen in nearly anaerobic water until the oxygen reaches undetectably low levels.

The vertical distribution of the species may perhaps be accounted for by pelagic, low-density eggs that rise into the upper waters before hatching. As the young mysids grow and become capable of better regulation, they seek the protection offered by the deeper, oxygen-low strata of water. Perhaps on attaining maturity they require more oxygen, become immobilized, and sink until they reach somewhat better oxygenated water below the oxygen minimum

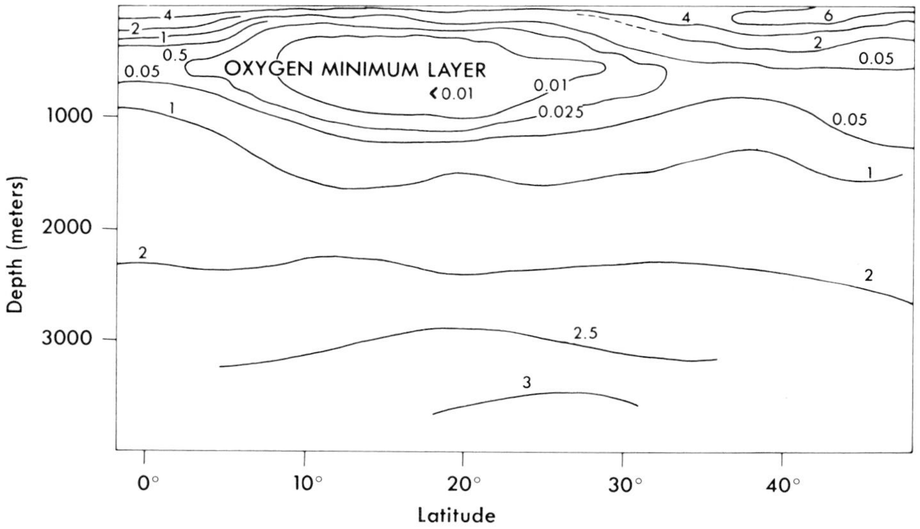

FIG. 1-6. Oxygen minimum layer in the northern Pacific. (After Sverdrup, H. U., M. W. Johnson, and R. H. Fleming. 1946. The oceans: their physics, chemistry, and general biology. Prentice-Hall, Inc., Englewood Cliffs, N.J.)

layer, where they are revived. Childress found that when subjected to oxygen pressures below 0.14 ml./liter, these mysids were soon immobilized, but that they would survive for several hours and revive when placed in water with slightly greater oxygen pressure.

In some regions occasional upwellings of water from the oxygen minimum layer into the surface layers may cause mass mortalities of fish and other animals in the water above. When such an event occurs, the decomposition of the excess organic matter may continue to deplete the oxygen for a time and slow down the recovery of the area.

Plant nutrients and circulation

It has recently been ascertained that much of the antarctic and arctic water from the sea floor eventually surfaces in the broad area of the North Pacific just south of the eastward-flowing North Pacific Current. Plant nutrients brought up with it enrich the North Pacific waters and are carried by the North Pacific Current to the area along the Canadian coast where the main body of the current turns south, as the California Current, which parallels the west coast of North America. A large gyre also turns north into the Gulf of Alaska, and then west along southern Alaska, some of it making its way into the Bering Sea and northward, and some turning south and reentering the northern edge of the North Pacific Current.

Furthermore, in the Arctic, and especially in the Antarctic, there is a continual upwelling of nutrient-rich water from intermediate depths, replacing the very cold surface waters that sink to the sea floor.

For all these reasons, far northern, and especially far southern, surface waters are particularly rich in plant nutrients, especially in the spring after the vertical exchange has been at its winter maximum, and plant growth at its winter minimum.

Plant nutrients brought south with the California Current from the North Pacific, and brought north by the Peru, or Humboldt, Current from antarctic regions, are chiefly responsible for the productivity of offshore and shelf waters along the west coasts of North and South America. Local wind-caused upwellings along these western coasts play a lesser role than was formerly assigned to them. Similar considerations apply also in the Atlantic Ocean.

Internal waves and plant nutrient renewal

Dr. John McGowan, of the Scripps Institution of Oceanography, has recently suggested another mechanism for plant-nutrient renewal in surface waters of the open ocean, through the action of internal waves. Internal waves of various amplitudes and periods, some of them very large, are moving in various directions through the deep water. Where they reach sufficient amplitude, or reinforce each other locally, or interact with mesopelagic horizontal water movements, mixing phenomena, analogous to that seen in surface waves when they form whitecaps, may be produced. Such mixing would detach localized masses of the colder, richer, lower water, bringing them into the warmer water above and producing temporary local temperature inversions in which cooler water overlies warmer water. Such inversions have been detected. As these blobs of deeper water mix with the surrounding water, they enrich it. This process would be maximal at depths where internal waves show maximum amplitude but would also occur to some extent throughout the water column, thus moving plant nutrients gradually up to the euphotic zone in a rather random, spotty manner.

For reasons not well understood, the depth at which internal wave action is maximal is not constant, but varies at least in some years. In the major North Pacific gyre, for example, it was much higher in the water in 1969 than in the preceding or following years. Plant nutrients were therefore being transported into the euphotic zone at a faster rate.

The result was an approximate doubling of the standing crop of zooplankton that year. Since phytoplankton production in the open ocean is nutrient limited (even with this increase), the extra nutrients are promptly removed from the water and incorporated into cell bodies. The resulting opportunity for more grazing by the zooplankton causes it to increase. The most dramatic observable effect was an approximate doubling of the standing crop of zooplankton that year.

The amounts of dissolved plant nutrients in the water did not show an appreciable increase because of their prompt removal by phytoplankton. The phytoplankton standing crop was either held down or brought down, by the time the survey was made, to a level that was not dramatically different from usual by the increased zooplankton grazing.

Had the increased upward transport persisted for several years, it seems possible that an increase in the standing crop of organisms still higher in the trophic hierarchy might have been the most dramatic observable effect, and that changes in species present, and in the proportions of the various species already typical for the region, might have occurred. Since, however, the change was so transitory, and even with the increase, the biotic community was still nutrient limited, the only observed effect was an approximate doubling of the entire zooplankton community, without dramatic changes in species composition or proportions.

Upward transport of nutrients through internal wave action is a very attractive hypothesis to explain the persistence of life in the surface waters of much of the open ocean. Other mechanisms invoked, such as the sinking of surface waters because of increased density either due to excessive evaporation or from winter cooling, in most parts of the world, do not bring about a turnover of water to sufficient depths to bring up water from below the nutrient-impoverished zone. Horizontal transport from distant regions with more enriched waters, such as the polar regions, is so indirect and takes so long that the nutrients carried in these surface waters would have been largely dissipated and lost to deeper water before getting to the regions of the central areas of the great oceanic gyres.

Deep circulation

Deep-water circulation in the Atlantic Ocean is much more complex and active than in the Pacific and Indian oceans. In all three the largest component of new, oxygenated bottom water is the cold antarctic water moving slowly northward following and spreading out from the lowest bottom contours.

In the Atlantic Ocean there is also a much larger volume of cold arctic water coming down from the north than is the case in the Pacific. In the Indian Ocean this component is entirely lacking.

Second, intense evaporation from the Mediterranean produces a highly saline and therefore denser water that sinks beneath the surface and flows out into the Atlantic in the Gibraltar countercurrent. Because of its density, it does not rise immediately and is gradually cooled, becoming even more dense as it moves out farther into the Atlantic. Some of it moves north, west of the British Isles. It can also be traced south as a deep layer, above the antarctic bottom water as far south as South Africa.

In the Pacific, the bottom topography, with the extensive east-west–trending archipelagoes in the western half, slows down, and interferes with, the northward movement of bottom water.

Thus in the Pacific and Indian oceans the bottom waters and other extremely deep waters are renewed much more slowly than in the Atlantic.

The Red Sea does not play a role in the Indian Ocean comparable to that of the Mediterranean water in the Atlantic partly because of its much smaller size, and partly because its highly saline waters are even warmer than those from the Mediterranean and do not sink as deeply.

CHEMISTRY OF SEAWATER

Despite the influx of approximately 3.64×10^4 km.^{3} of freshwater annually from rivers, plus an even larger amount falling directly on the oceans as unevenly distributed precipitation, and despite the annual evaporation of similar amounts of water (also unevenly and differently distributed) from the oceans' surface, the concentrations of major ions, the pH, and the osmotic properties of seawater are remarkably constant.

Precipitation is highest in the polar regions and causes some dilution of surface waters, especially in summer, when some of the ice is also melting. (The formation of ice in water, insofar as it is formed by the freezing of seawater, withdraws water from the surface, leaving the underlying water saltier than it was. However, since some of the ice and all the snow come from precipitation, the net effect is dilution.) Evaporation is greatest in tropical and subtropical latitudes, especially from seas surrounded by arid or semiarid lands, such as the Red Sea and the Mediterranean. Although some shallow tropical seas, such as the Java Sea, the Arafura Sea, and the Timor Sea, receive enough rain during the monsoon season to lower the salinities of their surface water, the general picture is that surface waters are somewhat higher in salinity and density in equatorial latitudes than they are in polar latitudes. Salinity and density of surface waters are also locally reduced where marine water is diluted by the runoff from large rivers.

The density of seawater increases as its temperature decreases or its salinity increases. In general, deep ocean waters are colder and more saline than surface waters and exhibit less variation in these respects.

Since cold water will hold more dissolved oxygen than does warmer water, the sinking of cold, highly oxygenated water in the polar regions is an important factor in maintaining circulation and oxygenation of abyssal depths.

On the average, seawater is composed of 96.52% water and 3.49% dissolved substances, mostly salts. The salts can be conveniently grouped as major constituents and minor constituents. All of the major constituents and several of the minor ones are found everywhere in the oceans in virtually the same proportions relative to each other. Such salts, found everywhere as a constant proportion of the total salts, are termed **conservative elements.** Other minor constituents, which show marked variations in relative concentration largely because of their selective removal from the water by living organisms, are termed **nonconservative elements.**

The major constituents, comprising 99.9% of all the dissolved salts, are the sodium, magnesium, calcium, and potassium cations and the chloride, sulfate, carbonate, bicarbonate, and bromide anions, together with boric acid mostly in the undissociated state (Table 1-3). A few of the other conservative elements, given in parts per million (ppm), are lithium (0.17), rubidium (0.12), barium (0.05), aluminum (0.01), uranium (0.003), and lead (0.0003). Others are found in still smaller traces.

The most important of the nonconservative constituents are the major plant nutri-

TABLE 1-3. Concentration of major components of seawater in moles per kilogram of seawater at salinity 35‰; for various ions the percentage that each makes up of total salts also given

Component	Moles/kg. seawater	Percent of total salts
Water (H_2O)	53.557	
Sodium (Na^+)	0.4680	30.4
Magnesium (Mg^{++})	0.532	3.7
Calcium (Ca^{++})	0.0103	1.16
Potassium (K^+)	0.0099	1.1
Strontium (Sr^{++})	0.0001	0.04
Chlorine (Cl^-)	0.5459	55.2
Sulfate (SO_4^-	0.0282	7.7
Carbonate and bicarbonate ($CO_3^=$ and HCO_3^-	0.0023	0.35
Bromine (Br^-)	0.008	0.19
Boric acid (mostly undissociated H_3BO_3)	0.0004	0.07
All others		0.09

ents, the phosphates and nitrates,* together with silicon, which is required by diatoms for construction of their frustules and by radiolarians for their skeletons. The extremely low solubility of silicon limits its availability in seawater in spite of its great abundance in the earth's crust.

These three essential nutrients tend to be in short supply in surface waters, where they are taken up by phytoplankton, but they increase rather abruptly below the euphotic zone as a result of their release from decomposing organic particles sinking from above and of the lack of functional photosynthetic organisms in the deeper water to utilize them.

Most of the remaining elements necessary for living organisms, although present only in what a chemist would term trace quantities, are nonetheless sufficiently abundant that they do not act as limiting factors. Most of them are not used for general tissue building or skeletal material, but rather enter into specific compounds, such as chlorophyll or respiratory pigments. Examples are magnesium in chlorophyll, iron in hemoglobin, copper in hemocyanin, iodine in thyroxine, and fluorine in enamel of teeth. Other essential plant nutrients needed in trace quantities include boron, zinc, manganese, cobalt, nickel, and molybdenum. Of these, cobalt (about 0.0005 ppm) and nickel (about 0.002 ppm) are in rather low abundance in the sea but have not been shown to act as limiting factors. Iron (about 0.008 ppm) and manganese (about 0.002 ppm), although equally present with, or more abundant than, nitrogen or carbon, probably do limit growth in parts of the sea because they are supplied principally in particulate form by erosion and drainage from the continents.

Organisms such as tunicates and some holothurians, which concentrate vanadium (about 0.002 ppm) or in one case niobium (about 0.00001 ppm), may well be limited in their abundance by the scarcity of these elements. Tunicate blood may be up to 10% vanadium, dry weight—50 million times its concentration in ambient seawater.

Salinity

These osmotic properties of seawater result from the total amount of dissolved salts. This is generally expressed in terms of sodium chloride equivalence, or **salinity**—the amount of sodium chloride that would give to distilled water the same osmotic properties. Thus a salinity of 32‰ would be equivalent, from an osmotic standpoint, to 32 gm. of NaCl/1000 ml. of solution, that is, a 3.2% solution of sodium chloride.

Since the ratios of major ions in seawater are constant, even though the total amount of salts may vary somewhat, the salinity can be calculated from a simple titration of the $Cl^- + Br^-$ ions with silver nitrate. This gives the **chlorinity,** which is defined as 0.3285 times the silver equivalent of seawater. The salinity can then be calculated by the following equation:

$$\text{Salinity} = 0.03 + (1.805 \times \text{Chlorinity})$$

The constant term 0.03 in the equation means that a water sample with zero chlorinity would still have a salinity of 0.03‰. This reflects the fact that low salinities in marine waters usually result from mixing with river water rather than from rain. River water contains much more sulfate and bicarbonate than chloride. Hence it may still have a detectable salinity even though the chloride value is extremely low or zero.

In the open sea, salinities commonly range from 32‰ to 37.5‰. The differences reflect local effects of evaporation, rain, freezing and melting of ice, or influx of river water. In estuaries and other brackish en-

*According to Sillén (1967) the most puzzling factor in the composition of both the sea and the atmosphere is the distribution of nitrogen. At the present pH and pE of seawater, one would expect much more of the nitrogen to be in the form of dissolved nitrates in the seawater rather than in the form of N_2 in the air. The known mechanisms for the return of nitrogen from the sea to the air—certain pockets of nitrate reduction—seem wholly insufficient to account for the balance actually observed. No large-scale, near-the-surface process of N_2 return from the sea has yet been discovered.

TABLE 1-4. Salinities of various types of water*

Salinity (0/00)	Type of water
0-0.5	Freshwater
0.5-3.0	Oligohaline brackish water
3.0-10	Mesohaline brackish water
10.0-17	Polyhaline brackish water
17-30	Oligohaline seawater
30-34	Mesohaline seawater
34-38	Polyhaline seawater
> 38	Brine

*From Välikangas, I. 1933. Über die Biologie der Ostsee als Brackwassergebiet. Verh. int. Verein. theor. angew. Limnol. **6**:1.

vironments, salinities may be much reduced and may vary according to location, tidal cycle, local runoff, etc. In isolated or semi-isolated seas in tropical or subtropical latitudes (the Red Sea or the Mediterranean), salinity may be higher than normal; in evaporate basins (some of the shallow lagoons along the edges of the Gulf of Mexico) and especially in isolated lakes with no runoff leaving them (the Great Salt Lake or the Salton Sea), salinities sometimes become very high, reaching saturation (Table 1-4).

The simplest and quickest way to measure salinities is to use a hydrometer calibrated to read directly in salinities at standard temperature, making a correction for the actual water temperature. This method is not as accurate as titration but is adequate for many purposes.

pH and E_h

The **hydrogen ion concentration** of seawater, expressed as pH, is rather constant, varying from 7.6 to 8.3. Buffering is largely the result of the carbon dioxide, carbonic acid, bicarbonate-carbonate equilibrium, the buffering effect of fine clay particles, and, to a lesser extent, boric acid. At higher pH values, precipitation of calcium as calcium carbonate is favored.

The **oxidation-reduction potential, redox potential,** or E_h, is expressed in millivolts and is related to the presence of molecular oxygen or of reduced compounds such as hydrogen sulfide, ferrous iron, and sulfhydryl groups. Positive E_h values indicate oxidative conditions, and negative values indicate anaerobic, or reducing, conditions. All organisms and also all chemical reactions of biological importance occur within certain pH, E_h, and temperature parameters. If these are known for a given environment, one has gone a long way toward characterizing that environment and can, to a considerable degree, predict which types of organisms and biological processes may be expected to occur there and which would be excluded. Similarly, if one knows the bounds of these parameters within which a given organism or a given chemical reaction can occur, one can predict in what types of environment it can be found.

The E_h values of seawater usually lie between +400 and +435 mV. at pH 7.6 to 8.3. The E_h is controlled by photosynthesis and by respiration. Photosynthesis acts on the system by utilizing CO_2 and HCO_3 and releasing oxygen, or removing hydrogen, thus tending to raise both the E_h and pH. Respiration has the reverse effect. The values in shallow or surface waters tend to rise by day and be depressed during the night.

The parameters of pH, E_h, and temperature within which most organisms can exist overlap considerably, enabling many organisms to live together in most environments. However, any shifts will tend to favor some and repress others, allowing different species or varieties to come to dominance. The activities of the organisms themselves are, of course, continually exerting such effects, changing these and other environmental parameters and setting the stage for further changes and fluctuations in the populations present.

The differences in salinity, pH, and E_h in the open sea are not sufficient to constitute major barriers to the distribution of organisms within the oceans, as do the differences in light and temperature. The constancy of conditions in the sea is of great biological significance. It permits, under pe-

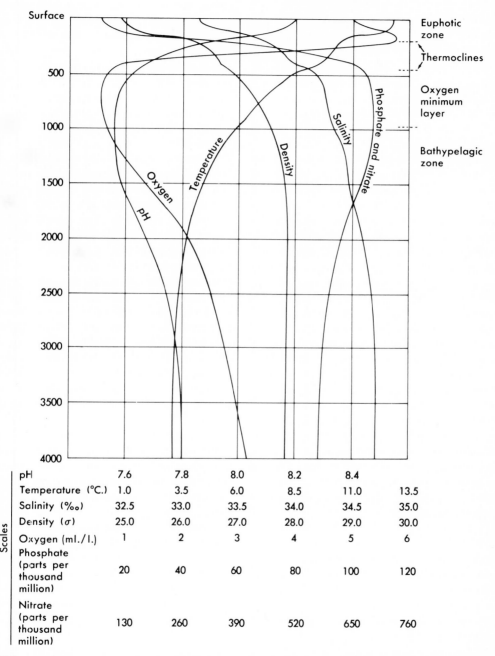

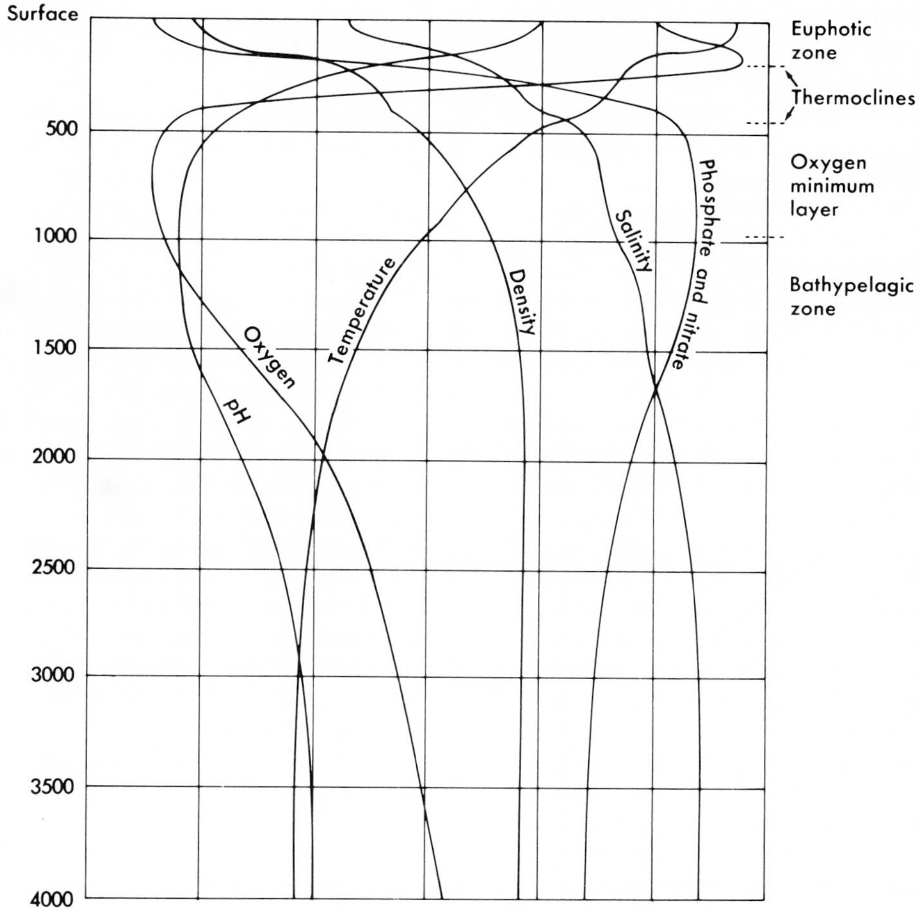

Scales						
pH	7.6	7.8	8.0	8.2	8.4	
Temperature (°C.)	1.0	3.5	6.0	8.5	11.0	13.5
Salinity (‰)	32.5	33.0	33.5	34.0	34.5	35.0
Density (σ)	25.0	26.0	27.0	28.0	29.0	30.0
Oxygen (ml./l.)	1	2	3	4	5	6
Phosphate (parts per thousand million)	20	40	60	80	100	120
Nitrate (parts per thousand million)	130	260	390	520	650	760

FIG. 1-7. Some physicochemical parameters, illustrated by a sampling off southern California.

lagic conditions, growth of embryos and of species of organisms that have not developed regulatory mechanisms.

At the pH of surface waters the concentration of bicarbonate carbon is about a hundred times that of carbon dioxide. Although carbon dioxide is the carbon source most readily transferred across plant cell membranes for photosynthesis, some plants—especially those habitually growing in alkaline environments—can also utilize bicarbonate. Hood and Park (1962) demonstrated that marine plants continue to take up C^{14} from bicarbonate under conditions of carbon dioxide purging by nitrogen or compressed air, whereas freshwater algae failed to do so. They concluded that utilization of bicarbonate as a carbon source in photosynthesis is probably universal among marine plants.

Organic compounds

Dissolved and particulate organic matter is found in small quantities in all parts of the oceans, about 1.2 to 2.8 mg. of organic carbon or 0.2 mg. of organic nitrogen per liter. Most of the amino acids have been demonstrated, usually less than 1 mg./m.^{3}, as well as carbohydrates and other organic compounds. Some of these substances may be directly available to certain organisms and made indirectly available to many more through agglomeration into larger particles or adsorption to the surfaces of larger particles that can be utilized by suspension feeders and adherent bacteria.

In connection with the estimate of dissolved organic compounds and other biologically active chemicals in the seawater, it should be borne in mind that chemical analysis of the seawater only measures the unconsumed residue at any time and does not necessarily give a clear picture of the importance of these substances in the marine ecosystem. Substances that are most readily utilized by living organisms may be taken up as fast as they are formed and hence may not appear in significant quantities in such analyses, even though they may be produced and utilized on a rather large scale. The less readily assimilated compounds will be those which tend to accumulate as a type of marine humus. The amount and variety of dissolved organic compounds in the water are much less and are much more uniform from about 300 meters on down than they are in the upper strata of water.

Two recent analytical advances open the way to much-improved understanding regarding the amounts and roles of organic materials in the sea.

By measuring the adenosine triphosphate (ATP), one can distinguish living organic matter from nonliving organic detritus, thereby making more precise the estimates of biomass. This is done by measuring the light produced when ATP is added to an enzyme mixture containing luciferin and luciferase extracted from firefly tails. By this technique it is possible to measure small amounts of ATP down to about 10^{-6} gm.

It has also been recently discovered that irradiation of seawater with high-energy ultraviolet light completely destroys all organic matter in it, producing carbon dioxide, orthophosphate ions, and nitrate and nitrite ions. The organic carbon is quantitatively recovered as carbon dioxide. This provides a powerful tool for determination of dissolved organic carbon, nitrogen, and phosphorus in seawater samples from any depth. Reliable carbon-nitrogen-phosphorus ratios are of prime importance in understanding the fluxes of organic matter in the sea.

This technique also makes it possible to obtain seawater totally free from unknown organic constituents, thus providing a highly reproducible medium for nutritional studies or for the study of effects of trace organic compounds important in the biological "conditioning" of water.

• • •

Fig. 1-7 brings together in a visual manner some of the factors discussed, for a representative sampling in the Pacific Ocean off southern California.

FIG. 2-1. Coastal waters, looking north along the coast of southern Oregon. This coastline is relatively young. The outlying rocks not yet wholly worn down indicate part of the former extent of the Pleistocene marine terrace represented by the flat forested area at the top of the present cliffs and currently being eroded back by the sea.

CHAPTER 2 COASTAL WATERS

CONTINENTAL SHELVES

The ocean basins usually have a narrow margin, or shelf region, surrounding the land masses and extending seaward a variable distance to a depth of about 100 to 200 meters. Being for the most part in the euphotic zone, continental shelves support abundant benthic fauna and flora—the shelf communities—that differ in many respects from the communities found in deeper water.

The water above the continental shelves and its fauna and flora are described as **neritic.** Neritic waters tend to be richer in plant nutrients and more productive than water of corresponding depths in the open sea. One reason for this is the greater mixing that occurs here as a result of turbulence, wave action, upwelling caused by offshore currents, winds, etc., bringing plant nutrients into all strata of the water. Also, additional nutrients are leached from the substrate and washed into the sea by rivers and streams from the adjacent land (Fig. 2-1).

Where the substrate is suitable for attachment, the shallow waters also support a rich growth of seaweed, turtle grass, eelgrass, and, in some areas, other plants.

The increased plant growth is reflected in an increase in animal life. Many species are found here that do not ordinarily occur far out at sea or in deeper water. Even the plankton, in addition to being more abundant, is of different character. Some of the oceanic species tend to be rare or absent in neritic plankton, or they occur only sporadically when oceanic water is swept in closer than usual. Many neritic species are limited to the shelf region and are seldom found far out at sea.

In addition to the **holoplankton**—species that are planktonic throughout their life—neritic waters contain much **meroplankton.** Meroplanktonic animals are those species found in the plankton only part of the time—for example, at night only. They seek shelter in the substrate the rest of the time, or they may be benthic forms that spend part of their life cycle in the plankton. Swarms of larval barnacles, decapods, echinoderms, and other nonplanktonic animals often make up an important part of the neritic plankton at certain times of the year.

The term neritic has often been applied to a species simply because it occurs in greatest numbers in neritic waters, even though the organism may be holopelagic, is not limited to the shelf in any way, and occurs regularly in high oceanic waters as well as in neritic waters. It is probably better to use the term in a more restricted sense for organisms that are bound to the shelf region in some way. Species that spend part of their time or go through any stages of their life cycle in or on the shelf substrate, as well as species that are tied to certain near-to-shore areas by salinity requirements, that feed on detritus washed from the land, or that for any other reason are bound to the shelf waters, are more properly termed neritic, even though some of them may at times or at certain stages of their life be found in truly oceanic waters.

TIDES

At its upper margin the continental shelf ends in the **intertidal zone.** This relatively narrow strip is exposed during the lowest low tides and either covered with water or splashed with waves during the highest high tides. The special conditions and stresses caused by alternating exposure and covering with water have resulted in the development of distinctive communities of intertidal animals and plants. The nature of the particular community at any given place is largely determined by the physical character of the substrate and the degree of its exposure to wave action. Differences in temperature are responsible for broader regional differences in fauna and flora.

Intertidal communities are more directly influenced by the local climate than are the subtidal communities. During periods of low tide they may be exposed directly to hot sunlight and drying, rain, cold winds and freezing, and predation from terrestrial animals and man. The tides operate on a lunar rhythm that is out of phase with the diurnal rhythm, resulting in a progression of tides through the lunar month, the high and low tides coming each day at different times and reaching different levels. That both the lunar tidal rhythms and the diurnal and seasonal rhythms of the earth-sun system exert profound effects on the lives of organisms is particularly evident in those forms which live part or all of their life in the intertidal zone.

There are usually four tides, of different heights, each day—two high tides and two low tides. The tides occurring during that part of each lunar month when the sun, moon, and earth are in line with each other—when the gravitational effects of the sun and moon reinforce each other—are strongest. Such tides, termed **spring tides,** exhibit the greatest difference in level between the highest high tide of the day and the lowest low tide. The opposite effect occurs when the sun and moon are in positions such that they partially cancel out each other's gravitational effects, resulting in weaker tides, which exhibit the least dif-

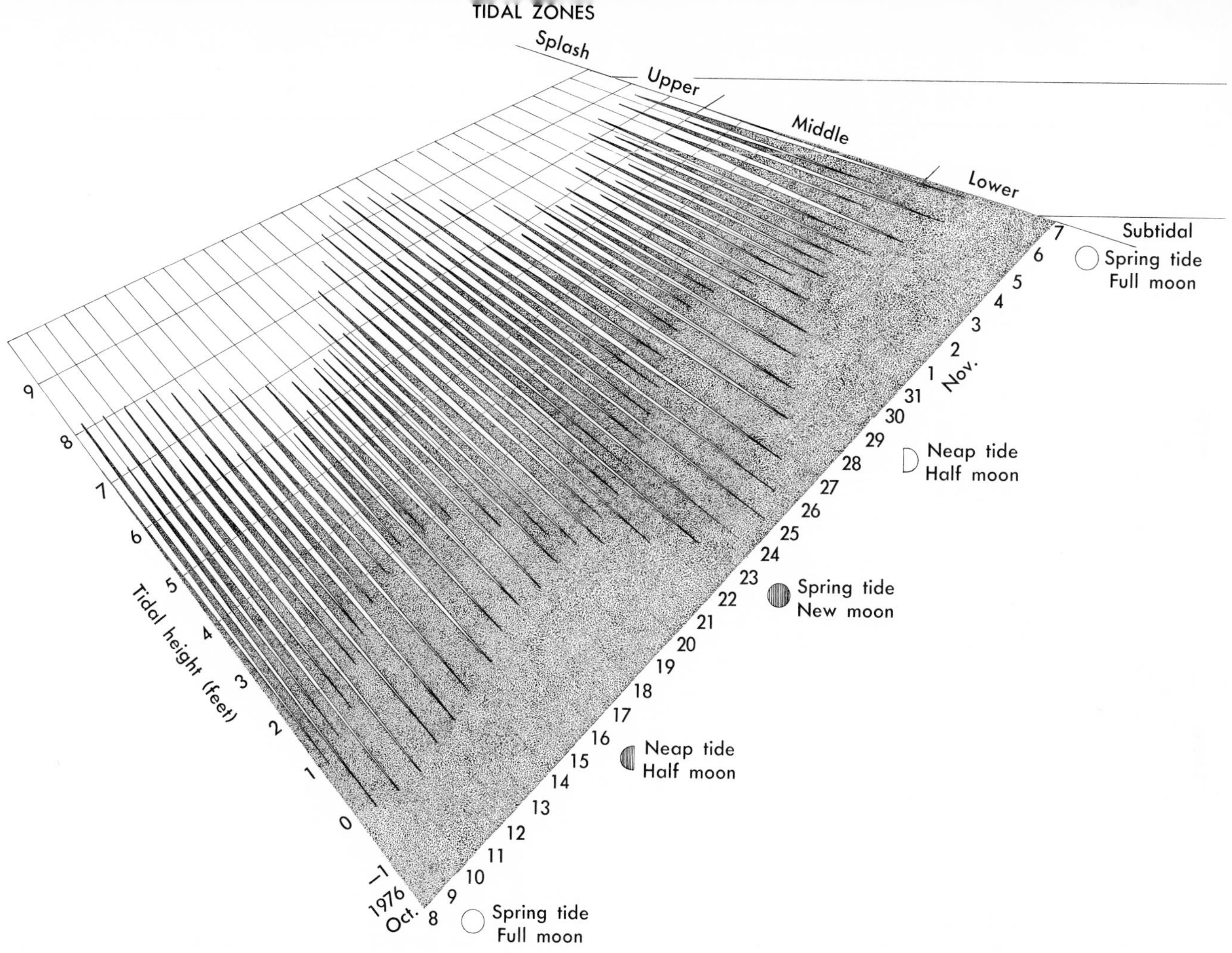

FIG. 2-2. The intertidal zone tides along the Oregon coast, October 8, 1976, to November 7, 1976; one full cycle. The shaded portion indicates levels of beach covered with water, and the correlation with phases of the moon is indicated. Note that the highest high tides are followed immediately by the lowest low tides. Similarly, the lowest high tides are followed by the highest low tides.

ference between the levels of high and low tide. Such tides are termed **neap tides.** Each lunar month has two series of spring tides alternating with two series of neap tides, with a few days of transitional, or intermediate, tides during each change. The actual time of occurrence and the level reached by a given tide at any particular place are influenced by many complicating factors, such as latitude, local and regional topography, currents, and winds.

The mean, or average, level attained by each of the four daily tides during the year in a given locality has been termed the **critical tide level.** The critical tide levels are the levels of the greatest average change in relative duration of exposure and submergence, and they have been correlated with major changes in the character of the intertidal communities.

Below the level of the lowest low tide, on open coastlines, there is a subtidal zone strongly influenced by wave action. Farther out the deeper waters over the shelf are much more quiet, showing only the circulation caused by currents, upwellings, etc.

VERTICAL ZONATION

As a result of the differences at various levels, the shelf communities tend to form as bands parallel to the coast. Intertidal communities are the most narrowly restricted and sharply distinguished, particularly where they occur on steep rock faces in areas with strong tides.

The intertidal zone is divided into three major and seven minor subzones (Fig. 4-5) by critical tide levels. The zonation is not everywhere the same. Local conditions may alter relative heights and times of tides in such a way that they eliminate one or more of the critical tide levels as a significant ecological factor. For example, the average levels of the low high spring tides and the high high neap tides may be sufficiently close together to constitute only one critical level. The same type of thing could happen at the low tide levels. In this case only five zones would be apparent in the intertidal.

Many areas, especially oceanic islands, have so slight a tidal rise and fall that local variations in weather, such as changes in wind direction and velocity, tend to obliterate the small differences in tidal levels, resulting in a narrow intertidal zone lacking well-marked subzones within it.

Since factors such as the degree of wave action, exposure to sun, wind, and rain, and extreme air temperatures also affect the vertical distributions of intertidal organisms locally, the boundaries of the intertidal biotic communities do not everywhere coincide precisely with particular critical tide levels. For an understanding of the biology of the seashore, the actual boundaries of the biotic associations are better indicators than are precise tide levels, and the zonation should be defined in terms of the distributions of animals and plants. Zonation is a result of the action of many factors, of which the tides are only one, albeit one of the most important.

Strong wave action, for example, tends to push the intertidal zones upward, allowing intertidal organisms to occur higher on the shoreline than they would in more protected places. If the coast has a gradual slope, this result of wave action may be evident only for the lower zones, the higher ones being, in effect, protected.

Direct sunlight and wind, by increasing temperature and evaporation, tend to restrict the upper levels various organisms can attain, especially in the upper intertidal zones, where exposure is longest and most frequent.

The frequency and degree of unusual weather conditions such as storms, unusually hot or cold weather, or exceptional rains also affect the zonation and character of intertidal communities. In polar regions, ice may scrape the intertidal rocks bare.

Interactions among various members of the biotic communities themselves also play an important part in determining vertical distributions. Especially important are the effects of various predators, the susceptibility of some species to being grown over and smothered by others below certain levels, the tendency of the larval forms of most sessile species to be extremely selective regarding the characteristics of the substrate on which they will settle, and particularly the strong tendency of sessile species in many cases to settle preferentially where adults or other larvae of the same species are already present.

Persons interested in more detailed discussions of intertidal zonation should consult the works of Stephenson and Stephenson (1947-1961) and the excellent book by Newell (1970).

For some distance below the level of the lowest low tide of the year, the inshore waters are strongly influenced by turbulence produced by wave action. In the deeper, calmer water beyond the zone of wave turbulence, changes in the benthic biota are less abrupt and less well marked than they are in the intertidal and shallow subtidal zones.

The subtidal benthic shelf communities tend to be much broader and less sharply defined than the intertidal communities. Their character is largely determined by the character of the substrate, the amount of illumination reaching the bottom (which in turn is a function of depth, turbidity of the overlying water, and latitude), and the temperature of the water. The zonation of the continental shelf region is summarized in Fig. 4-5.

There has been considerable difference in usage of the term **littoral.** Some writers use the term to describe the entire shelf, intertidal and subtidal; others restrict the term to the upper edge of the shelf, the intertidal area. In reading a particular work, one must note the manner in which the terms are being used. In this book the term will be used as indicated in Fig. 4-5, restricting the term littoral to the intertidal zones. The sublittoral areas are divisible into an upper marginal portion, strongly influenced by turbulence and wave action, and the deeper portions with quieter waters, grading into the archibenthal environment at its lower edge.

ROCKY AREAS

Second only to the great coral reefs in the abundance, variety, and beauty of the living communities they support are the rocky coasts, where headlands, reefs, and rock outcroppings of all kinds meet, extend into, and become submerged in the sea. Each such exposure is a unique microcosm, as individualized as a man's fingerprint. The extent, nature, and species composition of the communities occupying it depend on many complexly interacting factors.

In the rocky intertidal there is an alternating exposure and submergence by the tides, to a different extent and at a different time each day, with the lunar progression of the tidal cycles. At low tide there is direct exposure to the sunlight and rain, winds and desiccation, and predation from land animals. As the tides roll in, there may be exposure to tremendous power of the surf pushed by the incoming tide, especially during the fortnightly spring tides, in exposed situations on the open coast. There are abrupt changes in temperature, salinity, character of active predators, light, available food, etc. The degree of protection and opportunity for settlement vary with the nature and hardness of the rock, the direction and angles at which the rock strata meet the sea, the local tidal range and weather, and the direction and amplitude of currents, waves, and other water movements. Other important factors are whether sand or other loose material regularly, or ever, covers the rocks (and if so, how often and to what extent), the frequency of storms, and the force they attain locally. The geological age of the particular relationship between coast and sea—whether the land in geologically recent times has been rising and emerging from the sea or sinking beneath it, or whether there has been a long period of relatively stable relationship—is also important.

Zonation of biotic communities is more obvious in the rocky intertidal than elsewhere because the rocks provide attachment sites and protection for a luxuriant biota and because the vertical ranges are often more abrupt and compressed into much shorter horizontal distances than are those in areas of soft substrates (Fig. 2-3). The epifauna and flora of intertidal rocks

FIG. 2-3. Exposed seaward-facing rocky slope at Marblehead Neck, Massachusetts, showing zonation of intertidal communities. The lower edge of the barnacle zone shows at the extreme upper left. It is succeeded by a belt of rockweed *(Fucus)*. Below the *Fucus* is an approximately equally broad community dominated by the mussel *Mytilus edulis*. At the waterline, only partly uncovered, is the zone of Irish moss *(Chondrus crispus)*. (Helen P. Finlayson photograph.)

are more directly and more suddenly exposed to the alternating environmental stresses resulting from submersion and exposure than are the infaunal animals buried in soft substrates.

Biotic communities in rocky areas are much more complex than those in other areas because of the wide variety of ecological niches provided by the pools, channels, crevices, rock faces, etc. and their varying relationships to light, water movement, temperature changes, and other factors.

Ballantine (1961) and Lewis (1964) devised methods for assessing the degree of exposure of rocky coasts of the British Isles, using the intertidal organisms of the upper littoral and supralittoral zones—the combinations of species present, their relative abundances, and their vertical distributions. Ballantine used a scale of eight degrees of exposure ranging from extremely exposed to extremely sheltered, and Lewis divided the degrees of exposure into five categories.

Jones and Demetropoulos (1968) devised a simple dynamometer for measuring wave force and demonstrated that the zonation patterns of several intertidal organisms could be interpreted in terms of the readings of the dynamometer.

Subtidal rocks and reefs, within the euphotic zone, are likewise heavily colonized with a rich biota of algae, plankton-feeding animals, substrate scrapers, and others, including many species seldom or never found in the intertidal zones.

Pequegnat (1961) demonstrated that offshore submerged rocky reefs also show marked vertical zonation with respect to both the types and the abundance of organisms. They are most heavily colonized near the top. This is not a simple function of absolute depth, since the top of a submerged reef at the same depth as the bottom of another submerged reef closer inshore has a far richer biota than does the bottom portion of the latter. He believed that the degree of turbulence and resulting circulation of plankton and other nutrients around the various parts of a submerged reef is one of the most important factors determining the abundance and distribution of animals on such a reef.

Below the euphotic zone there are no functional producers—no macro or micro algae on the rock surfaces—and the plankton and seston in the water are also less abundant and varied. Such rock surfaces support much sparser populations of animals. Exposed rock surfaces in really deep water are for the most part either steep or vertical, the more level strata having been covered with sediment, and are the least well known of the biotic provinces of the sea due to the great difficulties in sampling or making direct observations of them.

SAND BEACHES

Most sand beaches are composed largely of quartz and feldspar, the most abundant and hardest particles left from the weathering of rocks in the mountains. In some areas corralline algae or calcareous remains from the coral reefs predominate. In others, darker minerals from volcanoes may cause beaches to be darker or appear blackish.

Sand beaches and sandy subtidal bottoms are limited to areas where strong water movements carry off the finer, lighter particles. The coarser particle size of the remaining sand permits percolation of water to greater depths and, in intertidal sands, **laminar flow** within the sand with the rise and fall of the tides. This oxygenates the surface layers to a much greater depth than is the case in quiet waters with finer sediments.

Coarse particle size also means that much less surface is available for absorption of dissolved and small-particle organic matter and bacteria. The total amount of organic matter and living biomass of open sand beaches is far less than in comparable sediments in quieter water. The smaller amount of organic matter and greater oxygenation permit aerobic decomposition of most of the organic matter, and the biome is dominated by aerobic organisms.

Since coarse sediments do not retain water well, the surface layers may drain and become dried out to a depth of many centimeters in the upper reaches of beaches exposed for long periods between high spring tides.

The coarseness of the sand and the slope of the beach are both increased by strong wave action. A typical open beach intercepts 20 to 200 metric tons of clean seawater per meter of beachfront per day. Since incoming waves have more force than receding waves, floating objects, such as logs or masses of seaweed torn from their holdfasts in rough weather, tend to be pushed to the upper edge of the beach and stranded there.

Beaches are far from static, particularly in the intertidal and wave-affected portions of the subtidal zones. Dry sands on the highest parts of a beach may be transported by wind. If the prevailing winds blow onshore, the result may be the formation of sand dunes in back of the beach.

Every wave on a sandy beach moves quantities of sand. In winter large waves of high spring tides may carry most of the sand away, depositing it as a bar a short distance offshore and lowering the level of the beach, sometimes exposing underlying rock. The smaller waves of summer tend to wash the sand back up on the beach, restoring its former level.

Since the prevailing direction of the waves is seldom precisely parallel to the beach, sand is steadily moved in the direction of the prevailing wave action as an **alongshore current** (commonly termed **longshore current**). If this alongshore current comes to the head of a submarine canyon, a great deal of sand may be cascaded down into the canyon, and the beaches on the far side of the canyon head may be largely stripped of sand.

If the source of new sand, such as a large river emptying into the ocean, is altered so that new sand no longer is available, alongshore transport may strip the region for progressively greater distances, leaving rocky or gravelly shoreline where there were formerly extensive sand beaches. Large dams and irrigation projects, although far inland, may produce such effects along the coasts on which the rivers empty.

When an alongshore current of sand comes to an embayment, the waves lose some of their force as they are refracted into the embayment. They drop part of their load of sand, so that a bar tends to build up across the mouth of the bay from the side from which the sand is coming. If this sand is not dredged and deposited on the far side of the bay so that alongshore transport can continue to carry it, it may eventually close off the embayment. If there is sufficient current into the embayment to maintain an opening, a long bar may form, narrowing the entrance and extending along the outer side of the channel.

Along the coast of the Pacific Northwest the direction of wave trains is seasonally reversed. Between headlands the beaches are largely fed by erosion from the sea cliffs. Here we find the longshore current piling up the sand at the south ends of the beaches during the winter, and at the north ends of the beaches in summer, with little net transport of sand along the coastline as a whole. Where there is an embayment in such an area, the sandbar may close one side of the mouth at one time, and the other side at another.

Sand also moves toward and away from the shoreline. Larger waves or storms that occur during the winter tend to cut away the beach and deposit the sand as a rise, or offshore sandbar. The smaller summer waves again push it shoreward. Some beaches may be stripped of sand in winter but revert to fine, sandy beaches in summer.

Even the difference between spring and neap tides makes a difference in areas where the tidal amplitude is marked. Spring tides tend to cut sand from the beach at about mean sea level and deposit it offshore, often heaping it into a sandbar rising well above mean sea level. The neap tides in summer and fall return sand to the beach. Where there is an annual change in mean sea level, the movements of the sand tend to be roughly correlated with the changes in mean sea level.

Strong onshore winds in areas with low land back of the shoreline may return much sand from the dry upper part of the beach to the land, forming extensive sand dunes.

On the shelf and continental slope beyond the zone of wave turbulence, the transporting forces are weaker and slower, but the sediments to be moved are commonly finer. Currents and internal waves also bring about the transport of deeper-water sediments.

Since smaller, finer particles are more readily transported, there is a continual sorting by size and density during transport, and the particles deposited at any one place are often uniform.

BAYS AND ESTUARIES

Embayments are of many types, depending on the recent geological history of the region; the size, water volume, and sediment load carried by rivers; the seasonal fluctuation in amount of water carried; the amplitude of the tidal fluctuations; the frequency, strength, and direction of storms; the amount of precipitation in the region; the direction or trend of rock strata relative to the shoreline; and other factors. If the sea level in recent geological times has risen relative to the land, many embayments may be in the form of large drowned river valleys, with numerous protected secondary embayments, and perhaps miles of relatively stable estuarine conditions, with a salinity gradient ranging from nearly that of open seawater at the mouths to that of freshwater at the extreme upper ends. Chesapeake Bay is a good example of this type of embayment. If, on the other hand, recent geological history has been one of a rising of the land relative to the sea level, rivers will tend to meet the sea more abruptly, with narrower and shorter estuaries in which changes in salinity and volume of water with tidal cycles are more marked.

In any case, when the relations between land and sea level are relatively constant for a long-enough period of time, rivers tend to build estuaries and deltas at their mouths. As tides push into river mouths, the water is slowed down, and the water level raised. Much of the silt carried may be deposited, tending to make the mouth of the river shallower and broader, which accentuates its tendency to deposit its load before reaching the ocean proper. The salinity may also be a factor causing agglomeration and settlement of particles. A sandbar is commonly built near the mouth, both by sand brought down by the river and by sand brought from the beaches of one side by alongshore transport. If the flow of water is sufficient throughout the year, channels to the sea will be opened and filled with deposits, and new channels will be made, the process resulting in a fanning out at the river mouth, with the formation of the deltas extending into the sea and the deposit of finer muds in the deeper water beyond the deltas. With

FIG. 2-4. South Slough—an arm of the Coos Bay estuary, Oregon. Looking north from near the upper end of the slough. Oregonians can be proud of the fact that through the efforts of Dr. Paul Rudy, of the Oregon Institute of Marine Biology, and others, South Slough has been designated as the first National Estuarine Research Sanctuary.

the slowing of flow near the mouth, more sediments are deposited back of the sandbar, and eventually mud flats or areas of mixed sand and mud are formed, through which the river passes in winding, ever-changing channels and over which the water rises during the high tides and may recede at low tide. Where alongshore transport from the beach is considerable, the sandbar may be built continuously, deflecting the mouth of the river to one side and forming a long estuary parallel to the coastline and separated from the ocean by a sandspit. Examples are the great southward-trending spits extending from the north sides of Delaware Bay, and Albemarle Sound. Perhaps the best examples of the development of such spits separating long estuarine areas from the main body of the oceanic water are those in the western end of the Gulf of Mexico, forming the outer side of Laguna Madre.

Where the water flow is not sufficient to maintain a channel at all times, such an estuarine area may be entirely cut off from the sea, either seasonally or semipermanently. Such an isolated estuary will become a freshwater lake eventually if the input of freshwater exceeds evaporation. If evaporation exceeds input, it may become even saltier than ocean water. If the inflow has significant seasonal variations, new channels may be cut during the wet season, restoring the connection with the sea, and the estuary may be closed off by sandbars at the mouth during the dry season.

Types of estuaries

Pritchard (1955) and Wood (1964) have classified estuaries into a few principal types, depending on the pattern of water circulation and flow:

1. The first category of estuaries has been termed **river-dominated, stratified,** or **two-layered estuaries.** This type is characterized by a high volume of freshwater runoff and relatively deep narrow contours. The freshwater tends to flow out at the surface, whereas the deep water comprises a wedge of saltwater with a net flow upstream, which may push upstream a considerable distance at high tide. The Mississippi estuary is an example of a stratified, or river-dominated, estuary. The flow of freshwater is immense, and the tidal amplitude slight, only about 6 inches. The deep saltwater wedge extends about 150 miles upstream. In much of this stretch it is possible to fish for demersal saltwater fish while boating in water fresh enough to drink. The margins and surface waters of such estuaries display essentially freshwater biota.

2. A second category has been termed **marine-dominated** or **vertically homogeneous estuaries.** In these estuaries the input of freshwater is relatively small and the tidal rise and fall considerable. These estuaries may be broad and shallow, with relatively small channels. The incoming marine water during rising tides pushes in over the freshwater and then, being more dense, mixes vertically, producing a homogeneous water column in which there is little salinity difference with depth at any given place, due to the **tidal overmixing.** At ebb tide much of this type of estuary may be drained, exposing extensive mud or sand flats. The flats in the upper portions tend to be of fine mud, and those in the lower parts, or where currents are stronger, more sandy. Coos Bay, Oregon, is a typical marine-dominated non-stratified estuary. In such estuaries the margins may get water of a more marine character than do the central channel areas because incoming marine water flows out over the mud flats, whereas the channels tend to be somewhat more dilute. These estuaries will have marine organisms such as barnacles, lugworms, and soft-shelled clams extending into their upper reaches, which are commonly bordered by salt marshes. Drowned valleys such as Coos Bay are commonly of this type.

Many estuaries are more evenly balanced or of mixed character. Some change in character with seasonal changes in the relative amount of freshwater runoff. Small estuaries are more subject to marked seasonal changes of this kind than are large ones. Estuaries subject to such seasonal changes of character tend to support a less richly diversified biota than do either the river-dominated or the marine-dominated estuaries because they do not remain suitable throughout the year for either type of biota.

3. A third category has been termed **evaporite estuaries.** These are more or less self-contained estuaries formed where rivers flow through low country into shallow coastal waters in areas of low rainfall. They form chains of shallow coastal lagoons with low offshore barrier islands. Examples are the Laguna Madre and others along the western and northern coasts of the Gulf of Mexico, the Gippsland Lakes of Victoria, and Lake Alexandrina of South Australia. During periods of low runoff such basins may become hypersaline.

Biota of estuaries

Estuarine waters mostly have positive E_h values (p. 41) ranging from +150 to +600 mV., although some basins with accumulated organic matter may develop negative E_h in the deeper water as well as in the sediment. Because of the higher E_h value and exposure of all depths to light during the day, photosynthesis is the main reaction, and the sulfur and iron cycles are unimportant, although some of the sulfur and iron bacteria may be present in and isolatable from the water. Photosynthesis causes a rise in the pH and in the carbonate-bicarbonate ratio. If the pH reaches 9.3, $CaCO_3$ precipitates out, reducing the bicarbonate available to plant cells and reducing photosynthesis so that the pH does not get beyond 9.4. The greatest pH fluctuations occur on shallow banks and channels well bedded with sea grasses, their epiphytes, and diatoms. Temperature fluctuations in such shallow areas may also be considerable.

The sediments largely control the ecosystems of estuaries because of the intense microbial activity in them (Chapter 15). In the sediments, changes in E_h become more significant than those in pH. Where the bottom deposits contain much organic matter and are not disturbed by vigorous water movements, the surface of the sediments has consistently low E_h values—about −130 to +160 mV.—and there is little or no oxidized surface layer.

Marine monocots such as *Zostera, Thalassia,* and *Posidonia* usually support a large biomass of epontic diatoms, algae, and small animals. *Ruppia* tends to have fewer epiphytic growths, probably because it occurs along the upper margins of the flats, where it is exposed longer between high tides.

The phytoplankton of estuaries comprises three principal components, the relative abundance and distributions of which depend on the nature of the estuary, the season, and other factors:

1. The **oceanic component**—stenohaline species that proliferate well only in oceanic waters but may be carried into the lower reaches of estuaries when oceanic water masses are swept inshore. *Climacodium frauenfeldianum* and *Ceratium candelabrium* are examples.

2. The **neritic component**—species that prefer marine-dominated inshore waters and proliferate well in the lower reaches of marine-dominated estuaries, as well as over adjacent continental shelf areas. The diatoms *Nitzschia serratta, Schroederella delicatula,* and *Asterionella japonica* are examples.

3. The **estuarine component**—euryhaline, eurythermal species preferring waters of rather high nutrient content and able to tolerate considerable fluctuations in salinity and temperature. Such species are nearly always found abundantly in the estuaries and much less abundantly outside. They often bloom in the upper, nonmarine-dominated parts of the estuary. Examples are *Coscinodiscus granii, Rhizosolenia robusta, Peridinium ovatum, P. subinerme,* and others.

Organisms of the nanoplankton and ultraplankton size ranges are far more abundant than the net protoplankton—by at least an order of magnitude. After the flowering and decay of the sea grasses, vast numbers of minute flagellates, other colorless protozoa, and bacteria are present in the water.

Seasonal changes

In temperate and boreal regions a seasonal sequence of biological activity is superimposed in the water and the surface layers of sediments. The normal sequence in the temperate zone is for a rapid growth of algae such as *Enteromorpha* and *Ectocarpus* to occur early in the spring. As this bloom begins to die off, sulfate reduction increases rapidly, and the surface layers of the sediments become deoxygenated sufficiently to allow purple sulfur bacteria to grow on top of the mud or the rotting algae. Where the algal growths are tapetic, forming mats or felts of algae on the mud, the E_h values at the lower surface of such mats or felts are negative, and the anaerobic photosynthetic sulfur bacteria grow there, utilizing complementary wavelengths of light that filter through the algal mat above. The green sulfur bacteria require lower E_h levels and more reduced sulfur compounds than do the purple ones.

The bloom of sulfur bacteria is commonly followed by a big bloom of blue-green algae such as *Lyngbya, Rivularia, Microcoleus, Phormidium,* and *Synechococcus*. Slightly later, pennate diatoms such as *Navicula, Pinnularia, Nitzschia, Hantzschia,* and others become abundant.

The succession is largely controlled by the hydrogen sulfide concentrations near the sediment surface. In the late stages, marked by heavy diatom populations, flagellates, ciliates, and aerobic bacteria also abound, and many small fish forage over the flats when the tide is in.

In places with considerable organic matter but with positive redox potential at the sediment surface due to the combined effects of oxygen in the water and photosynthesis of surface organisms, iron bacteria may grow profusely, forming surface films.

Because of the extensive surface areas of fine sediment for adsorption, slowing down of both marine and fresh waters as they meet and mix in the estuary, cyclic patterns of water flow, and intense microbial activity in the surface layers of the sediments, estuaries tend to act as nutrient traps, as well as to generate more organic nutrients on their own through photosynthesis and chemoautotrophy than do most areas of comparable size. For these reasons estuaries tend to be areas of unusually rich biological production. This, together with their freedom from some of the predators of truly marine or truly fresh water, enables them to support massive populations of some fish and invertebrates and to serve as spawning sites and nurseries for the young of many others.

MAN AND COASTAL ENVIRONMENTS

In recent times the activities of man have become a powerful force for change on a global scale. No place on earth is so remote or protected that it is free from these effects. The intertidal and shallow waters at the margins of the seas, being the most directly exposed and perhaps the most vulnerable, bear the brunt of this assault on nature. In all settled regions they have been and are being drastically changed. This is especially true of bay and estuarine environments.

It must be borne in mind in this connection that the intertidal and subtidal communities of animals and plants do not represent just the edge of a vast similar community extending far out to sea and almost infinitely renewable from the sea. Rather, they are thin communities, often measurable in a few inches or feet, extending parallel to the coastline as a series of narrow communities. Therefore they are particularly vulnerable to destruction from depredation and other changes occurring along the coastline.

The impact of man on these and other coastal environments will be considered in greater detail in Chapters 21 and 22.

FIG. 3-1. Sand dollar, *Echinarachnius parma,* off New England coast. (Chesher photograph.)

CHAPTER

3 THE FLOOR OF THE SEA

The ocean bottoms are, for the most part, comprised of layers of sediments that have been slowly forming for millions of years. The character of life in the overlying waters, the depth, the distance from land, and the currents all influence the nature of the sediments in any area. Over the ages, changes in climate, temperature, currents, volcanic activity, and isostasy, together with evolutionary changes in the fauna and flora of the seas, are reflected in the sediments and in the fossils preserved in them.

Marine sediments from coastal regions and former shallow seas now raised above sea level are directly available for study and have provided geologists and paleontologists with most of the information known about the history of the earth. The study of oceanic sediment cores, together with recent sonic, electronic, and magnetic methods for obtaining information about the floor of the sea, has added a new dimension to these studies. Submarine geology has become an exciting field in its own right.

TERRIGENOUS SEDIMENTS

Near land, the sediments and the rates at which they accumulate are more variable than in truly oceanic regions. These sediments, largely terrigenous in origin, consist of clay, sand, gravel, volcanic ash, and other materials washed into the adjacent sea from the land and often subsequently transported, sometimes great distances, by currents, internal waves, ordinary wind, and wave action along the shoreline. They are mixed to varying degrees with materials of oceanic origin. The patterns of sedimentation are strongly influenced by currents and other water movements, which also exert a marked sorting effect. Heavier objects and particles are deposited closer to their point of origin and transported, in general, shorter distances more slowly, whereas lighter, smaller particles may be carried great distances and in various directions, depending on the currents and water densities at various depths and the size, shape, and density of the particles. In some areas, larger, heavy terrigenous objects such as boulders or gravel may be incorporated into the ice from glaciers, which feed icebergs into the water, and be transported with the icebergs until melting releases them.

OCEANIC SEDIMENTS

Deep-water sediments are derived from many sources—weathered continental materials transported by rivers, glaciers, and winds; shells and skeletons of organisms sinking from the surface waters; volcanic materials from both terrestrial and submarine volcanoes; authigenic mineral deposits, or precipitates, formed from interacting compounds dissolved in the seawater; and even extraterrestrial particles derived from the continual rain of meteorites over the eons. Near land, especially off coasts with large rivers or glaciers, the sediments are primarily terrigenous in character.

It has long been thought that the deep-water sediments were almost entirely formed from materials drifting down from above. However, some of the materials, such as sponge or echinoderm spicules, calcareous or arenaceous foraminiferan shells, shell fragments, and radiolarian and even diatom frustules, may be largely autochthonous, the materials having been concentrated from dissolved substances in the water by organisms living in the sediment.

Even normally photosynthetic forms such as diatoms may live heterotrophically in deep sediments. Wood (1965) lists ten genera of diatoms taken at depths of 7000 to 10,000 meters and states that experimentally diatoms have been grown in the dark under pressures of 500 atmospheres, in the presence of glucose, and that they grow, reproduce, and continue to produce chlorophyll under these conditions. Microbial processes, with the exception of photosynthesis, can take place even at the greatest depths.

Truly oceanic sediments are of three principal categories—**calcareous oozes,** in which the shells of foraminiferans or pteropods and coccoliths commonly predominate; **siliceous oozes,** comprised largely of diatom frustules or radiolarian skeletons; and **red clay,** in which the remains of organisms are less conspicuous.

Calcareous oozes, often termed globigerina ooze in areas where the shells of the pelagic foraminiferan *Globigerina* are abundant in it, characterize especially the warmer and shallower oceanic regions. They cover approximately half of the oceanic part of the sea floor. Below 5000 meters they are largely replaced by red clay or by siliceous oozes because of the solubility of calcium carbonate at great depths.

Red clay is characteristic of the deepest ocean basins and carpets approximately 35% of the ocean floors, whereas siliceous sediments occupy about 14%. Siliceous sediments, largely of diatom shells, occur particularly well developed in the Antarctic Ocean and also in the far north just south of the Aleutian Islands. Siliceous sediments, largely radiolarian in character, occur in a relatively narrow band across the central Pacific north of the equator, from Central America to west of Hawaii, an area of the equatorial western Pacific, and an area southwest of Indonesia and northwest of Australia.

Andree (1920) classified the sediments into **littoral** (the sediments on the shelf), **hemipelagic,** and **eupelagic.** The eupelagic sediments are those deposited chiefly by pelagic organisms, such as globigerina ooze, together with an admixture of skeletal elements from benthic forms such as shell fragments, sponge spicules, and echinoderm spines. Since the term littoral is now usually restricted to the intertidal region rather than the entire continental shelf, perhaps the term **terrigenous** should be used to cover both Andree's littoral and hemipelagic sediments, which have a higher content of mineral matter of terrestrial origin and of organic matter than do the eupelagic sediments.

The accumulation of sediments on the sea floor, particularly the eupelagic sediments, is a slow process judged by our time standards, averaging perhaps about

0.015 mm. of new sediment per year. Kuenen (1941, 1950), using several independent methods, estimated that they should accumulate to an average depth of 3 km. every 2 billion years. The earth is known to be more than 4 billion years old, which should give an average depth of eupelagic sediments of at least 5 or 6 km.

In actual fact, however, there are only about 0.3 km. of sediments on the floor of the deep seas. This presents something of a paradox that has been resolved only by startling discoveries in marine geology during the last two decades (Chapter 20).

The sedimentation rate is not uniform over the various regions of the ocean, and it is not the same over a given region throughout the ages. It is fastest in shallow seas and shelf areas and slowest in abyssal depths beneath the open ocean.

Many animals live at or close to the interface between the bottom sediments and the cold, slowly moving water above. Some burrow or dig themselves into the oozes; others are attached by rootlike or rhizoidal processes; many crawl or creep on or through the ooze, feeding on the organic humus or on other animals they find there; and many demersal fishes and some cephalopods move over the sea floor, feeding on benthic animals. The character of the animal population is largely determined by that of the sea floor. Where there are rocky exposures, cliffs, or other hard substrate, the benthic fauna is much different from that where the bottom is carpeted with soft marine oozes. By far the greater part of the ocean floor is covered with such ooze, although in many areas it may be somewhat compacted, or lithified, rather than soft and fluffy.

The ocean floor is populated with animals at all depths. However, the quantity of animal life decreases markedly with great depth. Some groups disappear altogether, whereas others, not found in shallow waters, form a characteristic part of the deep-sea assemblage. For example, Zenkevitch (1963), reporting the results of a Russian deep-sea expedition in the northwest Pacific, indicates that between 1000 and 5000 meters there is a sharp decline in the relative importance of sponges. Between 4000 and 6000 meters, starfish, sea anemones, and polychaete worms became much less prominent. Decapod crustaceans were not found deeper than 5000 meters, but holothurians became relatively much more important. Below 8000 meters there were no more starfish, and gorgonians, pennatulids, amphipods, and isopods became rare as did also anemones and molluscs at 9000 meters. Below this depth only a few polychaete and echiuroid worms, pogonophores, and holothurians were taken, the latter constituting 90% of the catch by weight.

At 1000 meters a single haul contained 120 species. At depths between 8000 and 9000 meters three hauls yielded 6, 7, and 17 species, respectively. The weight of living organisms taken per square meter of deep-sea floor was 2 to 5 gm. at depths of 4000 to 5000 meters but decreased to 20 to 30 mg. below 9000 meters.

An important reason for the great decline in biomass with depth is the paucity of available food and the relative uniformity of its character. Organic materials produced in the euphotic zone are largely cycled in the upper strata of water—eaten by the animals or decomposed by microorganisms long before reaching the bottom. The organic material that finally drifts to the bottom is both small in amount and consists largely of the less digestible residues—the so-called **marine humus,** together with skeletal elements such as foraminiferan and radiolarian shells and diatom frustules.

Eupelagic sediments are markedly poor in organic material. Terrigenous sediments on or near the continental shelves are characterized by a higher content of terrigenous minerals and a much higher content of organic matter, derived in part from the terrestrial fauna and flora of the shelf, and in part from the rich life in the neritic waters above the shelf. A large proportion of this material reaches the bottom here, before decomposition is completed, and supports a large population of mud eaters and seston feeders. Much of this sediment is swept out past the edges of the continental shelves so that archibenthic and even some abyssal sediments near the shore are commonly satisfactory from a nutritional standpoint and can support a rather large population of benthic animals. These partly terrigenous sediments swept out beyond the continental shelves constitute the **hemipelagic sediments** of Andree. It is the nutritional content, rather than the mineral composition or the origin of various sediments, that is of greatest biological significance. The line between eupelagic and hemipelagic sediments is, of course, vague. It usually lies some distance beyond the rim of the continental shelves. Typical samples of each, however, differ sharply in organic content and nutritional value.

FORMATION AND CHARACTER OF SOFT SUBSTRATES ALONG SHORES

Materials eroded from the continental land masses are continually being transported by water, sorted, worn down, partially dissolved, and eventually brought to the sea. Here, some of it forms mudflats, sandbanks, beaches, deltas, and other formations at the edge of the sea. Some of the finer particles may be carried far out to sea, becoming part of the fine, particulate matter that eventually settles to the floor of the abyss. Also, tidal action, waves, and activities of countless boring and scraping organisms, plus the ordinary erosive factors of wind, rain, frost, etc., continually wear down the rocky areas of the intertidal zone. The world's rivers transport approximately 8 billion tons of sediment to the sea each year. These sediments are deposited most rapidly at the mouths of rivers because of the slowing down of the river as it meets the sea and because of the precipitating action of the sea salts.

In addition to the visible deltas built up at their mouths, large rivers may deposit layers of underwater mud in the form of large fans extending far out to sea. In Fig. 19-1 is shown, for example, such sea-floor fans of alluvial deposits brought to the northern edge of the Indian Ocean on either side of India by the Ganges and Brahmaputra river systems on the east and the Indus and Narbada rivers on the west. These fans extend out into the Indian Ocean the full length of the Indian subcontinent on the west and for twice this distance on the east.

Long-term changes in sea level or isostatic changes in the level of bordering lands also cause extensive marginal regions either to rise, during which time deposits of terrestrial origin may come to overlie older marine deposits in a seaward direction (a phenomenon termed **offlap**), or to sink relative to the sea level, in which case new marine deposits may come to overlie earlier terrestrial ones in the landward direction (a phenomenon termed **onlap**).

At present North America is largely emergent, but ancient marine terraces built up during the last 100 million years, since the Lower Cretaceous, show the extent of former shallow seas on present continental areas. The terraces of the Atlantic coast reach inland to the "fall line" in the eastern Appalachians. Those from the Gulf of Mexico are far more extensive, stretching all the way up the Mississippi Valley to Cairo, Illinois, which was the high-water mark of the shallow Cretaceous seas occupying much of the central portion of the present North American continent. The present continental shelves represent submerged terraces, emergent during periods of lowest sea level in the past. These sediment platforms formed as stratified wedges, with the thicker edge on the seaward side. At the present shoreline of Louisiana and Texas, the sedimentary platform is about 4000 feet thick. The bottom of the present Gulf of Mexico is largely muddy because the great muddy Mississippi and Rio Grande rivers empty into the Gulf. The rivers of the Atlantic and Pacific seaboards are much clearer and freer of silt. Much of the silt they have is trapped in large estuaries (Chesapeake Bay) before getting to the ocean. Swifter bottom currents along the open sea margins and stronger tidal action keep the finest particles fairly well swept out to sea, so that the near-shore deposits tend to be sandy or gravelly, and the shelf is relatively clean. Since fine particles are swept farther out to sea than are coarser, heavier ones, offshore deposits are usually finer than those close to shore. During onlap, if the invasion of land areas by sea is sufficiently great, an overlapping of shale on sandstone may occur. During offlap coarser, new deposits may overlap the finer, older mud laid earlier in deeper water.

The marine terrace along the Atlantic coast is only about one fourth as thick as that along the Gulf coast and is cut across by numerous steep submarine canyons, some of which may represent former river canyons. Others were probably formed by turbidity currents, or slides.

PARTICLE SIZE OF DEPOSITS

The particle size of a deposit exercises marked biological effects through differences in water retention and movement, organic content, compactness, and the ease with which burrows can be excavated and maintained. Table 3-1 gives a résumé of particle sizes, together with names applied to sediments having particles in certain size ranges and the Phi scale, based on a logarithmic relationship, which has been found convenient in quantitative work with sediments.

TABLE 3-1. Particle sizes of sediment

Size (mm.)	Name*	Phi scale†
256 and larger	Boulder	−8
64-256	Cobble	−6, −7
32-64	Pebble	−5
16-32	Coarse gravel	−4
8-16	Medium gravel	−3
4-8	Fine gravel	−2
2-4	Granule	−1
1-2	Very coarse sand	0
0.5-1	Coarse sand	1
0.25-0.5	Medium sand	2
0.125-0.25	Fine sand	3
0.0625-0.125	Very fine sand	4
0.0039-0.0625	Silt	5, 6, 7, 8
0.0039 or less	Clay	9 or more

*In part from Wentworth, C. K. 1926. Methods of mechanical analysis of sediments. Univ. Iowa Stud. Natur. Hist. **11**(11):3-52.

†$-\log_2$ of diameter in millimeters. (Krumbein, W. C. 1936. Application of logarithmic moments to size-frequency distributions of sediments. J. Sed. Petrology **6**:33-47.)

Mechanical analysis of sands and sediments has commonly been done by sieving them through a series of standard mesh screens and calculating the percentage of each size class of particle as a percentage of the whole. Emery (1938) developed and standardized a rapid method especially for shipboard use, in which a tall column of water is used to measure settling velocities. The coarser particles settle faster. For most sands complete analysis takes only about 5 minutes, and data can be taken for a continous frequency curve, giving a more complete picture of the sorting by water. The results more closely relate to the actual conditions of deposition, since the rate of deposition depends on settling velocities rather than mere grain size. This method is not suitable for sediments with particle sizes less than 0.5 mm. In mixed sediments the sand fraction can be obtained by decanting the finer particles after mixing the sediment with water and allowing it to briefly settle several times.

In any sediment the smaller the average particle size, the greater is the exposed surface, the capillarity and water-holding capacity, and, given the same organic content, the bacterial flora.

Fine sand may give a **capillary lift** ten times as high as does coarse sand. This is of major importance to organisms in the intertidal zone, where retention of water during the low tide period may be vital. The **porosity,** or proportion of space between grains, would be 25.95% regardless of grain size if the grains were all perfect spheres. Actually, small grain size increases porosity, so that there is not only more surface to exert holding power for water but also more space for water in fine-grain sediment than in coarse-grain sediment.

Where shallow marine and estuarine sediments and beach sands are not stabilized by the growth of plants, they are continually being shifted about by the movements of water, covered during high tides and exposed during low tides, or, if they are below the low tide level, swept back and forth by tidal movements, currents, and internal waves. They do not offer the firm anchoring sites of the rocky areas. Survival here depends largely on the ability to burrow into the wet substrate and the possession of adaptations enabling the animal to breathe, feed, and reproduce while lying below the surface, where it is out of the reach of the surf and largely protected from desiccation and predators. A mud flat or sand beach may appear barren and lifeless to the casual observer, even though it is teaming with countless living things ranging from microscopic bacteria and protozoa to large worms, ghost shrimps, and clams. The finer sediments, deposited where the water is quieter and often containing considerable organic matter, have a slower circulation of water. They are soon rendered anaerobic below the surface by the action of bacteria. Such sediments are blackened because of the accumulation of hydrogen sulfide and iron sulfhydryl compounds $\left(Fe\begin{smallmatrix}SH\\SH\end{smallmatrix}\right)$. Animals living in such sediments must have some kind of direct communication with the surface for respiration, and in the intertidal zone they must be able to survive until the next tide brings oxygenated water over the flats.

In sediments with mean particle size of medium sand or less, the percentage of organic nitrogen and carbon increases logarithmically with decrease in particle size. The organic nitrogen has been found to correspond directly to the quantity of adsorbed microorganisms, representing the protein in the heterotrophic populations present. The carbon represents the living biomass plus nonliving organic debris such as cellulose, chitin, and lignin, which may be present in varying amounts. In a general way the number of deposit feeders present shows a good correlation with the amount of organic nitrogen present, demonstrating the importance of the microorganisms, chiefly bacteria, to these animals.

Extremely thixotrophic semifluid muds with suitable organic content have fewer large deposit feeders because of the difficulty of maintaining contact with the oxygenated surface layer by tubes or burrows. Extremely small, light animals such as the amphipod *Corophium*, the small gastropod *Hydrobia*, or extremely small bivalves such as some species of *Macoma* may still be there.

Fine deposits in areas of low salinity tend to have both lower organic nitrogen content (microbial population) and lower populations of deposit feeders.

THE ANAEROBIC BIOME AND RPD LAYER

All sediments in aquatic systems, with the exception of surf-swept sand beaches, tend to develop anaerobic reducing conditions in their deeper layers. This is especially true of finer sediments with high organic content, where such layers may reach the surface or even extend above the sediment surface into quiet waters.

In general, the anaerobic marine sediment biome exists in all bottom sediments and in gentle intertidal slopes where the sand or mud is of small-enough mean grain size to give it high water-holding capacity. Such beaches do not drain well when the tide recedes, and thus a constant anaerobiosis prevails. The fine grain prevents oxygenation by laminar flow with the incoming tide except in the upper few millimeters or centimeters. In general, mixed sands or muds show lower porosity than do well-sorted sediments with particles of uniform size. Addition of a small amount of silt to well-sorted sand markedly reduces its porosity.

Small grain size also greatly increases the surface area available for adsorption of molecular and colloidal organic compounds from the water, making them more readily available to bacteria and promoting a higher bacterial content.

Where oxygenated surface layers are present, they almost invariably rest on an anaerobic foundation. E_h **values (oxidation-reduction potential,** measured in millivolts) are positive in the oxygenated layers, decreasing with depth, and negative in the anaerobic layers. The transitional zone is usually relatively thin and has been termed the **RPD (redox potential–discontinuity)** layer. The location of the RPD layer occurs where the oxygen input becomes insufficient to keep pace with the oxygen demand created by decomposition of organic matter. Typically, oxygenated layers in fine sediments tend to be yellowish, due to the presence of ferric compounds, the RPD layer gray, and the anaerobic layers black. Most of the H_2S produced by decomposition of organic matter is bound to ferrous ions in the unstable ferrous sulfhydryl compound $Fe(SH)_2$ that gives the sediment its black color. The bleached color at the surface is due to partial oxidation to $Fe\genfrac{}{}{0pt}{}{\diagup SH}{\diagdown OH}$ (hydrotroilite) and further to FeS_2 (pyrite). In Fig. 3-2, modified from Fenchel (1969), there is a schematic summary of some of the changes in a vertical section of a typical sediment.

Sediments should not be thought of as static, especially those in shallow waters where seasonal changes may cause rising or lowering of RPD layers due to changes in the rates of deposition and of decomposition of organic matter. Storms and other strong water movements causing mixing and oxygenation of surface layers drive the RPD layer down. In shallow waters where the RPD layer is only a few millimeters from the surface, photosynthesis by organisms in the oxygenated layer may depress it slightly during the day.

The complex dynamics of the sediment layers result from the one-way system of oxygen and organic matter input (and, in shallow waters, of light). Light penetration is limited to usually less than 3 cm., even in well-lighted sediments, and is insignificant or absent in most sediments. Fenchel (1969) demonstrated that oxygen production by photosynthesis in any case does not serve to replace the oxygen requirements of the organisms present. Oxygenated surface layers are maintained by diffusion of oxygen from the water above, with the input from photosynthesis in well-lighted sediments of secondary importance.

Aerobic heterotrophic organisms in the oxygenated surface layers quickly deplete the oxygen levels, and since organic input exceeds the available oxygen for complete aerobic breakdown, most of the energy-yielding reactions are anaerobic. These involve many steps of degradation by series of organisms, each adapted for utilizing end products of previous organisms. The result is a general pattern of vertical distribution of organisms; each modifies its environment, providing conditions and materials acceptable to organisms of another trophic level.

Since sediments overlie consolidated, or hard, strata, the continual downward escape of accumulating metabolic end products from the anaerobic zone is prevented. Concentration gradients are built up that result in a net upward transport of the end products of anaerobic metabolism toward the surface. On reaching the RPD level, these products become important in stimulating a secondary production by chemoautotrophs, which are always abundant in and near the RPD layer.

Where the RPD layer is close to the surface, in shallow water sediments, both light and inorganic reduced metabolic products are available. Good growth occurs among photoautotrophic organisms such as the blue-green algae *Oscillatoria* and *Pinnularia,* sulfur bacteria, and others that can utilize the light as an energy source and take advantage of the reduced inorganic compounds as oxidizable hydrogen donors to carry out photoreduction of carbon dioxide.

Sorokin (1964, 1969) called attention not only to the importance of anaerobic decomposition in the sediments but also to the cycling and subsequent biological utilization of the reduced end products. The production resulting from their use is comparable to primary production in the sense that the energy source cannot be utilized directly by higher organisms.

The RPD layer is a site for convergence of materials and conditions from both the aerobic zone above and the anaerobic zone below. It is the site of maximum energy availability to sediment organisms, which is reflected in the facts that the highest rates of fermentation occur in this layer and the maximum biomass is developed there. Aphotic chemoautotrophs occur abundantly in the RPD layer all over the world, and, under photic conditions, are supplemented by photoautotrophs.

Wieser and Kanwisher (1961), Fenchel (1969), Riedl (1969), and others have noted that the distribution of ciliates, gnathostomulids, nematodes, and other microbenthic and meiobenthic organisms clearly shows not only characteristic vertical distributions of species in regard to E_h levels, oxygen and hydrogen sulfide concentrations, etc., with maxima in the RPD zone, but also that there are clear regional differences in these faunas. The ecosystem in the sediments is clearly far more complex than has previously been generally realized.

Many organisms capable of photosynthesis, or of aerobic metabolism, are also able to live heterotrophically, or under anaerobic conditions, for greater or lesser lengths of time and are found in the RPD layer or even beneath it, as well as in the oxygenated layers. This is especially true of blue-green algae. In the RPD layer Eubac-

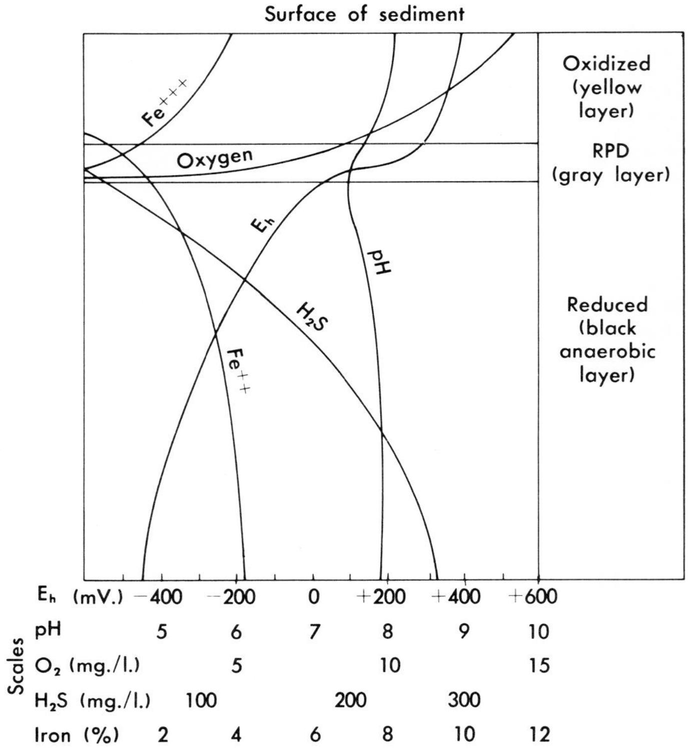

FIG. 3-2. Sediment profile. Other factors: nitrogen—nitrates present in the yellow layer, decreasing with depth, mostly reduced to nitrites by the time the RPD layer is reached, and to ammonia in and below the RPD layer; carbon—CO_2 maximum near bottom of RPD layer, decreasing both above and below this level. (Modified from Fenchel, T. 1969. Ophelia **6:**1-182.)

teriales and Spirochaetales are important links between the aerobic photosynthesizers and the anaerobic heterotrophs. Varying tolerance to oxygen and to hydrogen sulfide partially explains the correlation between vertical distributions and redox profile. Many ciliates and most gnathostomulids can live for long periods under completely anaerobic conditions and can endure or even require high concentrations of hydrogen sulfide, toxic to most aerobic forms.

The anaerobic and RPD layers may well constitute the most extensive biome on earth. It is found underlying the waters in all but coarse sediments overlaid by turbulent water. It is typical of the ocean bottoms, lakes, and estuaries, and many smaller bodies of water.

Organic input and oxygen are introduced at the top. Organic compounds are first acted on by aerobic heterotrophs, diminishing the oxygen and creating conditions favoring fermentation. As the products of fermentation reach deeper anaerobic layers, they are reduced by anaerobic bacteria to inorganic electron acceptors. Concentration gradients of these reduced compounds in the deep layers result in a net upward transport. On reaching the RPD zone, they are available for chemoautotrophs and serve as energy sources for reduction of carbon dioxide to particulate organic matter, which contributes to the substantial biomass of the RPD layer.

The basic vertical stratification of sediments in terms of E_h, pH, oxygen, and hydrogen sulfide is brought about and maintained by biological activity in a confining environment. If the RPD layer is at or near the surface in shallow waters, maximum production occurs there as the productivity of chemoautotrophs and photoautotrophs is merged.

Bacteria, diatoms, and flagellates contribute their energy mainly to ciliates, which are most numerous in the RPD layer. Micrometazoa, with the exception of gnathostomulids, are mostly found above the RPD layer and seem to occupy a higher trophic position, although nematodes range through all the upper layers. The mobility of many of the organisms functions to facilitate energy distribution and flow in these layers.

Fig. 3-2 summarizes in a schematic way some of the changes observed in a vertical section of sediment, with an oxygenated surface layer overlying a gray RPD layer and a black anaerobic layer.

SEA FLOOR SEDIMENTS AND LATITUDE

Although tropical waters in general tend to be less productive than those of the far north or south because the warm surface waters tend to stay at the top with little interchange between them and the colder, denser, nutrient-rich water below, both of the major westward-flowing currents (the North and South Equatorial currents) produce some upwelling along their equatorial margins, so that the equatorial zone itself between these currents is somewhat richer and more productive. Another factor that may influence final deposition of sediments derived from organisms is the fact that the thermoclines and density gradients slope slightly toward the equator. Thus many particles may accumulate along such gradients and be transported a greater or lesser distance toward the equator before eventually sinking to the bottom.

In any event sediments along the equator have accumulated more barium and calcite, derived from organisms living near the surface, than have the sediments farther north or south.

THE EARTH'S CRUST BELOW THE SEDIMENTS ON THE SEA FLOOR

Below the sea floor sediments the earth's crust is of very different character from the crust on the continents. It is much thinner and consists of basalt rather than the granitic rocks characteristic of continental platforms. Seismic studies indicate that it is only about 5 km. thick, whereas under continents it may be from 24 to 30 km. to the mantle.

PART TWO

THE BIOTA

CHAPTER

4 BASIC ECOLOGICAL CONCEPTS

Living things are of primary importance in determining the character of the earth's surface, atmosphere, and waters. Those portions of the earth's crust, together with its waters and atmosphere, which are permeated with living organisms or their products are collectively termed the **biosphere.** The volume of the biosphere represented by the oceans is approximately three hundred times that on land because on land only about 20 meters serves as effective biosphere. In the sea, living organisms occur at all depths.

The biosphere represents a vast seat of continual physical and chemical change enveloping the earth. The total amount of living matter present at any time is the **biomass.** This term is also used in a slightly different sense, as when one speaks of the biomass produced by a particular area or region or by a particular species or group of species over a given period of time.

All living things must feed, respire, synthesize new organic molecules, and eliminate unwanted metabolic products. There is a continual flow of materials into and out of every living thing. Furthermore, the organisms themselves are undergoing continual replacement and evolutionary change. The animals and plants of one geological period may be considerably different from those of another, and they exert different effects on their environment and on each other.

The manner in which an organism feeds and the quantity and quality of foods available to it determine in large measure not only its structure and mode of life and its relations with other organisms but also the total biomass of that species which can be produced.

The entire interacting system of organisms, together with the environmental factors with which they interact, is termed the **ecosystem.** With regard to their major roles in the ecosystem, that is, the overall result of their life and activities, organisms can be divided into three categories: **producers, consumers,** and **reducers.**

The producers are those which synthesize new organic matter from inorganic substrates such as carbon dioxide, water, and soluble salts. These are the photosynthetic forms—the plants. All other living things, except chemoautotrophic bacteria, depend directly or indirectly on the plants for their food and oxygen.

The consumers are organisms that ingest or absorb organic matter directly, assimilate what they need from it, and eliminate the rest. This group includes all animal life. Animals that feed directly on plants, the herbivores, are sometimes termed primary consumers, whereas the carnivores, scavengers, and parasites, which subsist on other animals, are termed secondary consumers. Many animals consume a mixed diet.

The reducers are mostly microorganisms, such as bacteria, molds, and actinomycetes, which cannot carry on photosynthesis or ingest particulate organic matter. They liberate exoenzymes into their environment to break down complex organic molecules into smaller units that can be absorbed and assimilated. To the extent that they assimilate and build into their own bodies the organic matter on which they live, they are, of course, consumers too, but the overall result of their activity is to break down the vast reservoirs of organic matter

A

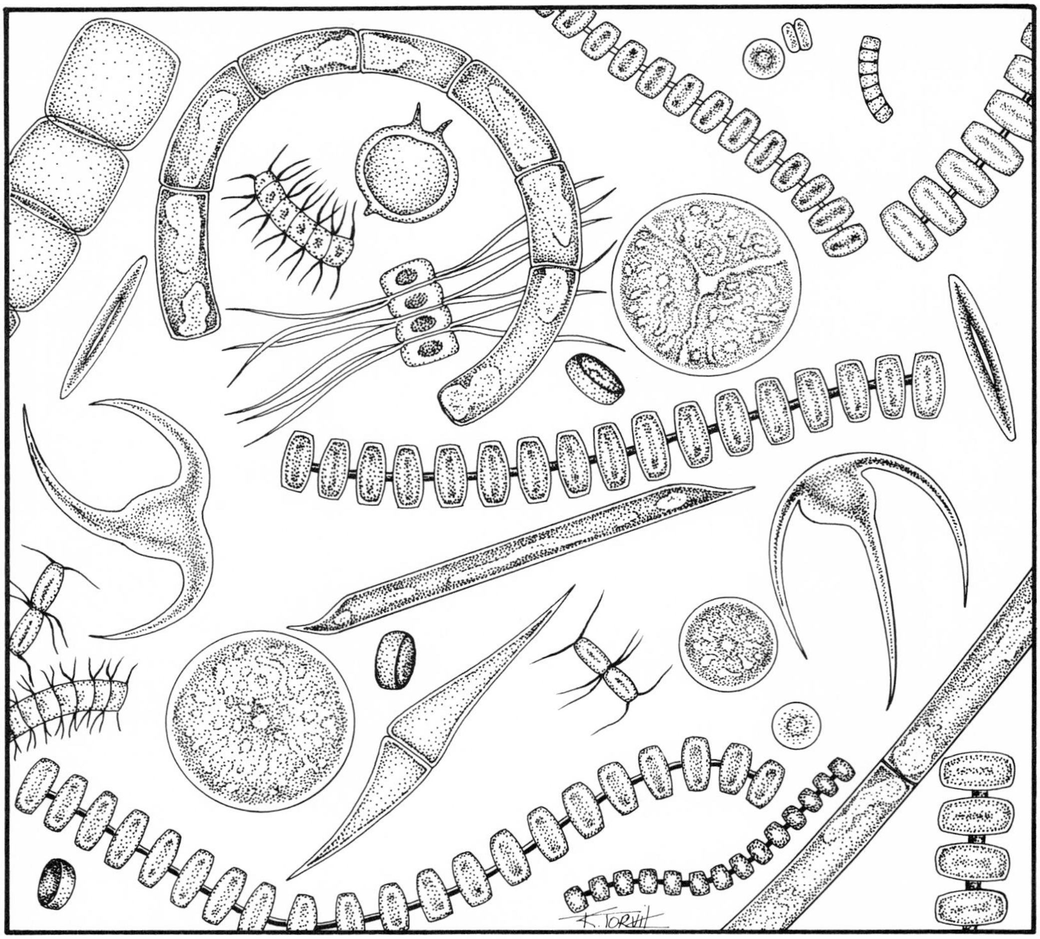

FIG. 4-1. A, Phytoplankton, the producers. Mostly minute microscopic plant cells floating in the waters of the euphotic zone.

represented by plant and animal bodies and residues, releasing the constituents in the form of simple ions and gases available once more to the plants. This process is called the **mineralization of organic matter.** It is represented by the various processes of decomposition, such as fermentation, decay, and putrefaction.

Through the combined activities of all these groups of organisms, a circulation of the chemicals necessary for life is maintained. This circulation enables life to flourish abundantly for millions of years in spite of the relatively limited amounts of necessary materials available on the earth's surface.

Since the efficiency of energy use in the conversion of organic matter from one organism to another is rather low, it follows that the total attainable production of consumers and reducers together must be less than that of the producers. Of necessity, a great deal more plant life than animal life is therefore produced. Likewise the production of herbivorous animals is far greater than that of carnivorous animals. This phenomenon is conveniently oversimplified in the "10% rule," which states that an animal assimilates only 10% or less of the food that it eats—the rest being used in maintaining its metabolism or being eliminated. In other words, in any food chain proceeding from

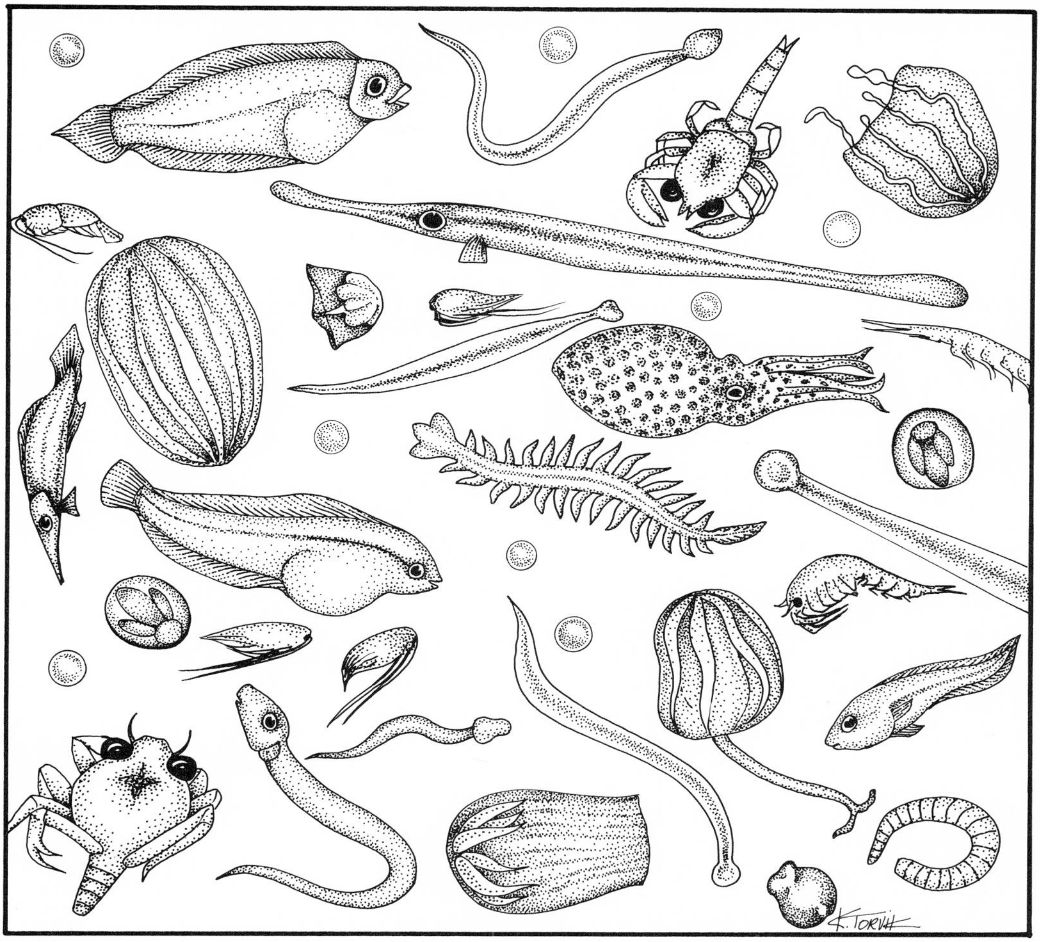

B

FIG. 4-1, cont'd. B, Zooplankton, the consumers. Mostly small-to-minute animals and immature stages of larger animals.

plants to herbivorous animals to carnivorous animals, there is a 90% reduction in total possible production attainable at each step.

This rule does not necessarily mean that the standing crop of producers and consumers in any given system must always be in a 10:1 ratio or more. In the ocean, where most of the producers are rapidly multiplying microorganisms and most of the consumers are more slowly growing larger animals, and where the producers are eaten about as fast as they are formed, situations in which the standing crop of consumers equals or exceeds that of the producers at any given time may actually be found. The necessary energy flow ratios appear, however, if the total biomasses, of the producers and of the consumers, produced over a long period of time are considered.

The farther removed from the plant base of the food chain and the more specialized or restricted the diet of an animal, the smaller will be the maximum possible total biomass of that species produced.

Producers, consumers, and reducers comprise a balanced interacting system. None could exist long without the others.

ENERGY FLOW

The concepts of energy flow and the efficiency of energy use in an ecosystem

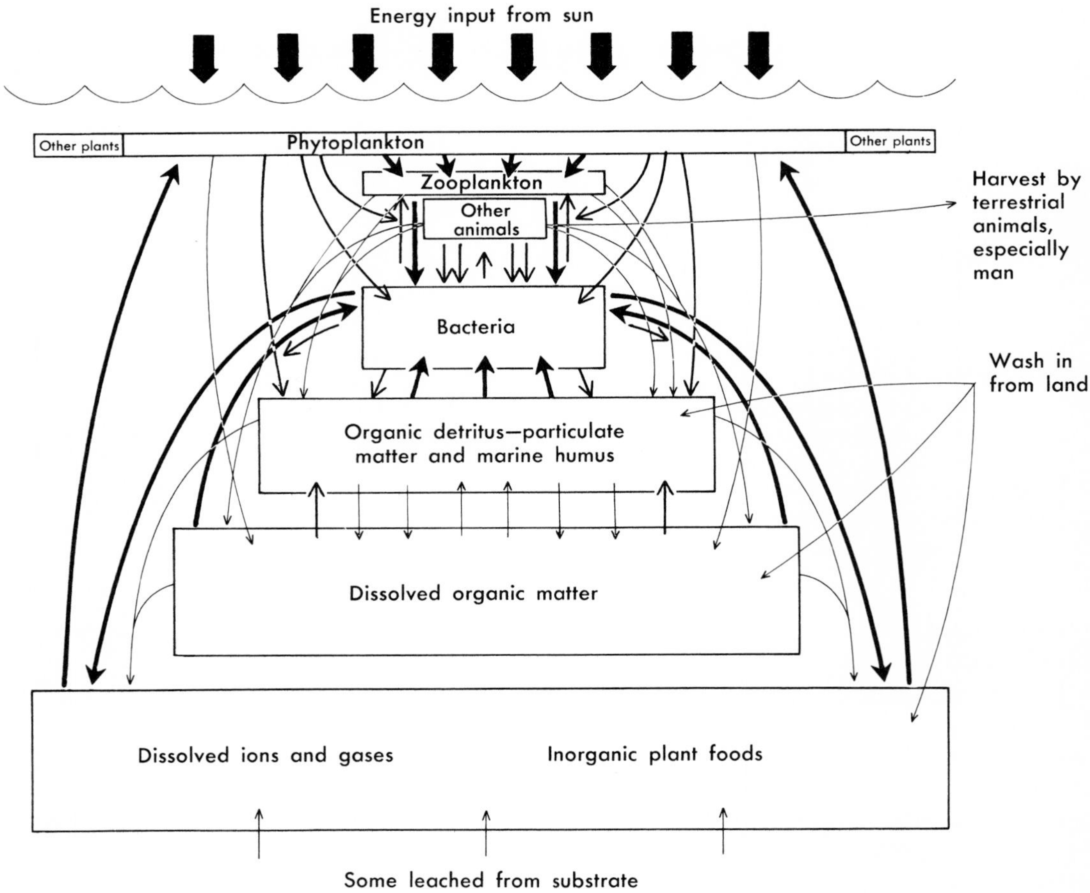

FIG. 4-2. Cycle of life in the sea. These items are not drawn to scale, but the relative sizes of the various rectangles and number and heaviness of the arrows are intended to convey some idea of which are the greater accumulations of matter and the more massive energy exchanges. Some organic residues are entrapped in the bottom sediments. Over time, some sediment exchange occurs between land and sea—wash-in from land, and emergence and submergence of coastal areas.

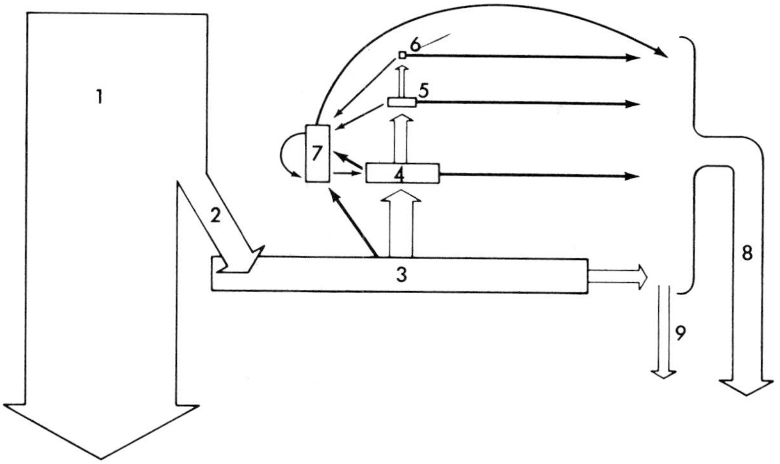

FIG. 4-3. Concept of energy flow through various trophic levels of an ecosystem (diagrammatic). **1,** Radiant energy from the sun intercepted; **2,** portion of radiant energy absorbed (the remainder is reflected back into space or converted directly into heat); **3,** the producers (photosynthetic organisms); **4,** primary consumers (herbivorous animals); **5** and **6,** different trophic levels of secondary consumers (carnivores, parasites, etc.); **7,** reducers, or decomposers (bacteria, fungi, actinomycetes, etc.); **8,** energy losses through respiratory metabolism; **9,** other energy losses. There are rather wide variations in the efficiency of transfer of energy from one trophic level to another by different kinds of organisms, efficiencies ranging from about 4% to more than 30%. In general, organisms lower in the trophic scale tend to have more efficient assimilation than do those at higher levels.

throw much light on the nature and evolution of ecosystems and the meaning of such terms as ecological niche.

Under relatively stable conditions, an ecosystem will tend to evolve toward the most efficient use of the energy captured by primary producers. It would be conceivable to have an ecosystem consisting only of producers and bacteria, but in such a system much of the energy captured by the producers would be wasted—that is, not used in building biomass—because of the inefficient use made by bacteria of various organic substrates.

Wherever there is an energy-use gap, or energy leak, of this type between one form of life and those next in the trophic chain, there is an opportunity for an organism able to move into an intermediate situation (ecological niche) to make use of some of this energy. As the energy gaps become smaller, the animals will become more and more diversified and specialized to take advantage of particular restricted opportunities for energy capture.

Thus the total biomass supported by a given amount of captured energy will increase. The ratio between the total energy captured by primary producers and the total biomass in an ecosystem is an index of the maturity and stability of the ecosystem.

The mature ecosystem will evolve to a point of near-maximal use of its captured energy and will be characterized by a complex hierarchy of specialized animals and plants, many of which are adapted for extremely narrow and particular channels of energy capture. More of the available nutrients will be tied up in living biomass. Production of new biomass will be largely controlled by the turnover rates of limiting plant nutrients. The average age reached by some of the principal organisms will be greater, and the rate of turnover and replacement slower.

TABLE 4-1. Estimates of gross and net organic production of various systems in grams of dry weight produced per square meter per day*

	Gross	Net
1. Mass outdoor *Chlorella* culture		
Maximum		28.0
Mean		12.4
2. Terrestrial crops in growing season (maxima)		
Sugar		18.4
Rice		9.1
Wheat		4.6
3. Terrestrial habitats		
Spartina (marsh grass)		9.0
Pine forest (best growing years)		6.0
Tall prairie		3.0
Short prairie		0.5
Desert		0.2
4. Marine (maxima for single days)		
Coral reef	24.0	(9.6)
Turtle grass flat	20.5	(11.3)
Grand Banks (April)	10.8	(6.5)
Continental shelf	6.1	(3.7)
Sargasso Sea (April)	4.0	(2.8)
5. Marine (annual averages)		
Long Island Sound	2.1	(0.9)
Continental shelf	0.74	(0.40)
Sargasso Sea	0.74	(0.35)

*Extracted from Ryther, J. H. 1959. Potential productivity of the sea. Science **130**:602-608.

TABLE 4-2. Solar energy available for photosynthesis in joules per year*

	Joules
1. Intercepted by earth	5.0×10^{24}
2. 40% reaches surface	2.0×10^{24}
3. 50% of (2) is infrared	1.0×10^{24}
4. 40% of visible light is reflected	6.0×10^{23}
5. Available in ocean, 75% of (4)	4.0×10^{23}
6. Assuming 2% efficiency in PS	8.0×10^{21}
7. Needed per gram of carbon assimilated	5.0×10^{4}
8. Maximum assimilation per year, tons of carbon	1.6×10^{11}

*From Vishniac, W. 1968. Autotrophy; energy availability in the sea and scope of activity. *In* C. H. Oppenheimer [ed.] Marine biology. Vol. IV. New York Academy of Sciences, New York.

As long as conditions remain essentially unchanged, such an ecosystem will be less subject to great fluctuations in numbers of individuals or of species than will a younger ecosystem with fewer and more generalized species. The former may, however, be more vulnerable to near-total catastrophe in the face of any basic or rapid changes because of the intricate chains of dependencies and the narrow abilities of many of its species to accommodate to any change in their highly specialized modes of energy capture.

Tropical ecosystems are in general more mature than are those of temperate, boreal, or polar regions. Conditions in the tropics

TABLE 4-3. Energy requirement for photoautotrophic growth*

Stage in biosynthesis	On basis of	Required per gram atomic weight of carbon: Cofactors	Energy in joules
CO_2 → Carbohydrate	Calvin-Benson pathway	3.0 ATP	1.2×10^5
		2.0 NADPH	3.6×10^5
Carbohydrate → Amino acids	Survey of known pathways	~0.5 ATP	0.2×10^5
Amino acids → Whole cell	10 gm. dry weight/ATP (Bauchop and Elsden)	2.4 ATP	1.0×10^5
			6.0×10^5†

*From Vishniac, W. 1968. Autotrophy; energy availability in the sea and scope of activity. *In* C. H. Oppenheimer [ed.] Marine biology. Vol. IV. New York Academy of Sciences, New York.
†Minimum energy requirment for photoautotrophic growth is therefore 5×10^4 joules per gram of carbon, that is, the amount required for 1 gm. atomic weight of carbon divided by the atomic weight of carbon (12).

TABLE 4-4. Ocean productivity*

Carbon assimilation per year	100-200 gm. C/m.^{2} (H. W. Harvey; G. A. Riley)
Area of oceans†	3.5×10^{14} m.^{2}
Annual productivity‡	5.3×10^{10} tons carbon
Maximum productivity	1.6×10^{11} tons carbon

*From Vishniac, W. 1968. Autotrophy; energy availability in the sea and scope of activity. *In* C. H. Oppenheimer [ed.] Marine biology. Vol. IV. New York Academy of Sciences, New York.
†Not counting ice-covered waters.
‡P and N are limiting, so that for this amount of organic matter all available P and N would have to turn over two to ten times each year.

are more stable throughout the year and from year to year, without drastic seasonal changes in available energy and in other conditions. Also, today's tropics represent the least changed, longest existent relicts from the past. Cold climates and marked seasonal changes are a relatively recent development, geologically speaking. They were caused by thermal isolation of the polar regions, resulting from continental drift, and by continental emergence, with draining of vast areas formerly covered by shallow seas. This emergence was due partly to deposition of ice in the cold, thermally isolated polar regions, especially Antarctica. lowering the general sea level. Vast continental areas are subject to more rigorous, more fluctuating climatic conditions than are marine or maritime regions because of the high heat-absorbing and heat-exchanging capacity of water, which exercises a marked moderating effect on climatic conditions.

TEMPERATURE RELATIONS

Stenothermal organisms are those which can tolerate a limited range of temperatures. **Eurythermal** organisms can live in a wide range of temperatures. Many organisms are reproductively stenothermal, or stenothermal during young stages even though the adult may be more eurythermal.

The maximum and minimum temperatures of temperate and boreal waters, (Table 4-5 and Fig. 4-4), with wide temperature fluctuations, may be limiting to stenothermal species from warmer or colder waters. However, many eurythermal species usually found in warmer or colder waters can live in these areas, even if they are somewhat stenothermal with respect to

TABLE 4-5. Surface temperatures of open waters in degrees centigrade

	Minimum	Maximum	Mean	Approximate percent of world oceans with that range
Tropical	20	30	24-27	31
Subtropical	16-18	25-27	20-22	16.5
Temperate	9-16	24-26	16-20	24
Boreal and antiboreal	1-9	10-17	8-14	16
Arctic and antarctic	−3 to +1	8-10	−1 to +5	12.5

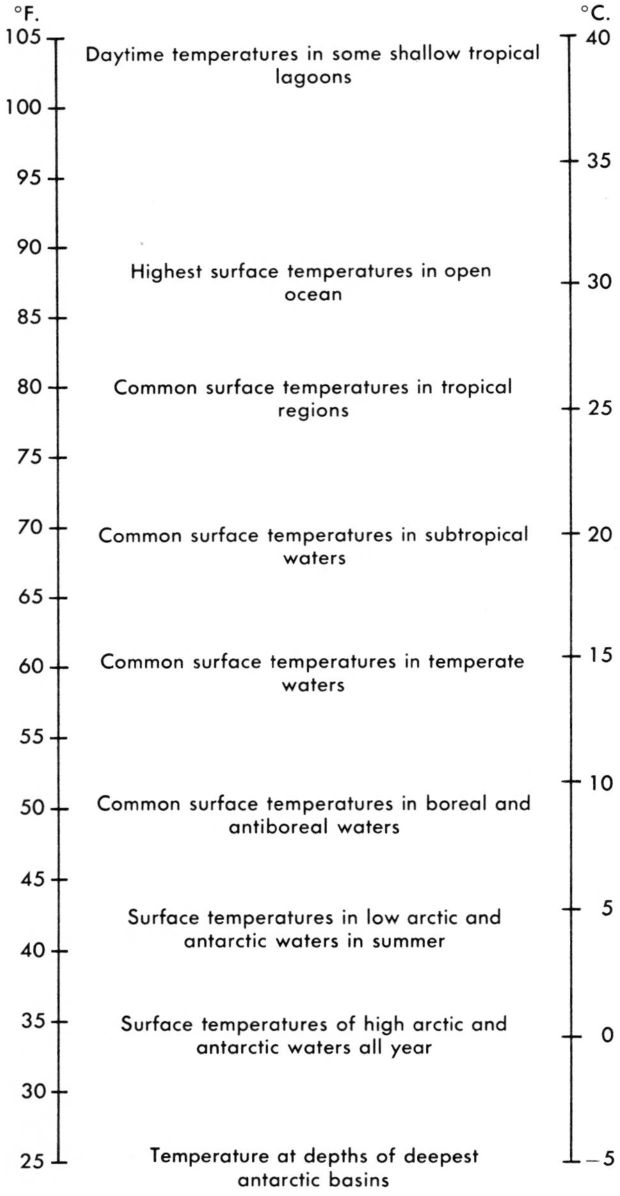

FIG. 4-4. Temperatures of marine waters.

TABLE 4-6. Examples of spawning season of some animals near Bergen, Norway*

Ecological region	Species	J	F	M	A	M	J	J	A	S	O	N	D
Arctic-boreal	*Strongylocentrotus dröbachiensis*	x	x	x									
	Cucumaria frondsoa		x	x	x								
	Dendronotus frondosus	x	x										
Boreal	*Pleuronectes platessa*			x	x	x							
	Mytilus edulis			x	x	x	x						
	Echinus esculentus			x	x	x	x	x					
	Asterias rubens			x	x	x	x						
Mediterranean-boreal	*Psammechinus miliaris*					x	x	x	x	x	x		
	Echinocyamus pusillus					x	x	x	x				
	Echinocardium flavescens					x	x	x	x	x			
	Echinocardium cordatum						x	x	x	x			
	Ciona intestinalis					x	x	x	x	x	x		

*From Sverdrup, H. U., M. W. Johnson, and R. H. Fleming. 1946. The oceans: their physics, chemistry, and general biology. Prentice-Hall, Inc., Englewood Cliffs, N.J.

TABLE 4-7. Mean annual temperatures at various depths and latitudes in the eastern Atlantic Ocean in degrees centigrade*

Depth (meters)	60° S.	40° S.	20° S.	0°	20° N.	40° N.	60° N.	80° N.
0	Below 0	15	17	27	20	16	9	2
200	Below 1	10	11	15	15	12	6	2
400	Below 1	10	9	9	12	12	9	1
800	Below 1	4	5	5	8	11	8	0
1000	Below 1	3	4	5	6	9	7	Below 0
2000	Below 0	2	3	3.3	4	4	3	−1

*From Ekman, S. 1953. Zoogeography of the seas. Sidgwick & Jackson, Ltd., London.

reproduction, by spawning during the part of the year that suits them best. Table 4-6 gives some examples.

Submergence

From Table 4-7 it is evident that for many stenothermal organisms from colder waters, suitable temperatures are found at greater depths as one progresses toward the equator. Cold-water stenothermal species that can tolerate a variety of depths may exhibit the phenomenon of submergence. For example, many species found in shallow areas on the continental shelf in arctic waters are found only at greater depths in the boreal waters of the northern Atlantic and Pacific oceans.

Submergence, together with changes in hydrographic conditions during the ages, accounts for cases of the appearance of the same or related species in temperate, boreal, or polar waters of both hemispheres, but not in the tropics. This phenomenon is called **bipolarity.** Some forms that show bipolar surface distributions may actually exist at greater depths in the tropics, having, then, a continuous distribution. Because of submergence, some cool-water species enjoy a greater geographical distribution than do species limited to warm water.

Surface distribution

In the surface waters, however, incursions of pelagic tropical, or warm-water, species into regions normally inhabited by cold-water forms are more frequent and more extensive than are occurrences of cold-water species in normally warm regions. One reason is that cold water tends to be more dense and sink beneath the warmer waters in regions where cold and warm waters meet. Another factor is that most organisms are living closer to the upper limits of their thermal tolerance than to the lower limits, probably because their biochemical processes are more efficient and rapid there. Unaccustomed warming is more quickly deleterious to most organisms than is cooling to the same extent. Cold-water species carried into warm areas die and disintegrate more quickly than do warm-water species carried into cool regions. Warm-water species may often survive a considerable time in cooler-than-normal water and even continue to grow, although they may be unable to reproduce and maintain a population there.

VERTICAL STRATIFICATION

Since virtually all the plant life of the oceans is confined to the euphotic zone and all animals depend on plants either directly or indirectly for food, most animal life of the oceans is also found in or near this zone. Vertically the oceans can be divided into four principal realms by depth and light penetration. (See Fig. 1-2.)

Photic zones

At the top is the thin **euphotic zone,** the upper 100 meters or so, containing all the functioning producers, as well as most of the consumers and a large portion of the reducers. It is important to bear in mind the diurnal, seasonal, and regional differences and periodicities in the illumination and temperature of the waters in this zone. Organisms of the euphotic zone are often described as **epipelagic.**

Below the euphotic zone are the **dysphotic,** or **mesopelagic, depths** extending down to about 1000 meters. These depths receive too little, and too brief, illumination for effective photosynthesis. Only shorter wavelengths of visible light in the blue end of the spectrum penetrate to these depths. No functional producers are present, but many animals occur here. Some of the animals are regular residents of these strata. Others perform daily or seasonal migrations into and out of them.

Aphotic zones

At their lower limits the mesopelagic depths grade insensibly into the **aphotic zone,** the waters that receive no light at all. This zone continues to the ocean bottom. The dysphotic plus the upper portion of the aphotic zone, from around 100 to 3000 or 4000 meters, is termed the **bathypelagic zone.** Below this zone, in the great depths, is the **abyssal,** or **abyssopelagic, zone.** The deepest portions of the oceans, near the bottom of the deepest ocean basins and trenches, are sometimes termed the **hadal zone.** The deeper zones are not sharply marked but grade insensibly into each other.

The terms bathypelagic and abyssal are not always used in precisely the same sense by various writers. Bathypelagic is sometimes restricted to include only the upper half of the aphotic zone and exclude the mesopelagic depths, and abyssal is commonly used in referring to the entire aphotic portion of the oceans.

Benthic organisms at corresponding depths are described as **archibenthic** if they are living in the upper bathypelagic and dysphotic depths at the outer edges of the continental shelves and on the continental slopes; they are described as **abyssobenthic** when they are living on the deep-sea floor.

The water in the abyssal region is cold, usually more saline than that in the layers above, and under immense hydrostatic pressure. Water movements are slower.

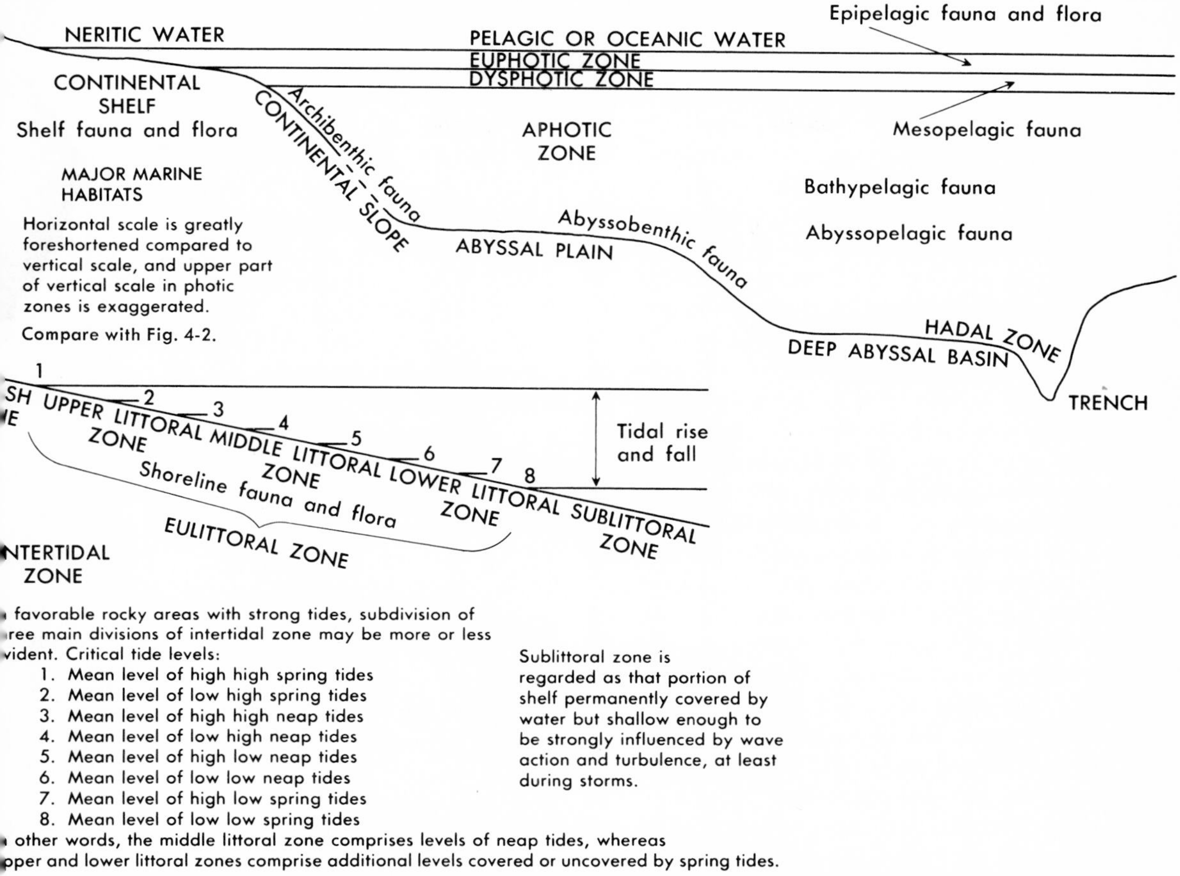

FIG. 4-5. Major marine habitats. In practice it is better to define the upper, middle, and lower littoral zones in terms of the zones actually occupied by organisms characteristic of them rather than in terms of absolute tide levels. Other factors—degree of wave action, exposure to or protection from direct sunlight, slope of the shore, etc.—also influence the vertical distribution of organisms.

Bathypelagic and abyssal animals depend ultimately on organic matter produced in the euphotic zone. There are no plants. Energy is transmitted to the deeper zones by settling of organic particles from the euphotic zone, by downward movements of water carrying organic materials with it, and by active transport by swimming animals that perform vertical movements (pp. 133 to 135).

Organisms limited to relatively narrow ranges of depth may be described as **stenobathic.** Those tolerant of great differences of depth, provided other conditions necessary for their life are met, are **eurybathic.** Organisms that spend their entire life in the water rather than on the substrate are spoken of as **holopelagic,** and **holoplanktonic** if their entire life is planktonic. Those which are planktonic only part of their life are termed **meroplanktonic.** The term deep sea commonly includes mesopelagic and abyssopelagic forms and both archibenthic and abyssobenthic forms. Pelagic animals, of course, face problems with respect to flotation, as mentioned in connection with the phytoplankton. Those living only at intermediate depths must be able to prevent excessive rising as well as sinking (Fig. 4-5).

USE OF DISSOLVED ORGANIC SUBSTANCES BY MARINE ANIMALS

In 1909 Putter suggested that many of the plankton feeders may also benefit by the uptake of dissolved organic materials in the water. Krogh (1931) reviewed the evidence available at that time and concluded that there was little evidence to support this view. However, it has been shown for some filter feeders such as mussels (Fox and co-workers, 1952, 1953) that the volume of phytoplankton removed from the water by them is insufficient to account for their growth and metabolism. These investigators also showed that *Mytilus* can remove fish hemoglobin from the water. MacGinitie (1939), using stained proteins, concluded that *Urechis* and *Chaetopterus* can remove particles of the order of 40 Å. Stepheus and Schiuske (1957) demonstrated that a wide variety of ciliary-mucus–feeding invertebrates remove amino acids from the water but that arthropods generally fail to do so. They suggested that whatever the mechanism, it is not direct filtration, which would imply a pore size too small to allow effective filtration of the volumes of water known to be filtered by such animals. They suggested adsorption of organic molecules onto the mucus as a possibility. Fox and co-workers suggested another mechanism by which such substances may be made available to filter feeders, that is, by their adsorption onto the surface of colloidal micelles or other particles suspended in the seawater that are large enough to be filtered. Polymerization of surface-active molecules into larger aggregates may also contribute to this process, especially where such molecules and other particles are concentrated from the water surfaces in small zones of convergence.

Analysis of seawater indicates that nonliving particulate organic matter greatly exceeds the amount of living organic matter and that the total amount of dissolved and colloidal organic matter is even greater.

For the oceans as a whole the values for dissolved organic matter run from about 3 to 7 mg./liter, with somewhat higher values near the surface, at the sediment-water interface at the bottom, and in most neritic coastal waters. According to Krogh (1931), Rakestraw (1936), Rakestraw and Carritt (1948), and others, there are from 1.2 to 2.8 mg. of organic carbon and about 0.2 mg. of organic nitrogen per liter.

The values obtained by chemical analysis of seawater may, however, be misleading. Marine biologists do not have good measures of the rates of production and consumption of dissolved and small-particle organic matter, but can only measure the unconsumed residue at any time. The classes of compounds most readily used by living organisms may be utilized by bacteria and other organisms as fast as they are produced, so that there may be a rather rapid production and use of such compounds even though chemical analysis of the water fails to show their presence in significant amounts. More refractory compounds would tend to accumulate and be the ones found in greatest abundance.

In line with this expectation, one finds that readily utilized compounds, such as monomers of sugars and amino acids, rarely exceed 20 to 30 μg/liter. Polymerized and macromolecular materials are at least an order of magnitude greater in amount, whereas the bulk of the marine humus consists of large molecular organic residues, containing little nitrogen or phosphorus and resistant to degradation by heterotrophs. The greater quantity of large molecular organic residues reflects the greater difficulty of their degradation and assimilation —not greater production of such compounds.

The rapidity with which smaller, more labile organic compounds are used and the fact that these may constitute a large fraction of the total dissolved organic material produced are indicated by the facts that dissolved organic matter and particulate organic detritus show little increase in the water during periods of high productivity and that at depths just below the euphotic zone the levels are almost independent of the productivity in the surface waters.

Perhaps the presence and nature of populations of heterotrophic microorganisms in the water are the best indicators of the production and probable amounts of labile compounds utilized in a given water mass. The more refractory compounds become available only through their adsorption to the surfaces of larger particles and their breakdown there by bacterial enzymes (Chapter 15).

If a 5% efficiency of removal is assumed, an organism would receive 0.15 to 0.35 mg. of organic matter per liter of water filtered, which would fall within calculated ranges of its need. However, if much of the dissolved matter consists of refractory compounds resistant to digestion, the significance of their uptake would be questionable.

Park and associates (1962) demonstrated the presence of seventeen dissolved amino acids in water from the central Gulf of Mexico. These ranged from more than 1 mg./m.^{3} for the four or five most abundant to less than 0.5 mg./m.^{3} for the seven least abundant. The concentrations were fairly uniform for all depths tested (10 to 3500 meters). They believed that dissolved organic matter may make a substantial contribution to the organic budget of the sea and is probably directly available to many animals through adsorption to the surfaces of suspended particles or sediments.

Electron microscope studies of the ultrastructure of marine animals reveal that many of them do have certain epithelial surfaces provided with microvilli and showing evidence of pinocytosis, together with high concentrations of alkaline phosphomonoesterases and mucopolysaccharides at the free cell borders. These are all indicative of a high level of active transport and furnish evidence for the utilization of dissolved and minute-particle organic matter by these organisms. Such areas have been found in the siphonal and mantle epithelium of *Tridacna,* in various other molluscs, in echinoderms, pogonophores, corals, cerianthids, other coelenterates, and numerous other animals (Fig. 12-5).

Near San Diego, California, sea urchins, which normally become scarce either through migrating away from or dying off in areas that they have completely denuded of kelps and other edible seaweeds, have been found to persist in localities influenced by municipal sewer outfalls. They apparently subsist, at least in part, on dissolved organic matter from the sewers and prevent normal reestablishment of the large seaweeds.

Even where dissolved organic matter may not be present in sufficient quantity to be responsible for initiating blooms of organisms capable of using it, once a bloom is under way, it is possible that excreted materials may have important effects on the growth and selection of the species to follow.

Refined analytic methods now make possible rather precise estimates of the amounts of living and nonliving organic matter.

The precise measurement of ATP enables living organic matter to be distinguished from nonliving, making possible better estimates of biomass of living cells. Measuring the light produced when ATP is added to an enzyme mixture containing luciferin and luciferase from firefly tails is the method used. Minute amounts of ATP, down to about 10^{-6} gm., can be detected in this manner. The validity of the method for estimating living biomass depends on the assumptions that ATP is found only in living cells and that the ratio of ATP to living cell protoplasm is constant.

ENVIRONMENTAL EXTREMES

Unusually high salinities (hypersalinity) are rare in marine environments, occurring only in shallow, isolated arms of the sea, where evaporation is high. Simons (1957) studied the ecology of Laguna Madre, Texas, where salinities vary from 27‰ to 78‰. he found that the higher the salinity above normal ranges, the fewer were the species present, but that the number of individuals of species that were present tended to increase. Fishes present in such environments were of larger-than-average size for their species, but invertebrates such as barnacles and crabs were smaller than average.

In most coastal areas and especially in bays, river mouths, and estuaries, salinities vary downward from the normal marine values as one ascends. Because of tidal movements of water, salinities may fluctuate considerably at any given station.

Where both temperature and salinity undergo great variation, often reaching extreme values, the number of both species and individuals is usually low.

Hydrostatic pressure increases by approximately 1 atmosphere (roughly 15 psi) for each 10 meters of depth. At 10,000 meters it is 15,000 psi, sufficient to slightly compress the water, increasing its density, and to alter the rates and character of various chemical processes of biological importance. The oceans are sufficiently deep to make this one of the important selective environmental factors.

Where hydrographic conditions are relatively constant but in extreme ranges, the number of species is usually low, but those species tolerant of the conditions may greatly increase in numbers of individuals, becoming overwhelmingly dominant in that environment. The deep-sea bottom is an exception. The hydrographic conditions there are both constant and extreme, but the scarcity of food reaching these depths imposes a low absolute limit on the numbers of any species that can be present. Thus populations never become as dense as in other extreme biotopes. The diversity, however, is relatively high.

Stable biotopes within normal ranges for marine environments, such as most tropical seas, tend to develop the maximum diversity—that is, the greatest number of species present in significant numbers—but usually do not show the overwhelming aggregates of any one species that commonly dominate more extreme biotopes.

LIMITING FACTORS

In any complex system involving utilization of many components, that essential component which is in least supply, in readily available form, in proportion to the need for it becomes a **limiting factor.** Its rate of turnover then determines the rates of operation for the entire system. This is the essence of the use of fertilizers to increase production—supplying those substances which, in a given situation, are acting as limiting factors.

The idea of limiting factors can be extended to include whatever physical, chemical, or biotic factors are in fact limiting either the total production of organisms, the production of certain categories of organisms, or the production of particular species in a given situation.

Total productivity

The amount of radiant energy reaching the earth's surface from the sun imposes overall theoretical limits on organic primary production in all latitudes. The total possible primary production in turn imposes theoretical limits on the possible secondary production of animals and microbes.

In practice, other limiting factors, such as the availability of plant nutrients, come into play well before the amount of light becomes limiting, so that production does not reach the theoretical limits calculated from only the available radiant energy. Light is not an important limiting factor except in those parts of the biosphere where it is deficient or lacking. In the aphotic zones of the oceans the absence of light precludes any primary production, and the populations of animals and bacteria there depend on production's occurring in the euphotic zone. The same is true for surface waters in polar regions permanently covered with ice. In other polar, high boreal, or antiboreal regions, production is confined to the short summer season, though it may be very intense during this period.

The effects of light and temperature are extremely important throughout the biosphere in determining what kinds of plants and animals can inhabit various regions or depths of water, and thus limit the various species rather than overall productivity.

Factor interaction

None of the physical, chemical, or biotic factors impinging on organisms are acting alone. A limiting factor that imposes some particular limit under one set of conditions may, under different conditions, though still acting as a limiting factor, impose a different limit or may not be a limiting factor at all. The presence or absence of one substance may markedly influence the degree or rate of utilization of a different substance, and thus modify the limits which the latter imposes. Since productivity is a result of many interacting factors, the situation is complex. Predicting the overall outcome for a given set of conditions involves the simultaneous solution of several rate equations.

The following are some of the more important factors to take into account in modeling an aquatic ecosystem and predicting production: solar radiation, transparency of the water, depth to the thermocline, NO_3 and PO_4 levels, temperature, and intensity of grazing. Models using these factors commonly give results that agree well (usually to within 25% or better) with actual observed production in carefully studied areas.

Life and the composition of seawater

About 55 elements are constantly present in seawater in amounts detectable by ordinary chemical means. Most of the remaining elements are probably also present in extreme dilutions. Most of the 55 detectable elements are "conservative" elements, always present in the same ratios with respect to each other. A few marked exceptions vary significantly from place to place and from time to time, and they are not found in a constant ratio with other dissolved elements or ions. These nonconservative elements all belong to that group of approximately 27 elements that make up the protoplasm, shells, or skeletons of living organisms. However, not all the elements or inorganic ions and compounds vital to living things are nonconservative—only those in which the amounts available in the water are small enough, and the amounts assimilated by living organisms are great enough, so that their uptake and release from living things makes a significant difference.

If we arrange the elements vital to living organisms in the order of the amounts present in all living things and on the basis of dry weight, and then consider this arrangement in relation to the amounts present in available form in seawater, it will become evident why some of these elements are commonly limiting factors and are nonconservative, whereas others are not.

There are five elements in which the major constituents are each found in protoplasm in amounts greater than 1% of the total weight. Another eight each make up from 1% to 0.05% of the biomass, whereas the remaining fourteen are present in very small to trace quantities. Arranged in this order they are as follows:

Series 1. C, O, H, N, P
Series 2. S, Cl, K, Na, Ca, Mg, Fe, Cu
Series 3. B, Mn, Zn, Si, Co, I, F
Series 4. Sr, Mo, Br, V, Ti, Al, Ga

The biological significance of some of the microconstituents in organisms depends on their roles as constituents of certain nearly universal or very widely spread enzymes or other critical molecules vital to certain metabolic processes, but needed only in very small amounts. Molybdenum, vanadium, and cobalt are components of enzymes involved in nitrogen metabolism. Vitamin B_{12} also contains cobalt. Manganese and silica are necessary for diatom growth. Boron is essential to plant cell growth. Silica is utilized in diatom frustules and in the skeletons of radiolarians.

The cycling of biologically important elements involves an alternation between their presence in seawater as inorganic ions or salts, and their assimilation into living organisms as constituents of organic molecules, a part of the biomass. The inorganic form is normally first taken up by plant cells, passed through food chains of varying length, and eventually returned to the environment in simple inorganic form as an end product of metabolism, respiration, or decomposition.

Carbon, oxygen, and hydrogen are present in vast amounts in available forms—water, carbon dioxide, and bicarbonate and carbonate ions—and are not limiting. The annual biological turnover of carbon in the oceans is only about 1% of that available. Assimilable forms of nitrogen and phosphorus, as nitrate and phosphate ions, respectively, are very limited—much less than the amounts used during the year. Thus, for productivity to be maintained at normal levels, the nitrates must be cycled from one to ten times annually and the phosphates from one to four times. The production of organisms then depends on the availability of these nutrients.

In extensive tropical and subtropical areas, where surface waters are separated from underlying nutrient-rich deep water by a permanent thermocline, these nutrients are replenished only at a slow rate by vertical mixing or turbulence caused by internal waves (p. 16) and by horizontal circulation from distant areas where upwelling near the polar regions and along the landward edges of major currents enriches the water. Since there is continual loss to deeper water by the sinking of some organic residues before decomposition is completed, and since much of the total amount of P and N present is tied up in the existing biomass, the productivity of such water is low. This accounts for the great transparency and deep-blue color of many tropical and subtropical waters. Biologically they are comparable to deserts. To some extent the lower productivity per unit of water in clear tropical waters, such as the Sargasso Sea, is offset by the fact that the nearly overhead position of the sun and the clarity of the water make the euphotic zone deeper in such regions so that much more water is involved in primary production. It also is productive all year rather than only seasonally. Thus the impact of the low productivity per unit of water, in the surface waters, on the fauna of deeper waters beneath is not as great as might at first be supposed.

The relative proportions of C:N:P in the phytoplankton, in the zooplankton, in larger animals, and in the water are relatively constant—about 41:7:1. Thus it seems that the phytoplankton assimilates these nutrients roughly in the ratio in which they occur in the water, and passes them along to other living things in about the same ratio; then they are eventually returned to the water in excreta and decomposition products in about the same ratio. In inshore waters the ratios are more variable than in the open ocean, and the ratio of available nitrogen to phosphorus may vary from 10:1 to 6:1. The ratios in organisms as well can and do vary to some extent with variations in the water.

The eight elements listed (p. 61) in the second group of elements found in all living things are usually present in sufficient amounts in most aquatic environments and are only rarely limiting (usually in freshwater situations). The levels in the water are rather low, however, and they are concentrated in phytoplankton and organisms beyond them in the food chain at higher levels than in the ambient medium.

The microconstituents—the elements of the last two groups—occur inorganically in available forms only in very low concentrations, and their availability frequently creates limiting conditions for productivity.

Limitations to particular categories, groups, and species of organisms

Now we are dealing with all the factors, both biotic and abiotic, that impinge on organisms, and with the complex interactions between them, as well as with the wide variety of response and control mechanisms evolved by organisms.

Abiotic, or physical, factors

On a worldwide basis three general factors—temperature, illumination, and the character of the ambient milieu—are the most important determinants of the distributions of organisms, and are responsible for the major biogeographic divisions. With respect to these factors, living things are divided into terrestrial, marine, and freshwater types. Within each of these categories are those adapted to warm, temperate, cool, or cold climates, or to marked seasonal changes.

Frequently the limits or extremes of climatic or other environmental factors, attained during their cyclic fluctuations, are more important than the average or mean conditions in determining whether certain species can establish themselves successfully in a given region.

Temperature. For surface waters the major faunal breaks occur between the subtropical and temperate zones—between areas where the surface waters are always 20° C. or warmer and areas where they are cooler at least part of the time. Secondary, less well-marked changes occur between the temperate and boreal or antiboreal zones, and between these zones and the subpolar regions.

With increasing depth, abrupt changes occur in both temperature and illumination, especially in regions with warm surface waters and a well-developed thermocline. Plant life is excluded below the euphotic zone, except for some types that can also exist heterotrophically—but in any case large-scale new production of organic matter from inorganic is precluded. Since most of the organic matter produced in the euphotic zone is cycled and recycled there many times, with only a small part of it sinking or being carried to greater depths, there is a marked decline in total biomass in deeper waters. The animals that do live in deep water show fantastic adaptations for existence in a world of perpetual darkness, cold temperature, low, intermittent food supply, and immense hydrostatic pressure. Many of these in the upper bathypelagic zone are small, are provided with bioluminescent organs, and have a relatively huge mouth with many sharp teeth, large, often upward directed eyes, a distensible stomach, and highly developed chemosensory organs—adapted for detecting and taking advantage of the rare discrete batches of food that become available. Others are detritus feeders with mechanisms for efficient concentrations of small particulate matter drifting down through the water.

Since cold water is continuous under much broader ocean areas than are the warmer surface waters above, deep-water communities, especially of the bathypelagic zone, may be very extensive. A continuous bathypelagic community may underlie several different regional surface water communities.

The extremely deep abyssopelagic and especially the abyssobenthic communities also often show very wide distributions but less so than those of intermediate depths, because here the bottom topography and character of the sediments exert influences that create barriers and regional differences in fauna.

Other abiotic factors. Other physical factors, such as differences in salinity, oxygen content of the water, or chemistry of the water, are more local in their effects and are particularly critical near the shore and in bay and estuarine environments. For infaunal animals, those living in the substrate, the physical and chemical nature of the substrate, the particle size of the sediments, etc. exercise marked effects in controlling distribution and size of populations, as do also the temperature and the chemical and biological characteristics of the overlying water.

Intertidal organisms are, in addition, subjected to stresses such as desiccation, temperatures above or below those found in the water, alternation between exposure and submergence, waves, wind, rain, and direct sunlight.

Biotic factors

Well-established, mature biotic communities themselves constitute effective limiting factors, both in the sense of maintaining limits and balance between populations of the species constituting the community, and also in the sense of preventing establishment of other organisms in the area occupied by the community or preventing their crossing through it. The partitioning of available energy and space resources between all members establishes very effective limits to the expansion of each species, and of each category of species, such as herbivores utilizing certain types of plants, or carnivores utilizing certain sizes and types of prey.

Pioneer vs. established species; succession. Wherever a biotic community is disrupted and some or all of the dominant species characterizing it are eliminated, the biotic environment is radically changed. There is an opportunity for certain species formerly constituting only a minor part of the community, and for certain outside species, to move in and colonize the disturbed area. Eventually, if the area remains undisturbed, the interaction between these new populations and the gradual reestablishment of dominant species best suited to the area produces a succession leading to reestablishment of a so-called climax, or mature, community in which the pioneer species may again be eliminated or relegated to minor positions.

When we consider the production of various categories, groups, or species within one of the general sorts of biotic communities delimited by the factors mentioned above, we find a host of additional limiting factors responsible for the relative abundance of varous species and types of organisms present. Perhaps chief among these limiting factors are the division of available resources, position in the trophic heirarchy, and the particular kinds of foods available to each species at various levels of abundance. We have seen that with each step in the trophic hierarchy there is approximately a 90% step-down in the producible biomass. For each species beyond the primary producers this means that the total biomass it produces cannot exceed about 10% of the total biomass of the particular kinds of organisms on which it feeds. The standing crops are not necessarily in this proportion to each other, however. Where the prey species are much smaller, faster reproducing, and faster growing than their predators, the biomass of the predators may greatly exceed 10% of that of their prey at any given time, but the total production over time of the two groups will bear the appropriate ratios to each other.

A given herbivore cannot feed on all plants, but only on a limited number of species of appropriate size range and characteristics for which it has become adapted to feed. Its total production is then a small fraction of the total production of these particular species. Anything that limits the production of these species of plants (including its own grazing) will also limit the production of that herbivore.

Competition with other herbivores that may also feed on some of the same plant species is a major additional limitation. The activities of any predators that feed on the herbivore in question or on its competitors also influence the population level attained.

Similar considerations apply to carnivores.

For a few exceptional animal communities, such as intertidal and subtidal plankton feeders, food is not a limiting problem but is all about them in the water that bathes them. Competition is here shifted from food to suitable sites from which to exploit it, and to protection from predators. Here we find many of the most abundant animals living superficially like plants—sessile, and crowded together in great numbers wherever there is suitable substrate for them.

As an animal develops from egg cell to adult, its habitat, food habits, and vulnerability to various external dangers may change radically. The reproductive potential of a species may in general be taken as an index of the average risk members of that species have faced in developing from egg cell to adult and successfully reproducing. If the chances of a given egg cell making it are one in a million, then each individual, on the average, must be responsible for a million offspring if the species is to hold its own.

Those animals which guard their young or take other measures that reduce the risk for the most vulnerable stages generally produce markedly fewer offspring than do those in which the eggs or young are simply cast out into the world to fend for themselves.

There is also a tendency to pass rapidly through those stages in the life cycle at which the risk is highest (mortality greatest) and to prolong those stages in which the organisms are relatively safe. High risk can serve as an effective selective factor for speed, since it puts a premium on individuals that pass through the most vulnerable stages quickly.

Whenever the risk factor for a given species changes, the result may be striking changes in the total population of that species developed in a few generations. In turn, there are changes in the feeding pressures it exerts on other species, as well as in the food supply of species that prey on or parasitize it. Adjustments throughout the biotic community may follow.

Often the effects are indirect—for example, a sudden decrease in one abundant prey species may cause a decrease in similar prey species rather than in its predators only, since the predators turn to alternate, less desired or formerly less readily available prey and intensify their hunting effort.

There is a damping off of the intensity of the effects as they become farther removed from the species that originally underwent a marked change in abundance. The reason is the netlike character of the food webs. The impact of the change spreads simultaneously in several directions and is not fully felt by any other single species. This damping effect is much more marked in mature, stable communities than in pioneer communities.

Pioneer communities and communities developing in difficult environments or regions with marked seasonal changes tend to be dominated by a relatively small number of species that have adapted to these conditions and that, having less competition for the available food, commonly develop huge populations. If a disease or other catastrophe selectively eliminates or reduces to insignificant numbers the population of one of these dominant species, the effects are drastic and felt at once by the entire community. Such communities generally show much greater fluctuations in numbers of the various species comprising them over a long period of time, than do more mature communities in more stable environments.

Mature communities in stable environments, because of the greater number of species they contain, the greater intricacy of the food webs, and the fact that the community is usually not so dominated by a few species, can better absorb a selective catastrophe to one or a few of its more abundant members without showing such drastic fluctuations. However, because of the greater number of very highly specialized species and their greater interdependence, any general change affecting the community as a whole, such as a slight change in climate, may lead to more drastic effects and perhaps the extinction of a much greater number of species.

The interactions between species in a biotic community thus tend to regulate the numbers of various types of organisms comprising it. An established community may itself constitute a limiting factor preventing establishment of species of other communities within the area occupied by it, or even their transit through it. On the other hand, it sometimes happens that an outside species introduced into a different community finds an unusual opportunity for expansion, free from its usual predators or other restraining influences, and multiplies enormously, with destructive effects on the structure of the indigenous community.

On land, plants are the largest, most numerous, most conspicuous organisms to be found, often growing in such thick masses that they form nearly impenetrable forests and jungles. Since their food is the carbon dioxide in the air around them, the water, and dissolved minerals in the earth, the most efficient manner for them to exist is as sessile organisms with two great absorptive surfaces—the photosynthetic surface, or leaves, which spread out in the air to receive light and carbon dioxide, and the roots and root hairs, which obtain water and dissolved minerals from the soil. Supporting and conducting tissues are required to hold the system up and to bring nutrients to all living cells. Where the soil is damp enough and rich enough in minerals to support an abundant plant growth, their competition is chiefly for light and space. Plants under such conditions tend to grow crowded together, evolving larger and larger forms and overtopping or growing on each other to reach the light until they arrive at the limits of size that can be efficiently maintained under the prevailing conditions.

In sharp contrast, when one goes out to sea he may well exclaim: "Where are the plants? Where are the producers in this system?" No obvious plants are to be seen except for the seaweed, marine grasses, and mangroves along the fringes of the continents and islands, or masses of small floating seaweed in exceptional areas such as the Sargasso Sea. These plants are so small in amount as to be totally unable to account for the great productivity of the oceans in fish and invertebrate animals, although they are, of course, important in the shoreline ecology.

The answer to this enigma, in fact our awareness that it is an enigma, has come only in the last century. If a finely meshed net capable of retaining microscopic objects is pulled through the upper layers of seawater, a mixture of microscopic plant cells and small animals is obtained. This is the **plankton.** Plankton is a term applied to the communities of floating, drifting organisms carried about primarily by movements of the water rather than by their own swimming activity. Those larger animals which control the direction and speed of their own movements, rather than drifting with the water, constitute the **nekton.** Organisms that live on or in the bottom constitute the **benthos. Phytoplankton** is the photosynthetic, or plant, portion of the plankton. Animals constitute the **zooplankton,** the nekton, and most of the benthos. **Demersal** animals are swimming forms that spend much of their time at or near the bottom. They may be considered as part of both the nekton and the benthos. Typical demersal animals include such forms as the flatfishes, octopuses, and many shrimps. Minute animals found in the surface film of the water are termed the **neuston.** Floating forms that protrude into the air, such as the Portuguese man-of-war or *Velella,* are sometimes termed **pleuston.**

The surface film contains a higher concentration of salts and organic matter than the water immediately below, as well as from ten to a thousand times as many bacteria, attracting certain protozoa, larval forms, and copepods. Floating fish eggs and larvae are also found here.

Certain oceanic birds, the skimmers, have beaks especially modified for skimming this layer. There are even some tropical bats that get their food from the neuston.

In the tropics the surface film also supports one of the few insects associated with the open sea. These water striders are of the genus *Halobates.* They dart about rapidly on the surface, eating small organisms found there and laying their eggs on bits of flotsam, such as floating feathers. They are air breathing and cannot survive long submergence.

Gooseneck barnacles of the genus *Lepas,* attached to bits of flotsam, to *Velella* floats, or to foamy floats secreted by themselves, are also part of the neuston community and have achieved wide oceanic distribution through ocean currents.

If one filters seawater through a fine filter or concentrates it with a centrifuge, he finds also a quantity of minute organisms —flagellates, blue-green algal cells, very small diatoms, and bacteria that are too small to be retained by a plankton net. Organisms in this size range have been collectively termed the **nanoplankton.** Nanoplankton is, then, simply a size designation that includes parts of both phytoplankton and zooplankton. Strictly speaking, the term nanoplankton is used for organisms ranging from 5 to 60 μm in size. Bacteria and other organisms smaller than 5 μm in length are termed **ultraplankton.**

Small particles of organic matter derived from the disintegration of the bodies, castings, or feces of organisms, and to an even greater extent from adsorption, polymerization, and aggregation of dissolved or colloidal moledules, together with inorganic particles of similar size, suspended in the seawater or sinking to the bottom, constitute **detritus,** or **seston.** Large organic molecules, macromolecules, or aggregations of organic molecules of colloidal size and minute particles of detritus dissolved or suspended in the water comprise the marine **leptopel.** The surfaces of detrital particles in the seawater, as well as all larger surfaces, become coated with dissolved, or colloidal, organic molecules. Some surface-active organic molecules are continually being swept to the sea surface by small bubbles and concentrated, along with other particles and small organisms, in small zones of convergence set up by Langmuir circulation (p. 13). Here they are polymerized into larger aggregates or adsorbed to the surfaces of larger particles, becoming, then, part of the detritus. Adsorption of leptopel to the surface of larger particles makes it available as food to filter-feeding animals. Fox and associates (1952) have made studies of the amount and distribution of the marine leptopel and of its significance for the nutrition of marine animals. Numerous animals known as **detritus feeders** ingest quantities of this material and digest the organic constituents, along with small organisms found in it.

The amount of nonliving particulate organic matter in the oceans exceeds the quantity of living matter by about an order of magnitude. The quantity of dissolved and colloidal organic matter is approximately another order of magnitude greater. Most of the organic matter is cycled and recycled in the upper layers of the ocean. Below about 300 meters the organic matter becomes much more uniform in character and less in total amount for a given volume of water, consisting largely of the less readily degradable residues—a type of **marine humus.**

Phytoplankton occurs in vast clouds in the upper 100 meters or so of the waters of all oceans except portions of the Arctic and Antarctic oceans permanently covered by ice. In the aggregate it comprises a mass of plant life even greater than that found on land.

Because of the necessity of light for photosynthesis, this plant life is restricted to the upper illuminated layers of water—the so-called **euphotic zone.** For each species of plant cell there is a certain depth below which it cannot carry on photosynthesis faster than it metabolizes organic molecules in its body; that is, below a certain depth, **anabolism,** the synthesis of new organic molecules and structure, just balances **catabolism,** the breakdown of these molecules. Under these conditions the species cannot grow or reproduce, although it may continue to live. This depth is known as the **compensation depth.** The precise depth for each species varies according to the turbidity of the water, amount and angle of incident light, and all factors that affect the rates of photosynthesis and metabolism.

In a general way, compensation depths vary with latitude, being greatest in clear, calm tropical waters in summer or fall, where they may exceed 100 meters. The more usual range seems to be from 10 to 60 meters. In some coastal areas during a dense bloom or when the turbidity is unusually great due to turbulence or influx of sediment from the land, the compensation depth may be as little as 4 meters.

Below the compensation depth the respiratory rate exceeds the rate of photosynthesis, and the plant cell eventually dies. Some photosynthetic forms apparently can also live heterotrophically. Many coccolithophores, some dinoflagellates, and others are regularly found at levels both well below and well above the compensation depth. Those in the deep waters must be enjoying heterotrophic nutrition.

CHAPTER

5 PLANT LIFE IN THE SEA
general considerations

FLOTATION AND VERTICAL DISTRIBUTION OF PHYTOPLANKTON

Protoplasm is, in general, slightly heavier than water and tends to sink. In a large organism there is far less surface in proportion to volume, and the tendency to sink is accentuated. Minute organisms, with a great deal of surface in proportion to volume, tend to sink far more slowly, especially if they are shaped to increase the relative amount of surface and to slow down any sinking that may occur by giving them a spiral or oblique path. Many marine organisms also adjust their specific gravity through the ionic composition of their body fluids, which contain a greater proportion of lighter ions than does seawater. Some have vacuoles containing less dense fluid or oil droplets. Some—for example, the dinoflagellates—can swim actively.

The necessity of remaining in the euphotic zone to grow and reproduce is one of the factors that helps to account for the small size and odd shapes of phytoplankton organisms compared with plants on land. Waters of the warm surface layers of tropical seas are both less dense and less viscous than are the cold waters of polar regions. Microplankton organisms from tropical waters tend to have more complicated and extended appendages or bristles, etc. for flotation. The formation of ribbons, chains, or rafts of many individuals may also serve as an aid in flotation, as well as help to protect them from grazing by small zooplankters.

FIG. 5-1. Young mangrove. (Chesher photograph.)

The absorption of essential nutrients is as

important as remaining in the euphotic zone. In addition to helping the cell stay in the euphotic zone, a large surface-to-volume ratio enables the plant cell to absorb the dilute nutrients from the surrounding water more efficiently, thus compensating for the fact that the nutrients are so dilute. However, when the phytoplankton is abundant, the nutrient supply in the upper stratum of water becomes impoverished. Some sinking then helps the cells to obtain nutrients. The sinking itself increases the depth of the euphotic zone by allowing more light penetration. In summer, maximum concentrations of chlorophyll are found near the bottom of the euphotic zone. At this time of year vertical diffusion is minimal and the quantity of surface nutrients is low. In winter, diffusion rates are higher. Growth, although slower, is greatest at the surface, and there is a linear decrease in chlorophyll with depth.

Increasing the temperature of the surface water decreases its density and viscosity and increases the tendency of phytoplankton cells to sink. This makes the water more transparent, and the euphotic zone becomes deeper. The interplay of these factors helps to account for the wide variation of shapes for chlorophyll curves observed in the ocean. Chlorophyll curves are graphs representing the total amount of chlorophyll per unit volume of water plotted against depth. The maxima would occur at those depths at which there is the greatest amount of plant life. Although many maxima are observed above the compensation depth, others are found just at or below it. Density differences in the water column are insufficient to fully explain the observed distributions.

The correspondence of chlorophyll maxima with an increase of nutrients in deeper strata suggests that a decreased sinking rate occurs as cells enter the darker, nutrient-rich water. Steele and Yentsch (1960) showed in cultures that sinking is increased with cell age but can be slowed down by the use of nutrient-rich water or by darkening the culture. Thus it appears that the general physiological condition of a cell is closely related to its buoyancy, as Gross and Zeuthen (1948) suggested. This factor is even more important than the changes in water density caused by changes in temperature for explaining differences in seasonal vertical distribution of phytoplankton.

In general, both the density changes and the changes in the buoyancy of cells with age, light, and nutrient environment tend to work in the same direction and reinforce each other in producing the observed seasonal differences in vertical distribution of phytoplankton in the water.

In early spring, when nutrients are abundant in the surface waters, there is rapid growth near the top, where cells tend to accumulate because of good nutrition and a low sinking rate. Later, in summer, the nutrients become depleted near the surface, the sinking rate increases, and cells begin to accumulate near the bottom of the euphotic zone, where better nutrition and darkness again revives them and retards their sinking. Also, at lower light intensities the rate of synthesis of chlorophyll increases.

According to Steemann-Nielsen (1954), maximum chlorophyll content of the euphotic zone is about 300 mg./m.^{2}, about one fourth to one fifth of the highest terrestrial concentrations. In water the quantity of organic matter produced is also controlled by cell density. When concentrations of cells reach the point at which the cells shade each other, growth becomes self-limiting.

At the surface, the light may often be too intense for many organisms, and the ultraviolet rays may injure them. The greatest bulk of phytoplankton occurs a little below the surface, from about 5 to 70 meters in depth. In northern and southern latitudes the euphotic zone tends to be shallower than that in the tropics because of less direct light, increased reflection resulting from the angle of incidence and greater wave action, and the greater turbidity of the water. In turbid coastal waters the euphotic zone may at times extend to a depth of only 5 meters or even less.

PHYTOPLANKTON BLOOMS

Since phytoplankton organisms are among the most rapidly growing elements of the ocean populations, some of them dividing several times a day under favorable conditions, the populations are very responsive to changes in the environment. This fact is displayed particularly well in the great spring outbursts of phytoplankton in boreal and polar regions. The cold of winter increases the density of the surface water, causing it to sink, which results in a turnover of waters from different depths. Some of the phytoplankton is carried below its compensation depth, which is at a high level in the winter due to the decreased illumination. Because of the decreased amount of phytoplankton in the euphotic zone, the decreased light, and the lower temperature, growth in winter is much slower, and the surface waters become enriched with dissolved plant nutrients from below. Better growing conditions in the spring cause a dramatic outburst of phytoplankton growth. Such outbursts are termed blooms. They may occur at any time and place where local conditions become more favorable for phytoplankton growth. Often there are several blooms between early spring and late fall. In some coastal situations special local conditions commonly produce winter blooms as well.

During a bloom the quantity of phytoplankton may double daily, producing great clouds of billions of cells, until impoverishment of plant nutrients and increased grazing by more slowly growing zooplankton populations slow down the rate of increase, bringing it into equilibrium or causing a sharp decline in total phytoplankton of the area and replacing it largely with zooplankton.

When a sudden dramatic bloom occurs in a given area, it tends to be dominated by a rather small number of species that happen to be most favored by the local conditions at the time or that happen to have a head start numerically when conditions favor a bloom. When a mixed population is subjected to a sudden increase in the carrying capacity of the environment, such as an increase in available nutrients together with better growing conditions in the spring, different species take full advantage of their respective rates of reproduction. These rates vary considerably from one species to another, and the fastest ones forge ahead. **Diversity**—in terms of numbers of species represented in roughly the same order of magnitude—decreases, and one or a few species become dominant. Some blooms are so dominated by one form that they are called **single-species blooms,** but more often several species are present in great numbers. Later, when the carrying capacity of the environment is approached, ecological succession sets in, and diversity increases until a somewhat more stable equilibrium is reached.

Burke and co-workers (1962) were able to select dominant species in mixed phytoplankton cultures by subjecting them to different light intensities. Selective grazing by zooplankton has also been shown to have far-reaching effects on the species composition, as well as on the total densities attained in mixed cultures. Both these factors are doubtless of great importance.

The nutritional requirements of different species of phytoplankton and the selective depletion of plant nutrients during a bloom also exert controlling influences on the course of the bloom, and of course the water temperatures are also of paramount importance in this regard.

It has been suggested that the great difference observed in seasonal phytoplankton distribution and blooming rates between the North Atlantic and the North Pacific, from about 45 degrees northward, may be largely due to the different species of the copepod *Calanus* dominant in these regions. In the North Atlantic, *Calanus finmarchicus* dominates the zooplankton. This species produces eggs in direct response to the availability of food, beginning to lay after the great spring phytoplankton blooms are already under way. In the North Pacific the dominant species, *C. cristatus* and *C. plumchrus,* store reserve materials in the bodies of the females and produce eggs early in the spring independently of the phytoplankton concentration. Thus the phytoplankton is grazed more intensively during the early stages of its blooms, largely preventing the massive spring outbursts that characterize the North Atlantic.

Where two great water masses meet and mix, the results may be startlingly different, depending on the character of the water masses. If each water mass supplies some complementary nutrient that was depleted in the other one, bloom conditions may prevail at the boundary region, resulting in a marked increase in productivity but a decrease in diversity. If, on the other hand, they do not furnish complementary nutrients, the populations may simply mix at the boundary, resulting in increased diversity.

Moving water masses carry with them the planktonic communities, or elements thereof, characteristic of the region of their origin, sometimes to great distances and into regions where these species are not characteristic. Divergences and convergences are of considerable biological importance. Water moving downward may carry with it phytoplankton into depths below the photic zone. Some of these organisms can live through long periods of such submergence, and if the water mass surfaces later in another region, they may produce blooms there if conditions are favorable.

Water masses frequently disappear under less dense water masses and later reappear on the other side, still containing much of the plant and microbial community characterizing the region of origin, although now separated from that region by a distinctively different community between them. Occasional upwelling of antarctic diatoms and dinoflagellates off the east coast of Australia is an example. These organisms must have been below the photic zone up to several years as this antarctic water passed below the tropical waters flowing across the Tasman Sea to New Zealand.

It is important to emphasize the dynamic and changing character of the plankton populations and the fact that their distributions in the water are not uniform. The ever-changing, drifting aggregations of planktonic organisms may well be compared to clouds in the atmosphere. Like the clouds, they occur at different levels, drift with the movements of their medium, and change in size, shape, depth, configuration, and composition; yet through all these changes they tend to display certain seasonal and regional regularities.

In any given place, marked changes in the plankton occur from time to time because of the varying seasonal conditions favoring different species in turn, because of interactions within the population itself, such as changes in the level of available nutrients and effects of grazing, and because water movements may bring plankton of a different character to the area and carry away those which were there.

All these factors, plus the difficulties of obtaining truly representative samples from the various depths and of knowing whether the samples obtained are representative, make the study of plankton both difficult and fascinating.

If plankton samples are taken regularly at any given station over a period of years, some of the broad trends and seasonal regularities become evident, and a type of plankton calendar for the area can be made, showing the principal seasonal changes and the species most likely to be dominant at different times of the year.

The continuous plankton recorder and the continuous, repeating water sampler-thermograph unit (pioneered by A. C. Hardy), which are towed behind commercial vessels on regular runs, as well as used with research ships, opened up a new dimension in marine research by making possible the regular plotting, on a wide scale, of marine conditions and the movements of water masses as reflected by the plankton organisms they contain. This procedure is making possible the prediction of marine conditions on a basis similar to the prediction of weather conditions based on widespread meteorological data. Since the movements of water masses are slower than those of the atmosphere, somewhat longer-range predictions can be made with the same degree of accuracy, which is of great importance to commercial fisheries. The continuous plankton sampler also helps greatly in assessing the general state of a fishery and the size of fish stocks, through its quantitative records of the numbers of fish eggs, as well as the quantities and types of fish food in the water. (See Hardy, 1926, 1936, 1939, 1958.)

STANDING CROP AND PRODUCTIVITY

Early estimates of the standing crop of marine phytoplankton and of productivity (Tables 4-1 and 4-4), based on net tows, gave misleading low figures because they neglected the nanoplankton. Even for the larger species of phytoplankton retained by the nets, such methods gave erroneous low estimates. Estimates based on other methods, such as the use of water-bottle samples and counting chambers, or dilution and culture techniques, have given significantly higher figures than estimates based on net phytoplankton taken at the same time and same station. The factor of increase is not constant but varies from one species to another.

Estimates of the standing crop at a given time and place are more easily made than are estimates of productivity. The standing crop can be estimated from direct counts or measurement of volumes or weights of samples taken from the seawater. Indirect methods, such as extracting and measuring the amount of chlorophyll from the plankton concentrated from a known volume of seawater and estimating from this the biomass of the phytoplankton that produced it, are also used.

Primary production is the production of new organic matter from inorganic substrates. It can be expressed either in terms of the energy utilized or in terms of the mass of new organic matter produced. Primary production is mostly carried on by photosynthesis. **Secondary production** is the production of nonphotosynthetic forms requiring organic substrates.

Gross productivity is determined by the total amount of energy absorbed. **Net productivity** represents the energy still available for organic synthesis after various energy losses—wastage, metabolic energy used in maintenance and respiration, etc.—are subtracted. **Gross production** is the total mass of organic matter, both plant and animal, actually produced in a given period of time. **Net production** represents the chemical energy in the gross product available for reuse.

The **standing crop** is the actual biomass of living things, or of particular organisms being considered, in existence at a given time. The **rate of turnover** is the rate at which the biomass of the standing crop is undergoing replacement. Rates of turnover vary widely for different species, as well as for different chemicals entering or being released from the biomass.

The **efficiency of production** is an expression relating the total available energy entering the system to the amount actually utilized in the production of organisms. The efficiency of production is highest for primary producers and diminishes rapidly as one goes to high trophic levels because of energy losses along the way. The efficiency of production for phytoplankton, for example, is of the order of 0.04% to 0.31%, whereas that for fishes is about 0.00005% to 0.00025%.

Primary production

To estimate productivity, it is necessary to measure directly the amount of new organic matter formed in a given time in a sample of seawater being held under the same conditions that obtain in the water mass for which productivity is being estimated. Alternatively, the productivity can be indirectly estimated from measurements of the rate of oxygen evolution, carbon dioxide incorporation, changes in pH, or some other physiological parameter correlated with the synthesis of new organic matter.

The rate of productivity in a given water mass changes with the diurnal and seasonal cycles; with changes in temperature, illumination, and depth; with the relative amounts of phytoplankton and zooplankton present in the water; and with the species present. There are so many variables and different types of factors that affect organic production, and production is so different from one area to another and from one time to another in the same area, that it is difficult to know to what degree the figures obtained can be extrapolated or generalized to give overall estimates of organic productivity in the oceans. It is also difficult to be sure to what extent the techniques used to measure production may have affected the rate of production in the samples.

One of the more difficult factors to evaluate is the relation between potential and actual photosynthesis in myxotrophic populations—populations of photosynthetic organisms that are also capable of obtaining part or all of their nutritional needs in a heterotrophic manner. It is difficult to determine the extent to which active photosynthesis is diminished by heterotrophy.

Further complications lie in the diurnal, seasonal, local, and irregular changes in activity and in rates of photosynthesis by plankton organisms. The maxima and minima of activity do not always occur at the same time of day or at the same time at different depths.

Another important factor, often difficult to assess, is the effect of grazing. Some calculations and experiments show that estimates of primary production, based on standing crops of phytoplankton, may be as much as an order of magnitude lower than actual production.

In addition to field and laboratory measurements, which supply most of the information, mathematical models of seasonal and local changes in plant biomass are a way of predicting changes and of showing where specific research is needed.

Methods of measuring production

Photosynthesis at sea is measured by the use of bottles containing either a given species of phytoplankton, a selected group of species, or a mixed random sample from the water being studied. The bottles are suspended at different depths in the euphotic zone or held aboard ship under conditions of light, temperature, etc. that simulate as closely as possible those of the water being studied. The **method of Gran** (1931, 1932) uses paired light and dark bottles, the dark bottle serving as a control to give a correction for the simultaneous respiration uncomplicated by photosynthesis.

Since 1952 the **incorporation of carbon-14** has been measured rather than oxygen formation. This more sensitive test was introduced by Steemann-Nielsen (1952). Solutions of $NaHC^{14}O_3$ are mixed with samples of seawater containing phytoplankton. After a suitable interval the phytoplankton is filtered out, and the amount of $C^{14}O_2$ taken up is measured by counting with a Geiger counter, usually the windowless type. Because the amount of plankton is usually small, it is assumed that there is no self-absorption of the beta particles under these counting conditions. It is, of course, necessary to know the amount of activity initially added, which must be measured under the same counting conditions, that is, with the same counting geometry and no self-absorption of beta particles. Since adequately thin planchets are nearly impossible to prepare, the zero-thickness activity is extrapolated from absorption curves obtained from thicker planchets. Jitts and Scott (1961), using liquid scintillation counters and new, thinner planchets, showed that this procedure is not accurate.

The possibility that some kinds of phytoplankton may excrete a portion of the C^{14} as soluble organic matter may also prove to be a limitation for this method. Appreciable amounts of C^{14} may at times appear in the form of the photosynthetic intermediate glycolic acid. The quantity of glycolic acid to appear seems unpredictable. Changes in respiration caused by the experimental technique may also cause differences in the results, as estimated by oxygen evolution and by C^{14} incorporations.

The two methods tend to agree fairly well in short-term experiments, but with more prolonged experiments, especially in tropical waters, the oxygen evolution method tends to give significantly higher results. In long-term experiments, depletion of nutrients in the light bottle may affect the results, as may the possible antibacterial effect of the light or of the actively photosynthesizing phytoplankton organisms in the light bottle. Also, since population changes in the light bottle and in the dark bottle are of a different character in prolonged experiments, the correction indicated by the dark bottle may not be valid for the light bottle.

Strickland and Terhune (1961) attempted to overcome the distortions introduced by the use of small containers, in which surface effects of the container walls exert noticeable influence on the results, by using large plastic bags, holding 32,000 gallons, submerged in situ. A raised, buoyed fiber-glass liner gave access to the bag for sampling. With such devices a less distorted picture of marine microbial interactions and ecology and of productivity can be obtained.

It is also difficult to be sure whether the C^{14} method is measuring net photosynthesis or total photosynthesis. It apparently measures net photosynthesis in long experiments and something between net and gross photosynthesis in short ones. Another factor of uncertainty lies in the possible importance of heterotrophic assimilation of dissolved carbon compounds. Healthy phytoplankton in the dark fixes about 1% of the amount of C^{14} fixed in the light, but under conditions of nutrient deficiency this amount may be increased to as much as 30% of the amount fixed in light.

Because there is a direct relationship in seawater between pH and total dissolved carbon dioxide, it is possible, with **the pH method,** to calculate the amount of carbon assimilated from changes in the pH of a sample during photosynthesis. For example, at 14° C. an increase of 0.15 in pH would be equivalent to the abstractions of 2.24 ml. (STP) of carbon dioxide, or 1.2 mg. of carbon per liter. If it is assumed that the plant cell converts all the carbon dioxide taken up into organic matter, a direct estimate can then be made of the amount of new organic matter produced during the period in which the measured rise in pH occurred (Raymont, 1963).

Discussion

In spite of all their limitations, these methods have given a great deal of valuable information. The vertical distribution of photosynthesis in a homogeneously distributed population under bright sunlight in the open ocean is given in Fig. 5-2. This type of curve is obtained by suspend-

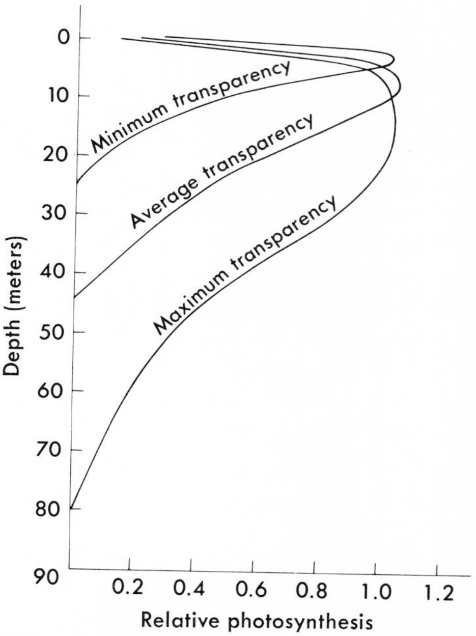

FIG. 5-2. Curves showing the vertical distribution of photosynthesis in a homogeneously distributed population under bright sunlight in the open ocean. (Modified from Yentsch, C. S. 1963. *In* H. Barnes [ed.] Oceanography and marine biology. Vol. 1. George Allen & Unwin, Ltd., London.)

ing a series of bottles, all containing the same amount and kind of phytoplankton, at different depths for the same amount of time. It must be realized, of course, that in the ocean the phytoplankton is not homogeneously distributed and that those species which normally occur near the top tend to have their maximum photosynthetic efficiency at higher levels of illumination than do those near the bottom. For species near the bottom of the euphotic zone, **light saturation**—the amount of light that gives maximum photosynthesis—occurs at lower intensities. The depth of the euphotic zone varies greatly, being a function of the amount and angle of incident light, water transparency, and the smoothness of the surface. On clear, bright days in the open ocean, the compensation depth may be approximately 85 meters. In more turbid oceanic waters, it is commonly between 25 and 85 meters, whereas in turbid coastal waters, it may run from 5 to 25 meters. The phytoplankton itself is somewhat self-regulating. As it becomes more abundant during a bloom, the water becomes less transparent, and the depth of the euphotic zone lessens.

The total amount of primary production per unit area of the ocean surface is best calculated by integrating the rates from several depths, and relating the total photosynthesis to light saturation (Pmax), to incident light (Io), and to a vertical gradient of intensity defined by the average attenuation coefficient for the euphotic zone (Ze), the base of which is the intensity at extinction (Ie). Talling (1961) calculated the relation between daily photosynthesis and that occurring at light saturation and gives the following formula for calculation of primary production per unit area:

$$\Sigma\, P = \frac{\text{Pmax}}{\text{Ze}}\left(\ln \frac{\text{Io}}{\text{Ie}}\right)$$

This formula does not wholly apply to

TABLE 5-1. Estimates of gross and net organic production of various systems in grams of dry weight produced per square meter per day*

	Gross	Net
1. Mass outdoor *Chlorella* culture		
Maximum		28.0
Mean		12.4
2. Terrestrial crops in growing season (maxima)		
Sugar		18.4
Rice		9.1
Wheat		4.6
3. Terrestrial habitats		
Spartina (marsh grass)		9.0
Pine forest (best growing years)		6.0
Tall prairie		3.0
Short prairie		0.5
Desert		0.2
4. Marine (maxima for single days)		
Coral reef	24.0	(9.6)
Turtle grass flat	20.5	(11.3)
Grand Banks (April)	10.8	(6.5)
Continental shelf	6.1	(3.7)
Sargasso Sea (April)	4.0	(2.8)
5. Marine (annual averages)		
Long Island Sound	2.1	(0.9)
Continental shelf	0.74	(0.40)
Sargasso Sea	0.74	(0.35)

*Extracted from Ryther, J. H. 1959. Potential productivity of the sea. Science **130:**602-608.

TABLE 5-2. Ocean productivity*

Carbon assimilation per year	100-200 gm. C/m.^{2} (H. W. Harvey; G. A. Riley)
Area of oceans†	3.5×10^{14} m.^{2}
Annual productivity‡	5.3×10^{10} tons carbon
Maximum productivity	1.6×10^{11} tons carbon

*From Vishniac, W. 1968. Autotrophy; energy availability in the sea and scope of activity. *In* C. H. Oppenheimer [ed.] Marine biology. Vol. IV. New York Academy of Sciences, New York.
†Not counting ice-covered waters.
‡P and N are limiting, so that for this amount of organic matter all available P and N would have to turn over two to ten times each year.

natural conditions, under which the rate of photosynthesis for each species of phytoplankton present also varies with the depth of the particular cells. Steeman (1962) suggests some modifying factors to compensate for light adaptation and efficiency of photosynthesis.

Ryther gives estimates of the gross and net organic production of various systems in grams of dry weight produced per square meter per day. Table 5-1 gives examples of these estimates.

It appears from these estimates that primary production in the ocean on a unit area basis is of the same order of magnitude as that on land or slightly below it. However, because of the greater extent of the seas, it is estimated that in their entirety, the oceans produce at least two to three times as much as all the land masses together.

A somewhat different approach may perhaps give more reliable estimates of overall productivity, or at least an indication of the upper limits of such productivity. Table 5-2 gives an estimate of the upper limit of ocean productivity based on what is known of the amount of solar energy available for photosynthesis and the amount of energy required for the fixation of a given amount of carbon. This estimate is then compared with an estimate of annual productivity.

CHAPTER

6 THE PHYTOPLANKTON

The most obvious elements of the phytoplankton are diatoms and dinoflagellates. Less conspicuous generally, but of local importance at some times and in certain regions, are other groups of photosynthetic flagellates and blue-green algae.

DIATOMS

Siliceous shell of a diatom

The diatoms are unicellular algae characterized by a skeleton, or capsule, the **frustule,** which is composed of two valves, the **epitheca** and the **hypotheca,** that fit together somewhat in the manner of the two halves of a Petri dish. These valves are impregnated with silica, which gives them a glasslike character. A broad, hooplike connecting band is attached at right angles to the incurved edges of each valve. Together, the two connecting bands form the **girdle.** As the diatom grows, the two halves are pushed farther apart and the connecting bands become broader.

Seen in lateral, or girdle, view, diatoms commonly look somewhat rectangular. Viewed from above in the top, or valve, view, they exhibit a great variety of forms: circular, triangular, linear, crescentic, wedge-shaped, boat-shaped, etc.

The valves are transparent and usually beautifully and symmetrically ornamented with a variety of markings. Often the marks consist of fine lines on the face of the valve, which in many cases, on very high magnification, can be resolved into rows of separate dots corresponding to minute concavities in the wall. Some genera show complex aerolated structures resulting from larger chambers within the wall. Motile naviculoid types have thickenings known as **nodules.** These complex structures, one at each end and one at the center, are generally connected by a long slit called the **raphe,** which is visible in the girdle view.

Most diatoms occur singly, but many species form characteristic chains or aggregates of individuals held together by protoplasmic junctions, mucilage, or interlocked spiny or hairlike projections from the frustules. The groups so formed have shapes that are characteristic for different genera and are often complex. Some of them look superficially like filamentous algae.

The cytoplasm forms a relatively thin lining along the inside walls of the valves, surrounding a vacuole filled with cell sap. The nucleus commonly occupies a more-or-less central position, with cytoplasmic strands extending from it to other parts of the cell (Fig. 6-2). The cytoplasm contains chloroplasts in which photosynthesis is carried out. The brownish color characteristic of most diatoms is caused by the pigment **diatomin** in the chloroplasts. Diatomin somewhat resembles the pigment of the brown algae, and masks the chlorophyll. Most diatoms store oil or fatty acids, rather than sugar, as the end product of photosynthesis. Under exceptional conditions a particularly rich growth of planktonic diatoms may produce sufficient oil to form an oily slick at the surface of the sea, up to several miles across. In the tropics somewhat similar slicks are commonly formed parallel to the coastline by the blue-green alga *Trichodesmium.*

Most of the figures for this chapter were redrawn from various sources, especially Lebour (1930), Gran (1912), Gross (1937), Marshall (1933), and Grassé (1952). Several have been somewhat simplified.

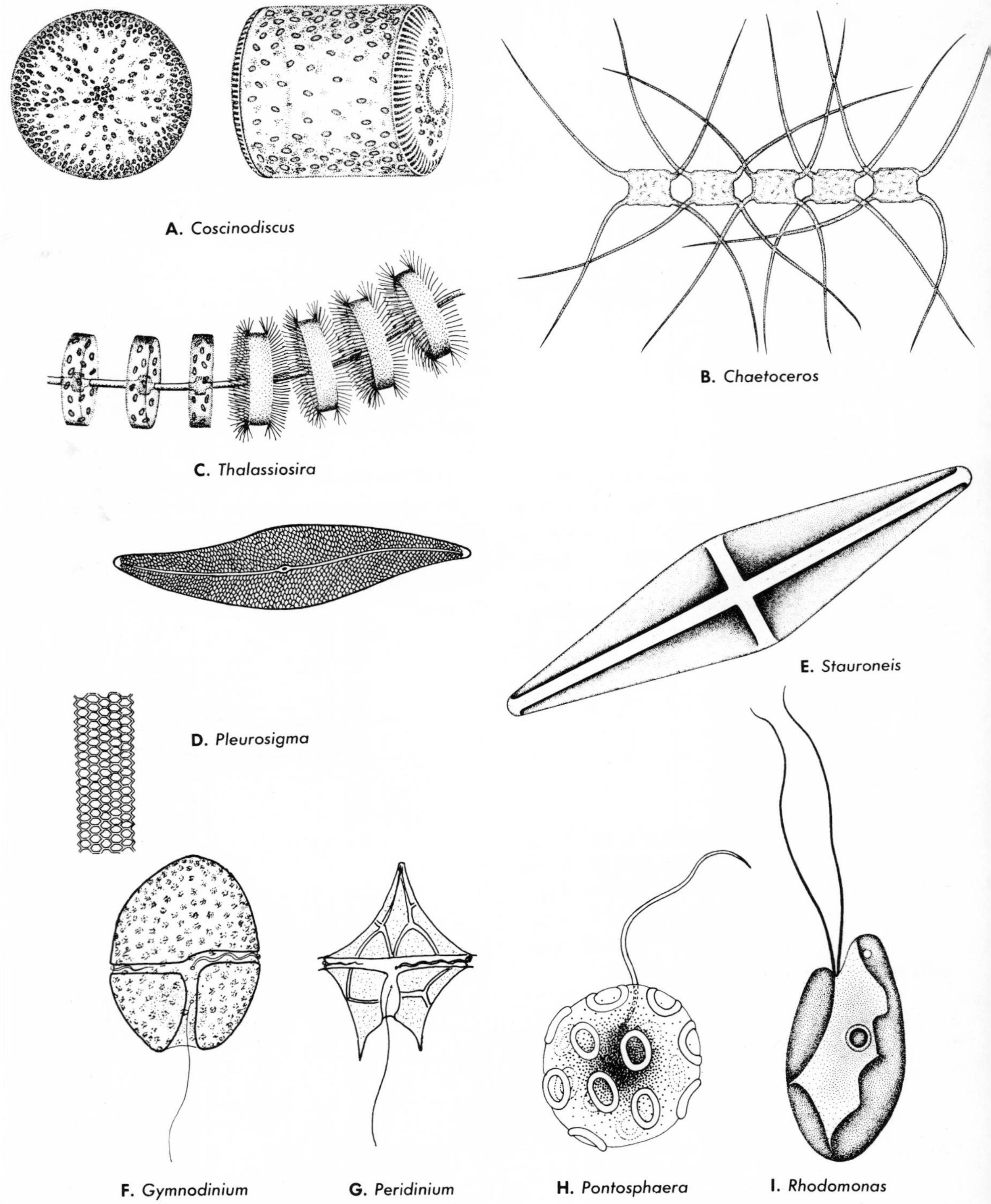

FIG. 6-1. Representatives of some groups of phytoplankton. **A-C,** Diatoms, centric types; **D-E,** diatoms, pennate types; **F,** naked dinoflagellate; **G,** thecate, or armored, dinoflagellate; **H,** Coccolithophore; **I-K,** others.

Diatoms range in size from less than 10 μm in length (or diameter, for small species) to about 1 mm. in very large ones. Even within the same species there may be considerable variation in size—up to thirtyfold or more in some cases. This variation is partly the result of the peculiar method of reproduction (Fig. 6-3). At each cell division one daughter cell retains the old epitheca, or larger half of the frustule. The other daughter cell retains the old hypotheca. Each daughter cell secretes a new hypotheca, so that the old hypotheca of the mother cell becomes the epitheca of one of the two daughter cells, which is therefore slightly smaller than the other daughter cell. Thus, as divisions proceed, part of the population is continually becoming smaller, resulting in a decrease in the average cell size for the population.

This process clearly cannot continue indefinitely for that part of the population approaching minimal size for the species. Maximum size is commonly restored by formation of so-called **auxospores.** The protoplast discards the old valves altogether, increases considerably in size, secretes a membrane with a little silica around

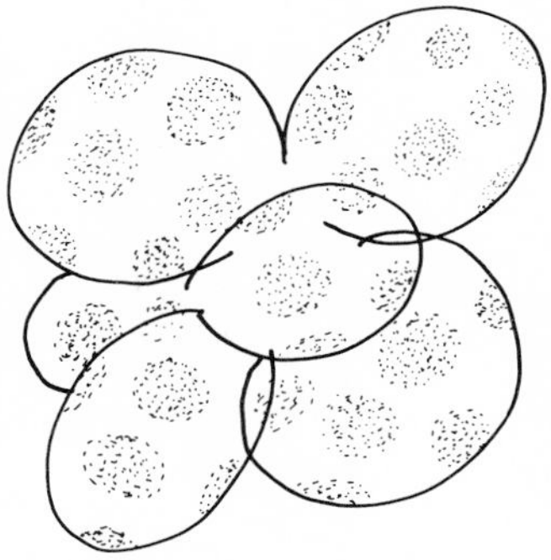

J. *Phaeocystis*

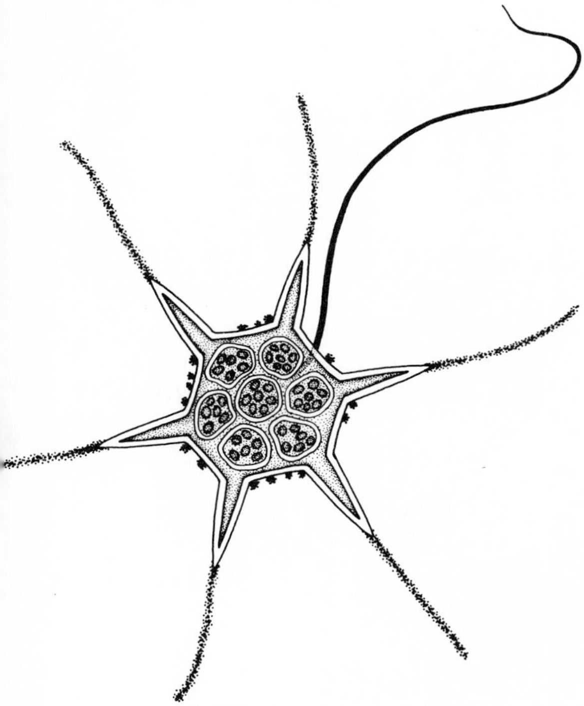

K. *Distephanus*

FIG. 6-1, cont'd. For legend see opposite page.

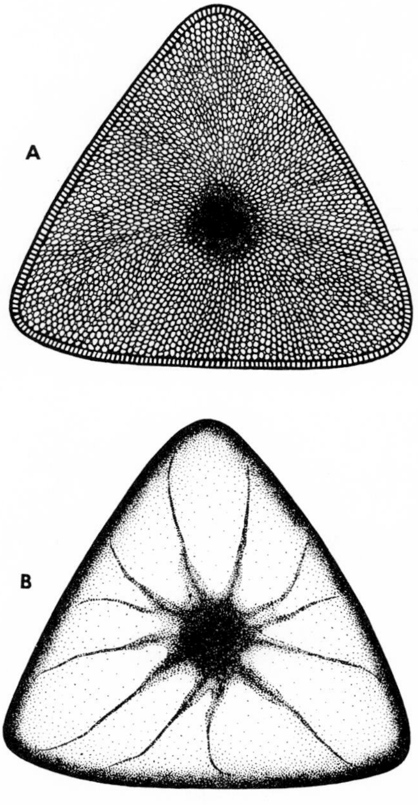

FIG. 6-2. *Triceratium,* showing, **A,** level of surface, and, **B,** nucleus and cytoplasmic strands.

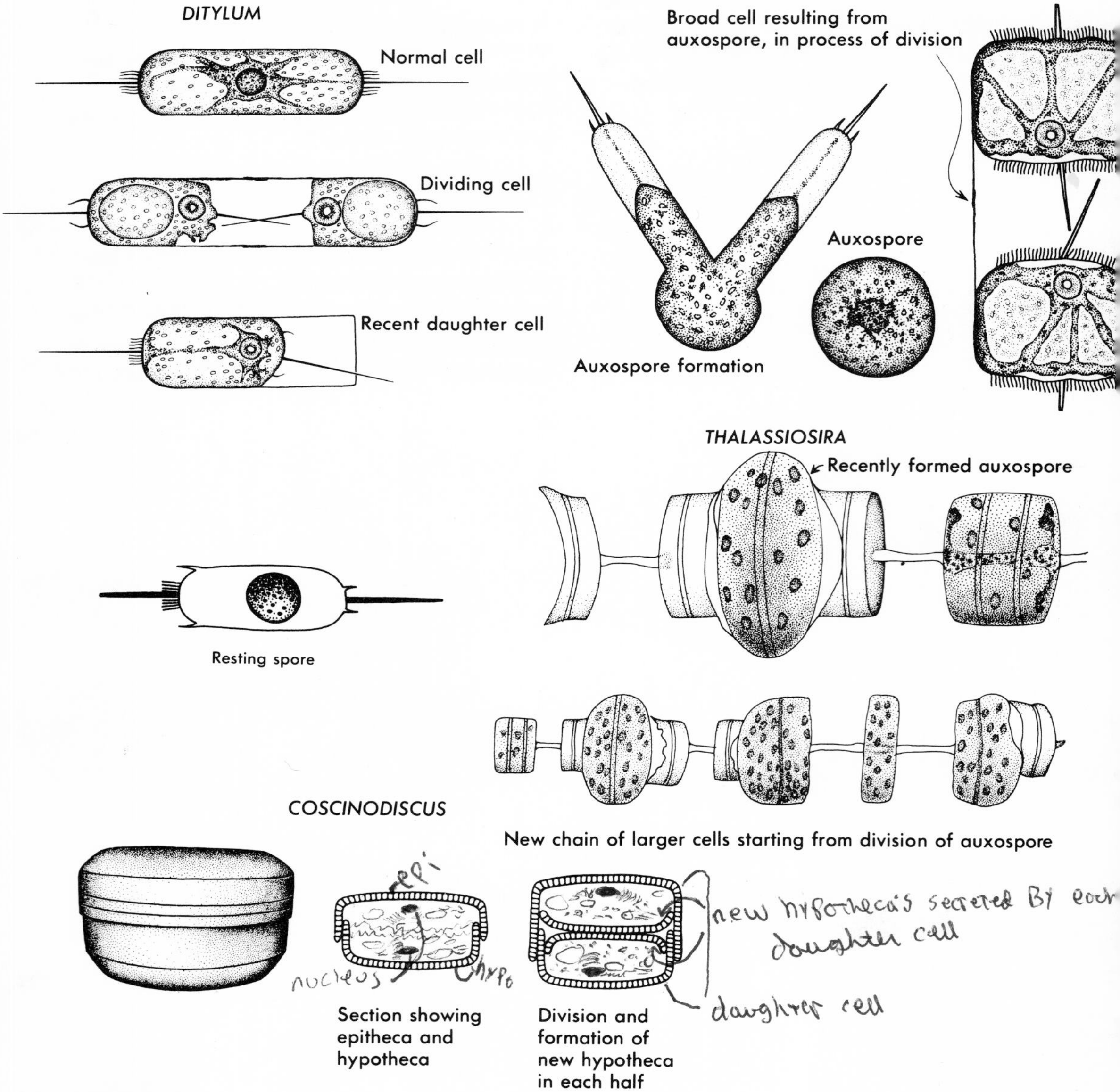

FIG. 6-3. Reproduction in diatoms.

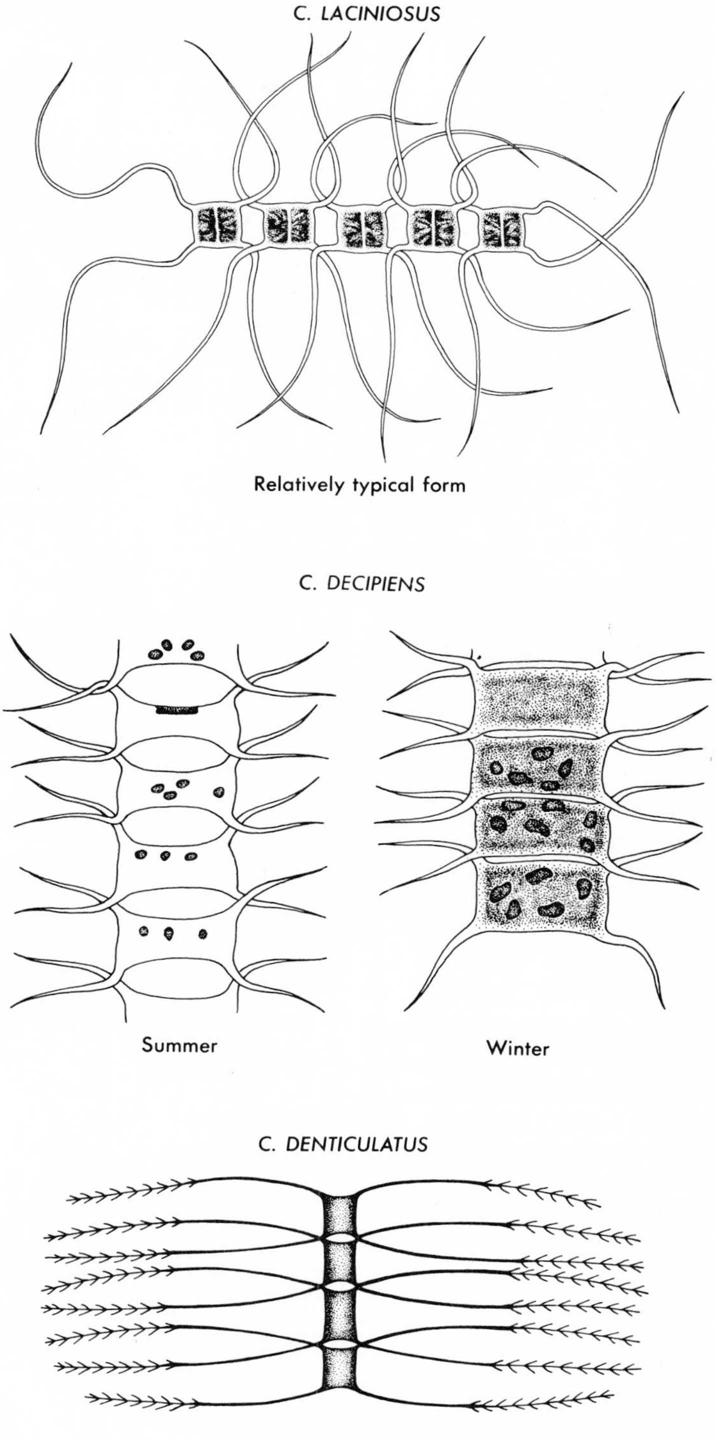

FIG. 6-4. *Chaetoceros,* illustrating variations within and between species.

Continued.

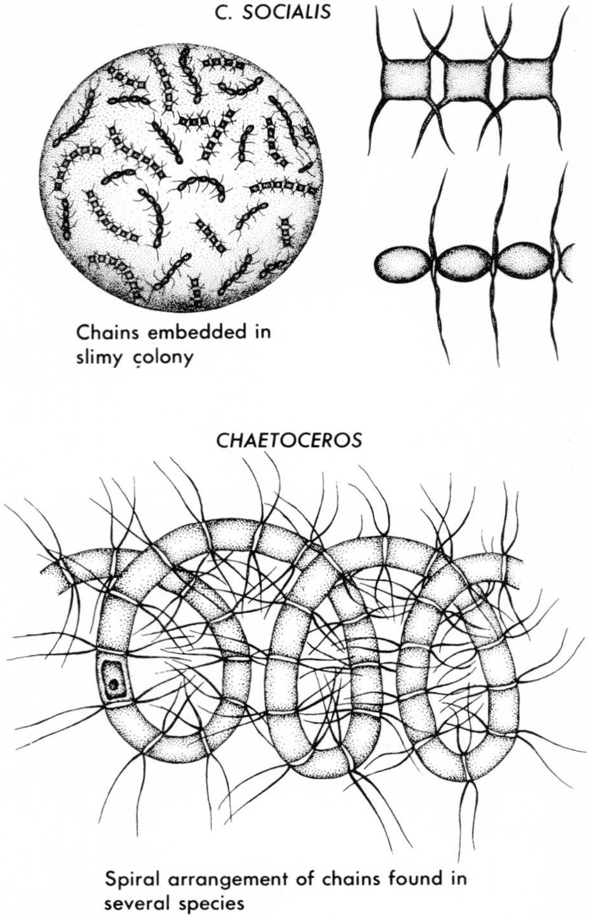

FIG. 6-4, cont'd. *Chaetoceros,* illustrating variations within and between species.

itself, and develops new valves of large size. The auxospore, then, is not a resting stage but a device for restoration of normal size. Auxospore formation may be influenced by external conditions such as light intensity and level of dissolved nutrients, and perhaps by the release of metabolites into the water, as well as by the absolute minimum size attained. There is evidence of formation of haploid sexual gametes prior to auxospore formation, the auxospore being formed by the fertilized diploid zygote.

Some diatoms, especially neritic species, also form resting spores, enabling them to survive severe winter conditions or other unfavorable periods. The formation of resting spores involves a considerable loss of cell sap and the rounding up of the relatively dense protoplast. Such denser resting spores commonly sink into deeper layers of water or to the bottom in shallow areas. They germinate again and begin reproducing when conditions are favorable once more. Low light intensity, low temperature, and depletion of nutrients in the water tend to favor the formation of resting spores. In cultures, overcrowding also promotes it.

In addition to the planktonic diatoms, all shallow waters where there is available substrate within the euphotic zone have large populations of benthic, attached, and

water over the continental shelf

living on bottom

upper strata of water receiving sufficient light for photosynthesis.

interstitial diatoms. These diatoms constitute an important element of the "algal film," forming diatom slicks on rocks and other submerged surfaces and providing a principal food for many grazing animals, including gastropods and chitons.

Vast numbers of diatoms die because of seasonal changes, local impoverishment in plant nutrients, grazing by zooplankton, or because they settle below the euphotic zone or are carried below it by vertical movements of the water. The result is an accumulation of great numbers of dead diatoms and their frustules at the sea bottom. In some areas they are a principal component of the marine ooze.

Where ancient marine sediments have been raised during geological time to become part of the land, fossil diatoms—along with the skeletons of Foraminifera, Radiolaria, and other organisms—may be studied to obtain clues to the conditions prevailing at the time the sediments were laid down. Such studies help in reconstructing and understanding the history of the earth. Where diatoms constitute the bulk of such sediments, they form the so-called diatomaceous earth, which has many industrial and scientific uses.

Distribution of diatoms

Diatoms are of worldwide distribution but constitute the dominant mass of the plankton only during the spring and sometimes the autumn in high latitudes and near the shore in the Pacific and Indian oceans. In these regions they sometimes reach densities exceeding 1 million/liter of water.

In high northern latitudes there are circumpolar deposits of diatomaceous earth. Most of the diatoms in these sediments are not the species that dominate the oceanic plankton today. They seem to have historical rather than contemporary significance, perhaps representing an ecosystem that no longer exists in the oceans. Many of them may have been deposited in animal feces or may have been heterotrophic forms living in the bottom sediments of shallow seas. Changes in climate and in the relative abundance of other groups such as various nanoplankton flagellates, bacteria, or blue-green algae may have contributed to their disappearance.

In general, diatoms are oceanic in cold waters of the polar regions, whereas in warm temperate and tropical areas they are most abundant as neritic and estuarine forms. Lohmann (1908) estimated that for the Northern Hemisphere as a whole, coastal waters are more than fifty times as productive of diatoms as are oceanic waters.

The fact that nitrate nitrogen is less abundant in warm waters, where it may at times be limiting, probably explains the greater relative abundance in the tropics of blooms of cyanophytes, which can utilize ammonia or free nitrogen.

It has been claimed that, in general, truly oceanic species and those living in warmer waters tend to have thinner, lighter frustules than neritic and cold-water species, and that truly plantonic forms have generally lighter, more delicate shells than bottom-living or attached forms. Some widely distributed species show distinct varieties differing in thickness of the frustules and in average size in polar and temperate regions.

Like flowering plants, diatoms appeared rather late in evolution and have come to occupy a dominant role in their environment, exerting far-reaching effects on the evolution of associated organisms and on the physical, chemical, geological, and biological processes of the earth's surface. Among the earliest and best-known fossil deposits of diatoms are those from the Moreno shales in central California, dating from the late Cretaceous period. Some records occur back into the Triassic period.

This period of the earth's history, the latter half of the Mesozoic era, when the great dinosaurs still ruled the earth, foreshadows some of the most striking changes that have ever taken place in the character of life on earth. It was during this period that many of the major groups of animals and plants

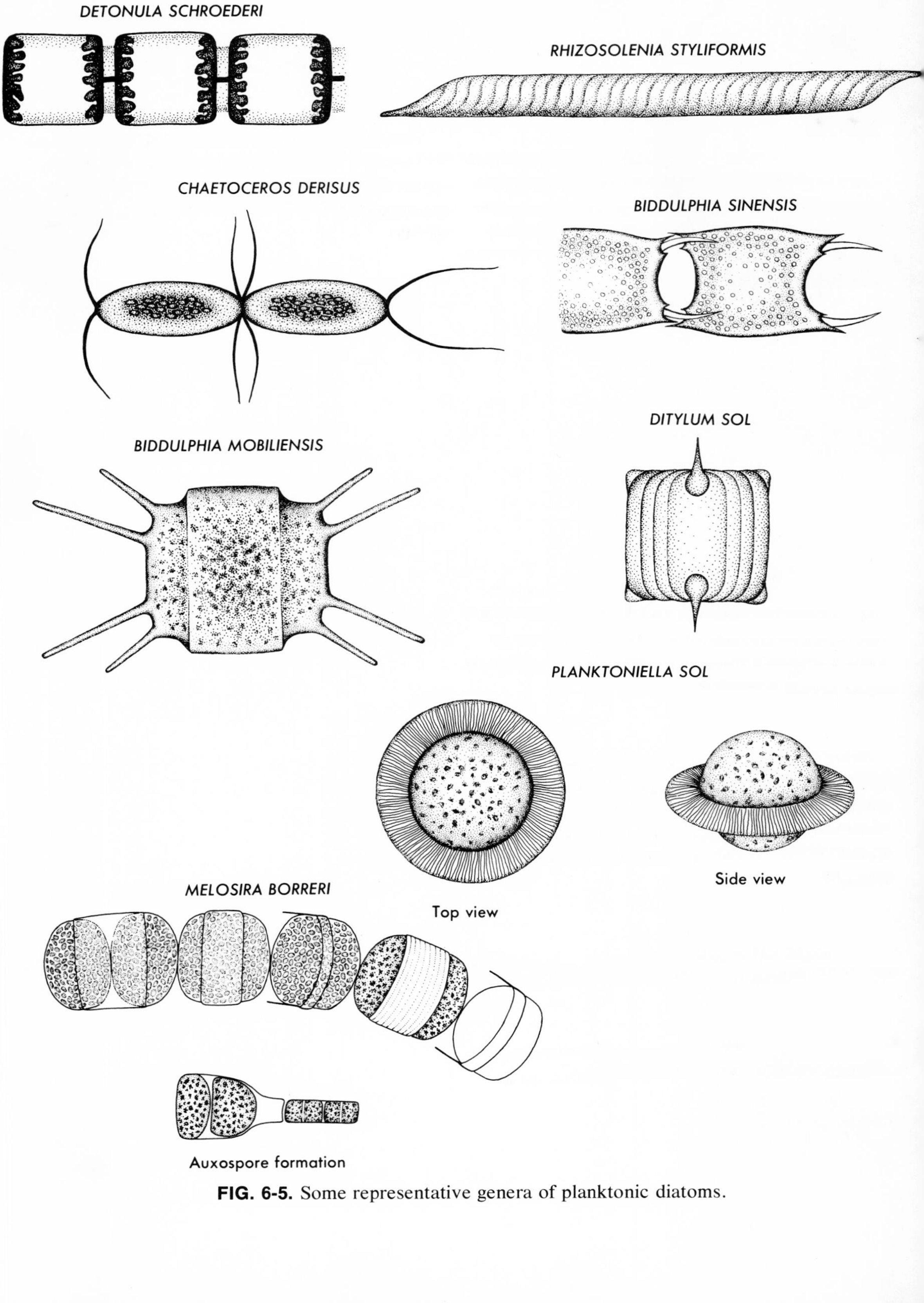

FIG. 6-5. Some representative genera of planktonic diatoms.

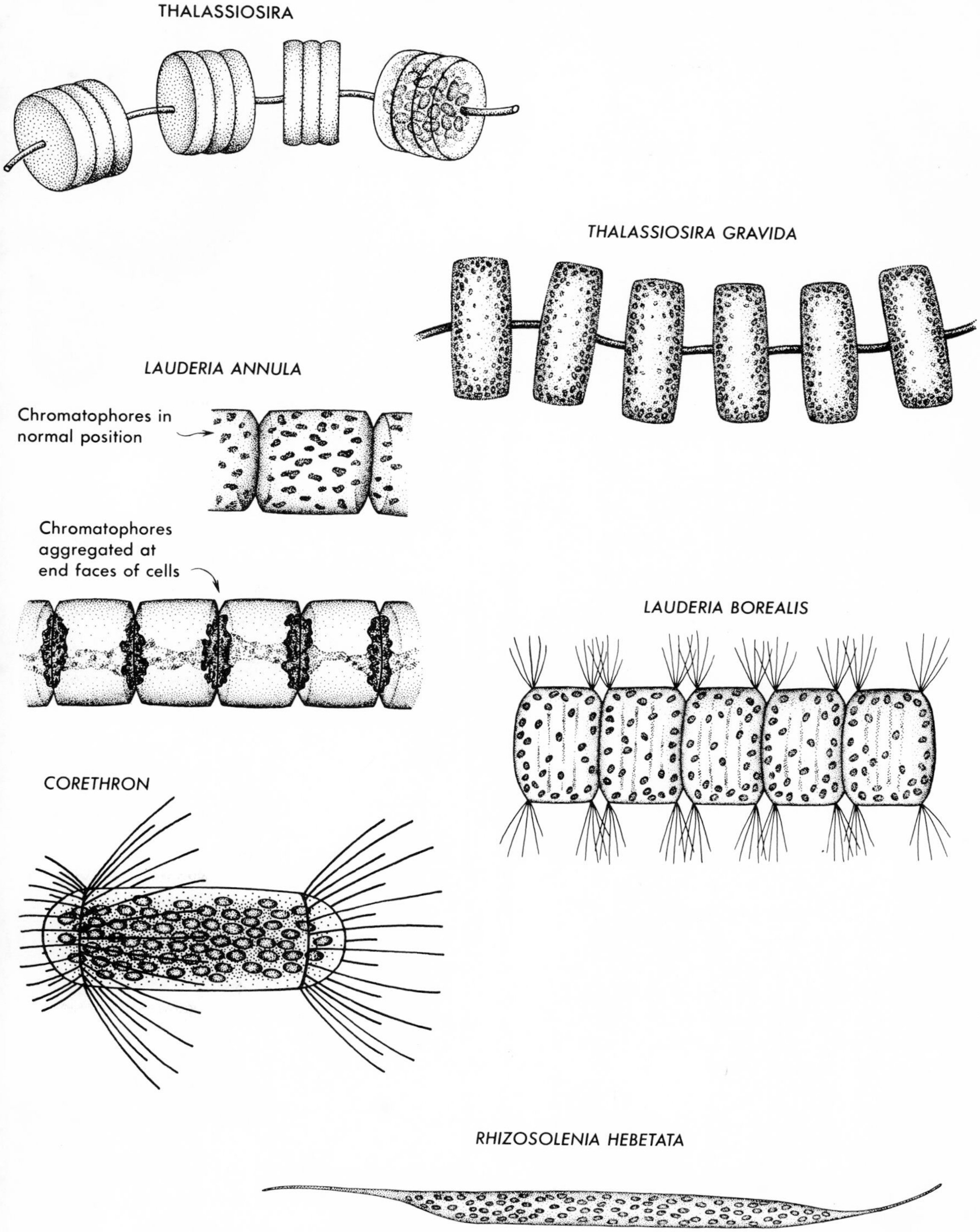

FIG. 6-5, cont'd. Some representative genera of planktonic diatoms.

Continued.

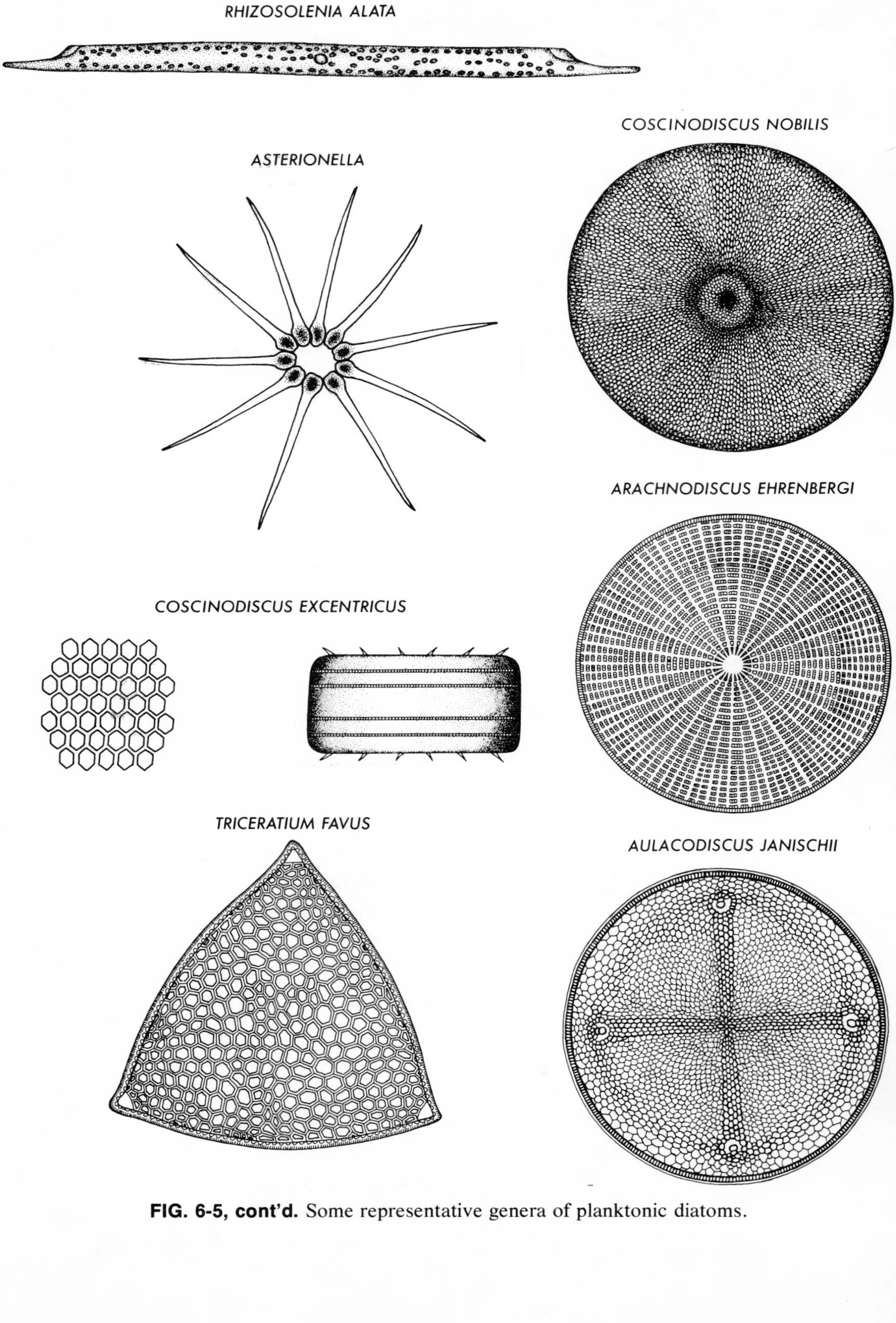

FIG. 6-5, cont'd. Some representative genera of planktonic diatoms.

that characterize life today—the diatoms, flowering plants, mammals, birds, and insects, to name a few of the more outstanding—had their origins. It was during this time that the great reptiles lost their long hold on the earth and disappeared forever, together with many of the archaic plants that characterized the landscapes of their time. It is hard for us to imagine what the world was like before these changes or to visualize the forces that culminated in such a sweeping renovation and alteration of life on earth.

Classification of diatoms

See taxonomic appendix.

DINOFLAGELLATA

The dinoflagellates (Fig. 6-1, *F* and *G*) constitute a second well-marked group of unicellular algae, the class Dinophyceae. Like diatoms, they are an abundant and important element in plankton of all seas.

Although diatoms usually dominate the phytoplankton in cold waters, dinoflagellates commonly outnumber them in tropical and subtropical waters. The genera *Peridinium, Ceratium, Prorocentrum, Goniaulax, Exuviella, Oxytoxum,* and *Gymnodinium* are especially abundant and in the seas of the southwest Pacific commonly comprise most of the phytoplankton.

The body of a dinoflagellate is enveloped in a cellulose membrane that in many forms also bears hardened plates variously shaped and sculptured.

The most typical and familiar dinoflagellates have a body surface provided with two grooves, each having a flagellum. These are the **annulus,** or **girdle,** a transverse groove passing around the body like a belt, and the **sulcus,** a longitudinal groove in the posterior part of the body from which a trailing flagellum arises. In the armored dinoflagellates the annulus may be covered by a perforated plate, the **cingulum.** The transverse flagellum is commonly somewhat flattened and ribbonlike. The action of these two flagella gives the dinoflagellates a characteristic swirling motion as they swim. The annulus divides the body into an anterior cone (the **epicone,** or **epitheca**) and a posterior portion (the **hypocone,** or **hypotheca**).

The nucleus is usually single and rather massive, with evenly distributed chromatin granules. Many planktonic dinoflagellates have numerous small chromatophores that are often dark yellowish brown or greenish and are located peripherally in the body, and often there is a light-sensitive red pigment spot called the **stigma.** In some groups the stigma is developed into a remarkable eyelike structure, the **ocellus,** with a myeloid lens and a dark-pigment ball (family Pouchetiaceae). There are no contractile vacuoles, but commonly there is a large vacuole, termed the **pusule,** filled with pinkish fluid and connected to the exterior by a canal. It may be concerned with digestion. Other vacuoles may also be present; sometimes there is one fairly large vacuole surrounded by a ring of small ones.

The typical dinoflagellates are divided into two orders—**Gymnodiniales,** the naked or unarmored dinoflagellates, and the **Peridiniales,** the armored dinoflagellates.

Red tides

Some species produce powerful poisons. When an extensive bloom of such a species occurs, it may produce a so-called "red tide," discoloring the water with the presence of millions of cells, killing great numbers of fish and other animals, and causing losses to the local fishing industry and great annoyance to residents in areas where large numbers of dead fish wash up onto the beaches or into the bays. People are sometimes poisoned by eating fish or shellfish that have concentrated such organisms with their filter-feeding mechanisms. *Goniaulax polyhedra* and *Gymnodinium breve* are among the better-known species causing red tides.

During blooms of these species the water begins to look discolored at concentrations of 200,000 to 500,000 cells per liter. At the peak of a red tide, bloom cell concentrations may reach from 1 to 6 million cells per liter. Cobalt, in the form of cyanocobalamin, seems to play a role in the production of red tides off Florida.

ZOOXANTHELLAE

Marine animals of diverse groups, including Radiolaria, many coelenterates, especially the corals, the giant clam *Tridacna,* and many others, carry in certain of their cells and tissues great numbers of small round algalike brownish or greenish brown plant cells—the **zooxanthellae.** These zooxanthellae may be of great importance to their hosts. In corals, for example, a vast amount of animal tissue is often concentrated in a relatively small amount of water—as on an extensive coral reef at low tide—and the respiration and excretory activity of so much animal tissue would quickly foul the water and kill the animals. However, the zooxanthellae use much of the carbon dioxide and nitrogenous wastes before these substances enter the water, and their photosynthetic activity liberates oxygen into the water and into the tissues of their hosts. Food materials synthesized by the zooxanthellae may also be shared with their hosts. Muscatine (1967) demonstrated that zooxanthellae carrying on photosynthesis in the presence of host tissue excrete up to 40% of the labeled photosynthate as glycerol into the host tissue. In the absence of host tissue, they carry on photosynthesis but produce little or no glycerol or other excreted photosynthate. In some cases, part of the zooxanthellae population may be regularly digested either in the host tissues or during times of stress, thus reducing the need to capture as great a quantity of zooplankton food as would otherwise be necessary.

The zooxanthellae at times produce a free-swimming, swarming stage, usually either small cryptomonads such as *Rhodomonas* or naked dinoflagellates, thus demonstrating the relationships of these symbiotic plant cells to parts of the plankton.

Some dinoflagellates have lost their photosynthetic capacity and exhibit saprozoic, holozoic, or parasitic modes of nutrition and life. Some of these have become so modified that they are difficult to recognize as dinoflagellates.

Classification of dinoflagellates

In addition to the typical dinoflagellates discussed above, there are two groups usually included in the class that differ considerably in structure—the adiniferids, or adinidans, and the Cystoflagellata.

The **Adinida,** or **Adiniferida,** do not possess the girdle and sulcus characteristic of other forms. The cellulose covering is commonly bivalved, the valves being regarded as lateral. The flagella are both at the anterior end, one extending forward, the other curving around more or less horizontally in the manner of the transverse flagellum of other dinoflagellates.

The **Cystoflagellata** are especially interesting. They are large, globular, rather gelatinous, hollow, fluid-filled cells that look almost like minute medusae. They have only the longitudinal groove, or sulcus, and a peculiar tentacle-like organ much thicker than a flagellum and slowly movable. The principal genus is *Noctiluca* —famed for its bioluminescence. *Noctiluca* gives off a flash of light when disturbed. When it is the dominant plankton, the surface waters may be dramatic at night—flashes and trails of light marking the movement of fish, boats, or nets through the water, or waves along the shore shining with a strange light as they break.

The classification of the dinoflagellates is summarized in the Taxonomic Appendix. Only a few of the principal families and genera are mentioned.

OTHER PHYTOPLANKTON

Some blue-green algae and some small flagellates belonging to the Chrysophyceae and Chlorophyceae sometimes occur in the plankton in sufficient numbers to be of at least temporary local importance.

Blue-green algae

Pelagic Cyanophyceae includes species of *Oscillatoria, Trichodesmium, Haliar-*

achne, and *Katagnymene.* In the tropics they may at times exceed all other phytoplankters in numbers.

Perhaps the best-known members of the blue-green algae in marine plankton are species of *Trichodesmium.* These slender cells, which float on the water surface, are rich in the pigment phycoerythrin, giving them a reddish color. When present in great numbers, they may produce extensive slicks and give off a characteristic odor of chlorine. *Trichodesmium* slicks are common in the Java Sea and Indian Ocean, sometimes paralleling the coast for many miles. The Red Sea was so named because of the periodic discoloration of the water by *Trichodesmium,* and the Gulf of California is sometimes called the Vermilion Sea for the same reason. In cold waters the only planktonic blue-green alga is *Trichodesmium rubescens.*

Mass deaths of coral have been reported when windrows of *Trichodesmium* driven ashore by wind pile up on the reef. This occurrence could be of geological importance, since it not only kills the existing reef but also results in lowered redox potential, favoring hydrogen sulfide production and mineralization, and severely interfering with further reef formation.

Some blue-green algae grow on and in reduced sediments in bays and estuaries at E_h values of -200 mV. or lower. In the East Indies and nearby tropical regions they constitute the principal element in the tambaks or marine fish culture ponds. Forms such as *Oscillatoria, Microcoleus,* and *Lyngbya* dominate the biocenosis, and *Phormidium* may form a skinlike growth over the mud. It is probable that the long-continued productivity of the tambaks, some of which have been producing fish for as long as 85 human generations without artificial fertilization, is dependent on blue-green algae. Some of these algae fix nitrogen, and some occur in metabolic association with bacteria that dissolve calcium and iron phosphates at pH values as

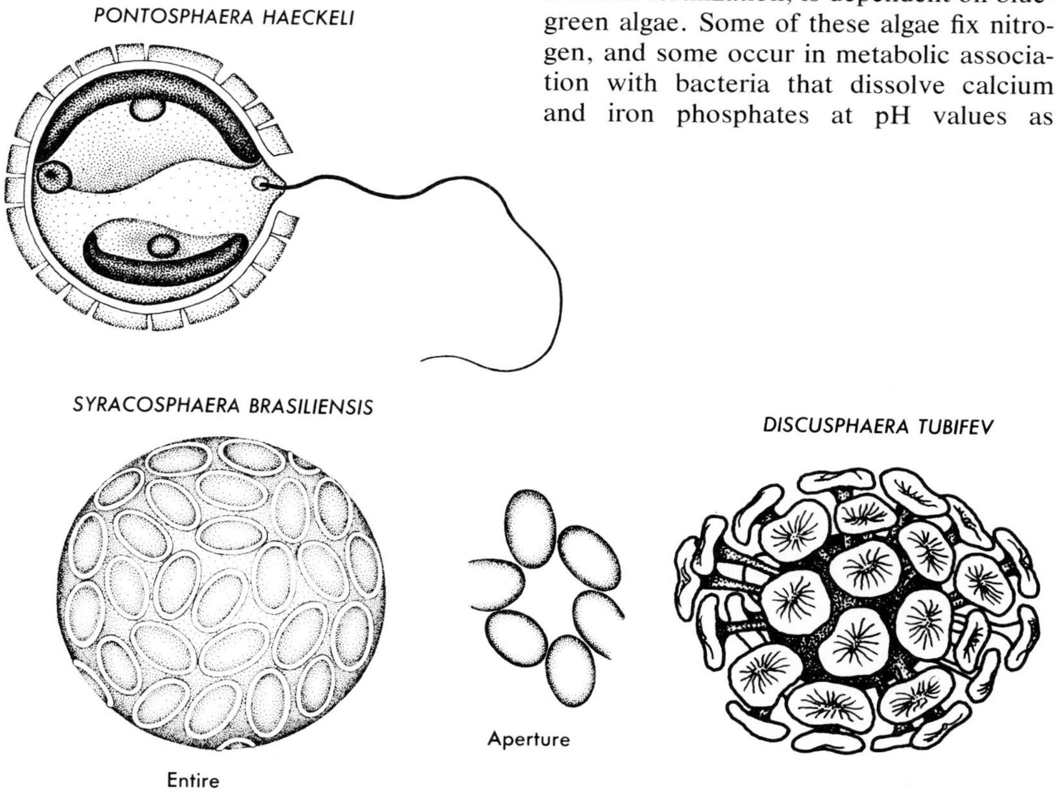

FIG. 6-6. Coccolithophores.

FIG. 6-6, cont'd. Coccolithophores.

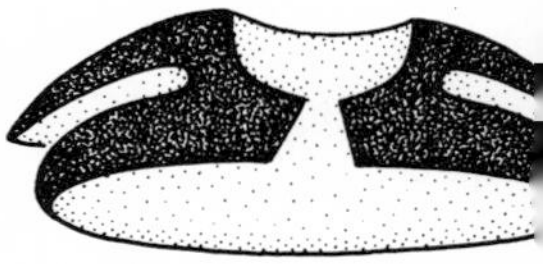

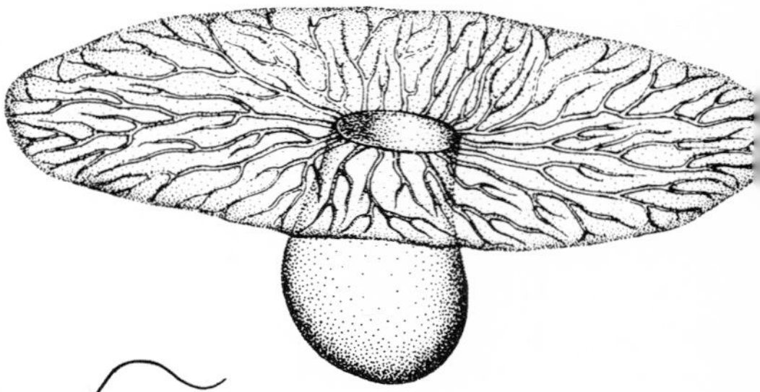

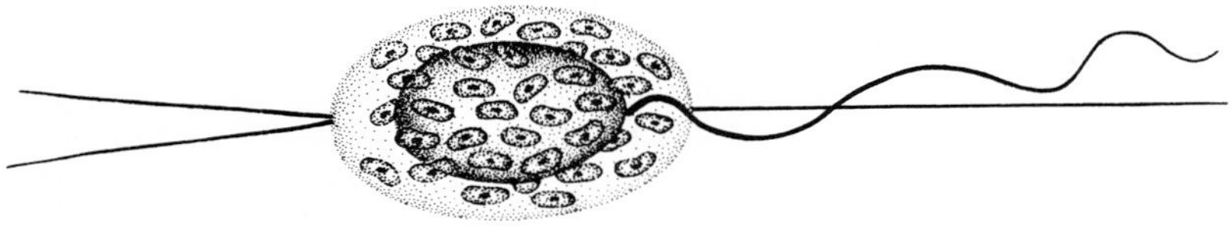

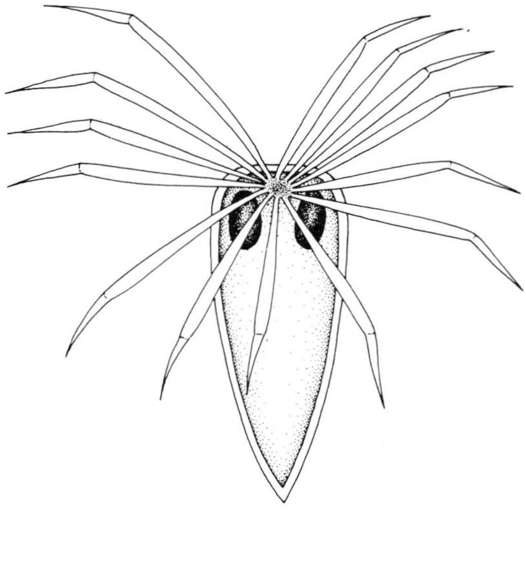

high as 8.5. Their planktonic hormogonia enter directly into the food cycle, and their decay enriches the mud. They give greater continuity of production than do the diatoms or green algae, which tend to occur irregularly or in bursts.

Tapetic species, forming felts on sediments, may aid in stabilizing the shore, preventing erosion and adding enrichment so that later settlement by *Spartina* and pickleweed can take place.

Species of *Anabaena,* a genus better known from freshwaters, occur in more northern brackish waters. One genus, *Richelia,* is found in diatoms of the genus *Rhizosolenia.*

Flagellates

Flagellates belonging to the group Chrysophyceae or Chrysomonadina are sometimes important. Among the larger forms, *Phaeocystis* forms gelatinous brownish colonies in which the individuals are rounded up in the **palmella phase.** In temperate latitudes, blooms of *Phaeocystis* often color the water brownish and may clog plankton nets. New colonies are developed from free-swimming flagellated zoospores. This genus is important to fishermen because herring avoid areas where *Phaeocystis* is blooming. Different species have been found to prefer different temperature ranges of the water and hence to characterize separate water masses. *Halosphaera* is a bright green unicellular flagellate sometimes occurring in great numbers.

One of the best-known groups is the family **Coccolithophoraceae** (Fig. 6-6). These small uninucleate flagellates usually have two yellowish chromatophores and two flagella. Discs of calcium carbonate, known as coccoliths, are on the surface of the cell. These coccoliths have been long known from marine sediments, where they sometimes make up a significant portion of the "Globigerina ooze," or calcareous sediment. The family is divided into two subfamilies on the basis of the structure of the coccoliths. Subfamily **Coccolithophorinae** has perforate (tremalith) coccoliths, whereas subfamily **Syracosphaerinae** has imperforate (discolith) coccoliths. The majority of species occur in warm seas, but some, such as *Pontosphaera huxleyi* and species of *Syracosphaera,* are abundant in colder waters.

In spite of the fact that Coccolithophoraceae are potentially photosynthetic, many of them are also found at considerable depths well below the euphotic zone, where they must be existing heterotrophically.

Another interesting group are the **Silicoflagellaceae** (also called Dictyochaceae). These have a siliceous skeleton. The cytoplasm contains small brownish yellow spherical bodies, and the nucleus is near the center of the body. One long flagellum is present. The latticed skeleton commonly consists of two circles of hollow siliceous rods joined by intercalary rods.

Other small flagellates without special skeletal structures, such as *Chromulina,* are also abundant at times.

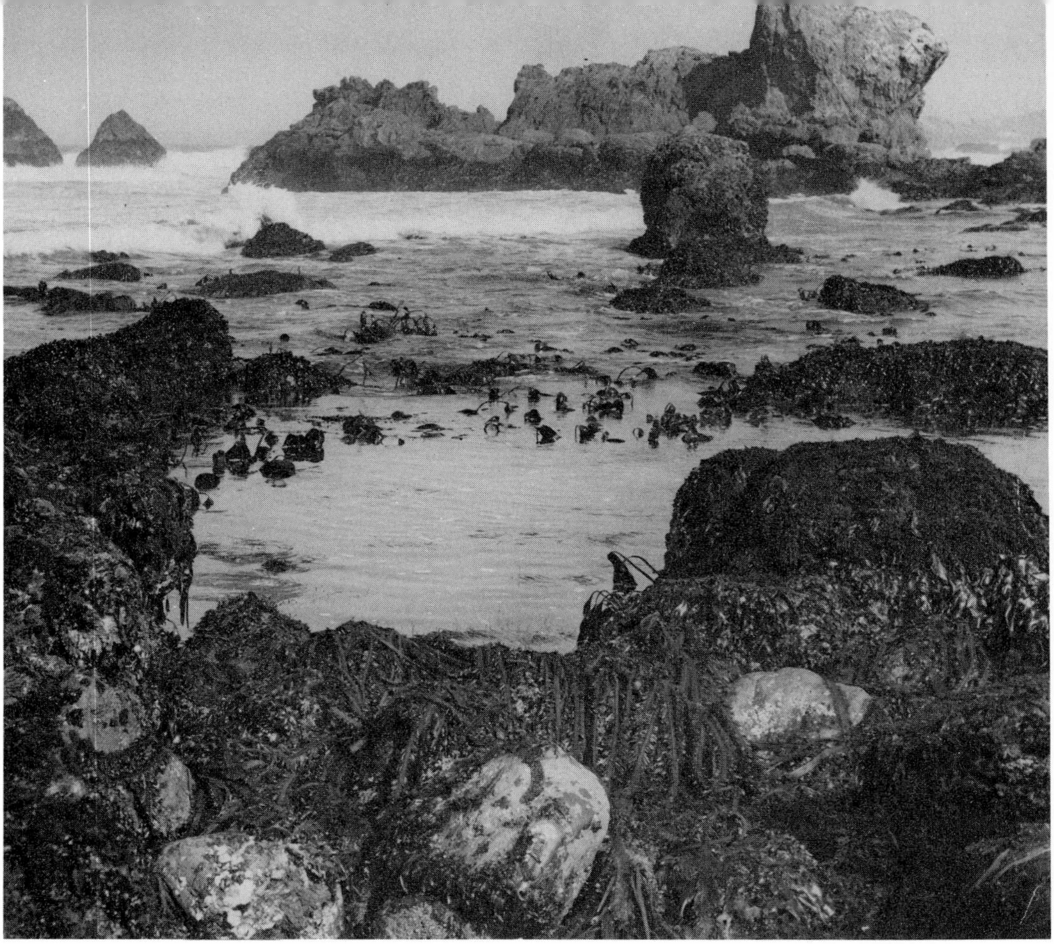

FIG. 7-1. Rocky area along California coast showing seaweeds exposed at low tide. Long straplike fringed seaweeds in foreground are *Egregia*. The broad blades on bent-over stems are *Laminaria*. Some eelgrass is visible in the quiet area of water near the rock. (M. W. Williams photograph.)

CHAPTER

7 LARGER MARINE PLANTS

Algae

THE SEAWEEDS

The great majority of larger marine plants are seaweeds. Seaweeds are for the most part fixed algae growing attached to firm substrates such as rocks, pilings, and shells. Attached seaweeds are of necessity confined to the fringes of continents and islands, or the tops of submarine banks where they can find attachment and enough light for photosynthesis. Larger floating plants are exceptional.

The most conspicuous exception is the subtropical *Sargassum,* or gulfweed. This small seaweed grows abundantly at the surface in the Atlantic south of Bermuda.

The peculiar patterns of weather and currents produce here a relatively stable calm region known as the Sargasso Sea. Gulfweed has accumulated in sufficient mass, forming a very special environment, over a long enough period of time to permit the evolution of a community of animals directly or indirectly dependent on the sargassum weed. Some of these animals exhibit amazing mimicry, imitating both the color and form of the *Sargassum.*

Sargassum, like other seaweeds, normally grows attached to the substrate in shallow water, but detached ones are trapped in the giant swirl that constitutes the Sargasso Sea, float in the surface water buoyed up by their air bladders, and continue to grow, although no longer reproducing sexually. Some of the fronds have apparently been floating for hundreds of years and have attained much larger size than those growing normally along shore. More old plants are found floating there than newly detached ones.

The mass of floating *Sargassum* averages approximately 1 to 1½ tons/km^2. The total mass is estimated at somewhere between 4 and 10 million tons. More than 50 species of animals, of many phyla, specifically adapted to life in the floating *Sargassum,* attest to the stability of this floating community over a long period of time.

Most seaweeds grow in shallow waters from the intertidal zone down to about 30 or 40 meters. In clear tropical or subtropical waters some of them may occur at ten times this depth. They are usually not abundant where there is nothing firm to which to attach—as on sandy or muddy bottoms. In the intertidal zone on suitable rock surfaces, these algae, together with the associated animals, display the phenomenon of zonation with dramatic sharpness.

Storms or rough water often detach considerable amounts of offshore seaweeds, washing them up onto nearby beaches in windrows, where they serve as food and protection to various scavengers of the beach, such as kelp flies and their larvae, amphipods, and small beetles.

Uses of seaweeds

Some seaweeds, such as Irish moss *(Chondrus crispus),* sea lettuce *(Ulva* spp.), and *Porphyra,* are widely used for food. Kelp meal is used as a dietary supplement and is sometimes incorporated as an additive in animal feeds.

Seaweeds cast up in windrows on beaches are sometimes gathered up and ground for use as fertilizer.

The most important commercial use of seaweeds is the extraction of phycocolloids used in a wide variety of products as gelling, emulsifying, suspending, and thickening agents. Examples are algin, carrageenin, iridophycin, and laminarin. Agar-agar, obtained from red algae such as *Gelidium* and *Gracilaria,* is the most widely used solidifying agent for microbial culture media and also has a variety of other uses.

The larger seaweeds also serve as sources of potash, iodine, potassium chloride, acetone, calcium acetate, and decolorizing carbon.

CALCAREOUS ALGAE

Green algae of the genus *Halimeda* (Fig 7-2) build coarse sediments that may become consolidated through overgrowth by encrusting coralline red algae such as *Lithothamnion* or *Porolithon. Halimeda* is the principal growth in the lagoons of some of the great Pacific atolls such as Kwajalein and Eniwetok. It can also grow below the level occupied by most of the reef corals. In Funafuti Lagoon, for example, down to depths of 30 meters, 80% to 95% of the sediment is comprised of *Halimeda.*

The coralline algae, red algae of the family Corallinaceae, are of worldwide distribution—not limited to the tropics, as are the reef corals. Both encrusting and erect nodulose types are common in lower intertidal and subtidal rocky areas in temperate and cool regions. Encrusting species commonly form a reddish or purplish band in the lower intertidal zone, where sea urchins are abundant enough to prevent other algae from covering the rock surfaces.

Encrusting and nodulose "nullipore"

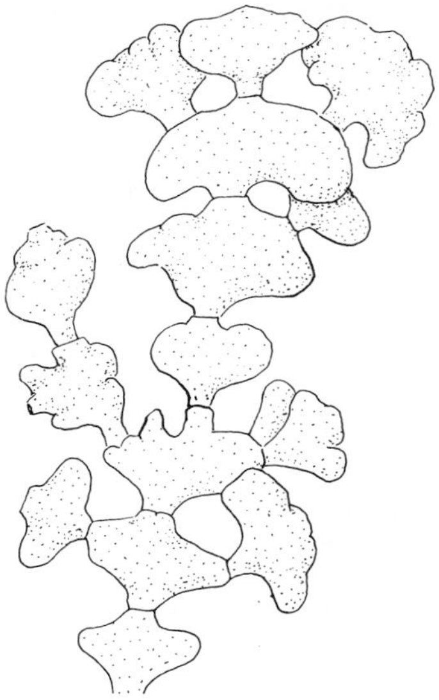

FIG. 7-2. *Halimeda.* The hard stony segments make up much of the calcareous debris in the lagoons of large tropical atolls.

coralline algae contribute to the formation and strengthening of coral reefs, especially on the exposed windward sides. They also form noncoral calcareous reefs in regions where true coral reefs do not occur. In the Indian Ocean several "coral reefs" contain no coral. As far north as Spitzbergen there are miles of calcareous banks formed chiefly by coralline algae. In Scotland some white sand beaches are composed largely of ground-up remains of offshore coralline algae.

CLASSIFICATION OF ALGAE

The larger algae include three great classes or subphyla of thallophytes: the **green algae (Chlorophyceae),** the **brown algae (Phaeophyceae),** and the **red algae (Rhodophyceae).** (The blue-green algae have been discussed on pp. 88 to 89.)

Green algae (Chlorophyceae or Chlorophyta)

The Chlorophyceae are clear green. The chlorophyll is not masked by other pigments. This group comprises practically all the more familiar types of freshwater algae and also contains interesting marine types. Morphologically, they range from simple rounded single cells through various kinds of filaments, plates, sheets, and cylinders of cells. There are also curious types without definite cell walls, but rather with a continuous cytoplasm containing many nuclei, a condition termed **coenocytic.** Some groups of unicellular algae are frequently classed as Protozoa (Phytomastignia). In the marine environment, they are less diversified and less important than the other two classes, the brown and the red algae.

Perhaps the best known and most familiar marine green algae are members of the so-called confervoid algae. They belong to a group, variously termed **Ulotrichales, Confervales,** or **Ulvales,** in which there is formed a multicellular thallus of simple or branched filaments, or a hollow flat or tubular thallus. They reproduce asexually by the production of flagellated zoospores and sexually with differentiated motile sperm cells and egg cells. There is an alternation of generations that are reproductively different but have a similar appearance.

Sea lettuce, *Ulva,* forms a large, flat thallus resembling lettuce leaves and comprised of one or two cell layers. It grows abundantly in protected places in the intertidal and upper subtidal zones. The species are edible and are dried and sold under the names green laver and Iceland sea grass. One species with narrow fronds resembling intestines is called water gut.

Another widespread genus is *Enteromorpha,* which forms large coarse complex branched filaments of tubular structure. it grows abundantly in bays and estuaries and as a waterline fouling plant on boats and pilings.

An interesting group is the **Codiales,** or **Siphonales.** The vegetative body of the alga is without cross walls, so that the cytoplasm is continuous throughout the plant. Such an arrangement is termed **coenocytic** or **syncytial.** The plants may become large and have a fleshy, complexly branched thallus, as in *Codium,* which has numerous

Text continued on p. 100.

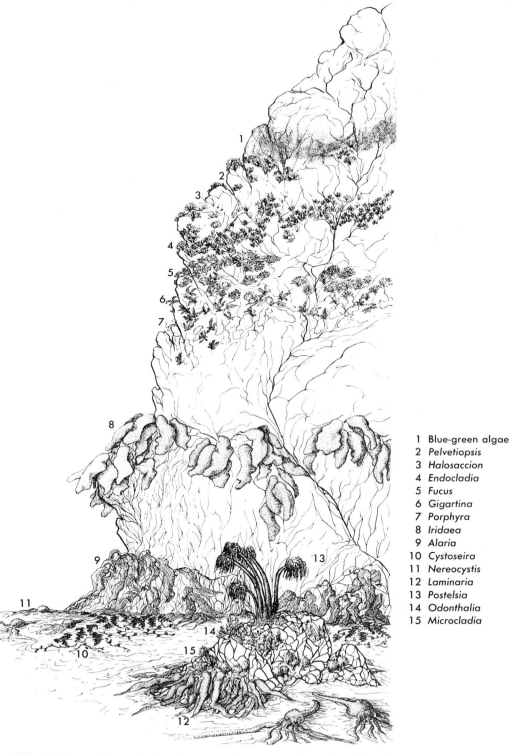

1 Blue-green algae
2 *Pelvetiopsis*
3 *Halosaccion*
4 *Endocladia*
5 *Fucus*
6 *Gigartina*
7 *Porphyra*
8 *Iridaea*
9 *Alaria*
10 *Cystoseira*
11 *Nereocystis*
12 *Laminaria*
13 *Postelsia*
14 *Odonthalia*
15 *Microcladia*

FIG. 7-3. Intertidal rock face in exposed area near Bastendorf Beach, Oregon, showing zonation of fifteen kinds of larger marine algae present.

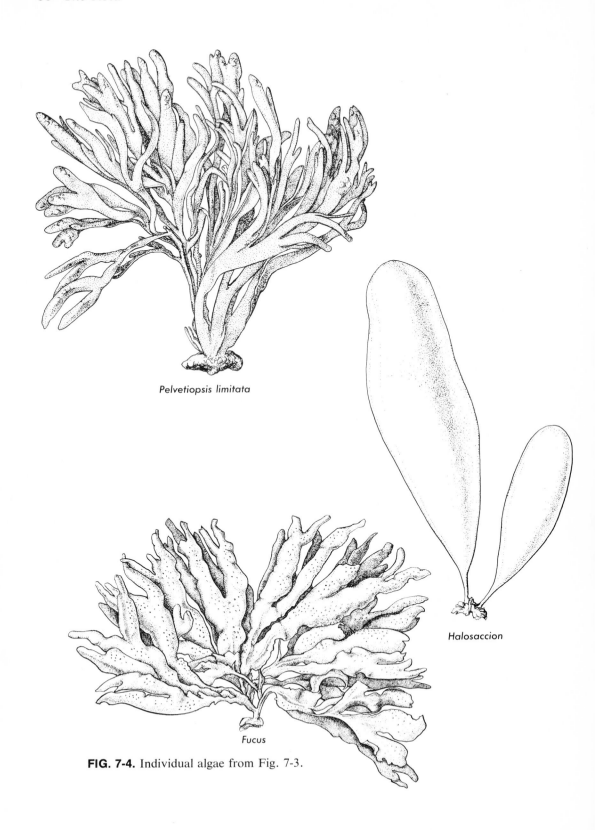

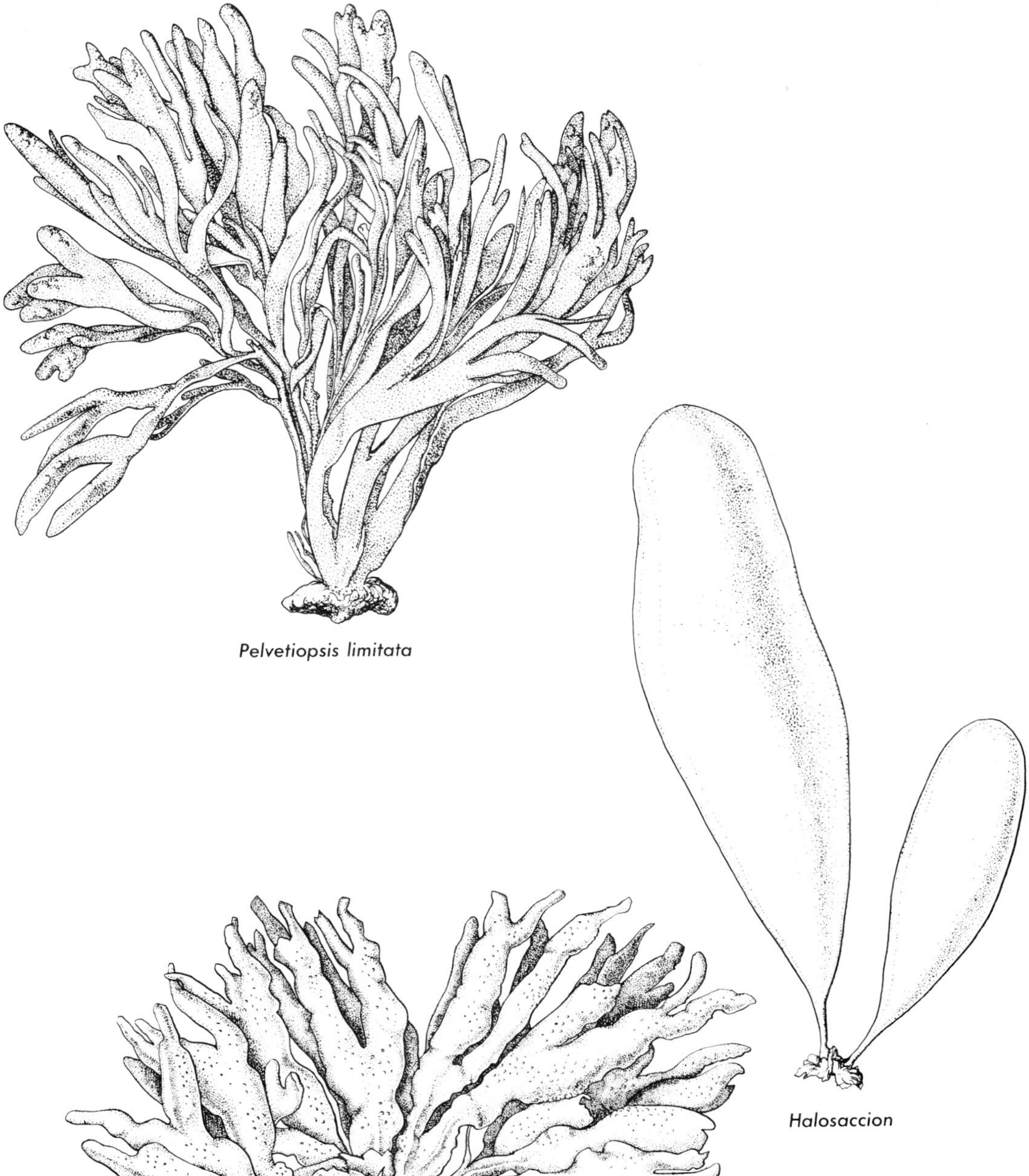

FIG. 7-4. Individual algae from Fig. 7-3.

Gigartina

Porphyra

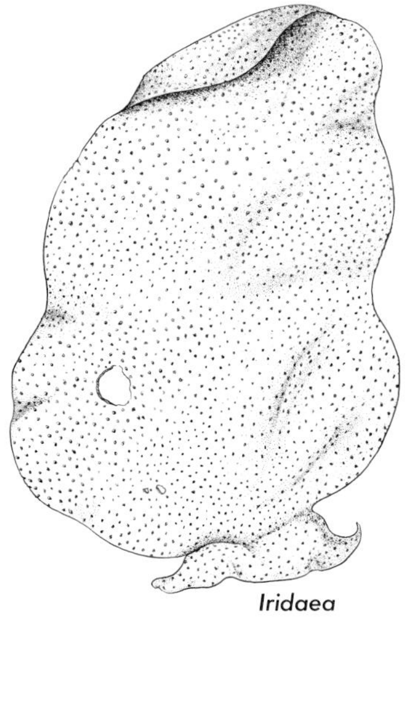

Iridaea

FIG. 7-4, cont'd. Individual algae from Fig. 7-3.

Continued.

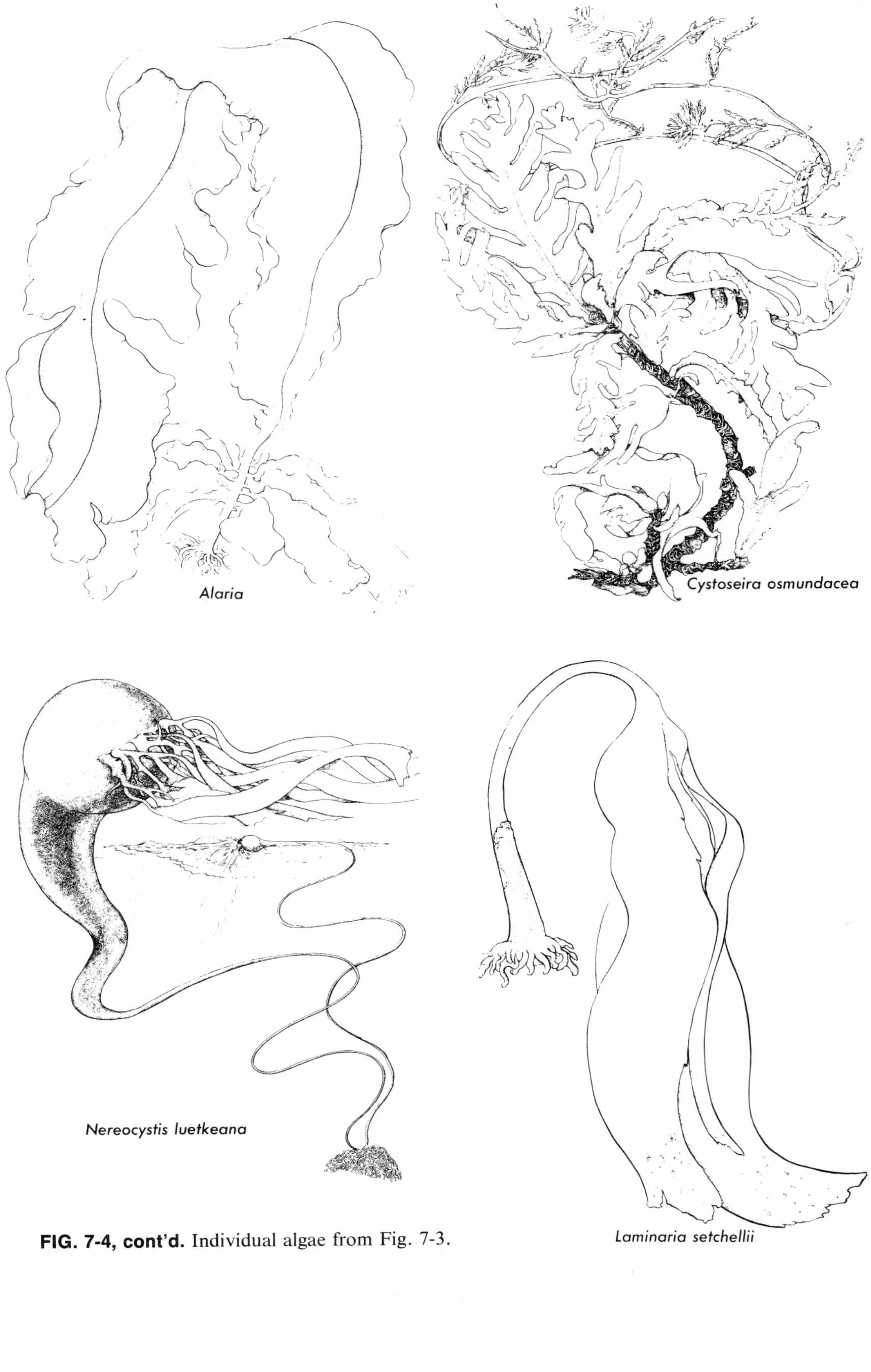

FIG. 7-4, cont'd. Individual algae from Fig. 7-3.

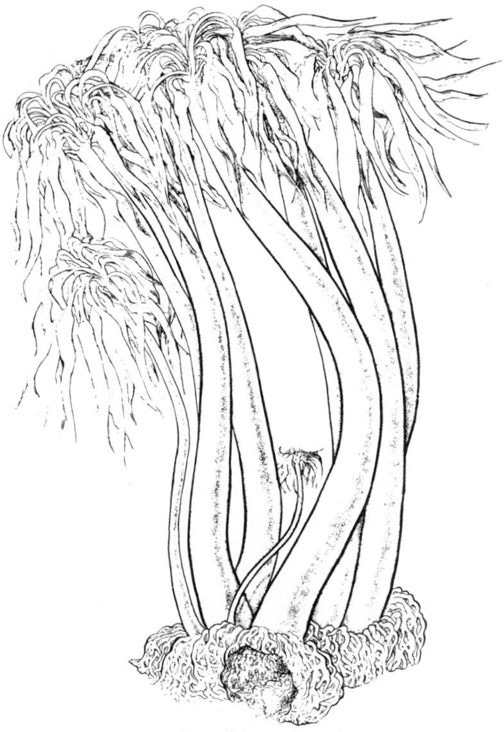

Postelsia palmaeformis

Odonthalia floccosa

Microcladia boreale

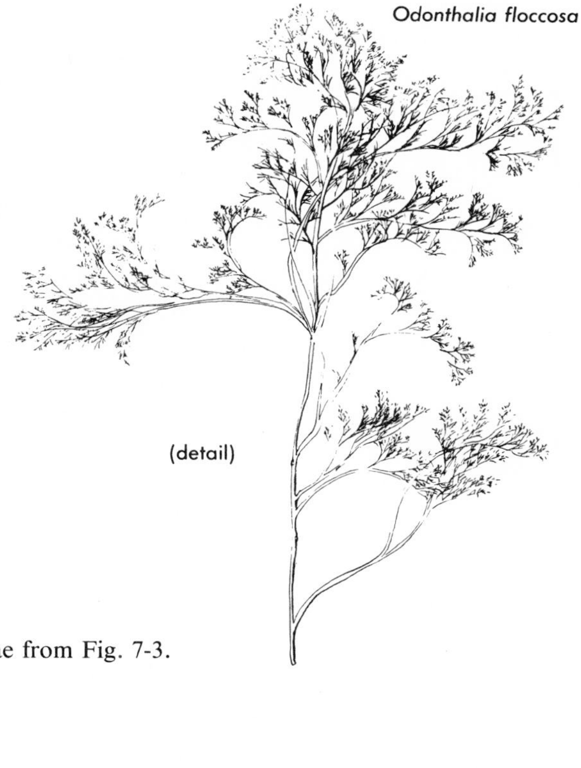

(detail)

FIG. 7-4, cont'd. Individual algae from Fig. 7-3.

interwoven, thick, cylindrical filaments with branching, club-shaped apexes.

Some are large hollow bladderlike single or multinucleate cells, attaining sizes up to an inch or more long, as in *Halicystis* or *Valonia* (Siphonocladales). Because of their large size, these cells have attracted the interest of biologists and have been much used in studies of penetration of ions and differential concentration of ions and other substances into the cell sap.

In *Caulerpa* there is a multinucleate creeping stemlike thallus that has rhizoids on the lower side and gives rise to foliose expansions above.

Classification of green algae

The classification of the green algae is summarized in the Taxonomic Appendix. Only the more important marine types are included.

Brown algae (Phaeophyceae or Phaeophyta)

Most of the familiar large seaweeds belong to the Phaeophyceae. They are brown or olive-brown in color, the chlorophyll in the chromatophores being masked by fucoxanthin or other carotenoids. Almost all are marine, anchored to the substrate by holdfasts. The primary food reserve is a

FIG. 7-5. *Ascophyllum* on rock wall at Marblehead Neck, Massachusetts. Zonation shows well here. Above the *Ascophyllum* is a heavy growth of barnacles. The *Ascopyhllum* itself is growing in a patchy manner over and among mussels *(Mytilus edulis)*. Along the upper edge of the *Ascophyllum*-mussel zone can be found numerous carnivorous gastropods (whelks), whereas at the lower edge of this zone and extending well up into it are numerous starfish *(Asterias)*. In the pool at the left can be seen a group of large sea anemones *(Metridium)*. (Helen P. Finlayson photograph.)

carbohydrate, laminarin, dissolved in the cell sap. The reproductive cells, both zoospores and gametes, are motile and have two laterally inserted, unequal flagella. All the Phaeophyceae are multicellular and are either filamentous or have a complex, thalluslike portion, the stipe, and a basal boldfast that is often rootlike in appearance. There are about 900 species, all but three of which are marine.

Brown algae occur in all oceans but are most characteristic of, and often predominate in, the algal flora of temperate-to-cold waters. For the most part they are found near shore in water not more than 20 meters deep.

Classification of brown algae

See taxonomic appendix.

Red algae (Rhodophyceae or Rhodophyta)

The red algae contain, in their plastids, in addition to chlorophyll, the red pigment phycoerythrin and sometimes the blue pigment phycocyanin. This group contains the most specialized and complex types of algae. The food reserve formed is an in-

FIG. 7-6. *Fucus* on intertidal rocks at Bolinas, California. (M. W. Williams photograph.)

soluble carbohydrate called floridean starch. Sexual reproduction is unique in that nonflagellated male gametes, spermatia, are passively transported to the female sex organs. There are no flagellated zoospores.

There are about 400 genera and 2500 species. The group is almost exclusively marine, about 12 genera and less than 100 species being found in freshwater. Red algae are present in all seas, including the polar seas, but are best developed in the tropics, where they often constitute the predominant benthic flora. Most of them grow attached to the substrate in shallow or intertidal areas. In the North Atlantic and North Pacific they rarely occur at depths greater than 35 meters, but in the tropics they have been dredged as deep as 170 meters.

The thallus is commonly a simple or complexly divided blade, or it may have a stemlike portion and separate blades. In some groups—the coralline algae—the body is heavily impregnated with lime. Coralline algae play an important role in the building of coral reefs and atolls.

The male sex organ, the **spermatangium,** is unicellular and produces one nonflagellated gamete, the **spermatium,** which upon liberation floats to and lodges against a hairlike prolongation of the female unicellular sex organ **(carpogonium)** at the apex of a carpogonial filament.

On fertilization, the zygote in some groups grows directly into a special filament, the **gonimoblast filament,** which bears sporangia **(carposporangia)** containing carpospores. In other groups the zygote produces a delicate tube, the **ooblast,** which grows from the carpogonium to another cell of the plant. The zygote nucleus migrates along the ooblast to that cell, and the gonimoblast fibers then grow from this **auxiliary cell.** The fruit of the sexual plant, consisting of the gonimoblast fibers and the carposporangia jointly, constitute the **cystocarp.** The cystocarp has been interpreted as a very small asexual generation growing on the sexual plant.

Each carposporangium liberates one spore. In primitive groups the reduction divisions occur in the zygote stage, and the carpospores then are haploid and develop directly into gametophytes. In more specialized forms the zygote nucleus does not undergo reductional divisions but produces 2N carpospores, which develop into 2N free-living asexual plants of the same size and general structure as the gametophytes. This asexual plant, the **tetrasporophyte,** bears **tetrasporangia** in which the single nucleus undergoes meiosis, producing **tetraspores.** On liberation, the tetraspores develop into gametophytes.

Classification of red algae

There are two subclasses, the **Bangioideae** and the **Florideae.** The taxonomy of the red algae is still in an unsatisfactory condition, particularly with regard to definitions of larger units such as subclasses and orders. (See taxonomic appendix for classification.)

Angiosperms

Few higher plants have become adapted to marine habitats, although there are a rather large number of estuarine and salt marsh types found around the edges of brackish lagoons and mud flats, and many terrestrial "maritime" plants occur only near the coast.

Two groups will be considered: (1) the eelgrasses and their relatives, which are the only angiosperms to have evolved a wholly marine way of life, and (2) the mangrove complex, which forms a unique and important habitat along many tropical shores and offshore islands.

TRULY MARINE ANGIOSPERMS

By "truly marine" angiosperms is meant those which tolerate the salinity of marine water, grow while wholly submerged, and have the capacity for underwater flowering and pollination. Furthermore, all of them have the ability to anchor themselves firmly enough with their roots to withstand the usual surge of waves and tides

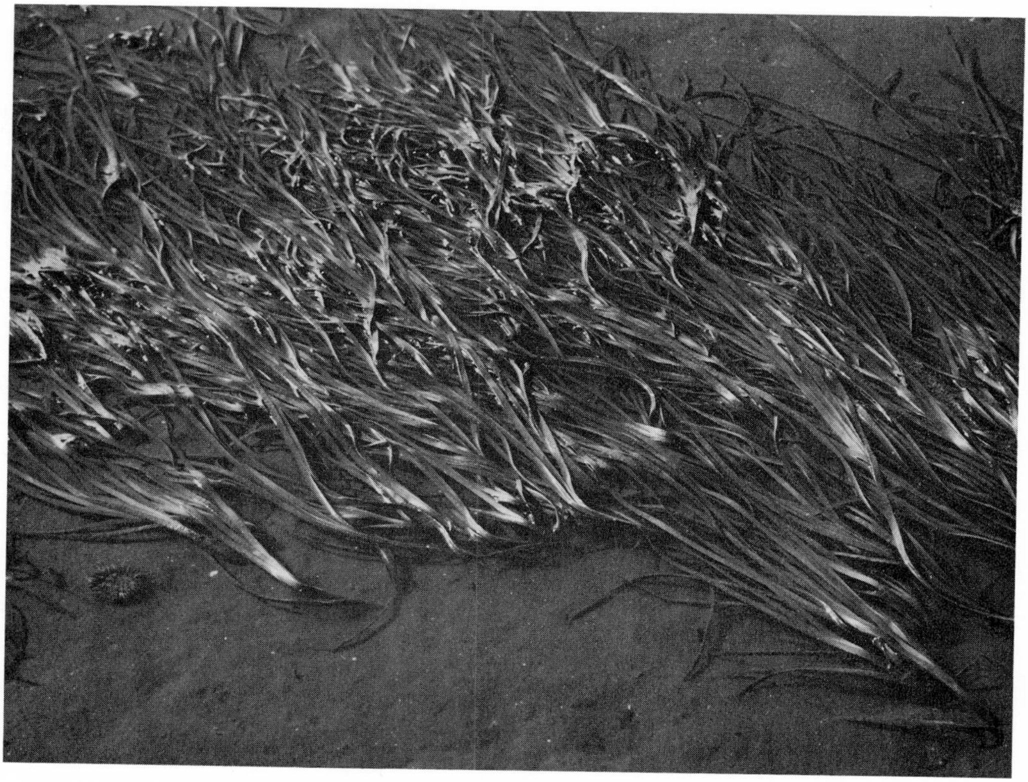

FIG. 7-7. Eelgrass in Tomales Bay, California. Open anemone shows that the area is covered by water. Streaming of eelgrass shows direction of tidal flow. (M. W. Williams photograph.)

in their particular habitats. None are planktonic. All have threadlike pollen, which drifts by and catches on the stylus-like stigma.

Five families of plants, all of which belong to order **Naiadales (Helobieae),** have evolved such species.

Ruppiaceae: *Ruppia*
Zosteraceae: *Zostera, Phyllospadix*
Posidoniaceae: *Posidonia*
Zannichelliaceae: *Zannichellia, Althenia, Syringodium, Amphibolis, Cymodocea, Halodule*
Hydrocharitaceae: *Halophilia, Enhalus, Thalassia*

Some authors include the first 4 families under **Potamogetonaceae.**

Naiadales is a group composed of mostly freshwater aquatic plants. The families that contain marine genera are widely represented in freshwater habitats, and some of their freshwater species had already evolved underwater flowering and pollination. The occurrence of such species as *Potamogeton pectinatus* and *Zannichellia palustris,* which inhabit both fresh and brackish waters, and typically brackish genera such as *Ruppia* and *Althenia,* shows how the transition from fresh to salt water may have occurred. Since freshwater Naiadales are themselves a highly specialized group evolved from terrestrial ancestors, it would appear that the marine representatives must be among the most recently evolved groups of angiosperms.

The tropical-subtropical species are distinct from those of the temperate and boreal regions and include the genera *Amphibolis, Cymodocea, Enhalus, Halodule, Halophila, Syringodium,* and *Thalas-*

sia. The species in turn may be grouped as Indo–West Pacific species, ranging in warm waters from the east coast of Africa to the Indo-Malay region and the Pacific islands, and the tropical American species centered in the Caribbean.

Tropical marine angiosperms

Indo–West Pacific group

Enhalus acoroides, Halodule pinifolia, Halodule tridentata, Thalassia hemprichii: Wide distribution from East Africa to the Indo-Malay region and the Pacific islands.

Halophila beccarii: Rather rare, scattered along Malaysian coasts and east to Borneo and the Philippines.

Halophila stipulacea: Red Sea and West Indian Ocean (spread to Mediterranean Sea through canal and now colonizing coasts of Greece).

Syringodium isoetifolium

Halodule uninervis

Tropical American group

Halodule bermudensis: Only in Bermuda Islands.

Halodule beaudettei: Amphi-American

Halodule ciliata: Caribbean coasts of Panama.

Halophila bacillonus

Halophila englemannii: In West Indies and later on Texas coast.

Syringodium filiforme

Thalassia testudinum (turtle grass): Florida and West Indies.

Halodule, Halophila, Syringodium, and *Thalassia* have their greatest species density in the Indo-Pacific region and probably originated there. All are absent from tropical west Africa except *Halodule wrightii,* which has an amphi-African distribution, and from the Pacific coast of America, except the amphi-American *Halodule beaudettei,* found in the Atlantic and Pacific coasts of Central America.

It is of interest that some of the Indo–West Pacific species find their nearest related species in the tropical Caribbean region. Such species form pairs of sibling, or twin, species. Examples of twin species are as follows:

Indo–West Pacific species	Tropical American counterpart
Thalassia hemiprichii	*Thalassia tistidinum*
Syringodium isoetifolium	*Syringodium filiforme*
Halodule uninervis	*Halodule beaudettei*

Temperate-boreal marine angiosperms

The temperate-boreal group is characterized by the genera *Althenia, Phyllospadix, Posidonia,* and *Zostera. Zostera* has the most extensive range, being found on both East and West coasts of North America, and on the coasts of Atlantic Europe, East Asia, Australia, and South Africa. One species of *Posidonia* is found in the Atlanto-Mediterranean region, another in Australia. *Phyllospadix* has an amphi-Pacific discontinuous distribution, being found along the shores of both eastern Asia and western North America, but not along the northern arc between. Since none of these plants is well adapted for long transport over open ocean, the discontinuous distributions of some of the genera and species have stimulated considerable speculation regarding the modes and routes of transport and the length of time involved.

Zostera, Phyllospadix, Cymodocea, and *Thalassia* are perhaps the best known and the most important genera from an ecological standpoint. They form submarine meadows, contributing a major part of the organic production in such areas, forming a protected habitat for characteristic assemblages of animals, stabilizing the bottom, and serving as a site for increased sediment depostion. *Zostera,* grass wrack, grows especially in protected areas in bays, estuaries, and river mouths, where the water is relatively quiet at least part of the time.

The ecological importance of these beds of eelgrass was dramatically demonstrated in 1931 when the beds of *Zostera marina* along the Atlantic seaboard of America and Europe suffered from a severe epidemic of infestation with a parasitic slime mold, *Labyrinthula,* which wiped out most of the beds. On disappearance of the *Zostera* the numbers of cod, shellfish, scallops, crabs, and other associated animals fell off sharply. The oyster industry of much of northern Europe and the Atlantic coast of America was ruined. Plant and animal life somewhat farther out from river mouths was overwhelmed by raw sewage, which formerly had been filtered out in the *Zostera* beds. Silt from erosion caused by lumbering and farming inland and silt eluted from the substrate of former *Zostera* beds spread farther than before, building new mud flats on formerly gravel bottoms. Smelt and other fish that spawned on these graveled areas could no longer do so and greatly diminished. Formerly richly productive bays and estuaries became wastelands. Brant and other birds that fed on the bivalves and other organisms in the eelgrass community began to die off. Brant decreased to one fifth their former number.

At Great Bay, New Hampshire, the bay became choked with sewage and silt. There were no oysters in Oyster River, no salmon at Salmon Falls. Other fish decreased drastically. The increased effects of pollution prevented eelgrass recovery. Former boatyards and shipping facilities were sludged up and deserted. The effects of the eelgrass catastrophe were still apparent and strongly felt in some areas as much as 20 to 30 years later. In this case the activities of man, causing excessive pollution and erosion, combined with a natural disaster to produce an overwhelming effect on the local biota.

On the other hand, in the Kattegat, where approximately 24 million tons of eelgrass thrived, its disappearance, while leading to the disappearance of animals and fish directly associated with *Zostera,* had surprisingly little effect on the infauna. These animals clearly did not depend directly or indirectly on the *Zostera* for their food, but on the plankton in the water.

Smaller eelgrass disasters have occurred in coastal lagoons of eastern Australia when flocks of hundreds of black swans descend upon the *Zostera* beds to eat the stolons. They sometimes denude whole areas of several lagoons. The result is failure of mullet and shrimp fisheries and unsuccessful spawning and settlement of oysters.

On removal of the *Zostera,* the redox potential of the sediments increases, and much of the silt fraction is washed out by increased water movement. Since *Zostera* roots prefer a reduced environment containing silt and free hydrogen sulfide, these changes seriously impede their return and slow the recovery of these areas.

In late summer *Zostera* flowers, then dies and rots. This rotting of old plants produces rapid growth of microorganisms, especially bacteria, flagellates, and ciliates, which, together with rising temperatures of late summer, stimulates spawning of the Japanese oyster and provides abundant food for its larvae.

THE MANGROVE COMPLEX—TIDAL WOODLANDS

Mangroves do not represent a particular taxonomic group of plants. The term is used for all the woody plants that invade shallow tidal areas, forming "mangrove swamps" or coastal woods extending into zones covered with seawater at high tide. Preeminent among the mangrove plants are members of the family Rhizophoraceae, with their characteristic branching, stilt-like roots holding the rest of the tree above the water. Species of *Rhizophora* are usually the outermost, or fringing, members of mangrove swamps and often constitute the only, or at least the dominant, growth in such areas.

Mangroves are found along shallow protected tropical humid coastlines around the globe, often forming impenetrable thickets. Through their role in stabilizing the substrate, holding sediments, and adding to the detritus, they help build soil and add to the land area of the coast, to form and extend low offshore islands.

Seawater is "physiologically dry." Terrestrial plants using it as a source of water are in danger of excess transpiration. Those which have become adapted to growth in the tidelands tend to have modifications somewhat like those of certain desert plants, such as thickened coriaceous, or waxy, leaves, to reduce transpiration. They are favored by conditions that reduce transpiration, such as humidity and cloudiness. Mangrove forests develop in parallel with tropical rain forests. Cloudiness is apparently an essential climatic condition for both.

Rhizophora is uniquely adapted for life in

A

FIG. 7-8. A, Mangrove. Same area as shown in Fig. 7-9. Closeup of roots and young plants exposed at low tide. (U.S. Forest Service photograph.)

shallow tidelands. The trees themselves send out adventitious roots from branches above the water, supporting them and extending the growth vegetatively out over the shallow water (Fig. 7-8). Long, pendant seedlings are produced. On being dropped from the trees, some of the seedlings penetrate the soft substrate, standing upright and taking root where they were dropped. Others are carried away by tides and currents; they may survive for months and be carried for great distances. If they lodge along a shallow shore, submerged bar, or fringe of an island, they take root and start to grow. In 20 to 30 years the new trees are fully mature and are creating a new mangrove swamp.

One of the greatest mangrove swamps of the world has developed along the southern coast of Florida, from the Florida Keys around the southern tip of the mainland and north along the gulf coast through the Ten Thousand Islands (Fig. 7-9). The thousands of mangrove islands here are for the most part shaped as elongated ovals, larger and with a lagoon situated at the southeast end. It seems probable that they developed along wave-created sand ridges or ripple marks in the shallow sea floor, the islands of today reflecting the shape and trend of the wave marks of yesterday. Over a period of decades the building of new land, the coalescence of small islands to form larger ones, and the merging of the shoreline with nearby offshore islets can be seen. Whether this entire area will eventually become low-lying land, as now seems to be indicated, or will be reclaimed by the sea, no one can say.

B

FIG. 7-8, cont'd. B, Young mangrove. (Chesher photograph).

A rise in sea level or isostatic changes in the earth's crust could reverse the direction of thousands of years of evolution of the coastline and shallow sea, turning it in new and unexpected directions.

In the Florida–West Indies mangrove swamps the roots of the mangrove are commonly thickly settled with mangrove oysters *(Ostrea frons),* which cling to the roots by curious, fingerlike extensions of the shell. In the mud among the roots and between mangrove thickets are innumerable burrows of fiddler crabs *(Uca pugilator).* Coffee bean shells *(Melampus bidentatus)* are abundant among the roots and along the upper edges of the swamp, where they climb on the stems of salt marsh grasses as the tide rises. Beautiful little shells of the rose tellin *(Tellina lineata)* may lie scattered on the mud, indicating the presence of a colony of these bivalves living nearby just beneath the mud surface.

Prominent among the predators in this community are the large crown conches *Melongena corona,* which prey on the oysters. At night on low tides, raccoons wander out into the mangrove swamp to feed on the oysters.

Curiously, as the mangrove plants have evolved as an assemblage of land plants invading the margins of the sea, the swamps they produce have been the site for the evolution of some striking cases in which marine animals have become adapted for life on the margin of the land. Mangrove periwinkles *(Littorina angulifera),* whose relatives live in the sea, have colonized the roots and branches above the tidal level. Land-dwelling crabs, the white land crab *(Cardisoma guanhumi)* and the purple-clawed hermit crab *(Coenobita clypeata),* which return to the sea only to spawn, live along the upper margins, out of the water.

As the mangrove swamp grows it is colonized with a characteristic assemblage of marine animals found nowhere else. It forms nesting and roosting sites for many large birds, which, with their guano, fertilize and help build the swamp.* The interior parts of the swamp become more terrestrial in character and may be colonized by other kinds of plants of the mangrove complex.

In the Indo-Pacific the curious "climbing fish," or mudskippers *(Periophthalmus),*

*In some cases small mangrove islands used as nesting sites are "burned out" by excess guano, which kills the trees and eventually destroys the island.

FIG. 7-9. Mangrove islands in the Ten Thousand Islands area off the Florida coast. (U.S. Forest Service photograph.)

may be seen at low tide, walking and jumping about on the mud or even climbing onto some of the mangrove roots by using their pectoral fins as limbs, and stopping at small puddles to moisten their gills. If frightened, they disappear into burrows in the mud, or if they must cross water, they do not enter it but skip across the top with a series of strong flexures of the body.*

*I once stopped for lunch with my family by a small mangrove swamp on the east end of Java. As we explored the swamp, dozens of curious mudskippers observed us, looking for all the world like some strange cross between frog and fish, with their large, raised eyes bugging out from the top of their heads, and their awkward jumping and walking movements. They seemed to be alert, intelligent creatures, displaying a genuine curiosity about us but staying well out of reach. We discovered that they liked bread crumbs—a new item in their diet. As we walked through the tangled mangrove, scattering bread crumbs, more and more of them followed us, until we felt indeed like the Walrus and the Carpenter.

Mangrove swamps of the Indo–West Pacific, from tropical east Africa to the Pacific islands, differ in species composition from those of the tropical Atlantic, found especially in the Caribbean–Gulf of Mexico region and to a lesser extent along the coast of tropical West Africa. Those of the Indo–West Pacific are best developed in parts of India and in the Malay Archipelago. They are composed of a rather wide assortment of plants, as the following list indicates. Those of the tropical Atlantic are best developed in the region of the Florida Keys and the adjacent gulf coast.

I. Indo–West pacific
  Family Rhizophoraceae
    Rhizophora mucronata Laup. (the most important and widespread species)
    R. conjugata Linn.
    Ceriops candolleana Arn.
    C. roxburghiana Arn.
    Kandelia rheedii W. & A.
    Bruguiera gymnorrhiza Lamk.

B. eriopetala W. & A.
B. caryophylloides Bl.
B. parviflora W. & A.
Family Combretaceae
Lumnitzera racemosa Willd.
L. coccinea W. & A.
Family Lythraceae
Sonneratia apetala Buch-Ham.
S. acida Linn.
S. alba Smith.
Family Maliaceae
Carapa moluccensis Lamk.
C. obovata Bl.
Family Myrsinaceae
Aegiceras majus Gärtn.
Family Rubiaceae
Scyphiphora hydrophyllaceae Gärtn.
Family Verbenaceae
Avicennia officinalis Linn. and var. *alba* Bl.
Family Acanthaceae
Acanthus ilicifolius L.
Family Palmae
Nipa fruticans Thumb.

II. Tropical Atlantic mangrove
Family Rhizophoraceae
Rhizophora mangle L.
Family Combretaceae
Laguncularia racemosa Gärtn.
Family Verbenaceae
Avicennia tomentosa Jacq.
A. nitida Jacq.

CHAPTER

8 ANIMAL LIFE OF THE SEA

general considerations

Marine plant life consists largely of simple unicellular forms, microscopic in size. The larger plants, confined to the edges and shallow areas, are far fewer in number of species, and for the most part are relatively simple and primitive, compared to the luxuriant variety and complexity of plant groups dominant on land. Animal life, on the other hand, is rich and diverse in the extreme. All phyla and major classes of animals are represented in the fauna of the oceans. Many are found only there.

PLANKTON FEEDERS

In the ocean, plankton feeders are in a position somewhat analogous to that of the plants on land. That is, their food is all about them in the ambient medium. They need methods of concentrating it. A large majority of the zooplankton organisms, the free-living microbenthic forms, and many of the larger animals are either plankton or detritus feeders. Many of them have ciliary or ciliary-mucus mechanisms for creating currents of water and retaining plankton; others have networks of bristles, hairs, or tentacles with which they strain or filter the water. Those which subsist primarily on zooplankton tend to have coarser filtering apparatus than do those which eat primarily phytoplankton or minute bits of detritus.

When we consider the benthic animals, where there is substrate to which they can attach or into which they can burrow, we see in the most dramatic way the effect of having a major food supply in the ambient medium. Here the majority of the most abundant and dominant animals, like plants on land, live as sessile organisms. Many, such as the bushy hydroid colonies, gorgonians, sea fans, and sea lilies, are even plantlike in their growth habits. Before biologists had a clear concept of the basic differences between plants and animals, many marine animals were classified as plants.

In shallow shelf areas and the lower intertidal zones, where plankton is most abundant, we find animals growing crowded together in dense masses, as close together as physically possible, sessile, and often growing on top of each other, much in the manner of land plants. Almost all phyla include forms so modified. A few examples are attached protozoa, sponges, coelenterates, bryozoans, tube worms, bivalve molluscs, barnacles, and tunicates.

SUBSTRATE GRAZERS

The surface of rocks or any submerged objects quickly becomes covered with a film of colloidal and molecular organic matter, bacteria, diatoms, other algae, and small attached animals. This "algal film" offers opportunity for another great group of animals that can pick or scrape it off from the rock surfaces efficiently enough to utilize it as food. Such grazers are particularly well represented by gastropod molluscs and chitons, which cling to and slowly creep over the surfaces, scraping off the algal film with a filelike tongue, the radula.

Where the substrate is sand or mud, the organic film in and on it can be utilized by

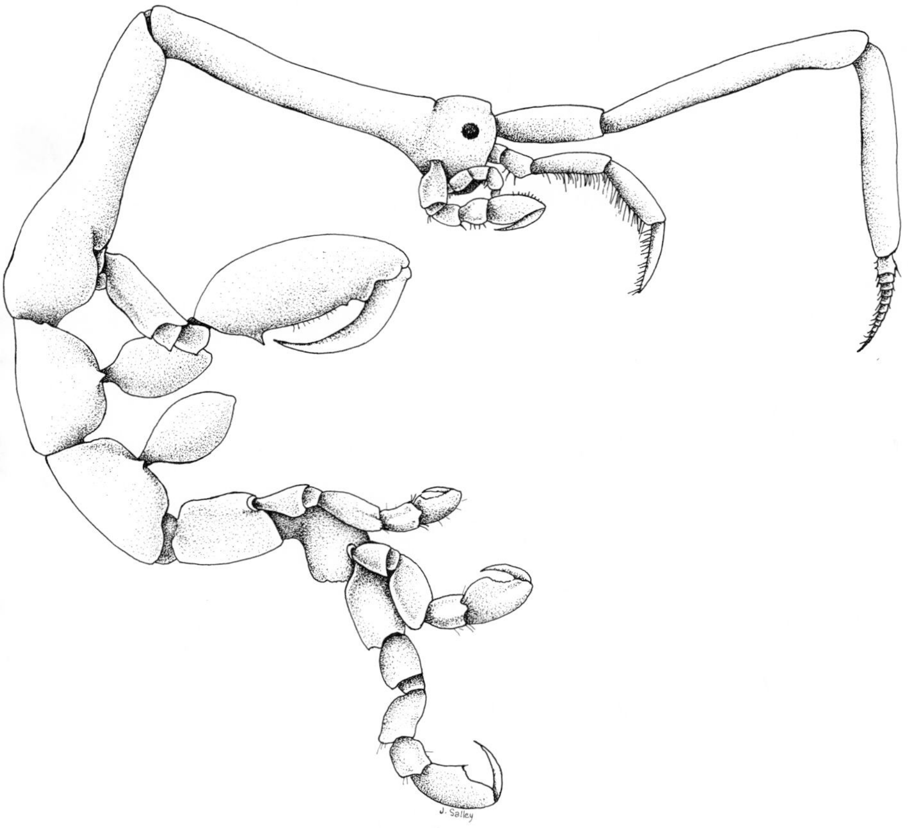

FIG. 8-1. *Caprella equilibria.* Caprellids are strangely modified amphipods found among hydroid colonies, under overhanging rock ledges, etc. (*In* Light, S. F. 1975. Light's manual; intertidal invertebrates of the California coast. 3d ed. revised and edited by R. I. Smith and J. T. Carlton. University of California Press, Berkeley.)

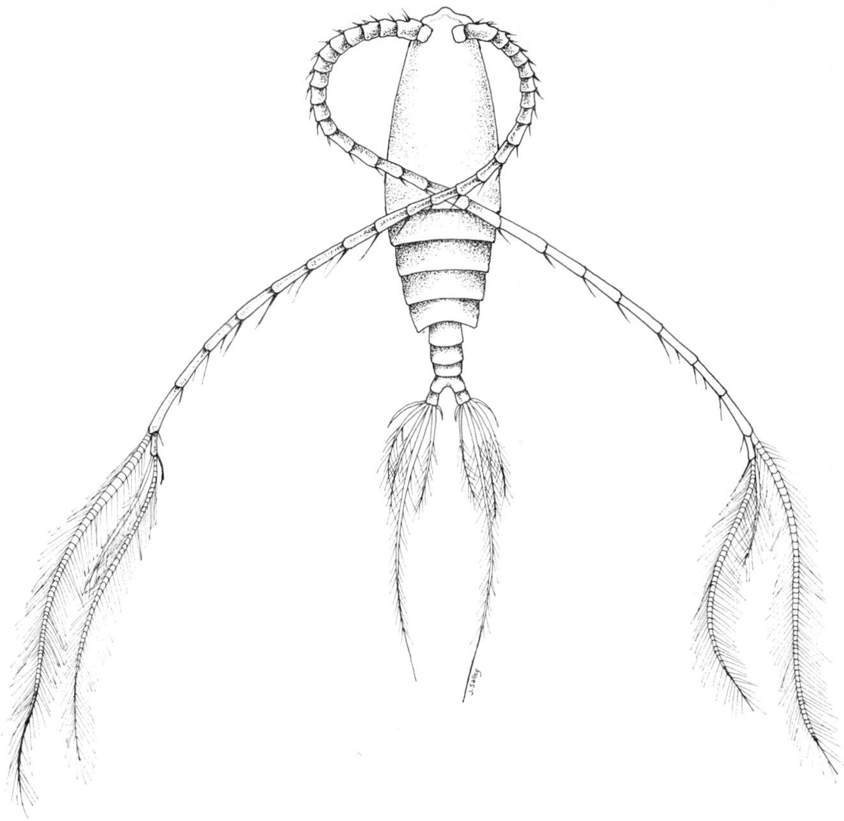

FIG. 8-2. *Calanus gracilis,* a phytoplankton-feeding member of the zooplankton. (After Brady, G. S. 1883. Challenger reports. Series 5, Vol. 8.)

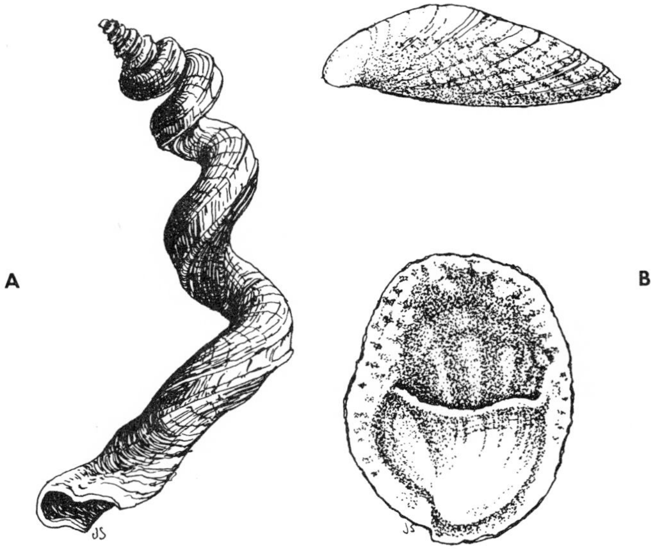

FIG. 8-3. A, *Vermicularia spirata;* **B,** *Crepidula fornicata.* (After Arnold, A. [Foote]. 1901. The sea beach at ebb tide. Century Co.; reprinted unabridged in 1968 by Dover Publications, New York.)

various burrowing or surface-crawling worms, sea cucumbers, balanoglossids, etc., which ingest the substrate, digesting the organic matter contained and ejecting the sand grains as the bulk of their feces.

PREDATORS AND SCAVENGERS

Anywhere a population of plants and animals exists, a portion of them will evolve as predators, scavengers, and parasites. These groups are found in all major environments. Since scavengers and predators depend on discrete sources of food (as do also primary consumers on land), rather than on food throughout their ambient medium, they must be able to move actively or trap their food. They do not differ so greatly or in such a uniform manner from one type of environment to another, in either structure or habits, as do the producers and primary consumers. Their modifications everywhere are more diverse and depend chiefly on the particular kind of food they eat and what they must do to obtain it, together with the physical character of their particular environment.

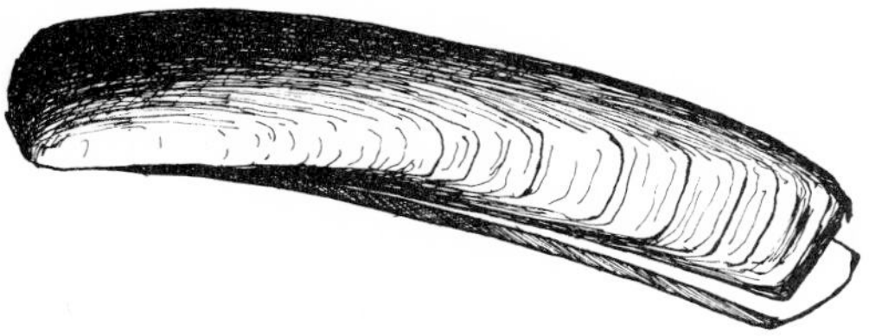

FIG. 8-4. *Ensis directa.* Soft substrate, buried plankton feeder.

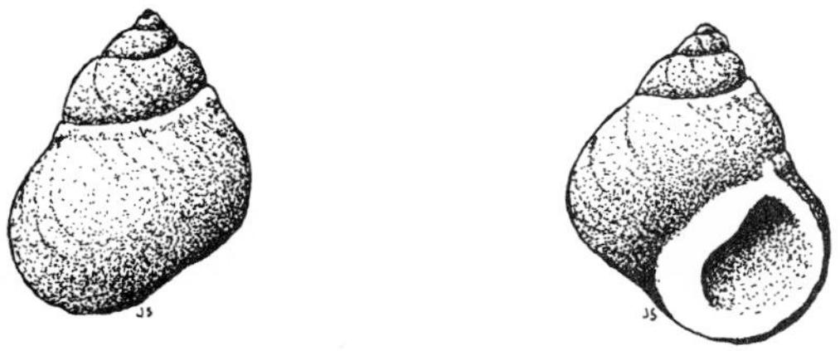

FIG. 8-5. *Littorina littorea,* an algal film scraper from rock surfaces.

Predator-prey relationships

The relationships between predators and prey are complex, being complicated by the interactions of both components with all the other physical and biotic influences of their environment. It is not to the advantage of the predator to overwhelm the prey population.

When a predator-prey pair of species is artificially isolated from the natural environment and left to interact only with each other in a homogeneous environ-

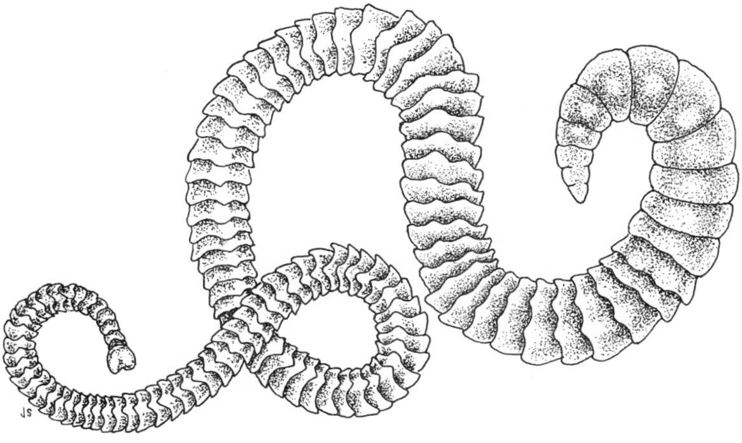

FIG. 8-6. *Capitella capitata,*, a substrate ingester, living buried in fine, soft mud or sandy mud. (After McIntosh, W. C. 1915. British marine annelids. Vol. III. Ray Society, London.)

ment, the prey population being provided with abundant food, a common result is that both undergo a period of increase followed by a period of decrease and finally extinction, the prey population being a phase ahead of that of the predator. If the environment is made heterogeneous in such a way that a significant portion of the prey population can successfully avoid predation, the situation more nearly approaches that found in nature and a periodic oscillation of predator and prey populations may be obtained.

In nature the interactions of the out-of-phase oscillations of predator-prey populations at times produce somewhat irregular long-term reinforcements or oscillations of greater amplitude. Calculations by Lotka (1925), Volterra (1926), and others indicate that predator-prey interaction in nature should ordinarily result in oscillatory patterns, rather than in extinction of the prey followed by that of the predator.

Natural selection will tend to increase or maintain the efficiency of the prey in escaping predation and of the predator in food capture. The net result is that the predator tends to live at the expense of the least fit, "most expendable" elements of the prey population, and the least fit predators are unsuccessful in maintaining themselves. Thus the predator-prey relationship may be important in keeping both populations in a state of fitness and removing the weak, sick, injured, and old from both.

Escape mechanisms are diverse but can be grouped into five major categories, often combined in various ways:

1. Speed or agility—sensing, outdistancing, or dodging a predator or in group action confusing to the predator
2. Concealment or hiding—including protective coloration, protective shells, burrows, tubes, etc. and behavior patterns that enable the prey to be overlooked or mistaken for something not wanted by the predator
3. Fighting back—self-defense or group defense
4. Chemical defenses—distastefulness or release of offensive or harmful substances when endangered
5. Timing of life cycles or of physiological and behavioral rhythms in such a way that critical sensitive stages coincide with periods of reduced activity or efficiency of the predator, or are especially well protected from predation

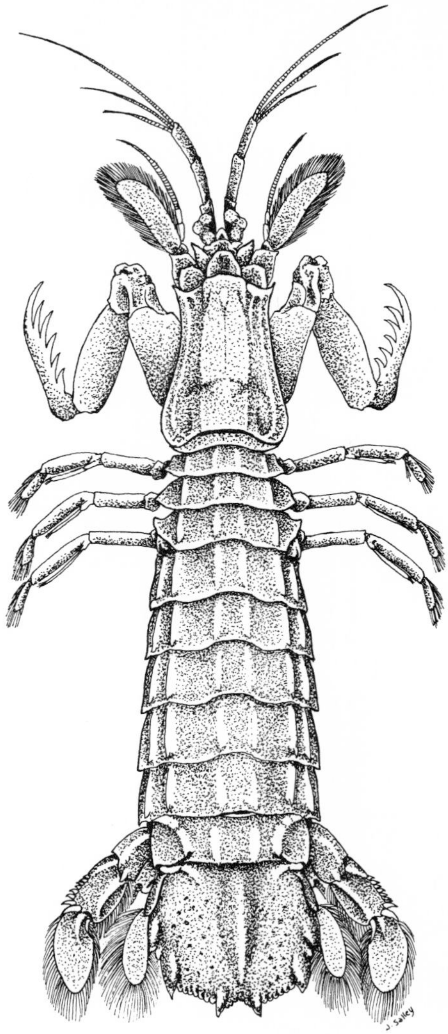

FIG. 8-7. *Squilla empusa,* a mantis shrimp, active predator taking rather large prey. (After Arnold, A. [Foote]. 1901. The sea beach at ebb tide. Century Co.; reprinted unabridged in 1968 by Dover Publications, New York.)

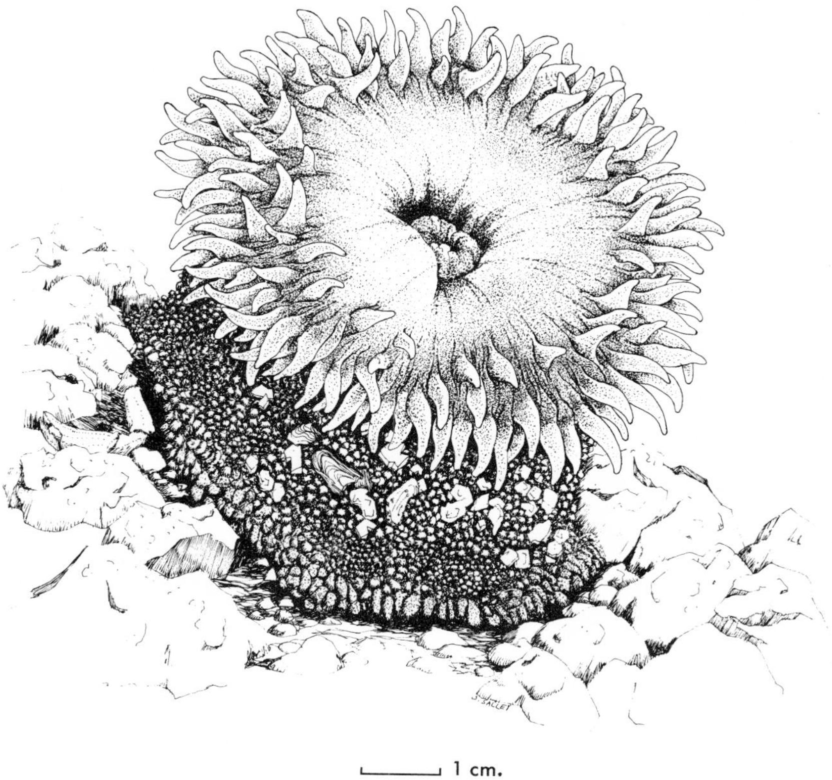

FIG. 8-8. *Anthopleura xanthogrammica,* a sessile predator taking rather large prey.

FIG. 8-9. Serpent stars dominating soft bottom surface. (Oregon Institute of Marine Biology photograph.)

Predators of course tend to evolve compensatory or complementary mechanisms enabling them to overcome to some extent the defensive adaptations of prey species.

Removal of predators may produce explosive growth of populations of prey species, sometimes to the point that they overreach their food supply and either migrate or undergo catastrophic decline, facing local extinction or existence at much lower than usual population levels until their food supply has been regenerated. Blooms of prey species may attract predators from surrounding areas to such an extent that the prey is greatly diminished, following which the predators decrease.

Survival of different species equally subject to predation by a given predator may be due to different mechanisms. For example, the starfish *Astropecten* can swallow both the bivalves *Venus* and *Spisula*. *Venus* has a heavier shell, lower metabolic rate, and much slower growth than does *Spisula*. If the starfish killed both at an equal rate, *Venus* would become depleted. However, with its slower metabolism and heavier shell, *Venus* can clamp shut when swallowed, remaining alive and resisting digestion as long as eighteen days, before which it has usually been ejected from the starfish stomach. *Spisula,* on the other hand, with its faster metabolism must open frequently to respire and is quickly digested. It maintains its population in the face of predation through faster growth rate and reproduction—replacing its population as fast as it is consumed.

Volterra also deduced that if a new factor moderately destructive to both predator and prey is introduced into the environment, the net result is often a decrease in the predator population and an increase in that of the prey. MacArthur and Connell (1966) reported an example in which introduction of DDT in certain citrus orchards for control of the scale insect *Icerya purchasi,* already controlled by ladybird beetles, eliminated the beetles and resulted in an increase of the scale insects.

In nature, predator-prey relationships are usually complicated by the competition of different predators for the same prey and by the fact that many predators carry on varying degrees of predation against a number of prey species and can shift emphasis from one to another, depending on availability. The resultant interactions often produce effects that would be difficult to predict and exert controlling influences over both the species composition and relative abundance of the organisms in a biotic community.

Often the life cycles, growth rates, and physiological and reproductive rhythms of predators and prey are adjusted to each other to the mutual benefit of both populations.

In soft-bottom communities dominated by the brittle stars such as *Amphiura* and *Ophiura,* the brittle stars are sometimes so numerous that every square centimeter of the bottom is searched daily for prey. How is it, then, that bivalves and other animals also occurring in these communities are able to establish themselves? Their newly settled young are in the size range utilized by these brittle stars.

The answer seems to lie in the fact that predatory species take little or no food during their own spawning period. In *Amphiura* the gonads may become so distended that the stomach is crowded or even histolized. There is a period of about two months when the *Amphiura* are feeding lightly or not at all. During this time many newly settled mollusc larvae will have grown beyond the size range preyed on by *Amphiura*. Species for which the area would otherwise be satisfactory but whose larvae and young happen to settle when the *Amphiura* are actively feeding are excluded. Similar relations have been established in other predator-prey situations.

This may also be of considerable advantage to the predator. For example, a young *Mytilus* settling at the beginning of the month during which *Asterias* is spawning may grow from about 0.35 mm. to about 10 mm. during this time, increasing its organic matter 2000-fold. More slowly growing forms will increase at least 500-fold. By eating only one prey animal a month after it has settled, the starfish can obtain as much food, expressed in calories, as it would have obtained by eating 500 to 2000 of the same brood a month earlier. The crop of prey animals can then support a much larger biomass of predators without becoming depleted.

A further conservation of prey species is effected by the fact that during the planktonic larval life of the predator species a further grace period of up to three weeks is given, during which many newly settled prey individuals grow to a size at which they are safe from the attack of newly settled, extremely voracious young of the predators.

Effects of predator-prey interactions on community structure

The physiological rhythms of predator and prey and their relative growth rates often play an important role in determining the species composition of animal communities. Since most communities include numerous different species of both predator and prey, and since all the species go through a series of critical size ranges during their growth, during which time they prey on different species and are subjected to predation by different species, these relationships can become very complex and produce fluctuations in the populations of various species not explainable by any simple change in the environmental conditions.

The predator-prey relationship extends beyond the species directly involved and strongly affects the entire community structure. Paine (1971), for example, has demonstrated that the presence, absence, size of the populations, and size of the individuals of the starfish *Pisaster* in rocky intertidal areas of the West Coast exert controlling influences over the nature of the entire biotic community of the lower littoral zone. In the absence of the starfish, horizontal and gently sloping or protected vertical rock surfaces tend to become covered with a monoculture of mussels, *Mytilis californianus,* and more

FIG. 8-10. *Pontobdella muricata,* a leech. (After Harding, 1910. *In* Mann, K. H. 1962. Leeches. Pergamon Press, Inc., Elmsford, N.Y.)

exposed vertical rock faces with gooseneck barnacles, *Pollicipes polymerus*. In New Zealand similar results were obtained even though all the species involved were different. The presence of starfish prevented the mussels or gooseneck barnacles from completely dominating the areas, permitting establishment of various algae and promoting a much higher degree of patchiness and diversity of the biota. In these situations, predation was found to be more important than competition in determining spacing and relative abundance of species in the community.

SYMBIOSIS—LIVING TOGETHER

Parasites are animals that have adopted the habit of living on or in other, usually larger, animals. Commonly they have become tied in an obligate manner to this association and are unable to live as free-living organisms. Often they have lost the ability to synthesize some vital nutritional factor or factors, which they can obtain directly from their hosts. Thus parasites are commonly deficient in biosynthetic abilities as compared with free-living relatives.

Many groups of parasites have become much modified for parasitic existence, showing reduction or loss of sensory and locomotor functions and organs, and sometimes of the digestive tract as well, and a great increase in reproductive potential. This latter reflects the difficulty of moving successfully from one host to the next. Commonly there is an exaggerated development of suckers, hooks, or rhizoidal attachment organs. The life cycle is often complex, involving asexual as well as sexual reproduction, adapting them to particular successful pathways from host to host and modes of access to their hosts.

Parasite-host relationships

Usually a mutual adjustment evolves between parasites and their hosts so that the latter are not overwhelmed by their guests, to the mutual destruction of both. In some instances mutual adjustment has evolved to such a degree that neither can survive alone without the other. Such a relationship is termed **mutualism.** There are all shades of relationship, from the strictly parasitic, in which the parasite lives at the expense of the host, inflicting injury upon it, to the completely mutualistic, such as exists between many cellulose-eating animals and the protozoan and bacterial fauna and flora of their gut. The term **symbiosis** (literally, living together) is used in its broad, literal sense by some biologists to cover all relationships here included in the term parasitism. Others have used it in the same sense as mutualism. Most parasites fall somewhere between these extremes; they live with their host, inflicting little noticeable damage and doing the host no apparent good. Such a relationship is termed **commensalism.** Some biologists prefer to restrict the term parasitism to cases in which the guest organism clearly lives at the expense of its host, and to use the terms commensalism, symbiosis, and mutualism in referring to the others. The distinction is not always sharp.

FIG. 8-11. Ectoparasitic arthropods. **A,** *Caligus clemensi.* **B,** *Ergasilus longipalpus.* **C,** *Pennella sagitta.* **D,** *Lernaea branchialis* (parasitic copepods). **E,** *Ione cornuta* (=*brevicauda* Bonnier) (parasitic isopod). (**A** after Parker and Margolis. 1964. *In* Light, S. F. 1975. Light's manual; intertidal invertebrates of the California coast, 3d ed. revised and edited by R. I. Smith and J. T. Carlton. University of California Press, Berkeley; **B** and **C** after Yamaguti, S. 1963. Parasitic copepods and Branchiura, John Wiley & Sons, Inc., New York; **D** after Smith, G. 1909. *In* S. F. Harmer and H. E. Shipley [ed.]. Cambridge natural history. Vol. 4. Macmillan & Co., Ltd., London; reprinted in 1923 ed.; **E** after Searle, H. Richardson. 1905. A monograph of the isopods of North America. Smithsonian Institution, U.S. Nat. Mus. Bull. no. 54, U.S. Government Printing Office, Washington, D.C.)

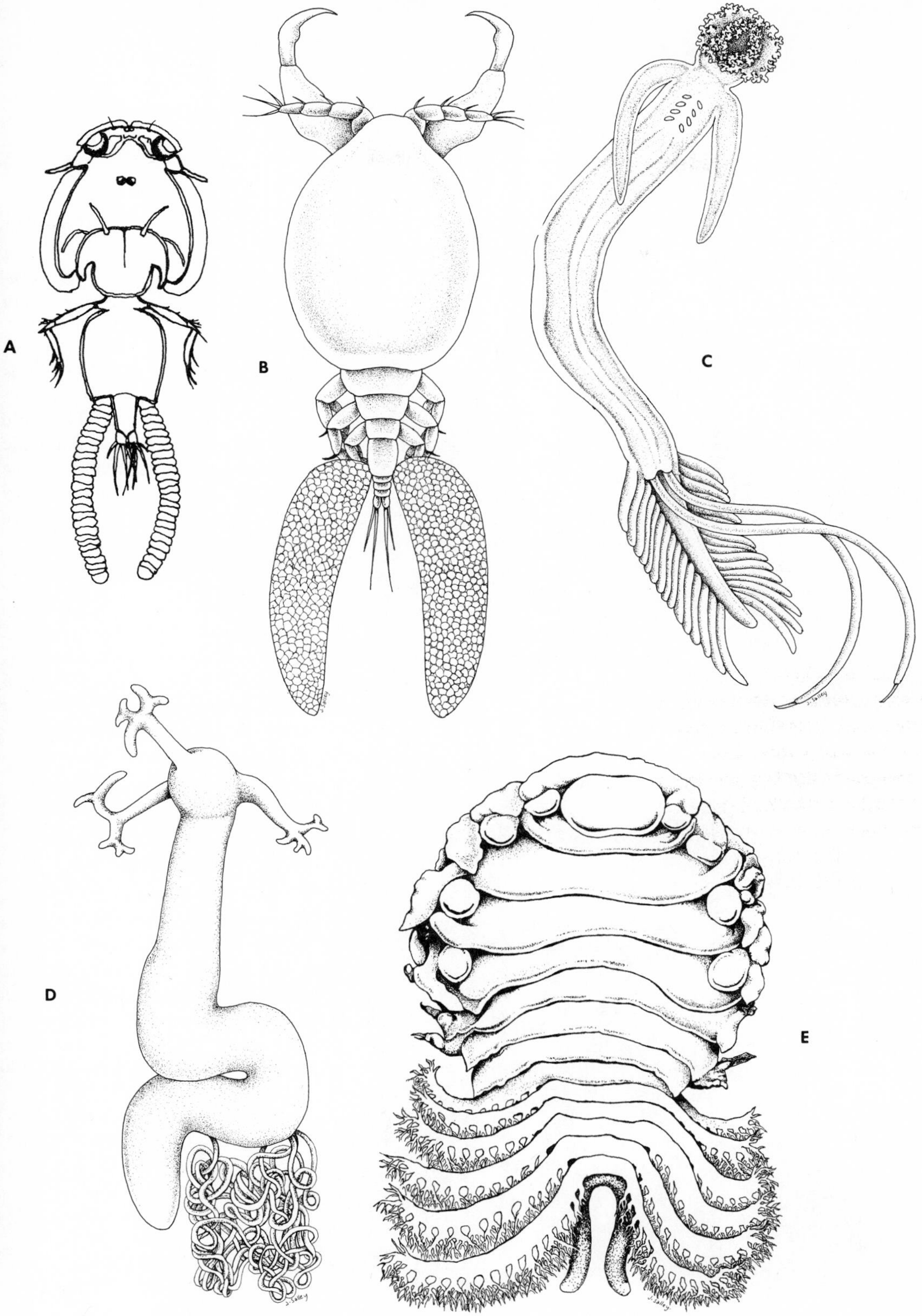

FIG. 8-11. For legend see opposite page.

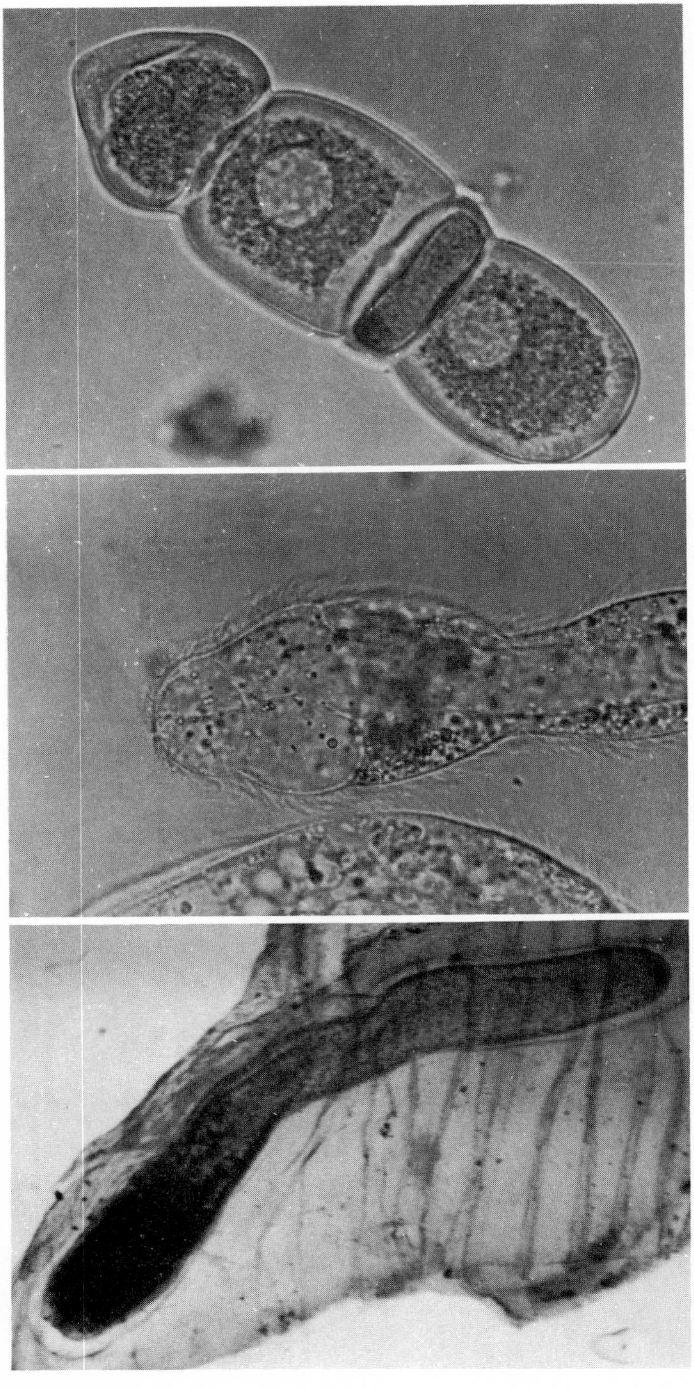

A

B

C

FIG. 8-12. Variety of internal parasites. **A,** *Gregarina valettei* from the barnacle *Pollicipes polymerus*. **B,** Callote of a dicyemid "mesozoan" from *Octopus*. Dicyemids have long been a controversial, little-understood group. Their life cycle, morphology, and systematic relationships have been given the most diverse interpretations by various zoologists who have studied them.

C, *Ouwensia catostyli,* a vermiform parasite of the rhizostome jellyfish *Catostylus,* discovered in 1965 by S. H. Moestafa in Jellyfish from near New Guinea. Its affinities are unknown. Probably it is a larval cestode. The same, or very similar, form was independently rediscovered by P. J. Phillips in 1969 in a related jellyfish, *Stomolophus,* in the Gulf of Mexico—a region thousands of miles away, and completely cut off, from New Guinea. This may indicate great antiquity for the relationship between these worms and the jellyfish hosts—perhaps going back to the Mesozoic era, when an uninterrupted tropical marine biota extended by way of the Tethys Sea from the Indo-Pacific to what is now the Gulf and Caribbean region.

(**A** and **B,** photographs by James L. Houk, Moss Landing Marine Laboratories; ×400.)

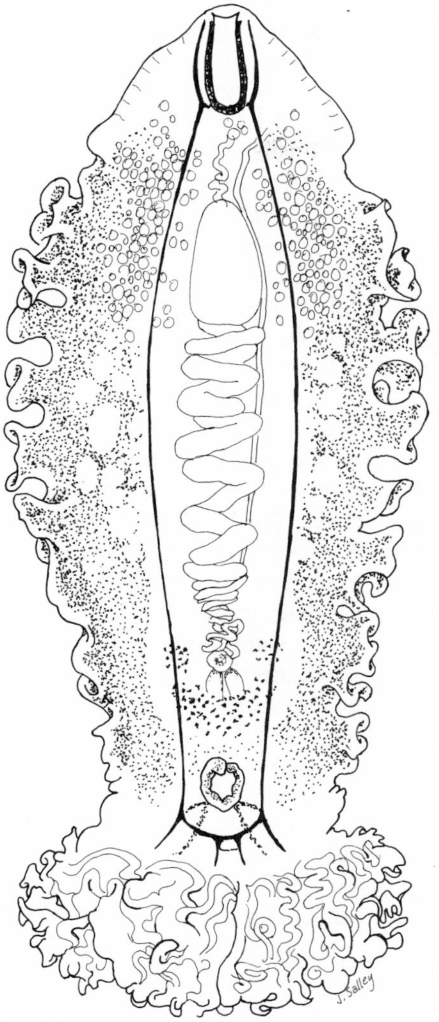

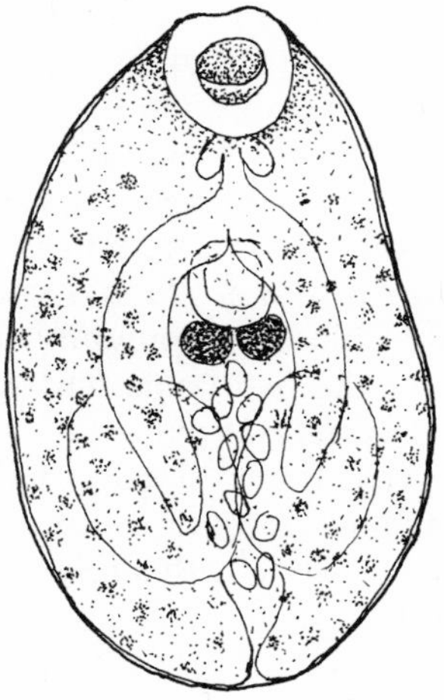

FIG. 8-13. A, *Gyrocotyle fimbriata,* a cestodarian cestode from the intestine of the ratfish (Elasmobranchii, family Chimaeridae). **B,** *Nanophyetus* (= *Troglotrema*) *salmincola,* a digenetic fluke, vector of *Neorickettsia helminthoeca,* a rickettsial (intracellular bacterium) organism causing "salmon poisoning" in dogs that eat infected salmon. (**A** after Lynch and Marshall, N. 1971. *In* Barnes, R. D. 1974. Invertebrate zoology. W. B. Saunders Co., Philadelphia; **B** after Faust and Russell, 1964.)

Organisms ordinarily living as commensals may at times be clearly injurious to their hosts, and conversely many pathogens are often found living in individuals of susceptible host species, doing them no harm, or living harmlessly in other host species. The term parasite will be used in its broader and more general sense to refer to any organism that lives in or on another (usually larger) animal.

Since most free-living animals harbor one or more kinds of parasites, often in great numbers, there are probably more kinds of parasites, and certainly many more individuals, than there are free-living animals. Because of their small size and hidden habitats, we often fail to realize this. We sometimes speak of three major habitats—the marine, the freshwater, and the terrestrial. The advent and proliferation of living things on earth provided a fourth habitat—the parasitic—of such importance that more than half of all the animals now living occupy it.

Parasite-host relationships often provide clues for better understanding of both the current biological scene and the historical processes by which it arose. For example, the abundance and variety of metazoan parasites of animals in a region is a fairly good indicator of the productivity and maturity of the biotic community of that region. Where the fish fauna harbors large numbers and variety of parasites, a survey will show a rich fauna of other animals—invertebrates and carnivores—and the biotic community is one with a relatively long and stable history.

The host range of parasites indicates something of the degree of evolutionary advancement (specialization) of the parasites. During early evolution of a parasitic group, utilization of an increasing number and variety of host species provides a measure of safety or insurance for the parasite against being eliminated accidentally because of host responses to the parasitism, fluctuations in host populations, etc. However, as specialization of both host and parasite populations proceeds, adapting the parasites more closely to their hosts, the adaptations tend to fit populations of parasites specifically to certain species or groups of species of hosts, dropping out less efficient hosts and intermediate hosts. The final result in long-established highly specialized groups is often a large array of genera and species each adapted to a rather narrow range of hosts.

Differences between host species—biochemical, behavioral, and ecological—tend to isolate from each other the parasitic populations inhabiting them and lead to separate evolution and specialization of the parasites.

As a rule, those parasites which have a wide range of host species tend to be more generalized and have not undergone as great a degree of evolutionary advancement or specialization as have those with more restricted host ranges. Commonly also they are parasites in which transmission is more or less direct by ingestion of contaminated food or water, mechanical transmission by flies, etc. and does not involve a complex life cycle with obligate periods in intermediate hosts.

Where a whole series of specialized related host groups is found, as among termites, each bearing its own special genera or species of related parasites, such as the various hypermastigote flagellates of termites, the conclusion is that the association has been one of long standing, with the evolution of both parasite and host proceeding together. The evolution in each group was influenced by the same isolating mechanisms that resulted in genetic isolation and evolutionary differentiation of the host populations. Thus the phylogenetic relationships of parasites and hosts in such groups tend in a general way to parallel or mirror each other, and the relationships of either group may sometimes be used to shed light on obscure facets in the relationships of the other.

Local variations in occurrence of parasites give useful information about movements and mixing of the host populations. On the Atlantic Seaboard, Sindermann (1957) found that the herring populations in the northern portion of their range around the Gulf of St. Lawrence commonly harbored a fungus parasite not present in the herring along the coast from Massachusetts south. The southern herring harbored larval cestodes, the adults of which occur in sharks. Larval cestodes were not found in the northern herring. These differences were interpreted as showing that the herring populations were, and had been for some time, distinct from each other, with little interchange or mixing.

Parasites requiring more than one kind of host during their life cycle are of course limited to areas of overlap in the distributions of appropriate hosts. The Australian fish *Siganus* along the barrier reef shows a high incidence of trematode infections including 10 species of trematodes, but populations of this fish around the sandy shores of Moreton Island, where the snail hosts are lacking, show none.

Animals occurring near the periphery of their distributions, where their occurrence is relatively sparse, usually lack parasites common in the more central parts of their range, presumably because the smaller number of hosts makes it more difficult for parasites to find them and successfully complete their cycles, and because of the smaller number of infections of intermediate hosts. Likewise, when animals become established in new areas, they frequently lose some of the parasites characteristically found in them in the regions from which they came.

The more than 1000 species of monogenetic trematodes and 2000 species of digenetic trematodes from fishes furnish good material for the study of parasite-host relationships, which may shed light on the history, relationships, and movements of populations. The association of trematodes with vertebrates has evidently been a long one, dating back to near the origins of both groups, and both groups have evolved together into a great variety of genera and species, commonly with specific and restricted host-parasite relationships.

The origins of populations of fish now isolated from other populations of their congeners can sometimes be inferred from the relationship of their parasites. For example, the Argentine hake *Merluccius hubbsi,* isolated from other hakes occurring now in the North Pacific and the North Atlantic, has parasites much more like those of the North Pacific hake than those of the North Atlantic hake and may be inferred to have been derived from the Pacific species.

The freshwater drum *Aplodinotus grunniens* has trematodes clearly related to marine groups rather than to those found in other freshwater fish, both the fish and its parasites showing evident relationship to marine rather than freshwater groups. It represents a relatively recent isolation of a marine fish in a freshwater habitat.

Some of the larger-scale geographic biotic affinities that are reflected in parasite populations, as well as similarities in the flora and fauna, are the connection in relatively recent geological times between Atlantic and Pacific sides of central America, the relationship between the Amazon region of South America and tropical Africa, and the relationship and former continuity, by way of the Tethys Sea, of the Caribbean and Indo-Pacific tropical marine biota now wholly isolated from each other.

For sexually reproducing organisms, frequent failure to find a mate could pose a serious problem, especially for internal parasites. It is noteworthy that the flatworms, the only animal phylum that is regularly (although not invariably) hermaphroditic, have been the most successful of the metazoan phyla in adapting to the parasitic mode of existence. Two of the major classes, the Trematoda and Cestoda, are exclusively parasitic, as are some species of Turbellaria as well. The most successful crustacean parasites belong to the Cirripedia, a group in which the adults are primitively sedentary and hermaphroditic. Not all hermaphroditic animals are capable of self-fertilization, but even in species with obligate cross fertilization, hermaphroditism reduces the problem of finding a mate by half.

Many other parasites in which the sexes are separate have evolved special behavioral and structural modifications for keeping the sexes together, such as reduction in size of the male relative to the female and his living as a parasite in or on the body of the female, as in many parasitic copepods and isopods. On the other hand, development of larger males with special structures for holding the female most of the time, as the muscular gynecophoral groove of male schistosomes, is occasionally seen. In the monogenean *Diplozoon,* although hermaphroditic, two individuals grow together during a late larval stage, forming a permanent tissue junction—a type of permanent copulation.

Another noteworthy feature in the lives of many parasites is the occurrence of asexual reproduction as a regular part of the life cycle, thus greatly increasing the biotic potential of the species. The biotic potential of a species is an accurate reflection of the average risk entailed, during the evolution of that species, in completing its life cycle, or to put it another way, of the probability that a given egg or larva will fail to complete the life cycle. This risk is much greater for most parasitic groups than for most free-living animals and is compensated for by development of excessive reproductive potential as compared with related free-living groups. The risk is especially great in forms in which the life cycle is complex, involving two or more different kinds of hosts for various stages, as compared with forms in which transmission is direct from one individual to another of the same species. Internal parasites tend to have greater biotic potential than ectoparasites, and ectoparasites tend to have greater biotic potential than related free-living organisms. Furthermore, in parasites with life cycles involving alternation of sexual and asexual reproduction in different hosts, the intensity of reproductive activity in each stage tends to reflect the risk involved in completing the next phase of the life cycle, whereas the length of time or the proportion of the life-span spent in a given stage of the life cycle tends, as in free-living species, to be a reflection of the relative safety and ease of maintenance of that stage. Stages that are relatively secure tend to be prolonged, whereas those entailing great risk tend to be shortened.

Pseudosymbiosis

A curious form of relationship obtains in a few instances in which particular structures or organelles evolved in one group of organisms are taken over by wholly different organisms, in which they remain functional for a time and useful to their new carriers. Examples are best known among opisthobranch molluscs. Some nudibranchs, for example, acquire nematocysts by grazing on hydroids. Undischarged nematocysts eventually lodge in the tips of the cerata, where they remain functional for some time, apparently exerting a protective influence for the nudibranch. Similar use of nematocysts have been reported in some flatworms and one ctenophore.

An even more remarkable instance seems to be the rule in several species of saccoglossans that feed on coenocytic algae of the order Siphonales, sucking out the protoplasmic contents of their large cells. Some of the chloroplasts are taken up by cells at the ends of the digestive ceca, in which they remain functional and continue to carry on photosynthesis, making a significant and apparently vital contribution to the nutrition of the mollusc. Many marine animals have symbiotic algae in the form of zooxanthellae in their tissues, but this is an instance not of a symbiotic alga, but only the chloroplasts.

The amounts and even the types of photosynthates released by either symbiotic algae or isolated chloroplasts may vary with different circumstances, being different in intact plant cells by themselves than is the case when either the plant cell or the isolated chloroplasts are in animal tissue. This is of considerable importance in assessing the significance of any particular symbiosis between plant cells and animals.

GENERAL EVOLUTIONARY IMPLICATIONS OF SPECIALIZATION

Morphological and physiological changes that can be classed as specializations adapting an organism for life in any particular restricted type of habitat tend to channel the further evolution of such an organism into a one-way bypath, making it increasingly specialized for, or adapted to, that particular type of habitat and increasingly unfit for life in other habitats. The explanation is twofold. The selective pressures on an organism in a given habitat favor any mutations that increase the probability of survival or of successful competition with other species in that habitat and select against other changes. Also, the tremendous complexity and number of genetic factors make of evolution a one-way process, rendering it improbable to the point of impossibility that a series of changes adapting an organism to one habitat could be undone to the point at which the organism would be liberated from that habitat and successful in the original one occupied by its ancestors. In those instances in which a return has been made, as in the case of terrestrial animals giving rise again to aquatic forms, it is accomplished by new structural and physiological adaptations—not by a return to the ancestral condition. This irreversible character of evolution has been termed Dollo's Law in honor of one of the first evolutionists to formulate it clearly.

Thus specialization for any restricted type of habitat takes organisms out of the mainstream of evolution onto a one-way side branch. Most organisms, especially in regions where the physical environmental conditions have been relatively stable for a long period, are on such one-way, blind-ended evolutionary branches. All taxa of organisms are continually giving rise to species increasingly specialized for life in progressively narrower, more restricted ecological niches. Such highly specialized forms seldom, if ever, give rise to new and different lines of evolution.

Perhaps for this reason, periods of unusually rapid change in the world—either geological, climatological, or biological—are marked by extinction of great numbers of formerly successful but extremely specialized animals and plants whose ecological niches are altered. Such periods are followed by relatively rapid evolution of new forms to fill new or unoccupied niches. The evolution of new forms in itself is a factor of change contributing to further diminution of the older, highly specialized groups.

The advent and evolution of life on earth in itself provided a whole new and continually changing array of possible specialized ecological niches. All major taxa of animals, and some plants, have responded by evolving species specialized to fill these niches. Parasitism, commensalism, and mutualism—all the various forms of symbiosis—have arisen repeatedly and independently in all major animal taxa.

Parasites, on the other hand, although probably a majority of species and individuals on earth, have seldom, if ever, given rise to new groups of free-living organisms or even to radically different groups of parasites.

EFFECTS OF CLIMATE AND SEASONAL CHANGES

Temperature is of paramount importance in determining the large-scale distributions of organisms and seasonal successions of plankton organisms in surface waters and shallow water. At all times, surface waters are cold in polar latitudes and become warmer as one approaches the equator. In general and especially in temperate and boreal latitudes, given surface isotherms move toward the equator in winter and toward the poles in summer. Since winter and summer occur at opposite times of the year in the two hemispheres, the surface isotherms of both hemispheres are in their more northerly positions from June to August and in more southerly positions from December to February.

These seasonal changes become much more marked the farther one goes from the equator, as do also seasonal differences in the amount, angle, and duration of light penetration into the ocean.

Tropical and subtropical climate

The overall result is that conditions in tropical and subtropical surface waters tend to be far more stable than is the case in higher latitudes. More energy is received from the sun. The euphotic zone is deeper. The thermocline is deeper in the water and more permanent. The warm surface waters are slightly less dense and less viscous and they hold less of the dissolved gases. Plant nutrients such as nitrates and phosphates, vitamins such as B_{12}, and very slightly soluble substances such as silica (used in forming diatom frustules and radiolarian skeletons) tend to become depleted except in areas of upwelling, since they are not replenished by annual overturn of surface waters and exchange between surface and deeper layers such as occur in winter in colder regions.

Under these conditions, nutrient limitations and rates of nutrient cycling tend to be critical factors determining overall biological productivity and species composition of the biotic community. Selection favors species requiring less of the limiting nutrients per unit of protoplasm, and forms that can fix atmospheric or dissolved nitrogen. Dinoflagellates, blue-green algal cells, and various small flagellates tend to be more prominent components of the phytoplankton, and diatoms a somewhat less prominent component than is the case in high latitudes. Planktonic diatoms with thinner, lighter frustules will be favored both because of limitations in available silica and because such frustules favor flotation in the warmer, less dense, less viscous tropical waters. In a given water mass, the rate of turnover of that vital nutritional factor which is in least supply relative to the demand for it determines the overall rate of biological production and exercises selective pressures favoring those species which require less of it or which can concentrate and utilize it best.

In areas with little vertical water movement the biomass per unit volume of water is relatively low because of nutrient depletion. The clear blue appearance of many tropical waters results from the relative poverty of life in them. Those tropical areas in which considerable upwelling occurs on a regular basis, as along the coast of Peru, tend to be among the most productive regions of the world because of the combination of nutrient replenishment and high energy input from the sun.

Under stabilized conditions, interactions among the organisms become the prime factors controlling species composition, succession, and diversity of the biota. The fact that the phytoplankton does not fluctuate tremendously in biomass with the seasons but is available all year tends to exert selective pressures favoring smaller plankton-feeding animals, which do not have to store much energy in the form of fats or other reserves, and tends to promote dispersal through swarms of small planktotrophic larvae rather than direct development or the production of lecithotrophic eggs and larvae. More prolonged planktonic larval life rather than very brief ones are also favored.

The fact that tropical ecosystems have existed continuously for a much longer period of time than have our present polar, boreal, and temperate ecosystems has allowed the evolution of the biota to proceed to the point at which practically all available ecologic niches are filled. Diversity has reached a maximum, and the communities tend to be characterized by great numbers of relatively well-marked species, none of which ordinarily dominate their communities in overwhelming numbers.

Temperate, boreal, and polar climates

In temperate and especially in boreal and polar regions the reverse situation obtains. The marked lowering of surface temperatures in winter increases the density of the surface waters, causing them to sink and be replaced by subsurface water rich in dissolved plant nutrients. However, phytoplankton growth is minimal during the winter because many of the cells are carried below the euphotic zone by the sinking of the surface waters. The euphotic zone itself is shallower because of the increased angle and lessened penetration of the sunlight, and less total light energy reaches the water and for a briefer period each day as winter sets in. The thermocline also disappears with the cooling and sinking of surface water.

By spring the water has become more or

less stabilized, there is a daily increase in the amount and duration of light, and the light penetrates more deeply into the water as the sun assumes a more nearly overhead position. Those species of phytoplankton which happen to be numerically ahead at the time locally, or which have the fastest growth rates, respond by producing massive blooms. A cloud of phytoplankton produced by such a bloom may reach densities of millions of cells per liter. Such massive clouds tend to be dominated by only those few species which were able to respond most rapidly to the favorable conditions. Once the bloom is well under way, products of the species dominating the bloom and selective depletion of certain plant nutrients by these species slow down the growth of the dominant population and exert controlling influences on subsequent species succession in the water mass concerned.

Since each species of grazing plankton-feeding animals is to some degree selective, feeding most efficiently on certain sizes and shapes of phytoplankton cells and at certain densities of these cells in the water, the dominant species composition of the bloom also exerts selective influences on the populations of the more slowly growing zooplankton to follow.

Massive phytoplankton blooms are of relatively short duration, and the growing and feeding seasons are progressively shorter nearer the polar regions. Selection here tends to favor larger species, which can store energy in the form of food reserves; forms with short planktonic larval life, timed to coincide with the phytoplankton blooms; or direct development of lecithotrophic larvae.

The extreme seasonal differences in conditions exclude species unable to tolerate or to avoid either of the extremes.

Furthermore, polar and boreal ecosystems, as they are known today, are much younger than the tropical ecosystems, having originated only since the end of the Mesozoic era (Chapter 19). The cold waters were populated by species derived from the biota of warmer waters, which became adapted to the cooler temperatures and are able to take advantage of the favorable seasonal availability of abundant food.

Under these conditions there is a tendency to develop huge populations of those species best able to exploit the brief seasonal food supply and adapt to the more stressful oscillation in physical conditions. The fact that these ecosystems are young compared to those of the tropics may also help account for their lower diversity. We may still be witnessing the filling up of these ecosystems, expressed in ongoing adaptive evolution of species to fill a growing number of possible niches as the populations evolve. Although temperate and boreal regions in general have far fewer species of organisms than do tropical areas, many of these species are less well marked—showing more tendency to produce varieties, local races, subspecies, hybrids, and intergrading series. The species, though fewer, are commonly harder to define or delimit.

Although the coldwater ecosystems are being filled with species originally derived from the warmwater biota, there is practically no return flow—species adapted to cold waters becoming adapted to and colonizing tropical or subtropical surface or shallow waters. There are several reasons for this phenomenon.

In the first place, the coldwater environments are relatively new compared to the well-filled warmwater environments. One would expect the predominant movement of species to be into the less-saturated environments until an equilibrium is reached.

Furthermore, the fact that most organisms are living much closer to the upper limits of their thermal tolerance than to the lower limits means that abnormal warmth is more quickly damaging or fatal to most organisms than is a comparable degree of cooling.

Coldwater organisms are also less frequently and less extensively carried into warm surface waters because where water masses of different temperatures meet, the denser cool water tends to sink beneath the warmer surface water, carrying its biota with it. Mobile animals that do find themselves in incomfortably warm water will actively seek cooler, deeper water.

MIGRATIONS

Movements of organisms over considerable distances may occur either actively or passively and either on a regular, usually annual, basis or fortuitously. The term migrations usually means fairly regular movements of populations, over more or less defined routes.

Migrations are usually bound up with seasonal changes in the abundance and location of suitable food supplies, seasonal changes in the suitability of the environment at given places, differences in environmental requirements of various stages of the life cycle, or safety for eggs or immature stages. Migrations also reflect the past history of the species and the changing climatic and geological conditions during its evolution.

Many pelagic marine animals tend in general to move north and south with the seasonal movement of surface isotherms, thus remaining throughout the year in water of the temperature range to which they are best adapted. By moving toward the equator with the onset of winter and toward the pole in spring, animals not only avoid the unfavorably warm or cold temperatures they would experience if they remained in the surface waters of one region but also stay in water with adequate food production, since, in general, biological productivity is sharply and progressively curtailed from the pole toward the subtropics with the onset of winter and increases progressively from the subtropics toward the pole with the coming of spring.

The tern and the petrel

The arctic tern and Wilson's petrel perform remarkable migrations. The arctic tern breeds in the Arctic but in winter migrates south to an area south of the Antarctic Circle. Wilson's petrel, on the other hand, breeds on barren islands of the Antarctic Ocean but migrates north to spend the summer along the Gulf Stream, going as far north as Labrador, and in winter returning to the Southern Hemisphere for the southern summer. In both instances the tern and the petrel travel 32,000 km. or more each year, mostly over open ocean out of sight of land. Because of their migrations, they utilize the massive blooms of food organisms during the short summers in the high latitudes of both hemispheres; they utilize the barren, relatively predator-free islands or tundras of polar latitudes for breeding during the brief but luxuriant growing season when food is abundant there; and they do not compete with each other for the limited nesting sites, although both species are in the Northern and Southern hemispheres at the same time.

The salmon and the eel

The spawning migrations of anadromous fish have also long been a source of amazement to all who have studied them. The Pacific salmon, for example, spawn in the shallow gravelly areas of streams, often hundreds or even a thousand or more miles upstream from the ocean. The young fingerlings migrate downstream to the ocean in the same manner. In the ocean they follow an extended migration, some species traveling most of the way across the North Pacific and back, over a two- to three-year period. Then they return unerringly to the same river from which they entered the ocean and perform a long, difficult, and dangerous upstream journey, involving correct choices of numerous branchings and tributaries, to the same spawning grounds they left as small fingerlings two or three years previously. Because they have not seen or traversed those waters since their initial journey downstream, this return would seem to represent an altogether incredible feat of memory. This ability has been shown to depend on the sense of smell, which demonstrates the fact that various tributaries and streams must be much more individualized in a chemical sense than one would suppose, since their dilute traces in the larger river systems of which they become a part and in the offshore ocean waters are sufficient to guide the salmon back.

The tendency of each population to return to the same place from which it came, to spawn, together with differences in timing between species and populations, doubtless plays a role in keeping the various races and species genetically distinct. Those populations which make the longest upstream journeys as a rule begin their run earlier in the season than do those making shorter runs.

Salmon and other anadromous fish are threatened by overfishing during their spawning runs, when they are highly vulnerable to commercial fishing methods. An even greater threat is the rapidly increasing, nearly universal pollution of rivers and streams and their alteration by dams and impoundments for flood control and hydroelectric projects.

The complex interrelationships of salmon with their environment are illustrated by the fluctuations in Karluk Lake on Kodiak Island, one of the world's most important spawning grounds for the red salmon. Nelson and Edmondson (1955) showed that the decomposition of 1 million spawned-out salmon contributes the equivalent of 87 tons of commercial phosphate fertilizer and 420 tons of nitrate fertilizer (worth about $28,000) to the lake annually. Salmon runs formerly were of the order of 3 million or more, or equivalent to about 1300 tons of commercial fertilizers. This enabled the lake to support rich phytoplankton production, which in turn supported the vast swarms of zooplankton needed to support the young fry. The great reduction of escapement into the lake because of commercial fisheries resulted in a great decline in the fertilization of the lake and hence indirectly to a further decline in the numbers of salmon beyond the direct effect of the fisheries.

American and European eels of the genus *Anguilla* are equally remarkable. They spend most of their adult life in ponds and marshes or other bodies of freshwater. When ready to spawn, they enter streams, swim down the rivers to the ocean, and migrate to their spawning grounds in the Sargasso Sea, hundreds or thousands of miles from their home ponds.

As they enter upon this migration, they change from a nondescript dirty yellowish color to glistening silver. The eyes enlarge greatly (Fig. 8-14). Those changes, together with the fact that migrating eels are never found in surface waters between the coastal streams and the Sargasso spawning grounds, indicate that the migrations are done at considerable depths.

The eggs are pelagic and hatch into

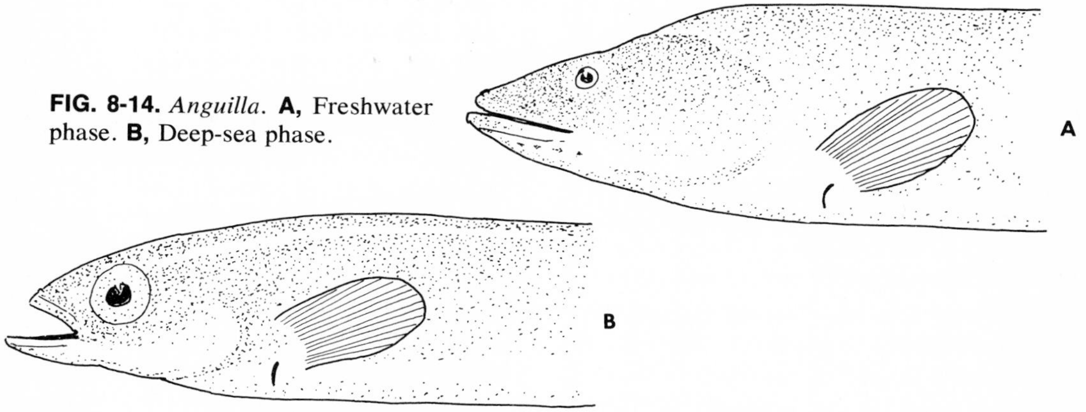

FIG. 8-14. *Anguilla*. **A,** Freshwater phase. **B,** Deep-sea phase.

small transparent flattened forms that were given the generic name *Leptocephalus* before their relationships were known. They are now termed leptocephalus larvae. The larvae come to the surface waters and float as part of the plankton for one or two years.

The American eel *(A. rostrata = A. chrysypa)* matures in a year, just as the Gulf Stream has brought the larvae alongside the North American coast. The European eel *(A. vulgaris = A. anguilla)* matures a year later, when the Gulf Stream has carried the larvae across the Atlantic to the European shores. As they enter inshore neritic waters, the leptocephali transform into young eels, called elvers, and begin their migration up rivers and streams. The males remain in the lower reaches of the rivers or upper ends of estuaries. The females often migrate upstream far inland to take up their residence in suitable ponds or marshes.

The eel is the most remarkable example known of adaptation of life cycles to the great oceanic currents. Knowledge of the migration was brought to light through the tireless efforts of the Danish scientist Johannes Schmidt (1925), whose researches constitute a classic in fisheries biology.

It is probable that a growing stenothermy—requirement of water temperature at 16° to 17° C. for reproduction—plus negative phototaxis guides the eels to their spawning ground, since the area between Bermuda and Puerto Rico is the only region where this particular narrow temperature range prevails at suitable depths. The newly hatched larvae, on the other hand, show positive phototropism, which brings them to the surface, where the great surface currents determine their subsequent distribution.

How did the precise differential arise in the length of time needed for development between the European and American eels? It seems probable that during the ice ages the belt of uniform temperature of 16° to 17° C. at a depth about 500 meters included most of the central Atlantic. The American eel probably spawned in the western edge and the European eel in the eastern edge of this zone, each having a migration of approximately the same length. During subsequent geological time, the 17° C. temperature zone shrank, especially toward the west, with the result that the European eels had a longer and longer trip to make. The increasing time needed for the larvae to float back to the European coast placed a selective value on longer larval life, those which metamorphosed too soon perishing at sea. Eventually the shrinking of suitable breeding grounds brought the two together in a very restricted region, but the European eel had already evolved a longer larval life.

The eel and the salmon, with their similar yet opposite biological histories—one spawning in inland waters far from the sea but living most of its life in the ocean, the other spawning far out in the deep sea but living most of its life in inland waters far from the sea—present fascinating problems in the origin and evolution of behavioral patterns, the guidance mechanisms in animal migrations, and the adaptive changes involved in alternating between freshwater and saltwater twice during the life cycle, once as immature individuals and once as mature ones ready to spawn and die. They also shed some light on the probable biological and geological history of the world during the evolution of these patterns of life.

Antarctic krill

The movements of the antarctic krill *Euphausia superba* are a good example of passive migratory movements, with the life cycle adapted to take advantage of the different directional movements of water at different depths.

The adults appear in antarctic surface waters in dense shoals near the continent of Antarctica in September and October and remain in the surface waters feeding and spawning, drifting north and northeast throughout the antarctic summer, November to February. As they approach the

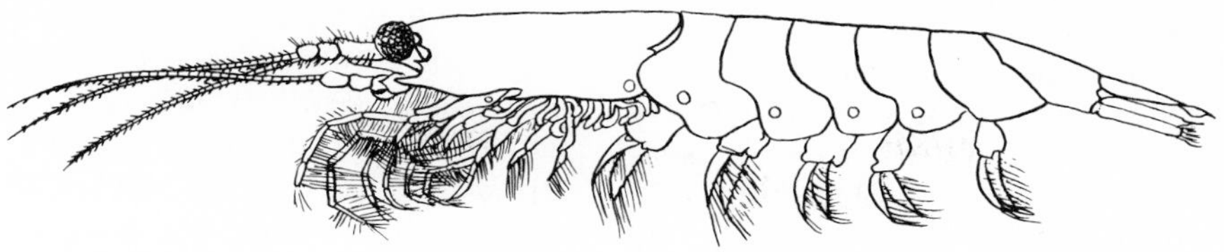

FIG. 8-15. Euphausiid shrimp, *Euphausia superba*. The antarctic krill, or whale food. (After Bargmann, H. *In* S. Ekman. 1953. Zoogeography of the seas. Sidgwick & Jackson, Ltd., London.)

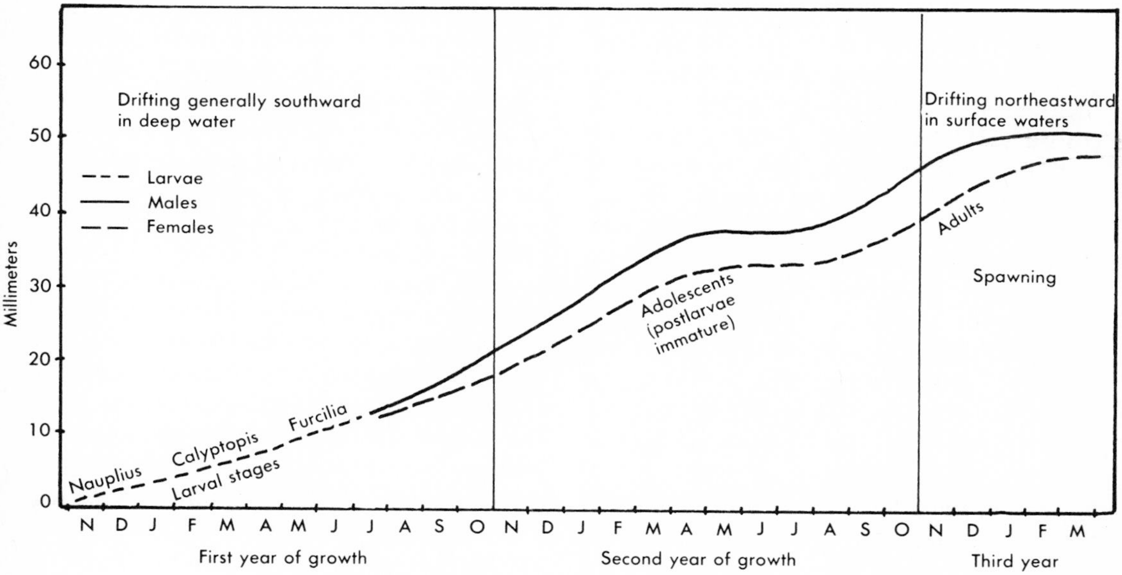

FIG. 8-16. Growth of *Euphausia superba*.

Antarctic Convergence, they sink into deeper water, the eggs sinking farthest.

Here they are caught in the southerly drift of the water of intermediate depths and are gradually returned southward as juvenile stages. Although the calyptopis and furcilia stages show diurnal vertical migrations, the balance of time at various depths is such that the net southward drift is continued. The later stages tend to accentuate this drift by remaining in deeper water more of the time. As the krill finally approach the Antarctic continent two years later, they are carried up with the upwelling near Antarctica, molt to the adult stage, and begin the feeding, spawning, and northward drift.

This shrimp is an excellent illustration of the adaptive timing of a life cycle, with different parts of it spent at different depths, enabling the species to remain in the same general region in spite of the continuous drift of surface waters out of that region.

Marine turtles

During their spawning season marine turtles come up onto sandy beaches after dark and lay their eggs in excavations in the sand above the high-water line in tropical regions. They perform extended migrations between these beaches and their feeding areas.

By far the most abundant and best known are the herbivorous green turtles *(Chelonia),* which feed on shallow-water meadows of turtle grass *(Thalassia).*

Populations of these turtles have been shown to return to the same beaches on which they are spawned, sometimes after traveling more than 2200 km. at sea. Careful studies of the migrations and habits of sea turtles have been made by Carr (1954, 1961, 1965) for the turtles of the tropical American region and by Bustard and Tognetti (1969) for turtles of the Great Barrier Reef, Australia. Apparently individuals also establish particular feeding areas to which they can return after spawning, and are also able to return to them even if removed and liberated at a considerable distance in areas presumably completely unfamiliar to them.

Adults spend more than a year at sea before spawning. On the Great Barrier Reef marked individuals returned to spawn after a four-year interval. Thus turtles spawning on a particular beach in successive years are not the same individuals.

Where turtle population densities are high and the populations relatively free from predation, the final check that brings the population into equilibrium was found to be the increasing destruction of some of the nesting sites by other nesting turtles digging into them, when the spawning beaches became very crowded—a strictly density-dependent phenomenon. This effect is negligible at lower population densities. Certain protected beaches along the Great Barrier Reef may be the last in which it is observable. Most of the world's green turtle populations have been drastically reduced through predation by man, especially at the spawning beaches.

Other species of marine turtles, which are carnivorous and can therefore attain only markedly smaller populations, being at a higher trophic level at which less food is available to them, are limited primarily by food availability, rather than by crowding and egg destruction on the spawning grounds.

Demersal fish

Demersal fish, for example, plaice and others that spend much of their time on rich feeding grounds (shallow banks of the North Sea, or those off Newfoundland), commonly perform spawning migrations to other areas such that water movements bring their pelagic larvae into the feeding areas at the time they are ready to assume a demersal life, rather than sweeping them out of the feeding areas to be lost at sea in less favorable regions. Presumably also the eggs and young larvae are less subject to predation than would be the case if they were released in the richer feeding grounds.

THE STAY-AT-HOMES

Temperate, boreal, or polar animals such as infaunal organisms, sessile attached forms, flightless birds (penguins), and small planktonic, demersal, or benthic organisms unable to perform extended seasonal migrations respond to the seasonal changes by feeding voraciously during the season when food is available, storing as much energy as they can; by reducing the duration of planktotrophic larval stages or substituting lecithotrophic larvae or direct development; and by greatly reducing activity and metabolic rate during the most unfavorable portions of the year. In many species of rapidly reproducing small organisms, populations are rapidly built up during the summer, whereas the winter is passed in the form of resistant eggs or cysts with low metabolic activity. Populations are considerably decimated during the winter.

Planktonic and free-swimming forms often seek deeper water in winter, whereas mobile benthic inshore forms may do the same by moving farther offshore.

LOCAL SURFACE DISTRIBUTIONS

It is commonly observed that the surface distributions of phytoplankton and of zooplankton are often patchy in character, the maxima of zooplankton being found not where the phytoplankton clouds are thickest but near their edges, or in waters between the great phytoplankton clouds. Hardy and Gunther (1935) suggested the

animal exclusion theory to account for this distribution. According to this theory, a dense cloud of phytoplankton may impede the upward movement of many zooplankton species through it, so that they may be prevented from completing their upward migration until the most dense portions of the phytoplankton cloud have drifted past, at which time they rise to higher levels, thus creating a discontinuity between the maxima of phytoplankton and of zooplankton. However, as Bainbridge (1949) found, for *Neomysis vulgaris,* at least some zooplankters will actively swim from diatom-poor waters into diatom-rich waters, especially if they are partially starved. Harvey (1934) proposed that great swarms of zooplankton may consume the phytoplankton to such an extent that they thin it out locally, producing the observed discontinuities. Steemann-Nielsen (1937*b*) noted that differential rates of multiplication may produce similar effects, the fast-growing phytoplankton producing dense blooms where they are locally more abundant in the water at a favorable time. Slower-growing zooplankton may graze on this bloom, and by the time their population became heavy, the bloom of phytoplankton would have passed its peak because of exhaustion of surface nutrients.

It seems probable that all these mechanisms may play a role in producing the discontinuities at various times and places.

Hardy and co-workers (1936*c*) devised a simple plankton indicator for herring fishermen to aid them in locating areas where zooplankton, especially *Calanus,* abound. It had been observed that herring are usually found where there are clouds of *Calanus* and tend to avoid areas of dense phytoplankton bloom. The indicator consisted of a cloth disc that was towed behind the fishing boat for a given length of time and then examined. If it showed a reddish color as a result of having retained many of the red copepods, the area was judged to be favorable for fishing, but if it was green, from retaining mostly phytoplankton, the area was judged to be less favorable. Boats making use of the plankton indicator reported consistently higher catches than those which did not use it.

DIURNAL VERTICAL MIGRATION

Many animals find the twilight zone, where there is a bare minimum of light, a favorable depth during the day. Here it is more difficult for predators to see them, and yet they have not gone so deep that they cannot swim up into the richer waters nearer the surface to feed at night. **Diurnal vertical migration** is a striking feature of the life of many marine animals. The usual tendency is to go down during the daylight hours and come up as the light decreases. Some organisms follow an isolume rather closely—that is, they adjust their depth during the day in such a way that they remain throughout the day in water with a particular degree of intensity of illumination. The assumption is, of course, that they would descend to deeper water during the morning and ascend again late in the afternoon, or during cloudy weather. However, this will not serve as a general explanation of diurnal vertical migration, since most plankters do not scatter out at night, as may be expected, and many of them begin their downward migration well before dawn and their upward migration after sunset, at hours when little or no change of illumination is occurring (Fig. 8-17). Nor is the concept that hunger combines with negative phototropism a wholly satisfactory explanation, since in areas where the bulk of the phytoplankton is at some depth in the water, many zooplankters migrate through the most dense strata of phytoplankton to the less dense surface waters.

Many other animals live at different depths at different times of the year or during different stages of their life cycle.

Such periodic changes in depth are often important in other ways than in feeding and avoiding predators. For example, water at different depths may be drifting in different directions. This drifting may play a

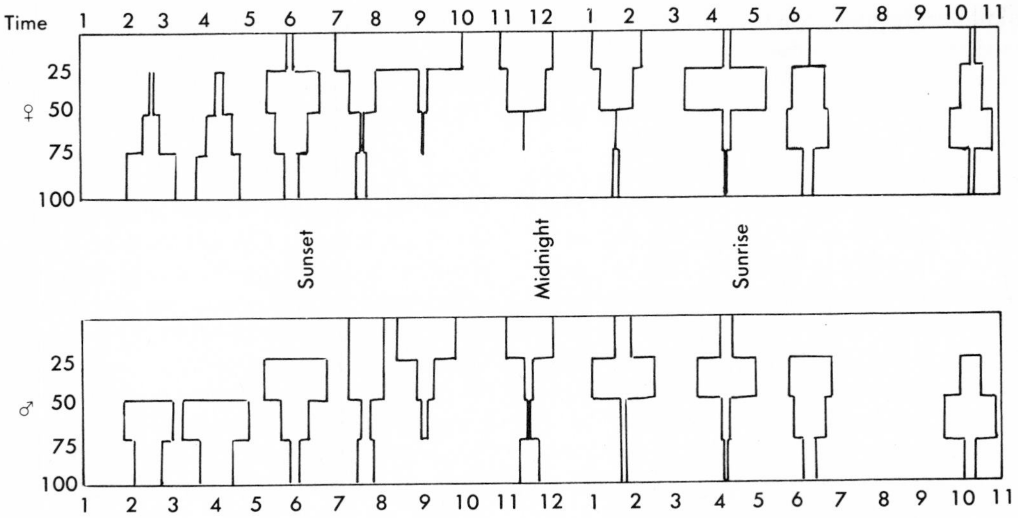

FIG. 8-17. Diurnal vertical distribution of male and female *Calanus finmarchicus*. South of Ireland, August, 1932. (After Farran, 1947.)

role in keeping a population of animals in a given region in spite of the movements of the water. In other cases, animals may take advantage of the interfaces between different water masses moving in different directions. By concentrating at such a level they can avoid the full effects of drift in either direction.

Because a large proportion of the animals in the euphotic zone descend to lower depths during the day and rise toward the surface at night, and because the phytoplankton remains nearer the surface whereas waters of different depths move at different rates or even in different directions, there is a continual redistribution of the animals in the phytoplankton. A given cloud, or bloom, of phytoplankton does not necessarily carry the same animals with it from day to day as it drifts across the face of the ocean. Animals finding themselves in phytoplankton-exhausted surface waters one night may find themselves in richer water on a subsequent ascent.

For small animals such as copepods, daily vertical migrations of any extent mean that they must spend most of their time swimming up or down. Hardy and Bainbridge (1954) devised an ingenious plankton wheel in which planktonic animals could be introduced into a large rotating circular glass wheel-shaped tube filled with seawater. The speed of rotation was adjusted to the speed of upward or downward swimming. Adjustment was made at various times and under controlled conditions of lighting so that the speed and reactions of the animals could be readily observed and studied. They clocked upward movement of the copepod *Calanus* at 15 meters/hour, *Centropages* at nearly 30 meters/hour, *Centropages* at nearly 30 meters/hour, and the small copepod *Acartia* at 9 meters/hour. The euphausiid *Meganyctiphanes* achieved 135 meters/hour, and the polychaete *Tomopteris* attained over 200 meters/hour. Downward speeds often exceeded the upward speed of the same species by a considerable amount. *Calanus*, for ample, can descend 47 meters in an hour and is capable of short bursts of speed at rates exceeding 100 meters/hour. The results of these experiments suggested that the observed daily migrations in the sea were indeed feasible for these small animals.

Enright and Hammer (1967), in carefully controlled, large-scale experimentation with mixed zooplankton just pumped in from the offshore saltwater line at the Scripps Institute of Oceanography, demonstrated that for some species there is an endogenous physiological timing mechanism, or **biological clock,** that causes them to retain the timing pattern of their vertical migration behavior for some time under conditions of constant illumination and temperature. With others this was not the case, and exogeneous stimuli were required to elicit the vertical movements. They conclude that "although vertical migration of zooplankton may appear to be a single phenomenon in an ecological context, conceivably evolved on the basis of a common selective pressure, the mechanisms underlying the field behavior are by no means uniform."

The extensive vertical migrations of animals play an important role in the transport of energy from the surface layers to the deeper layers of the ocean. The descent of the animals is much faster than settling rates. Animals remaining in deeper water prey on those descending from above. Some of these in turn migrate to still deeper levels.

Much organic matter that, if merely settling, would be consumed and recycled in the surface layers is thus carried directly to deeper strata.

Deep-scattering layer

When great numbers of animals migrate downward during the day, they form one or more strata at intermediate depths sufficiently dense to reflect an echo from echo-sounding equipment, giving a record of a "false bottom," the so-called deep-scattering layer. As night approaches, the deep-scattering layer rises and becomes more diffuse or even disappears, and then forms again at dawn. The daily increase of animals at such intermediate depths may induce a type of reverse diurnal migration of certain predators living still deeper. These predators may rise to intermediate levels to feed during the day and return to their murky depths at night.

Large schools of fish at intermediate depths will also produce echoes, often characteristic for particular species. Echo-sounding equipment has become a valuable aid for commercial fishermen in locating and trawling for such schools, in addition to its value as a navigational aid and for mapping the bottom.

Commonly, three fairly well-defined deep-scattering layers are formed—the deepest at about 520 to 580 meters, sometimes descending to around 700 meters, composed of small myctophid fishes; the second at about 430 meters, mostly of euphausiids; and an upper stratum, at around 230 to 270 meters, of sergestid shrimp. At night the three layers rise and merge into a single diffuse band up to 150 meters thick, but before dawn the three bands begin to sort themselves out again and descend to their characteristic levels, the animals of the lowest strata beginning their descent first. The layers never cross. Other kinds of fish and crustacea, as well as siphonophores and squid, may also be associated with deep-scattering layers.

The animals in the deep-scattering layers are not as profuse and crowded as had been anticipated from early studies of their echograms. Observations by divers in bathyscaphes and small research submarines, and net hauls from deep-scattering layer levels indicate that, more often than not, the animals of these strata are at some distance from each other, and the deep-scattering layers are not likely to support any spectacular new commercial fishery.

Animals such as schooling fish with swim bladders and siphonophores with small gas floats produce the best echoes.

LIFE BELOW THE EUPHOTIC ZONE

Below the euphotic zone, life is less abundant than in the surface layers and depends for its nourishment entirely on the drifting or carrying of organic materials down from above or on the ability of animals to rise up into the euphotic zone at times to feed. There are no producers.

Conditions here are far more uniform over most of the world than is the case near the surface, and from region to region there is a greater uniformity of animal life. There is darkness, cold, and immense hydrostatic pressure. Currents are slower than at the surface. Seasonal changes are slight or absent. It is not surprising that animals inhabiting such an environment should have evolved what to us appear to be fantastic forms and adaptations.

Pelagic fish from the upper strata of water tend to be silvery or light-colored ventrally and on the sides, resembling the bright reflective air-water interface when seen from below, and darker, bluish, or greenish above, resembling the darker water below when seen from above. The darker color dorsally also serves as obliterative countershading, making the shadow on the ventral portion less conspicuous. Crustacea from upper strata are commonly nondescript gray, whitish, or somewhat transparent. At intermediate depths many of the fish are much more uniformly silvery and tend to have larger eyes. Here the light intensity is about the same in all directions, and obliterative countershading would have no function. Demersal fish and crustacea at these depths often are partly red, with subcuticular red chromatophores or oil droplets. In deep water the fish tend to be entirely black (except for bioluminescent organs), and the crustacea are commonly either uniformly red, with the red pigment incorporated into the cuticle, or black. In deep waters the eyes are usually small and in some cases absent.

Numerous other animals, such as most hydromedusae, ctenophores, siphonophores, salps, and occasional species from other groups, are so transparent that they are difficult to see at whatever level they are found. Deep-sea medusae, however, also tend to be red or purplish.

In the bathypelagic and abyssal zones, food is relatively scarce and hard to find. Many animals have evolved fantastic adaptations for obtaining it and the ability to ingest large amounts when it is found. Some can swallow objects larger than themselves. Often they have a fierce, dragonlike appearance, with large mouths filled with long, sharp teeth. They often have long tactile processes for probing the dark water about them, large eyes (especially animals that live where there is only dim light), and commonly, photophores. They have evolved as peerers, gropers, and gulpers.

Bioluminescence

The luminescent organs serve several purposes. They may make species recognition possible, playing a role in keeping schools of the same species together, which is important in mating and reproduction. They may play a role in courtship activity in some species. Some fish use them to attract other food organisms. A few have evolved a dorsal fin ray resembling a fishing pole. It can be extended forward just above and in front of the mouth, dangling a luminescent organ at the end to attract other organisms within easy reach. Luminescence also serves as a defense against predators. A sudden flash of light when a fish is disturbed may momentarily blind or scare a predator, enabling the prey to escape. Some deep-water squid have evolved luminescent ink, which has the same functions in defense and escape in the zones of darkness that the usual black ink of their shallow-living relatives has in better-lighted zones.

Species recognition, attraction of food, and defense, then, are some of the functions of bioluminescence.

It is more difficult to see any selective advantage of bioluminescence in some of the microorganisms such as continuously luminescing free-living bacteria. The light in the photophores of higher animals is often made by luminescent bacteria living symbiotically in the photophore.

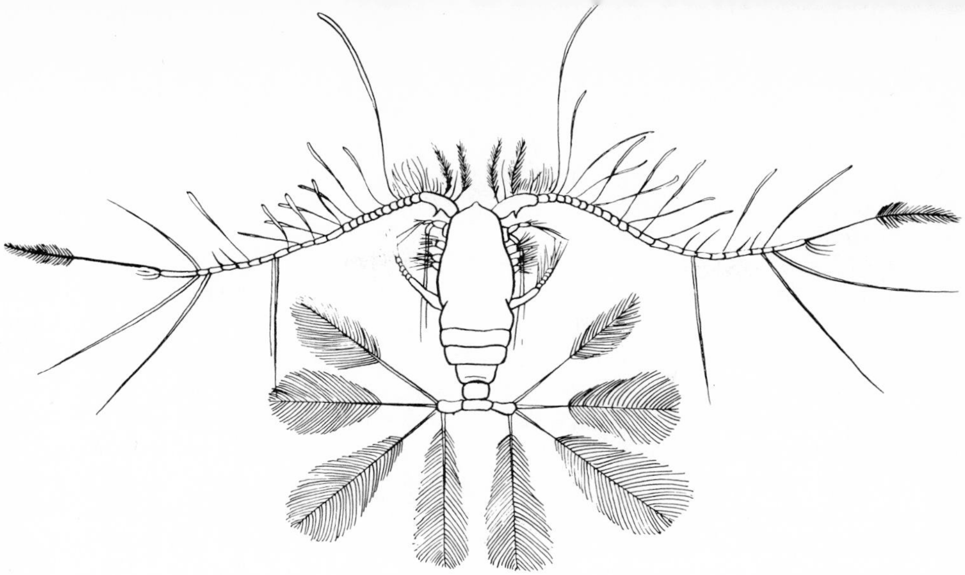

FIG. 9-1. Copepod, *Calocalanus pavo*.

CHAPTER

9 THE ZOOPLANKTON

Protozoa

There are three groups of protozoa that are almost exclusively marine and are found in the plankton of all oceans and seas: the orders Foraminifera and Radiolaria and the ciliate family Tintinnidae.

CLASS SARCODINA
Order Foraminifera

The Foraminifera (Fig 9-2) are shell-bearing rhizopods. They are primarily benthic or attached, but there are a number of important pelagic species. The cytoplasm is rather homogeneous and streams out through one or more apertures in the shell, and in perforate forms through the many small pores as well, forming long, fine, branching, anastomosing, radiating, or net-like pseudopodia, the so-called **reticulopodia** characteristic of this group. In living specimens these pseudopodia are in constant activity, extending, contracting, fusing with others, etc. Minute granules can be seen moving along them, those in the inner core of a pseudopodium going, in general, in an outward direction and those in the outer sheath of the pseudopodium passing inward. The cytoplasm may contain brown granules that seem to be waste matter and in some are extruded prior to the formation of a new shell or new chambers of a shell. In some species the cytoplasm is colored. No contractile vacuoles are present.

The test, or shell, varies greatly in form and color. In most species it is less than 1 mm. in diameter or length, although a few attain several millimeters. It may be almost wholly or partly arenaceous, with foreign particles such as sand grains incorporated into it, or calcareous—either perforate (pierced by many small pores in addition to the main aperture), or imperforate and porcelaneous. The young of all species with calcareous tests are perforate, only later depositing a porcelaneous, smooth layer over

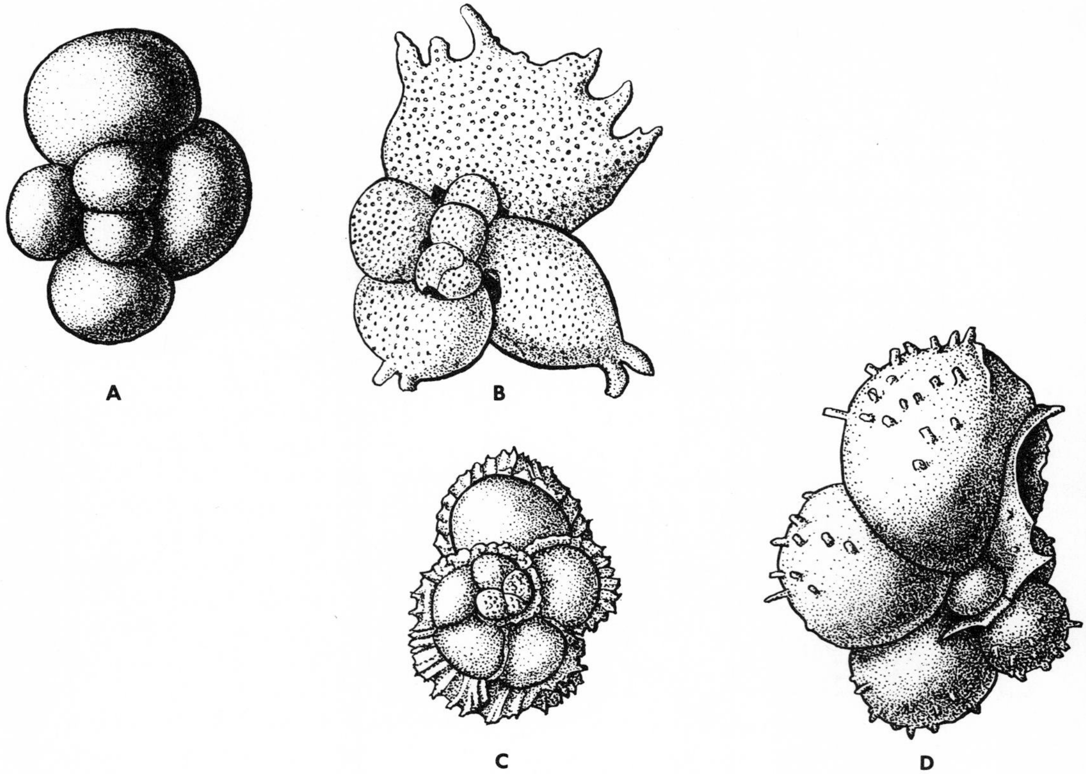

FIG. 9-2. Foraminifera. **A,** *Globigerina bulloides* d'Orbigny, dorsal view (after H. B. Brady). **B,** *Globigerinoides sacculifera*. **C,** *Globorotalia fimbriata* (after H. B. Brady). **D,** *Hastigerina pelagica* (after H. B. Brady). (After Cushman, J. A. 1959. Foraminifera: their classification and economic use. Harvard University Press, Cambridge, Mass.)

the outside of the test. The majority of species remain perforate.

Polythalamous species, with tests of more than one chamber, first form a single chamber, the **proloculum,** to which other chambers are added later during growth. In some species the chambers themselves may become subdivided into chamberlets. The septa between chambers are perforated by one or more pores called stolon canals, through which the protoplasm is continuous to all chambers. The last chamber bears one or more apertures, which are the main openings of the shell.

Foraminifera, for the most part, feed on diatoms and algae, but pelagic species may also catch other protozoans and microcrustacea. Some of them carry symbiotic zooxanthellae in their cytoplasm.

Most Foraminifera exhibit dimorphism, occurring in two forms that differ from each other in the size of the first chamber, the proloculum, and commonly in other details as well. The **megalospheric** form possesses a large proloculum. The organism has a single nucleus, usually found in the middle chamber and shifting its position to maintain this relationship as new chambers are added. When full growth has been attained, the nucleus divides rapidly many times and isogametes are formed that leave the parent test and fuse in pairs to form zygotes.

The zygote secretes a **microspheric** proloculum around itself. As growth occurs and new chambers are added, the nucleus continues to divide, forming a multinucleate individual. The nuclei in the various chambers vary in size proportionately with the size of the chambers.

Eventually, after full growth is attained, the cytoplasm around each nucleus becomes individualized, and the young mononucleate cells desert the parent shell, secrete a megalospheric proloculum around themselves, and begin the cycle again.

This reproductive sequence may be variously modified in some species, especially by the interposition of more than one asexually produced megalospheric generation before gamete formation, so that megalospheric individuals tend to be more abundant than microspheric ones. For some species, no microspheric generation is known.

In a general sense, in dimorphic species, the microspheric phase may be said to be the more conservative and the more likely to include stages indicative of the phylogeny of the species. The megalospheric form is more likely to be specialized and to skip over such early primitive stages in its development. In fossil Foraminifera, therefore, study of early microspheric and mature megalospheric stages of a species shows something of its ancestry and of its relationship to later descendants, respectively.

Movement is slow in most species, up to 6 cm./hour, but usually only 1 cm. or less per hour. Some species have been observed to abandon their test, move about for a time as naked protoplasts, and then make a new adult test. In species that incorporate foreign materials into their tests, these materials are first ingested and then later concentrated at the surface from within and held in place with a secretion of cementing substance. This explains the growth of one-chambered forms and also the predominance of adult shells in collections.

The earliest, most primitive type of test is chitinous. Next in order are the arenaceous tests, which are developed by adding foreign material to the exterior of chitinous tests with various kinds of cementing materials. The cement may be chitinous, ferruginous, calcareous, or siliceous. As the cement becomes a more dominant feature, its secretion renders foreign material unnecessary, and a smooth test results. Those with ferruginous cement are usually yellowish or brown, and most are arenaceous. Those with calcareous cement constitute the majority of species. They are all perforate when young, but some species later become covered with a smooth, imperforate layer. The change from chitinous-arenaceous test to purely calcareous test can sometimes be seen in one individual, the early chambers being arenaceous and later ones being purely calcareous. Among arenaceous types, more primitive species show less selectivity of materials, whereas advanced types often show remarkable selection and arrangement of foreign materials. Species with arenaceous tests are benthic, and below 2000 meters they are usually the only ones present.

Some holoplanktonic types, such as species of *Globigerina,* are so abundant that the bottom ooze for many square miles may be composed largely of their calcareous shells, which have sunk to the floor of the seas, accumulating over the centuries as the so-called **globigerina ooze.** Such sediments may, in later periods, become consolidated and made into calcareous rock or chalk, as in the famous white cliffs of Dover. Globigerina ooze, such as that found in large areas in the North Atlantic, commonly contains approximately three-fourths Foraminifera shells and one-fourth coccolithophores, by weight. The most abundant species are the foraminiferan *Globigerina bulloides* and the coccolithophore *Pontosphaera huxleyi.*

Faunal groups can be recognized by cooccurrences of species in particular water masses. There are some 28 species of abundant, widely distributed pelagic Foraminifera, 22 of them occurring in tropical and subtropical waters, 4 in boreal or cool temperate waters, and 1 each characteristic of transitional and of polar waters. Most species spend their early stages in the upper part of the euphotic zone and then mature at somewhat deeper levels below the euphotic zone, but for the most part in the upper 200 meters. Bé (1968) determined the **porosity** of the shells (a function of the number of pores per unit area, and pore size) for many of the species. He found that in addition to the fact that Foraminifera from cold waters tend to make, on the average, smaller, more compact shells than do those from warm waters, they also have smaller, more closely spaced pores. Porosity (the proportionate area occupied by pores) is lower, however. There is, then, a correlation between the temperature at which the shells are formed and their porosity and thus, secondarily, between latitude and porosity. In this connection he also called attention to the work of Wiles (1960), who found that in deep Atlantic cores, the concentration of pores in 1 species, *Globigerina eggeri* Rhumbler (= *G.* dutertrei d'Orb.), varied at different levels in the core and that changes in pore concentration could be correlated with Pleistocene climatic changes. High pore counts appeared in the interglacial sediments when an assemblage of warmer-water species was present, and low counts appeared in sediments corresponding to glacial epochs. Although Wiles did not determine the shell porosities, Bé points out that if the size of the pores remained fairly constant in this species, the results would be in line with the general differences found between cold-water and warmwater species.

According to Waller and Polski (1959), salinities of less than 27‰ are in general limiting to the survival of planktonic Foraminifera. Benthic forms from inshore shelf regions tend to be more euryhaline than do strictly pelagic species, since the former commonly encounter more variable salinity.

Most of the planktonic Foraminifera are in the upper 200 meters of water, and none of them is below 2000 meters. At great depths the calcareous shells begin to dissolve. Shells of planktonic species are absent from sediments at depths of 5000 meters or more. Berger (1967) suspended samples of washed shells of mixed species from eastern Pacific core samples at various depths for a month. He noted not simply that the rate of dissolution increased with depth but also that those species with lighter or thinner tests disappeared first. Thus the depth at which a sediment is formed not only will determine whether or not tests of pelagic Foraminifera will be preserved in it but also, in those containing Foraminifera, will strongly influence the species composition.

Classification of Foraminifera

Some authorities divide the Foraminifera into more than 300 genera distributed among 45 or more families. For details, special monographs of the group should be consulted.

Order Radiolaria

The Radiolaria (Figs. 9-3 and 9-4) are mostly rather large, spherical, pelagic rhizopods from 50 μm to several millimeters in diameter, sometimes forming colonial aggregations up to several centimeters across. The pseudopodia are radiating axopodia or filopodia, and in many there are also radiating skeletal elements as well. The cytoplasm is divided into a central portion surrounded by a capsule and containing the nucleus or nuclei and an outer, or extracapsular, portion. The nucleus, when single, is sometimes large and may be polymorphic. Some radiolarians have many smaller nuclei instead.

The **central capsule** is a tectinous membrane with pores through which there is a continuity of intracapsular and extracapsular cytoplasm. The pores may be uniformly distributed or arranged in one to three **pore fields.** The extracapsular cytoplasm, the **calymma,** is commonly rather stiff and gelatinous and filled with large vacuoles containing a watery fluid, giving it a frothy appearance. There are no contractile vacuoles. In several groups of Radiolaria the calymma also contains

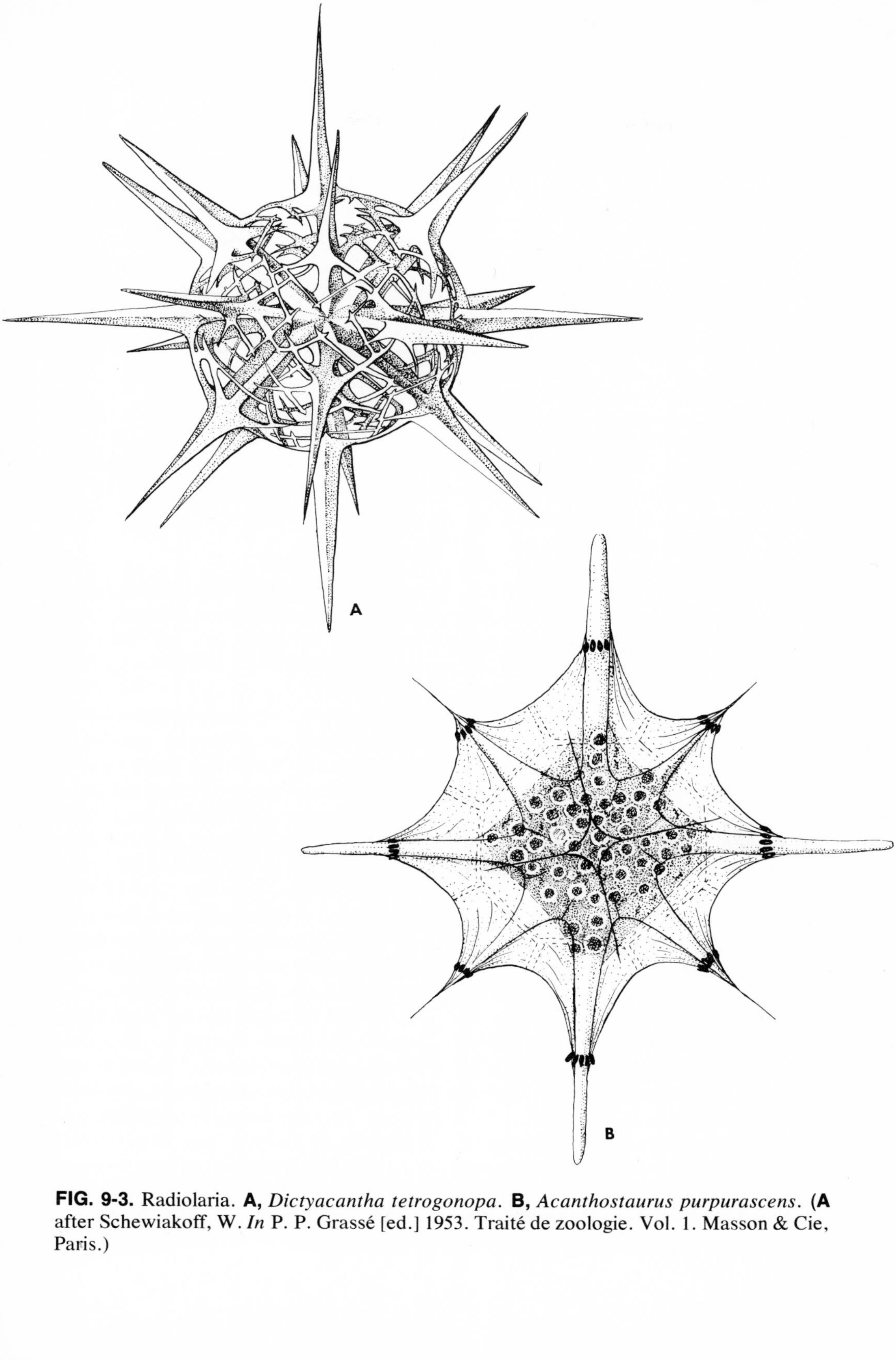

FIG. 9-3. Radiolaria. **A,** *Dictyacantha tetrogonopa.* **B,** *Acanthostaurus purpurascens.* (**A** after Schewiakoff, W. *In* P. P. Grassé [ed.] 1953. Traité de zoologie. Vol. 1. Masson & Cie, Paris.)

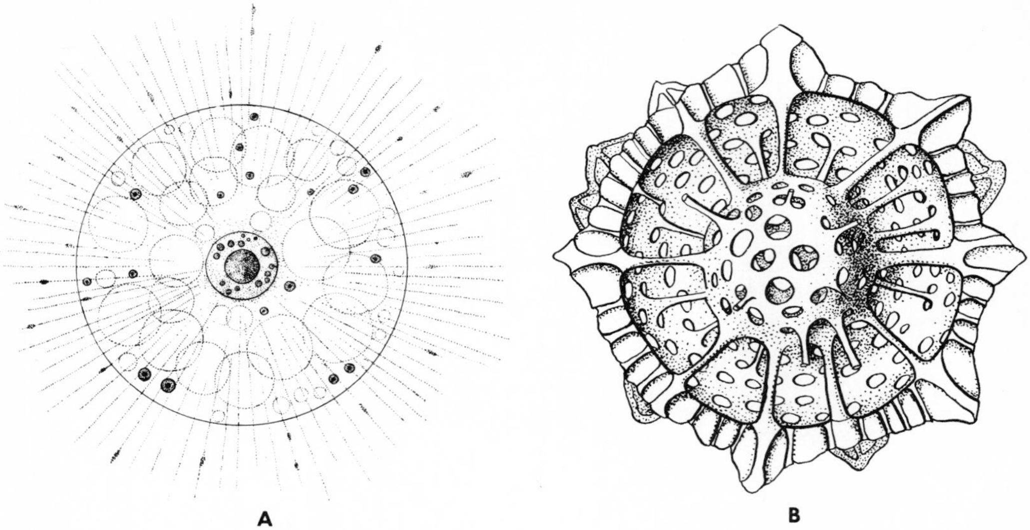

FIG. 9-4. Radiolaria. **A,** *Thalassicolla pellucida*. **B,** *Carposphaera nodosa*. (After Haeckel. *In* P. P. Grassé [ed.] 1953. Traité de zoologie. Vol. 1. Masson & Cie, Paris.)

rounded yellowish bodies, symbiotic zooxanthellae.

Other cytoplasmic inclusions such as fat droplets, pigment granules, and crystals are sometimes present. The pseudopodia continue as denser rays of cytoplasm through the calymma to the central capsule.

The skeleton, usually of silica except in the Acantharia, which possess skeletons of strontium or calcium aluminum sulfate, may be radiate, concentric, or of mixed type. Many radiolarians form extremely intricate and beautiful lattice-type skeletons. Radiolarians are holozoic, feeding on diatoms and small zooplankton caught among their pseudopodia. Those which have zooxanthellae may also obtain some nutrients from them.

Like the calcareous skeletons of Foraminifera, radiolarian skeletons accumulate at the bottom of the seas and are readily fossilized. The oldest known Precambrian fossils are radiolarian skeletons. In some of the deeper places in the oceans subject to immense hydrostatic pressure, the calcareous shells of Foraminifera may be dissolved, but the glassy skeletons of Radiolaria persist and come to characterize the bottom deposit so that a radiolarian, rather than a globigerina, ooze may be formed.

CLASS CILIATA: FAMILY TINTINNIDAE

Tintinnids are a large family of mostly marine ciliates belonging to the order Spirotricha. Characteristically, the Tintinnidae secrete a chitinous or pseudochitinous **lorica** about themselves, within which they live. The lorica tends to be characteristic for a given species, and since it is commonly the only part that can be studied satisfactorily in ordinary plankton samples, it has been extensively used as a basis for the description and classification of species. The lorica is commonly shaped like a vase or bowl, and in many species it also incorporates foreign particles such as sand grains, coccoliths, or fecal detritus. Both sessile and free pelagic species occur. Some are also found in brackish or fresh waters.

Kofoid and Campbell (1929) described over 300 species, which they grouped into 12 families (here regarded as subfamilies) and 51 genera.

The body is somewhat trumpet-shaped and is attached aborally at the narrow end to the base of the lorica. Strong myonemes enable it to contract rapidly back into the lorica. It bears a conspicuous aboral zone of sixteen to twenty-two membranes, sometimes plumose, or feathery, in appearance. There may also be four or more longitudinal rows of shorter cilia on at least the anterior portion of the body. A cytopharynx with neurofibrils is also present.

There are one to four (commonly two) macronuclei and micronuclei.

Sometimes a rounded cell bearing a large nucleus is seen in the cytoplasm. This parasitic dinoflagellate, *Duboscquella,* is of the family Blastodiniaceae.

Nongelatinous Metazoa

CRUSTACEA

By far the most numerous and important members of the zooplankton are the Crustacea, which swarm in the seas as insects do on land. Wherever zooplankton is particularly rich in crustaceans, it attracts and supports a great population of fish and other predators.

As with other arthropods, the body of a crustacean is encased in a jointed cuticular **exoskeleton** secreted by the epithelial cells, in arthropods commonly termed the **hypodermis.** This exoskeleton is shed at intervals throughout the crustacean's life, allowing for growth and change in form and dividing its life into a succession of **instars** marked by **ecdysis,** or molting, at the end of each instar.

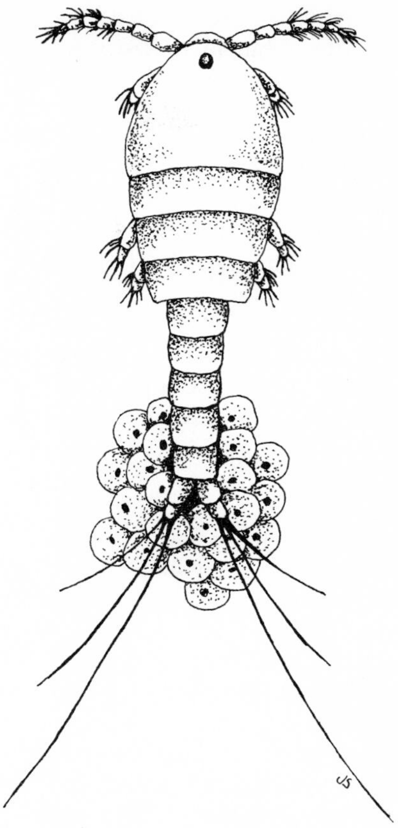

FIG. 9-5. *Tigriopus californicus,* an inhabitant of high upper littoral tide pools. (*In* Light, S. F. 1975. Light's manual; intertidal invertebrates of the California coast. 3d ed. revised and edited by R. I. Smith and J. T. Carlton. University of California Press, Berkeley; after McCain, J. C., 1968; Parker and Margoles, 1964; J. H. Hedgpeth, 1962; and R. Bolin, 1944, respectively.)

Crustacea are characterized by five pairs of head appendages: two pairs of **antennae,** one pair of **mandibles,** and two pairs of **maxillae.** The first and second pairs of antennae are often termed **antennules** and **antennae,** respectively. The first and second pairs of maxillae are commonly called **maxillulae** and **maxillae.** These are followed by a variable number of thoracic and abdominal appendages. The thorax is often termed the **pereion,** and its appendages, the **pereiopods.** The abdomen is termed the **pleon** and its appendages, **pleopods.** The abdomen ends in the **telson,** which bears the anus, but it has no appendages and is not regarded as a somite. The number of thoracic and abdominal somites varies greatly in different groups of Crustacea. The number of apparent segments of the body, as seen from above, is even more variable because of the frequent occurrence of coalescence, or fusion, of two or more adjacent somites to form one apparent segment.

The Crustacea are divided into five subclasses: **Branchiopoda, Ostracoda, Copepoda, Cirripedia,** and **Malacostraca.** The first four subclasses are often collectively termed the **Entomostraca.** Entomostracans are usually smaller than most malacostracans and are regarded as more primitive. Most of them hatch from the eggs as nauplial larvae. The **nauplius** is a minute oval free-swimming larval form with a single median eye near the anterior end, and, in the first instar at least, only three pairs of appendages, all pediform, representing the two pairs of antennae and the mandible.

Copepods

More than 4500 species of copepods have been described, of which about 90% are marine. Approximately 3000 are free-living, and the remainder are parasitic. Copepods comprise the bulk of the zooplankton. In both numbers of individuals and numbers of species they exceed all the rest of the metazoan plankton combined. They are a key group in the economy of the seas. The free-living copepods are usually small, ranging from 0.2 mm. to about 2 cm. in length.

After a brief discussion of the morphology and biology of copepods, *Calanus finmarchicus,* one of the most abundant, important, and most studied species, will be considered in more detail as an example.

Morphology

The body of most copepods is divided into a broader oval convex anterior region of several somites, the **prosoma,** in which the segments are firmly united and immovable, and a narrower posterior movable portion, the **urosome.** The articulation between these two regions of the body is termed the **hinge joint,** or **major articulation.** Some authorities define the urosome as the **genital somite** plus all somites behind it. It seems preferable to define it functionally as that part of the body behind the major articulation. The major articulation does not fall between the same somites in all groups of copepods. This fact, together with the differences in apparent segmentation caused by various combinations of fusion between somites in different groups of copepods, has caused some confusion in terminology and in the interpretation of copepod morphology. When one is reading about copepods, it is important to determine how the author is using the terms. The terms will be used in the sense outlined in the tabulation below.

prosoma The anterior part of the body in front of the major articulation (= metasome in Davis' usage and that of some other writers).

cephalosome The apparent first segment when this comprises the head plus the first thoracic somite.

cephalothorax The apparent first segment when it comprises the head plus two or more thoracic somites.

metasoma The thoracic segments behind the cephalosome or cephalothorax, and in front of the major articulation.

urosome The part of the body behind the major articulation. This includes the abdominal somites and anal segment, or telson, with the caudal furca, and in some copepods it also includes the posterior thoracic segment.

Those copepods in which the major articulation occurs between the last pedigerous somite and the genital somite, and in which, therefore, the urosome bears no appendages, have been classed as the **Gymnoplea.** They comprise the suborder **Calanoida.** In the other suborders the major articulation occurs anterior to this, so that the urosome includes one thoracic somite in addition to the genital and abdominal somites, and hence may bear one pair of appendages or rudiments thereof. These have been termed the **Podoplea.**

Copepods have fifteen **somites.** The first five comprise the head, or **cephalon.** These somites bear the two pairs of antennae, the mandibles, and the two pairs of maxillae. The thorax is composed of six somites, which bear the maxillipeds and the five pairs of swimming legs. The genital somite occurs behind the hinge joint, or major articulation. In females it is usually united with the first abdominal somite, forming a compound genital somite as the last thoracic somite (some researchers consider it to be the first abdominal somite). In any case, it is part of the urosome and bears either no appendages or rudimentary ones. The abdomen consists of the last three somites plus the anal segment, or telson. In copepods these somites do not bear appendages. Behind the abdominal somites is the telson (anal segment of Davis and regarded as a true segment), which bears the caudal rami. In some free-living copepods the fifth pair of legs is much modified, reduced, or absent. Parasitic copepods in some groups, especially the females, show such great modification that they are hardly recognizable as copepods in the adult condition.

The number of apparent segments, as seen from above, is fewer than fifteen because of coalescence, or fusion, of certain somites. The somites of the head plus one or two thoracic somites are fused together into a complex that looks like a single anterior segment. In some groups the last two thoracic somites are fused. In females the genital and first abdominal somites are usually fused together as the "genital segment." In some the last two abdominal somites are fused.

Locomotion and feeding

Locomotion is jerky and rapid and produced by strong, synchronous beating of the swimming legs, during which the long antennules lie back along the sides of the body. When the beating stops the antennules are extended laterally, slowing down sinking. Bursts of active swimming alternate with short periods of gliding or resting. The gliding movements of forms such as *Calanus* and *Diaptomus* are caused by the currents set up by the anterior feeding appendages. Some gliders can reverse the currents and glide backwards as well as forward.

Diatoms constitute the principal food of most planktonic copepods. In many, as in *Diaptomus* and *Calanus,* the appendages around the mouth (second antennae, mandibular palps, and first maxillae) are clothed with setae and are rapidly vibrated (between 600 and 2640 times a minute), creating swirls of water that pass along both sides of the body. The resulting eddy is sucked forward into a median filter chamber by the maxillipeds and out of the chamber laterally by the first maxillae. There is a dense screen of setae on the second maxillae. The food particles caught in these setae are scraped off by the endites of the first maxillae and by special setae of the base of the maxillipeds and passed to the mandibles and mouth.

In others such as *Acartia,* instead of acting as passive filters, the maxillae are modified to act as scoop nets. Some copepods capture larger organisms in addition to filtering for phytoplankton or detritus, and others are wholly predaceous.

Filtering efficiency

The filtering efficiency, or filtering rate, is calculated by allowing known numbers of a given species of copepod or other filter feeder to graze in a definite volume of water containing a known number of suitable phytoplankton cells per milliliter for a given length of time. From the decrease in phytoplankton cells the amount of water filtered per copepod in a unit of time can be calculated, using an equation that takes into account the increasing volume of water that must be filtered to obtain the same number of phytoplankton cells as the phytoplankton decreases.

Experiments by Harvey (1937) indicate that the larger calanoids can filter up to 70 ml. of water per day, and that they take about an hour to fill or empty the gut. In the presence of normal amounts of mixed phytoplankton, they would therefore receive the amount calculated as necessary to maintain their metabolism. However, in the case of species such as *Calanus finmarchicus,* which migrate to depths below most of the phytoplankton for part of the day, their hourly filtering rates while actively feeding must often be greater.

Harvey also showed that the calculated filtering rate increased with increasing size

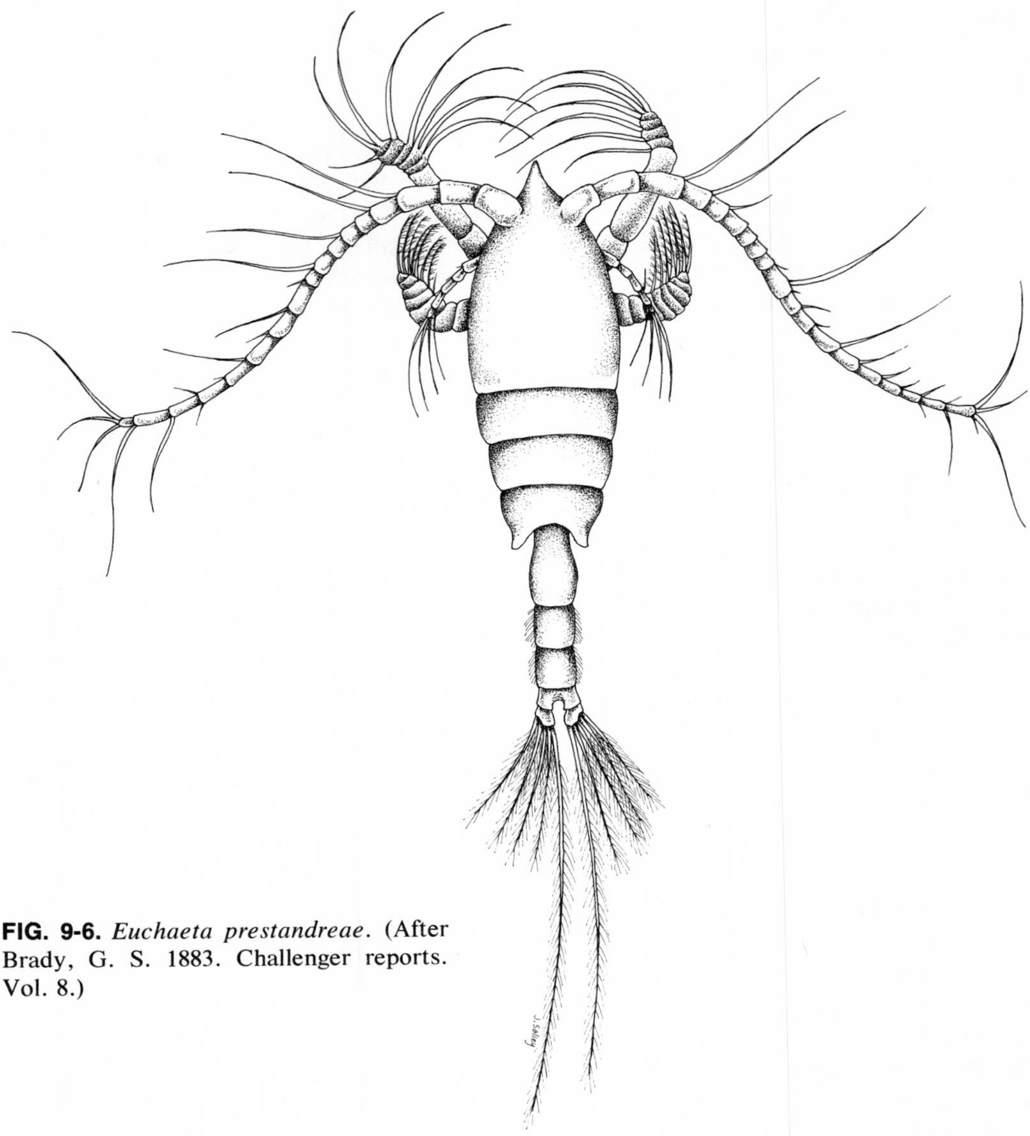

FIG. 9-6. *Euchaeta prestandreae.* (After Brady, G. S. 1883. Challenger reports. Vol. 8.)

of the diatoms being grazed. When fed larger diatoms they appeared to filter more water in a given unit of time. This may reflect a greater efficiency in filtering rather than a faster rate—the larger diatoms being retained by the setae of the filtering apparatus more effectively.

Reproduction

Copepods are the only entomostracans that form **spermatophores.** The male grasps the female and deposits spermatophores on the genital segment of the female at or near the opening of the seminal receptacles, by means of a special adhesive cement.

When the eggs are laid, they are surrounded by an **ovisac** produced from secretions of the oviduct so that they stay attached to the female in the ovisacs, which serve as brood chambers. Each sac contains from a few to about forty eggs, and clutches may be produced frequently.

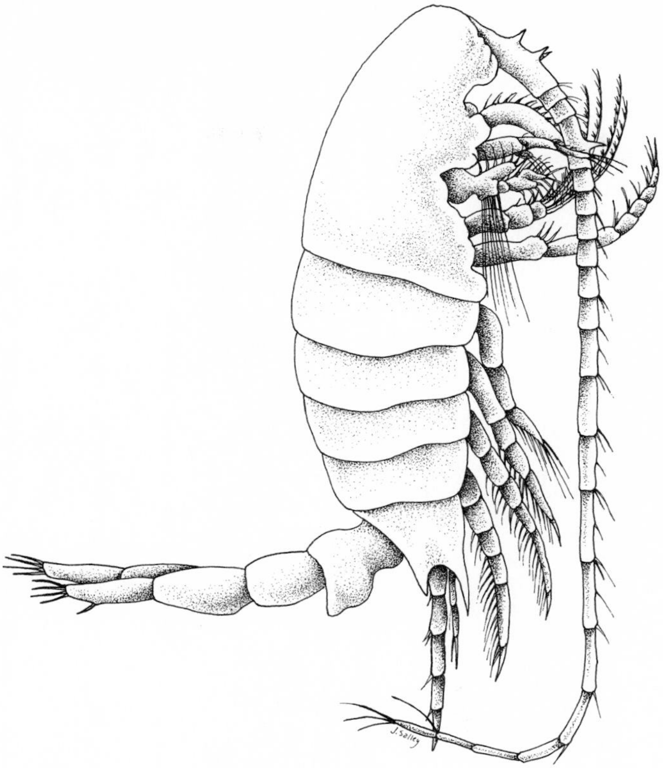

FIG. 9-7. *Centropages furcatus.* (After Brady, G. S. 1883. Challenger reports. Vol. 8.)

Some species set the ovisacs free in the water rather than retaining them attached to the female. Many calanoids simply shed the eggs singly into the water. The egg hatches into a free-swimming planktonic nauplius, a small oval form with three pairs of appendages and a single median eye. The appendages represent the first antennae, second antennae, and mandible. The latter two in this stage are biramous pediform appendages. No trunk segmentation is evident. Successive molts gradually add more appendages, beginning with those at the anterior end. There are commonly six nauplius instars (five molts), after which the body form changes to one similar to the adult form. Immature stages of this kind are termed **copepodids.** Usually there are five or more molts, the mature individuals constituting the sixth instar after attainment of the copepodid form.

Color of copepods

In copepods, which are colored red, orange, black, or white with some orange, the pigment is always astaxanthin, produced from yellow xanthophyll obtained from the phytoplankton. The bright blue color of the eggs of such species as *Euchaeta japonica* is also astaxanthin. Astaxanthin produces different colors when associated with different proteins. When it is extracted with an organic solvent such as chloroform, the solution is always reddish orange.

Metabolic fuel storage

Many copepods synthesize and store waxes and fats as reserve metabolic fuel for use during periods when there is little or no food available, or during nonfeeding stages in the life cycle. The lipids are obtained from phytoplankton. Calanoids inhabiting

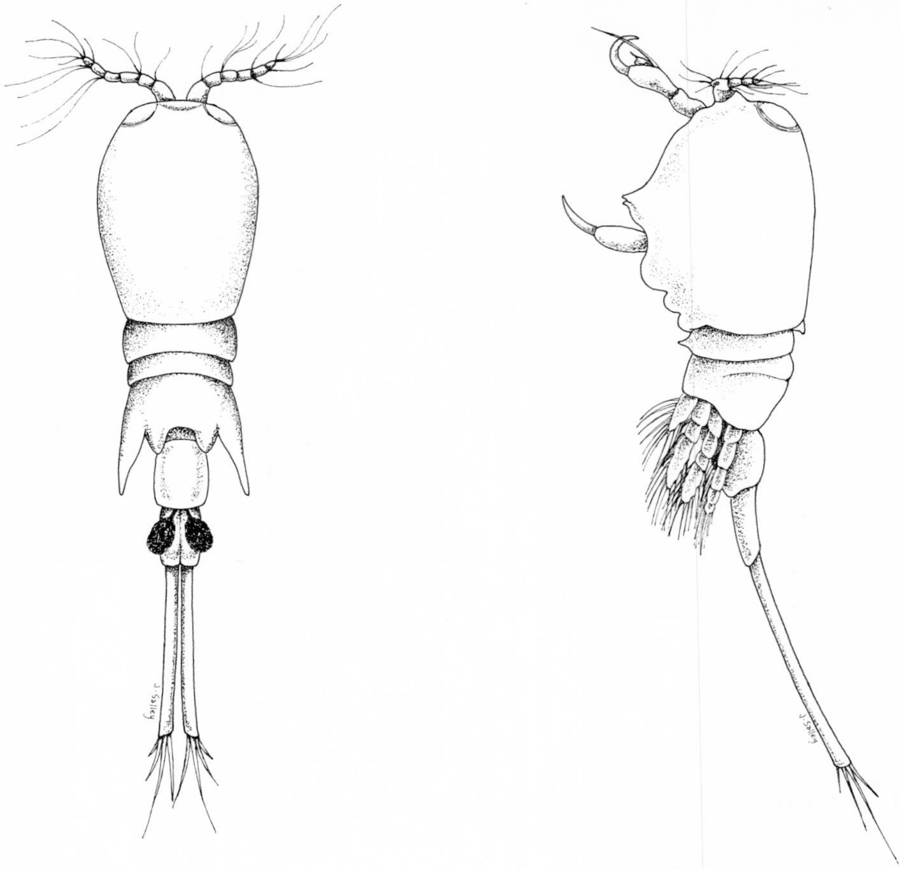

FIG. 9-8. *Corycaeus varius.* (After Brady, G. S. 1883. Challenger reports. Vol. 8.)

cool to cold waters have a large sac in which wax is stored. If starved, they utilize their fat reserves first, conserving the more valuable polyunsaturated fatty acids that are in the wax.

In *Calanus plumchrus,* for example, the predominant copepod in the Straits of Georgia, British Columbia, the adults are nonfeeding. They live from large stores of pure wax, constituting up to 50% of the dry weight of the body, and accumulated during the copepodid stages. These adults live in relatively deep water at about 400 meters throughout the winter. Most of the reserves of the female go into the production of eggs, which are shed in late winter or early spring. The eggs, which contain oil droplets, rise toward the surface and hatch in two or three days. The first-stage nauplii still have sufficient wax reserves for their needs and do not feed until after their first molt. By this time the spring phytoplankton bloom, fertilized by runoff from the Fraser River, is under way, and the succeeding nauplius and copepodid stages feed well, through the spring and summer, accumulating sufficient stored wax to meet their energy needs for the last seven months of their life. In late summer or fall they cease feeding, migrate to deeper water, and transform to the nonfeeding adult stage.

The storage of high-energy fat and wax

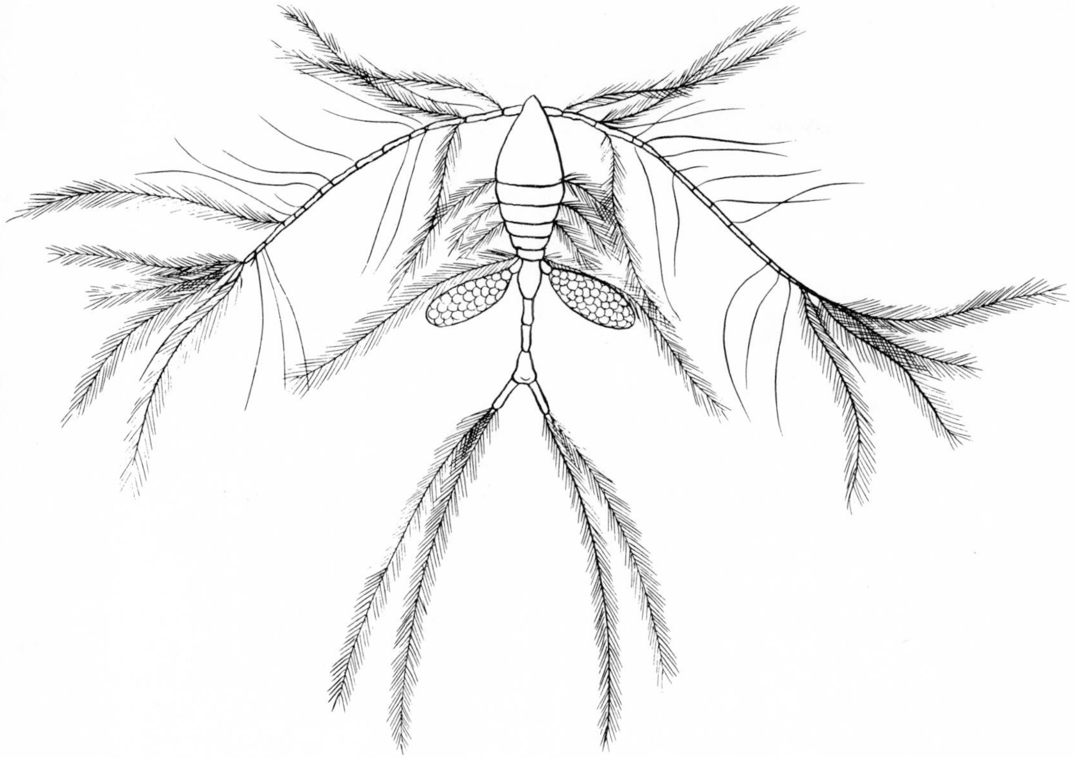

FIG. 9-9. Copepod, *Oithona* sp. (After Sverdrup, H. U., M. W. Johnson, and R. H. Fleming. 1942. The oceans: their physics, chemistry and general biology. Prentice-Hall Inc., Englewood Cliffs, N.J.)

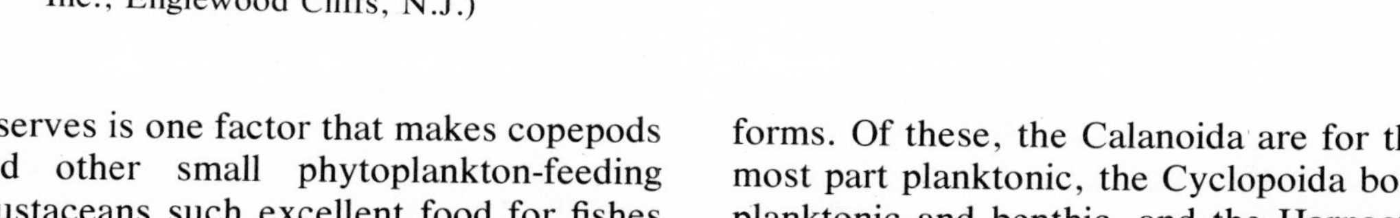

FIG. 9-10. Copepod, *Aegisthus* sp. (From Sverdrup, H. U., M. W. Johnson, and R. H. Fleming. 1942. The oceans: their physics, chemistry and general biology. Prentice-Hall, Inc., Englewood Cliffs, N.J.)

reserves is one factor that makes copepods and other small phytoplankton-feeding crustaceans such excellent food for fishes and other animals at the next trophic level.

Classification of copepods

The copepods are divided into 7 suborders, of which 3—Calanoida, Cyclopoida, and Harpacticoida, each with more than 1000 described species—contain free-living forms. Of these, the Calanoida are for the most part planktonic, the Cyclopoida both planktonic and benthic, and the Harpacticoida largely benthic in habit, though some of them are also found in the plankton. (See taxonomic appendix for more details.)

Calanus finmarchicus

C. finmarchicus (Fig. 9-11), often termed red feed, herring feed, or krill, occurs in

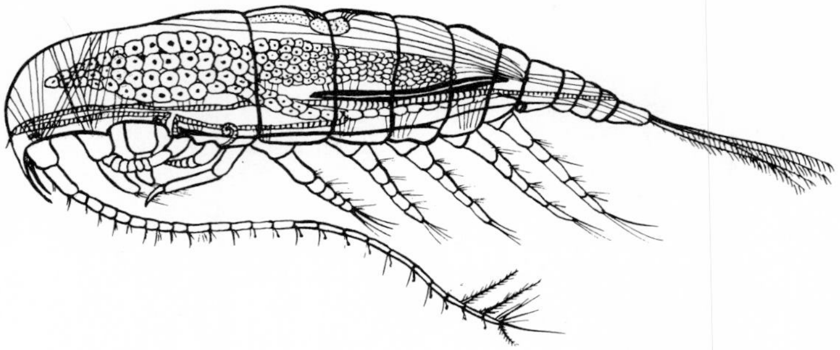

FIG. 9-11. Copepod, *Calanus finmarchicus.*

such great swarms in the colder waters of the North Atlantic that it gives the water a reddish tinge for several miles. It extends through the polar sea to the North Pacific and is a major food for several fish, including such commercially important species as herring and mackerel, as well as for baleen whales such as *Balaenoptera borealis* and *B. physalis*.

C. finmarchicus exhibits a rather wide bathymetrical distribution, dominating both the surface and subsurface communities of planktonic organisms of some regions. It exhibits well-marked diurnal vertical migrations in some areas, as well as the phenomenon of submergence in warmer latitudes. The diurnal migratory behavior differs regionally and seasonally, and in a given region the two sexes and various immature stages may differ considerably in their vertical distribution and movements.

Breeding of *C. finmarchicus* commonly begins at the time of the great early spring blooms of phytoplankton, with the population maxima coming somewhat later. The eggs are planktonic. Early nauplii are so small that they pass through all but the finest plankton nets. They are usually found in the upper strata of water. Older copepodid stages commonly inhabit the deeper strata of water occupied by the species. Where the species occurs submerged in warmer regions, it does not occur in dense swarms or with such marked seasonal fluctuations. It reappears nearer the surface in cold-temperate or antiboreal southern waters, but not in the great swarms characteristic of the North Atlantic. Since it is not present in such great numbers in regions where it is submerged beneath warmer waters, the suggestion has been made that such living conditions are unfavorable for it. This may not be true, however, since even in these warmer regions, it is just as numerous as some of the species characteristic of and limited to the warm regions. A similar argument can be made with regard to its occurrence in neritic waters. Even though the greatest swarms are commonly found in cold neritic waters, the species is not tied to the shelf in any way and is of regular occurrence as a holopelagic high oceanic species.

If anesthetized, *Calanus* sinks fairly rapidly. Living individuals maintain their level in the water by almost constant upward swimming. Experiments with the Hardy-Bainbridge plankton wheel indicate that *Calanus* can climb 15 meters in an hour, and can descend at rates up to 48 meters/hour. Maintenance of stocks in some areas has been correlated with the vertical movements of the copepods. Off the coast of Norway they descend into deep water during the winter, rise to near the surface in early spring, and reproduce actively through the summer while drifting mainly north and northeast toward arctic waters. Near the end of summer many of them descend into deeper water, 200 to 300 meters deep, drifting back toward their original location to rise again and reproduce the stocks next spring.

Metabolism during the breeding season

is rapid. Individuals seem to need about 2% to 4% of their body weight in food each day. Experiments placing them alternately in water containing abundant phytoplankton and in clear water indicate that they require about an hour to fill or empty the gut. They may actively seize diatoms with their antennae as well as obtain them through their filtering mechanism.

During the winter, when much of the population spends all the time in deeper water, starvation may be a real problem, even though the metabolic rates and food requirements are lower. The number falls to a minimum about February or March in the Clyde area. Here the population consists largely of stage V copepodids in December, which begin molting to the adult stage in January, and the small early breeding population begins shedding eggs by late February and early March. Soon a high percentage of nauplii appear. They mature about early April, when a second and larger spawning occurs, followed by a sudden rise in the overall population and augmented again by a third spawning about June and July. A few may have a fourth spawning in late summer, but most are in about the fifth copepodid stage in late autumn, in which stage they overwinter. Males seem to have a distinctly shorter life-span than do females and tend to feed less extensively. The different generations during the breeding season may show differences in the mean size of the individuals, the most robust individuals appearing in the spring brood, when the population of copepods is relatively small and the phytoplankton is abundant. In colder waters of the far north, the number of generations per year is reduced and the individual life-span increased up to 2 years.

With their rich, oily bodies and tendency to gregariousness, they commmonly form a great floating mass that does not mix readily with other water masses of pure oceanic water. The line of separation between a swarm of *C. finmarchicus* in one water mass, and another water mass of pure Atlantic water, is often like the interface between two fluids of different viscosity that have a slight tendency to mix—the difference in viscosity in this case being at least partly created by the presence of the swarm of copepods. The fact that *Calanus* is a principal food of herring and that herring tend to congregate in areas where crustacean zooplankton predominates was employed by Hardy (1936) in devising a simple plankton indicator for fishermen. It consisted of a disc of bolting cloth to be towed behind the fishing boat for a given length of time. If it came up reddish, a predominance of calanoid plankton was indicated, the degree of redness roughly indicating the richness of the zooplankton. If it came up green, showing a predominance of phytoplankton, the area was judged less favorable. Fishermen using this indicator obtained consistently better catches than those not using it.

Since the species is both euryhaline and eurythermal to a greater extent than many other common plankton organisms, it is not a reliable indicator of hydrographic conditions, currents, or water stratification.

Marshall and Orr (1953), who have done a great deal of research on this species, brought together the results of their long and careful studies, together with the other available information on the biology of the species, in a fine book that should be consulted by everyone interested in this copepod or in the biology of copepods generally.

Euphausiids

Second only to the copepods as important elements of the zooplankton are malacostracans of the order Euphausiacea. These small shrimplike crustaceans are 1 to 6 cm. in length, with prominent compound stalk-eyes. The eight thoracic somites are covered by a common carapace, which does not, however, form lateral gill chambers. Five abdominal somites bear pleopods, and the sixth bears the uropods, which together with the telson make up the tail fan. The telson bears a subapical spine on each side. All the thoracic limbs

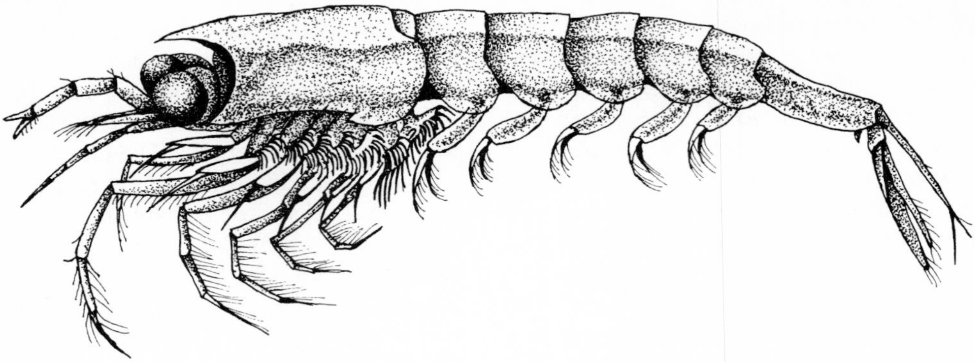

FIG. 9-12. *Thysanoëssa macrura*. (After Sars, G. O. 1885. Challenger reports. Vol. 13.)

are biramous, the second to eighth bearing natatorial setae, and a gill is formed from a branched epipodite. The gills become larger and more branched posteriorly. The anterior pair is not modified into maxillipeds. The last one or two pairs of thoracic legs may be reduced or vestigial. In males the first two pairs of pleopods are modified for copulation. The sperm are transferred in spermatophores, as in copepods. The female does not have oostegites but carries the eggs in a basket formed by setae on the thoracic limbs, or in ovisacs cemented to the ventral surface of the thorax. In some species the two pleopods of the same somite are coupled together by a median process, the **appendix interna,** or **stylambis.** Most euphausiids exhibit bioluminescence, the photophores located on the margins of the eyestalks. The body is nearly transparent in many species, whereas in others it is bright red.

Euphausiids tend to aggregate at times in great swarms or to become stratified at certain levels in the water. In many regions they perform diurnal vertical migrations. Their abundance and tendency to swarm, plus the fact that their bodies are rich in protein, make them a prime food for many predaceous fish, cephalopods, birds, and whalebone whales.

The eggs hatch as nauplii, but the larvae go through more complex metamorphoses than do copepods, including **nauplius, metanauplius, calyptopis** 1-3, **furcilia** 1-6, and finally the adult stage.

Euphausia superba (Fig. 8-15) is a bright-red dominant and characteristic surface-dwelling zooplankter of antarctic waters, the krill on which baleen whales, many fish, penguins, etc. largely depend for food during the antarctic summer.

Blue whales and other animals that feed largely on krill have an unusually short, efficient food chain, only one step removed from the phytoplankton. Hence they are able to produce a much greater biomass than would otherwise be the case (Pequegnat, 1958).

Herbivorous animals preyed on by selective predators commonly evolve highly developed escape mechanisms, such as speed, keen senses, or social defense. It has been claimed that nonselective predators—such as the blue whale, which simply swims through swarms of krill, straining out tons of them as they go—would not tend to exercise any selective genetic effect of evolutionary significance on the population of the prey. This may not be entirely true, however. For example, they may give a selective advantage to rapid reproduction—those individuals which have already spawned before being eaten having the best chance to transmit their characteristics to future generations.

The relatively long submergence of the juvenile stages and the short, single-season adult life illustrate another evolutionary tendency. In any organism the relative and absolute lengths of time spent in particular stages of the life cycle tend to reflect the degrees of risk to which the organism is subjected in the various stages. Animals tend to hurry through or abbreviate and telescope those stages of the life cycle in which they are subject to the highest mortality, and to prolong the safer stages. The swarming adults in surface waters during the brief antarctic summer are subject to a far higher intensity of predation than are the larvae drifting southward in deeper, darker water.

The reproductive potential of a species reflects a summation of all the risks involved in completing the life cycle. Species suffering the highest mortalities before the cycle is completed have the greatest number of potential offspring. If changed conditions of any kind significantly alter the risk a given species runs at any stage of its life from that under which the species evolved, the result may be dramatic increases or decreases in the population of the species in question, which in turn alters the conditions for all species dependent on or associated with that species, setting off a chain reaction of adjustments, with what often seem surprising and unpredictable results.

Classification of euphausiids

Three families are commonly recognized: Euphausiidae, Nematoscelidae, and Bentheuphausiidae. These families, as well as some representative genera, are given in the taxonomic appendix.

Other planktonic Crustacea

Several additional groups of Crustacea contain at least some important planktonic members, even though in many of these groups the majority of members may be nonplanktonic.

Order Mysidacea

The mysids (Fig. 9-13), or opossum shrimps, superficially resemble the euphausiids, with which they have sometimes been grouped in the order Schizopoda. Most of them live in shallow waters, burying themselves in the substrate part of the time and appearing only temporarily, usually at night, in the plankton. They are often abundant around river mouths and in brackish water. A few are truly planktonic, and some are bathypelagic or abyssal. As in the Euphausiacea, the thoracic limbs are biramous. They differ, however, in many important respects. The carapace covers most or all of the thorax and forms lateral gill chambers. The first thoracic limbs are modified as maxillipeds and are shorter than the following thoracic limbs. The female has a ventral thoracic brood pouch composed of **oostegites** (ventral plates representing epipodites from the thoracic limbs are expanded, flattened and directed medially beneath the thorax), in which the young develop to an advanced stage. The heart is

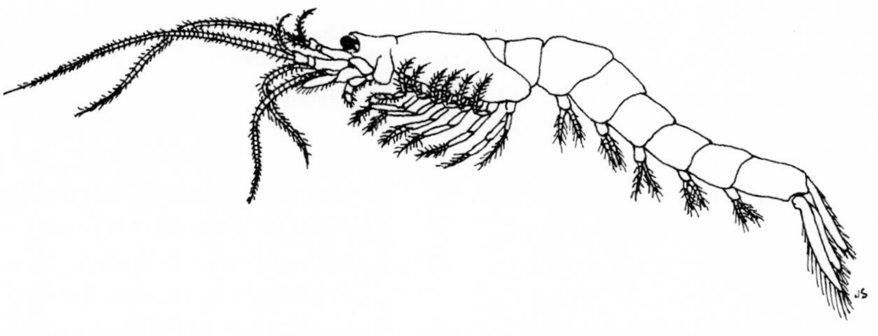

FIG. 9-13. *Mysis stenolepis.* (After Arnold, A. [Foote]. 1901. The sea beach at ebb tide. Century Co.; reprinted unabridged in 1968 by Dover Publications, New York.)

tubular. The telson lacks the subapical spines found on that of euphausiids. Gills, when present, are highly subdivided, each apparently representing one epipodite. Gills are lacking, however, in the family Mysidae, in which respiration takes place through the thin lining of the underside of the carapace. A current of water is drawn under the carapace by action of the epipodites of the maxillipeds. The carapace is often marked by a cervical sulcus, or groove, and the last abdominal somite is usually elongated and marked by a transverse groove. Deep-sea forms sometimes have oddly modified eyestalks, which in some cases extend beyond the eye itself. A well-developed antennal scale is present. Pleopods are always present in males but commonly reduced or absent in females.

Mysids have a worldwide distribution. There are about 450 species, mostly marine. They are usually 2 to 3 cm. long, although some much smaller forms, only 3 mm. in length, are known, and a few of the largest reach 15 cm. Both the antennules and antennae usually have long flagella. Representative genera are given in the taxonomic appendix.

Order Cladocera

The cladocerans, water fleas and their relatives, are primarily a freshwater group of entomostracans in which the carapace encloses the body but not the head, as a bivalved but not hinged structure open ventrally. The second antennae are enlarged and used for swimming. A single large compound eye and sometimes an ocellus are present. Some brackish-water species and a few truly marine forms are found, especially in neritic plankton. In suborder Calyptomera the carapace encloses the whole body and limbs. In suborder Gymnomera it is somewhat reduced, serving mainly as a dorsal brood sac.

Order Ostracoda

The ostracods are small crustaceans, laterally compressed and wholly enclosed within a bivalved movable, or hinged, carapace that gives them, superficially, the appearance of minute bivalve molluscs. They are ubiquitous in both freshwater and marine environments, many of them living in or near the substrate. A number of truly planktonic forms occur (Fig. 9-15). The appendages are reduced in number more than in any other group of Crustacea; usually only seven pairs are present. They are (1) first antennae—uniramous; (2) second antennae—biramous; (3) mandibles, usually with a large biramous palp; (4) first maxillae, serving as jaws; (5) second maxillae, their function varying in different groups, serving as jaws, maxillipeds, or legs; and (6 and 7) trunk appendages, legs, sometimes absent. The thoracic appendages have lost the functions of swimming, respiration, and feeding, which they serve in most other entomostracans. Both pairs of antennae are used in swimming. Many ostracods carry the eggs in a brood pouch or space between the dorsal shield and the body. An unpaired nauplius eye consisting of ventral median and two lateral elements is present. These elements may be united in the midline, or they may be separate. The abdomen is short, ending in a caudal furca. The development of marine forms is usually rather direct, there being no free larval stages or metamorphosis.

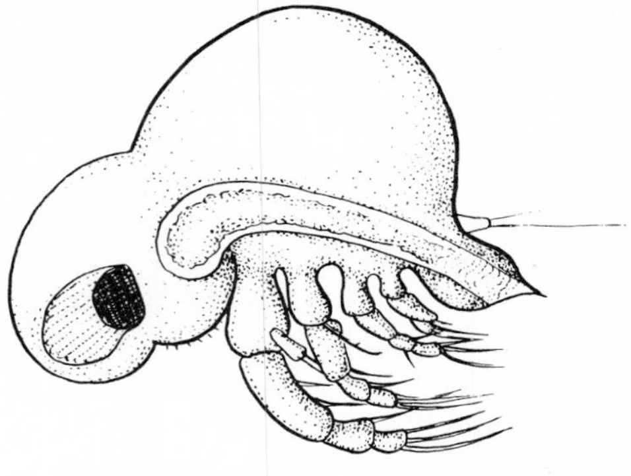

FIG. 9-14. Cladocera, *Podon* sp.

Fossil ostracods occur in nearly all geological strata. They abounded during the Silurian period, at which time some attained a length of 9 cm. The largest known

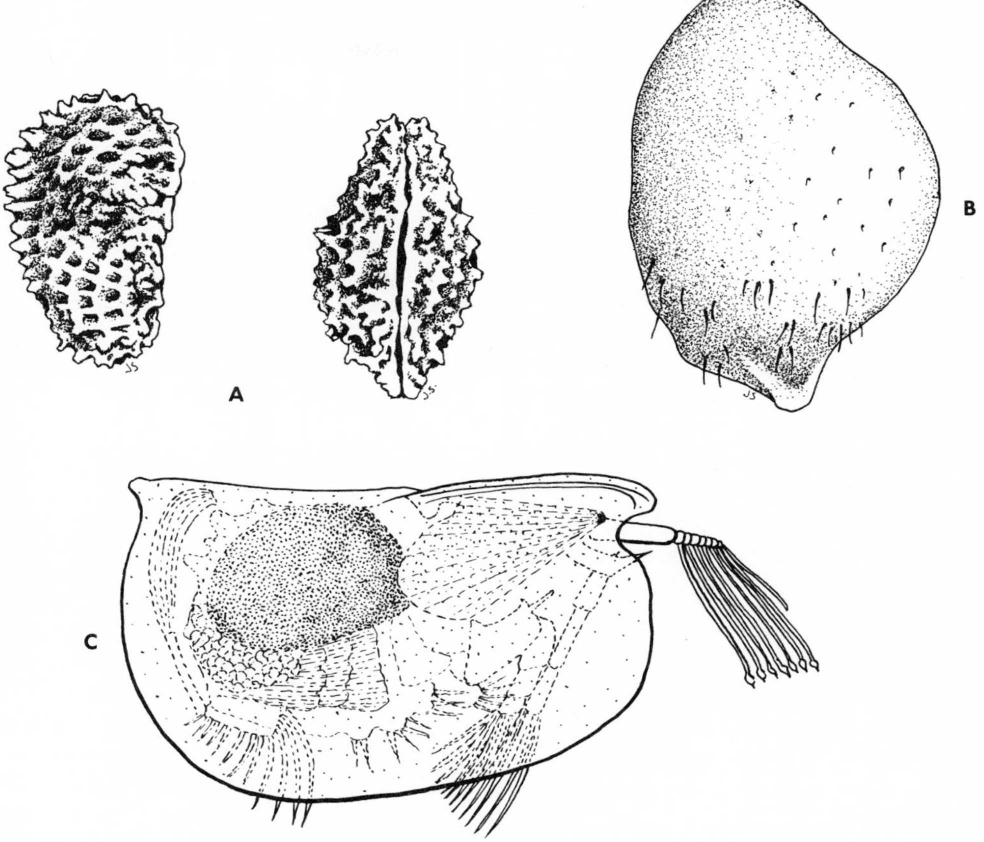

FIG. 9-15. Three ostracods. **A,** *Cythere radula.* **B,** *Bairdia victrix.* **C,** *Conchoecia* sp. (**A** and **B** after Brady, G. S. 1880. Challenger reports. Vol. 1; **C** after Davis, C. C. 1955. Marine and freshwater plankton. The Michigan State University Press, East Lansing.)

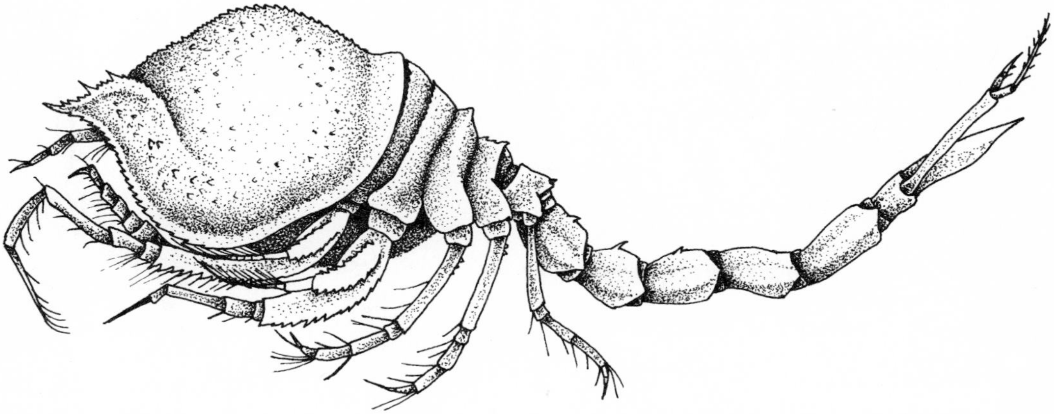

FIG 9-16. Cumacean *Diastylis mysracina.* (After Sars, G. O. 1887. Challenger reports. Vol. 1.)

living forms are deep-sea species of *Gigantocypris,* which grow to 23 mm. in length.

Order Cumacea

The small crustaceans of order Cumacea live in mud or sand, mostly on or near the bottom. Occasionally they appear in plankton in shallow waters or in hauls from near the bottom.

Orders Amphipoda and Isopoda

The orders Amphipoda and Isopoda together constitute the group **Arthrostraca,** or **Edriophthalma.** They are mostly benthic forms with no carapace, the bulk of the body being comprised of free thoracic segments. They have seven pairs of thoracic legs, which are uniramous, there being no exopodites. The eyes are sessile. These groups will be considered in more detail in connection with animals of the substrate. A few, however, are typically elements of the macroplankton. Amphipods tend to be laterally flattened and to have the thoracic limbs in two groups, the four anterior pairs directed forward and the three posterior pairs directed backward and upward. In Isopoda all seven pairs of thoracic limbs are more alike, usually shorter, and fitted for crawling or clinging. The body is usually dorsoventrally flattened. Both are large groups (Fig. 9-17).

Planktonic amphipods, particularly the family Hyperiidae, are often parasitic on or associated with medusae or salps. Many of the hyperiids have large heads, comprised for the most part of huge eyes.

The genera *Euthemisto, Hyperia* (Fig. 9-18), and *Parathemisto* are of worldwide occurence and at times are important members of the zooplankton community.

Other genera of the Amphipoda, includ-

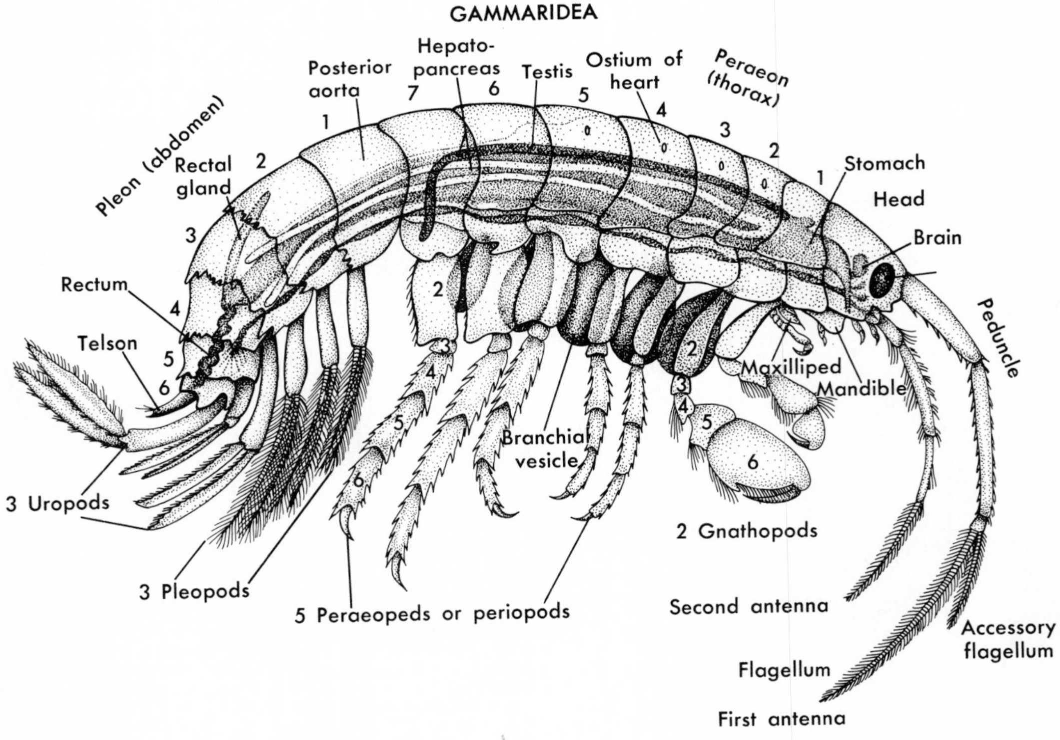

FIG. 9-17. Amphipod Gammaridea labeled with terms applied to major parts of anatomy of malacostracan Crustacea. Many Malacostraca have five pairs of pleopods and only one pair of uropods.

ing holopelagic species, are *Cuphocaris, Eurythenes, Brachyscelus, Euprimno, Phrosina, Phronima, Themisto, Lanceola, Scina,* and *Vibilia. Themisto libellula* (also known as *Euthemisto libellula*) is important in arctic waters. Its biology has been investigated by Dunbar, 1946. Several genera occur in the antarctic macroplankton widespread in the antarctic waters, including species of *Cyllopus, Eusiris, Vibilia,* and *Primno.*

Only two genera of isopods, *Munnopsis* and *Eurydice,* are commonly planktonic, and even these genera also have benthic as well as pelagic species.

Many amphipods and isopods that spend the day in or on the substrate may enter the plankton of shallow neritic waters at night, types such as *Nototropis* and *Apherusa* often being the most conspicuous elements of such plankton at certain hours. Crustacea with this habit include especially mysids, cumaceans, amphipods and a few isopods, and some euphausiids.

Order Decapoda

The order Decapoda includes most of the larger familiar conspicuous crustaceans, such as shrimps, prawns, lobsters, hermit crabs, and crabs, and are generally benthic in the adult state. Most of them have pelagic larvae, and some of those which are pelagic as adults are strong-enough swimmers to be classed as nekton rather than plankton. *Lucifer* is the genus most highly modified for planktonic existence.

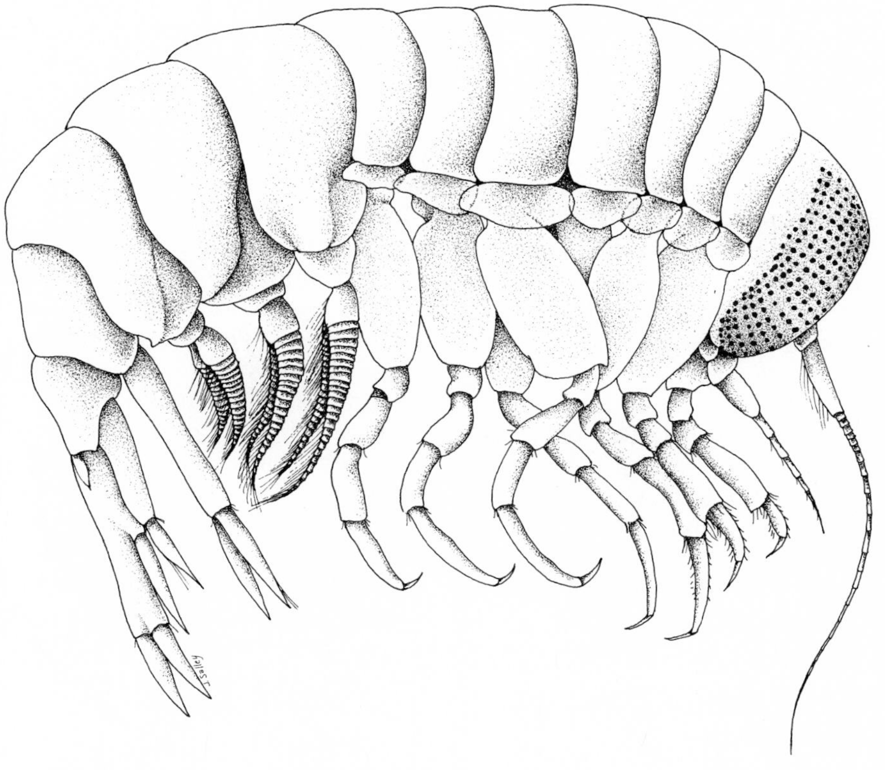

FIG. 9-18. *Hyperia gaudichaudii.* Hyperiid amphipods commonly inhabit the jelly of gelatinous animals such as medusae and salps, and usually show exaggerated development of the head and eyes. (After Stebbing, T. R. R. 1888. Challenger reports. Vol. 29.)

Some of the deep-sea prawns, such as *Gennadas, Sergestes, Hymenodora, Acanthophyra,* and *Systellaspis,* are weak swimmers and therefore to be included in the macroplankton.

Decapod planktonic larvae will be considered in the section on larval forms in the meroplankton. Gurney (1939, 1942) gives a good treatment and bibliographies of larval decapods.

As has happened in many animals in which the larval stages are much different both structurally and ecologically from the adult stages, some of the larvae were discovered and independently named before their relationships were understood. Some of these invalid generic names such as zoea and megalops have continued in use as designations for those stages in the life cycle for which they were originally proposed. The usage is generally broadened to include the same stages in the life cycle of all closely related crustaceans having a similar stage. Others have simply dropped out of use in synonymy.

A few decapods (*Lucifer* spp.) and some peneid shrimps hatch as nauplii resembling those of most entomostracans, but most decapods pass through the nauplial stage within the egg before hatching and then hatch at a more advanced stage.

Numerous shrimps of the family Sergestidae are planktonic, often highly modified, with long slender bodies, and transparent or nearly so in some species. The rostrum is short or absent. The antennules bear one long and one short flagellum. The short flagellum in males may be bifurcated, forming a short prehensile organ. In *Lucifer,* perhaps the most modified genus for planktonic life, the cephalic region and the eyestalks are much elongated, giving the animal a strange appearance.

OTHER NONGELATINOUS METAZOA

Phylum Chaetognatha

Aside from the Crustacea, the arrowworms, or glass worms, are perhaps the most widespread and characteristic nongelatinous element of the plankton. The phylum is a small one; only about 50 species have been described. They are, however, in all oceans and often in sufficient abundance to form a significant part of the zooplankton.

Chaetognaths are small, slender, bilaterally symmetrical, transparent, wormlike organisms usually about an inch or less in length. The body is straight and divided into three segments, head, trunk, and tail, each containing a pair of coelomic spaces of enterocoelous origin. The head coelom is nearly obliterated in the adult by the development of a complicated mass of muscles that move the lateral head setae. These strong, curved setae are used in the manner of jaws to seize prey. The head and "jaws" can be retracted into a hoodlike fold of the body wall and protruded to seize prey. The mouth lies subventrally between the jaws. The body has one or two pairs of lateral fins and a terminal horizontal tail fin, which are used as stabilizers in swimming. Propulsion is achieved by alternate contractions of the dorsal and ventral bands of longitudinal muscles, which give the animals a rapid darting and gliding movement. They are actively predaceous, feeding on copepods, small fish larvae, other arrowworms, larval decapods, etc. In turn, they are eaten by fish, medusae, pelagic worms, etc.

The digestive tract is a simple straight tube, or in some it possesses a pair of diverticula in the anterior region. A large ventral ganglion connected to the brain by two large connectives is found in the trunk segment, usually in the anterior half. All chaetognaths are hermaphroditic, the ovaries lying one on each side along the posterior part of the trunk coelom and the paired testes lying behind them in the tail segment. Mature sperm are stored in lateral seminal vesicles, which are conspicuous in mature arrowworms.

Because different species of arrowworms prefer water masses of different character, they are good indicators of hydrographic conditions or of the origin of waters in which they occur. The geographic and vertical distributions of chaetognaths are of considerable interest.

All chaetognaths are marine, and all but 2 or 3 species are strictly holopelagic and planktonic. The 1 or 2 presumed benthic species of *Spadella* are sometimes classified as a separate genus, *Bathyspadella*.

Because of their delicate body, which is without hard parts except for the jawlike setae of the head, and because of their pelagic habit, chaetognaths are rarely fossilized. They are, however, apparently a very old group. The only fossil known from mid-Cambrian Burgess shale has been given the generic name *Amiskwia*.

Mollusca and Annelida

Although the molluscs and annelids are both major phyla of animals and have successfully adapted themselves to freshwater and terrestrial habitats as well as marine, the great majority have remained benthic in habit. Only a few small bizarre groups of gastropods and a few scattered genera of annelids have adopted a fully planktonic mode of existence. Aside from these, the only pelagic forms are the squids. They rival or exceed fish in their ability to swim and are to be counted among the nekton rather than the plankton. Most molluscs and annelids do have pelagic larvae, which are commonly abundant especially in neritic plankton.

Mollusca

In the molluscs the principal modifications involved in the change from benthic shelled ancestral forms to planktonic habits involved reduction or loss of the shell and shell gland, reduction of the dense muscular connective tissue that gives the body of most molluscs its firmness and weight, and the development of special means of flotation.

Two groups—the Pteropoda, a suborder of Opisthobranchia, and the Heteropoda, a

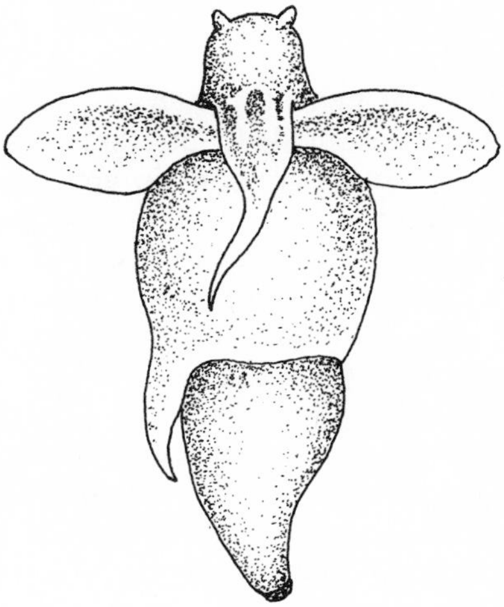

FIG. 9-19. *Dexobranchia ciliata*. (After Pelseneer, P. 1887, Challenger reports. Vol. 19.)

TABLE 9-1. Chaetognatha arranged according to type of water preferred

Cosmopolitan, oceanic, warm to temperate waters	Cosmopolitan, tropical-subtropical, neritic	Cosmopolitan, cold oceanic or deep in mid latitudes (submergence)	Costal, temperate to cool, neritic
Eukrohnia subtilis *Krohnitta pacifica* *Krohnitta subtilis*	*Sagitta bedoti* *Sagitta neglecta* *Sagitta oceania* *Sagitta tropica*	*Eukrohnia hamata* *Sagitta arctica* *Sagitta elegans*	*Sagitta setosa*
Bathypelagic	**Restricted to intermediate waters**	**Arctic**	**Deep, or benthic**
Eukrohnia fowleri	*Sagitta serratodentata*	*Sagitta arctica*	*Spadella cephaloptera*

suborder of Streptoneura—have become wholly planktonic, as have a few genera belonging to other groups.

In **Pteropoda** the foot is expanded into two large flaps, or fins, that function in both locomotion and flotation. The shell is thin and delicate, or lacking; it is seldom coiled when present. Pteropods are hermaphroditic, usually without gills. Pteropods are epipelagic—usually found in surface waters. They sometimes occur in great swarms, particularly *Spiratella (Limacina)* and *Clione*. These genera are important as whale food in northern waters. The shelled pteropods are largely herbivorous, whereas the naked forms are voracious predators.

Sediment beneath waters in which vast numbers of shelled, or thecosomatous, pteropods live may contain great numbers of the empty shells, which form a significant fraction of the deposit. Such pteropod ooze is especially well developed in areas of the North Atlantic of moderate depth, around 2000 meters. It is not found in the Pacific and Indian oceans, although thecosomatous pteropods are common in the plankton of these oceans as well.

The **Heteropoda** usually have a pair of tentacles and a pair of large, well-developed eyes. As in pteropods, the foot is greatly modified to form an organ of flotation, usually forming from one to three fins. It often also bears a sucker, and in some species, an operculum. The head and foot are large compared to the small visceral sac (nucleus). The body and shell are rather transparent in most species. Best known genera: *Atlanta, Carinaria,* and *Pterotrachaea*. The shell is absent in some. Heteropods are confined to warm waters of tropical and subtropical regions.

Aside from these two groups, two other genera of gastropods are regularly encountered in the plankton. *Janthina* is a thin-shelled streptoneuran that relies on a mass of bubbles for flotation at the surface and whose foot is not greatly modified. *Janthina* is a beautiful snail with violet- or lilac-colored shell and bright red body. It feeds on the floating siphonophore *Velella* and on medusae. The second, more rarely encountered, is the nudibranch *Phyllirhoë,* another striking tropical form. These genera are tabulated in the Appendix.

Fossil record. The fossil record is fairly good for thecosomatous forms, indicating an extremely early origin for the group. The genus *Cornularia,* from the Paleozoic and Mesozoic, possessed a larger shell than existing pteropods, the shell tapering, four-angled in cross section, and having a narrowed aperture. A related genus, *Hyolithes,* is known from the Cambrian. *Tentaculites* are small conical shells abundant in some Paleozoic rocks and thought to be pteropods.

Annelida

Few annelids are holopelagic, although most polychaetes have planktonic larval stages. A number of species enter the plankton for brief periods, as for spawning. Pelagic annelids are never extremely abundant, although they occur with con-

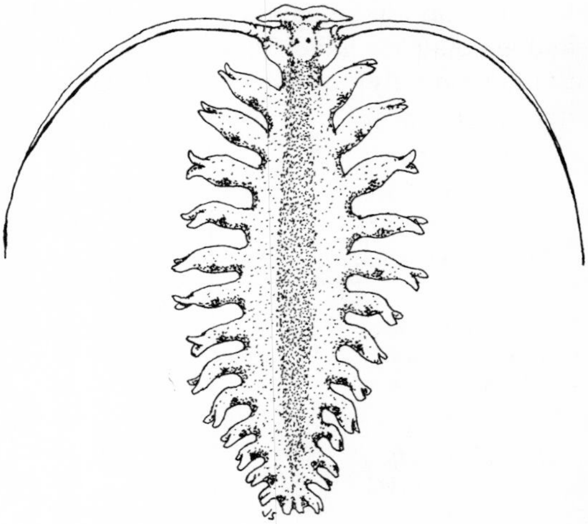

FIG. 9-20. *Tomopteris rolasi,* a pelagic polychaete annelid. (After Benham, W. B. 1901. *In* S. F. Harmer and H. E. Shepley [eds.]. Cambridge natural history. Vol. 2. Macmillan & Co., Ltd., London; reprinted in 1923 ed.)

siderable regularity in some regions. They belong to the families **Phyllodocidae, Alciopidae, Typhloscolecidae,** and **Tomopteridae.** The most common genera can be distinguished as follows:

Calizona: Eyes prominent. Parapodia with one appendage.
Greefia: Eyes prominent. Parapodia with two appendages.
Pelagobia: Eyes small or lacking. Many setae near aciculae.
Tomopteris: Eyes small or lacking. Few or no setae.

Nemertea

The nemerteans are another rather large group of benthic worms, mostly with pelagic larvae. The few holopelagic species that have been described are all deep-water forms. Bathypelagic nemerteans are all greatly modified for a planktonic mode of living and superficially do not look at all like ordinary nemerteans, the body being relatively shorter and flatter, and expanded into a pair of lateroposterior fins and a caudal fin that aid in swimming. Some show sexual dimorphism, with the male bearing a pair of strong tentacles near the anterior end, apparently for use in holding the female during copulation. *Nectonemertes, Pelagonemertes,* and *Planktonemertes* are the most frequently found genera.

Rotifera

A few planktonic rotifers are found in marine waters, especially neritic or somewhat brackish waters. Most of these species are derived from euryhaline freshwater forms that have migrated into marginal, somewhat brackish waters.

Brachionus: Has well-developed lorica, usually bearing spines on anterior border. Foot posterior, with toes of equal size.
Keratella (= Anuraea): Anterior edge of lorica has six spines; dorsal part of lorica reticulate. Gut complete, with anus. No feet. *K. cochlearis* is found in open sea in such waters as gulfs of Bothnia and of Finland, which are much less saline than is typical oceanic water.

Gelatinous Metazoa

Gelatinous zooplankton includes those animals which either have gelatinous bodies or are embedded in a gelatinous matrix, and which are either holoplanktonic or have an adult sexual stage that is planktonic. Three groups of animals contribute most of the gelatinous zooplankton: **Cnidaria (Coelenterata), Ctenophora,** and **Thaliacea.** The Larvacea, or appendicularians, are also included here with some justification.

Where gelatinous zooplankton predominates, the plankton is not as nutritious for larger predators and does not support as good a fishery as does a predominantly crustacean plankton. Larger medusae, because of their bulk and their stinging cells, occasionally constitute a real nuisance to fishermen using nets and are a hazard to swimmers.

CNIDARIA
Medusae

Medusae of 2 of the 3 classes of coelenterates are a conspicuous element of the macroplankton. They are saucer- or bowl-shaped organisms. The upper, usually convex, surface is termed the **exumbrella,** and the lower concave surface, the **subumbrella.** The body is comprised chiefly of thick, jellylike **mesogloea,** lined both externally and internally with a thin epithelium. Subepithelial muscle fibers form bands beneath the epithelium responsible for the contractions of the bell by which locomotion is effected. The mouth is at the open end of the **manubrium,** suspended from the center of the subumbrella, and opens into the stomach or **coelenteron,** which is the only internal cavity. The coelenteron is continued radially as a series of **radial canals** extending through the mesoglea at the edge of the bell, where they usually communicate with a **circular canal** going around the circumference. Most medusae have a series of hollow or solid (filled with a core of endodermal cells) muscular tentacles extending from the margin of the bell and armed with **nematocysts.** In addition, some species also have oral tentacles arising from the manubrium or from the extended lips of the mouth, or the mouth may be drawn out into oral arms. Marginal sense organs—**statocysts, ocelli, rhopalia,** etc.—are also characteristic of most medusae. There is no central nervous system, but the nerve net may be concentrated around the bell edge, forming what amounts to nerves along the tentacles and also around the mouth. Like all coelenterates, the medusae are actively carnivorous, capturing prey with the aid of the long, highly contractile, nematocyst-armed tentacles or, in some, with the aid of the oral arms. Because of the great elasticity of the body, medusae are able to ingest surprisingly large objects. Nematocysts are characteristic of, and unique to, all coelenterates. There are many different types characteristic of various groups.

In the class **Hydrozoa** there is usually an alternation of generations. Gametes produced by the medusa give rise to small ciliated solid-bodied **planula** larvae, which settle to the substrate and give rise to a sessile **polypoid generation.** When mature, the polypoid generation gives rise to **medusae** by asexual budding, and the cycle is completed.

Because of this sessile generation, most hydroid medusae are meroplanktonic and tied to neritic waters. Hydroid medusae are usually rather small, large individuals being only a few centimeters in length or diameter. They are most easily recognized by the presence of a **velum,** a small shelf of tissue growing inward horizontally from the edge of the bell, somewhat restricting the subumbrellar orifice, and probably, in those species in which it is well developed, aiding in swimming (Fig 9-21). Such medusae are called **craspedote medusae,** in contrast to the **acraspedote**—scyphozoan medusae that lack a velum. The stomach in hydromedusae is simple, not divided into radial chambers by mesenteries and not bearing gastric filaments. The marginal sense organs are statocysts or ocelli or both. Most coelenterates are active carnivores. The prey is stunned and immobilized or killed by nematocysts and swallowed into the large coelenteron, where it is attacked and rapidly broken up by extracellular enzymes. The resulting small particles and dissolved organic material are ingested or absorbed by gastrodermal cells. In medusae much of this material is distributed into the radial canals before being taken up by the gastrodermal cells.

A

B

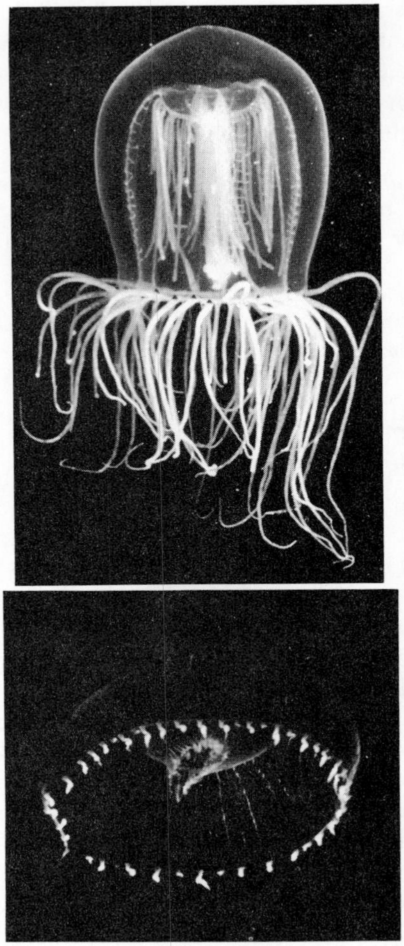

FIG. 9-21. Hydrozoan medusae. **A,** *Polyorchis.* **B,** *Aequorea.*

Hydrozoa show all gradations, from types such as *Polyorchis* and *Aequorea,* in which the medusa is large and well developed and is the major stage in the life of the organism, to types such as *Tubularia* and *Plumularia,* in which the polyp or hydranth is highly developed, whereas the medusa may be reduced to a mere budlike appendage of the hydranth, which never becomes free. This is doubtless the outcome of the relative success of the two stages in the particular lines of hydrozoans. Selective pressures are different in the two stages. If the habits and environment of a particular type of hydrozoan favor one stage more than the other, the long-term evolutionary result will tend to be a shortening and reduction of the stage less favored and a flowering and great development of the more successful portion of the life. Many polyps remain active most of the year and may live more than one year, whereas their medusae are released only seasonally and have a brief life-span. Such medusae are little more than a means for dissemination of the species. Since zygotes and planula larvae may be carried about by currents and tides almost as effectively as small medusae, sessile medusae are just as effective and may be better protected than free medusae for these species.

Because selective pressures are so different in the two stages of the life cycle, evolutionary changes in polyps and in medusae have not necessarily proceeded at the same rate or in directions that are closely correlated with each other. Thus it is difficult to correlate a given medusa with a given polyp unless the production of medusae can actually be observed. Furthermore, the methods of collection differ widely, the medusae most commonly being taken in plankton tows and the hydranths collected by low-tide shoreline collecting and dredging, and therefore they are not usually taken in the same collection. The result has been the growth of a double taxonomy for hydroids, one system based on the study of the polypoid generation and the other based on study of the medusae. This has caused considerable confusion in the taxonomy of the group, especially when the two stages have been discovered independently, assigned different generic and specific names—sometimes even put in different families—and only much later found to be stages in the life of the same species.

Class Hydrozoa
Order Hydroidea

The polyps of hydroids are grouped into two suborders based on the presence or absence of a **hydrotheca** (a protective cup of secreted perisarc into which the hydranth can withdraw for protection). These suborders are termed the **Gymnoblastea,** or naked hydroids, which are without such a structure, and the **Calyptoblastea,** which possess a hydrotheca. The medusae to which these hydroids give rise differ in characteristic ways and have been termed **Anthomedusae** and **Leptomedusae,** respectively. In general, the Anthomedusae are taller and always retain a tetraradiate form, and many of them have the gonads on the manubrium. They usually have marginal ocelli. Leptomedusae tend to be flatter and more saucer-shaped; some of them show branching of the radial canals or an increase in their number, which obscures the tetraradiate symmetry, and most of them have statocysts but not ocelli on the bell margin. The gonads are commonly borne on the radial canals rather than on the manubrium.

In addition, there are three more groups of craspedote medusae—the Narcomedusae, the Trachymedusae, and the Pteromedusae, often collectively termed the trachyline medusae—which differ in several respects from both Anthomedusae and Leptodmedusae and for which polyp stages do not occur.

In the **Narcomedusae** the tentacles pass through a portion of the bell to emerge from the exumbrellar surface above the bell margin. The margin is scalloped by the tentacle bases. The radial canals are absent or, if present, are in the form of flat radial gastric pouches. The gonads lie in the floor of the stomach. The planula larva develops tentacles, becoming an **actinula,** which either transforms directly into the medusa or buds off medusae. The actinulae are commonly parasitic on other medusae. The nematocysts are all of one kind, the simplest

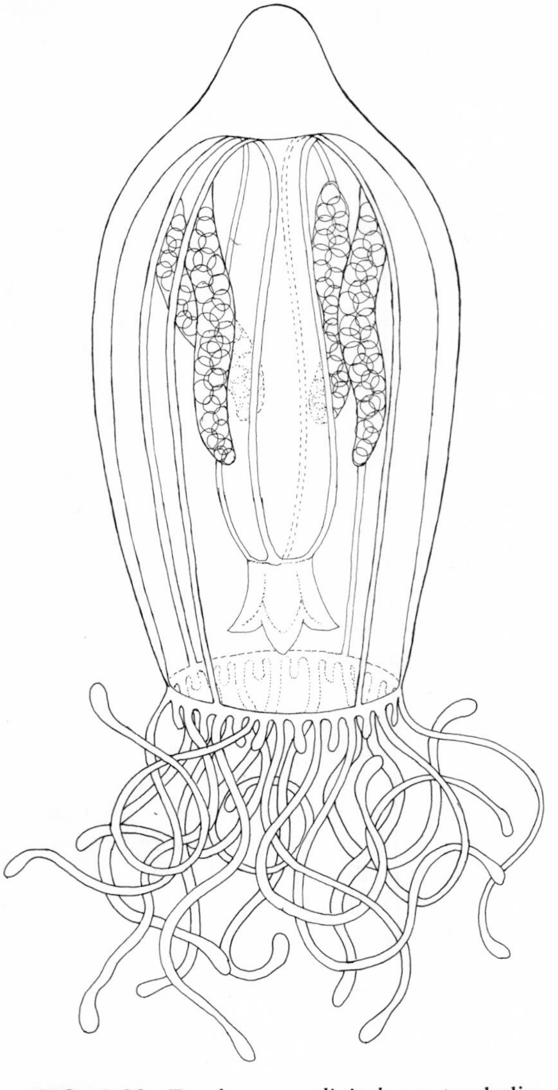

FIG. 9-22. *Trachynema digitale,* a trachyline medusa with tentacles inserted above bell margin. Polypoid generation reduced. (After Arnold, A. [Foote]. 1901. The sea beach at ebb tide. Century Co.; reprinted unabridged in 1968 by Dover Publications, New York.)

type—atrichous isorhiza. **Otoporpae** ("auditory tentacles") with strips of ectodermal tissue termed peronia may be present along the bell margin between the main tentacles.

The **Trachymedusae** resemble ordinary hydroid medusae more closely. The stomach is not lobed. The tentacles arise from the margin of the bell, are solid, and have an endodermal core. There are four, six, or eight radial canals, and in some species there often are blind centripetal canals between the radial canals. Ectodermal gonads lie beneath the radial canals. A band of nematocysts around the edge of the umbrella forms a thickened margin. The nematocysts are of only one kind—heterotrichous microbasic euryteles. Development is direct, the eggs giving rise to a swimming planula that becomes an actinula by formation of a mouth and tentacles and transforms directly into a medusa.

The **Pteromedusae** are represented by the genus *Tetraplatia,* a peculiar medusa found in oceanic waters. The body is bipyramidal, with an equatorial groove from which four swimming lappets extend, each bearing two statocysts. These lappets are thought to represent the velum. The upper portion of the body is regarded as equivalent to the bell of other medusae and the lower part as equivalent to the manubrium, with the mouth at the lower pole.

Trachyline medusae, with their direct development, are of considerable phylogenetic interest. Some authorities regard the medusa stage as the primitive one and believe that polypoid stages were derived through prolongation and modification of the actinula larval stages. If the view of Weyl (1968) regarding the Precambrian origins of animal groups in the tropical thermocline is correct, this phylogeny would appear to be the most likely one.

Order Siphonophora

Siphonophora are pelagic colonial hydroids in which several types of modified polypoid and medusoid individuals are united into a colony or higher-grade individual in which the individual components serve as organs. All are connected by common coenosarc and gastrovascular space. Some forms simply float at the surface of the sea, and others swim actively below the surface. Larger floating

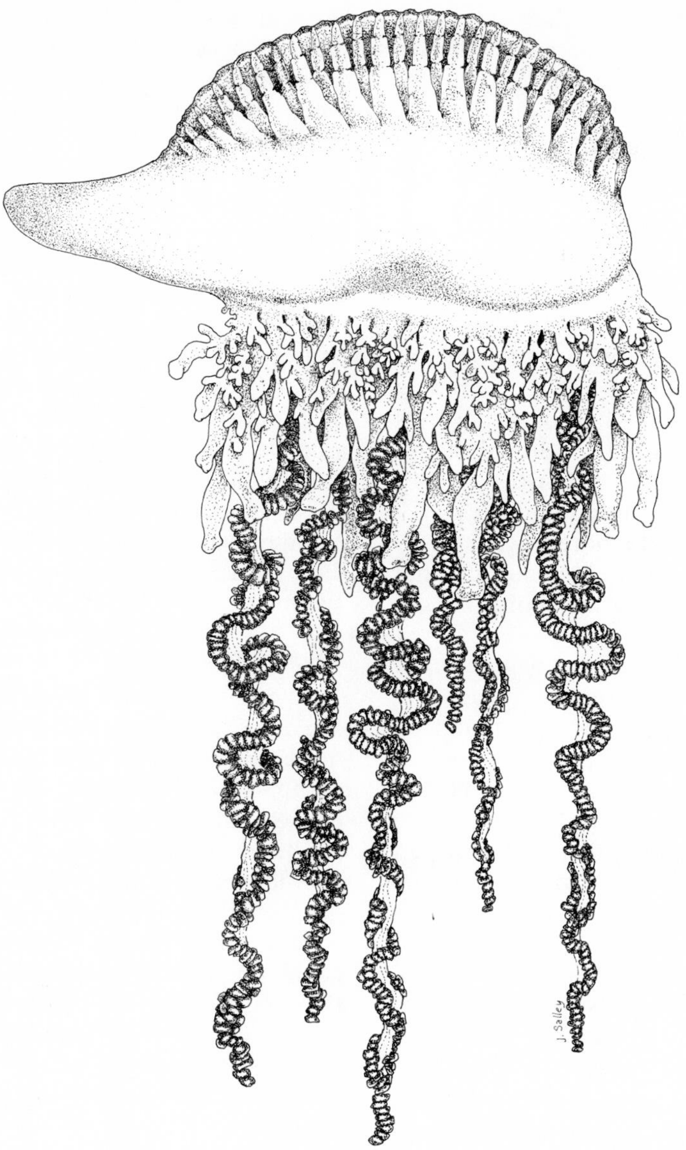

FIG. 9-23. *Physalia physalis,* the Portuguese man-of-war, a compound colonial hydrozoan.

forms, such as the Portuguese man-of-war, have powerful nematocysts with which they can kill fish and inflict serious injury on man. Siphonophorans are most abundant in tropical waters, to which many are limited, but some occur in all oceans.

The components of siphonophoran colonies are as follows:

1. **Gastrozooids.** Usually possess one hollow tentacle armed with nematocysts for capture of food, a mouth, and a coelenteron.
2. **Palpons.** Simpler than gastrozooids, without a mouth, and serve as accessory digestive organs.
3. **Dactylozooids.** May be long tubular tentacle-like zooids without a mouth, or shorter zooids armed with a single

long tentacle. Dactylozooids are armed with powerful nematocysts and are the principal organs of the capture of prey.

4. **Gonozooids.** Short zooids, with or without a mouth, bearing clusters of gonophores.
5. **Gonophores.** Medusoid zooids bearing sex cells on gonodendrons.
6. **Swimming bells** (nectophores, or nematocalyces). Used for locomotion.
7. **Bracts.** Thick gelatinous structures of various shapes that aid in swimming and protection of the colony. Bracts have been regarded as altered medusoids, but recent authorities regard them merely as altered tentacles.
8. **Floats** (pneumatophores). Also regarded as an altered medusoid, and now thought to be an aboral invagination.
9. **Stem.** Structure along which other components are strung.

Not all these components are found on any one siphonophoran. Siphonophora in which an elongate, stemlike coenosarc trails behind a nectophore, or swimming bell, have the various individuals arranged along this stem in definite groups, or **cormidia.** Each cormidium consists of a definite number and arrangement of individuals. New cormidia form at the proximal growing end of the coenosarc in the nectophore, and they become larger, more individualized, and more mature distally.

Many siphonophorans are delicate and extremely transparent in the water. Forms in which many individuals, bracts, etc. are strung out along a long stem are difficult to collect in perfect condition, only disconnected parts commonly being found in the usual plankton haul. Some confusion has thus arisen, since detached cormidia have often been described as separate species.

A type of nematocyst found only in siphonophorans is the rhopaloneme.

It is thought that Siphonophora may have evolved from trachyline hydroids with a swimming actinuloid stage and an adult medusoid stage—that the merger of these two stages, remaining together and differentiating into the various modified polypoid and medusoid forms found in the colonies, produced the complex forms found today. By using whole individuals as organs, the siphonophorans have achieved an organ grade of construction in spite of their simple diploblastic tissue-grade organization of the individuals. The tendency of colonial animals to specialize nonreproductive members of the colony to form organs is one of the strongest lines of support for the theory that metazoan animals orginated from colonial Protozoa. A colonial protozoan that developed along these same lines, forming a highly complex colony, with various cells performing different functions, would be defined as a metazoan.

Class Scyphozoa

Scyphozoans, or true jellyfish (Fig. 9-24; see also the taxonomic appendix), seem to represent a line in which medusa success has been so marked that it has led to the evolution of a group in which the medusae are usually larger, more complex, and longer lived than those of most hydroids and in which the polypoid generation is reduced or wholly suppressed. Scyphozoans retain tetramerous symmetry in both the medusa and the polypoid generations. The stomach is divided into four radial chambers containing **gastric filaments.** The medusa is **acraspedote,** that is, it does not possess a true velum, although in some species the edge of the bell is turned in, constituting a type of pseudovelum, the **velarium**. The marginal sense organs are typically **sense pits** and **rhopalia.** The bulk of the mesogloea is essentially free of cells, although a well-developed layer of circular muscle may grow beneath the epithelium of the subumbrella and the tentacles and oral arms may have well-developed muscle tissue. There is no stomadaeum such as that found in most hydromedusae. The corners of the mouth are drawn out into oral arms.

There are only some 200 species of true

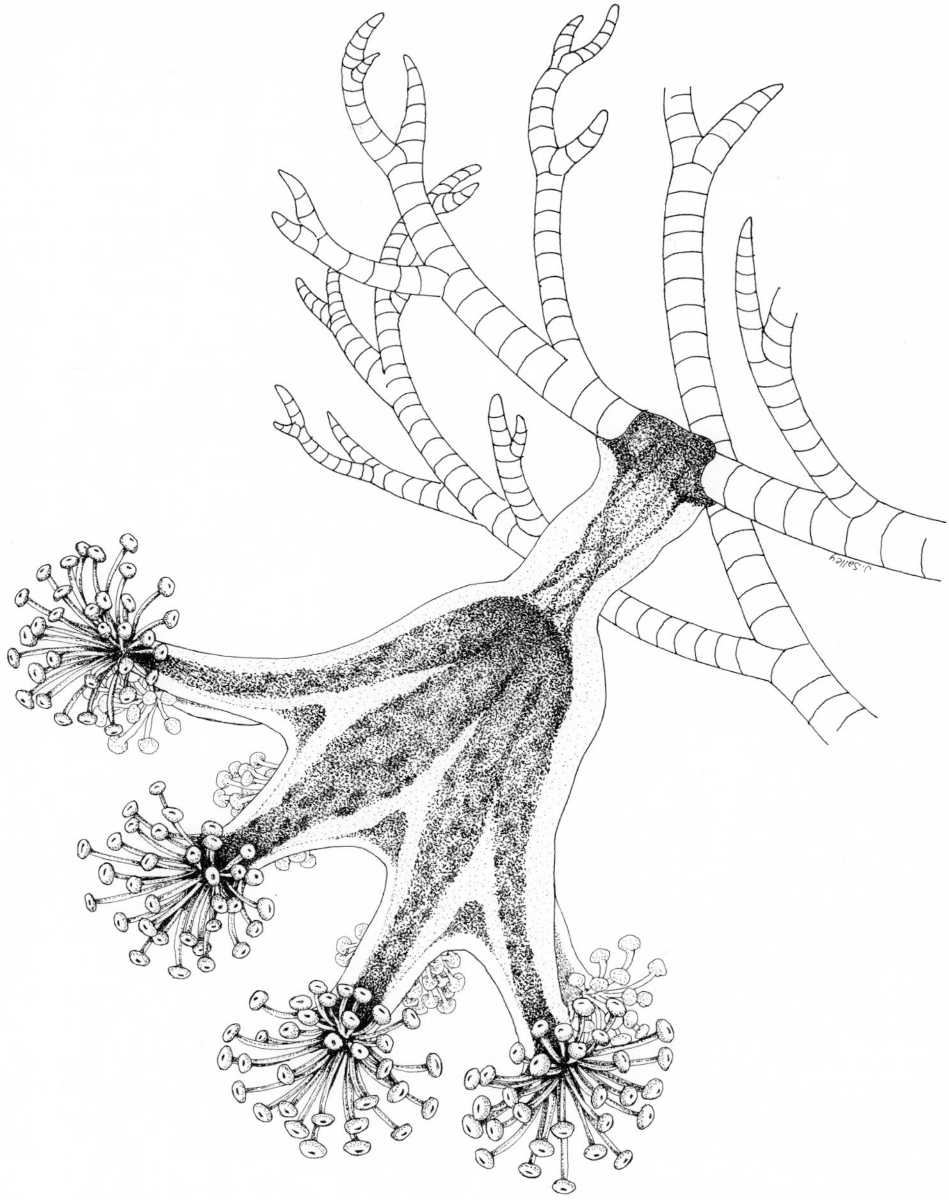

FIG. 9-24. *Haliclystus auricula,* order Stauromedusae, the only true jellyfishes to live as sessile medusae attached by a dorsal extension of the bell, forming an adhesive disk. (Redrawn from Barnes, R. D. 1974; after Marshall, N., 1971.)

jellyfish, but they are so abundant and widely distributed that they are an ecologically important group. They are found in all oceans and range from surface-swimming forms down to 3000 meters' depth or more.

Scyphozoa are divided into five well-marked orders, four of which are free-swimming and are to be regarded as part of the plankton or nekton. Most of them are much larger than the majority of organisms usually thought of as constituting

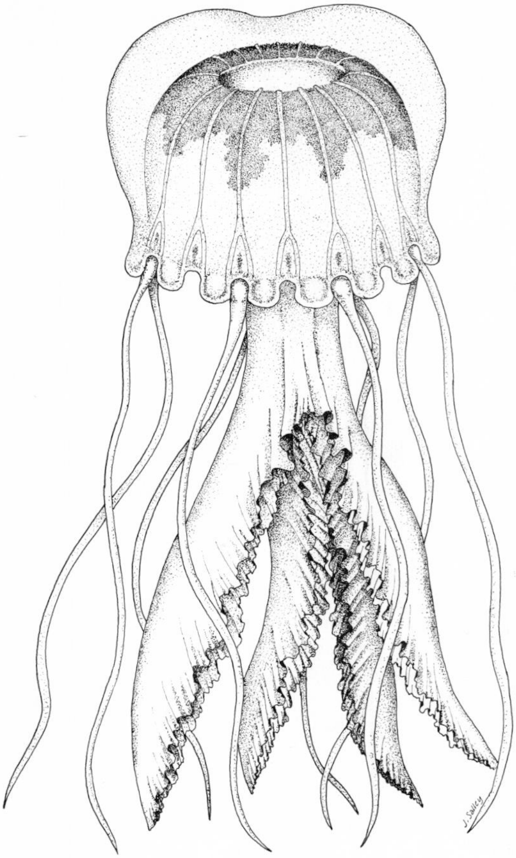

FIG. 9-25. *Pelagia cyanella,* a large pelagic semaeostome jellyfish. (After Arnold, A. [Foote]. 1901. The sea beach at ebb tide. Century Co.; reprinted unabridged in 1968 by Dover Publications, New York.)

the plankton, and many of them swim as actively as, or more actively than, some other animals generally regarded as nekton. The largest jellyfish known is *Cyanea arctica,* which sometimes attains a weight of nearly a ton and may have tentacles more than 30 meters long.

Order Semaeostomeae

The most familiar jellyfish and the only ones usually seen in temperate and cold waters are in this order. In this group the bell is bowl- or saucer-shaped, without any furrow. The tentacles arise from the margin or from the subumbrellar surface near the margin, are hollow, and do not arise from pedalia. The corners of the mouth are prolonged into four long, frilly oral arms. The margin of the bell is scalloped into eight to many lappets. There are eight to sixteen rhopalia borne in marginal notches between special rhopalar lappets. Most semaeostomes have a fixed larval stage known as a **scyphistoma** and are hence neritic forms more or less bound to coastal waters. Scyphistomae undergo a peculiar transverse constriction to bud off immature medusae known as **ephyrae.** *Pelagia,* however, lacks such a stage and is found in high oceanic as well as neritic waters. Scyphistomae of jellyfish have two kinds of nematocysts (atrichous isorhiza and microbasic heterotrichous eurytele), whereas the medusa has also a third kind (holotrichous isorhiza).

The remaining orders of jellyfish are largely tropical or subtropical in their distributions or are swept farther north in the warm currents along eastern North America and eastern Asia.

Order Cubomedusae

Members of this group are readily recognized by the squarish shape of the bell and the fact that the tentacles are borne on **pedalia** at the corners of the bell so that there are either four tentacles or four groups of tentacles. The margins of the bell are turned in, constituting a velarium, or false velum, with septa. There are four rhopalia, occurring along the umbrella margin midway between the bases of the pedalia. At night they are strongly phototropic and are a nuisance as they crowd around fisheries' lights. These jellyfish are strong swimmers and voracious predators, eating mostly fish, and have powerful nematocysts that inflict a painful sting, giving this group the name sea wasp. They are commonly dull yellowish or livid in color. They seem to stay near the bottom much of the time until fully mature, and then they swarm to the surface to shed eggs and sperm. The very serious reaction of some persons to the sting of these

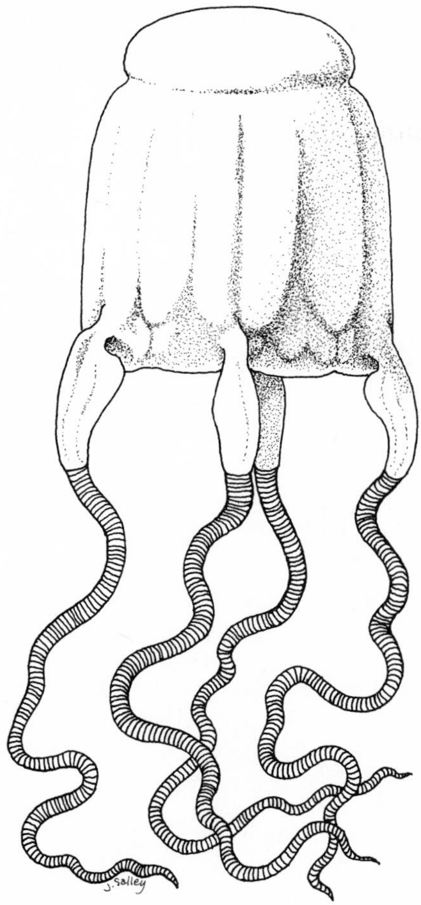

FIG. 9-26. *Charybdea murrayana,* a cubomedusan jellyfish. (After Haeckel, E. 1882. Challenger reports. Vol. 4.)

jellyfish and of certain other coelenterates may be largely due to hypersensitivity induced by previous exposure to the sting.

Order Coronatae

The members of this cosmopolitan order are mostly small- to medium-sized jellyfish, the bell marked with a circular coronal constriction, or groove, that separates the central domelike portion from the marginal scalloped portion. The solid tentacles are borne on pedalia alternating with the lappets. Several are deep-water bathypelagic species, dark purple or reddish brown in color.

Order Rhizostomae

Rhizostomae is a distinctive order. It is particularly abundant in shallow tropical and subtropical waters of the Indo-Pacific. The bell is firm and bowl- or saucer-shaped, and it often bears nematocyst warts on the exumbrella. There are no marginal tentacles. The bell margin is scalloped and bears eight or more rhopalia. The upper ends of the oral arms are fused, obliterating the mouth, which runs as branching channels through the oral arms and ends in hundreds of small mouth openings in the lower inner portion of the eight mouth arms. The subgenital pits are, in many of the genera, so large that they form a continuous cavity, or **porticus,** separating the bell from the oral arms, which are then connected to the bell by four pillars through which run the canallike extensions of the stomach into the mouth arms. The oral arms are provided with filamentous appendages loaded with nematocysts. Subumbrellar musculature is usually strong. In some genera the oral arms terminate in long pendant clublike appendages of unknown function.

PHYLUM CTENOPHORA

The ctenophores comprise a small, sharply defined phylum of usually delicate transparent pelagic gelatinous animals, often of great beauty. The common name, comb jelly, refers to the unique mechanism of locomotion, which consists of eight meridional rows of strongly flagellated comb plates running from near the aboral pole of the body, over the surface, and toward the oral pole. The body symmetry is biradial around the oral-aboral axis, one plane, the tentacular plane being defined by the presence of two tentacle pouches, from each of which a long tentacle, bearing short branches armed with **colloblasts,** or sticky cells, can be protruded. The other plane, the sagittal plane, lies at right angles to the tentacular plane, is situated at the apical pole, and is marked by the **polar**

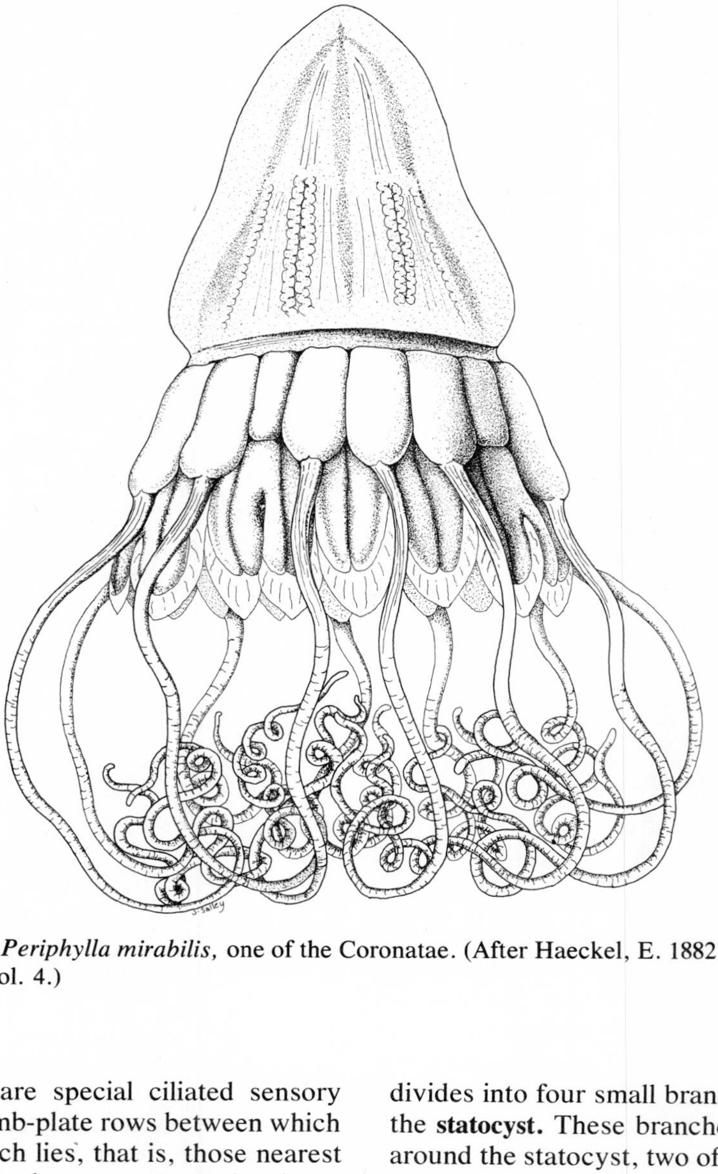

FIG. 9-27. *Periphylla mirabilis,* one of the Coronatae. (After Haeckel, E. 1882. Challenger reports. Vol. 4.)

fields, which are special ciliated sensory areas. The comb-plate rows between which a tentacle pouch lies, that is, those nearest the tentacular plane, are termed subtentacular comb rows; those nearest the sagittal plane, are termed subsagittal comb rows. The digestive system consists of a long, heavily ciliated pharynx extending about halfway or more from the mouth toward the aboral pole. At the upper end the pharynx is dilated and glandular, serving as a stomach. The ramifications of the digestive tract are thin, transparent canals. The intestine, or aboral canal, continues straight toward the aboral pole, where it divides into four small branches just below the **statocyst.** These branches pass upward around the statocyst, two of them ending in anal pores. At the junction of the stomach and intestine, two **radial canals** are given off and pass toward the tentacle pouches. Just before or soon after reaching the tentacle pouches, they give off two oblique, lateral branches, the **interradial canals.** The interradial canals, in turn, branch dichotomously into **adradial canals.** Each adradial canal terminates in a **meridional canal** lying just beneath one of the comb rows. The meridional canals bear the gonads, which are arranged in a character-

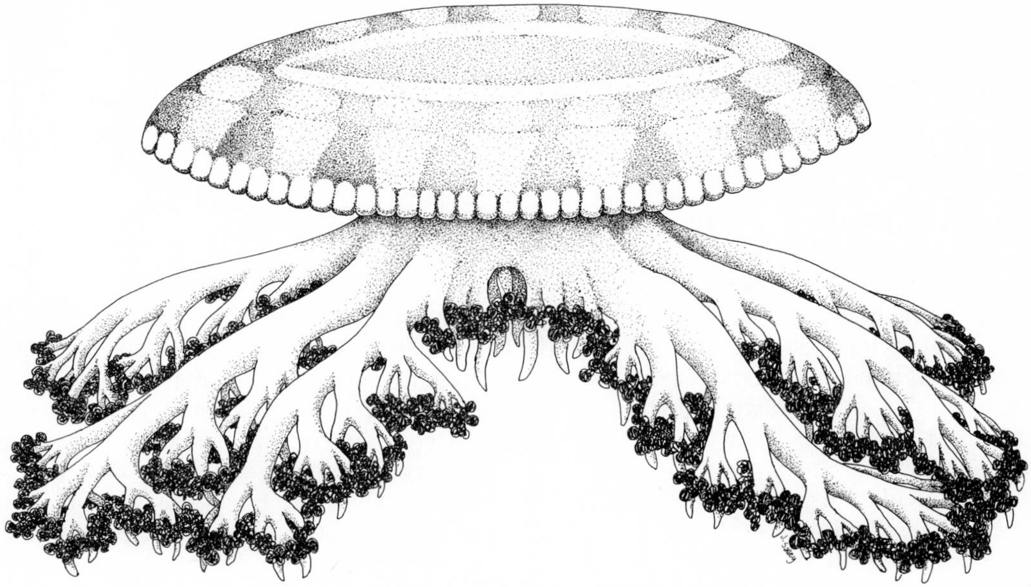

FIG. 9-28. *Cassiopea amachana,* a rhizostome jellyfish that spends much time in shallow tropical lagoons upside down with the exumbrella appressed to the substrate. (After Marshall, N. 1971. Pocket encyclopedia of ocean life. The Macmillan Co., New York.)

istic manner, the ovaries lying along the perradial wall of the canal and the spermaries lying along the interradial wall. The radial canals also give off a pair of paragastric canals that pass down along each side of the pharynx toward the mouth in the tentacular plane.

The above characterization applies primarily to the Cydippida, which are regarded as the most typical, or the least modified, of the ctenophores. Larvae from other groups resemble the cydippids for a brief time and are known as cydippid larvae.

Ctenophora are divided into two classes and five orders. The members of all but one of the orders are planktonic. These groups, together with some of the more familiar genera, are briefly characterized in the taxonomic appendix.

PHYLUM CHORDATA
Subphylum Urochordata
Class Ascidiacea

The ascidians, or true tunicates, are mostly benthic throughout their adult life. The pyrosomids, however, are a group of colonial forms in which the entire colony behaves as a free-swimming pelagic organism.

Suborder Pyrosomida: Most tunicates are benthic throughout adult life. Pyrosomids, however, are a group of colonial forms in which entire colony behaves as a free-swimming pelagic organism. Colonies cylindrical, and tapering or flattened, with one end open. Zooids arranged in one layer in walls of gelatinous cylinder, perpendicular to surface, with their incurrent openings on outside and their atrial openings on inner surface. In some species outer surface of common tunic may be produced into long fingerlike processes. As zooids pass water through their branchial chambers and out atrial openings for feeding and respiration, water streaming out from open end of colony causes colony as a whole to swim with closed end forward. At fore end of branchial sac of each zooid are paired luminescent organs. Colonies noted for brilliant bioluminescence they display when disturbed. They float horizontally in sea and have only weak powers of movement. Colonies vary from a few inches to over 4 feet in length. *Pyrosoma:* Has characters of the suborder.

Class Larvacea or Copelata

The family **Appendiculariidae** really fits into both categories (gelatinous and nongelatinous zooplankton), since most of them secrete a temporary gelatinous **"house"** about themselves, which aids in filter feeding and locomotion. They often abandon this house and secrete a new one, and in plankton samples are commonly free from a house, perhaps because of disintegration of such a fragile structure or because the house has been shaken free in handling.

Appendicularians are much like the larvae of sessile tunicates, from which they differ primarily in the possession of functional gonads and of an anus. They also lack the closely investing tunic of larval tunicates. The body is divided into two portions, an anterior globular part containing the organs and variously termed the "head," the "trunk," or the "body," and a flat tail attached to the ventral side of the head near the middle. The head is more sharply differentiated from the tail than it is in larval tunicates, and the tail has a series of ganglia. It is probable that the appendicularians are an example of **paedogenesis,** in which the formation of gonads has occurred in a developmental stage corresponding to the larval stage of the ancestral forms. The result has been the loss of the ancestral sessile habit, freeing the organisms for a pelagic existence.

The house is a remarkable adaptation. On the upper side is a grilled filtering surface, the porosity of which allows the smaller plankton organisms (nanoplankton) to pass but excludes larger particles. A current of water is set up by activity of the animal's tail. Water entering through the coarse filter is swept posteriorly, around through a finer internal filter, and then finally into the large ciliated pharynx of the appendicularian. Here food particles are retained, the water leaving through the paired branchial openings, one on each side. When sufficient pressure has built up inside the house through the intake of water, a trapdoor at the posterior end suddenly opens and the water rushes out, propelling the house, with its contained owner, forward by jet propulsion to new feeding grounds. After several hours, if the filters become clogged or the house is damaged, the appendicularian escapes through another trapdoor, or escape hatch, at the front and soon secretes a new house. Appendicularians are the only group of filter-feeding animals in which essential parts of the filtering apparatus are not part of the structure of the animal itself, although it is true that many animals do use a secretion of slime or mucus as an aid in the entrapment and transport of planktonic or detrital particles brought to them by a feeding current.

It is of interest that although coccolithophores had been known to exist for some time because of the coccoliths found in geological deposits, living ones were first discovered by members of the *Challenger* Expedition in the filters of the houses of appendicularians, coccolithophores being too small to be retained by the plankton nets in use.

Some of the principal genera are given in the taxonomic appendix.

Class Thaliacea

The salps are pelagic transparent tunicates, rather cylindrical, with a large oral opening at one end of the body and a large cloacal opening at the other. The body wall is a thick elastic resilient cellulose **tunic,** the elasticity of which restores the body shape after muscular contraction. The **mantle tissue** immediately below the tunic has several conspicuous ring-shaped muscle bands, the contraction of which forces water out through the **cloaca,** propelling the animal forward. The body cavity is divided by a respiratory partition, which may vary from a complete partition bearing slits, or **stigmata,** to a single ciliated bar. This partition divides the body into an anterior pharyngeal chamber and a posterior cloacal chamber. An **endostyle** and **dorsal lamina** are present in the mantle on the ventral side of the body below the cloacal chamber. In the dorsal wall of the pharyngeal chamber lies a ganglion, which in some species has an eyespot and subneural gland associated with it.

Salps alternate between solitary and aggregated generations, which in the past often received separate generic names. A common practice has been to use both names, the oldest one first, with a hyphen between them.

Salps are ciliary-mucus feeders, and they remove phytoplankton and other fine particulate matter from the water. Since this food is also used by many larval and some adult plankton-feeding fish, the occurrence of large numbers of small salps may be restrictive to the development of successful stocks of plankton-feeding fish or of fish with plankton-feeding larvae. The salps themselves are a poor food, not much utilized by fish. Salps are characteristic of warm waters. One species occurs in the Antarctic, and there are none in the high northern latitudes.

Two principal groups are recognized, the **Cyclomyaria,** with the large family **Doliolidae,** and the **Hemimyaria,** of which the chief family is the family **Salpidae.**

Suborder Cyclomyaria, family Doliolidae: Body barrel-shaped, with a series of usually complete muscular rings around it. Respiratory partition platelike, with two rows of stigmata. Oral and cloacal openings, at opposite ends of body, are lobed. Three types of individuals produced asexually, and one type hatches from fertilized eggs. A tailed larva always developed.

Doliolum: Sexual generation has eight muscle bands, twelve oral lobes, and ten cloacal lobes. First asexual generation has nine muscle bands, and ten oral and twelve cloacal lobes. From the hinder end grows a dorsal protuberance and a ventral stolon. Buds originating on stolon migrate to tubercle, which becomes long and taillike and bears buds in five rows. The median row becomes detached and develops into next asexual generation. Asexual generations are similar to sexual generation except that the former lack gonads and produce a ventral stolon, the buds of which detach and grow into sexual generation. (Buds in lateral rows on stolon of first sexual form are called gastrozooids. They develop into doliolum-like zooids with only rudimentary gonads that soon atrophy, but the gut is well developed, and zooids actively take nourishment and serve for nutrition of growing stolon.)

Suborder Hemimyaria, family Salpidae: In this group, tunic thicker and body often somewhat flattened dorsoventrally. Oral and cloacal openings may be either terminal or subterminal. Respiratory partition reduced to a single ciliated bar running diagonally from dorsal wall to esophagus in ventral wall. Life cycle includes alternating solitary asexual and aggregated sexual types of individuals. No tailed larva formed. Gametes produced by sexual form develop into asexual individuals that produce chains of sexual individuals. Four to twenty (usually six to nine) muscle bands, continuous dorsally but usually not ventrally, except for atrial sphincter. Aggregate sexual phase may differ greatly in size, shape, and other characters from solitary phase. Commonly they develop only one ovum per zooid. Muscles are less numerous and less well developed. Each sexual zooid has eight pairs of adhesive papillae on ventrolateral surfaces, by which it is attached to next zooids in line. If zooids become isolated due to breaking of chain, they continue to exist alone, and adhesive papillae disappear.

Salpa: A typical "nucleus" is present, or if not, muscle strands are numerous. Straight, double chain of aggregate sexual individuals. Digestive tube usually bent on itself and contained in nucleus.

Cyclosalpa: No "nucleus" is present. Intestine not twisted but passes forward along endostyle in sexual form and along dorsal lamina in asexual form, to open anteriorly into atrial cavity. Salp chain often circular, each chain, or wheel, containing about thirteen individuals of aggregate generations.

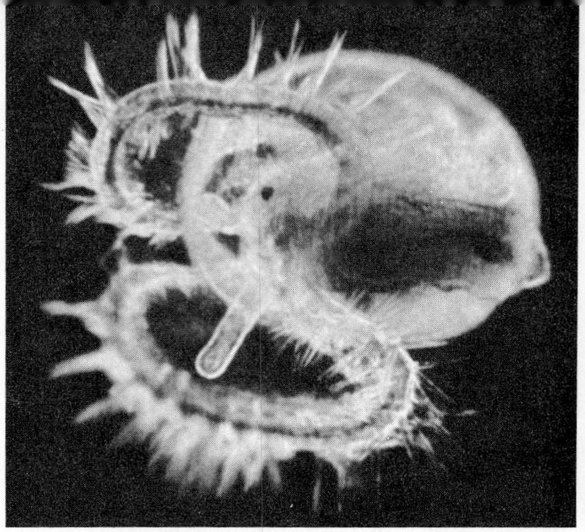

FIG. 10-1. Veliger larva of a gastropod mollusc. (Photographed alive by Douglas P. Wilson.)

CHAPTER

10 PLANKTONIC LARVAL FORMS

Because much of the shelf fauna and some of the pelagic nekton and deep-water benthos have larval forms that are planktonic for a time, and because these larvae are in most cases extremely different from the adults, from both a morphological and an ecological point of view, and are most commonly taken in plankton tows, a number of representative planktonic larval types will be considered together here, rather than separately in connection with the corresponding adult groups.

A short description of larval forms in general will be given before the various larvae are described individually. Ecologically they may be thought of as **planktotrophic,** that is, eating other planktonic organisms or detritus, most commonly elements of the phytoplankton, or as **lecithotrophic,** subsisting during part or all of their planktonic existence on stored food from the egg. The latter types usually develop from larger, yolkier eggs. Some of them may subsist for half a month or more, about as long as most planktotrophic forms. Lecithotrophic types grade insensibly into forms in which an increasing portion of their development takes place in the egg stage, to types that have no planktonic larvae. Thorson (1950) called attention to an interesting correlation of latitude with the proportions of species in various groups of benthic animals that have planktonic larvae. The highest proportion of species with planktotrophic larvae are found in tropical and subtropical waters, but as one goes toward higher latitudes, or to great depths, the proportion with lecithotrophic larvae or with direct development without planktonic larvae increases. This phenomenon is thought to be correlated with the highly seasonal character of the outbursts of phytoplankton in the spring in the higher latitudes and with the paucity of suitable planktonic food at great depths. Swarms of planktotrophic larvae might face starvation if reproduction were not timed nicely to periods when there is plenty of phytoplankton. It has been shown repeatedly that in northern latitudes, species with long-lived planktotrophic larvae show much greater fluctuation from year to year in success of settlement than do species that have a brief period of existence as planktonic larvae, or none at all.

For sessile animals long-lived planktonic stages give a better chance for successful dispersal to more distant localities with suitable habitat, but there is a greater risk that the larvae may not survive at all. The balance of risk and advantage shifts in favor of planktotrophic larvae in the tropics, where the phytoplankton supply does not fluctuate so markedly, and in favor of direct development in the high latitudes, though in neither case is the shift so drastic that it has led to the elimination of one or the other mode for all the animals. Direct development in the high latitudes seems also to be correlated with a longer average life-span for the sessile stage or stages. It is obviously possible for an animal with a given amount of biomass to produce a great many more planktotrophic larvae than either lecithotrophic larvae or directly developing forms, which must depend on stored food. Where the chances for their survival are good, this becomes the method of choice.

Having a planktonic larval stage means that such larvae are at the mercy of the water currents of the region. The water currents thus may have a considerable influence on the ability of such benthic species to live and replace themselves in a given locality. Johnson (1939), reporting on the occurrence of nauplii of the sand crab *Emerita* in offshore waters far from land, found that these larvae are reliable indicators of waters of inshore origin that have been swept out to sea. Such larvae are, of course, doomed, the loss of many larvae being a price paid by the species for the potential of wide dispersal. If the prevailing currents at a particular place are such that they carry away any planktonic larvae liberated, benthic species that have such larvae may be excluded from the area, even though in other respects it may be entirely suitable for them. In some such cases, settlement may occur successfully if larvae at a stage ready to settle are also swept in from other areas.

Local rotary drifts and eddies over banks and along shores play an important part in keeping enough planktonic larvae over such areas until they are ready to settle, enabling benthic populations to be maintained in places that might otherwise be swept bare by the major currents of the region (Fish, 1935).

The great rotary eddies, such as the Sargasso Sea and the eddy in the Gulf of Mexico, probably maintain endemic plankters from which the surrounding currents are continually seeded. Redfield (1941) studied the great eddy in the Gulf of Maine and found that much cold, relatively barren water enters from the north during the winter and spring, but that as it swings counterclockwise around the gulf, the biomass of plankton is greatly increased. Thus the water escaping over Geroges Bank is relatively rich in plankton, which probably accounts for the favorable fishing conditions found there.

SPONGE LARVAE

Amphiblastulae of sponges are blastula-like larvae with large, yolk-laden, smooth cells in one hemisphere and small, flagellated cells in the other. They have only a brief planktonic life before settling down to the benthic life of a sponge. The flagellated cells become the choanocytes of the adult sponge (Fig. 10-2). Other cells of the sponge body are derived from the large, smooth cells of the posterior pole. These larvae are seldom numerous in the plankton, and since they do not feed during this stage, they do not exert much influence on other animals there.

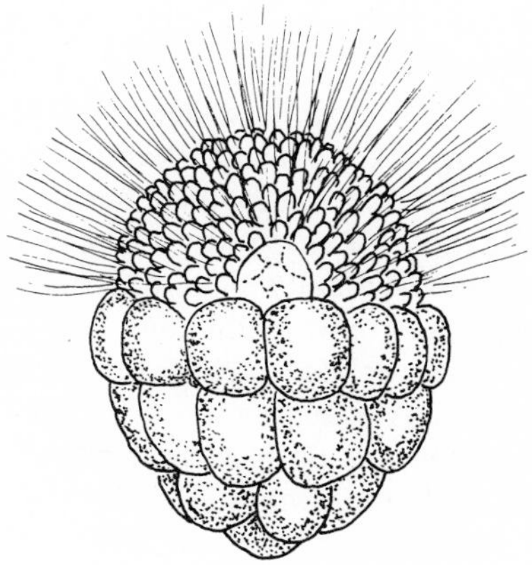

FIG. 10-2. Amphiblastula larva, *Sycandra*.

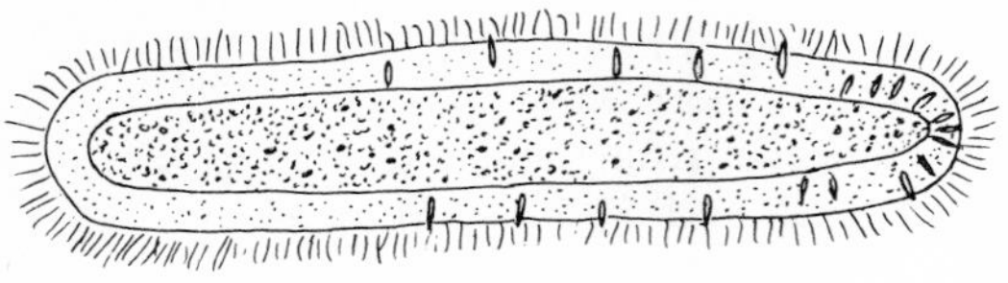

FIG. 10-3. Planula larva of *Polyorchis* drawn from life. Nematocysts appear as refractile bodies in the ciliated epithelium in the anterior two thirds of the larva.

PLANULA LARVAE OF COELENTERATES

The product of sexual reproduction is a short-lived solid ciliated larva, the **planula,** resembling an acoel flatworm, but without a mouth or digestive tract. In coelenterates with a sessile polypoid generation, the planula (Fig. 10-3) attaches to a suitable place on the substrate and transforms, by way of an actinula-like stage, into the polypoid. In pelagic medusae without a polypoid generation, the planula generally becomes an **actinula larva** that either transforms into, or buds off, medusae. The actinulae of some of the trachyline medusae are parasitic on other species of medusae. Young medusae of scyphozoans, termed **ephyra,** budded off by the **scyphistoma,** may also be considered a larval stage before they have attained the size and proportions of the adult medusa. The planula, a short-lived, nonfeeding form is seldom abundant in the plankton and exerts little effect there.

PLATYHELMINTHES

Small free-living flatworms, particularly of the groups Acoela and Rhabdocoela, are occasionally encountered in neritic plankton, but most of them are, at best, tychopelagic, not typically plankton organisms. Larval stages, **miracidia** and **cercaria,** of digenetic flukes also have a brief plankton existence, as do the free-swimming ciliated larvae of monogenetic trematodes and the coracidia of pseudophyllidean tapeworms. All these are nonfeeding larvae that are merely en route from one host to another. They do not constitute an important element in the plankton and are infrequently encountered in plankton hauls.

NEMERTEA

Many benthic nemertean worms, order **Heteronemertea,** have a unique planktonic larval form for dispersal known as the **pilidium larva,** which looks somewhat like a small swimming football helmet with large earlobes. It is derived by gastrulation and the downgrowth of large oral lobes on either side of the blastopore. The invagination produces the foregut and midgut regions, which can be considered endodermal. Ectodermal invaginations produce paired discs that detach and move into the interior, uniting around the gut to form a larva. These discs are the cephalic, cerebral, and trunk discs, together with unpaired dorsal and proboscis discs. When the larva has been fully formed inside the pilidium, it escapes, sinks to the substrate, and grows into an adult worm. The empty pilidium may continue to swim for a time before it undergoes disaggregation or is eaten.

ANNELIDA

Archiannelida and **Polychaeta** usually produce **trochophore larvae.** These larvae are formed through spiral determinate cleavage in which the various blastomeres each give rise to particular parts of the larva. During late cleavage stages a characteristic annelid cross is formed. It is interradial in position and has an apical annelid rosette of four cells. As the embryo grows, it develops into a hollow biconical organism with a beltlike band of large cilia in front of the mouth. The band, called the **prototroch,** is equatorial in position. A second band of cilia, situated postorally, is termed the **metatroch,** and there may be one around the anus, the **telotroch.** There is a complete digestive tract, with a mouth near the midlevel just below the prototroch, a curved gut that includes an enlarged stomach, and a short intestine passing to the anus at the lower pole. There is an apical ganglion and a sense organ with a patch of long cilia at the

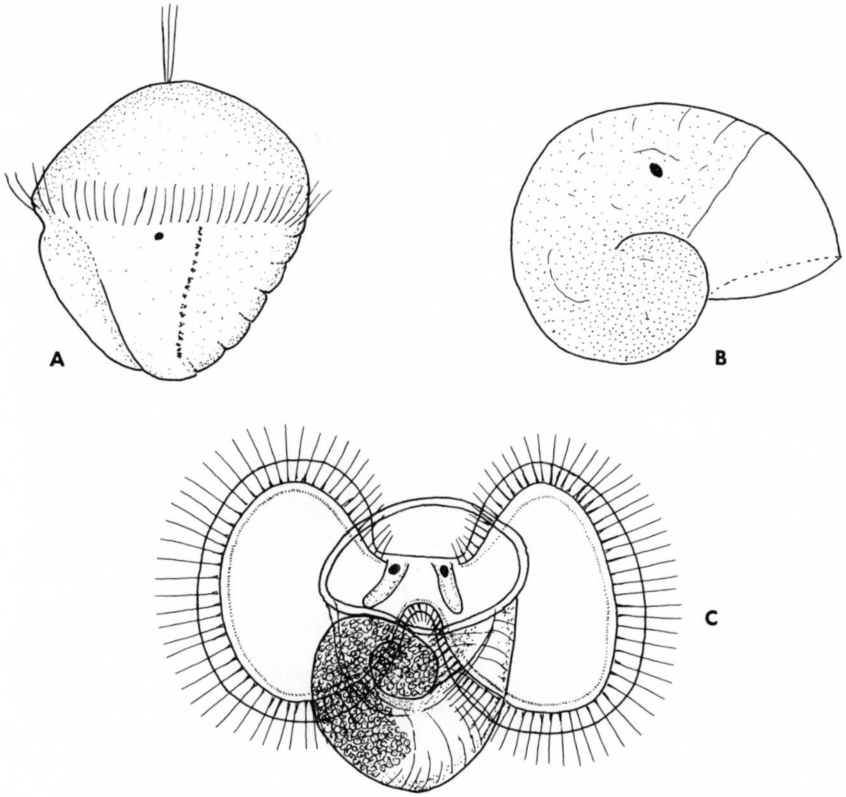

FIG. 10-4. A, Late trochophore larva of a chiton *(Ischnochiton)*. Differentiation has proceeded to the point where the foot, the locations of most of the future shells, and the edge of the mantle can be seen. **B,** Gastropod veliger larva as it usually appears in preserved plankton sample. Soft parts withdrawn into shell. **C,** Veliger larva, *Eulimella nitidissima* Montagu, with velum expanded. (**B** drawn after Davis, C. C. 1955. Marine and freshwater plankton. The Michigan State University Press, East Lansing.)

top of the upper hemisphere, a larval nephridium, sometimes a statocyst, and mesodermal bands derived from a pair of posterior teloblasts. In some there is also a pigmented eyespot. Some species such as *Eupomatus* have only a prototroch. Others such as *Chaetopterus* have only the metatroch. Still others have several rings of cilia and may be described as polytroch *(Ophryotrocha)*. Some tube worms have an even more modified trochophore, bearing two bundles of long setae, known as the **metraria larva** *(Myriochele danielssoni)*.

Transformation into the adult worm is direct, the lower hemisphere elongating, becoming segmented, and developing segmental setae groups and parapodia. The metatrochophore finally settles to the substrate and develops into an adult individual. The anterior hemisphere becomes the preoral elements of the head.

MOLLUSCA

Many molluscs also have a similar **trochophore larva** (Fig. 10-4, *A*), which differs from that of annelids in detail of its development and differs greatly in the manner of its transformation into an adult. The cross of cells formed in late cleavage, known as the molluscan cross, is composed of blastomeres different from those occurring in annelids, and it is radial in position. The anterior hemisphere of the trochophore becomes the head of the mollusc. The ven-

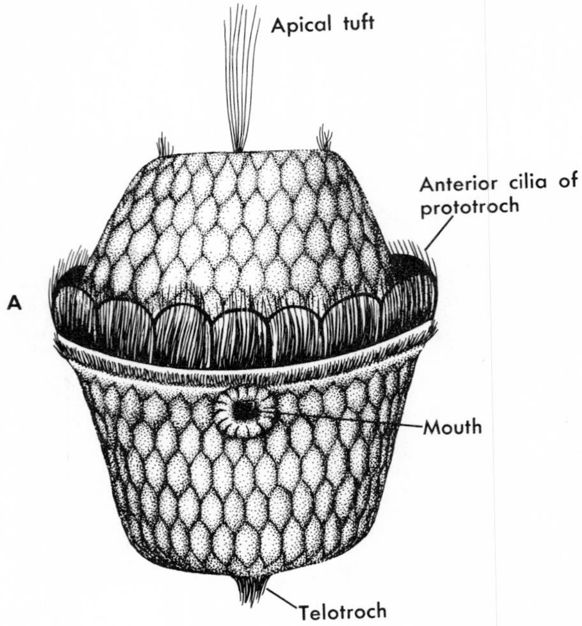

FIG. 10-5. Muscle morphogenesis in a primitive gastropod, *Patella vulgata,* and its relationship to torsion. **A,** Trochopore larva; **B,** veliger before torsion; and **C,** veliger after torsion. (After Crofts, 1955. Proc. Zool. Soc. London **125:**711-750.)

tral portion of the lower hemisphere becomes the foot, and the dorsal part becomes the mantle. In gastropods there is a rotation of the visceral hump, termed torsion (Fig. 10-5), which brings the mantle cavity and gills to a forward position over the head and gives the viscera a characteristic twist. For a time the young larvae continue to swim, developing a pair of large, ciliated swimming lobes, collectively termed the velum, from the old prototroch. These lobes grow and gradually shift to a more dorsal position, forming a large bilobed funnellike velum with prominently ciliated margins. In this stage the larva is known as a **veliger.** Eventually the veliger settles down, the velum is resorbed, and the individual assumes the form characteristic for its species. Gastropods other than Archaeogastropoda tend to pass through the trochophore stage or its equivalent in the egg and to hatch as veligers. In the case of land snails and their derivative freshwater pond snails, the Pulmonata, the veliger stage, too, is passed through in the egg membranes, and the young snail hatches fully formed. Bivalves and scaphopods also hatch as a free-swimming form called a trochophore, which transforms into a bilaterally symmetrical veliger of brief duration, during which time the main organs of the adult are laid down.

In the case of Cephalopoda, the eggs are yolky and development is direct. However, in many instances young cephalopods are still small, still dependent on the yolk for nourishment when they are liberated from the egg, and spend some time in the plankton as a planktonic larva before the yolk is entirely absorbed. In the case of squids, the young remain pelagic, becoming part of the nekton, whereas the octopuses are mostly benthic.

BRYOZOA

Adult bryozoans are sessile benthic animals, but the peculiar **cyphonautes** larvae are characteristic members of the meroplankton in neritic waters. The cyphonautes are flat, triangular larvae encased in a rigid chitinous shell open at the oral end of the

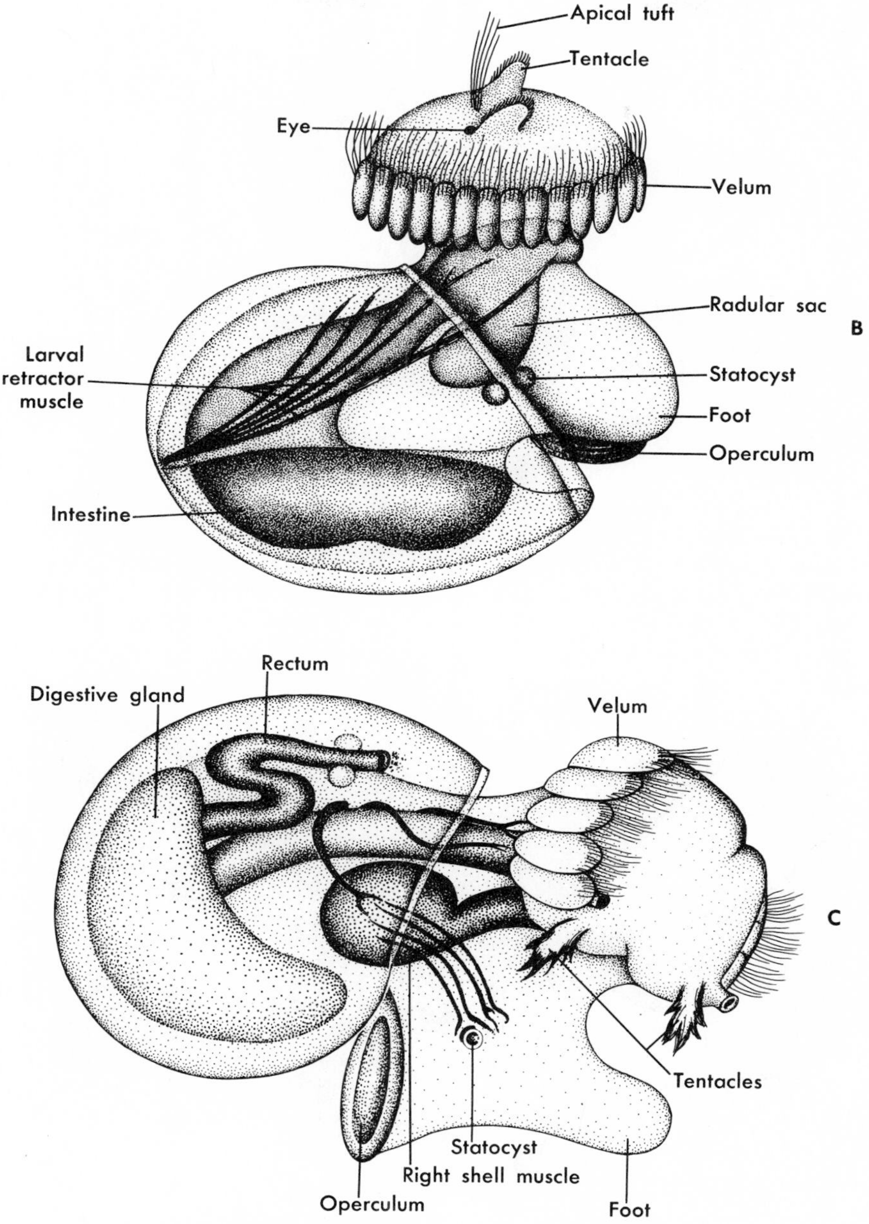

FIG. 10-5, cont'd. For legend see opposite page.

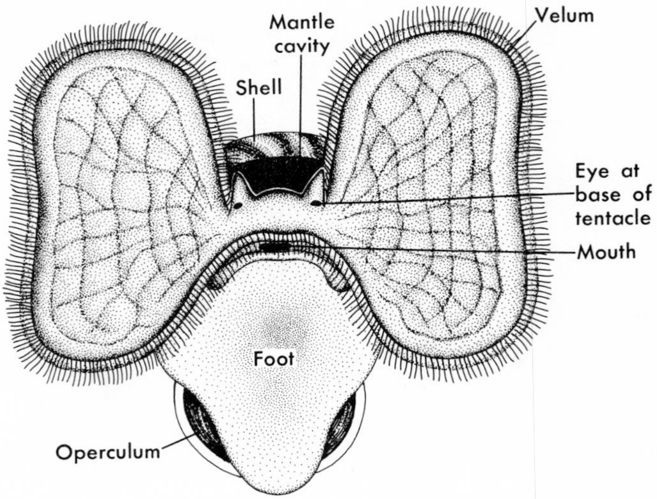

FIG. 10-6. *Littorina littorea,* ventral view, swimming stage veliger larva. (After Fretter and Graham. 1962. British Prosobranch Molluscs, Ray Society, London.)

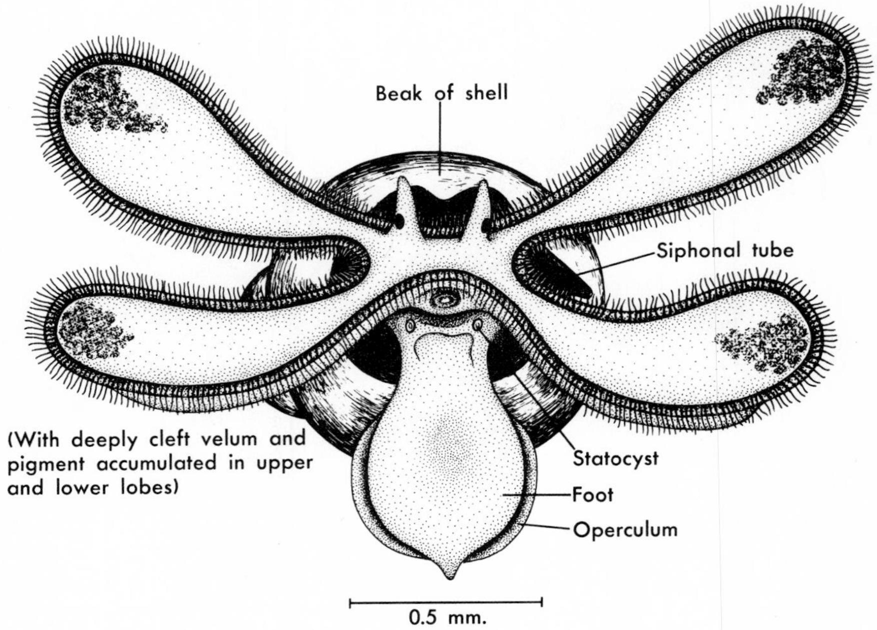

FIG. 10-7. *Nassarius* veliger. (After Fretter and Graham. 1962. British Prosobranch Molluscs, Ray Society, London.)

animal. It is thought by some to represent a highly modified trochophore.

BRACHIOPODA

Pelagic larvae of brachiopods are only rarely encountered in the plankton. This phylum, too, occurs in neritic waters. The adults are all fixed benthic forms.

PHORONIDA

These wormlike, tube-building, lophophore-bearing forms have a planktonic larva, the **actinotrocha,** which can be considered to be a modified trochophore in which the metatroch has developed into a series of tentacles.

ECHINODERMATA

Among the most bizarre and characteristic ciliated meroplanktonic larvae that superficially resemble trochophores are the planktonic echinoderm larvae. They are all derivable from the pluteus type of larva. *Pluteus* was a generic name proposed for certain larvae that were later found to be ophiuroids. Similarly structured larvae in

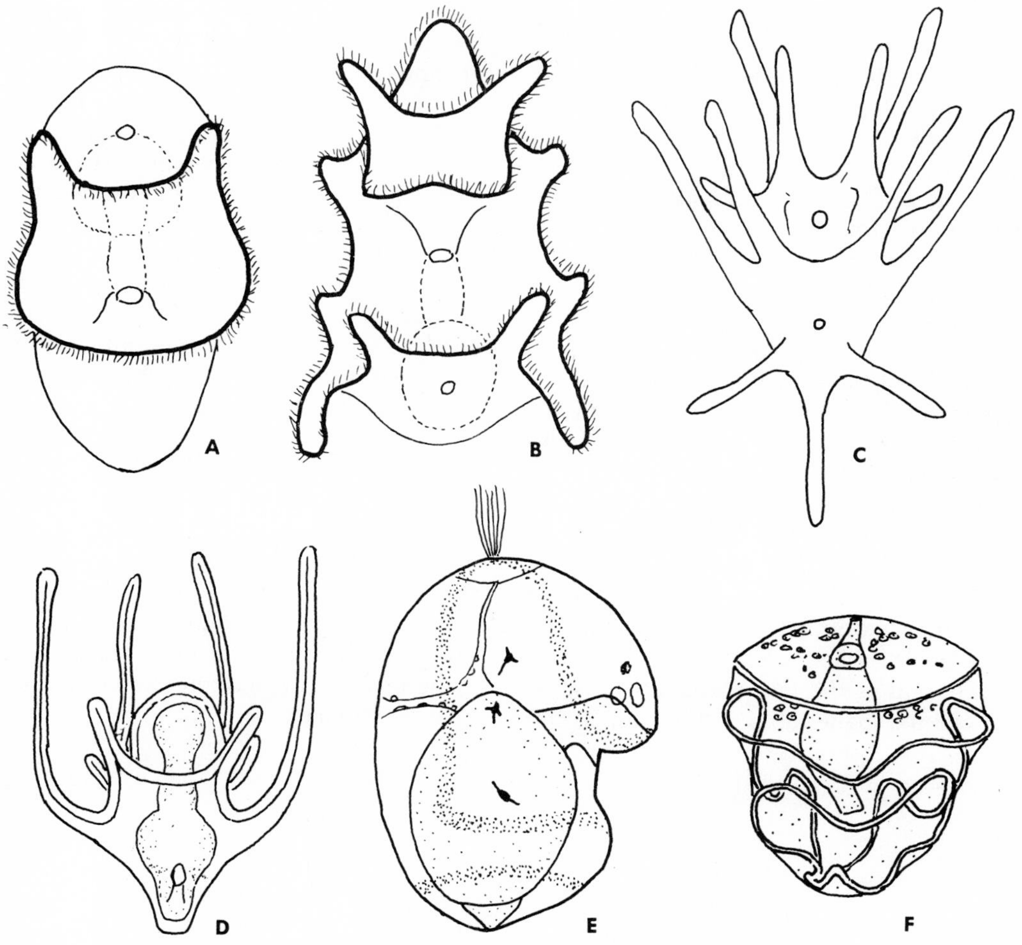

FIG. 10-8. Echinoderm and hemichordate larvae. **A,** Early bipinnaria stage. **B,** Brachiolaria stage. **C,** Spatangidae, echinopluteus larva. **D,** Ophiopluteus larva. **E,** Early tornaria larva of *Balanoglossus clavigerus* Delle Chiaje, showing general similarity to echinoderm larva. **F,** Late tornaria, *Balanoglossus clavigerus* Delle Chiaje. (**D** after Davis, C. C. 1955. Marine and freshwater plankton. The Michigan State University Press, East Lansing.)

other echinoderms and in certain lower chordates are referred to this general type, used now as a descriptive name for such larvae (Fig. 10-8).

The coelom in echinoderms and chordate larvae is enterocoelous in origin and undergoes a complex development in the case of echinoderms, thus differing greatly from the schizocoel developed in the trochophores of annelids and molluscs.

In its simplest early form the **pluteus** can be thought of as a ciliated larva with a complete digestive tract somewhat like that of a trochophore but bent ventrally so that both the oral and anal openings are toward the concave ventral side. A large ciliated band that may be thought of as corresponding to the metatroch passes essentially around the mouth, but is produced into lobes at the anterior and posterior lateral angles. In later stages these lobes are variously divided and extended to form larval arms, and in asteroids there is a meeting and fusion across and in front of the anterior lateral lobes in such a manner as to cut off a circle of the ciliated band in front of the mouth, separating it from the remainder of the band.

The plutei of ophiuroids and echinoids are rather typical and are termed **ophiopluteus** and **echinopluteus,** respectively. They differ chiefly in details of body proportions and in the number, branching, and orientation of the larval arms. Larvae of holothurians usually have earlike lobes rather than extended larval arms and are called **auricularias,** whereas the larvae of asteroids, after the cutting off of the separate ciliated band above the mouth, are termed **bipinnarias.** The auricularian larva often changes to form a so-called "pupa," or **doliolarian larva,** before leaving the plankton. The asteroidal larva changes to form the **brachiolarian larva,** which develops additional long tentacles, not involving the ciliated band, that later function to provide temporary attachment to the substrate while the larva undergoes metamorphosis.

The larval arms of plutei are not muscular and are used only indirectly for locomotion through the activity of the cilia on the portions of the ciliated bands that traverse them. They seem to be mainly organs of flotation, and are supported by rigid skeletal spicules that would make movement impossible.

Crinoids are mostly deep-sea animals, and they have direct development. Some of the shallower-water tropical forms such as *Antedon* produce a short-lived doliolarian type of larva that has only a blind buccal depression representing the mouth and lives off the yolk of the egg during its few days of swimming existence. It then attaches itself by a small adhesive pit near the apical tuft of cilia and metamorphoses into a stalked sessile form, about 3 mm. long, known as the pentacrinoid stage.

As one goes from the tropics toward the high latitudes, a larger proportion of the echinoderms have yolkier eggs, pass through a greater part of their development in the eggs, and have more abbreviated, less fully developed planktonic larval stages, some of them developing directly, with no planktonic larvae.

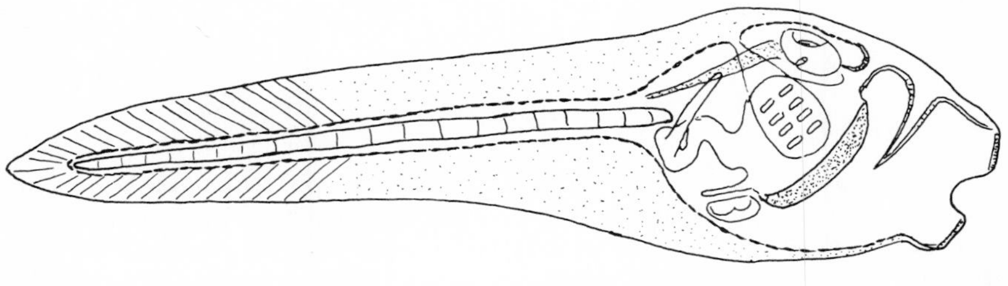

FIG. 10-9. Free-swimming tunicate larva, *Clavelina.*

CHORDATA

Hemichordata

The wormlike enteropneustans, or balanoglossids, have a ciliated, pluteus type of larva, the **tornaria,** named after the old larval genus *Tornaria* (Fig. 10-8, *F*), to which unidentified tornaria are sitll referred. They are so similar to the bipinnaria larvae of starfish that they have been mistaken for them at times. The similarities in structure and development of the larvae provide the strongest arguments for postulating a relationship between the echinoderms and the chordates. The tornaria even produce such characteristic developmental echinoderm features as a hydrocoel, a pore canal, and a water pore on the left side. In some of the tornaria the ciliated band becomes more complicated than that in echinoderm larvae, but it remains essentially a band around the mouth.

Urochordata

The larvae of the tunicates, or sessile Urochordata, are similar to the appendicularians (already described under holoplanktonic forms), except for the lack of gonads and an anus, a more developed tunic, a slightly less differentiated nervous system, and a less sharp differentiation of head from tail (Fig. 10-9). The "tadpole" stage of tunicates is a briefly planktonic nonfeeding stage that soon settles down; undergoes regression and absorption of the tail and notochord, a great enlargement and increase in the complexity of the pharynx, and an enlargement of the atrium and the enveloping tunic; and becomes a typical sessile tunicate. Tunicates, with their well-developed larval notochard, seem much more closely related to the higher chordates, the Cephalochordata and vetebrates, than to the Hemichordata, and their larvae do not show any great resemblance to those of echinoderms.

Cephalochordata

The Cephalochordata are benthic in habit but are capable of swimming, and young ones are occasionally taken in the hypoplankton near the bottom, especially at night.

Vertebrata

Many marine fish shed floating planktonic eggs. Young fry of both these and many species with sessile or demersal eggs also spend some time in the plankton. The eggs and young fry of many fish are difficult to recognize in regard to species, but as they grow older the fry become more and more recognizable and begin to behave as part of the nekton. Few older fish continue to be planktonic in habit, possible exceptions being the huge *Mola mola,* or ocean sunfish, which has such limited powers of swimming that it may be considered planktonic (Fig. 10-10, *C*). If the ocean sunfish, the basking shark, and the huge arctic jellyfish *Cyanea arctica* are considered to be part of the plankton, they are by far the largest of the plankters. Perhaps the sargassum fish may also be considered planktonic.

ARTHROPODA (NONCILIATED PLANKTONIC LARVAE)

The larval stages of innumerable crustaceans, both holoplanktonic types such as copepods, and benthic types, form a large and characteristic portion of the plankton. Most malacostracans pass through the nauplius stage within the egg membranes and hatch at a later stage. Most of the so-called Entomostraca, as well as a few Malacostraca, hatch as nauplius larvae. **Nauplii** are commonly more or less oval and have a single median eye and three pairs of appendages, at least in their first instar. The appendages represent the first and second antennae and the mandibles, all of which are pediform in this stage. With successive molts, additional appendages are added, beginning with the more anterior ones. Later nauplius-life stages, but with more than three pairs of appendages, are sometimes termed metanauplii. The nauplii of the Cirripedia are distinguished by the fact that the anterolateral corners of the carapace are drawn out into lateral spines.

Postnaupliar stages are greatly varied in

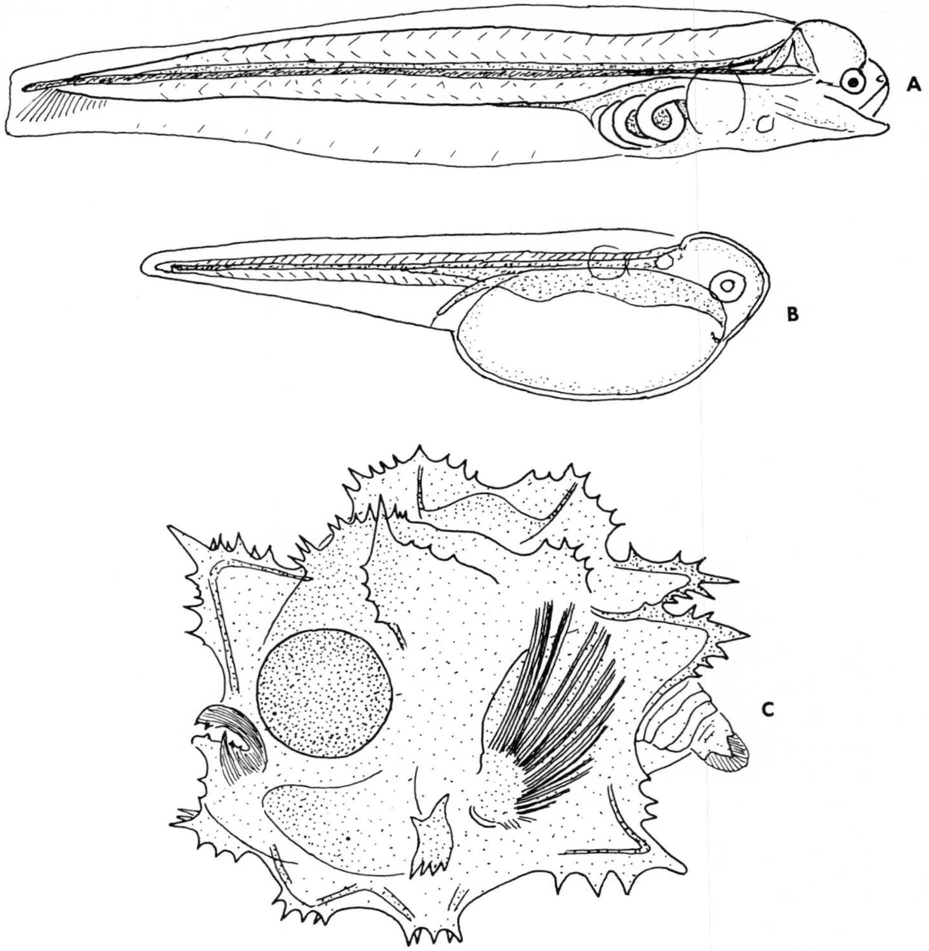

FIG. 10-10. A, Planktonic larva of *Pleuronectes platessa* (15 mm.). **B,** *Pleuronectes platessa,* just hatched (6 mm.). **C,** Larva of *Mola mola,* newly hatched (1 mm.). (**C** after Davis, C. C. 1955. Marine and freshwater plankton. The Michigan State University Press, East Lansing.)

the different crustacean groups. Some of the more commonly encountered ones will be mentioned.

Copepoda

Copepods pass through six instars as **nauplii** (Fig. 10-11), after which they transform into **copepodids.** These look and behave much like adult copepods but are sexually immature and may differ in details of the number and exact form of appendages, and of the apparent body segmentation. The sixth instar, after becoming a copepodid, is the adult stage.

Cirripedia

Swarms of larval barnacles are, at times, a conspicuous element in neritic plankton. The characteristic **nauplii** are readily recognized by the anterolateral spines of the carapace. After a few nauplier instars, they transform into a different-appearing larva, the **cypris** larva, which resembles a small ostracod, having the carapace developed

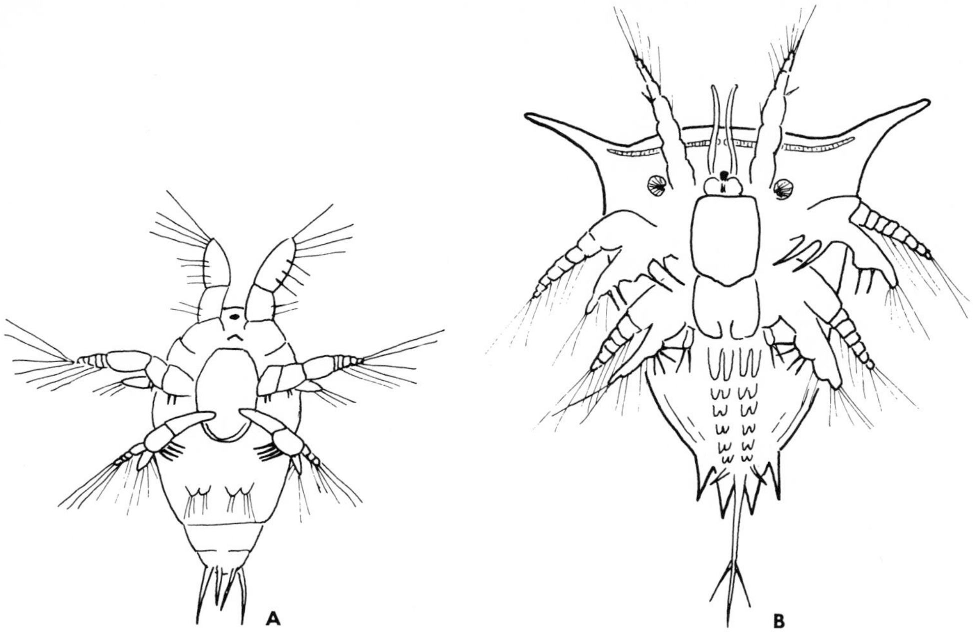

FIG. 10-11. A, Copepod nauplius larva *(Eurytemora hirundoides)* (fourth instar). Ventral view showing the three pairs of swimming appendages (uniramous antennules, biramous antennae, pediform mandibles) and the single naupliar eye, characteristic of nauplii. **B,** Late nauplius (metanauplius) of a barnacle, *Balanus*. Characteristic lateral extensions of forepart of carapace as "frontal horns" identify nauplii of barnacles. Note presence of both paired compound eyes and the unpaired nauplius eye.

into a bivalved shell enveloping the body. The cypris larva typically has both compound eyes and a median nauplious eye. The antennules have four segments, and a disc at the end of the second segment on which the duct of the cement gland opens. The cement glands occupy much of the anterior end of the larva. The second antennae have disappeared, and the mandibles and maxillae are rudimentary. Posteriorly, there are six pairs of biramous, setose swimming appendages. A short abdomen ends in a caudal furca. The mouth lies behind the middle of the ventral region and is directed posteriorly. Two strong apodemes project into the body at the base of the antennae for attachment of the antennary muscles. A transverse adductor muscle closes the valve of the shell. After a very short period of rapidly swimming and crawling about on the substrate, the cypris selects a suitable spot and glues itself firmly to it with its antennules by means of the secretion from the cement glands. It then undergoes a remarkable metamorphosis to become an adult barnacle.

Malacostraca

Euphausiacea. Most euphausiids simply shed their eggs into the water. A few hold them temporarily attached to the thoracic legs. The egg hatches into a **nauplius.** Afterward there is a complicated series of changes. After three or four naupliar instars the larva changes to a **calyptopis** larva (Fig. 10-13), which has three pairs of appendages behind the mandibles, and in which the thoracic segments are clearly differentiated and the telson is more fully developed. The calyptopis instar also has a pair of uropods. Soon the calyptopis undergoes another marked change, becom-

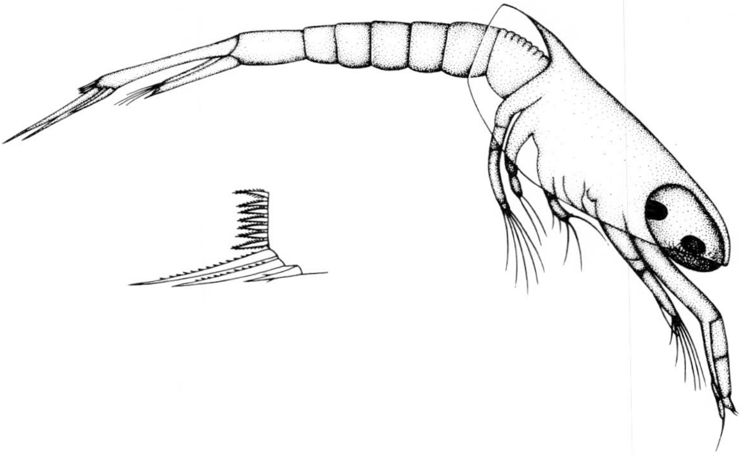

FIG. 10-12. Euphausiid protozoea and detail of spination of telson.

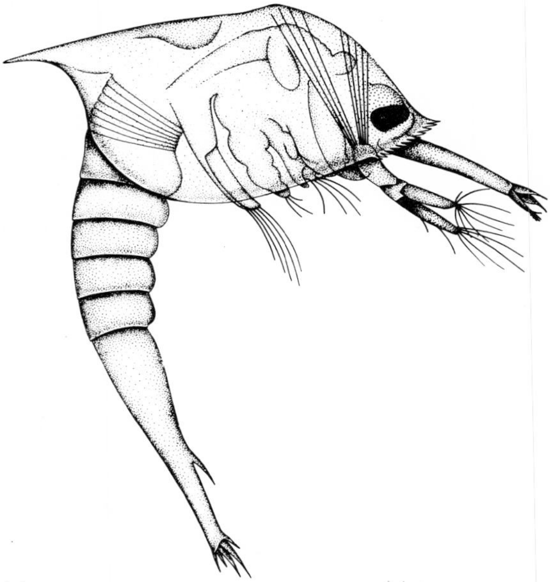

FIG. 10-13. *Euphausia brevis* protozoea (calyptopis).

ing a **furcilia** (Figs. 10-14 to 10-16). The furcilia has stalked eyes and more thoracic and abdominal appendages. The young euphausiid in a postlarval stage is called a **cyrtopia.** The changes from cyrtopia to adult are so gradual and small that it seems proper to consider the cyrtopia a postlarval rather than a larval stage.

Stomatopoda. The eggs of stomatopods hatch at a more advanced stage than the nauplius. Two types of larvae hatch: from *Squilla* (Fig. 10-17) and related genera, a **pseudozoea;** from *Lysiosquilla* and other genera, an **antizoea.** They are not comparable to or derived from the zoea stage of decapod crustaceans.

The carapace of the pseudozoea is produced into a rostrum over the head and bears two anterolateral spines and two posterolateral spines. Only two thoracic appendages are present. Both are uniramous, and the second is large and raptorial. The abdomen bears five pairs of biramous appendages.

The antizoea has a large carapace covering much of the body but attached only at the head. It bears a large rostrum and a pair of large posterolateral spines, in addition to which there may be other, smaller spines. The thorax and abdomen are not sharply differentiated, but there are five pairs of simple biramous thoracic appendages.

All antizoeae and some pseudozoeae metamorphose into a second larval stage known as the **erichthus.** The rest of the pseudozoeae become **alima** larvae. The erichthus and alima stages are much more alike than the pseudozoea and antizoea stages.

Decapoda

Most benthic and nektonic marine decapods have planktonic larvae. These larvae are of great variety and are so numerous that they constitute an ecologically important fraction of the zooplankton. Many of them were discovered and given generic and specific names before their relationships were known. Some of these names survive as descriptive names for various stages in the life cycle. Nonuniformity in usage of such names has led to some confusion in earlier literature in the designation of stages of larval decapods. Gurney (1942) has done much to clarify relationships. He divided the life cycle into **naupliar, protozoean, zoeal,** and **postlarval** stages. In many instances there is no marked change between the protozoean and the zoeal stages, and it seems preferable simply to number these stages without attempting to make a hard-and-fast distinction between them. As a rule, the first zoeal stage does not yet have the thoracic legs

Text continued on p. 194.

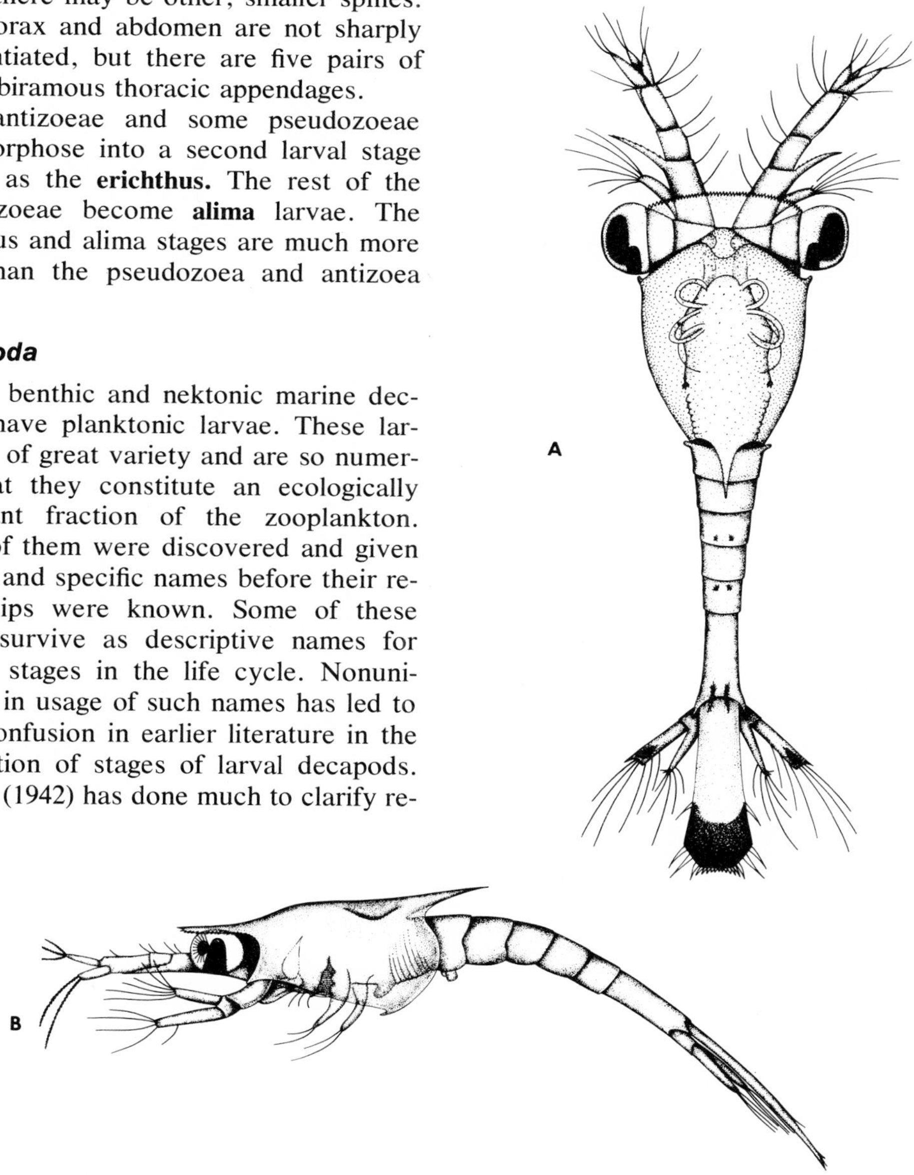

FIG. 10-14. *Euphausia brevis* furcilia. **A,** Dorsal view. **B,** Lateral view.

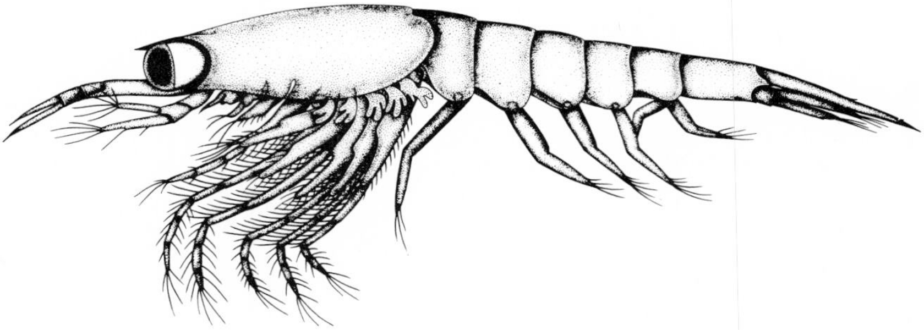

FIG. 10-15. *Euphausia superba* furcilia.

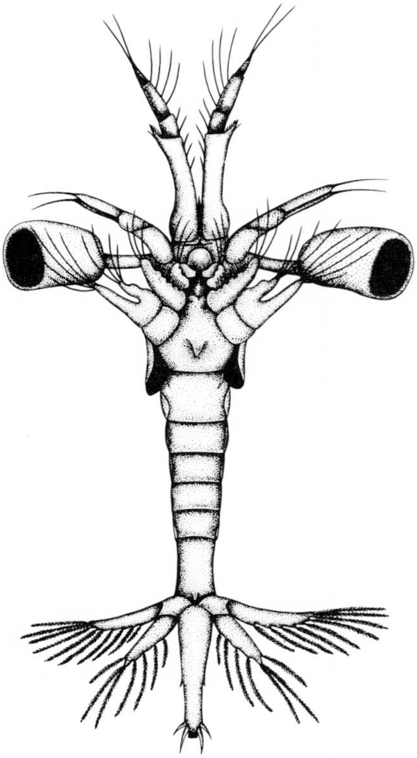

FIG. 10-16. Euphausiid furcilia with stalked eyes.

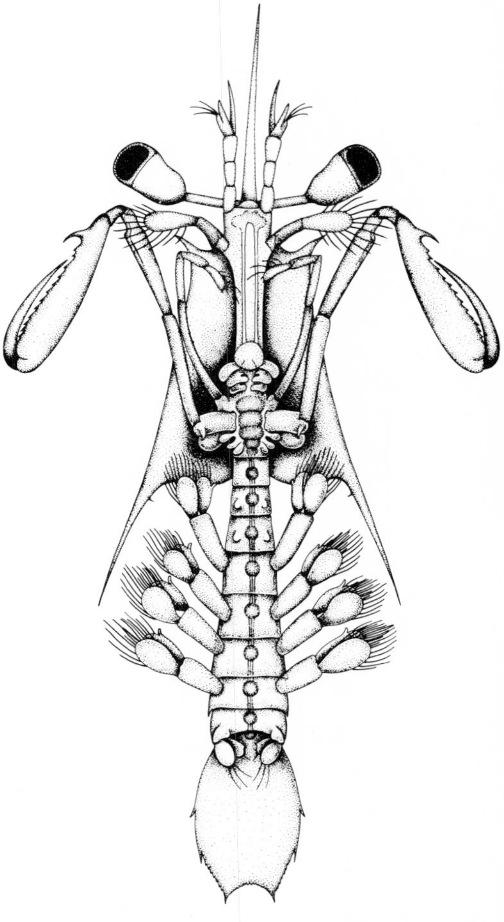

FIG. 10-17. *Squilla mantis* pseudozoea larva.

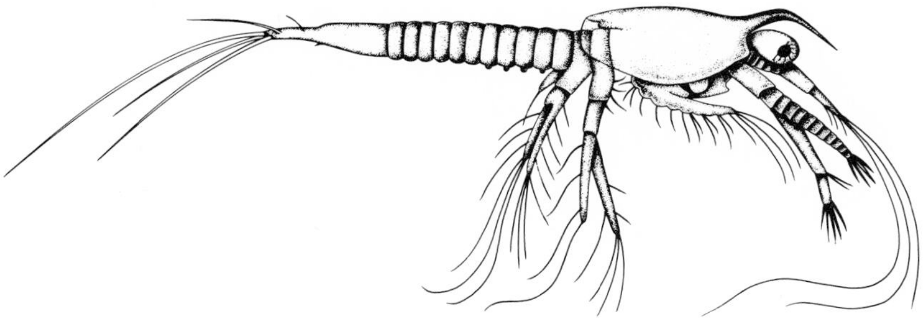

FIG. 10-18. Shrimp larva, *Gennadas* protozoea.

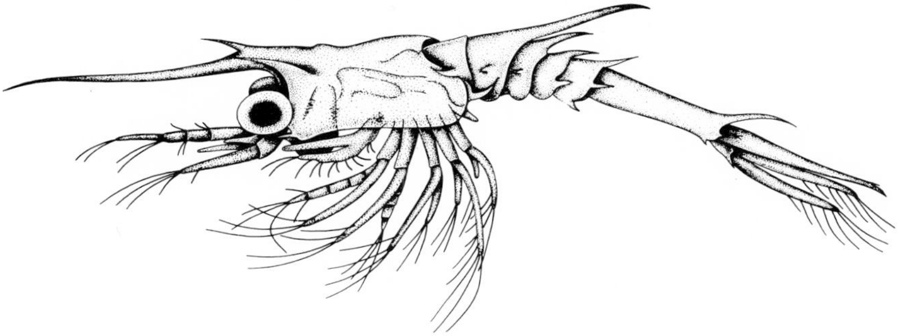

FIG. 10-19. Shrimp larva, *Gennadas* zoea.

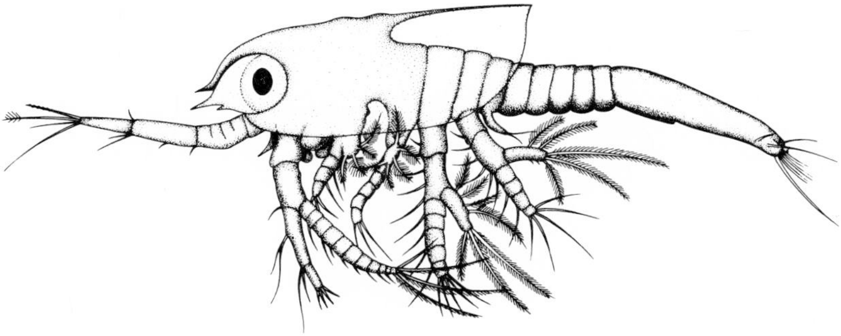

FIG. 10-20. Shrimp larva, *Penaeopsis* protozoea.

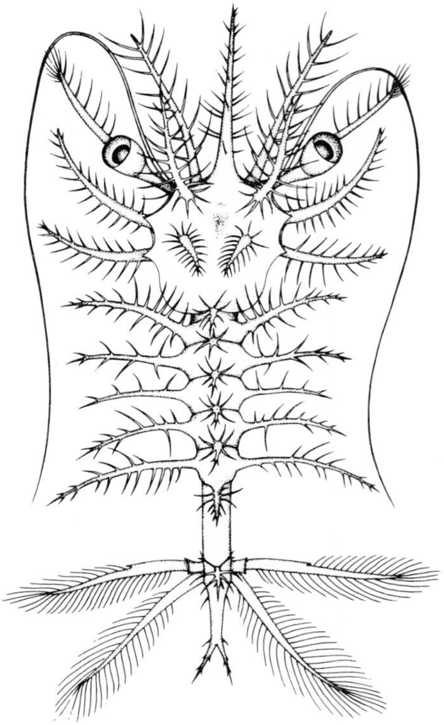

FIG. 10-21. Shrimp larva, acanthosoma.

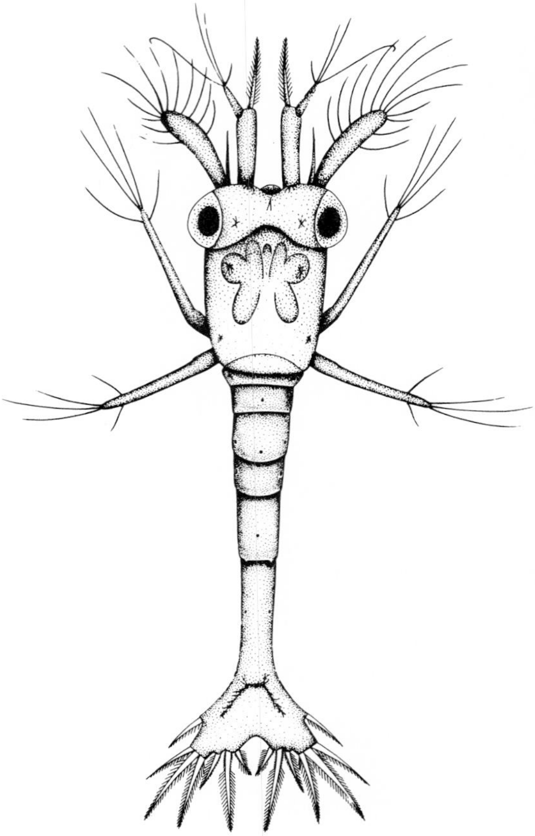

FIG. 10-22. *Processa edulis,* protozoea stage.

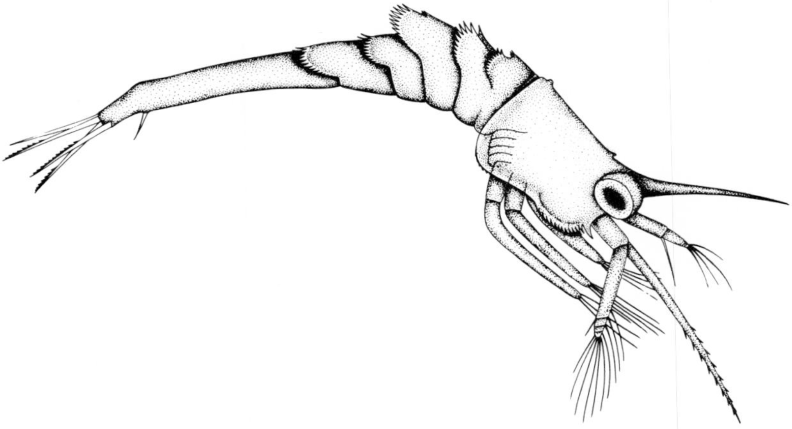

FIG. 10-23. Caridean shrimp. *Pandalus steriolepis* protozoea stage 1.

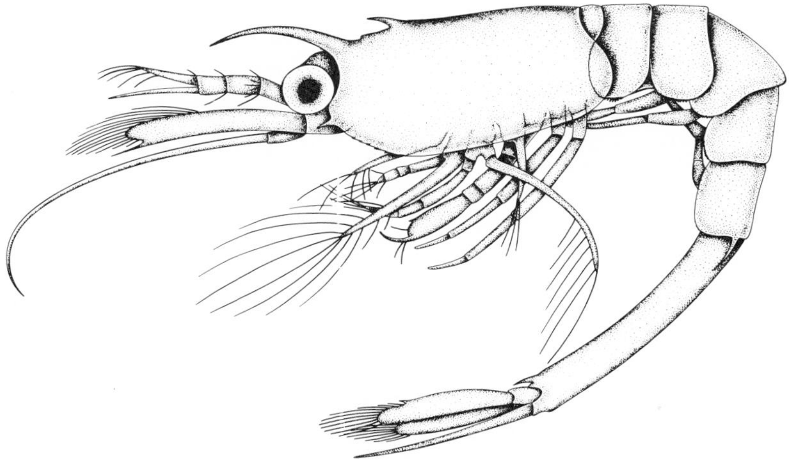

FIG. 10-24. Caridean shrimp larva, *Crangon antarctias,* side view.

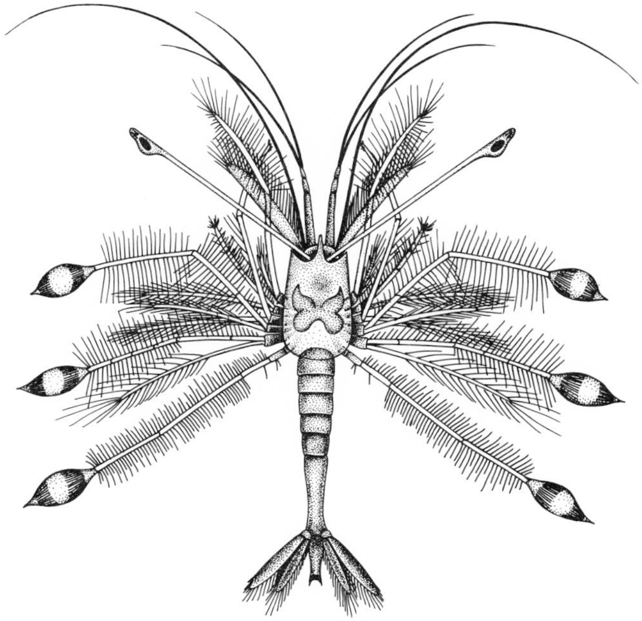

FIG. 10-25. Caridean shrimp, eretmocaris larva of Hippolyaceae.

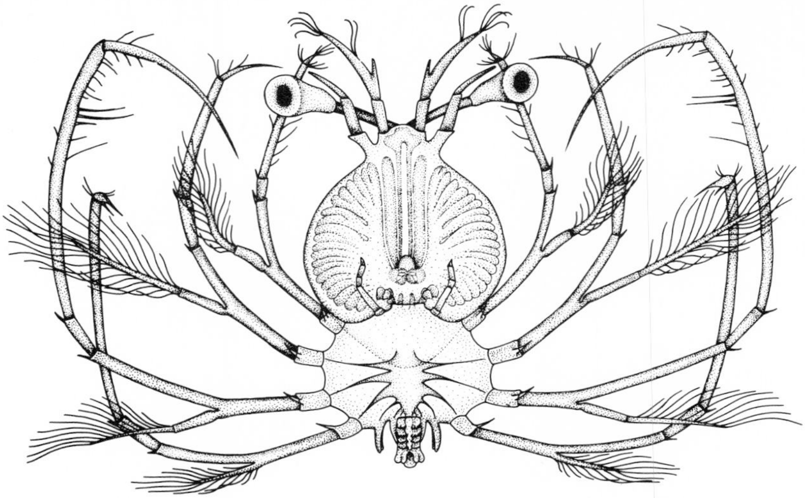

FIG. 10-26. Palinurid (spiny lobster) phyllosoma larva of *Palinurus vulgaris*.

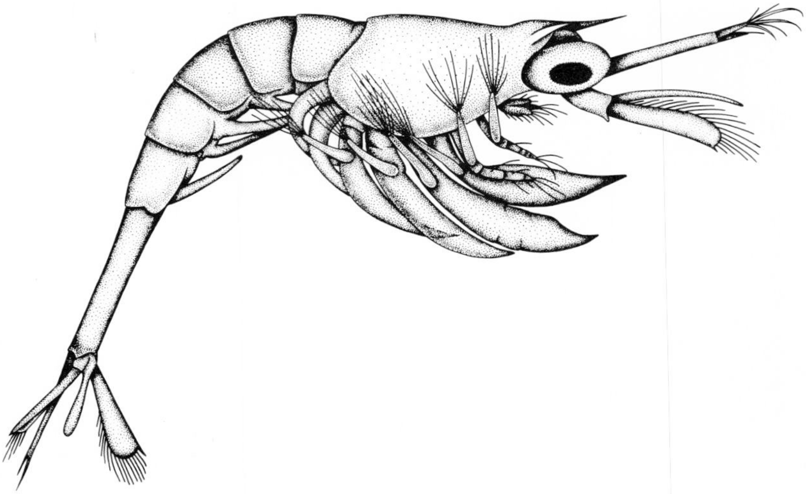

FIG. 10-27. Anomuran larva, *Upogebia deltaura*.

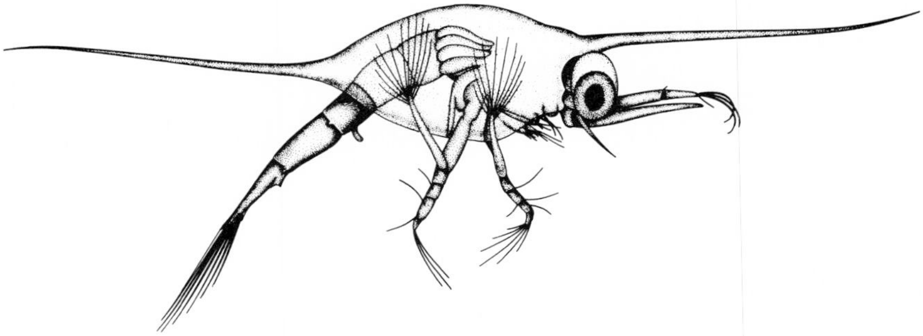

FIG. 10-28. Anomuran larva, the flat crab, *Porcellana* zoea.

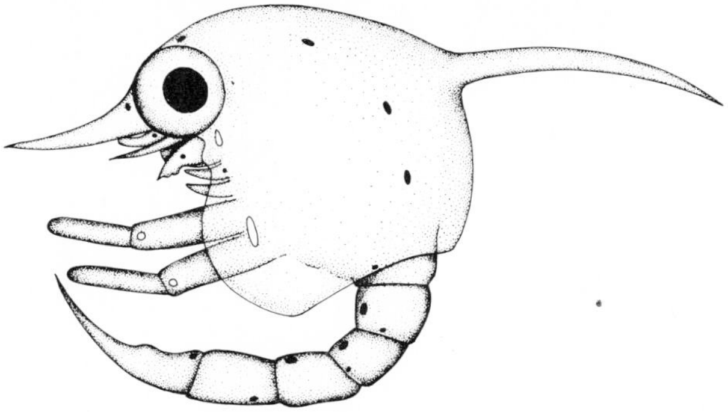

FIG. 10-29. Brachyura, true crab, *Carcinides maenas,* first-stage zoea.

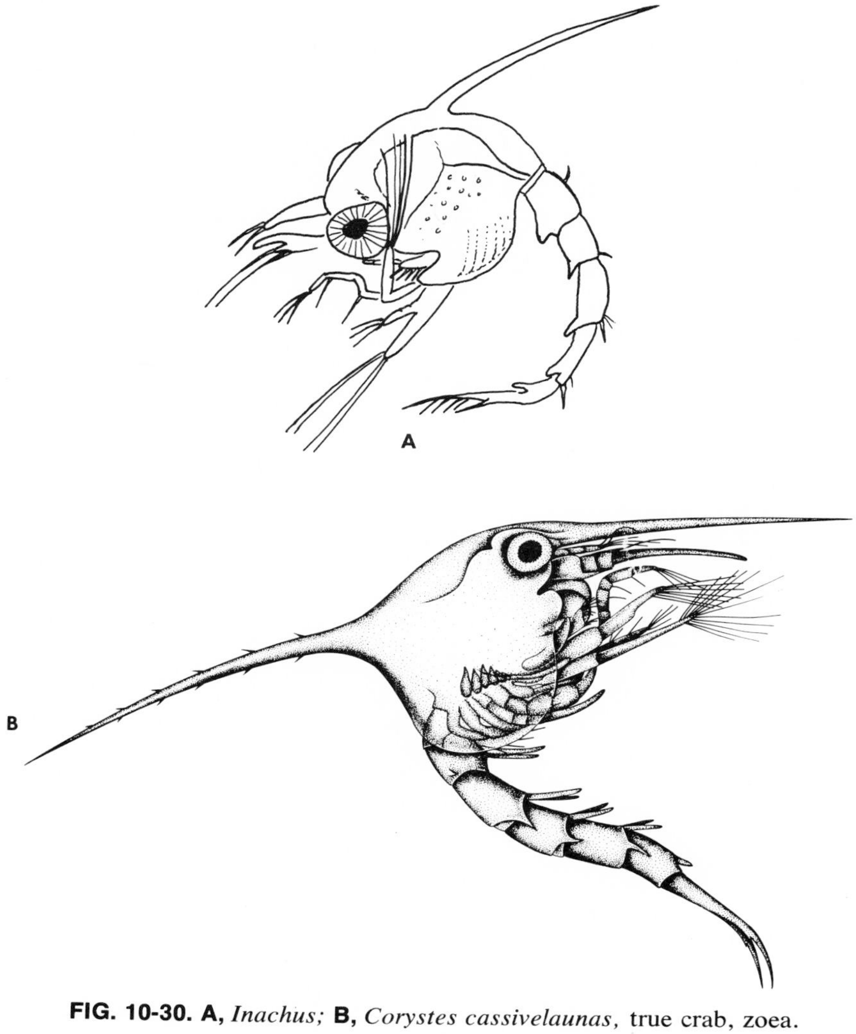

FIG. 10-30. A, *Inachus;* **B,** *Corystes cassivelaunas,* true crab, zoea.

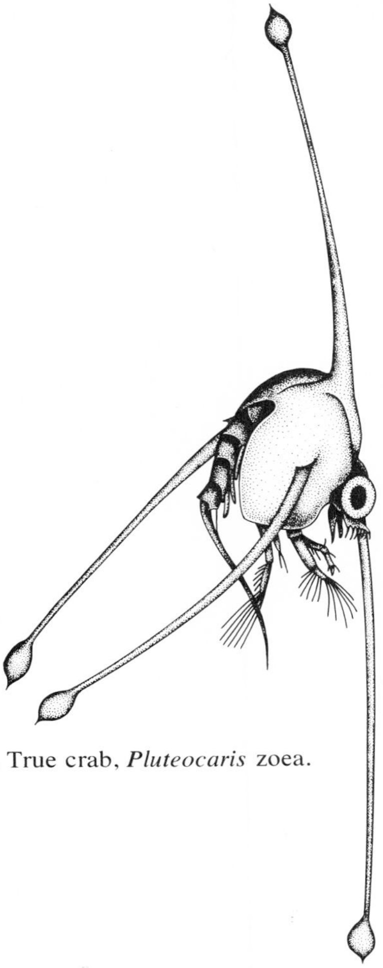

FIG. 10-31. True crab, *Pluteocaris* zoea.

developed, and often the spines and rostrum of the carapace are weakly developed or lacking. Such larvae would be called protozoeae in Gurney's scheme. After the zoeal stages there is a sudden metamorphosis, and the zoea is transformed into a postlarva in which the appendages have, in general, adult characteristics. Subsequent molts result in a more or less gradual metamorphosis into the adult body form. In the Brachyura the postlarva is termed a **megalops.** It has the general appearance of a crab but still possesses a well-formed abdomen not tightly folded beneath the cephalothorax, as in the adult. The postlarvae of Macrura and Anomura are more varied and still go under a variety of names for various types, such as **mysis stage** (an undesirable term, since it implies a relationship that does not exist), **eretmocaris,** with very long, sometimes segmented eyestalks (a convenient designation of certain postlarvae, but it does not represent a particular taxonomic entity), and **phyllosoma** (larvae of Palinuraceae).

FIG. 11-1. Mixed herd of herbivorous fish, tangs, and parrot fish, picking over rocky substrate off Guam. These fishes have been shown by Bakus to be important in preventing establishment of beds of larger marine algae in many tropical areas. (Chesher photograph.)

CHAPTER

11 THE NEKTON

There is no sharp line between pelagic animals regarded as plankton and those constituting the nekton. Most free-swimming invertebrates are generally treated as part of the plankton, even though many of them, such as many crustaceans, arrowworms, and others, have powers of swimming equal to those of some of the animals usually treated as part of the nekton. Many members of the nekton also exist as plankton during early stages of their life.

Two groups of animals—the cephalopod molluscs and the fish—have, however, evolved numerous large, actively swimming species that constitute the bulk of the nekton as generally understood.

Nor is there any sharp division between the nekton and the benthos. Demersal fish and octopods, for example, may just as well be treated as part of the benthic community. They are included here with the nekton because many of them do swim about extensively and because their closest relatives constitute the bulk of the nekton.

Cephalopods

The class Cephalopoda of today represents a remnant of an ancient and once-dominant group of molluscs that differentiated early from primitive molluscan stocks; became immensely abundant and diversified during the late Paleozoic and Mesozoic eras, numbering more than 10,000 species; and then dwindled in numbers of species to the approximately 600 living today. Some of the major groups became wholly extinct, whereas others have only a few living representatives. However, several of the recent cephalopods are successful, being enormously abundant and widely distributed, and it seems likely that the octopods are at present undergoing active evolutionary diversification.

In contrast with the sedentary benthic habit of most members of the phylum, cephalopods evolved as active carnivores and scavengers. The earliest cephalopods were benthic, retaining the primitive bilateral symmetry of presumed molluscan ancestors with a posterior mantle cavity but differing in the great growth, dorsally, of the visceral hump, so that the dorsoventral axis became the functional longitudinal body axis. The greatly stretched-out anterior and posterior surfaces now serve as the functional dorsal and ventral sides, respectively. The foot became crowded forward and more or less fused with the head, forming an oral crown of long tentacles around the mouth. A portion of the foot became modified as a ventral siphon through which exhalant currents from the mantle cavity are expelled. The tentacles developed suckers and in some cases chitinous hooks as well. The elongated, now "ventral" mantle cavity contained the paired gills—two pairs in the more primitive cephalopods, one pair in most recent species—and water circulation was effected by means of muscular movements of the mantle tissue. Water is drawn in ventrolaterally along the entire open edge of the mantle cavity, and this slitlike orifice is closed and

FIG. 11-2. Octopus captures crab by surrounding it with tentacles and the connecting web. (M. W. Williams photograph.)

the water expelled through the siphon. The mantle tissue lost the cilia found in the mantle tissue of many molluscs, and became increasingly muscular, with sufficient power that the exhalant jets from the siphon were used for locomotion more effectively than the tentacles, which pulled the animal over the substrate.

Early cephalopods, like most other molluscs, had large external shells, which in many species became variously coiled, and as the animal grew, divided into successive chambers. The cephalopod lived in the last and largest chamber, the posterior chambers being filled with gas to offset their weight, and all chambers were connected to the posterior part of the body of the cephalopod by the membranous tube, the siphuncle. The four living species of *Nautilus* are the only living representatives of the once-great group of Tetrabranchiata, or four-gilled, shelled cephalopods. They are also primitive in having numerous isopodous tentacles without suckers, imperfectly developed eyes without lenses, no ink sac, four auricles, no branchial hearts, and other features setting them apart from most modern cephalopods.

FIG. 11-3. A young specimen of a common southern octopus, *Octopus bimaculoides,* crawling across the aquarium glass. (M. W. Williams photograph.)

FIG. 11-4. *Benthoctopus hokkaidensis,* small, purple species from about a 100-meter depth off the Oregon coast.

Most recent cephalopods have much reduced shells, completely buried in the dorsal mantle tissue, or none at all. The mantle is much more muscular than that of nautiloids, having developed into an efficient swimming organ. The nervous system and sense organs are exquisitely developed: the cerebral, pleural, pedal, propedal, and visceral ganglia are fused together in what amounts to a complex brain encased in a cartilaginous "skull." The remarkable cephalopod eye, analogous in every respect to the vertebrate eye—with lens, anterior and posterior chambers, retina, and optic tract—has excited the admiration and wonder of biologists over the years. That such a remarkable structure as the camera eye should have evolved independently twice, in such different evolutionary lines as the molluscs and the vertebrates, is one of the most extraordinary of evolutionary coincidences or cases of convergent evolution. In some ways the cephalopod eye seems even more logically structured than that of the vertebrate. The sensory cells are in front of, instead of behind the nerves that connect them to the optic tract, and there is no blind spot in the retina for the passage of the optic nerve.

Many cephalopods have chromatophores of two or three different colors under nervous control, enabling them to effect the most remarkable and rapid changes. When the animal is observed closely, a constant play of color change is seen, even though the overall average effect may remain the same. This play of color in the skin may continue for several hours after the death of the cephalopod.

The mouth of cephalopods is equipped with a horny beak somewhat resembling that of a parrot. A radula is also present. Large salivary glands are associated with the mouth, and the bite of some species, at least, is poisonous and sometimes fatal. The general fear of octopuses on the part of the public, however, is not justified, since they are secretive and will avoid people whenever possible. Even when handled, they seldom bite. I have caught hundreds of octopuses bare-handed, ranging in size from small, immature specimens weighing less than a gram to large ones weighing over 80 pounds, without ever having been bitten. Squid seem to show more of a tendency to bite if given the opportunity when captured. However, they are completely helpless out of the water and can be handled safely if one is careful to keep his fingers away from the mouth.

The characteristic defensive mechanism

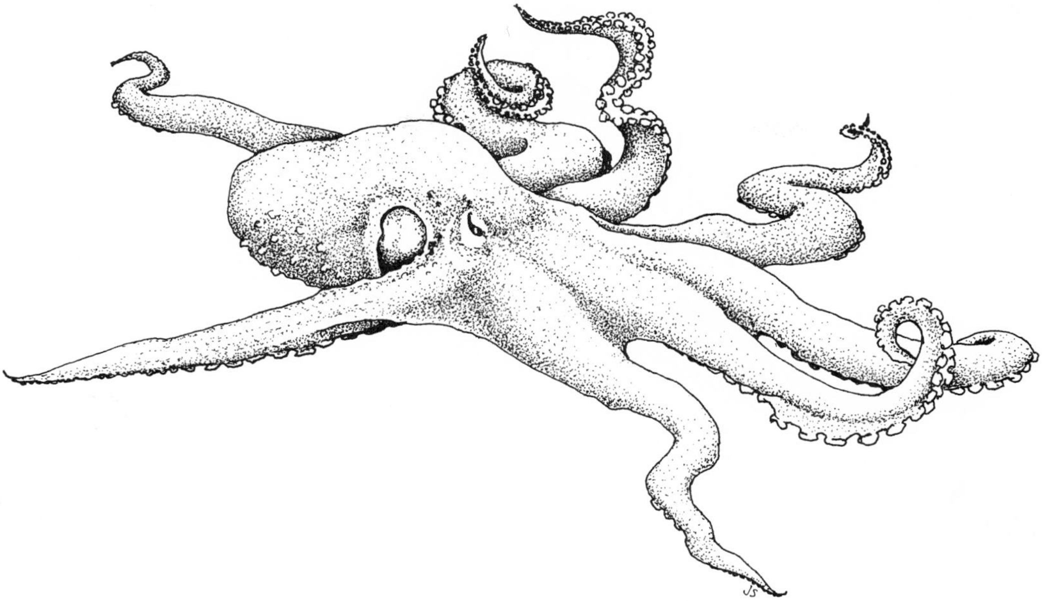

FIG. 11-5. *Octopus dofleini.*

of most cephalopods is a jet of ink from an ink sac lying on, or embedded in, the ventral surface of the liver and discharging along with a jet of water through the siphon. Most people regard this action as primarily setting up a smoke screen behind which the cephalopod can escape. More often the ink probably serves as a dummy in the water to confuse the predator, the small blobs of ink in the water usually being not much different in size and shape from the cephalopod that made them, and often retaining their shape for several seconds. The rich brown ink of cuttlefish, sepia, has long been used by artists. Some deep-water bathypelagic squid have luminescent ink, rather than the black or deep brown ink of their relatives living in well-lighted waters.

Many squid, particularly various oegopsids living in deeper-water strata, have a large posterior coelomic space filled with a jellylike material that serves as a hydrostatic organ in much the same manner as do the large containers of gasoline of the bathyscaphe. This space has an advantage over the swim bladder of fish in that it does not change markedly in volume with rapid changes of water level. Such cephalopods often characteristically hang in the water head downward when at rest.

A curious and characteristic modification connected with the mode of locomotion is a pair of flaps, or valves, on the end of the intestine, which opens in the mantle cavity back of the siphon. When the mantle contracts suddenly, creating the considerable pressure in the mantle cavity necessary for jet propulsion, these flaps close the orifice, thus preventing the cephalopod from administering a strong enema to itself whenever it goes somewhere in a hurry.

Some of the squid rival or excell most fish in the speed and grace of their swimming. They can direct the siphon either way, swimming forward or backward. Some of them have even been observed to jump from the water and glide considerable distances in the air, much in the manner of flying fish.

The courtship and mating of cephalopods are also distinctive and varied. Some species exhibit peculiar courtship activities involving rapid changes in color. The

A

B

FIG. 11-6. A, Squid over patch reef, Florida Keys. **B,** Same squid as seen 32 seconds later. Note color and pattern change. (Chesher photographs.)

sperm of the male are contained in large spermatophores, which the male transfers to the mantle cavity of the female with one of his arms, which is specially modified for the purpose. In *Argonauta* the end of the male arm, including several suckers holding the spermatophores, is detached and left in the mantle cavity of the female. This was mistaken for a peculiar, many-suckered parasitic worm by Cuvier and given the generic name *Hectocotylus*. Since then the modified male arm has been known as the hectocotylized arm and the modified tip of the arm as the hectocotylus. Most cephalopods do not leave part of the arm in the female mantle cavity but simply use it to transfer the spermatophore.

The spermatophore itself is a remarkable structure. One must watch the discharge of a spermatophore with a stereoscopic microscope to appreciate its complexity and activity. A ripe spermatophore will usually discharge if placed in a little seawater on a slide. It consists of a swollen, saclike portion containing the sperm, and a long, tubular, necklike portion with a distal cap and filament. Within the neck is a complex, spirally wound, springlike mechanism, which, when the spermatophore is exposed to seawater, suddenly releases and begins

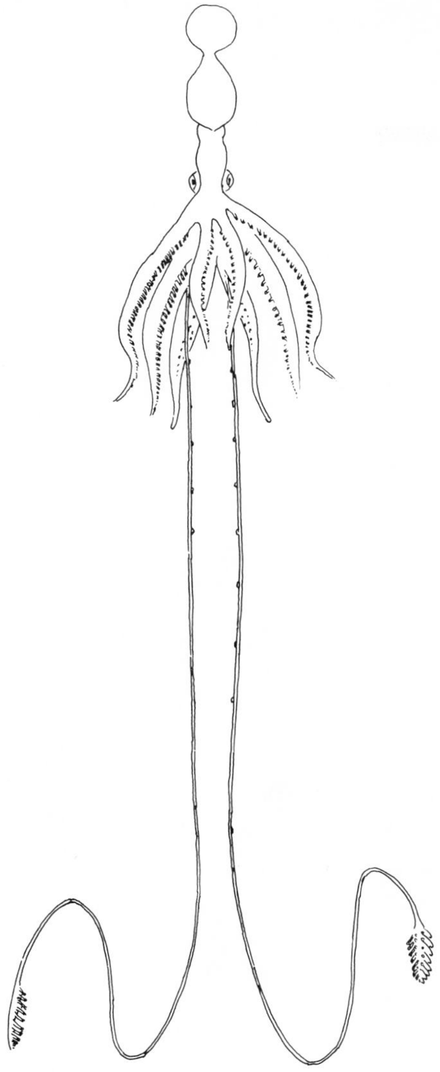

FIG. 11-7. *Chiroteuthis veranyi.* A deep-sea squid that hangs head downward in the water, extending the long arms below it. The arms are equipped near the ends with luminescent organs, sharp hooks, and suckers. Prey, attracted to the lights, is snagged by sudden jerks of the tentacles and then brought to the mouth.

to uncoil stretching out the tubular portion of the spermatophore to a much greater length. The sperm become active and come boiling up the tube as if under pressure, to escape from the end of the tube or from ruptures along the way. In some cephalopods—for example, *Rossia*—the end of the spermatophore tube, as it elongates, penetrates the tissues of the female, inside the mantle cavity, and then blows up like a small balloon, anchoring the spermatophore firmly in place so that it looks like a parasitic worm with its head buried in the tissue. In large octopuses the spermatophores may be over a foot long.

Some of the gregarious schooling squid such as *Loligo* may occur in immense numbers and constitute a valuable fishery. The muscular mantle tissue and parts of the arms are excellent eating. Squid are ordinarily not especially numerous in northern waters, but in 1958 great numbers appeared in the fjords of Norway, resulting in a commercial catch of 8500 tons, valued at half a million dollars.

Some of the giant squid such as *Architeuthis* and *Ommastrephes* constitute a principal food of the sperm whale, which is itself so abundant that it provides an annual catch of about 18,000 animals. The fact that these whales are voracious predators with rapid digestion indicates that giant squid must be far more abundant than had formerly been supposed on the basis of infrequent sightings or capture and occasional dead ones washed up on the beaches.

Modern cephalopods are so sharply set apart from other molluscs that they seem out of place in that phylum and are in many respects more like vertebrates. The size range resembles that of vertebrates, the largest cephalopods attaining lengths of more than 12 meters if the two tentacular arms are included, which is much larger and bulkier than any other invertebrates. The body tissue is muscular and has considerable connective tissue resembling that of vertebrates and even containing true cartilage. The shell remnant, or pen, of squid provides a stiffening element in a position similar to the spinal column of vertebrates. Unlike other molluscs, they have a closed circulatory system with capillary beds. Their activity, intelligence, and neuromuscular development rival those of fish. Above all, their eyes give them a pe-

A

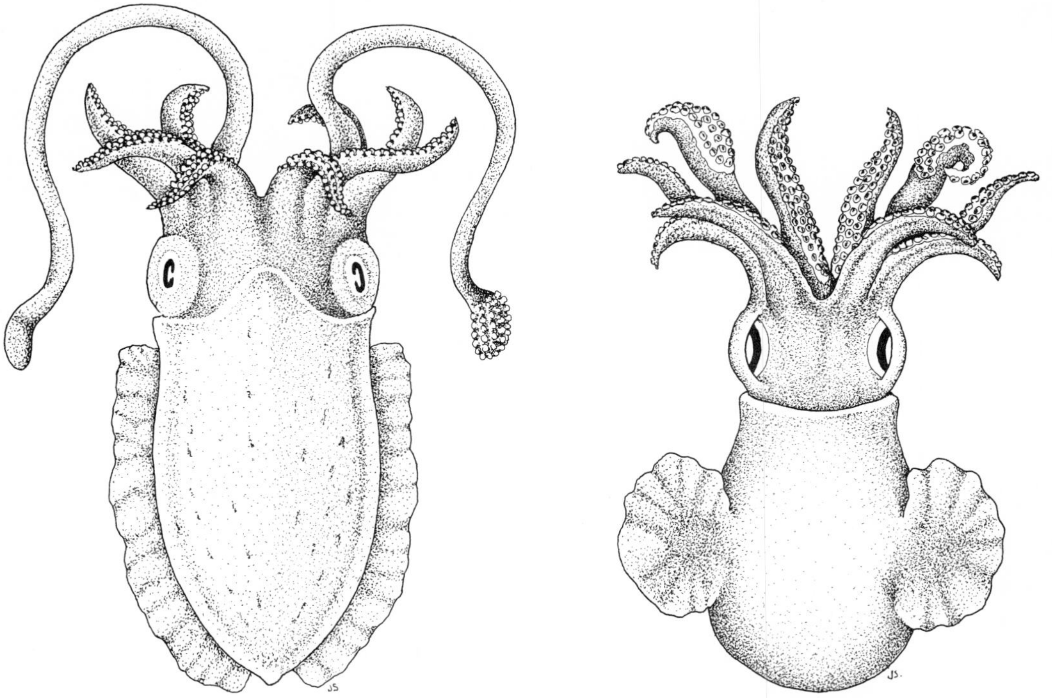

B

FIG. 11-8. A, *Sepia cultrata,* the cuttlefish, and **B,** *Rossia oweni* are examples of benthic squids. (After Hoyle, W. E. 1886. Challenger reports. Vol. 16.)

culiarly vertebrate appearance. It is difficult to look an octopus in the eye and believe that you are not seeing something closer to us—something more nearly human than a mollusc.

Fish

Fish are by far the most abundant of all vertebrates, in numbers of both species and individuals. It is estimated that there are more than 30,000 different species in the class Pisces. They range in size from a species no more than half an inch long when mature to such giants as the whale shark, which is said to reach 70 feet in length and a weight of several tons. Fish continue to grow throughout their life, although at a much diminished rate after attaining maturity, so that the size of mature individuals of a given species is, at least in many cases, less rigidly fixed than in most other animals.

Being the principal free-swimming aquatic animals, without serious competition from other groups, fish have dominated aquatic environments everywhere. Only the squid (molluscs) and the cetaceans (mammals) have successfully invaded their realm. Competition between species, the balancing of one against another, has led to evolutionary divergence in all directions and their expansion into almost every conceivable ecological niche. The diversity among fish is much greater than that found in any other class of vertebrates, making them a particularly fertile field for the student of evolution. The vastness of the aquatic environment also contributes to this diversity by making possible the survival of various types that have been superseded by different groups in centers of abundance and competition but that may continue to exist in other regions, in isolated or deep waters.

FIG. 11-9. Clown fish, *Amphiprion percula,* a brightly colored coral fish. (M. W. Williams photograph.)

The abundance of fish life is a reflection of the abundance of plankton in localized areas, as well as a reflection of the type of plankton. Since all the phytoplankton and a large portion of the zooplankton are limited to the upper strata of water in or near the euphotic zone, most fish are also found there. Just as there is more plankton in regions where great water masses meet and mix than in monotonous stretches covered by a water mass of relatively uniform character, so too there are more fish and other marine animals in such areas. Neritic waters, which receive materials washed out from the land and which also are enriched by increased turbulence, upwelling, etc., are particularly rich in fish. Such waters provide a diversity of ecological habitats far greater than that found in the open ocean, and they also contain the greatest diversity of fish species.

Coloration

The coloration of fish is of almost infinite variety and beauty and is of great biological importance. The most widespread features are simulation of background color and obliterative countershading, which have survival value to both predator and prey species. Most species that swim near the ocean surface, such as herring, tuna, mackerel, and flying fish, are greenish or bluish above, and whitish, silvery, or light colored below. Those living in inshore waters are more commonly green, and those in offshore waters, blue. Thus they match the waters in which they live when they are seen from above, and their light undersides blend with the sky when they are seen from below, matching the surface film that shines from below like a mirror.

Demersal fish, which live on the bottom, tend to be uniformly colored, speckled, or mottled, with neutral colors more or less matching the type of bottom on which they are found. Many of them, particularly the flatfish, can rapidly change both color and pattern to match their background.

For such fish the albedo, or reflecting power of their background surroundings, is more important in determining the chromatophore response than is the intensity of the overhead illumination. It is more important for them to match the background, whatever the incident light, than to respond simply to light intensity.

The upper and lower halves of the visual field influence the response in opposite ways. Light reflected from the substrate mostly strikes the upper portion of the retina. That coming from above strikes the lower part of the retina. The ratio between the two is important. Increasing the illumination of the upper half of the retina, relative to the lower, leads to paleing, the converse of darkening.

If the substrate is illumined from a source of light invisible to the fish, the usual ratio of illumination of the upper and lower portions of the retina no longer obtains. The fish may assume a very different coloration than it normally does on the same substrate illumined with the same intensity with light from above.

Tide pool fish living among the vari-colored plants, coralline algae, and animals of the tide pools are brightly and variously colored, often matching the rocks, coralline algae, or other features of their environment in a striking way.

Obliterative countershading is the rule among fish living in well-lighted waters, the upper part being dark enough to offset the shadow on the lower part, giving the whole fish a rather flat, indistinct image. Some notable exceptions to this pattern confirm its purpose, such as the Nile catfish *Synodontis,* which habitually swims upside down and has the ventral side darkened, and the remoras, which attach to sharks by means of a dorsal sucking disc—the dorsal surface, which is in the shark's shadow, being light, and the ventral surface, countershaded.

Sharply contrasting bands, stripes, or blotches, although they make a fish that is out of its environment extremely conspicuous, may also serve a concealing and protective function in that they break up the outline of the fish. Such patterns commonly include a dark bar across the eye, making it less conspicuous, and sometimes ocellar, or eyelike, markings on other parts of the body, which may serve to confuse a predator. Some fish mimic particular seaweeds with which they are associated in a most striking manner, involving not only color

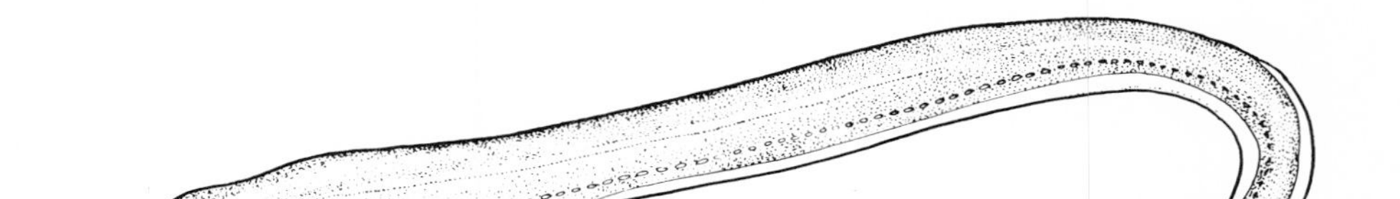

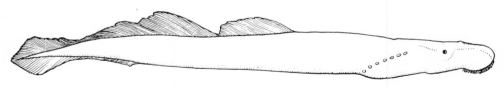

FIG. 11-10. Lamprey, *Petromyzon marinus.*

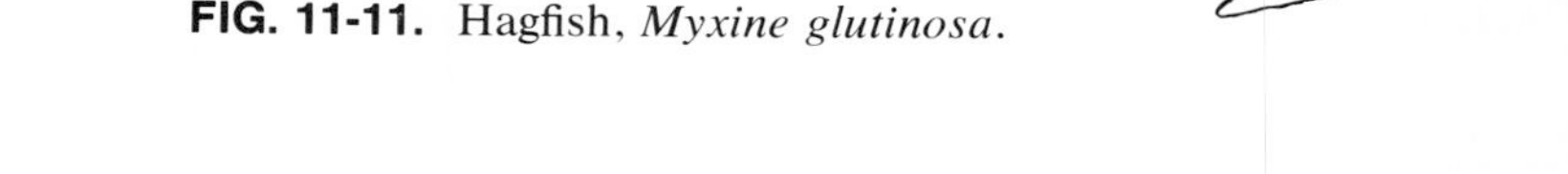

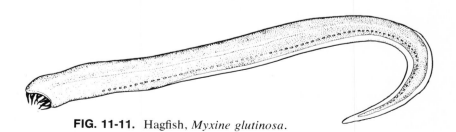

FIG. 11-11. Hagfish, *Myxine glutinosa.*

but orientation and shape of the body and outgrowths from the body surface that resemble seaweed in shape.

Fish living most of the time in the mesopelagic dysphotic depths are commonly red when seen in white light, but in their own environment, below the depth of penetration of the red light, they tend to absorb the greenish rays of light reaching them and, there being no red rays to be reflected by their red pigments, are inconspicuous. Others tend to be colorless, silvery all over, or more or less transparent.

Below the mesopelagic depths, in the deep waters of the bathypelagic and abyssal zones, where there is no light at all, most of the fish are uniformly black, except for luminescent organs.

Some of the pigmentation of fish probably has a protective function in shielding vital organs from excessive light. The concentrations of melanin, commonly centered over the brain, probably have such a function, and the black peritoneal lining of the body cavity of many fish may protect the viscera or certain of the digestive enzymes.

Some species that are aggressively territorial and fight off intruders, such as the garibaldi of southern California, or that engage in strong rivalry for mates show conspicuous warning coloration that may serve both to frighten rivals and to aid in sex recognition. Through selective breeding, aquarists have come up with numerous brilliantly colored "tropical fish" and goldfish strains. In some the male becomes more

FIG. 11-12. A, Basking shark, *Cetorhinus maximus*. **B,** Whale shark, *Rhineodon typus*.

intensely or deeply pigmented during the spawning season. Species that select mates and that have a complex courtship or breeding behavior tend to show greater difference between the sexes than do the majority of fish, in which the sexes look alike. The so-called "cleaning fish" are also usually conspicuously colored, advertising their services.

Colors of reef fish and other coral reef animals. The striking colors and strange shapes and patterns of many reef fish and other animals associated with coral reefs have long excited the admiration and wonder of visitors. One might suppose that such brilliant coloration would invite disaster where hungry predators are never far away. If this were the case, the bright colors and patterns would long since have been eliminated through natural selection. It must be assumed they serve useful functions for their bearers that at least counterbalance any such adverse effects. There is probably no single pat answer to the problem, but rather each case must be examined individually. However, some generalizations can be made that perhaps will shed some light on the question.

The reefs themselves constitute a colorful background, against which some of these fish may not be as conspicuous as they appear in aquariums and in other waters. Also, a coral reef provides endless small channels, nooks, and crannies in which the coral fish can hide when alarmed, reducing the selective premium on being overlooked by a predator.

Sharply contrasting blotches and bars of color may also serve to break up the outline of the fish so that it is not so easily recognized as such by a predator. In some instances the color may serve as a warning, advertising distasteful flesh or poisonous spines.

Conspicuous coloration of species with poisonous flesh, foul taste, or formidable defensive stings does not appear to have the same importance among fish as it is supposed to have among insects. Many of these species seem to be just as concealedly colored as their innocuous relatives, and there is not as much mimicry of such species by harmless species as occurs among insects.

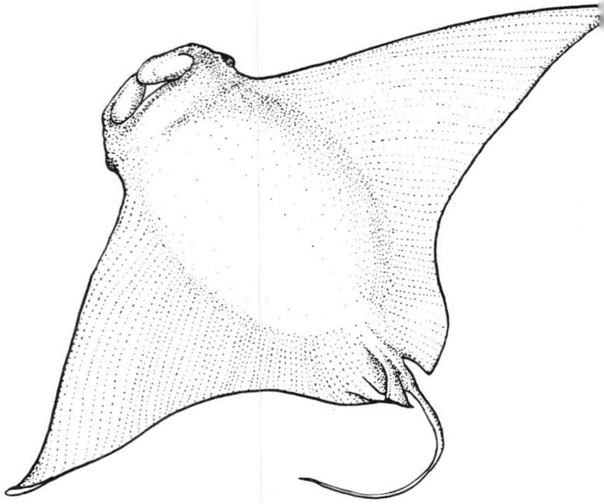

FIG. 11-13. Manta ray, *Manta hamiltonia*.

The reason may be that aquatic vegetation is less diversified than that on land. Numerous groups of terrestrial plants have evolved chemical repellents that protect them from grazing by all except special insects. These insects have become tolerant to the chemical from a particular protected plant group and, in turn, derive protection from their predators through ingestion of the repellent chemical. Such insects commonly evolve conspicuous warning coloration—and may even serve as models mimicked by other insects not sharing their chemical protection.

The cleaning symbionts seem to use their bright colors and characteristic motions to advertise their presence and attract larger fish to the cleaning stations.

Also, it must be remembered that these fish are observed through human eyes and brains. What is seen does not necessarily appear the same as it does to another fish or squid.

On the positive side, coloration can play important roles in species recognition, in helping keep schools of fish together, in courtship, and perhaps in other ways of which we are not aware.

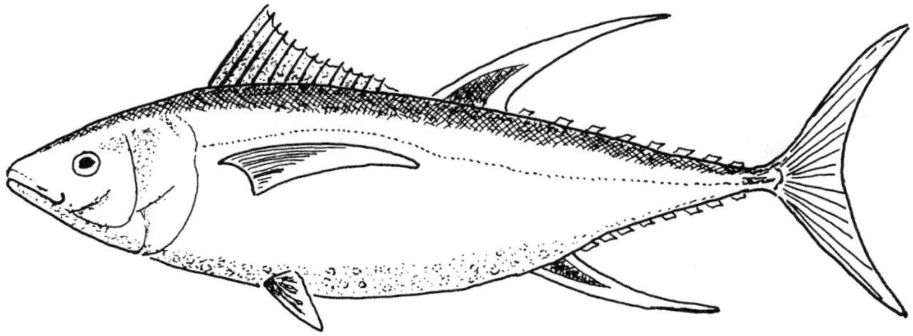

FIG. 11-14. Yellowfin tuna, *Neothunnus macropterus.*

Adaptations for swimming

The fins of fish are thought to have arisen through division and modification of fin folds such as those found on *Amphioxus,* which has paired lateroventral metapleural folds running along the anterior part of the body and uniting behind the atriopore into a single median fold that runs back around the posterior end of the body and continues forward along the middorsal line as a dorsal fin fold. A division in the paired lateral folds gave rise to two pairs of ventrolateral fins, a forward pectoral pair and a posterior pelvic pair. These fins foreshadow the paired limbs of all higher vertebrates. The unpaired fold has given rise to the caudal fin or tail, a varying number of dorsal fins and finlets, and the ventral or anal fin and finlets behind the anus. In swimming, propulsion is achieved in most fish through strong lateral flexures of the hind part of the body, causing the caudal fin to push strongly against the water. The dorsal and anal fins serve primarily as rudders, and the paired fins as balancing and braking organs. The heavy front end acts as a fulcrum for the tail, which acts as a flexible lever.

Sustained swimming is seldom or never faster than about 32 km./hour, although short bursts may in some cases exceed this rate. In the majority of fish, swimming is considerably slower. Maximum speed for small fish is about ten times body length per second for short bursts. Fast-swimming species show streamlining of the body, which minimizes the water resistance. It has been found that the swimming movements are such that the drag caused by water turbulence behind a swimming fish is less than that on a model of such a fish being propelled or pulled through the water; thus the energy required in swimming is less than one might suppose from a study of such models.

Of all the large fish, the tuna and their relatives are uniquely specialized for rapid long-distance swimming. The body is elegantly streamlined. Alone among fish, they have red flesh—a sign of high muscular activity. There is a band of dark red, highly vascularized myoglobin-containing muscle along each side. This area is maintained at a temperature of 3° to 12° C. above that of the ambient water, and higher than the temperature of the rest of the body.

In huge experimental tanks, tuna have been clocked at cruising speeds up to 230 km./day. Tagged individuals have been captured near Bergen, Norway, less than four months after having been released in the Bahamas, indicating average cruising speeds of 65 to 162 km./day in these cases.

Since tuna are not protected against heat loss by layers of fat under the skin as are some marine mammals and birds, the maintenance of higher-than-ambient temperature and intense muscular activity entail great energy cost and require a high metabolic rate. Young tuna must eat as much as one fourth of their own weight daily for normal growth.

FIG. 11-15. Opaleye. The glowing ring around the eye of *Girella nigricans* led to the common name of "opaleye." As a juvenile, this fish is common in the tide pools of southern California. Later, the adults take to offshore waters. (M. W. Williams photograph.)

They seem to occupy a specialized high-energy ecological niche. They are forced to range widely over the seas in pursuit of the shoals of other fish to sustain the high metabolic rate needed to carry out this wide-ranging pursuit.

Flotation and maintenance of level

In all except the demersal spieces of fish, which prefer to live on or near the bottom, the problem of flotation is more acute than it is in most of the invertebrates and plankton organisms discussed earlier. There is a much smaller surface-to-volume ratio, and the musculature and bones necessary for efficient swimming of a larger organism make the fish proportionately denser. Most teleost fish with heavy skeletons and muscles compensate for this density with an air bladder, which, like the lungs of higher animals, is derived from an outgrowth of the gullet. The fact that only the teleosts possess an air bladder lends some support to the hypothesis that both lungs and air bladder presumably arose from a simple pouch of the gullet that originally had a respiratory function. The pouch probably occurred in fishes inhabiting freshwater where low oxygen concentrations are more frequently encountered, and oxygen deficiency was compensated for by gulping air. In teleosts the air bladder lies dorsal to the intestine and serves as a hydrostatic organ. The gas is obtained by secretion from a gas-secreting gland and is a mixture of gases different from that present in the atmosphere or water. A fish can control its depth in the water by adjusting the volume of gas in the air sac. Control is not a voluntary muscular action

FIG. 11-16. Viviparous black perch school swims gracefully in the aquarium at the Scripps Institution of Oceanography, La Jolla, Calif. (M. W. Williams photograph.)

but rather a slower process of secretion or absorption of gas into the bladder. The fish then balances its overall density to match that of the water at the level in which it is living, becoming, in effect, a Cartesian diver. Strong-swimming predatory fish, such as mackerel, some tuna, and sharks, which swim at various levels in pursuit of their prey and often rapidly traverse a considerable range of depths, do not have a swim bladder. They sacrifice buoyancy for speed and facility in going up and down. They have to swim to maintain their level. Demersal species with air bladders, brought suddenly to the surface in trawls, are commonly dead on arrival, with the air bladders much distended, sometimes even protruding from the mouth, and vital organs ruptured.

Stretch receptors associated with the air bladder have also been thought to play a role in the perception of changes of hydrostatic pressure, which is particularly useful to fish inhabiting intermediate levels in the water. However, many kinds of fish that do not have air bladders or other compressible gas vesicles also show definite responses to changes in hydrostatic pressure.

Morris and Kittleman (1967) found, during two summers of trapping fish and diving in McMurdo Sound in Antarctica, that there is a well-defined depth stratification of several species of fish, none of which has a gas bladder. Furthermore, there is in these waters no temperature gradient that may serve to keep species at given levels. The entire water column has a uniform temperature of about −1.9° C.

It was known that pressure changes, as well as discrimination of frequencies of sound waves, involved, at least in part, the ear, but the exact mechanism had never been defined. It occurred to Morris and Kittleman that the otoliths might serve both these functions admirably if they were piezoelectric. Examination of otoliths of the antarctic notothemid *Trematomus bernacchii* and the North Pacific demersal flatfish *Parophrys vetulus* revealed that in both cases the otoliths are indeed strongly piezoelectric, oscillating over frequency ranges of 300 to 8000 cycles per second in *Trematomus* and from 1 to 15,000 cycles per second in *Parophrys*. Not all species of fish examined possessed otoliths with piezoelectric properties, but it appears that this may account for discrimination of sound frequency and perception of depth through hydrostatic pressure in some groups of fish. If this proves to be the case, it will be the first instance of a piezoreceptor system reported for animals and the first really new sensory receptor system reported for some time.

Olfactory and chemical senses

In many fish the olfactory lobes are the dominant part of the brain, and organs of chemical sensing are not only concentrated in the nasal region but spread out over the head and body as a complex series of lateral line organs. Chemical or olfactory sensing is used not only to sense food at far greater distances than the fish can see in the water—especially in dark or somewhat turbid waters—but also for communication.

Sexual identification, regulation of courtship and territorial behavior, and maintenance of social status in dominance hierarchies have all been shown in some fish to be mediated by pheromones. Such substances, as well as enhanced chemical senses, are also known to be important in many other groups of animals, such as crustaceans and insects.

Feeding habits

Perhaps the factor that most profoundly influences the evolutionary diversification

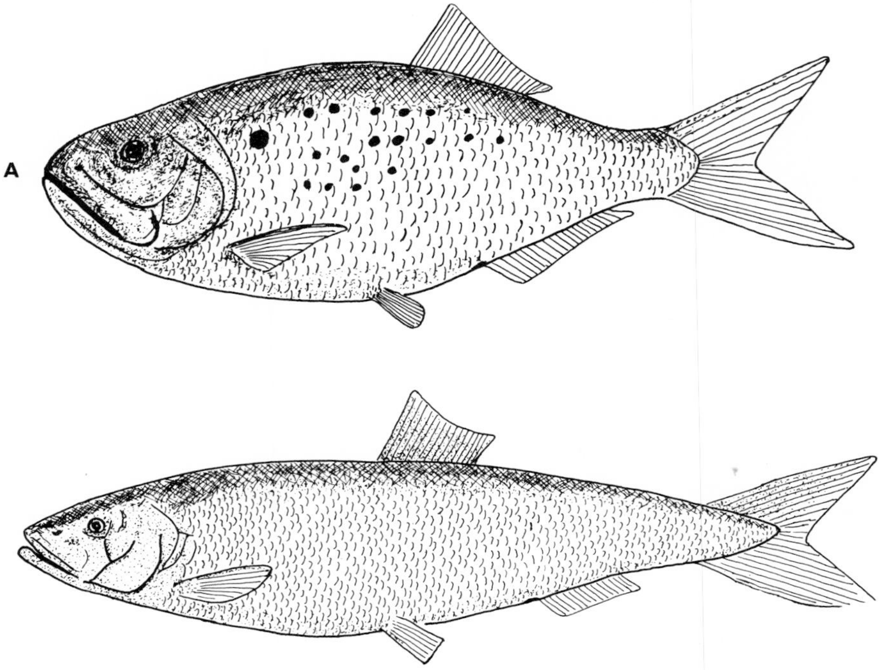

FIG. 11-17. Clupeoid fish. **A,** Menhaden, *Brevoortia tyrannus.* **B,** Herring, *Clupea harengus.*

and modification of fish is feeding habits. On this basis fish can be divided broadly into plankton feeders, nekton feeders, and benthos feeders. The lines between these categories, especially the last two, are not sharp. Of course, it is also evident that for most fish, the feeding habits must change radically during the growth of the young fry until they attain maturity. The young fry of most fish feed on smaller elements of the zooplankton or benthic organisms of similar size that they can pick off. As they grow, they begin to take larger organisms and, finally, whatever food is characteristic for the adults of their species.

Plankton-feeding fish and others that eat small-to-minute animals generally have well-developed gill rakers on one or more of the gill arches, on the inner concave side, opposite the gill filaments. As the fish passes water through the mouth and out over the gills for respiration, plankton organisms of a particular size range are filtered out and retained in the throat by the gill rakers until they are brought against the opening of the gullet and swallowed. The gullet opening is kept closed most of the time, except for the swallowing of food, so that the fish does not swallow too much water. In only a few kinds of fish are the gill rakers numerous enough and fine enough to retain much of the phytoplankton. However, some fish, such as the menhaden, feed directly on diatoms. The majority of plankton-feeding fish feed on zooplankton. The coarseness and number of gill rakers give some indication of the principal food of such fish: those feeding on the macroplankton have, in general, fewer and coarser gill rakers than do those that feed on the smaller elements of the zooplankton, whereas those feeding on phytoplankton have a great many fine gill rakers. At the opposite end of the scale, fish that prey on other fish commonly have only rudimentary rakers or none at all. Fish that feed on macroplankton begin to select and snap up individual macroplankters as well as simply to filter them out of the water. From this point it is a small step to typically predaceous habits, in which all the food items are thus selected and larger and larger organisms are utilized. The herring is a good example of a typical zooplankton feeder, feeding chiefly in clouds of copepods such as *Calanus finmarchicus*. Even at times just prior to spawning, when it is not feeding, the herring tends to remain associated with areas rich in copepods rather than with areas of dense phytoplankton blooming.

Since plankton is by far the major food source in the oceans, plankton-feeding fish can build up by far the greatest biomass of fish. They form the principal food of all the larger fish-eating carnivorous fish, birds, seals, etc. and the basis for our most productive fisheries. Some familiar examples, in addition to those already mentioned, are sardines, pilchards, *Rastrelliger,* and shad.

Schooling

Most species of abundant fish that feed on plankton or on a variety of small invertebrates and small fish, in more or less open waters, show the phenomenon of schooling. By day they tend to swim in relatively compact groups, the fish within the group all being nearly the same size and spaced at rather constant intervals from each other. Such schools often act almost as if they were individual organisms, so precisely do the fish swim and turn together. Other species sometimes school shortly before spawning and migrate to the spawning grounds as a school, although they may not remain so organized during ordinary feeding. Some of the schools may attain great size, including hundreds of thousands or millions of individuals. In the North Sea, schools of herring 15 to 17 km. long by about 5 km. wide (about 9 by 3 miles) have been observed. The individual fish in large schools do not remain in the same relative positions but are continually shifting. The school does not have definite leaders. If it turns suddenly, those fish which are on one side of the school suddenly are in front, whereas the former front members now find themselves on one side. Fish from the edges

FIG. 11-18. A Hawaiian moray eel in the Steinhart Aquarium, San Francisco. (M. W. Williams photograph.)

of the school tend to work in toward the middle, leaving others on the exposed edges.

Schooling fish have wide-angle eyes with broad lateral fields, helping to keep the school together (Fig. 11-9). Each fish can easily keep nearby fish both ahead of it and behind it in view. It also is difficult for a predator to approach them unseen.

Since the fore fins are usually relatively fixed and immovable, the fish cannot stop, reverse, or hover like solitary bottom fish with movable fore fins, but must keep swimming. They can swerve or veer, however. In true schools the fish are spaced so that there is swimming room but not turning room between them. One fish within the school cannot turn around without colliding with a companion.

Schooling is probably a protection to the individual fish in the school, which would be at the mercy of any larger, faster predator if found alone in the open water. The individual fish, particularly if it is in the central part of a school, may also take evasive and confusing action when approached by a predator, darting rapidly about in a manner that makes it difficult for the predator to select and capture a particular individual. The ability to make a sudden spurt of rapid swimming seems to be more important in evading predators than sustained speed. Most predators are bigger and faster anyway, so that sustained speed would not help much in escaping. Some small fish attain maximum speed within 1/20 second in response to an appropriate stimulus.

When prey fish are gathered together in large schools, there is a smaller chance that a predator will find the school than that it would find some of the individuals of the prey species if they were scattered widely and more evenly through wide areas of the ocean. Oddly enough, the ability of the predator to take all the prey it wants when it does encounter a school may also in the long run contribute to the safety of the prey. A predator can eat only so many prey fish at a feeding. The chance that any particular fish in a school will be eaten in an encounter between the school and a predator would be 1 in x/a if a is the number of prey individuals the predator can eat at a time, and x is the number of individuals in the school. The larger the school, the less the chance that any particular individual will be consumed in a given encounter with a predator. If, on the other hand, the predator encountered individuals fairly regularly at widely spaced intervals, it would never become satiated, and every fish encountered would be consumed. It is true that large schools of fish also attract more predators to the area than would otherwise be there and make it easier for at least some of the predators to obtain their fill, but the balance of the advantage lies with schooling. Otherwise, it could not have become such a widespread hereditary trait in so many diverse fish as well as in some squid and other animals.

Several kinds of small schooling pelagic fish such as anchovies have deciduous scales, which gives them a slightly better chance to escape when seized by a predator.

The appearance of man on the scene, with efficient mechanized equipment for taking advantage of the schooling of fish and taking them in hitherto unheard-of numbers, may well shift the balance of advantage and bring disaster to a number of species, as apparently happened in the case of the sardine fishery off the California coast shortly after World War II. It is possible that such an event will be even harder on the larger predators than on the schooling prey, both because man also fishes intensively for them and they are far fewer in number to begin with, and because their food supply in a region may be reduced to below the level at which either man or the large predator species could fish efficiently. In such a situation the large predators may be wholly eliminated, but vestiges of the population of plankton-grazing species would survive, from which, now relieved of most of the predation pressure, a population even larger than before may be reconstituted. Since most of these species produce tremendous numbers of eggs and mature faster than do large fish, population comeback can be relatively fast, given favorable opportunity. A much more ominous possibility facing many species is the destruction of their spawning grounds, particularly when these grounds are in coastal waters and estuaries, through alteration and pollution.

The schooling of large pelagic migratory species such as tuna may primarily serve to keep the fish together for reproduction, rather than serving as a protection from still larger predators. This function would, of course, also be served, in addition to whatever protection schooling may give, in the case of the smaller fish discussed previously.

Most of the remaining fish are active predators, feeding on other fish or on invertebrates of comparable size. Those which eat other fish or squid in open waters are, in general, fairly large, fast swimmers and are often well armed with sharp-edged or needlelike teeth. Those of the mesopelagic and bathypelagic depths tend to be smaller than those of the epipelagic waters and to have relatively larger heads and mouths and more exaggerated teeth, giving some of them a fierce, dragonlike appearance. Predatory fish swallow their prey whole, or if attacking something too big to be swallowed in this manner, tear off bite-sized pieces that are swallowed whole. Many of them have a keen sense of smell and can detect the presence of suitable food in the water far beyond the distance at which they can see it. Some also seem to be attracted to disturbances in the water, such as those created by the swimming of a school of fish or a human swimmer.

In shelf areas where there are suitable concealment places among rocks or seaweed, or on the bottom, numerous species have adopted the habit of lurking in hiding until suitable prey comes within close range and then darting forth to catch it. Fish of this latter group that hide on or near the bottom and habitually lunge upward to catch swimming fish or invertebrates above them usually have large heads and mouths, and the mouth and eyes are directed upward. Bottom-dwelling fish, such as the flatfish, skates, and numerous scorpaenids, usually are sluggish. They have lower metabolic rates, reflected in lower rates of oxygen consumption, and smaller ratios of gill surface to body weight than do more actively swimming types. On the other hand, fish such as the cod, which habitually picks up molluscs and other benthic invertebrates, commonly have the mouth and associated sense organs (nose, barbels, etc.) directed straight ahead, or even on the ventral side of the body, as in rays. Fish that eat molluscs and crabs commonly have rather broad, flat teeth for crushing the shells rather than the sharp, needlelike teeth of fish-eating types. Of special interest are some of the more specialized flatfish such as the soles, in which the jaws of the eyed side are toothless. Predaceous fish are the objects of all sport fishing and commercial bait fishing.

Osmotic relationships

Elasmobranchs and marine cyclostomes keep the tissues isotonic with the surrounding water by the resorption of urea through the renal tubules until the concentration in the blood reaches 200 to 2500 mg./100 ml. The tissue osmotic pressure is thus raised above that of seawater, enabling the fish to absorb water through the gill and oral membranes in small amounts adequate for the formation of urine. This system requires that the embryo, prior to developing its own regulatory mechanisms, must be either in a tough-shelled egg or inside the female. Internal fertilization effected by copulation is therefore necessary and has led to the evolution of claspers in the male. This system apparently arose with the evolution of elasmobranchs and cyclostomes as far back as the late Silurian or early Devonian.

Teleosts, on the other hand, underwent a different evolutionary history. They appear to have evolved in brackish estuaries or freshwater rivers where the problem was not getting enough water into the tissues, but rather excluding or eliminating excess water without at the same time losing valuable salts needed in the tissues and body fluids. The water tended to be absorbed because of the hypertonicity of the tissues relative to freshwater. The glomerular kidney was evolved in response to this need. Each nephron consists of a tuft of capillaries, the glomerulus in the mouth of an excretory tubule, combined with an enzymatic transport system for resorbing most of the salts from the renal tubule before the excess water reaches the exterior as urine. Thus the glomerulus and tube constitute a system for pumping out excess water without undue loss of needed salts. The urine is dilute and copious. When the teleost returns to the sea, the osmotic conditions are reversed, and the tissues tend to lose too much water and gain too much salt. Apparently the return of teleosts to the sea came too late in

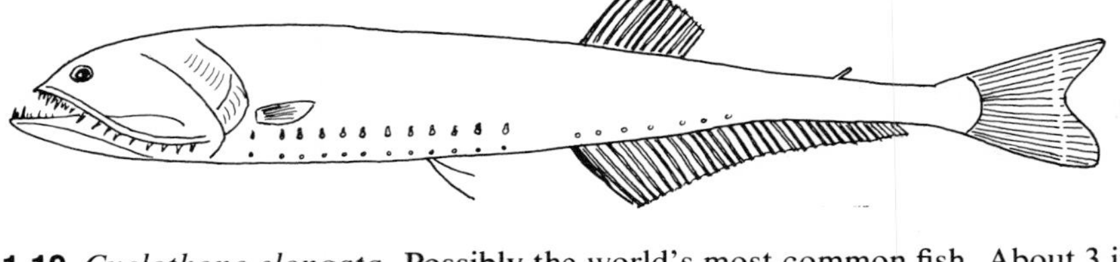

FIG. 11-19. *Cyclothone elongata.* Possibly the world's most common fish. About 3 inches long. Family Gonostomidae.

their evolution for them to change the basic pattern of their protein metabolism or structure of their kidney. The problem was met in a new manner through the evolution of special salt pumps—patches of cells in the gills or mouth that actively excrete excess salt from the blood, leaving the water with which it had been absorbed in the first place free for urine formation. Also, most marine teleosts make much less urine than do freshwater fish. The more urine they make, the more they must drink and the more salt must be excreted. They therefore tend to keep urine formation at the lowest possible level consistent with elimination of their soluble nitrogenous waste products. The glomeruli tend to degenerate, and the urine is much more concentrated than is the case with freshwater fish.

FIG. 11-20. Pipefish. *Syngnathus californiensis* resembles in design, color, and pattern the eelgrass in which it lives. Horizontal eelgrass blade is covered with diatoms. These relatives of the sea horse are common in the eelgrass beds of California bays. (M. W. Williams photograph.)

Classification of Pisces

Authorities differ rather widely in the constitution, ranking, and names applied to some of the higher taxa of fishes. The scheme used in this book does not follow any one source wholly but is a compilation from several, with an effort to adopt what appear to be the simplest, most practical arrangements. Because of their great numbers and diversity in today's oceans, the elasmobranchs and teleost fish are accorded a much more detailed treatment than are the other groups, but even here many interesting groups at various levels have of necessity been omitted. The term Pisces is here used in its broadest sense and includes some groups often recognized as independent classes by many specialists, who restrict the name to bony fish. In this discussion the class Pisces is composed of aquatic vertebrates breathing by means of gills throughout life. They do not possess tetrapod-type limbs. The homologues of these limbs, when present, are paired fins. The following outline will show the division of the class Pisces into major groups used here:

- Subclass Agnatha
  - Order Ostracodermi
  - Order Cyclostomata
- Subclass Placodermi
  - Order Acanthodii
  - Order Cladoselachii
  - Order Coccosteomorphi
- Subclass Chondrichthyes
  - Order Elasmobranchii
  - Order Holocephali
- Subclass Osteichthyes
  - Grade Choanichthyes
    - Order Crossopterygii
    - Order Dipnoi
  - Grade Actinopterygii
    - Order Polypterini
    - Order Chondrostei
    - Order Holostei
    - Order Teleostei

The taxonomic appendix gives a somewhat expanded classification, together with some of the most used synonyms for various group names.

FIG. 12-1. *Acanthaster planci,* a large, spiny starfish that is destructive to coral reefs in many parts of the tropical Pacific, crawling over a massive meandriform coral head. (From Chesher, R. H. 1969. Destruction of Pacific corals by the sea star *Acanthaster planci.* Science **165:**280-283. Copyright, 1969, by the American Association for the Advancement of Science.)

CHAPTER

12 BENTHIC COMMUNITIES ON HARD SUBSTRATES

The general character of the benthic fauna at any given place is largely controlled by the physical character of the substrates. Substrates may be divided into two quite different series—the rocky, or hard, substrates, and the soft substrates such as sand, mud, and marine oozes. Climatic differences, physical barriers to the spread of particular species, interactions between species, and other factors impress further differences. Thus the species composition of similar substrate communities from one region to another may be different, even though the overall character of the communities and the roles played by ecologically equivalent species may be similar. In the consideration of representative types of benthic communities, each will be seen from the standpoint of the most important food supplies, the types of animals exploiting the food supplies, and the special adaptations impressed on the animals by the physical characteristics of their environment. We will begin with hard substrates in the littoral, or intertidal, zone; consider the character of shallow-water, intermediate-

A

B

C

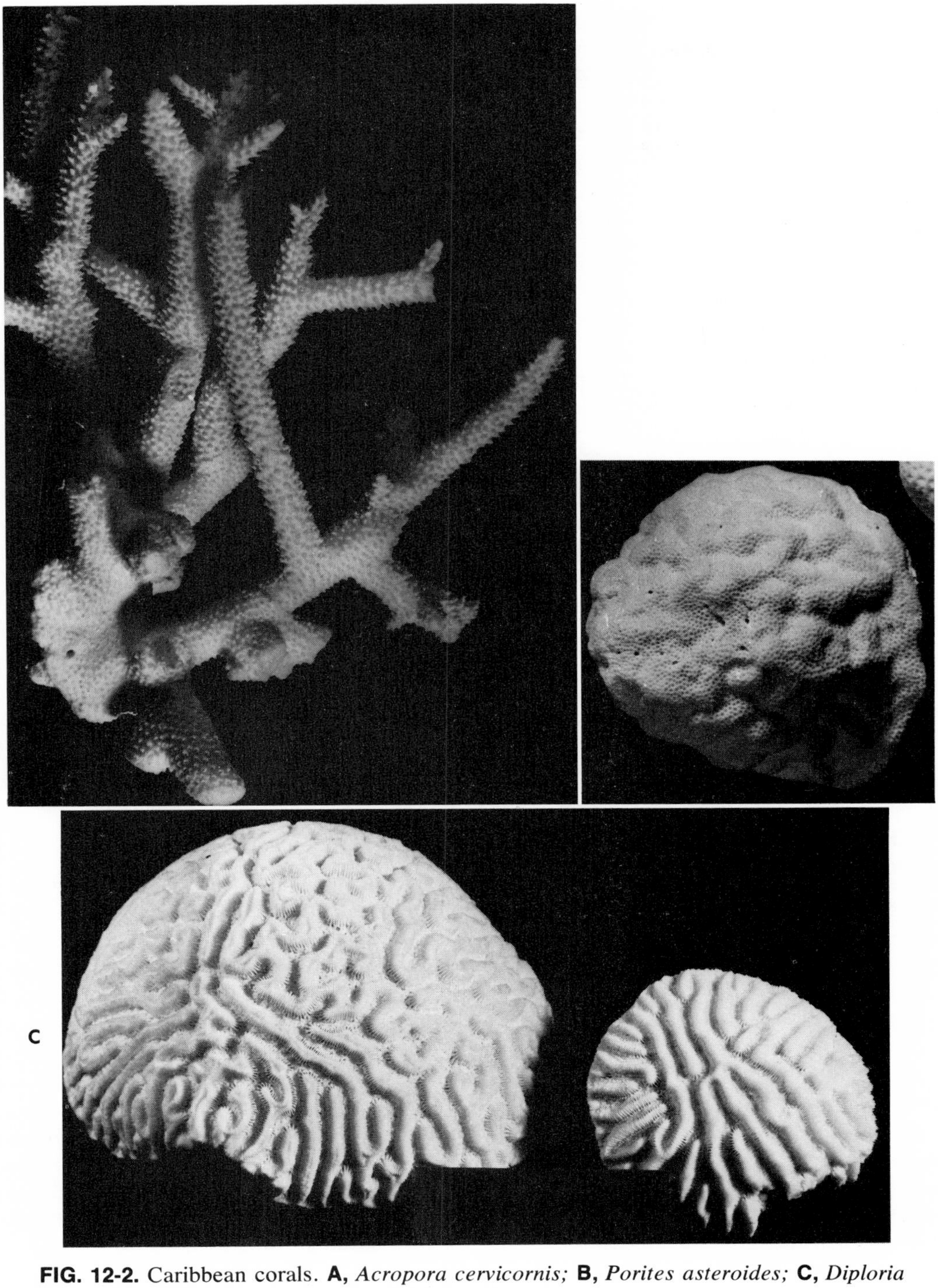

FIG. 12-2. Caribbean corals. **A,** *Acropora cervicornis;* **B,** *Porites asteroides;* **C,** *Diploria labyrinthiformis* and *D. strigosa*. (R. Anderson photographs.)

depth, and deep communities; and then follow a similar scheme for soft substrate communities.

The overriding feature of almost all shallow-water, marine shelf communities is the presence of relatively abundant plankton and seston in the water, constituting the major available supply of food. This permits the development of animal communities dominated by sessile animals equipped with the means of concentrating plankton from the water and by substrate grazers specialized for obtaining their food from the substrate. Such attached or slow-moving animals commonly shed their gametes into the water and achieve dispersal through planktonic larval stages. In communities where there is enough available food in the ambient medium to support large animal populations, other factors—space for attachment, oxygen content, wave action, protection from predation, salinity, degree of exposure by tides, and nature of the substrate—become the limiting factors, determining the species that can live there and the biomass of animals produced in a given place. In favorable locations, plankton-feeding animals sometimes become crowded, growing on each other even to the extent of smothering the lower members until the mass becomes too unwieldy and is detached from the substrate.

SHELF COMMUNITIES

The coral reef

Coral reefs are a dominant feature of shallow tropical marine waters where the mean annual temperature is at least 23.5° C. and where the water does not fall below 20° C. They are so characteristic of the tropics that the northern and southern extremes of coral reef formation may be taken as the boundaries of the tropical zone. Even in the tropics, coral reef development may be prevented by cool water temperatures in areas where offshore winds blow surface water away from the land, causing upwelling of cooler water, as in Western Australia, or where cold currents occur near shore, as off parts of tropical western South America.

The stony corals, order Madreporaria or Scleractinia, are hexamerously symmetrical anthozoan polyps related to sea anemones, from which they differ in lacking a siphonoglyph, in secreting an exoskeleton of calcium carbonate, and in being predominantly colonial through rapid asexual budding.

Typically each **polyp** sits in a skeletal cup, the **corallite,** the outer portion of which is the **theca.** In the center of the bottom of the cup there is usually a column, or spine-like process, the **columella.** Radiating from the columella to the theca are a series of radially arranged **septa,** which in a general way alternate in position with the mesenteries of the polyp. It is almost as if the polyp were sitting on a tack in the middle of a cup with radial knife blades passing from the tack to the wall of the cup (Fig. 12-3). Throughout its life the polyp continues to build up the theca, the columella, and the septa. From time to time the polyp withdraws from the bottom of the corallite, secretes a horizontal partition, or **tabula,** as a new bottom for the corallite. Thus a coral head in vertical section may show many long, parallel tubes partitioned by tabulae. Former postions of the corallites may be seen, and new tubules, beginning at various levels between the longer ones, show intercalated or budding new polyps established as the coral grows.

Variations in the pattern of budding of new mouths, in the degree of separation attained, and in the size of corallites result in a considerable variety of skeletal patterns among corals. In types with extratentacular budding, in which new polyps are formed entirely from tissues outside the ring of tentacles of preexisting polyps, in which there is a complete separation of the new individual, even if it budded from an intratentacular position, skeletons form in which each polyp has an individualized cup, or corallite, and is an easily recognizable unit. In other corals new mouths arise within the ring of tentacles, and the polyp shape alters to accommodate them. This process may be so highly developed that it leads to the for-

FIG. 12-3. *Fungia fungia,* large solitary coral polyps. (Chesher photograph.)

mation of one or more polyps, each with a number of mouths arranged in a row. The resulting skeleton is a series of winding ridges alternating with valleys containing septa—the meandriform corals. In the Fungidae we find the development of much larger than usual single polyps, which form a type of mushroom-shaped skeleton expanding from a short stalk, with the sharp septa radiating from the center of the top of the mushroom. Larger specimens are commonly detached from the stalk portion.

The Indo-Pacific region is the world's richest in coral species and in the development of coral reefs. Here there are some 80 genera and 700 species of corals. In the tropical Atlantic, on the other hand, there are only about 26 genera and 35 species. This is probably the result of climatic changes (Ice Ages) since the end of the Mesozoic era, which have severely reduced the former tropical marine faunas of the Atlantic but not of the Indo-Pacific. In Both regions the genera *Acropora* and *Porites* are among the most important and heaviest contributors to reef formation. In the West Indies there are 3 species of each of these genera. In the Indo-Pacific there are about

150 species of *Acropora* and 30 of *Porites*.

Three suborders commonly recognized are the Perforata, Aporosa, and Fungacea.

In the perforate corals, **Perforata,** the zooids are usually small, often crowded, and usually individualized, and the corallum is porous or reticulate. The septa are relatively few in number, not more than twelve, and sometimes indistinct, and in many of them the columella is absent. Most of the branching corals such as the staghorn coral belong to this group. Many important reef-building corals are included.

The imperforate corals, **Aporosa,** possess solid coralla, usually with numerous septa, commonly in multiples of six, without crossbars **(synapticula).** The disc may have one or more mouths, and there is usually a mesentery in each interseptal locus. Many of the more massive corals and meandriform types such as the brain corals belong here.

Fungoid corals, **Fungacea** or **Fungida,** have solid, imperforate skeletons; septa are usually numerous and with crossbars (synapticula). The discs are not circumscribed and in colonial forms may be confluent. The tentacles are scattered soft lobelike structures, not covered when contracted, and sometimes obsolete.

Despite the striking differences in coral skeletal patterns, the group seems to be a relatively homogeneous one, differing from anemones not only in growth pattern, skeleton formation, and absence of a siphonoglyph in the pharynx but also in having septal filaments consisting only of a cnidoglandular band and in having rather simple, unspecialized nematocysts of only three kinds.

Spawning of corals commonly displays a marked lunar periodicity. In many species, development of the ova proceeds to the planula stage before they are discharged into the water. The planulae already contain zooxanthellae when discharged.

Roles of symbiotic algae in corals

All the reef-building corals have abundant symbiotic zooxanthellae (small, rounded algal cells belonging to the groups Cryptomonadida and Dinoflagellida) in their ectodermal cells. There may be as many as 30,000 zooxanthellae cells per $mm.^3$ of tissue. In addition to the zooxanthellae, there are an even greater number of skeletal algae forming bands just beneath the polyp zone and in contact with the living tissue through minute pores in the calcareous skeleton. Together, these skeletal algae and the zooxanthellae comprise a mass of tissue about three times as great by dry weight as the animal tissue of the reef. The zooxanthellae and skeletal algae are of vital importance to their hosts, although their role in the alimentation of the host has long been disputed. When grown in culture separately, they do not liberate photosynthates into the water in significant amounts. For this reason it has been thought that their primary importance was in merely providing the coral with oxygen during the daylight hours and perhaps utilizing nitrogenous and other metabolic products of the coral so that these products are not excreted in quantities that would become toxic to the coral.

The amount of animal tissue in an extensive coral reef such as the Great Barrier Reef of Australia is so great that during low tide, one may otherwise expect that the corals would be killed through oxygen depletion and the accumulation of metabolic wastes. Indeed, this phenomenon does occur to some extent on the inner, or protected, side of coral reefs, where many dead corals are found. The corals grow most vigorously at the outer, exposed edges of the reef, and the less favorable growing conditions on the protected side are perhaps responsible in part for the formation of channels and lagoons between the reef and adjacent shorelines. Thus fringing reefs are usually at a little distance from the shoreline.

Recent studies using labeled carbon dioxide have brought to the fore once more the idea that the symbiotic algae may also play an important role in the alimentation of their hosts. Labeled carbon compounds can be found in the host tissue as well as in the zooxanthellae themselves after a period of photosynthesis. Furthermore, calculations made by the Odums at Eniwetok Atoll indicate that the open sea there does not provide enough zooplankton to account for the productivity of the reef. To arrive at a reasonable food budget, or trophic structure, it was necessary to count the algae as major producers for the reef as a whole. The reef efficiently conserves nutrients, obtaining only a small part of them from the surrounding ocean and losing only a small portion to the lagoon.

The seeming contradiction between these findings and the failure to find photosynthates in the water after illumination of cultures of zooxanthellae has been cleared up at least in part by Muscatine (1967), who confirmed that if the algae are cultured in seawater separately from the coral, they fix labeled CO_2 but excrete little soluble radioactive organic photosynthate. However, he found that if they are cultured in the presence of some tissue homogenate from the coral, they secrete as much as 40% of the total carbon fixed, mostly as glycerol, or with long incubation, as glycolic acid. The stimulative property of host tissue demonstrated a previously unsuspected role of the host in the regulation of specific metabolic activities of the symbiont, perhaps in this case through activation of a glycerol permease, enabling the glycerol to be excreted from the zooxanthellae and made directly available to the coral.

Thus it now appears that the algae may supplement the plankton diet of the corals in the clear plankton-thin tropical waters, enabling a greater mass of animal tissue to be maintained than would otherwise be the case and perhaps supplying necessary nutrient components not present in the plankton diet.

It should not be concluded, however, that the reef corals are essentially autotrophic in the same sense that a few alcyonaceans and xoanthids are. Some of the latter have evolved complete nutritional dependence on zooxanthellae and direct use of dissolved substances from the water. They no longer feed at all. The typical cnidarian feeding structures are absent or rudimentary. They show no response to food stimuli nor do they even digest food if it is injected into their coelenteron.

Reef corals, on the other hand, are efficient, voracious carnivores that take in almost any kind of particulate animal food. The feeding mechanisms of different reef coral species vary considerably in detail, including use of tentacles (as in sea anemones) by some, entrapment of particles in ciliary-mucus tracts by others, and even entrapment and digestion by mesenterial filaments extruded through the mouth or other openings throughout the body. Furthermore, corals exhibit great sensitivity to chemical food stimuli and even open up in anticipation as a cloud of zooplankton approaches the reef, no doubt sensing the dissolved metabolites that diffusely surround plankton swarms.

Oceanic reef systems seem to be, on the whole, autotrophic units operating at high levels of productivity, fast turnover rates, and great efficiency in recycling nutrients. External energy loss is reduced to a minimum, and local nutrient levels, are maintained at a high, steady state.

The water circulating within the reef system, termed boundary layer water by Goreau and co-workers (1971), has much higher levels of zooplankton, organic detritus, and dissolved nutrients than does the surrounding open ocean. Thus nutrients available to the corals in the water actually bathing them may be as much as one or two orders of magnitude greater than those available in surrounding oceanic water.

Furthermore, corals have highly developed structural modifications for taking in colloidal and dissolved nutrients directly from seawater. Electron micrographs of exposed epithelial surfaces show that they possess microvilli, greatly increasing the absorptive surface; have high concentrations of alkaline phosphomonesterases associated with neutral mucopolysac-

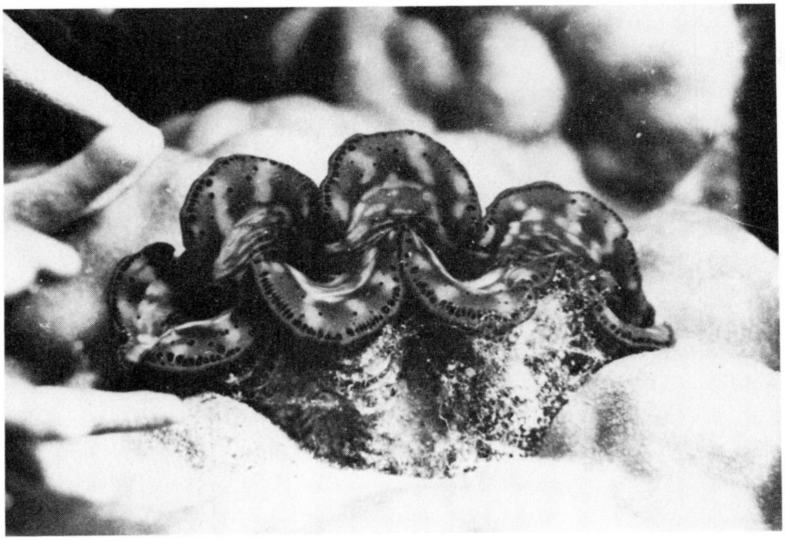

FIG. 12-4. Mantle and hypertrophied siphon tissue of *Tridacna,* in its natural habitat. (Fankboner photograph.)

charides at the free cell border; and can actively transport dissolved organic substances into the cell by pinocytosis against the concentration gradient. In these respects they are comparable to such absorptive surfaces as the duodenal epithelium or the renal tubules of mammals.

Thus, although it is clear that photosynthates do pass directly from zooxanthellae to the coral, it cannot be concluded that the coral depends wholly or even principally on them for its nutrition. The spectacular size of coral reefs tends to give an exaggerated idea of the amount of actual living tissue present and the amount of food required.

The zooxanthellae may be important in other ways than merely providing nutrients. Their products have been shown in the giant clam *Tridacna* (Figs. 12-4 and 12-5) to contribute the byssus formation, the crystalline style, and certain other specialized structures. In corals their photosynthetic activity has been shown to enhance calcification at the tips of the branches, where it is most active, the photosynthate being transported in the coral tissue to the sites of the most active calcium deposition.

Neither corals nor *Tridacna* seems to depend on digesting zooxanthellae cells directly. In both, older and dying zooxanthellae cells are engulfed and partly digested by amoebocytes. In corals they are continuously eliminated to the outside in strands of mucus, and in *Tridacna* they are transported to the digestive gland and digested, but in neither case do the hosts seem to be using these cells as a major source of nutrients. Rather, the process seems to be one of phagocytic culling out of those cells in the zooxanthellae population which are no longer fulfilling a useful function, since only the older, dying, less nutritious cells are taken.

Symbiotic ties between the corals and their zooxanthellae and the need of the latter for adequate illumination to carry on photosynthesis help explain why reef-building species of corals are limited to shallow waters.

Other components of coral reefs

Corals are not the only, and in some cases not even the major, components of calcareous "coral reefs." Various coralline algae are also important, as are the calcare-

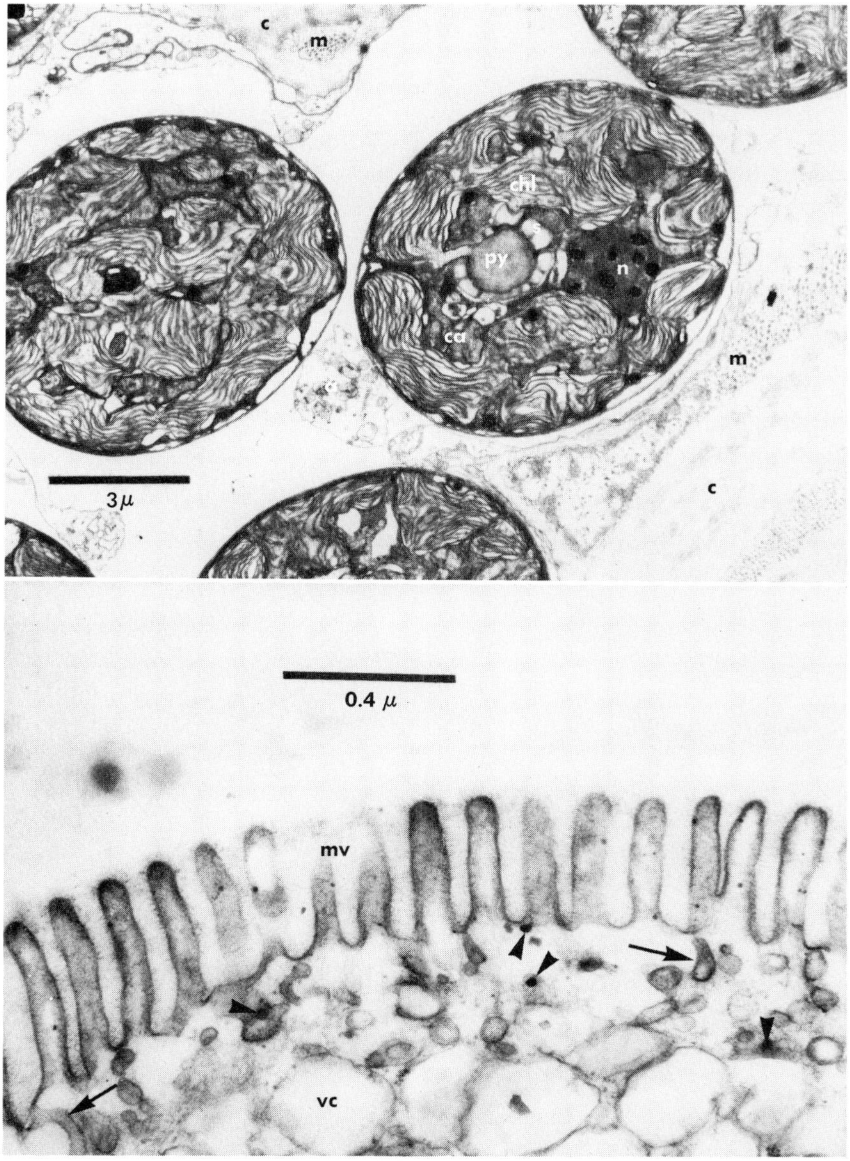

A

B

FIG. 12-5. A, Electron micrograph of symbiotic zooxanthellae in siphonal tissue of *Tridacna.* **a,** Part of an amebocyte; **c,** connective tissue; **chl,** chloroplast; **m,** muscle strand; **n,** nucleus; **py,** pyrenoid; and **s,** starch cap of pyrenoid. **B,** Electron micrograph of a portion of the surface of an epithelial cell from the siphonal tissue of *Tridacna,* showing microvilli, **mv,** and pinocytosis vesicles. **vc,** Clear vacuole; short arrows, pinocytosis of particulate matter; and long arrows, pinocytosis of fluid matter or channels filled with fluid. (Fankboner photographs.)

ous skeletons of the Hydrocorallina, and in some cases even the skeletons of Foraminifera. Shells of molluscs, calcareous tubes of tubeworms, echinoderm spines, and calcareous remnants of bryozoan colonies and other organisms also contribute.

Coral reefs generally have a conspicuous crest of *Lithothamnion* on the windward, actively growing edge, which serves to cement the looser coral together and give the reef the strength to withstand the buffeting of wind and waves.

Structure and origin of coral reefs and coral atolls

Perhaps the most characteristic coral formation is the **fringing reef.** Here the growth of corals in shallow waters produces a shelf of coral growth, the top of which is at about the height of the mean low-tide level. If this shelf attains any considerable width, the inner side—protected from wave action and containing quieter water already strained past the outer portion of the reef, as well as perhaps less of the oxygen and plankton and more of the metabolic products—becomes less favorable for coral growth. Here the corals tend to grow more slowly or to die, and eventually a lagoon may be formed between the outer reef and the shore. Darwin was the first to study carefully the formation of coral reefs and to correlate the formation of fringing reefs, isostatic changes in the level of the sea floor, or changes in sea level with the formation of barrier reefs and coral atolls—the isolated rings of coral islands often many miles across and best developed in the tropical western Pacific. He postulated that where there are a series of fringing reefs surrounding an island, either a slow subsidence of the island or a slow rise in sea level, at such a rate that the growth of coral can keep pace with it, will first produce a ring of coral islands surrounding the central island at an increasingly greater distance from it as more and more of the central island is flooded. Finally, only the

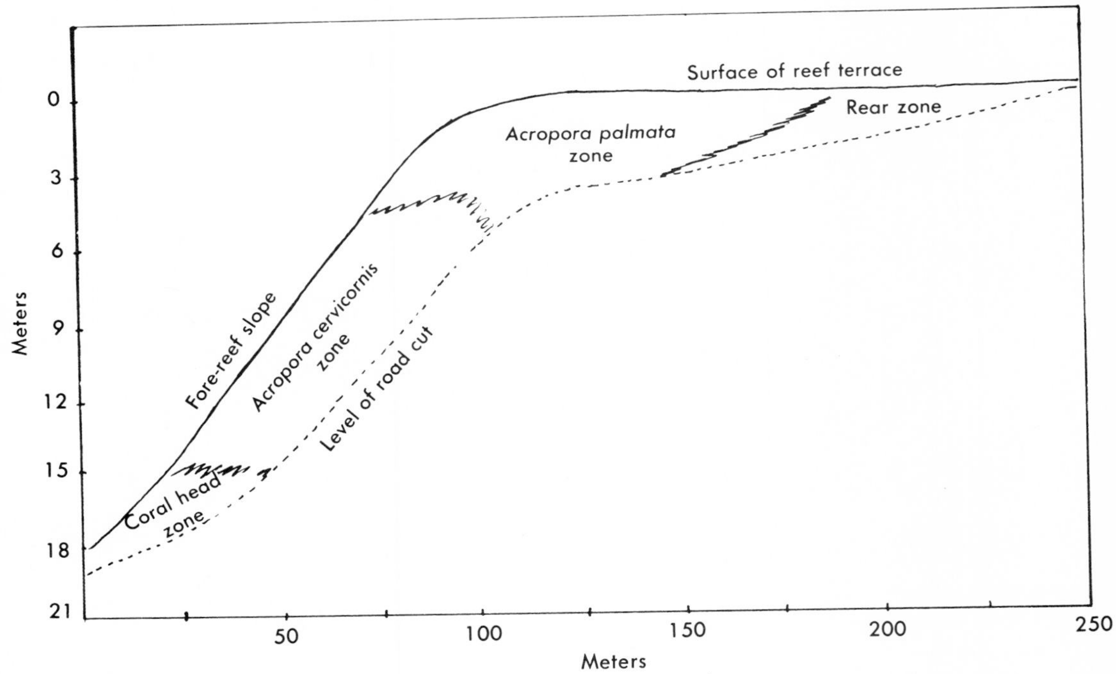

FIG. 12-6. Section of exposed coral reef to show general structure of West Indies coral reef.

ring of coral islands will be left. The coral islands comprising an atoll would then consist of dead coral remains to a considerable depth, only the top portion in the illuminated zone being alive.

After the *Challenger* Expedition, John Murray (1895), having failed to find convincing evidence of subsidence in the coral reef areas investigated by him, proposed an alternative theory—that piles of zooplankton shells and other debris eventually build up a high-enough platform for living corals to establish themselves. Since the corals at the outer edge of the platform grow best, a lagoon gradually forms—and there is the atoll without the necessity for all the subsidence. This would especially be the case where submerged mountains already furnish a beginning platform within reasonable depths from the surface.

Alexander Agassiz (1906) also reinvestigated this question and concluded that there is probably no single simple explanation for coral reef formation—some form on banks and shoals built up from erosion of land, some on plateaus built up ac-

FIG. 12-7. Portion of Barbados Island. Aerial view to show present fringing reef, and five successively older Pleisotocene reefs that have been raised, forming marine terraces. (Mesolella photograph.)

FIG. 12-8. Portion of road cut through a raised coral reef on Barbados Island. *Acropora palmata* zone of the reef. (Mesolella photograph.)

cumulation of sediments until the proper, height is reached, some on the peaks of submerged mountains, and some in the manner described by Darwin.

Subsequent borings in Pacific coral islands have confirmed Darwin's hypothesis in its general outlines. In some instances several hundred meters consist of the same kinds of coral as those now living at the top of the reef. They must have been near the surface at the time of their growth, proving the hypothesis of slow subsidence in such cases. Darwin's hypothesis first brought the principal reef forms—the fringing reef, the barrier reef, and the atoll—into relationship with each other.

Accumulation of organic debris and eventual development of vegetation on some of the larger low coral islands have created many habitable islands in the South Pacific.

If barrier reefs and atolls are, in fact, formed by slow subsidence with which the growth of coral has kept pace, instances where the rate of subsidence exceeded the growth rate of the coral, taking it to greater depths than it could tolerate, can be expected, and in fact such drowned reefs have been discovered. The best known of these reefs comprise the Chagos Archipelago in the Indian Ocean.

Later students of the problem have supplemented Darwin's observations and theories. Particularly important is the glacial control theory of Daly (1910), who showed, at least for the West Indian coral reefs, that during the Quarternary glacial period an interruption of earlier coral reef

FIG. 12-9. Soft coral, Alcyonacea, with polyps expanded. (Chesher photograph.)

development took place as a result of lowering of the sea level through the formation of the great continental ice caps. This caused an emergence of the reefs and, of course, stopped their growth during the periods of emergence. Both the climatic changes of the time and the lowering of the sea level would operate to stop growth during these periods.

Another new concept is the possibility that the reef-building organisms themselves may be a contributory factor in causing subsidence. Limestone debris broken from and washed off the coral reef settles as a sediment in their vicinity, accumulating through the ages in considerable quantities. It has been calculated that at the present rate of limestone production in a group of coral islands such as the Maldives, sufficient limestone would accumulate to overcome the solidity of the earth's crust and cause isostatic subsidence in a period of time ranging from about 400,000 to nearly 2 million years, which is not a particularly long time in geological terms.

Other methods of coral island formation have been suggested, such as growth of a reef on the top of a submerged mountain.

Some areas, such as Barbados, the outermost island of the West Indies, have undergone a series of changes of elevation throughout Pleistocene times, resulting in the formation of a series of fringing and barrier reefs subsequently raised as part of the land, the structure of which may now be readily studied along road cuts. Raised reefs, together with the development of scuba diving, have enabled marine biol-

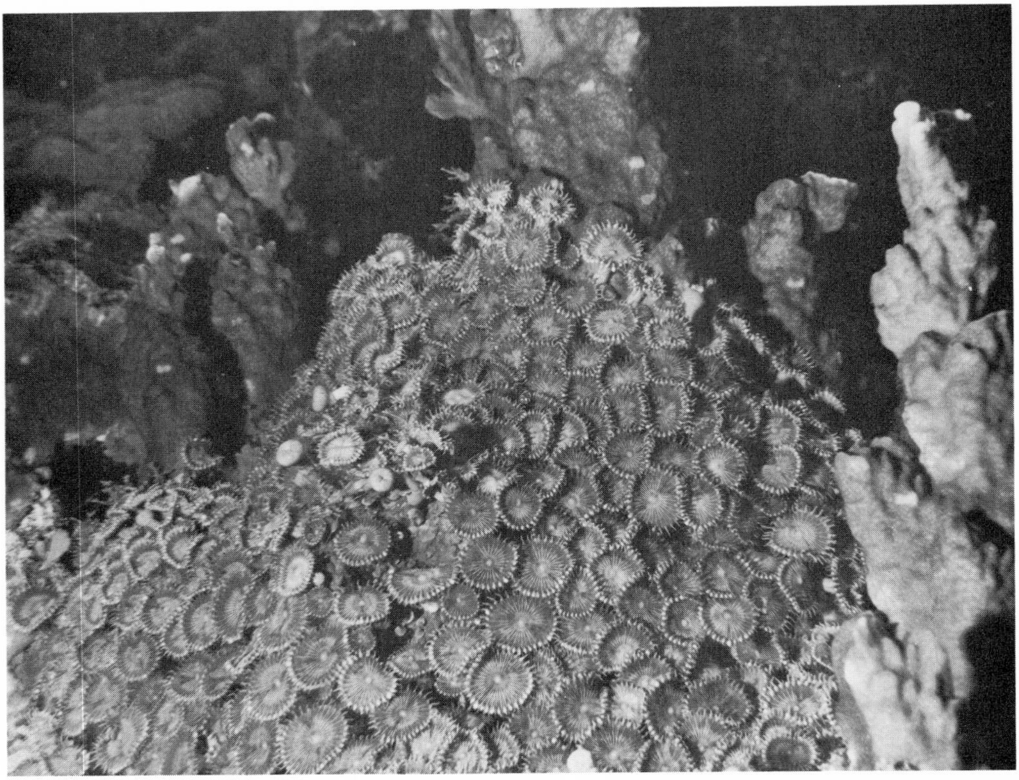

FIG. 12-10. Zoanthids, a type of anthozoan frequently found associated with coral reefs. (Chesher photograph.)

FIG. 12-11. Serpulid worm extending from its tube in a coral reef, radioles and operculum showing. (Chesher photograph.)

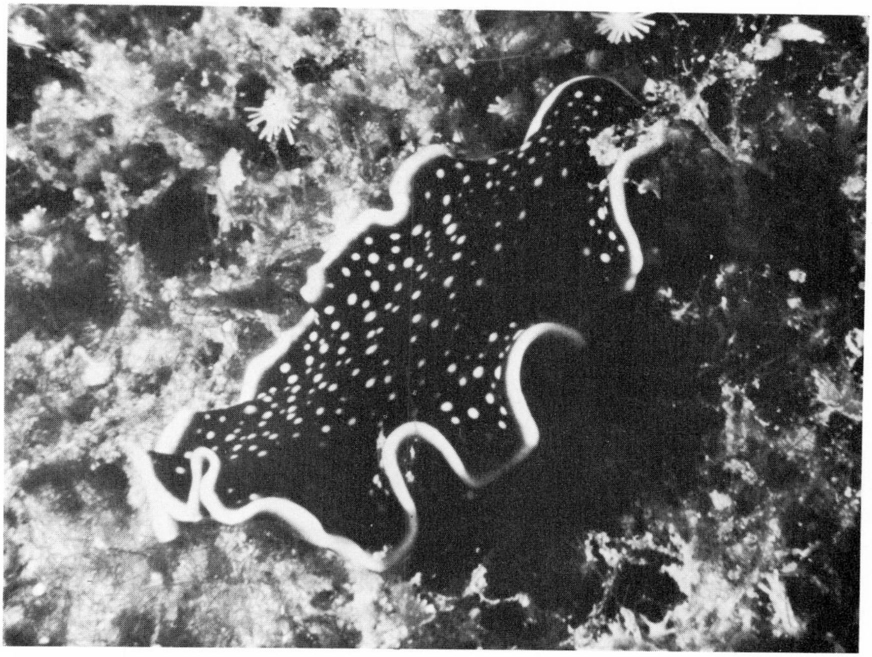

FIG. 12-12. Highly colored flatworm, and stellate foraminiferan. Guam. (Chesher photograph.)

A

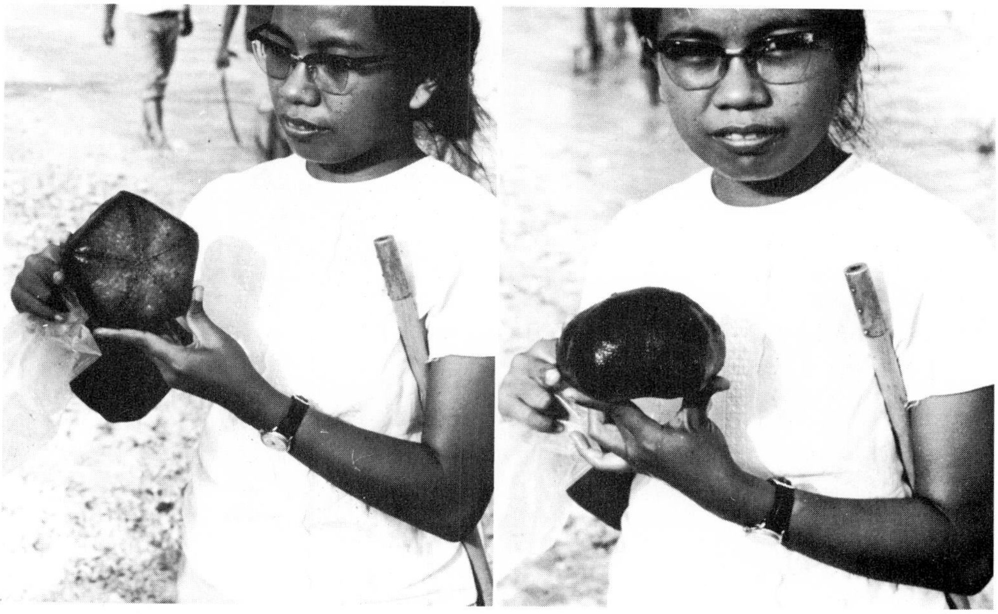

B

FIG. 12-13. Young Indonesian zoologist with large black armless starfish, *Culcita* sp., on the coral islands in the Java Sea off western Java. Ventral view, **A,** and side view, **B,** of the thick, fleshy starfish.

ogists to obtain a clearer idea of the structure and species succession of typical coral reefs. Coral reef morphology has been studied by Mesolella (1967) on Barbados, where a series of eighteen major reefs formed from time to time, the oldest ones being found at the highest elevations and most central positions on the island.

In Figs. 12-6 to 12-8 a typical West Indian coral reef, comprised of four major zones, is shown.

The outer, deepest part of the reef has been termed the **coral head zone** and is comprised largely of species such as *Montastrea annularis,* which form massive, solid-coral heads. Next is the **slope zone,** in the West Indies comprised largely of *Acropora cervicornis,* a typical branching form. Where the slope approaches the mean low-tide level and flattens to form the main platform of the reef, we have the **crest zone,** characterized by the large expanded flattened heads of *Acropora palmata.* Finally, there is the **rear zone,** or platform back of the crest, which contains a considerable variety of species, including recurrences of numerous species found in the other zones.

The corals in a typical West Indies coral reef can then be tabulated as in Table 12-1.

Several of the same species of corals found in the great coral reefs also occur beyond the tropical zone as individual coral growths, but without forming large reefs, especially along western and southern Africa and in Japan. There are also some deeper-water coral species without zooxanthellae, some of which are found as far north as the Norwegian fjords.

Coral reef associations

Corals are important not only because they comprise gigantic biocenoses of their own, characteristic of a distinct zoogeographical region, the tropical zone, but also because they provide the habitat for

TABLE 12-1. Typical components of a West Indies coral reef*

Major elements	Minor elements
Coral head zone	
Montastrea annularis	*Porites astreoides*
Siderastrea siderea	*Agaricia agaricites*
Siderastrea radians	*Favia fragrum*
Diploria strigosa	*Maeandrina maeandrites*
Diploria labyrinthiformis	*Maeandrina brazilensis*
	Colpophyllia natans
Slope zone	*Montastrea cavernosa*
Acropora cervicornis, together with scattered *Montastrea annularis, Siderastrea* spp., and *Diploria* spp.	*Porites porites*
	Eusmilia fastigiata
	Madracis sp.
Crest zone	
Acropora palmata	
Rear zone	
Montastrea annularis	*Maeandrina maeandrites*
Acropora cervicornis	*Colpophyllia natans*
Diploria sp.	*Porites astreoides*
Porites porites	*Montastrea cavernosa*
Agarica agaricites	*Occulina diffusa*
Eusmilia fastigiata	*Madracis* sp.
Favia fragrum	Also hydrocoral *Millepora albicornis*

*From Mesolella, K. J. 1967. Zonation of uplifted Pleistocene coral reefs on Barbados, West Indies. Science **156:**638-640. Copyright, 1967, by The American Association for the Advancement of Science.

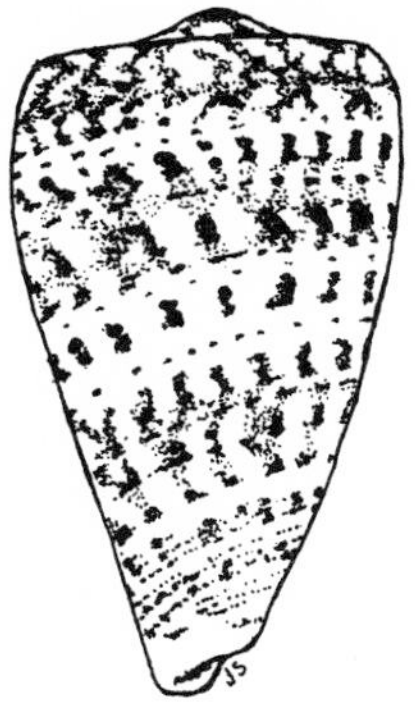

FIG. 12-14. *Conus spirus.* Cone shells are carnivorous gastropods, a prominent and beautiful element of tropical biota.

a special fauna of surpassing variety, beauty, and zoogeographical importance. No other animal associations can compare in these respects with those of the coral reefs.

Some coelenterates belonging to other groups also contribute to the calcium carbonate structure of the reefs. Among them are, especially, the athecate colonial hydroid *Millepora,* which produces massive branching skeletal structures superficially resembling corals. Some West Indies reefs are chiefly *Millepora albicornis* rather than true coral. This organism is called stinging coral or fire coral because of its strong nematocysts. The calcium carbonate skeleton of *Millepora* is much more massive than that of the true corals. Unlike the corals, *Millepora* produces minute free-swimming medusae. The medusae also have strong stinging cells and may cause severe pain to swimmers when they swarm in great numbers. The colony form varies considerably—even in the same species, depending on the ecology—from rather lumpy massive to much-branched growth. Certain alcyonarians such as the organ-pipe coral, *Tubipora,* as well as *Heliopora,* also come in this category.

Some bryozoans also produce calcareous colonial skeletons; these so closely resemble the skeletons of corals that they are almost invariably mistaken for corals by persons who have not studied them carefully. Many of the Paleozoic limestones in America are particularly rich in bryozoan remains. More than 1500 species have been described. Equally rich deposits of bryozoan calcareous rock of Mesozoic age are known in Europe. Bryozoan reefs were formed from the Silurian through the late Tertiary. The contributions of calcareous algae and the many benthic foraminiferans found on coral reefs have already been mentioned.

The coral reef provides a firm base for attachment, hiding places, calcareous rock that can be readily penetrated by boring or corrosion, and abundant calcareous sediment into which animals can burrow. They are inhabited by such a variety and abundance of animals that they constitute rich hunting grounds for a variety of predatory animals especially adapted to taking certain of the inhabitants.

Some of the largest, most beautiful anemones known from shallow waters, such as *Discosoma* and *Actinodendron,* are found here. Fleshy alcyonarians of the group Octocorallia, genera *Lobophytum, Sarcophyton,* and *Sinularia,* and others can sometimes be found carpeting considerable stretches.

Numerous different types of animals have become adapted for burrowing into and living inside the coral heads. Among them are such bivalves as *Lithophaga* and *Lithodomus,* which burrow by means of acid corrosion. The beautiful sinuous many-colored iridescent mantle lobes of the giant clam *Tridacna* can be seen along the surface of stones in which they lie buried. Crabs of the family **Haplocarcinidae** live in galls that they cause to develop on the branches of corals and have become so modified by their peculiar mode of life that it is difficult to determine their systematic relationships. A prawn, *Paratypton,* has adopted a similar mode of life. A peculiar gastropod, *Magilus antiquus,* lives inside the solid *Maeandrina* corals, the last convolution of its shell straightened out during growth and keeping pace with the growth of the coral.

Some crustaceans characteristic of coral reefs are the common widespread crabs *Leptodius exaratus* of the Indo-Pacific and *Trapezia* spp., both of the family **Xanthidae.** Several pinnotherids are also found here, as well as porecellanids and swimming crabs of the family **Portunidae.**

A variety of echinoderms are also characteristic, especially sea urchins (such as the long-spined *Diadema* spp., *Eucidaris,* and others); many sea cucumbers (such as *Holothuria* and *Stichopus*), which are found in abundance in loose coral sand under coral heads, etc., and the worldwide circumtropical *Holothuria atra* (the large, tough black sea cucumber); and some of the most striking and beautiful of the starfish are found in open patches of coral sand between reefs. In shallow lagoons of coral islands one can occasionally find numerous long colorful sea cucumbers of the family **Leptosynaptidae,** sometimes 0.6 or 0.9 meter in length and striped like a king snake, with such thin body walls that they stretch out like a long balloon full of water when picked up and may even break from the weight of water and body fluids. Occasionally one also finds comatulid crinoids actively crawling about like highly ornamented giant spiders.

In the West Indies region the small, short-spined sea urchin *Echinometra lacunter* is found in holes and depressions in the coral rock at the level of the low-tide line. This genus is cosmopolitan in warm seas, and the sea urchins are always found only at this particular tide level—never wandering about over the flats.

These same lagoons may at times be almost carpeted by large medusae, *Cassiopea,* lying with the exumbrellar surface flattened against the substrate and the oral arms, with their many appendages, appearing like colorful vegetable growths.

Many striking and beautiful species of molluscs, such as the cone shells (family **Conidae**), cowries **(Cypraeidae),** conch shells **(Cassididae),** strombids **(Strombidae),** and many others, occur in these waters.

Coral reefs provide abundant hiding places and hunting grounds for such predators as octopuses, eels, and blennies.

The so-called coral fish comprise a great variety of colorful, often peculiarly shaped fish found around coral reefs. Among the most characteristic are the butterfly fish **(Chaetodontidae),** with their vivid coloring, deep, compressed bodies, high dorsal fin, and elongated snout with sharp, bristle-like teeth adapted for pulling small animals from holes in the coral rock. Other prominent members of the coral fish assemblage are the gorgeously colored scarids (*Scarus, Sparisoma,* etc.), numerous members of the family **Labridae,** the Moorish idols **(Zanclidae),** surgeonfish **(Acanthuridae),** trunkfish **(Ostraciontidae),** filefish and triggerfish **(Balistidae),** and the peculiar shrimpfish **(Centriscidae),** which swim about in small schools in a vertical position, head downward.

Cleaning symbionts. Cleaning symbionts constitute a curious and important element of the reef fauna. They are small, colorful fish and certain highly colored shrimp that live by cleaning off parasites and infected tissue from larger fish. Many fish deliberately come to the stations, where these cleaning symbionts live, to be cleaned, and allow the cleaning symbionts to pick away at parasites or growths on them, even entering the mouth or gill chambers in their search. Conrad Limbaugh (1961) found that when the cleaning symbionts were removed from a reef, the reef was visited by a markedly smaller number of larger fish and that the indigenous fish living in the area of the reef tended to become "ratty" and less healthy in appearance (Feder, 1966).

Certain other species have evolved, closely resembling the true cleaning symbionts, but when fish come to be cleaned by them, they not only take parasites and infected tissue but commonly take a bite of healthy tissue from their unsuspecting customer. Older fish learn to tell the difference and go to the right shrimps. Only the inexperienced young and perhaps a few of the more stupid among the older fish continue to make the same mistake. In the Caribbean region the best-known cleaners are the Pederson shrimp *(Periclimes pedersoni)* and the neon goby *(Elacantinus oceanops)*, which sometimes work together at the same cleaning station. The shrimp usually locates next to the anemone *Bartholomea*, thus making a type of triple symbiosis.

The rocky intertidal

SUPRALITTORAL ROCKS

Below the margin of typical terrestrial flora, on the seaward-facing rocky slopes, species of rock tripe, the lichen *Umbilicaria*, are the last of the larger terrestrial flora. During the heat of the day they become blackened, thin, and brittle, giving the rock the appearance of scaling of thin layers. When wet with rain or fog they are like thin, pliable sheets of greenish leather. Below the *Umbilicaria*, to the upper extremity of the highest level of the spring tides, the rocks are commonly barren, with only a few insects or other terrestrial animals temporarily resting on them.

The junction between sea and land, at the lower margin of the high areas, where the rocks are dry almost all the time is usually marked by a darkened zone, which looks more like a stain on the rocks than

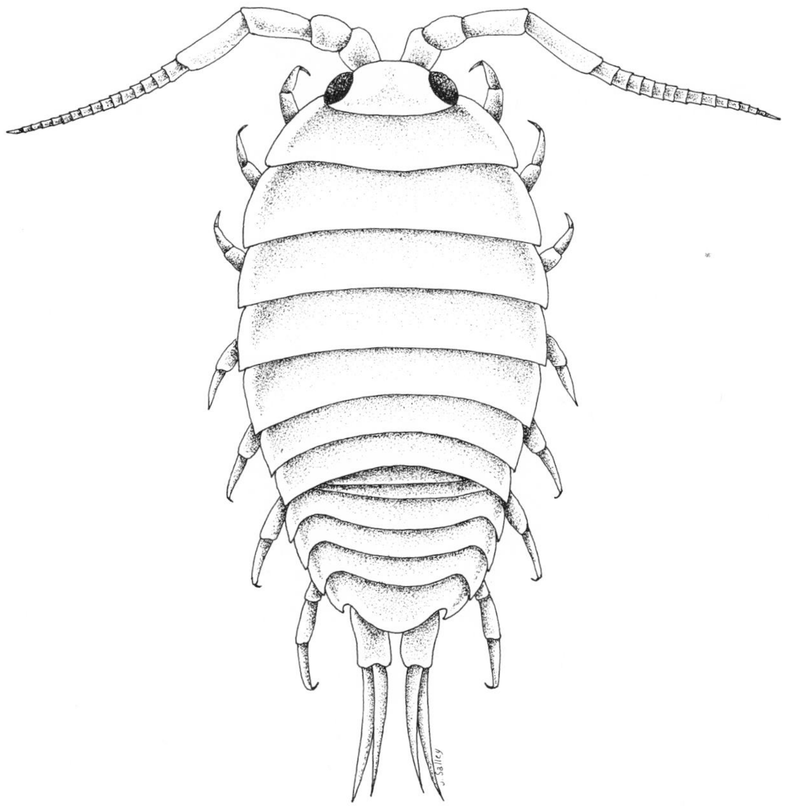

FIG. 12-15. *Ligia oceanica*. The *Ligia* are large, fast-moving isopods found on the supralittoral rocks. (After Richardson, H., 1905. Monograph on the Isopoda of North America, U.S. Nat. Mus. Bull. No. 54, Smithsonian Institution, U.S. Government Printing Office, Washington, D.C.)

like a band of living organisms. This **black zone** is formed by a growth of innumerable microscopic plants, most of them too small to be seen with the naked eye, and many encased in gelatinous, or slimy, sheaths. These sheaths protect them from drying during the long exposures to sun and air and make the rock surface of the black zone extremely slippery when wet. This zone of microscopic plants at the extreme upper margin of the intertidal area is perhaps the most universal mark of the transition from land to sea. Above it the environment is essentially terrestrial. Below it the world belongs to the sea. Most of the microscopic plants of the black zone are primitive blue-green algae. A few green algae and minute lichens are often present among them. They are seldom submerged, but receive the splash and spray from the large waves of the fortnightly spring tides.

In some regions small tide pools in the uppermost edge of the intertidal zone, which only occasionally receive water from the sea and are subject to rain and evaporation, when containing calcareous shell fragments that render them alkaline, become blood red with the growth of *Sphaerella,* an alga (Volvocales) related to species that sometimes produce "red snow" farther inland.

Rock pools in the upper littoral and splash zones are subject to many stresses —dilution by rain, concentration by evaporation, thermal changes, and changes brought about by the activities of organisms in them, such as changes in dissolved oxygen, pH, and dissolved organic materials.

Nonetheless, they often contain a considerable flora and fauna, generally decreasing in numbers of species and in biomass per given volume of pool water as one goes higher up the rocks, and with increasing terrestrial components.

Oxygen content in pools containing appreciable quantities of algae tends to rise during the day as a result of photosynthesis and to drop during the night. Common night levels are 60% to 80% of air saturation. During a bright day the level may rise to as much as 260% air saturation. The pH levels tend to rise and fall along with the levels of dissolved oxygen, both reaching maximal levels around noon or shortly thereafter. During the night and early morning the pH may be just slightly alkaline, but may rise to a strongly alkaline pH of 9.0 or even higher during the day. These effects are more pronounced in higher pools with longer periods of exposure and are also strongly affected by the particular times of submersion or splash. The highest pools in the splash zone may receive water only during high spring tides every two weeks or during storms. There are seasonal as well as daily variations, and cloud or fog cover markedly reduces the amplitude of daily changes.

Temperature fluctuation seems to be of overriding importance in controlling the biota of rock pools.

LITTORAL ROCKS

Attached plankton and seston feeders

The dominant animals of the rocky intertidal are usually attached, or sessile, plankton and detritus feeders that draw their nourishment directly from particles suspended in the water. Some, such as many sponges, bryozoans, some coelenterates, tunicates, and others, have developed the encrusting habit—flat growths firmly adherent to, and following, the contours of the surfaces on which they are attached. Others are raised more from the rock surface but are firmly attached and so shaped that they shed the water readily and offer minimum resistance to the impact of waves. Animals growing in crevices, under ledges, among the holdfasts of marine algae, or in relatively quiet waters, wherever they can obtain enough protection, may assume other shapes and growth habits, such as the many bushy, erect hydroid and bryozoan colonies, globular tunicates and sponges, etc. Still others construct firmly cemented calcareous tubes in which they live or burrow into the substrate, exposing their plankton-gathering apparatus only when the burrow is covered with water. Some of the more conspicuous of these plankton or detritus feeders will be considered group by group.

Barnacles (class Crustacea, subclass Cirripedia)

Barnacles comprise one of the most remarkable, widespread, and characteristic groups of animals to be found in the intertidal and shallow marine shelf waters. In rocky coastal areas certain species of acorn barnacles or sessile barnacles, especially genus *Balanus*, are commonly found among the highest of the truly marine animals on the rocks, extending from the upper limits of the highest tides or even into the splash zone down to about the level of the mean high neap tides. They often occur in such numbers that they form a white band marking the uppermost tidal levels. It is visible from a considerable distance on the darker rock, and in places may grow so crowded together that the shells are distorted—much taller and more cylindrical than normal. Lower on the rocks, these barnacles are replaced by such species as the clumps of stalked, or gooseneck, barnacles, at the beginning of the zone of mussels, and often by some much larger species occurring in smaller numbers in the lowest part of the lower littoral zone and subtidally.

It seems remarkable that a sessile plankton-feeding animal should be among those to occur in greatest numbers and most successfully in the highest part of the intertidal zone, where they have the briefest periods of submergence for feeding and respiration, where they must endure the longest periods of exposure to the air, with the dangers of desiccation, overheating or chilling, and exposure to rain, and where they often grow in exposed situations, enduring the impact of the largest waves during the spring tides or storms.

Some barnacles can tolerate long periods out of the water. *Balanus balanoides* has been kept out of water as long as six weeks without apparent ill effects. During prolonged exposure, air is admitted by a minute opening, small enough to avoid serious moisture loss. *Chthamalus stellatus* has been kept on a desk top for three years with only brief submergence one or two days a month.

Unlike most animals of the intertidal zone, most barnacles are hermaphroditic cross-fertilizing forms, with internal fertilization requiring copulation. Thus if a barnacle of this type happens to be located more than a few millimeters from another individual, it may be unable to reproduce. Some species are capable of self-fertilization. The eggs undergo development within the mantle cavity, and the young are discharged into the water as nauplii.

Within the barnacle zone on the rocks one can often observe that the individuals at the highest levels are all much smaller than those somewhat lower, near the lower edge of the zone they occupy. The reason is probably that the former receive the briefest period of submergence and cannot obtain enough food to support normal growth. Other aspects of their metabolism may also be interfered with by the prolonged periods between high tides, during which they must remain closed up and inactive.

When the tide ebbs, the full active rate of oxygen consumption falls to a much lower quiescent level. Aerial respiration is sufficient to meet the needs during quiescence, but with the onset of excessive desiccation, barnacles are forced to shut down. If this continues, their combined oxygen is eventually exhausted, and they switch to anaerobic respiration, which may continue until the accumulation of acid products reaches lethal levels or may be partly reoxidized by periodic air gaping.

Reduction of metabolic level during periods of exposure not only decreases the oxygen demand but also helps compensate for the shortened feeding time.

Measurements of cirral activity of submerged barnacles (number of sweeps of the cirri per unit of time) plotted against the dry weight of the barnacles' soft parts

FIG. 12-16. Acorn barnacles *(Balanus glandula)* and limpets *(Acmaea digitalis)* on rock face in upper littoral zone, California. (M.

show that younger, smaller individuals are more active—sweep more often per unit weight of tissue—than larger, older ones and that the temperature at which they are most active tends to be correlated with the average water temperatures of the water in which they are living. Temperature acclimation at various latitudes results in somewhat greater activity for those living in colder, more northern latitudes.

Similar considerations apply to numerous other intertidal animals that reduce their activity and close up or seek protected areas during ebb tide.

Adult acorn barnacles are enclosed in a calcareous shell secreted by the expanded carapace, sometimes termed the mantle because of its peculiar character and its surrounding of the body. The volcano-shaped shell consists of a basal part cemented to the rock surface; a parapet, or wall, composed of six nonmovable plates (a dorsa carina, two carinolaterals, two rostrolaterals, and a ventral rostrum); and one or two pairs of movable plates (the scutum and the tergum) near the aperture of the parapet, which allow the barnacle to close itself from the outside environment during periods of stress or alarm, but which can be separated to allow protrusion of the six pairs of thoracic limbs, the cirri, for feeding (Figs. 12-16 and 12-17). Various genera of barnacles are characterized, among other ways, by the varying relationships of the plates of the parapet to each other and in some cases by fusion or elimination of certain of the plates.

A surprising characteristic of some barnacles, discovered by Hoyle and Smyth (1963), is that the scuto-tergal adductor muscles and the depressor muscles are composed of giant cells. In the large barnacle *Balanus nubilus* Darwin, these cells are so large that conventional nerve-muscle preparations can be made, using single muscle cells, and microelectrodes can be readily inserted to obtain recordings from

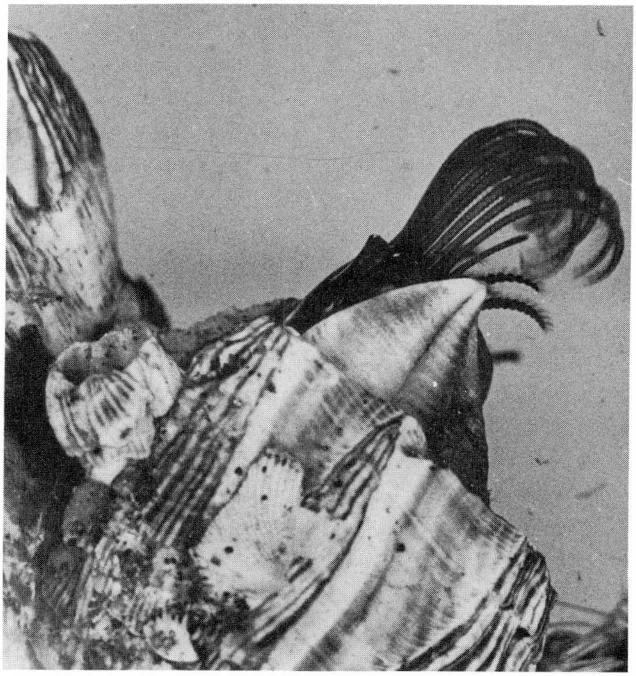

FIG. 12-17. Acorn barnacle *(Balanus tintinnabulum),* feeding. Southern California. (M. W. Williams photograph.)

FIG. 12-18. Stalked barnacles *(Mitella polymerus)* grow in dense clumps in upper part of the mussel zone. La Jolla, California. (M. W. Williams photograph.)

individual muscle cells. They are therefore useful for certain kinds of research in neuromuscular physiology.

In the stalked, or gooseneck, barnacles (Fig. 12-18) the body is enclosed in a soft mantle, the capitulum, commonly protected by calcareous plates. The capitulum includes all the body except the preoral portion from which the stalk was developed. The calcareous plates usually include two scuta, posterior in position, two lateral terga, and an anterior keellike carina. Transverse adductor muscles enable the barnacle to draw the edges of the terga and scuta together for protection.

Barnacles undergo a remarkable metamorphosis during their development. The young are liberated as characteristic nauplii, with the anterolateral corners of the carapace drawn out into lateral spines, giving them a triangular shape when seen from above or below. The nauplii are at first nourished by large fat globules, which also keep them near the water surface. As these are metabolized, the nauplii swim lower in the water.

After the naupliar instars they transform into a cypris stage, with the body enclosed in a bivalved carapace. At this time they resemble ostracods somewhat. When the cypris is ready to settle down and metamorphose into the adult stage, it begins to explore the substrate actively. The cypris larvae in general prefer clean, rough, pitted rocks to smooth surfaces, and are repelled by diatom slicks or algal film. They seem especially to prefer sites on which older barnacles have already settled and grown. They may spend an hour or more exploring rock surfaces before making a choice and settling down. On selecting a suitable spot, the larva attaches to the surface by a small disc on each of the antennules and secretes the basal attachment from the cement glands, which occupy much of the preoral region. A dorsal groove appears and enlarges, nearly separating the portion of the preoral part of the body containing the cement glands and antennules from the rest of the body proper, whereas the bulk of the body turns a half somersault within the carapace. Thus the barnacle ends up lying on its back at right angles to the long axis of the carapace in such a position that the six pairs of thoracic limbs point upward. These become the cirri, fringed with setae. When protruded and spread out, the cirri form a net with which the barnacle sweeps the water immediately above the orifice of its shell. Upon retraction the cirri come together, enclosing any food particles encountered and bringing them into the mantle cavity, where they are scraped off and transferred to the mouth by the maxillae. Small barnacles that have been exposed out of water for some time become active upon being immersed, and the feeding reaction is readily observable.

It has been experimentally shown that the barnacles characteristic of the higher parts of the rocks in the intertidal zone do not require the periods of exposure above water. In fact, they grow faster and larger if maintained continually submerged, where they can feed continuously. It is probable that they are limited to the upper fringe of the intertidal zone by competition for space with algae and encrusting organisms that require somewhat longer periods of submersion or that cannot stand such long exposure. In the lower parts of the barnacle zone, where the barnacles grow intermingled with small, attached seaweeds such as *Fucus* or *Ascophyllum,* it can often be seen that the sweeping motion of the algae over the rock face with the surge and ebb of waves commonly keeps the rock surface swept clean. Predation by carnivorous gastropods and starfish may also play a role in limiting some species of barnacles to the uppermost zones at which plankton feeding is possible.

Barnacles are a prominent element of the "fouling community," organisms that grow on ship bottoms, underwater pilings, pipes, etc.

Bivalve molluscs

Below the barnacle zone the most abundant and conspicuous filter feeders are various bivalve molluscs. The great majority of bivalves feed by catching small particles of plankton or organic detritus and bacteria on mucus of the gills as the respiratory current passes through them. The mucus, with the contained food particles, is swept to the lower margin of the gills by special tracts of strong cilia and then brought anteriorly along the food groove at the gill margin to the vicinity of the mouth, where the palps transfer it to the mouth.

In the rocky intertidal the bivalves can be grouped into three categories on the basis of their relationship to the substrate.

In the first category are the most conspicuous bivalves such as mussels, which live on the surface of the rock and are held in place by strong byssal threads secreted from a byssal gland near the base of the foot. Mussels commonly occur in great numbers, as closely spaced as possible, forming mats, or beds, of mussels sometimes covering much of the exposed rock surface between the lower margin of the barnacle zone and the upper edge of the zone of permanent seaweeds. Some species such as *Mytilus californianus,* the California sea mussel, prefer open, exposed coastline where they meet the full brunt of the waves breaking over them on incoming tides. Others, such as the widely distributed *Mytilus edulis,* often called the bay mussel, prefer somewhat quieter, more protected waters. It is probable that massing together, giving each other additional support, is essential in the case of mussels growing in exposed situations, enabling them better to withstand the wave shock. The mussel bed itself provides protection for many other animals, including worms, gastropods, small crabs and shrimps, and other small bivalves and often serves as a substrate for settlement of barnacles, bryozoans, tunicates, sponges, and small mussels. Eventually, the mass of animals may become so thick and heavy that big chunks break away during periods of unusually heavy waves, leaving patches of bear rock, where the cycle of colonization may begin again.

The principal predators of mussels are starfish, carnivorous gastropods, and man. Of course, if any of the mussels are crushed or damaged, as by heavy objects being cast on them by waves or by people walking on them, the injured individuals are promptly disposed of by scavengers belonging to many groups.

Below the beds of mussels exposed at low tide, beginning at the lower edge of the *Chondrus crispus* (Irish moss) zone where the sweep and backwash of the waves is still heavy, and occasionally in deep crevices or pools in the higher zones are found the larger horse mussels *(Volsella modiolus),* which are up to 5 inches long, with heavy, bulging shells and strong byssal threads attaching them in all directions. They often also attach byssal threads to shell fragments and pebbles brought into their zone by waves. The mat of byssal threads entangled debris etc. creates a protected habitat for a variety of small organisms, worms, crustaceans, echoinoderms, etc., and for the young of numerous others that would otherwise be unable to live in this rocky zone where the surge of the waves is so strong.

In somewhat deeper water or in places in warm shallow calm waters, a relative of the mussel, the large, fragile pen shell (*Atrina* spp.), sometimes occurs in great numbers, often where a few centimeters of loose material overlie something firmer to which they can attach. They occur offshore in places from Cape Hatteras south along the East Coast, becoming especially abundant along the Gulf coast of Florida. Millions of them are sometimes cast up on beaches by storms, second only to the arc shells in abundance on the beaches. They are of worldwide occurrence in tropical waters. The byssal threads of pen shells have a beautiful golden sheen and fine texture. In the Mediterranean region they have been used for the production of an especially fine-textured, soft, highly valued "cloth of gold."

The second category of intertidal bivalves consists of such forms as oysters, rock oysters, and some others that cement part or all of one of the valves firmly to the substrate as they grow. In such forms the valve that is cemented to the rock assumes the irregularities in shape of the surface to which it is fastened, and the upper valve becomes somewhat irregular to fit snugly.

The third category of bivalves finds protection from the waves and from predators by boring into the substrate, extending only the tips of the siphons to the surface for feeding. Where the rock is relatively soft sandstone, or calcareous, the surface few inches may be riddled with the burrows of pholads and date-nut shells. They have thus become a factor of some geological importance in hastening the disintegration of exposed rock. The boring habit has been assumed independently by different unrelated groups of bivalves. Still others, which do not themselves make burrows, make use of those made by other organisms. Such species may be termed nestlers.

Boring organisms fall into two categories depending on the method used in making their burrows—the chemical borers, limited to calcareous rocks, and the mechanical borers, in which a portion of the shell is especially modified for scraping out the burrow. The burrows are started soon after transformation of the swimming larval stage and continue to be enlarged throughout the growth of the occupant, with the result that the opening is constricted and the occupant is permanently imprisoned in the burrow.

Chemical borers dissolve calcareous rock by means of acid secreted against the rock as the flattened surface of the foot is pressed against the wall of the burrow. The acid is neutralized there before gaining access to the burrow generally. Further protection of their own calcareous shells is afforded by a thick horny outer periostracum.

One of the most characteristic and widely distributed groups of rock-boring clams are the pholads, or piddocks, family **Pholadidae.** In these clams the shell is an elongate oval, without ligament and hinge. There are specialized, filelike rasping areas on the anterior part of the shell. The animal clings to the wall of its burrow with its short, broad foot and rasps with the shell, changing its position from time to time. Piddocks burrow into relatively soft rocks such as consolidated clay, marl, shale, peat exposures, and some sandstones or conglomerates. The boring action can be readily observed by putting young animals in a test tube of the same bore as their burrow, immersed in an aquarium. Piddocks have the mantle cavity closed for most of its length. The siphons are long and united nearly their whole length, the posterior end of the shell permanently gaping to permit their passage. There are three adductor muscles.

One of the best known is the angel wing shell *(Pholas [Barnea] costata)* of the East Coast from Massachusetts to the West Indies, which burrows deeply into peat exposures, clay, etc. It grows up to 7 inches long and has a beautiful, fragile, pure-white shell. It is most abundant from Virginia southward.

One genus, *Martesia,* the wood-boring piddock, has taken to boring into wood rather than rock. Unlike the shipworm, the piddock's burrows are superficial, since it excavates only nestling sites and does not continue to bore deeply through the wood.

Perhaps the largest is the rough piddock, *Zirfaea pilsbryi,* which burrows over a foot deep in soft shale along the shores of the Pacific Northwest.

Another characteristic mechanical borer is the rednose, *Hiatella,* a rock-boring clam recognized by its red siphons, also found among crusts of coralline algae and in kelp holdfasts.

The giant clam *Tridacna* is exceptional in that it grows embedded in hard substrates in a supine position; it grows with its large mantle lobes containing symbiotic zooxanthellae exposed, rather than the ends of the siphons. It may be that in this case the clam is too large and massive to support itself solely by plankton feeding and requires the zooxanthellae to supplement its diet.

Among the most remarkable of the mechanical borers are members of the family **Teredinidae,** which includes shipworms. These molluscs have become adapted for burrowing in wood. They grow as long, wormlike organisms, with the shell restricted to two small scraping valves at the anterior end and most of the body being the long, tubular mantle. Instead of enclosing the body with shell, they line their burrows with shell material. Since the opening of the burrow, made by newly transformed larval shipworms, is scarcely larger than a pinhead, the only external sign that a piece of wood is under attack by shipworms is the presence of minute holes in the surface, which are easily overlooked, even though the wood itself may be riddled by the burrows. Shipworms cause immense damage to wooden ship bottoms and structures such as pilings, etc. in shallow marine waters. In San Francisco Bay in the early part of this century, they caused millions of dollars' worth of damage before anyone was aware of their presence. The problem was recognized only when piers, wharves, and other structures supported by pilings began to cave in.

Tunicates (order Ascidiacea)

In the lower intertidal and subtidal zones where there is firm substrate to which to attach, solitary and colonial tunicates are another conspicuous group of filter feeders, with a different variation in the ciliary-mucus method of gathering food from the water.

The solitary tunicates (Fig. 12-19) are globular or elongated saclike animals attached to the substrate at one end and with two openings, an incurrent oral opening and an excurrent atrial opening, usually

FIG. 12-19. Solitary tunicate *(Styela montereyensis),* with attached small mussel *(Mytilus californianus)* and hydroid *(Sertularia* sp.).

FIG. 12-20. Compound tunicate *Distaplia.* (M. W. Williams photograph.)

rather close to each other at the free end. The body is invested with a tough protective cuticular covering, the tunic, made largely of cellulose secreted by the underlying epithelial cells and containing mesenchymal cells that have wandered into it. Most of the body is occupied by a greatly expanded pharynx, or branchial basket, which communicates with the surrounding atrial space by numerous stigmata, or gill slits. The combined respiratory-feeding current created by the cilia of the pharynx enters the oral opening and passes through the stigmata into the atrial cavity and out the atrial opening. Food particles are retained by the mucus lining the pharynx and passed by special tracts of cilia to the endostyle, a longitudinal, glandular ventral groove, and back into the esophagus. The digestive tract consists of esophagus, stomach, and intestine, which is twisted into a loop so that it comes forward to empty into the atrial cavity, as do also the ducts from the gonads. The circulatory system of tunicates is unique in that the direction of circulation is reversible, the heart pumping the blood alternately in one direction into the pharyngeal vessels and then reversing itself and pumping it into the body and the viscera. Tunicates are hermaphroditic, in most cases shedding the gametes into the water, where they develop into a tailed, tadpolelike stage that possesses a notochord in the tail. Larvae thus display the chordate relationships of tunicates more clearly than do the adults.

Some tunicates can also reproduce asexually, budding off new individuals and forming colonies of more or less closely associated individuals. This highly developed capacity has led to the formation of so-called compound tunicates (Fig. 12-20), in which a great many small individuals or zooids are embedded in a common gelatinous matrix. Different patterns of budding and colony formation and differences in relative amounts of gelatinous material between the zooids in different groups of compound tunicates have led to the development of an amazing variety of colorful forms, ranging from thin flat encrusting growths closely adherent to the substrate to large fleshy or lumpy colonies such as *Amaroucium* (often misspelled *Amaroecium*).

In some places, such as certain subtidal areas in the Bay of Naples, rock faces have become so coated with compound tunicates and sponges that they form spectacular submarine gardens.

Sponges

Among the most ubiquitous and conspicuous components of lower intertidal and shelf fauna, wherever the substrate presents suitable attachment and the water is reasonably clean, are the sponges. They grow in a great variety of forms ranging from thin encrustations (Fig. 12-21) to massive growths and erect (Fig. 12-22) branching types. All sponges are sessile animals with a highly porous body permeated with channels and spaces permitting the flow of water through them, a type of living sieve. The water enters through minute incurrent openings, the **ostia;** passes through one or more chambers lined with special flagellated collar cells, the **choanocytes;** and leaves either directly or by way of a series of larger anastomosing channels, through one or more larger openings, the **oscula.** The choanocytes maintain the flow of water through the sponge and capture minute food particles as it passes. Oxygen is brought directly to all the cells by the circulating water.

Most functions are carried out at the cellular level, there being no organs and organ systems as we know them in higher animals. The low degree of organization of sponges is shown by the fact that if they are cut into pieces, each piece continues to function as if nothing had happened. Some have even been completely disaggregated by forcing bits of them through bolting cloth, and the cells have been observed to reaggregate into small sponges of the same type as the parent individual. Growth is plantlike, and in many species asexual

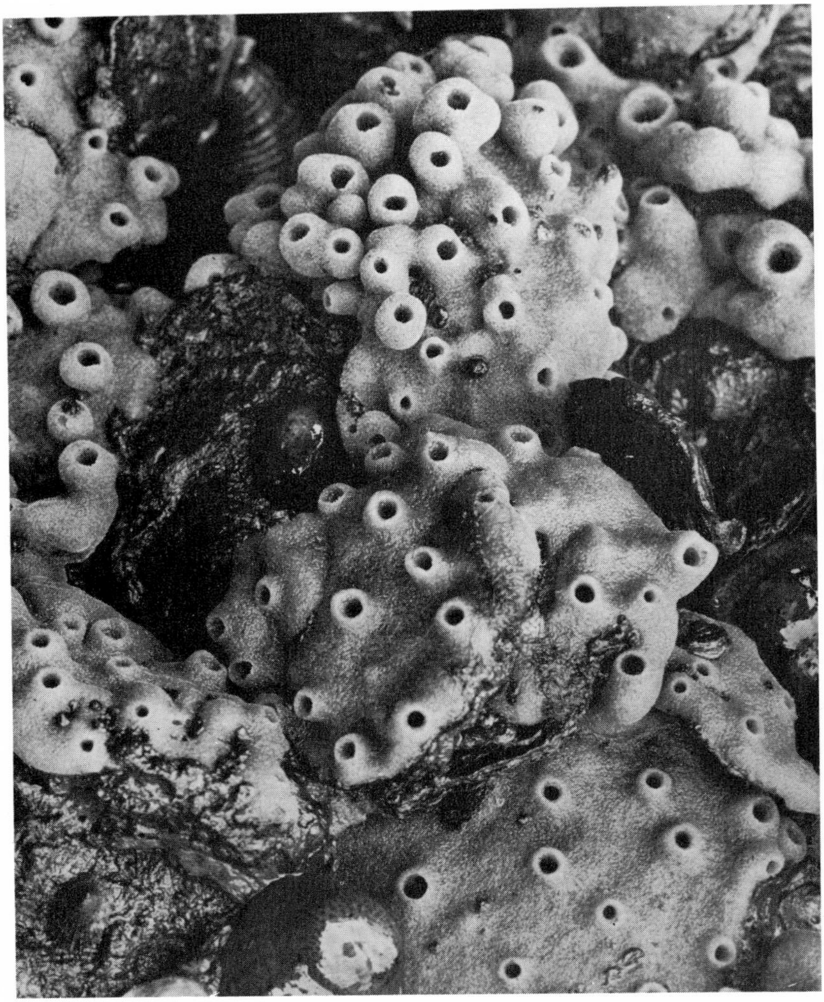

FIG. 12-21. Crumb-of-bread sponge, *Halichondria panicea,* a common West Coast encrusting species with raised, volcano-like oscula. (M. W. Williams photograph.)

buds of various types are produced. It is difficult to say just what constitutes an individual in such an animal. Perhaps the question is meaningless at this level. The most obvious functional unit of most sponges would be one osculum, together with the most directly associated flagellated chambers and water channels. In the Huxleyan sense, an individual would be all the growth that proceeds from one fertilized egg cell, whether or not it remains together.

The body of the sponge, or mesenchyme, contains several different types of cells, many of which are amoeboid and some of which secrete the characteristic skeletal elements—spicules or fibers of spongin.

Sexual reproduction in sponges is by gametes developed in the mesenchymal tissue. Male gametes liberated from one sponge may enter another along with the incurrent streams of water and fertilize ripe egg cells there. The fertilized cells commonly undergo part of their development in place and leave the parent sponge as peculiar, nearly spherical larvae, known as

FIG. 12-22. Dead-man's-fingers, *Chalina* sp., an erect tubular sponge. (Oregon Institute of Marine Biology photograph.)

amphiblastulae, containing small flagellated cells in one hemisphere and large yolky nonflagellated cells in the other. The amphiblastulae are carried about for a brief time in the plankton. If successful in finding suitable substrate when ready to settle, they affix themselves to it by the flagellated pole and develop directly into small sponges. The flagellated cells give rise to the choanocytes, and the rest of the sponge body is derived from the large nonflagellated cells. The amphiblastula should be regarded as a nonfeeding larval stage, not as a homologue of the embryonic blastula stage of higher animals. Its subsequent development in no way resembles that of an ordinary blastula, and there is little to suggest that the various cells of sponges can be homologized with the ectoderm, endoderm, or mesoderm of the embryos of higher animals.

The classification of sponges is based largely on their skeletal structures, since only they can be studied in ordinary museum specimens. A formidable terminology has grown up describing the various types of spicules and their arrangements. In spite of their primitive degree of organization, the various species of sponges maintain distinctive and recognizable growth habits. Some of them are highly colored—yellow, red, purple, or greenish.

Sponges are preyed on by relatively few animals. The prickly skeleton of spicules of most of them and probably chemical distastefulness in many seem to give good protection, although some molluscs, notably various nudibranchs and a few other animals, do graze on certain species of sponge. Many small invertebrates are commonly associated with larger sponges, nestling in or under them for protection.

FIG. 12-23. *Astrosclera willeyana* Lister, from Guam. This sponge represents a new class of sponges, the Sclerospongiae, described by Hartman and Goreau (1970). (Chesher photograph.)

FIG. 12-24. *Aglaophenia* sp. Ostrich-plume hydroid colony. (Oregon Institute of Marine Biology photograph.)

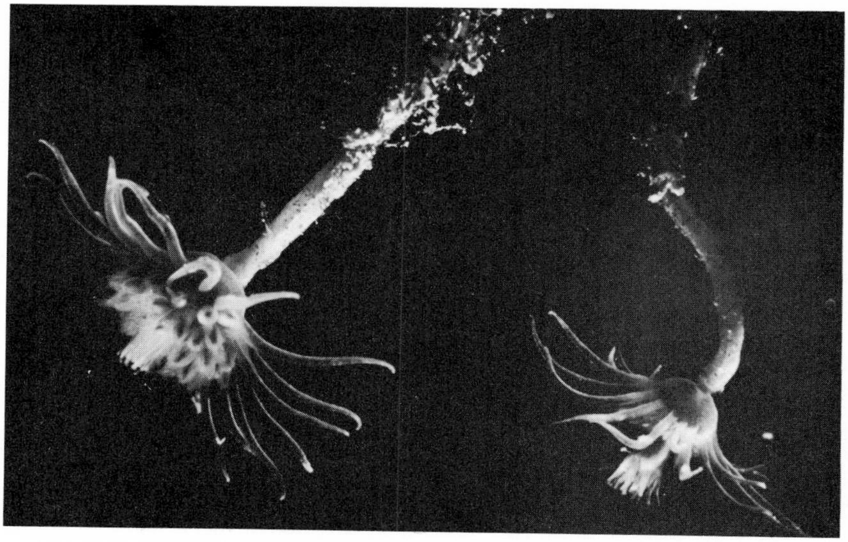

FIG. 12-25. *Tubularia,* with sessile medusoids between the two rings of tentacles. (Oregon Institute of Marine Biology photograph.)

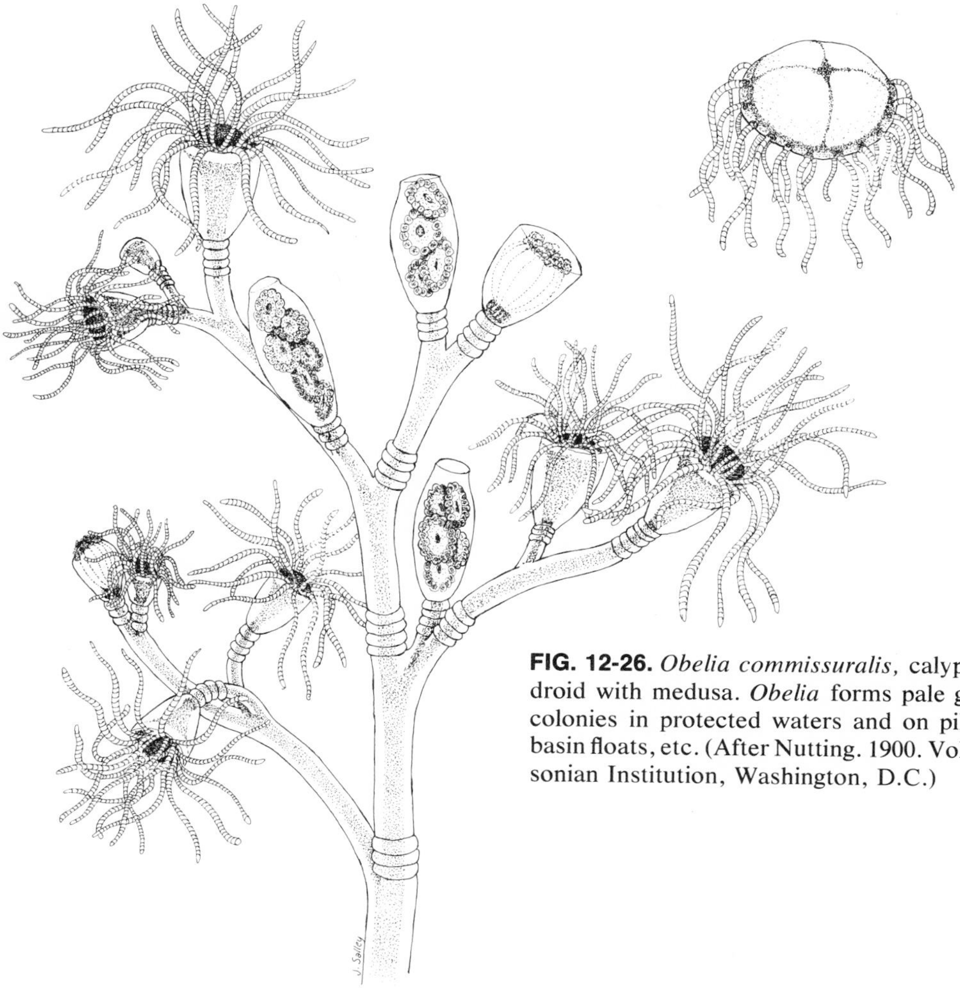

FIG. 12-26. *Obelia commissuralis,* calyptoblasthydroid with medusa. *Obelia* forms pale gray bushy colonies in protected waters and on pilings, boat basin floats, etc. (After Nutting. 1900. Vol. 1. Smithsonian Institution, Washington, D.C.)

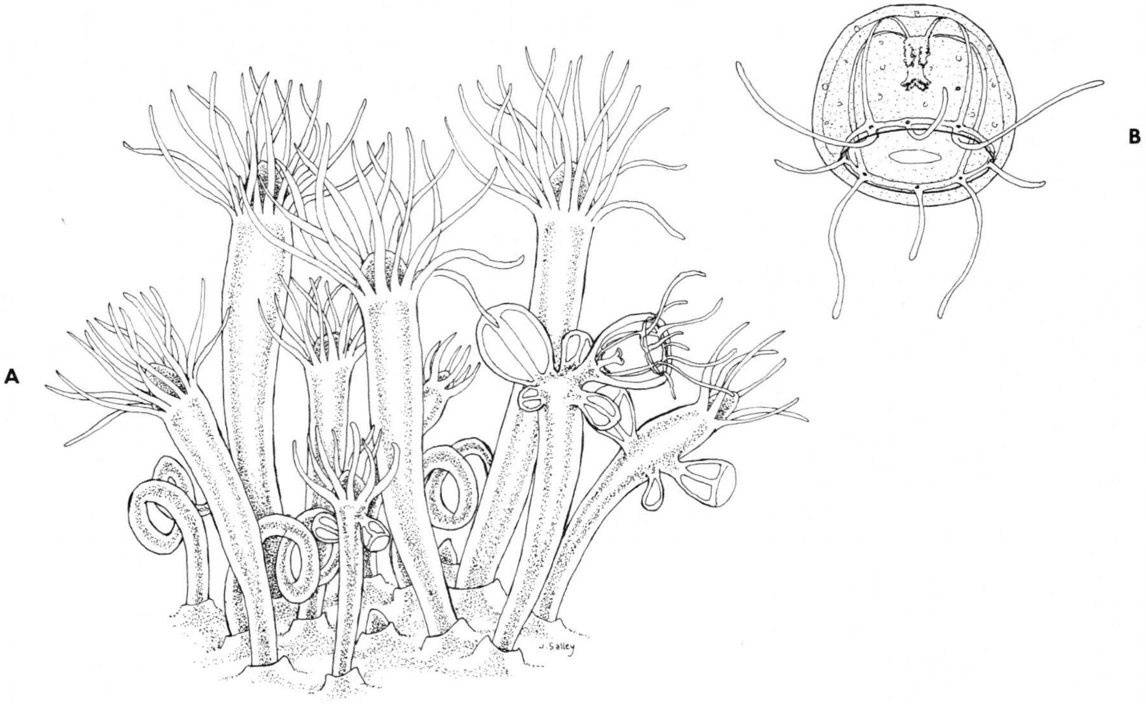

FIG. 12-27. *Podocoryne carnea*—another hydroid of protected waters. **A,** Colony; **B,** medusa. (**A** after Fraser, C. M. 1944. Hydroids of the Atlantic coast of North America. University of Toronto Press; **B** after Russell, F. S. 1954. Medusae of the British Isles. Vol. 2. Cambridge University Press, New York.)

Hydroids

Hydroids represent the polypoid generation of Hydrozoa. The medusoid generation has been discussed in the section on gelatinous zooplankton. The majority of the hydroids belong to intertidal and shallow shelf fauna. Most are colonial, constituting variously branched bushy or feathery (Fig. 12-24) growths in which the body layers and gastrovascular cavities of the polyps or hydranths are continuous throughout the colony, so that the entire colony is nourished by whatever any of the polyps feed on.

Each polyp is provided with a crown of nematocyst-bearing tentacles around the mouth, with which they capture and immobilize zooplankton organisms. A bushy hydroid colony containing dozens or hundreds of such polyps is a formidable peril for any zooplankter swimming near it or carried through it by water movements.

Most hydroids are protected by a noncellular secreted **perisarc,** which covers the stems and stolons (or **hydrorhiza**), giving stiffness, elasticity, and support to the colony. In many it also forms a cup, the **hydrotheca,** which extends up around each polyp and into which the polyp can contract if alarmed, and a protective cover, the **gonangium** or **gonotheca,** which surrounds reproductive polyps.

The **coenosarc,** or living portion of the stems and stolons, is composed of the same two layers of cells of which the polyps are comprised, the outer epithelium and inner gastrothelium. These layers have commonly been termed ectoderm and endoderm in coelenterates, but it seems preferable to reserve those terms for their embry-

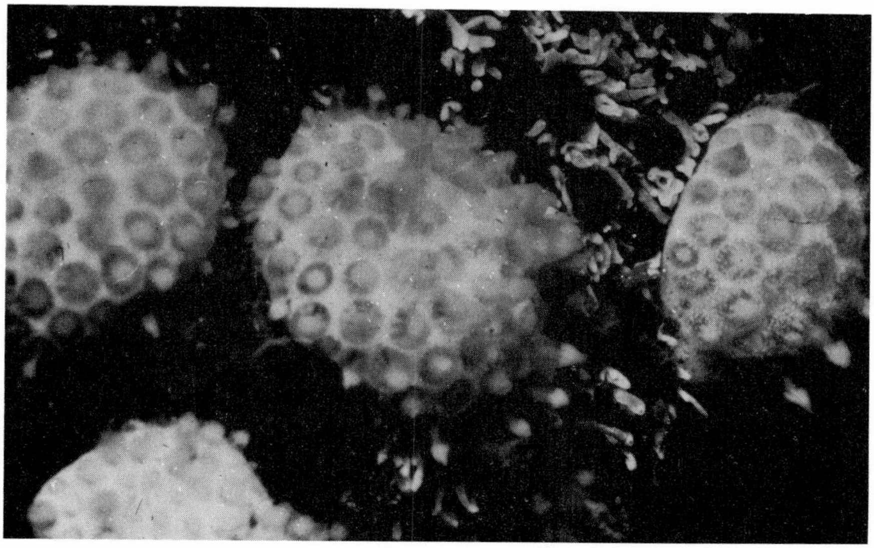

FIG. 12-28. Soft coral, *Alcyonium* sp., from lower littoral of Oregon. (Oregon Institute of Marine Biology photograph.)

ological usage and to describe the adult tissues of coelenterates in the same manner employed for similar tissues in other phyla, although recognizing their ectodermal and endodermal origin.

In many hydroids more than one kind of polyp have differentiated. The two most frequent are the **gastrozooids,** or feeding polyps, and the **gonozooids,** or **gonophores,** the reproductive polyps, which bud off medusae asexually. In addition, some species have developed **dactylozooids,** rodlike structures heavily armed with nematocysts and without tentacles or mouth.

CLASSIFICATION OF HYDROIDS

The order Hydroidea is divided into two suborders. In suborder **Gymnoblastea** the hydranths and reproductive zooids are devoid of perisarc, that is, there is no hydrotheca or gonotheca, although the stems and hydrorhiza may bear a perisarc. Members of this group are variously termed athecate hydroids or tubularian hydroids (Fig. 12-25). In those that give rise to free-swimming medusae, the medusae are usually tall and bell-like and bear marginal ocelli but not statocysts, and their gonads are usually borne on the manubrium. Such medusae are termed **Anthomedusae.**

The suborder **Calyptoblastea** contains species in which the perisarc forms hydrothecae and gonothecae. Free medusae of hydroids in this group are commonly flatter and more saucer-shaped, possess marginal statocysts, and bear the gonads on their radial canals. They are termed **Leptomedusae.**

Alcyonaria

In addition to the hydroids many anthozoans of the group Alcyonaria, also called Octocorallia, form characteristic growths on subtidal rocky areas and coral reefs. Some of them, especially small colonial "soft corals" (Fig. 12-28), may also be found in the low intertidal rocks, under ledges, in crevices, etc.

Alcyonaria are all colonial. The polyps are readily recognized by their eight tapering pinnate tentacles with pinnules. Anatomically they are also a homogeneous group, the polyps each possessing eight complete septa, with the longitudinal muscles located on the sulcal side of each septum (the side of the animal bearing the siphonoglyph) and with the gonads developing on six of the septa (all but the sulcal pair).

The alcyonarian colony is composed of a

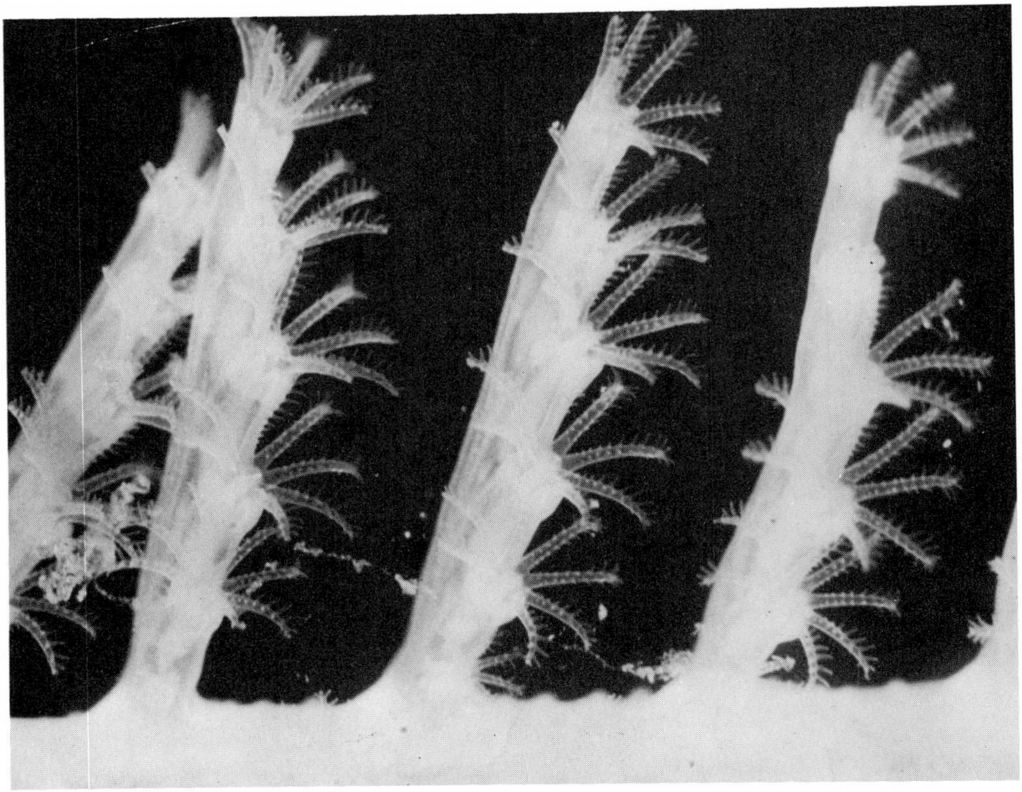

FIG. 12-29. Gorgonian, small portion showing a few side branches of the colony, with polyps extended so that the eight pinnate tentacles are visible. (M. W. Williams photograph.)

fleshy **coenchyma** representing the thickened cellular mesogloea, containing amoebocytes of various kinds and secreted skeletal elements. **Solenia,** or endodermal tubules, which are continuous with the gastrodermis of the polyps, run through it, forming a continuous gastric network throughout the coenchyma, connecting all the polyps. All the nematocysts are small, simple atrichous isorhiza (simple tubular types without spines).

The order **Gorgonacea** in particular, which includes gorgonians (Fig. 12-29) (sea whips, sea fans, etc.), includes some of the most colorful and characteristic animals of submarine gardens. The colony possesses an axial horny or calcareous skeleton and grows as bushy or fanlike growths. Sea fans grow in relatively quiet water at depths and locations where they are not subject to excessive surge or turbulence from wave action. They develop with the plane of growth of the colony at right angles to the direction of prevailing water movement so that the water strains through the fan, giving its hundreds of polyps maximal chance to capture small plankton organisms.

The genus *Corallium* of the suborder **Scleraxonia** includes the red corals, or "precious coral" of commerce. Another colorful member of the Alcyonaria, found in warm waters, is the organ-pipe coral, a rather aberrant member of the order **Stolonifera,** which also forms a red calcareous skeleton colored by iron salts incorporated into it and somewhat resembling tiers of organ pipes.

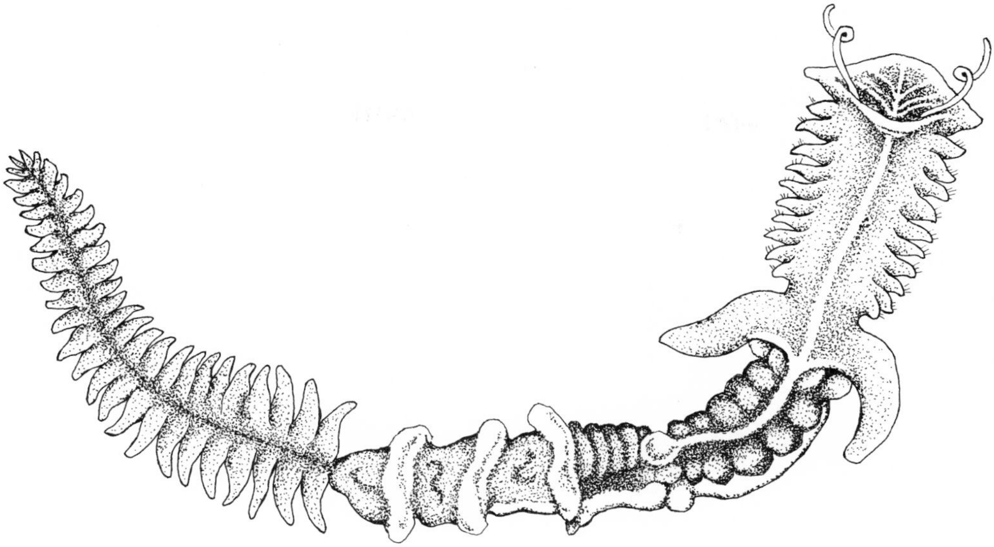

FIG. 12-30. *Chaetopterus variopedatus.* This animal constructs a mucus funnel in anterior portion of tube, circulates water through the tube by means of modified paddlelike parapodia, and from time to time eats the mucus funnel with its retained organisms and organic detritus. (After McIntosh, W. C. 1915. British marine annelids. Vol. III. Ray Society, London.)

Tube worms

An amazing variety of polychaete worms have adopted the sessile, tubicolous habit, together with elaborate specializations of the body enabling them to gather plankton directly from the water or to obtain particles of organic detritus from the neighborhood of their tube. Among those found on hard substrates, perhaps the most characteristic are the fan worms, families **Serpulidae** and **Sabellidae,** and certain members of the **Cirratulidae, Terebellidae,** and others, which also have many members in softer substrates. Many terebellids, chaetopterids, and others actually characteristic of softer substrates are commonly found under rocks on sandy mud or in crevices in rocky areas where detritus collects.

Many serpulids construct calcareous tubes, ranging in size from a few millimeters to several centimeters in length, strongly cemented to the rock surface. The anterior end of the worm bears two large groups of stiff food-gathering tentacles, the **radioles,** which, when extended, form a beautiful, often highly colored funnel-shaped crown. The cilia on these tentacles create currents that bring food particles to the mouth. If alarmed, serpulids instantly snap back into their tubes, closing the orifice with a specially modified group of tentacles that form an **operculum.** Serpulids possess an anterior collar and a thoracic membrane formed by fusion of dorsal and ventral cirri of some of the thoracic segments and used in molding the tube as calcium carbonate is secreted from glands beneath the collar. Serpulids and other worms of this type, which have a rapid withdrawal reaction in the presence of danger, have giant nerve fibers running through the nerve chain without synaptic interruption at each segmental ganglion.

Among the most familiar and ubiquitous of the serpulids are members of the complex usually called *Spirorbis* (Fig. 12-31). These small worms build flat, spiral-coiled tubes encrusting on seaweeds,

FIG. 12-31. *Spirorbis,* minute serpulid tube worms. (Oregon Institute of Marine Biology photograph.)

FIG. 12-32. *Eudistylia,* a large sabellid tube worm, sometimes called the feather-duster worm. May attain a length of 2 feet. (M. W. Williams photograph.)

shells, rocks, pipes, etc. These white tubes, forming little coils only 2 or 3 mm. across, may be so numerous that they practically cover the surfaces on which they have settled and may become a nuisance in salt-water plumbing systems. The larvae tend to settle gregariously and to prefer to settle near adults of their own species. This tendency may be an advantage in that it ensures selection of a location where conditions are suitable for survival as well as for attachment, since the adults have survived there. Some authorities maintain that there are sufficient differences between those species that form sinistrally coiled tubes and those with dextrally coiled tubes to warrant splitting the group into two genera —*Laeospira* and *Dexiospira*.

Among the most spectacular tube worms are various large species of Sabellidae, such as *Eudistylia* (Fig. 12-32), some of which construct parchmentlike tubes more than a foot long, usually in somewhat protected places such as the undersides of docks and on pilings in harbors, but sometimes clumps of them can also be found in intertidal rocks. These so-called plume worms, or feather-duster worms, have a big cluster of colorful tentacular radioles at the anterior end but no operculum, and the body differs from that of a serpulid, among other ways, in having a longitudinal ciliated "fecal groove," which carries feces deposited in the back of the tube forward and crosses and interrupts the lines of thoracic notosetae in a curious manner. In some sabellids such as *Schizobranchia*, the tentacular radioles may be branched several times. Many serpulids and sabellids bear eyes either at the base of, or at intervals along, the radioles.

The family Cirratulidae comprises worms in which anterior segments bear long, filamentous outgrowths continuing to arise, in

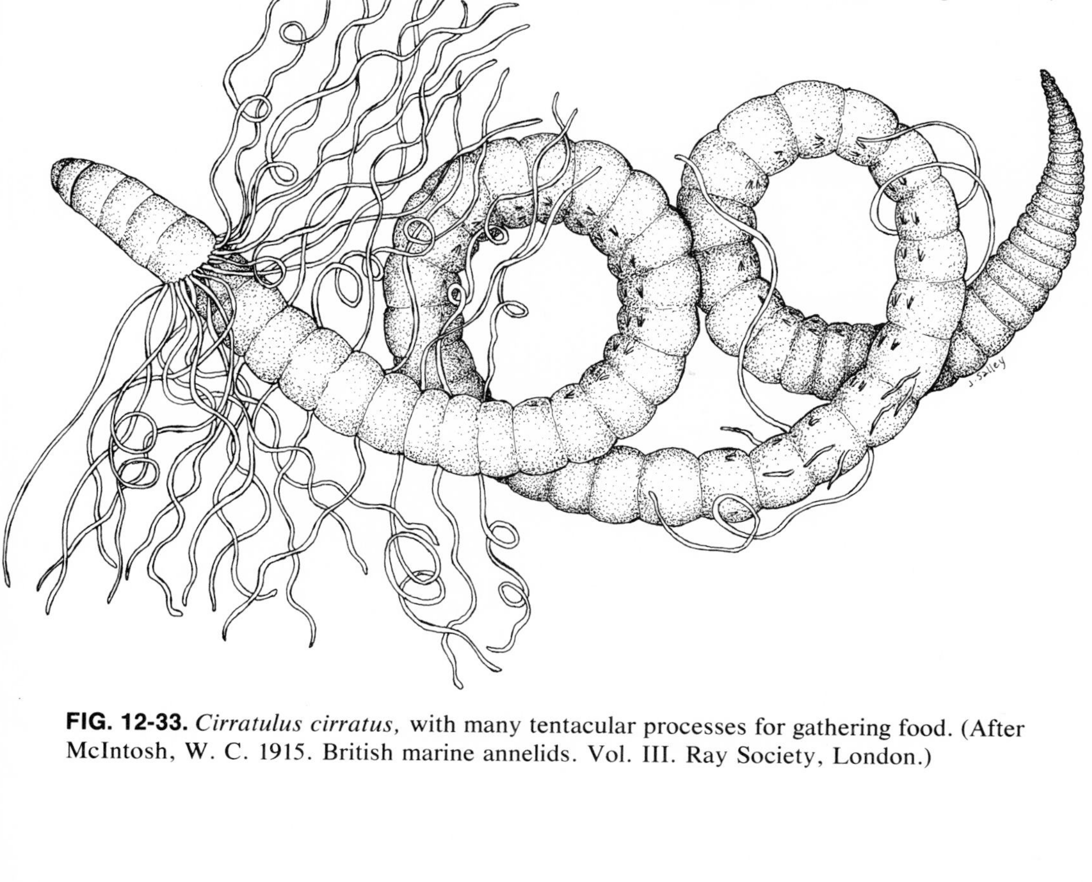

FIG. 12-33. *Cirratulus cirratus,* with many tentacular processes for gathering food. (After McIntosh, W. C. 1915. British marine annelids. Vol. III. Ray Society, London.)

many of them, along the sides for some distance back from the anterior end. The body is not clearly divided into two distinct regions. One genus, *Dodecaceria,* contains rather small, blackish worms with only a few filamentous processes. *Dodecaceria* constructs hard tubes of calcium carbonate. Through asexual reproduction large colonies may be built up, producing thick, firmly attached, rocklike masses comprised of the agglomerated tubes of hundreds or thousands of the worms and sometimes termed "worm rock." *Dodecaceria* is common in the lower littoral zone in rocky areas along the western coast of North America, as are also various larger cirratulids that do not construct calcareous tubes.

The lophophore bearers

Bryozoa. The Bryozoa, or moss animals **(Ectoprocta),** comprise a group of small colonial animals attached to rocks, shells, seaweeds, etc. large bryozoan colonies may contain thousands of individual zooids produced asexually by budding. Each zooid is housed in a boxlike or tubular **zooecium** of hardened, often calcified cuticle secreted by the epidermal cells. The body cavity is extensive and lined by a peritoneum. The **lophophore** is a circular, crescentic, or double spirally coiled ridge, bearing hollow ciliated tentacles that surround the mouth. The gut is a broad ciliated U-tube, looped in such a way that the anus lies close to the mouth, outside the lophophore. A muscular band, the **funiculus,** runs from the stomach, near the curve of the looped gut, to the aboral end of the zooecium, enabling the bryozoan to retract the lophophore region into the zooecium when alarmed or during periods of exposure at low tide. Most bryozoans are hermaphroditic, the testes developing from peritoneum lining the funiculus and the ovaries developing from peritoneum lining the lateral walls of the zooecium. Ova and sperm are shed into the coelom, and development of the larvae may proceed there. In those forms in which the zooecium is hard and imperforate, the body wall is reflected back under the anterior edge of the wall of the zooecium, forming a sacklike hydrostatic organ, the **compensatorium,** from which water is expelled on retraction of the zooid, making room for it in the zooecium.

The form, patterns, and color of bryozoan colonies vary greatly. Many species form thin flat encrusting growths closely following the contours of the substrate and tightly adherent to it. Others form fleshy thickened or branching colonies, and some form erect bushy dendritic colonies. Some bryozoan colonies such as *Flustrella* form fleshy growths commonly mistaken for brown algae. Others somewhat resemble certain corals.

Approximately 4000 recent and 15,000 fossil species of bryozoans have been described. Recent marine bryozoans fall into the subclass **Gymnolaemata,** in which the lophophore is more or less circular. Freshwater species are classified as subclass **Phylactolaemata.** They possess a horseshoe-shaped oval or double-spiral lophophore and an epistome dorsal to the mouth, give rise to statoblasts, and display other features clearly distinctive of their group.

Endoprocta. Another group of small animals formerly included in the Bryozoa are the **Endoprocta,** commonly called nodding-heads. They are often found especially on the underside of rocks in the lower intertidal and subtidal and sometimes on mollusc shells and other hard substrates. They are not nearly as common as the ectoproctans. In the endoproctans the anus is inside the circle of tentacles; the tentacles are folded in toward the center, as in hydroids, when the zooid withdraws; and the body cavity is a pseudocoelom rather than a true coelom and is secondarily filled with parenchymatous tissue and obliterated. The zooids of a colony are connected only basally through stolons and do not have zooecia. The stalks of the zooids are muscular, and they exhibit a characteristic nodding movement, which gives them their popular name.

FIG. 12-34. Brachiopods dredged near Santa Catalina Island. (M. W. Williams photograph.)

FIG. 12-35. Brachiopod from Coos Bay, Oregon, to show peduncular orifice of the ventral valve. (Oregon Institute of Marine Biology photograph.)

Brachiopoda. Brachiopods (Fig. 12-34), or lamp shells, are small bivalved animals, most of them superficially resembling bivalved molluscs. The two valves are dorsal and ventral rather than lateral in position. They usually are unequal and taper toward the posterior end where a stout, muscular peduncle attaching the shell to the rock passes out, between the two valves in some, or through a hole in the posterior end of the ventral valve in others. Some brachiopods cement the ventral valve to the rock as they grow and have no peduncle. The mantle cavity between the valves is mostly occupied by the large double-spiraled lophophore, which is sometimes supported by calcareous extensions of shell from the dorsal valve. Each lophophore bears a set of ciliated tentacles alongside a ciliated groove in which small particles are swept back to the mouth. Larger particles are rejected. The mouth is simple, without special jaws, lips, or palps. It leads into a short esophagus, stomach, and intestine. Saclike digestive glands open into the stomach. Sexes are separate. The shells are composed of calcite rather than the aragonite usually found in mollusc shells.

Brachiopods have a long fossil history and have been much used in stratigraphical correlation.

Mobile plankton feeders

Although plankton feeding by benthic organisms is usually associated with the

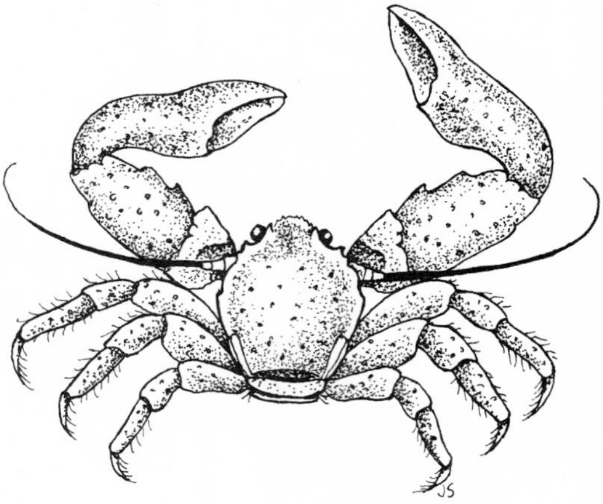

FIG. 12-36. *Porcellana serratifrons,* an anomuran. One of the few crabs that are plankton eating. (After Henderson, J. R. 1888. Challenger reports. Vol. 27.)

FIG. 12-37. Mossy chiton, *Mopalia mucosa*. (R. Buchsbaum photograph.)

sessile habit, a few of the plankton feeders of the rocky intertidal are mobile. The most conspicuous group are probably the flat crabs or porcelain crabs, a group of anomuran crabs found under loose stones and boulders (Fig. 12-36).

ALGAL-FILM GRAZERS

A second major source of food in the intertidal and subtidal rocky substrate communities is the film of growth and of molecular and colloidal organic matter that coats the surface of all rocks and other objects exposed to marine waters. This film is a mixture of diatoms, minute algae, bacteria, early stages in the growth of larger algae, and minute animals, together with adsorbed and settled particles and molecules of organic matter—collectively termed the "algal film."

Two groups of molluscs, the chitons and the gastropods, have evolved as algal-film grazers. Both groups are adapted for clinging to the rock surface by means of a broad flat muscular foot and slowly creeping about, scraping off the algal film with a filelike tongue, the **radula.** The tenacity with which some of them can cling is truly remarkable, enabling them to withstand the heavy surf and surge of incoming tides on exposed coasts without being swept away. Both groups have developed a calcareous shell for protection from predators and environmental stresses. Basically they have rather similar life cycles, involving a planktonic trochophore stage that later undergoes changes leading to the familiar mollusc form. Both are ancient groups, going all the way back to Cambrian times. The gastropods have evolved into the largest, most varied group of molluscs, and one of the major animal groups on the earth. They have colonized most marine habitats, have evolved plankton-feeding and carnivorous groups, and have extensively colonized terrestrial and freshwater habitats as well.

The chitons, on the other hand, have been an extremely conservative group from an evolutionary standpoint. From their beginning they seem to have been primarily confined to coastal waters, many of them to intertidal or shallow subtidal waters on rocky substrate. Chitons are an example of a group of animals that appears to have become well adapted to a particular type of ecological situation early in its evolution and to have found conditions sufficiently constant to enable it to persist through the ages in this situation with relatively little change.

Perhaps the difference in evolutionary history between the two groups is due in part to a difference that appears during larval life. In the case of the chitons (Fig. 12-38), the primitive bilateral symmetry is retained, the chief features of their evolution apparently having been an elongation and dorsoventral flattening of the body that mostly obliterated the large posterior mantle cavity, a growth of the mantle down around the rest of the body as a tough, protective girdle that comes into contact with the substrate on all sides, and the secretion of the shell as eight separate plates partly embedded in the dorsal mantle tissue, giving the body as a whole enough flexibility to fit itself to the contours of the substrate or curl up for protection when pried loose. With the body and head covered and protected by the girdle, most of the sensory functions usually associated with the head either degenerated or were transferred to the edge of the mantle, which bears innumerable minute sensory spines and which in some is adorned with variously formed sensory "hairs." Young chitons may in some species even bear light-sensitive organs, called aesthetes, at the ends of small channels in the shells on their back. There has also been, in chitons, a secondary multiplication of the ctenidia, which are found in a row on either side of the body in the pallial groove between the foot and the girdle. Water is circulated over them by ciliary action, the chiton lifting the girdle in the anterolateral areas to admit the water and discharging it posteriorly, often through a permanent notch in the posterior edge of the girdle.

A

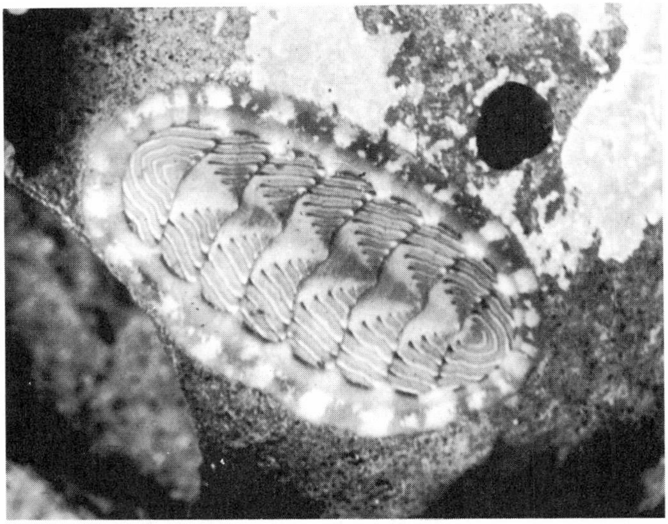

B

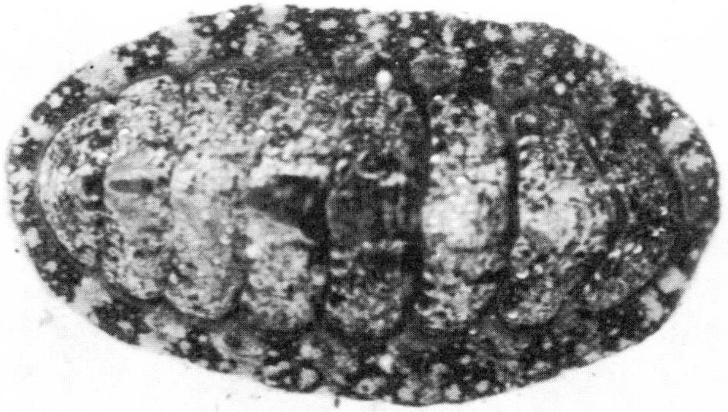

FIG. 12-38. A, *Tonicella lineata,* the lined chiton, a pretty, small species common along the western coast of North America. **B,** Another small chiton from the Oregon coast, about 1 cm. long or less, probably *Nuttallina.* (Oregon Institute of Marine Biology photographs.)

In gastropods, on the other hand, the visceral hump is not flattened but during larval life undergoes a remarkable rotation, termed **torsion,** effected by contraction of the strong right velar muscle of the larva. In some groups torsion is effected more slowly as a growth phenomenon rather than by a sudden contraction of the velar muscle. Torsion rotates the entire visceral hump 180 degrees, bringing the mantle cavity, gills, anus, and excretory and reproductive openings to the front directly over the head and reversing the direction of all organs in the affected part. The result is a twist between the foot and the visceral mass that bends the nervous system into a figure eight (the **streptoneurous** condition) and puts morphologically right structures on the left side and morphologically left structures on the right in the affected parts. Torsion should not be confused with spiral coiling of the visceral mass, which is another phenomenon, probably an adaptation for better compacting and balancing of the visceral mass. Spiral coiling is brought about by differentials in growth rates on the two sides of the animal.

Although enabling the gastropod to pro-

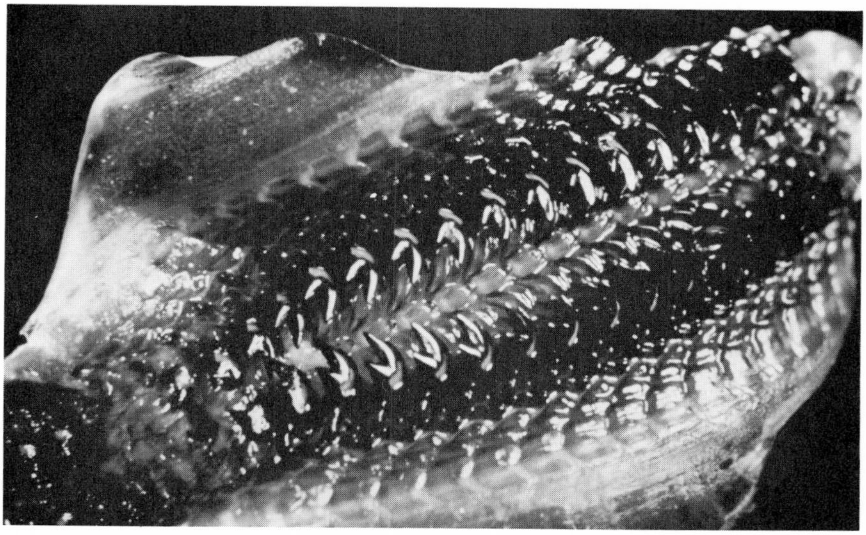

FIG. 12-39. Distal end of the radula of *Cryptochiton stelleri*.

tect its head by withdrawing it into the shell-protected mantle cavity, torsion also brings several problems. The discharge of deoxygenated water, feces, and gametes from the mantle cavity now occurs directly in front of the mantle opening, over the head of the gastropod, directly in the path of the animal's progression instead of at its posterior end. These waste products and any debris stirred up by the advancing edge of the foot on the substrate or by feeding activity are likely to enter the mantle cavity and contribute to possible fouling of the gills. Many of the features evolved by various groups of gastropods seem to be in large part adaptations to meet these problems, as well as the problem of balancing the high, twisted visceral mass.

Four principal directions of adaptation to the various problems created by torsion have been successfully explored by different groups of gastropods.

Perhaps the most primitive adaptation, found in the pleurotomariids today and in certain ancient gastropod fossils, is a groove, or notch, in the leading edge of the shell. It allows the water entering the mantle cavity from the front and sides to escape dorsally over the back rather than directly in the path of the animal. Refinements on this plan consist of closing the leading edge of the notch with subsequent growth of the shell so that it becomes a dorsal hole leading from the mantle cavity. The abalones and various kinds of keyhole limpets, or volcano shells, represent culminations of this line of adaptation.

The most successful, widely found adaptation is a change in direction of the flow of water through the mantle chamber so that the water comes in from the left side, passes through the mantle chamber, and is discharged on the right. With this basic change, various secondary modifications have been evolved in various groups, making it more efficient: shifting the anus and excretory and reproductive openings to the right, reduction or loss of the right gill with compensating development of the left gill, and development of a portion of the mantle edge as a siphon, which better directs the water flow into the mantle chamber. In burrowing forms or forms that live in soft substrate, such as olive snails, this siphon may become a rather long, wormlike snorkle that can be extended up into the water even though the mollusc is largely or wholly buried. In siphonate gastropods, many of which live in muddy or turbid waters, the gill axis is usually fused to the top of the mantle chamber and has only a single row of filaments **(monopectinate condition),** which is less easily fouled than the large, free bipectinate gills of more primitive gastropods.

Terrestrial gastropods, the Pulmonata, evolved from forms with this left-to-right adaptation of the mantle cavity and its contents. Adaptation to terrestrial life involved loss of gills altogether, wrinkling and vascularization of the walls of the mantle cavity, converting it into a type of lung, and restriction of the opening to prevent desiccation. The anal and reproductive openings are usually on the right side and wholly outside the mantle cavity. The life cycle also required modification, with elimination of planktonic larvae.

Finally, in the opisthobranch gastropods are forms that meet the problems of torsion by simply undoing it through subsequent detorsion. In most opisthobranchs the shell is reduced, sometimes enveloped in mantle tissue, or wholly absent. The sea slugs, or nudibranchs, not only undergo detorsion but have altogether lost the shell, mantle cavity, and gills and have developed secondary branchial structures, not homologous with the lost ctenidia, on the dorsal surface of the body. In the dorid type of nudibranchs these structures consist of a circle of retractile "gills" around the anus, which is in a middorsal position near the posterior part of the body. The aeolid type of nudibranchs, on the other hand, have no such anal gills but develop a variety of colorful nonretractile cirri, or processes, from the dorsal surface of the body. In many of them diverticula from the gut may extend up into these processes.

The Gastropoda are the largest, most varied group of molluscs. In addition to algal-film grazers, numerous groups have evolved other feeding habits, ranging all the way from plankton feeding to predatory hunting. The slipper shells (*Crepidula* spp.) have adopted the ciliary-mucus method of plankton feeding, using the gill as a collecting organ much in the manner of bivalves. The Vermetidae form tubular shells firmly fixed to the substrate and looking more like worm tubes than gastropod shells. They extend the gill tips and sweep the water for minute plankton organisms in much the way that a barnacle sweeps the water with its cirri—an adaptation unique among gastropods.

FIG. 12-40. *Thais lapillus,* one of the whelks, or carnivorous gastropods, to be found in the barnacle-mussel zone.

In carnivorous gastropods (Fig. 12-40) the radula may be modified as a poison dart, as in cone shells, or as a drill for making holes in other shelled animals such as oysters, mussels, or barnacles, as in the oyster drills, whelks, etc. Others, with less change of structure and habit, have taken to grazing on sessile animals such as bryozoans, sponges, hydroids, and gorgonians instead of the algal film. The many beautiful nudibranches are an example of a whole major group that has changed to the carnivorous habit without making a radical change in the method of feeding, but even here some surprising individual developments are found. For example, the nudibranch *Melibe* has a large, thin frontal membrane dorsal to and partly surrounding the mouth. It uses this membrane

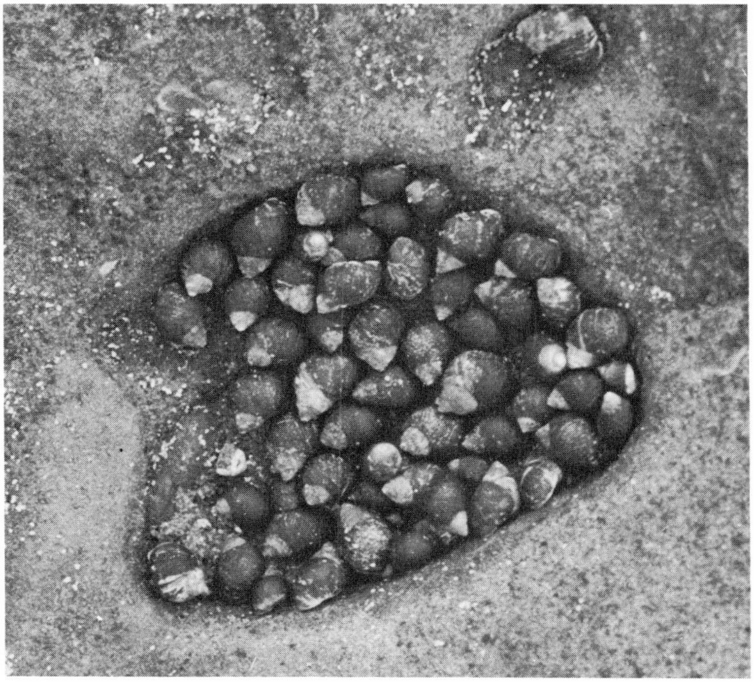

FIG. 12-41. Periwinkles, *Littorina planaxis,* nestled in crevices or small pockets in rocks between high tides. (M. W. Williams photograph.)

in the manner of a butterfly net, spreading it out in front of the head and capturing passing copepods or other small zooplankters.

Perhaps the great evolutionary success and diversity attained by the gastropods, as compared with chitons, are bound up with the problems created by torsion and by the high, unwieldy visceral mass. These continued challenges may have led to selection of diverse lines that improved on the imperfect basic pattern in a variety of ways. The chitons are, in a way, the victims of their own early success, having become so well adapted for a particular mode of life in a particular type of habitat, early in the development of the group, that they have never been forced to improve or change. There has been enough similarity in shoreline conditions throughout the ages to enable them to persist as a small, moderately successful group.

Representative intertidal gastropods

In the upper fringe of the intertidal zone, in and extending a little above the barnacle zone, the rocks may be peppered with small periwinkles, snails of the genus *Littorina* (Fig. 12-41). They show an interesting evolutionary series of species adapted to different degrees of exposure, ranging from species that occur only in the lower intertidal, where they are covered with water most of the time, to forms occurring so high on the rocks that only the highest spring tides and the spray from the waves reach them. The periwinkles seem to be midway in an evolutionary course such as that which in earlier times brought forth the present terrestrial snails out of the sea and onto the land. The three common species along the New England coast illustrate different stages of this progression. The smooth periwinkle *(L. obtusata)* is found in the lower intertidal zone, where it is submerged most of the time. It can tolerate only relatively short periods of exposure. The mantle cavity and gill are characteristic of marine gastropods generally. The eggs are laid in small gelatinous masses adherent to fucoid seaweeds, and the young enter the air bladders of these plants for protection. The adults remain among wet seaweeds at low tide. The common periwinkle *(L. littorea),* which lives higher on the rocks, where only the high tides cover it, can stand relatively long periods of exposure, and has the wall of the mantle cavity more wrinkled and vascularized and the gill somewhat reduced. The rough periwinkle *(L. saxatilis)* is almost a land snail. The mantle cavity is highly vascularized and functions much like a lung. Continued submersion is fatal to it. It can tolerate a month of exposure to the air before becoming too desiccated. Furthermore, it has freed itself of the reproductive tie to the ocean by becoming viviparous. It is most active during fortnightly high spring tides and during the interval between these tides becomes more sluggish and slightly desiccated. Curiously enough, the rhythm of the tides is somehow deeply impressed in its being. If kept in the laboratory under constant conditions, its activity pattern continues to reflect the rise and fall of the tides over its native rocks for several months. A similar series of species occurs along the Pacific coast and the coasts of Europe.

I have kept several specimens of a related periwinkle *(Tectarius muricatus)* from the West Indies on my desk top as a paperweight for nine months and had them promptly revive and start crawling about when put into seawater again. According to Moore (1958), they have survived in a dry state for at least seventeen months.

The second group of gastropods particularly characteristic of the intertidal zone, and with some species extending into the highest parts of the intertidal, are the limpets (Fig. 12-42), families **Acmaeidae** and **Patellidae.** In limpets the shell is conical rather than spiral, some species having an appearance that has given them the name "Chinaman's hat." The shell apex is usually closer to the anterior end of the shell, so that the anterior slope is usually somewhat shorter and steeper than the posterior slope, and the apex itself may be in-

FIG. 12-42. Limpets, *Acmaea limatula,* common on intertidal rocks, at a lower level in the intertidal than *A. digitalis.* (Oregon Institute of Marine Biology photograph.)

clined anteriorly. In the Acmaeidae there is a large bipectinate gill in the mantle chamber. The Patellidae have lost the gill and developed a number of platelike pallial branchiae, or pseudogills, in the mantle groove lateral to the foot. Acmaeids are of worldwide distribution and are the most common limpets in most of the world. Patellids are, for the most part, Old World forms, found along the coasts of Europe and Africa.

Limpets move about and feed when covered with water or while the rocks are still wet. Between tides they tend to cling tightly to the rock, preventing desiccation and predation by terrestrial animals. Many limpets seem to adopt a particular spot on the rock to which they return after feeding excursions. This spot may be worn so that the resident limpet just fits snugly into it. This "homing instinct" has been investigated a number of times ever since Aristotle first reported the habit, but the precise mechanism, the physical basis for the homing ability, and the organs and senses involved are still uncertain.

Constant scraping on rock surfaces subjects the radular teeth to considerable wear. As they are worn down or lost, they are replaced by new portions of the radula pushed up from behind. In both limpets and littorinids, the radula is long relative to the size of the animal. The continual scraping, year in and year out over the centuries, of countless millions of small radulae also has an appreciable wearing effect on the rock, cutting it away grain by grain. In one tide pool carefully observed for sixteen years, periwinkles lowered the floor about ⅜ inch.

Among the species of limpets living in the lower parts of the intertidal zone, some have taken on the habit of scraping the cortical cells of seaweeds, rather than the algal film on rocks. Some of them have become so specialized that the foot and shell are specifically shaped to fit the particular plants on which they live. On the Pacific coast of North America the common large brown alga *Egregia* often has characteristic holes in the beltlike stipe made by *Acmaea insessa*, or it may have numerous depressions, each containing one limpet, where they have not eaten entirely through the stipe. The most highly modified of these plant-eating limpets are probably species such as *A. depicta* and *A. paleacea*, which live on eelgrass and have become shaped to fit the narrow blades of this plant.

In the lower intertidal and subtidal zones are also found other, more primitive limpet-like gastropods, the keyhole limpets, in which both gills are retained and the excurrent water from the mantle cavity is discharged through a hole at the apex of the shell. The resemblance of some of these conical shells to volcanoes has given rise to the popular name "volcano shells."

The group of gastropods to which these keyhole limpets belong, the **Zygobranchiata,** are the most primitive of living Gastropoda. Others in this group are abalones, which have a series of holes in the shell for the excurrent current, and the deeper-water pleurotomariids, which have a slit in the leading edge of the shell, rather than a hole, for this purpose.

Going into the lower intertidal and subtidal zones, out onto the shelf, numerous other groups of gastropods are found, in addition to those which have been mentioned.

The algal film itself is better developed and more lush where it is submerged most or all of the time. It supports a considerable fauna of microcrustaceans, nematodes, flatworms, small annelids, protozoans, and other small animals that live in or derive their sustenance from it. Many larger scavengers and predators such as crabs and small fish also pick at it and derive a significant part of their nourishment from elements in the algal film. The young of many fish, and other animals that later change their diet, may also pick small organisms from algal film as well as from plankton.

In tropical coral reef areas certain of the coral fish such as filefish and triggerfish

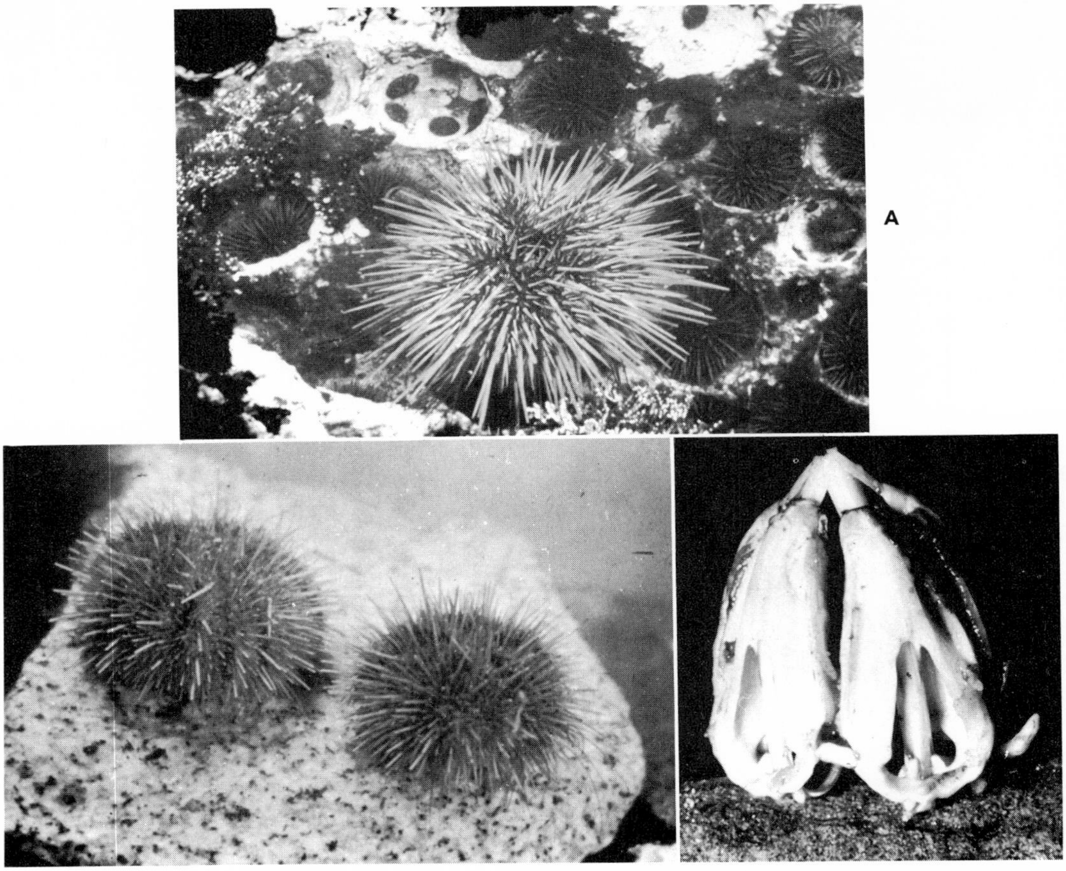

FIG. 12-43. A, The two common intertidal and subtidal West Coast sea urchins: *Strongylocentrotus franciscanus,* the large one in the center, and *Strongylocentrotus purpuratus.* **B,** *Strongylocentrotus dröbachiensis,* the green sea urchin. **C,** Aristotle's lantern, the mouth parts of a sea urchin. Perhaps this is the most complicated mouth of any animal. Five long, chisellike teeth are maneuvered by a complex of over sixty-five calcareous and muscular parts. (Oregon Institute of Marine Biology photographs.)

have broad, chisellike plates, something like broad, curved incisor teeth, at the front of the jaws, with which they can scrape the surfaces of the calcareous rocks and coral heads, scraping off the algal film and leaving recognizable scars on the rock face. These fish have been shown by Bakus (1964, 1966, 1969) to be an important element in limiting the establishment and growth of larger algae in these areas and to exert a marked effect on the evolution of the shallow-water benthic invertebrate fauna.

Herbivores of the rocky coast

The oceans have relatively few animals that browse on larger plants as terrestrial herbivores do because of the paucity of such plants in the sea, although, as mentioned, some algal-film grazers, especially certain limpets, have taken to scraping the stipes of larger marine plants instead of the film on rock surfaces. A few fish browse on certain species of algae. Sea urchins eat a variety of organic material; some are largely carnivorous, some scavenge almost anything they come across, and others eat, for the most part, pieces of the larger seaweeds. Such are the common purple sea urchins of the Pacific coast, *Strongylocentrotus purpuratus* (Fig. 12-43), and the larger red urchin, *S. franciscanus*. On opening them, one commonly finds the capacious intestine full of fragments of the larger algae. It has recently been found that these urchins sometimes become so numerous subtidally that they clean out most of the available seaweed fragments that might sustain them, whereupon great numbers of them leave their holes in the rocks and crevices and go foraging. Such groups of foraging sea urchins may cut off growing kelp plants near the holdfast, with the result that the kelp is washed up on shore in windrows and the kelp bed denuded. It has been shown that one *S. franciscanus,* ten *S. purpuratus,* or ten *Lytechinus anamensis* per square meter are sufficient to prevent kelp from developing. There may follow a period when, without enough food, the sea urchins either leave the area or die off until the numbers are so reduced that young seaweeds can reestablish themselves, eventually building back the kelp beds and establishing a temporary equilibrium between the available seaweeds and the population of sea urchins. In the area affected by sewage outfalls in and near San Diego Bay, it was found that the sea urchins did not disappear after denudation of the kelp beds but persisted, apparently obtaining some nourishment from dissolved organic materials in the water, perhaps through the tube feet or respiratory structures. In any case, the kelp was prevented from reestablishing itself until enough lime was dumped into the area of the former kelp beds to kill off the urchins, whereupon the kelp soon reestablished itself in sufficient quantity for the profitable kelp industry to resume operations there.

Carnivores and scavengers

The great abundance of animal life in the rocky intertidal and subtidal areas, made possible by the supply of food present as plankton and algal film as well as the fact that much of this life is sessile or slow moving, has left its mark on the carnivorous and scavenging elements of the community. Here it is possible for a large proportion of the carnivores to browse, much in the manner of herbivores on land. Also, entrapment and ambush techniques become much more feasible methods of capturing prey. A large portion of the more numerous and conspicuous carnivores are either nearly sessile, as are the sea anemones, or slow moving like the starfish, nudibranches, and whelks, moving about in a leisurely manner and having only to find and select their prey from among the lush growths of sessile or nearly sessile animals. Among the groups that are traditionally more active, numerous species also resort to somewhat similar modes of hunting, waiting at the mouths of burrows or in hiding places for their prey to come close enough to be taken with a swift lunge or thrust.

Among the coelenterates there is a whole gamut of forms, from the bushy hydroid colonies, the polyps of which are so small that they capture only smaller elements of the zooplankton, to the large sea anemones such as *Discosoma* and *Actinodendron* of the coral reefs, large species of *Metridium,* and others found offshore or sometimes in bays in cooler waters as well, which are capable of capturing and eating rather larger fish and crustaceans.

Along the Pacific coast of the United States and Canada the beautiful large green anemone *Anthopleura xanthogrammica* (Fig. 12-44), colored by symbiotic algae in its tissue, is characteristic of rocky pools and crevices everywhere that the rocks are in exposed localities but afford some protective pools, channels, and ledges. The related *Anthopleura elegantissima* is found somewhat higher in the intertidal and also often in more protected waters, commonly in dense masses formed largely by a peculiar kind of asexual reproduction in which the anemone slowly creeps in opposite directions until it literally pulls itself in half, each half rounding up, healing, and

A

B

FIG. 12-44. A, The common large, green intertidal sea anemone of the West Coast, *Anthopleura xanthogrammica*. B, The even more abundant aggregate sea anemone of the same region, occurring higher in the intertidal, *Anthopleura elegantissima*. (M. W. Williams photographs.)

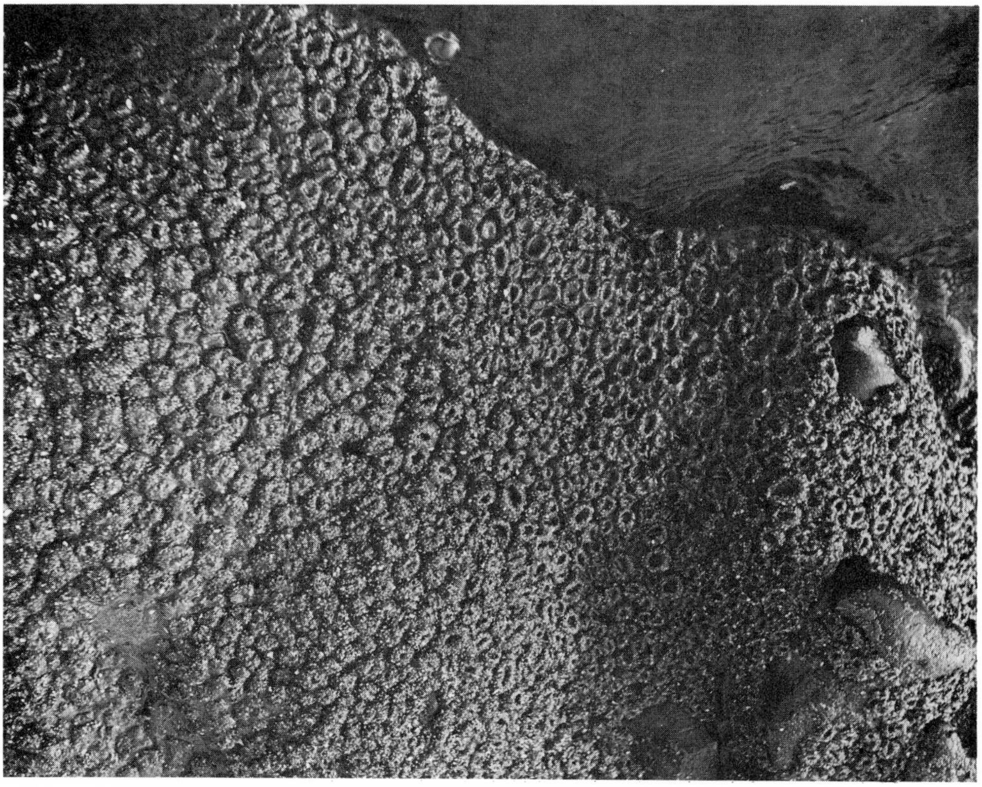

FIG. 12-45. A mat of the aggregate anemone exposed by low tide. Mats of this type are produced by a peculiar asexual reproduction. An anemone simply creeps on its pedal disc in opposite directions at the same time, literally pulling itself in half, each portion healing and regenerating to produce a new individual. Sexual reproduction, as in other coelenterates, results in the liberation of free-swimming planula larvae, which seek new locations and start new colonies of the anemone. (M. W. Williams photograph.)

FIG. 12-46. A small, colorful anemone, *Epiactis prolifera,* common from northern California to Japan. The young, when liberated, attach to the column of the mother, where they remain until about one fifth grown, when the creep off by themselves. (Oregon Institute of Marine Biology photograph.)

FIG. 12-47. *Metridium senile,* a large anemone, with the entire oral disc covered with tentacles. Species of *Metridium* occur along both the Atlantic and Pacific coasts of North America. Along the New England coast, *Metridium* occupies much the same ecological position as does *Anthopleura* on the West Coast. (M. W. Williams photograph.)

regenerating to form a new individual (Fig. 12-45). This species also has the habit of affixing sand grains, small shell fragments, etc. to the tubercles that line its column. Another common anemone of the rocky shores of this region is the smaller *Epiactis prolifera* (Fig. 12-46), which occurs in an arc capping the North Pacific from Japan to California. It is surprisingly variable in color and pattern and in summer usually has one or more younger anemones of the same species affixed near the base of the column of the parent. The eggs are apparently discharged from the female, and some stick to the column mucus. The young remain there, looking like asexually produced buds, until ready to fend for themselves. Species of the genus *Tealia,* with red or scarlet body and sometimes variegated with green patches as in the common West Coast *Tealia crassicornis,* are among the more striking species fairly common in the lower intertidal and subtidal. The plumose anemone *Metridium dianthus* occupies a

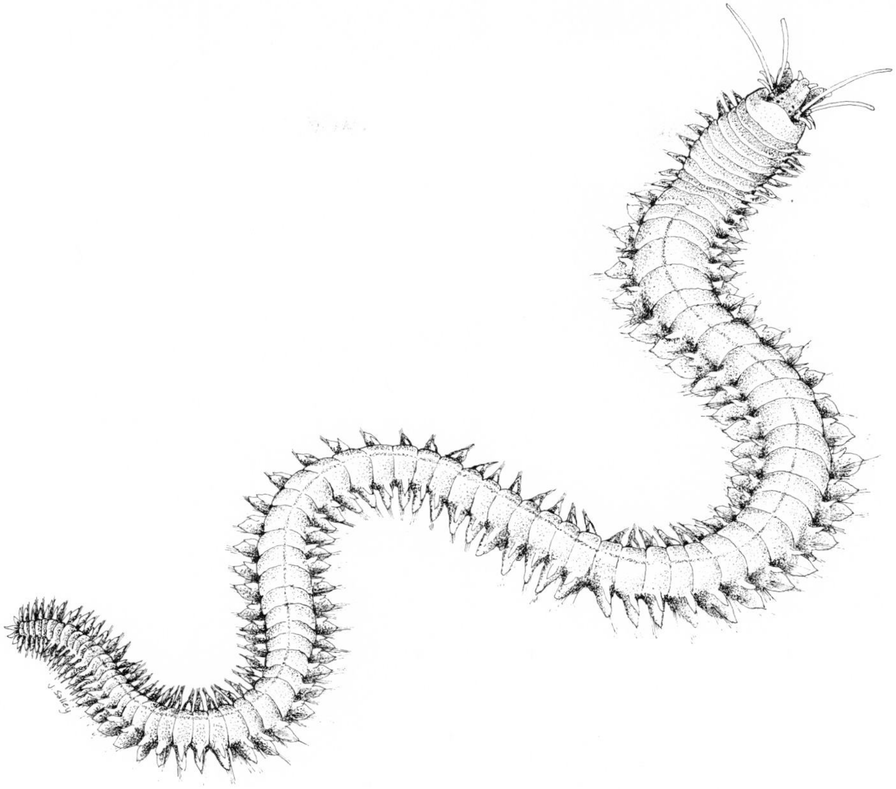

FIG. 12-48. *Nereis virens,* predatory polychaete often found in mussel beds.

somewhat similar position on the northeastern coast (Fig. 12-47) of the United States to that of the large green anemone of the West Coast. Metridia are commonly pure white, but lavender and rose-colored specimens also occur. When well expanded on their tall columns in a large aquarium, a group of large metridia makes a striking display.

In spite of the delicate flowerlike appearance and exquisite beauty of expanded sea anemones, they are formidable predators capable of paralyzing and devouring whole fish, crustaceans, or other small animals that brush against their nematocyst-armed tentacles.

Numerous predaceous worms, nemerteans and annelids, occur in rocky areas. Nereids (Fig. 12-48) are common in mussel beds as well as in sandy mud habitats. Cirratulids (Fig. 12-49) use their tentacles for prey capture. Nemertean worms have a remarkable eversible proboscis, the rhynchodaeum, which can be suddenly everted and used for prey capture. In many of them it is armed with a piercing stylet. Their body is very distensible, and many of them can swallow prey of considerably greater diameter than their own.

Many predatory and scavenging arthropods such as crabs, hermit crabs, small shrimps, certain amphipods and isopods, and mysids are also found here.

Perhaps the animals that more than any others have come to symbolize the seacoast are the starfish and sea urchins, members of the only major phylum of animals wholly confined to marine waters. Although related to the lines of animals from which the chordates arose, the development of a

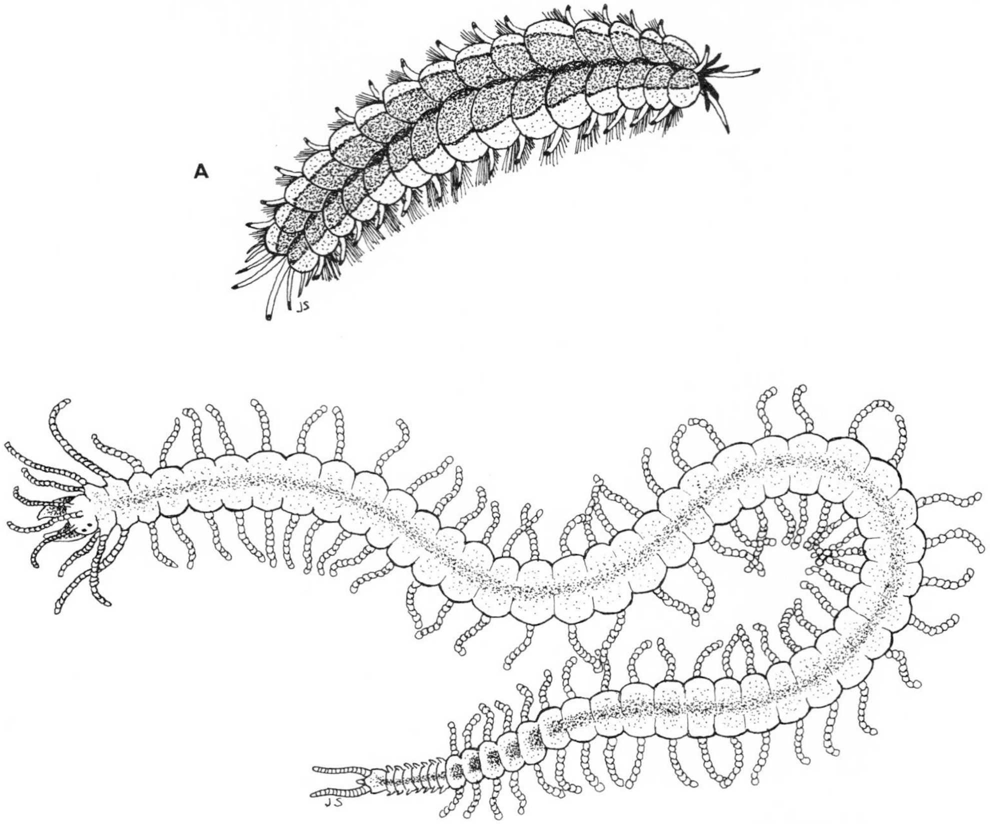

FIG. 12-49. A, *Harmothoe imbricata,* one of the scale worms, and **B,** *Syllis gracilis.* (**B** after McIntosh, W. C. 1915. British marine annelids. Vol. II. Ray Society, London.)

secondary radial symmetry, the unique water vascular system, the dermal calcareous skeleton, and the extensive and complicated coelom set them apart so sharply from all other animals that their affinities were for a long time obscure. Early zoologists classed them together with Coelenterata and Ctenophora in a phylum termed Radiata.

A few species are characteristic of the lower intertidal zones in rocky areas. On the eastern coast of North America two species of *Asterias*—the common starfish, *A. vulgaris,* and the northern starfish, *A. forbesi*—are the most abundant and familiar. Both occur in several color variations. Where they occur together, as in the Massachusetts area, the easiest, most reliable way to distinguish them is by the color of the madreporite plate—bright orange-red in *A. forbesi,* and light yellow in *A. vulgaris.* On the western coast of North America the same niche is occupied by the colorful, abundant, and widely distributed *Pisaster ochraceus* (Fig. 12-50), which likewise comes in several colors—yellow, orange, reddish, purplish, and chocolate. Sometimes all the colors are found together in the population at a single locality.

These starfish are voracious predators, subsisting on a diet chiefly of bivalve molluscs, especially mussels, but also eating other animals that either cannot get out of their way or fail to do so.

Numerous animals of the intertidal zone

FIG. 12-50. Typical rocky lower midlittoral zone assemblage exposed at low tide along West Coast of United States. Two starfish *(Pisaster ochraceus),* the upper one in humped feeding position and eating a California sea mussel *(Mytilus californianus),* some of which can be seen at the right, encrusted with acorn barnacles *(Balanus glandula).* At left is part of a clump of stalked barnacle *Mitella polymerus.* (M. W. Williams photograph.)

exhibit great alarm and make every effort to avoid a starfish that moves near them (Feder, 1963, 1967). They are warned by a chemical exuded into the water by the starfish. Even some animals ordinarily regarded as sessile will move out of their way. For example, in Puget Sound, sea anemones of the genera *Stomphia* and *Actinostola* will detach their pedal discs from the substrate, begin to writhe, and remove themselves from the immediate vicinity by a series of contortions somersaulting them through the water when a starfish is held near them. *Actinostola* (Fig. 12-51) also reacts to *Stomphia* if the latter is held close to it.

One of the most spectacular and voracious starfish of the West Coast is the large twenty-ray starfish, *Pycnopodia helianthoides* (Fig. 12-52), found in the lowest parts of the intertidal zone and the upper subtidal areas. This species is so large and soft bodied that it makes a rather sad, droopy specimen when removed from the

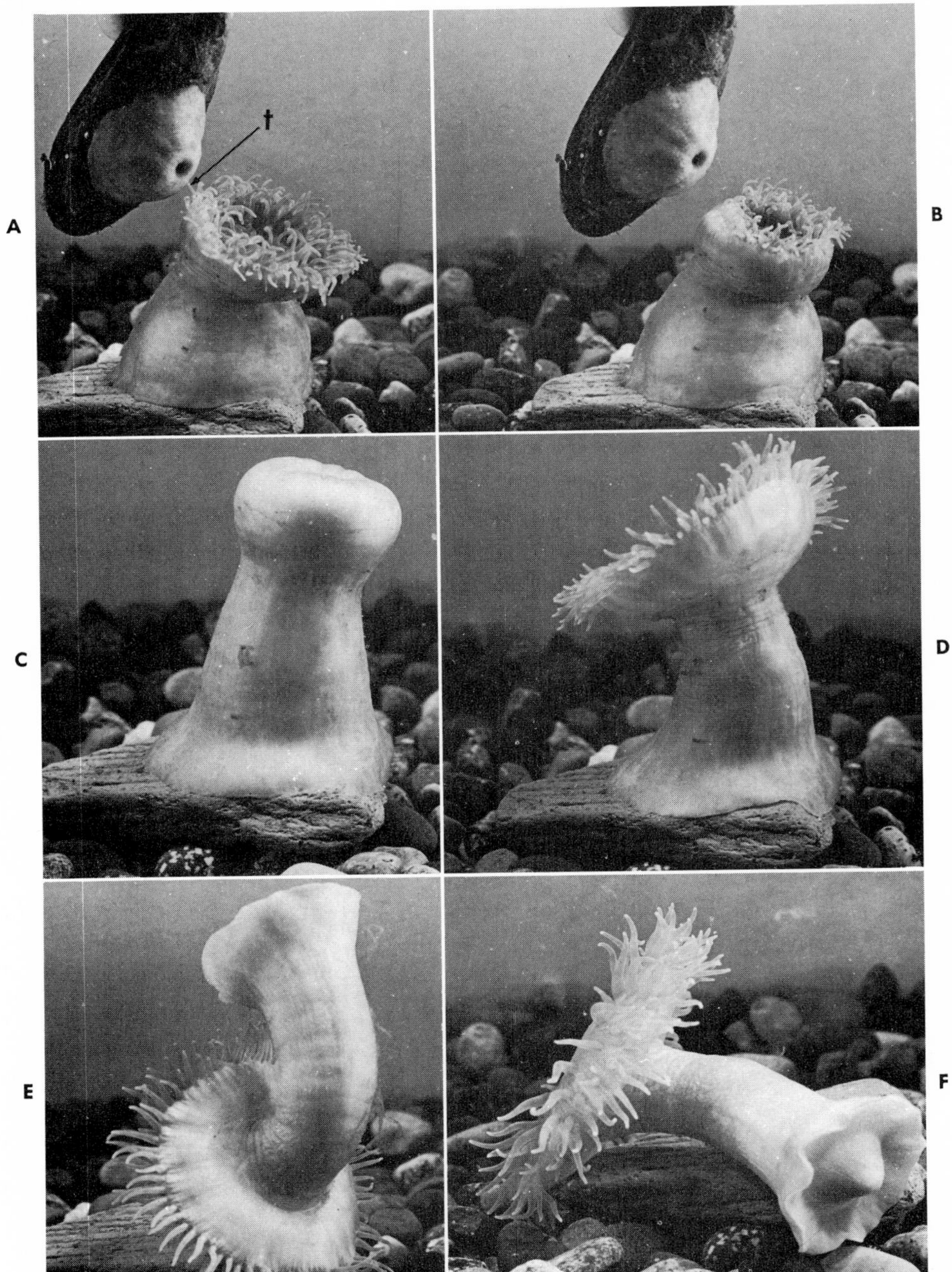

FIG. 12-51. A "swimming anemone" from Puget Sound. Anemones of the genus *Actinostola,* when excited by chemical substances (t) given off by echinoderms or, in this case, by species of *Stomphia,* become very excited, at first withdraw, then begin to expand and writhe, and eventually detach their grip on the substrate and somersault through the water by a series of writhing movements to land some distance from the offending stimulus. (Photographs by D. M. Ross. In Sutton, L. 1967. Science **155:**1419-1420. Copyright, 1967, by the American Association for the Advancement of Science.)

FIG. 12-52. The large, twenty-ray starfish, *Pycnopodia helianthoides,* of the West Coast, turned oral side upward to show the many tube feet. (M. W. Williams photograph.)

FIG. 12-53. The sun star, *Solaster dawsoni,* a West Coast species. (Oregon Institute of Marine Biology photograph.)

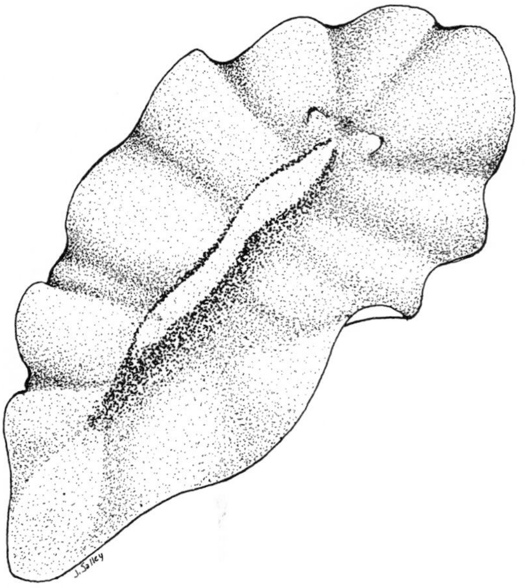

FIG. 12-54. *Notoplana,* a polyclad flatworm.

water and is a poor dried specimen, but in its natural habitat few animals can rival it in beauty and interest. Its diet is also surprising. I have found large, lumpy-looking specimens that, on examination, proved to have two or even three fully grown sea urchins *(Strongylocentrotus purpuratus)* in the stomach. How such a soft-bodied animal with a delicate thin-walled stomach can accommodate itself to such an unlikely, spiny diet is hard to understand.

Among the more active predators, the most conspicuous groups are the fish, cephalopods, errant polychaete worms, nemertean worms, polyclad flatworms (Fig. 12-54), and various groups of larger crustaceans (Fig. 12-55). These groups are all best developed in the subtidal shelf waters, but characteristic representatives occur also in the intertidal or may invade this zone during high tides.

Various species of tide pool cottids (Fig. 12-56) are found in pools and channels well

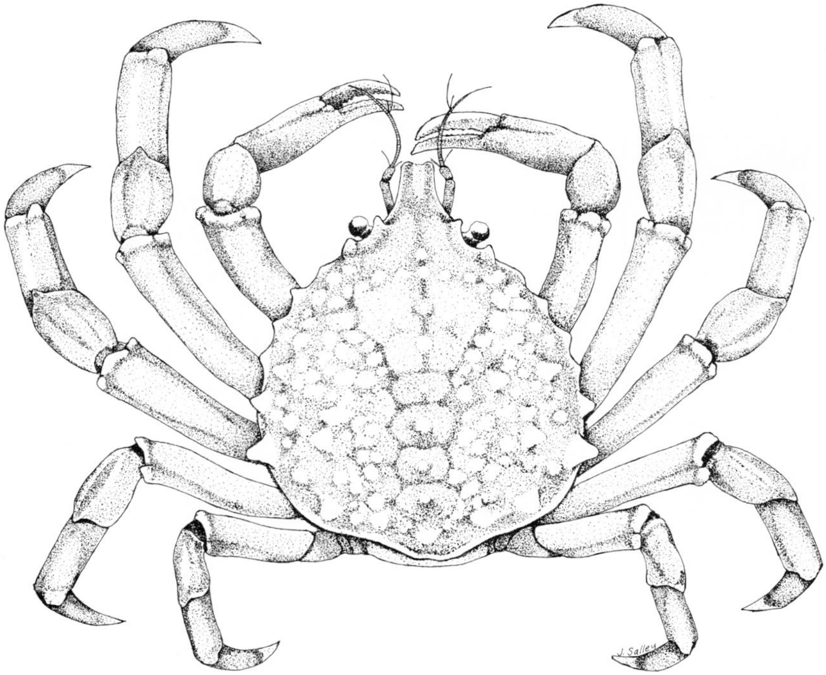

FIG. 12-55. *Libinia emarginata,* American spider crab.

up into the middle and upper parts of the intertidal zone. They are commonly patterned and colored to match the variegated coralline algae and rock surfaces among which they live. Habitually they spend much time resting on the bottom of the pools or along the sides, using the pectoral fins almost as limbs, and swimming and gliding in short, jumplike spurts from one resting place to another. They have large heads and mouths and are voracious eaters. Other fish commonly found in the intertidal include the young of several species found offshore when mature. Eel blennies, with elongate, slippery, almost snakelike bodies, are common in the lower parts of the intertidal under rocks or protecting seaweed, as are also clingfish, which cling to the undersides of rocks with their peculiar, modified suckerlike ventral fins. Some species of *Octopus,* such as *O. bimaculoides* of the southern and Lower California coast, characteristically occur in the rocky intertidal, but most species are subtidal, and only the fringe of the population occurs in the lower parts of the intertidal.

Most of the crabs, as well as hermit crabs, amphipods, lobsters, spiny lobsters, shrimp, and other crustaceans of the rocky shelf communities, are predators, scavengers, or both. In the deeper, dimly lit or nearly dark portions of the shelf, the fish and crustaceans characteristically take on red hues and patterns.

The larvae of sessile rock-dwelling animals may drift for hundreds or even thousands of miles in the plankton, ready to settle on any favorable substrate that presents itself. This is dramatically shown by the settlement of relatively isolated rocky outcroppings, or artificial reefs, in regions dominated by otherwise unfavorable substrate.

The eastern coast of the United States from southern New England to the tip of

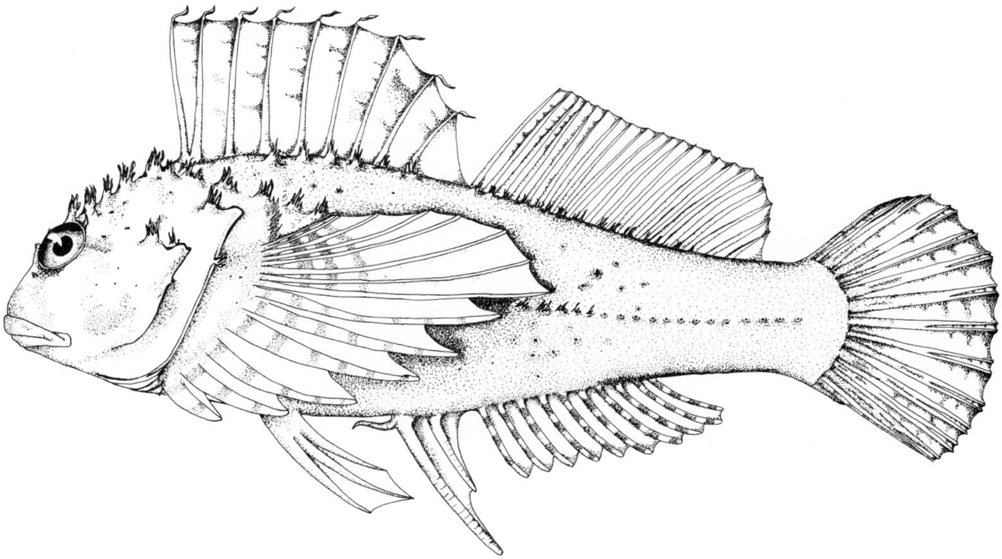

FIG. 12-56. *Oligocottus snyderi,* a tide pool cottid. (In Light, S. F. 1975. Light's manual; intertidal invertebrates of the California coast. 3d ed. revised and edited by R. I. Smith and J. T. Carlton, University of California Press, Berkeley; after McCain, J. C. 1968. Parker and Margoles, 1964; J. H. Hedgpeth, 1962; and R. Bolin, 1944, respectively.)

Florida is for the most part sand beaches, and a sandy bottom extends out over the continental shelf. However, off the coast of the Carolinas, here and there, are broken chains of submerged reefs and ledges—outcroppings known as "black rocks" to fishermen because blackfish congregate there. They are sometimes also called coral reefs, although they are not composed of coral but rather of relatively soft, fine-grained marl of the Miocene age.

These sunken reefs support forests of *Sargassum* and are in places covered by masses of calcareous tubes of tube worms and tube-building gastropods, or gardens of sponges, tunicates, gorgonians, hydroids, and bryozoans. Where sufficiently free from encrusting coralline algae or tube worms, they are riddled with the burrows of piddocks and date-nut shells.

When storms create deep turbulence reaching down to these submerged reefs, many of these rock-living animals, foreign to the prevailing sandy bottoms of the region, are dislodged and cast ashore, giving mute evidence of the unseen offshore reefs.

Because of the greater difficulty in sampling and observing and because of the greater inherent variety of ecological niches on rocky substrates and their close approximation and interdigitation, the communities of organisms occupying them, except for those in the intertidal zone, long remained much less well defined and understood than those of soft substrates. Only in recent years have adequate techniques for sampling and direct observation become available. For an understanding of the communities occupying such substrates, rather detailed data are necessary with collections. These data should include

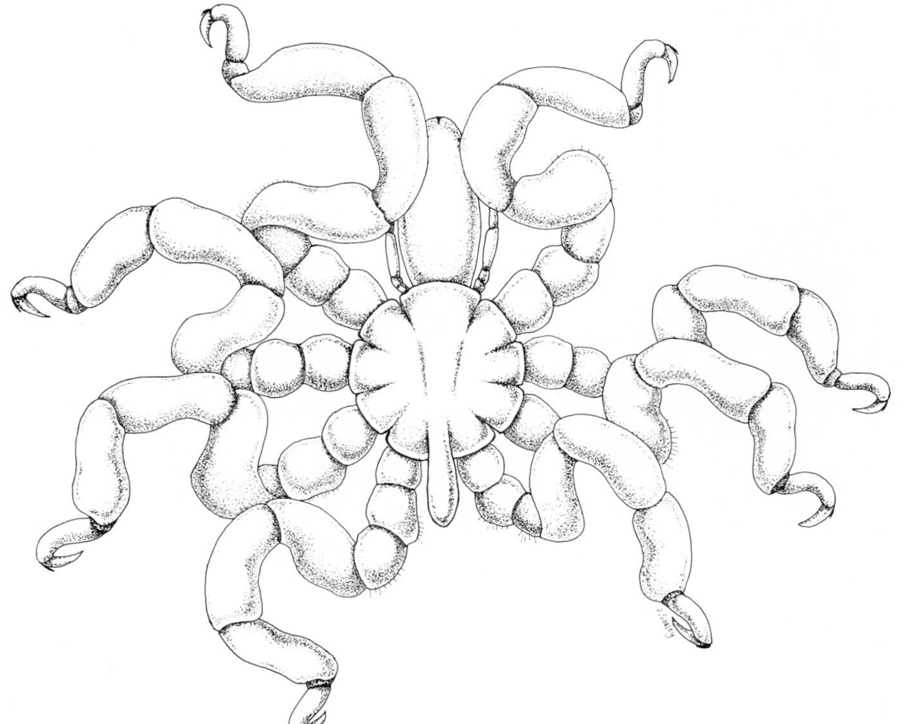

FIG. 12-57. *Discoarachne brevipes,* one of the Pycnogonida. (After Hoek, P. P. C. 1881. Challenger reports. Vol. 3.)

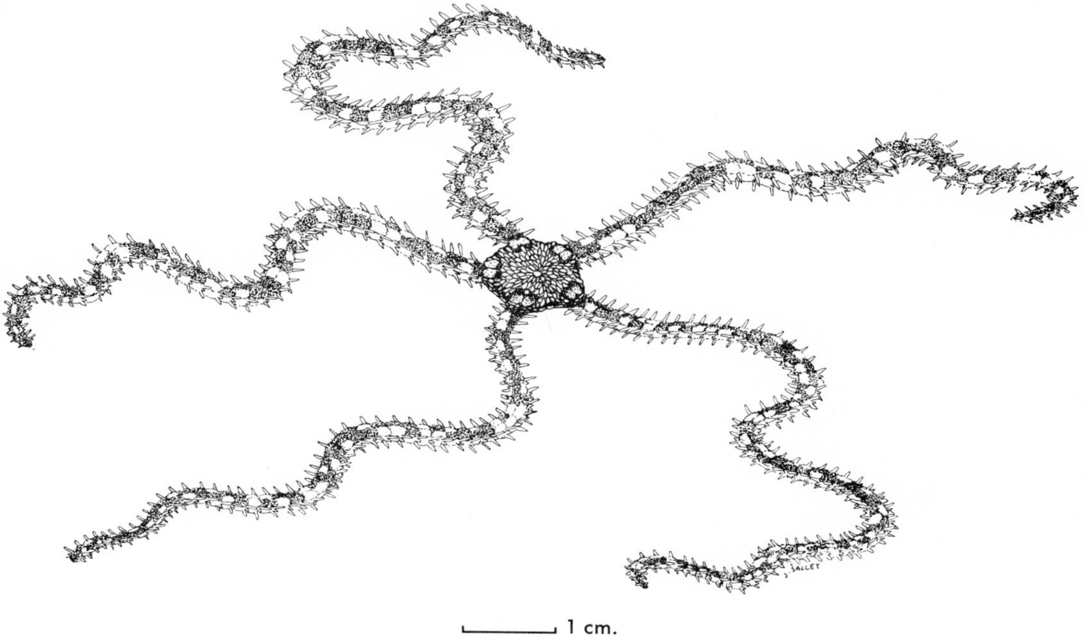

FIG. 12-58. *Amphioda occidentalis.*

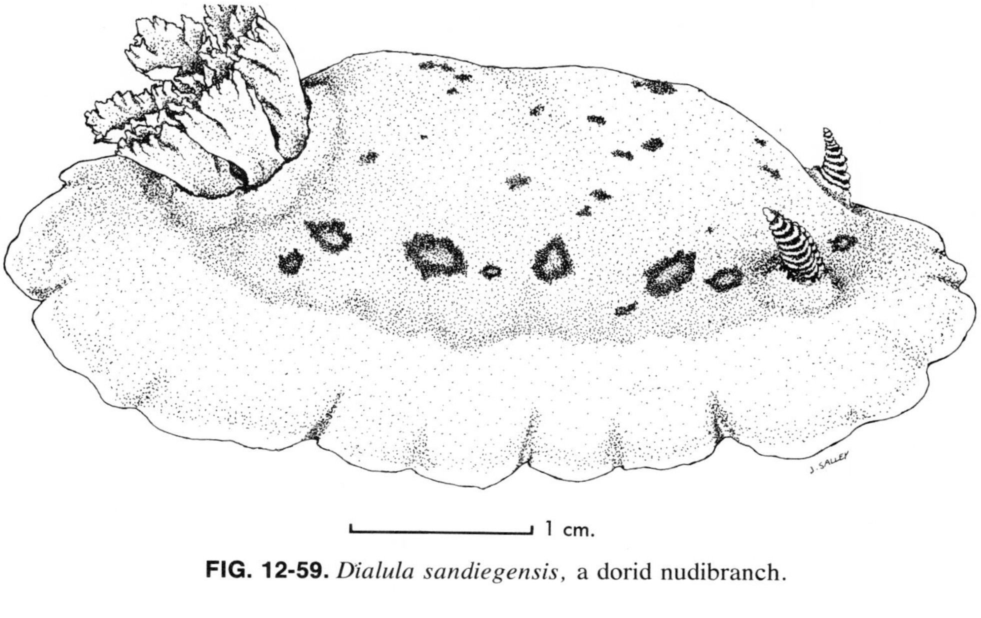

FIG. 12-59. *Dialula sandiegensis,* a dorid nudibranch.

depth, amount, and duration of light reaching the actual spot where the specimen was taken; temperature; whether on seaward, sheltered, or lateral rock face; water movement; associated organisms; and the nature of the rock (granitic, conglomerate, calcareous, etc.). With regard to position, eight categories are commonly recognized: (1) horizontal, with less than a 20-degree slope; (2) moderately inclined, with a 20- to 60-degree slope; (3) sharply inclined, with a 60- to 85-degree slope; (4) vertical, with an 85- to 90-degree slope; (5) slightly overhanging; (6) strongly overhanging, as the roofs of small caves; (7) in crevices; and (8) on the underside of individual rocks. The presence, type, and abundance of seaweeds and their relation to the collected specimens should also be noted.

ROCKY AREAS OF APHOTIC DEPTHS

The very deep, bathypelagic and abyssal, rocky areas are the least well known of all the marine habitats because of the difficulties of sampling and making observations. Most of the level or nearly level regions of the sea floor are soft-bottomed and covered with sediments. For the most part, the exposed, rocky areas either are the steep sides of submarine canyons, mountains, or escarpments or are relatively recently formed lava flows such as are found along some of the midocean ridges. Below the euphotic zone, there are, of course, no seaweeds. The attached growths are entirely animal in character—gorgonians, hydroids, sponges, bryozoans, branchiopods, crinoids, etc. Most of the groups of animals represented in the subtidal rocky shelf areas have representatives extending down into deep regions, but the populations are sparser because of the diminished quantity of planktonic food and the lack of a producing algal film.

CHAPTER

13 BENTHIC COMMUNITIES OF SOFT SUBSTRATES

SHALLOW-WATER BENTHIC COMMUNITIES

Since the pioneering work of Petersen (1913, 1915, 1918) and Petersen and Boysen-Jensen (1911) in the shallow Danish fjords, it has been recognized that the benthic animals of soft substrates are not randomly distributed with respect to each other or with respect to the environment but tend to occur in more or less well-defined associations, or **communities,** that can be identified or defined in terms of their most abundant and prominent members. Animals living in soft fine-grained substrates such as sand or mud, Petersen termed **infauna,** in contrast to those living on the surface of hard substrates, which he termed **epifauna.** Through extensive quantitative sampling and statistical analysis, Petersen demonstrated in the Baltic region several distinctive infaunal communities. Such communities could usually be correlated with certain combinations of ecological factors, of which the exact nature of the substrate (particle size, stability, compaction, organic content, etc.), the average and extreme temperatures and their seasonal fluctuation, illumination, salinity, and the movements of the overlying water are important. There are also a number of other, more subtle factors that have far-reaching influence on biotic communities. Often the composition of the communities themselves is a more reliable and readily available key to such parameters than is an examination and analysis of the environment itself.

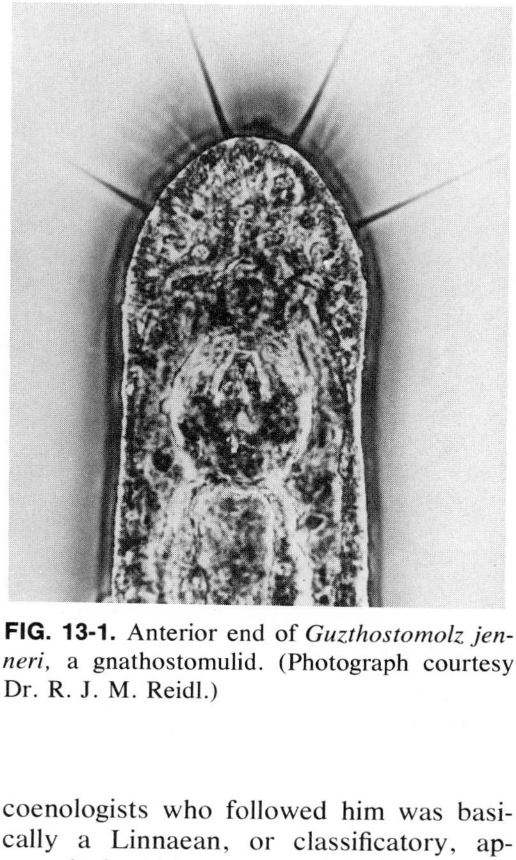

FIG. 13-1. Anterior end of *Guzthostomolz jenneri,* a gnathostomulid. (Photograph courtesy Dr. R. J. M. Reidl.)

The work of Petersen and of the biocoenologists who followed him was basically a Linnaean, or classificatory, approach, in which they attempted to divide and classify the various types of environment into units and subunits in a manner analogous to the approach of a taxonomist to animal and plant groups. A good example of the culmination of this approach is the work of Gislén (1930), who attempted to exhaustively divide and classify all the communities in a small Swedish coastal region into a system of faunal units, subunits, and microunits.

This was termed the **particulate approach** by Wieser (1953, 1954, 1957, 1959). It is believed by some that, carried to its extreme, this approach results in a bewildering patchwork of units and subunits in which the content of general information is eventually lost and the labor involved in analyzing and defining subunits of the environment becomes out of proportion to the return in an understanding of the factors underlying the observed distributions and associations. Jones (1950) and Thorson (1958) have presented good summaries of progress in the study of marine benthic communities.

The other approach to synecology is more analogous to the approach of a geneticist or physiologist to organisms and has been termed the **comparative approach.** In this approach the trends of faunal change in different types of habitat are compared, or the effects of a variable environmental parameter over a range of different situations are studied. The work of Wieser (1953, 1954, 1957, 1959) on free-living marine nematodes from different types of substrate exemplifies the comparative approach.

The two approaches are not, of course, mutually exclusive, any more than are taxonomy and physiology. Both, judiciously employed, are necessary to a proper understanding of biotic communities.

Communities are more than chance aggregations of species. They are relatively permanent associations. Not only are the communities of a given region fairly well defined and the order of importance of the species within a given community rather constant, but parallel communities occur in different latitudes or in widely separated regions where appropriate combinations of environmental conditions occur. Such parallel communities commonly show similar structure and species composition but with a number of the species of one replaced by ecologically equivalent species in the other, sometimes belonging to the same genera.

Because of different growth rates, lifespans, differences in success in settlement of immature individuals from year to year, and fluctuations in the environment that do not affect all members of the community in the same way, a given community may show wide fluctuations in relative abundance of some of the species from year to year or over a longer period of time.

Communities, like individual species, may also show the phenomenon of submergence. A community that occupies shallow inshore water in the far north tends to follow the isotherm into more southern latitudes, being found there in deeper offshore water, whereas warmer-water communities replace them near shore.

Mare (1942) divided the benthos into three categories—the **macrobenthos,** including all organisms too large to pass through a 1 mm. mesh sieve; the **meiobenthos,** including those which pass through a 1 mm. mesh but are retained by a 0.1 mm. mesh; and the **microbenthos,** those forms that pass through a 0.1 mm. mesh.

The macrobenthos contains all the obvious animals and plants included in most surveys of communities. The meiobenthos includes many small copepods, nematodes, flatworms, the most minute species and immature stages of various molluscs, worms, and other animals, rotifers, gastrotrichs, gnathostomulids, the larger foraminiferans, and others. Individually, such forms commonly range from 0.05 to less than 0.001 mg. in weight. The microbenthos is composed primarily of small protozoans, especially flagellates, ciliates, and amoebae, and of bacteria.

Until recently it was thought that the upper few centimeters at the immediate surface of muddy or fine-sand substrates comprised by far the richest layer for all groups of the meiobenthos and most groups of the microbenthos. This does indeed seem to be true for most groups. It is now apparent, however, that in the case of the gnathostomulids, which dominate the anaerobic layers in fine sands, the fauna becomes richer in individuals and more diverse in species below this surface layer. This situation is somewhat analogous to the recent finding that diversity of benthic foraminiferans of the sea bottom increases with greater depth and distance from shore.

In recent years careful attention to the microfauna, to the precise nature and organic content of the sediments, to the effects of organisms such as predators belonging to the nekton, to the effects of plankton blooms and of the biological properties of the water masses passing over the communities, and to numerous other factors has added greatly to our understanding of the dynamics of infaunal communities, as well as to our understanding of the changes observed in them from time to time and from place to place.

In most communities the bulk of the standing crop, or biomass, is present as macrobenthos. However, the number of individuals and the rates of metabolism and reproduction among the meiobenthos and microbenthos are so much greater that it is these elements of the community which may often be responsible for the greater turnover of organic matter and the greater total production over a period of time.

The dominant members of the macrobenthos are plankton feeders and detritus or substrate ingesters. Although always far less in total biomass, the carnivores are not necessarily fewer in species and may even exceed in this respect some of the more dominant categories.

Areas of extremely cold water, such as those found in the arctic or antarctic regions, tend to be characterized by communities of fewer species, but sometimes there is a large standing crop of those species that are present. The individuals are commonly larger, longer lived, and slower growing, and a greater proportion of them show direct development. They are communities in which there is a high biomass of the dominant species but a relatively slow rate of production. Changes of the composition of such communities tend to be slow. The overall result is that such communities tend to be uniform.

Productivity of benthic communities

The standing crop of benthic communities can be estimated in various ways. Estimates may be based on the numbers of animals of the dominant groups and on the wet weight of these animals or the weight of dry organic matter present per unit of area. The latter gives the most reliable comparisons between different communities, since the great differences in size between different species, the differences in the proportion of shell material and of water content of the tissues, and other differences make comparisons based on numbers or wet weights difficult and often misleading.

Productivity of the community is even more difficult to assess because of the great differences in the life-span among different members of the community. The microorganisms present, although not representing the bulk of the biomass, may, because of their more rapid multiplication, be responsible for the greater part of the organic production and turnover. A community with the greatest biomass, or standing crop, is not necessarily the most productive. A community in which the greatest part of the biomass is comprised of relatively large, long-lived animals may have a large standing crop but a relatively small rate of turnover. Such communities tend to be relatively stable. It must also be remembered that in benthic communities most of the actual organic production does not take place in the benthic community itself but rather in the plankton of the waters that pass over the community. What marine biologists see in production of the benthos is largely a matter of the extent and efficiency of the conversion of organic materials derived from the plankton, rather than primary production.

For practical purposes, reliable estimates of the production of economically significant species, rather than total productivity of the community, are desired.

Petersen (1915, 1918) and Petersen and Boysen-Jensen (1911) estimated that in the Kattegat area of Denmark the benthic fauna contained about 1 million tons of "fish food" (bivalves, gastropods, polychaetes, and crustaceans), about 5000 tons of flounders, and about 75,000 tons of invertebrate predators such as starfish and brittle stars. Assuming a ratio of consumption proportional to the weights of the predators, they concluded that only about 6% or 7% of the fish food was actually consumed by fish. The rest was consumed by the invertebrate predators. It has subsequently become clear that in reality the invertebrate predators eat about four times as much in a given period of time in proportion to their weight as do the flounders, so that the fish are actually receiving only about 1% or 2% of the fish food. Furthermore, the fish cannot readily take advantage of increases in the standing crop of "fish food," since the invertebrates, with their faster reproduction, can respond to such an increase long before the fish can.

It is often found that the production, or standing crop, of given species or of groups of species in benthic communities on seemingly similar sea floors sometimes not very distant from each other, varies to a surprising degree. In some cases the reason is fairly obvious, but other instances are found in which no ready explanation is apparent why one area should be poor in bottom fauna or in certain elements thereof, whereas another, comparable area is rich.

In particularly rich beds the bivalves alone in soft substrates sometimes reach a weight of as much as 300 gm. of dry organic matter per square meter. Usually the values are much less, ranging from only a few grams up to approximately a hundred. Dense mussel beds on rocky stretches may attain more than 3 kg./m.^{2}.

Part of the difficulty in interpreting the differences between areas may lie in the lack of information on seasonal and other long-term fluctuations of the biomass. The majority of quantitative investigations cover relatively short periods such as a few weeks or one summer. Until the magnitude of the changes from year to year and something of their causes are known, comparisons of different areas based only on short examinations of each are bound to contain large elements of uncertainty. The role of predators, expecially of mobile ones, not themselves part of the benthos but able to range over wide areas of sea bottom, is also difficult to assess.

The settlement of larvae of different species occupying the same water mass and therefore subject to the same gross environmental conditions may nonetheless vary greatly and in directions inconsistent with each other or with any general, overall environmental influence. For example, Loosanoff and associates (1955) found that in Long Island Sound, over a period of eighteen seasons, there was no consistent agreement between the number of American oyster larvae and the number of starfish larvae settling on collectors placed in the water.

Influence of the biological properties of the water masses

Subtle properties of water masses may strongly influence the biology of organisms over which these water masses pass. For example, Wilson (1951*a*), Wilson and Armstrong (1958), and subsequently numerous other investigators have found that apparently **external metabolites,** or **exocrines,** of organisms in the water in high-dilutions exert far-reaching influence. Wilson and co-workers demonstrated that off Plymouth, when water masses containing a certain characteristic assemblage of plankters (termed *Sagitta elegans* water) were present, the water was eminently satisfactory for use in culturing the larvae of certain benthic forms. When, on the other hand, other water masses with a somewhat different assemblage (termed *Sagitta setosa* water) were present, these same larvae could not be nearly as readily cultured. The addition of some *S. elegans* water to batches of *S. setosa* water rendered them satisfactory. The two water masses could not be differentiated on the basis of the physiochemical properties tested by conventional hydrographic techniques. Areas consistently characterized by the presence of *S. setosa* show distinctly different bottom fauna from areas in which *S. elegans* is of frequent occurrence. From the air the two types of water mass can be distinguished by a difference in color.

In the North Sea and the English Channel, *S. elegans* water indicates at least some admixture of open Atlantic water with the coastal water masses. This mixed water, for some reason, supports a population of *S. elegans* and certain other plankters not found in unmixed waters of either the open sea or coastal water masses and is intimately bound up with the success of the larvae of various benthic forms of the region. Since not all benthic species are affected the same way or to the same extent by the biological properties of the water masses passing over them, the effects may be highly selective.

The biological condition of the water masses is, of course, of great importance to the plankton and nekton in these water masses as well. The mixed waters indicated by the presence of such forms as *S. elegans* and *Clione limacina* also seem to be correlated with the shoals of herring, whereas *S. setosa* water is correlated with poor fishing conditions for herring. Thus the questions of biological conditioning of the water take on great practical as well as theoretical importance. Since the herring remain associated with this "mixed water" even during the nonfeeding period immediately prior to spawning, it has been suggested that perhaps it is not simply the presence of swarms of *Calanus,* their principal food, that brings the herring but that perhaps both herring and *Calanus* are drawn to this water because of its biological properties.

Temperature and the shallow benthos

That portion of the sessile or slow-moving epifauna exposed by high tides develops great resistance to the temperature extremes of the region.

Tropical periwinkles close the aperture with their horny operculum and, even if the sun-heated rocks reach high temperatures, can survive out of the water for a long time. In the laboratory some have been kept out of water almost two years and found to be still alive and active when resubmerged. Evaporative cooling and the ability to tolerate a greater degree of water loss than can animals lower in the intertidal zone enable many upper littoral forms to tolerate periods of exposure.

In northern waters, freezing may be a problem. Many of the more active animals move out to deeper water in winter, but less mobile forms must endure the cold. Various cockles, clams, and lug worms can survive frozen sand or mud for a month or more. The clam *Mya* has survived freezing at −4° C. for seven weeks. Nonetheless, especially long hard winters can be fatal for much of the intertidal benthos. Ice, if covering sand or mud for prolonged periods, may suffocate the animals beneath it.

Speciation in infaunal and epifaunal groups

Most epifaunal groups, such as prosobranch and nudibranch gastropods, crabs, amphipods, etc., show large increases in the number of species as one proceeds from polar regions toward the tropics, whereas most pronounced infaunal groups, such as level-bottom bivalves, cumaceans, brittle stars, and holothurians, do not. In the far north and along the shores of Antarctica and the antarctic islands, the epifauna of the intertidal and upper subtidal depths is severely limited or eliminated by ice scouring. The more stable temperature conditions as one approaches the tropics permits greater diversification to occur, whereas the seasonal changes in temperature in the temperate and boreal regions tend to restrict the numbers of species because of the increased range of environmental extremes with which they must cope. The overall result among epifaunal groups is the development of epifauna consisting of fewer species but more individuals of those species which can live in the northern and southern regions and of more species but in less massive growths of most of the species in tropical regions.

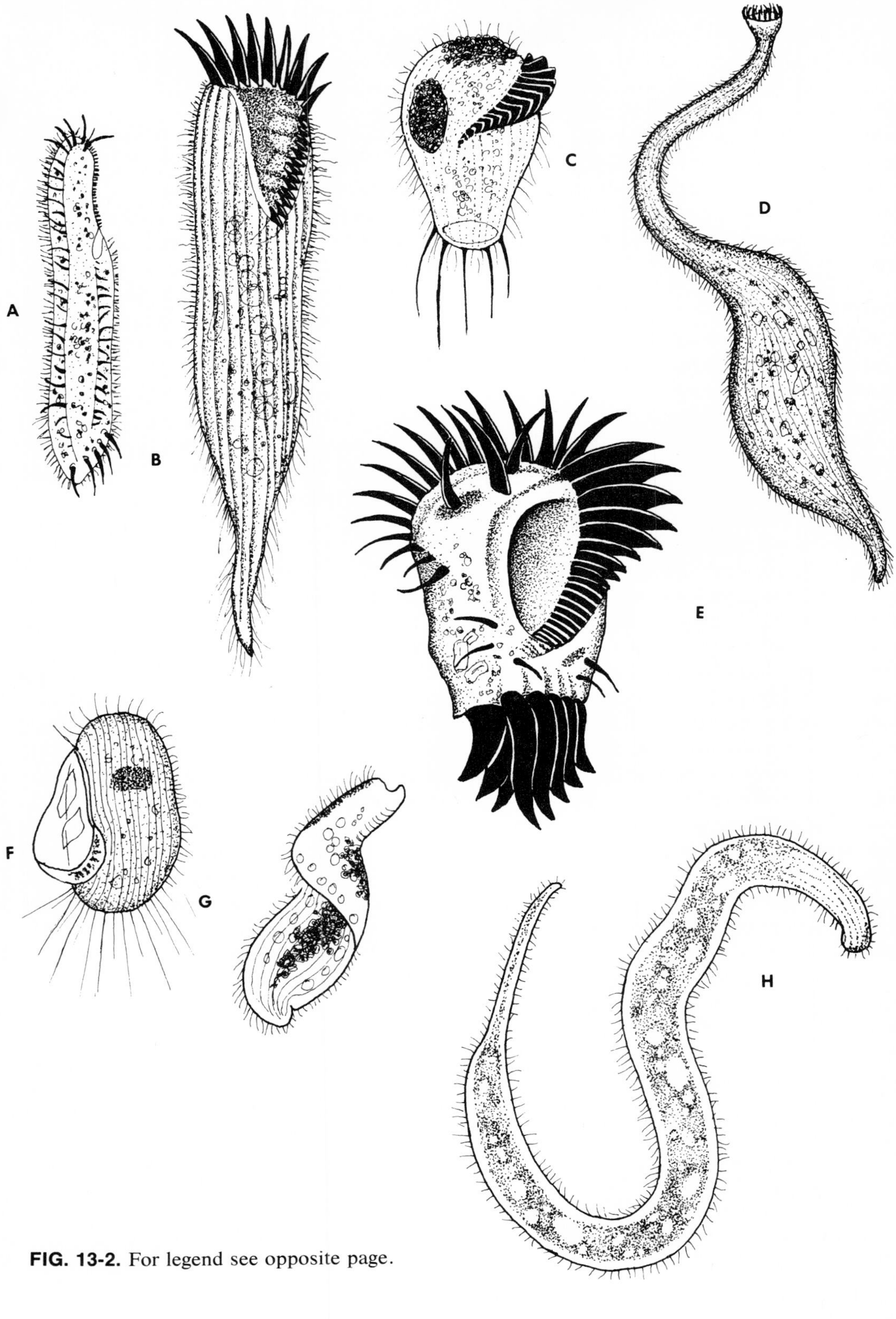

FIG. 13-2. For legend see opposite page.

In the case of the level-bottom infauna, these conditions are much more constant in almost all regions, and the number of types of microlandscapes that may provide habitats for new species and absorb them is comparatively restricted. Thus for infauna, the differences in conditions that lead either to restriction of the number of species or to diversification are much less between different latitudes than is the case for the epifauna.

The endopsammon

Wet or submerged sands usually contain a rather varied microfauna, the endopsammon, living between or on the sand grains. The composition of this microfauna varies according to the particle size of the sediment, its location relative to tidal or subtidal depth, the salinity and temperature of the water, and other factors. In coarse, aerated sediments typical members include many microcrustaceans (harpacticoid copepods and ostracods), small turbellarian flatworms (Acoela, Macrostomida, Otoplanida, and others), protozoans (especially ciliates and foraminiferans), small nematodes, gastrotrichs, rotifers, and tardigrades. In fine sand deposited in sounds and estuaries, the deeper sand below the surface centimeter or so is usually anaerobic and black because of the production of hydrogen sulfide from organic matter by bacteria. Many groups of typically endopsammal species disappear in such sediments. These fine-grained anaerobic muddy sands are commonly dominated by gnathostomulids, a group of lower worms almost wholly overlooked until recently because they are small, delicate, and recognizable only when alive. They are strongly **haptic,** that is, they stick to the sand grains. They are resistant to bad water conditions and emerge from deteriorating, stinking samples of substrate only after most other endopsammon have left. The samples would ordinarily be discarded before they come out (Fig. 13-3).

OPEN SANDY BEACHES AND SANDY SUBTIDAL COMMUNITIES

Sandy beaches facing directly into the ocean without the protection of headlands or offshore islands are areas of constant change and motion. Every wave lifts and moves quantities of the sand. The breakers pound the shoreline ceaselessly. The tides move up and down the beach in daily, monthly, and annual rhythms and progressions. Exposed sands are subjected to direct sunlight, drying, wind transport, rain, and freezing. Floating objects, masses of seaweed (Fig. 13-4), and subtidal animals torn from their anchorages offshore by the turbulence of storms are cast ashore and pushed to the upper part of the beach by the waves of the highest tides. Each level of the beach is submerged and exposed for different amounts of time and at different times of day or night each successive day. High tides bring with them certain predators from the fringe of the sea, and low tides bring another set of hungry predators from the land.

Only specialized animals can live successfully and maintain themselves in this turbulent world, where even the sand itself displays something of the fluidity of water. Most of the animals found here are amazingly proficient burrowers and are relatively small in size.

Plankton feeders

Mole crabs

One of the most successful animals of the surf zone is the mole crab, or sand crab. These are not true crabs but mem-

FIG. 13-2. Eight genera of ciliate protozoans commonly found in and above the RPD layer of sediments: **A,** *Trachelostyla pediculiformis;* **B,** *Condylostoma remanei;* **C,** *Metopus;* **D,** *Tracheloraphis* Kahli; **E,** *Diophrys scutum;* **F,** *Pleuronema coronatum;* **G,** *Loxophyllum;* **H,** *Galeia orbis.* (Redrawn from Fenchel, T. 1969. Ecology of marine microbenthos IV. Ophelia. **6:**1-182, July.)

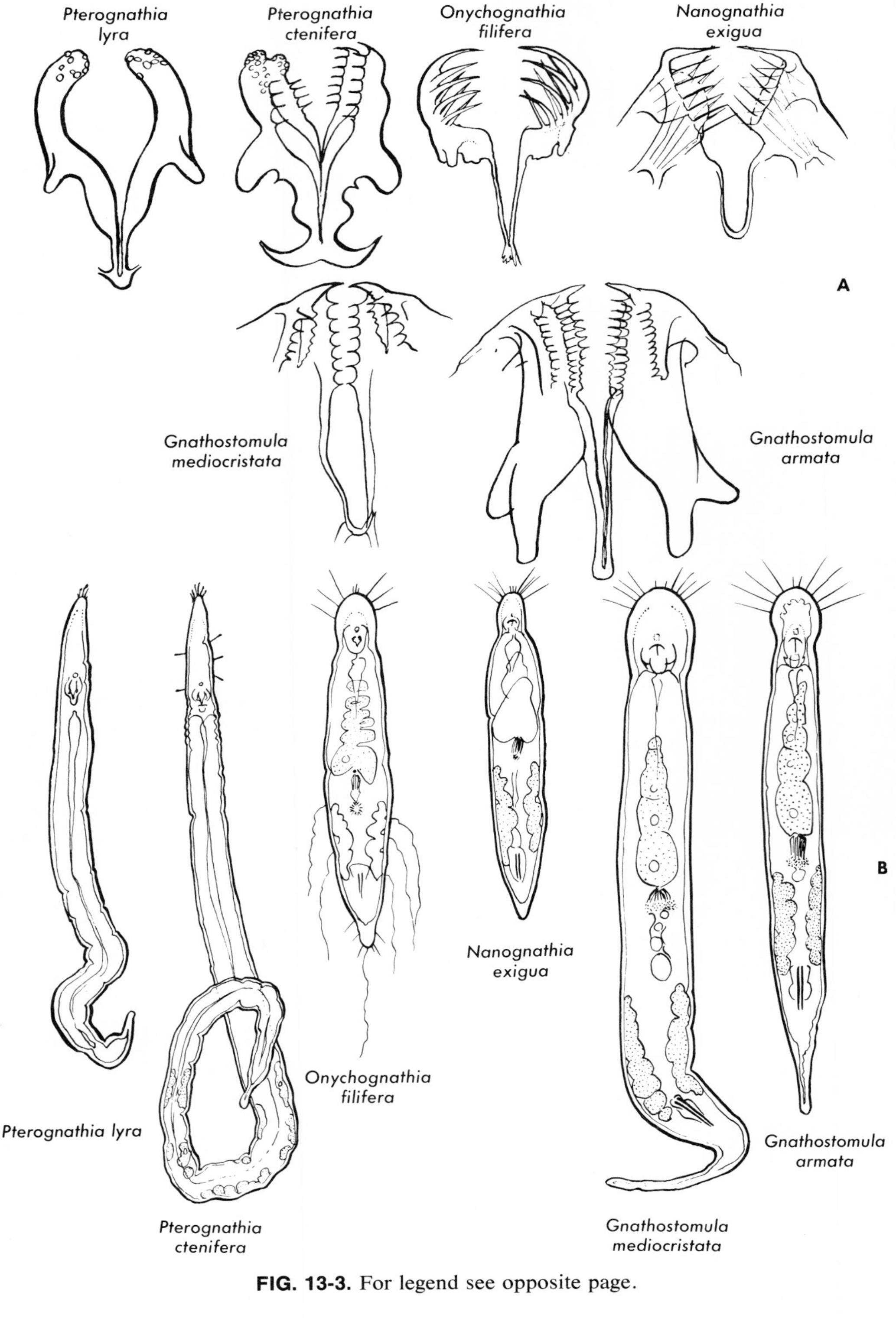

FIG. 13-3. For legend see opposite page.

FIG. 13-4. Sandy beach, Drake's Bay, California. Piles of large kelp torn from offshore kelp beds during rough weather are cast up on the beach. The beach itself has been somewhat eroded, exposing rocks that are frequently covered with sand, as indicated by the lack of attached animals. (M. W. Williams photograph.)

FIG. 13-3. A group of the recently discovered phylum Gnathostomulida, a dominant element of the endopsammon in fine sands that are anaerobic below the surface layer. **A,** Jaws and basal plates. **B,** Series of species showing increasing differentiation of the head and sensory cilia. Only the sensory cilia are drawn. There has been a suggestion that perhaps at least some of the enigmatic microfossils known as conodonts may be gnathostomulid jaws or basal plates. (From Riedl, R. J. M. 1969. Science **163**:445-452. Copyright, 1969, by the American Association for the Advancement of Science.)

bers of the curious group **Anomura,** which is intermediate in many ways between the macrural, or large-abdomened, decapod crustaceans such as shrimp and lobsters, and the true crabs, or **Brachyura,** in which the abdomen is small and carried tucked out of sight beneath the cephalothorax. Other familiar examples of Anomura include such diverse forms as hermit crabs, porcellanid crabs, ghost shrimp, galatheids, and lithodids, ranging from shrimplike forms to crablike types. The crablike members of the group can be readily distinguished from true crabs by their long antennae and by the fact that the last pair of walking legs is usually greatly reduced and often held up against or beneath the posterior edge of the carapace so that the crab appears to be missing one pair of legs.

The mole crabs, however, resemble neither shrimps, hermit crabs, nor true crabs. They have become so modified for existence in the wave-swept sands that one would hardly recognize them as decapod crustaceans at all without careful examination. The body is a smooth, oval shape and is largely covered by the hard carapace. The appendages can all be tucked in, and they fit neatly beneath and at the sides of the body so that the general streamlined form is maintained. They are flattened and shaped as efficient swimming and digging organs.

Like most other animals dominant in marine environments, the mole crabs are plankton feeders. Great numbers of them burrow backward into the soft sands facing the sea, where every wave sweeps across the wet beach and recedes as a thin sheet of water. As a wave recedes they extend their plumed antennae, capturing minute planktonic organisms carried in the surf, and then draw the antennae through the mouth parts, scraping off the food they have captured. As the tide moves up and down the beach, they move with it in stages, suddenly emerging from the sand and allowing themselves to be carried either up or down, depending on the direction of the tide, and promptly burying themselves at a more favorable level to continue their feeding. Any that are left behind by a falling tide bury themselves a few inches into the wet sand and await the return of the water.

In the summer the females carry bright orange egg masses firmly attached beneath the body. As the time for hatching approaches, they cease feeding and remain lower in the water. The young hatch as zoeae—minute, grotesquely spined planktonic creatures that are sometimes carried great distances along the coast by currents. Sometimes they are swept out to sea and lost. Off California they have been taken as far as 200 miles or more from shore. By the end of the summer they have undergone their larval molts and are ready to assume a benthic existence. The second-stage larvae seek out sandy bottom near the shore, where they undergo another molt, bringing transformation to the adult form and feeding behavior.

In the northern parts of their range, where the sand becomes cold and ice may be forming on the beaches, the mole crabs move out beneath slightly deeper water.

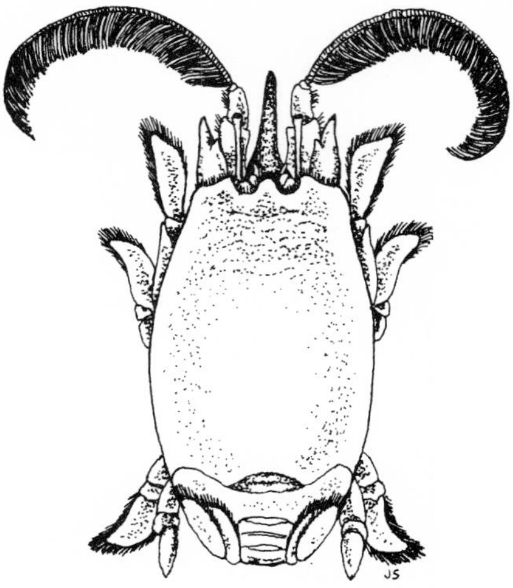

FIG. 13.5. *Emerita talpoida,* mole crab or sand crab, east coast, United States. (After Arnold, A. [Foote]. 1901. The sea-beach at ebb-tide. Century Co.; reprinted unabridged in 1968 by Dover Publications, New York.)

A

B

FIG. 13-6. A, Mole crab, or sand crab, *Emerita analoga,* from the Oregon coast, dug out of the sand to show general appearance. **B,** Same, buried in sand beneath water, with only the eyes and antennae protruding. (**A,** Oregon Institute of Marine Biology photograph; **B,** M. W. Williams photograph.)

A B

FIG. 13-7. A, *Nucula;* **B,** *Yoldia.* (After Keen, A. M., and E. Coan. 1974. Marine molluscan genera of western North America: an illustrated key. 2d ed. Stanford University Press, Stanford, Calif.)

In warmer areas they may remain active in the intertidal zone throughout the year. In spring, mating occurs. The males soon die, and the females carry their eggs through much of the summer and then die as winter approaches.

The common mole crab along Atlantic shores is *Emerita talpoida* (Fig. 13-5) and that of the Pacific coast, *E. analoga* (Fig. 13-6). A somewhat larger, more grotesque species, the spiny sand crab *(Blepharipoda occidentalis),* occurs lower in the intertidal and upper subtidal sands.

Bivalve molluscs

Some bivalve molluscs have also become adapted to life in the open sandy beaches between the tide lines or just below the low tide level. These, too, are plankton feeders.

On many open Atlantic beaches countless thousands of small coquina clams *(Donax variabilis)* are found in the surf zone. As waves wash them out of the sand, they promptly burrow back beneath the surface with their muscular foot, to approximately the depth of one shell length, and then extend their siphons to the surface of the sand to bring the life-giving current that bears their food and oxygen and carries away their waste products. Coquinas are not permanent residents of any particular beach. They may be present by the thousands at one time and absent at another. Perhaps, like the sand they live in, communities of coquinas are moved along the shore with the alongshore current of sand.

Donax also move up and down the beach with the tide. At low tide, as the amplitude of the waves begins to increase with the incoming tide, they emerge from the sand as the waves break and are carried up the beach, where they rebury themselves as the wave slackens. By repeating this maneuver, they maintain their position in the wave zone. As the tide begins to ebb, they display the opposite behavior, emerging from the sand only as each wave recedes, and are thus carried back again.

Species of *Donax* also occur in great numbers on some Pacific beaches. In southern California, *Donax* communities about 5 meters wide may extend several kilometers along the beach, with about 20,000 individuals per square meter.

On many beaches other clams, too, find a living. Most of these clams are somewhat larger and burrow more deeply into the sand where there is a greater degree of stability. Preeminent among rapidly digging, sand-dwelling clams are the razor clams, much sought after by gourmets on the lowest of spring tides.

A large, heavy-shelled clam, the pismo clam (Fig. 13-8), occurs on California beaches at the lowest intertidal and subtidal levels. They lie barely buried in the sand, facing the sea, with only the umbo of the shell reaching the level of the sand

FIG. 13-8. Pismo clam, *Tivela stultorum,* on a California beach. (M. W. Williams photograph.)

surface. The intensity of the rock-dwelling animals' competition for firm substrate to which to attach can be seen in the fact that the shell apexes of the pismo clams, hardly reaching the sand surface, commonly become sites of attachment for tufts of hydroid colonies. Clam diggers often locate the clams by these small feathery hydroids that stick up out of the sand.

Detritus feeders

On many beaches where the sand is of the proper porosity—medium to fine sands, with sufficient organic matter in it—peculiar polychaetes of the family **Opheliidae,** *Ophelia, Thoracophelia* (= *Euzonus*), or others, known as bloodworms because of their bright red color and their red hemoglobin-containing blood, live in dense mats, or colonies, just below the surface of the sand, and the colony may extend the full length of a beach. These worms, having adopted a permanent burrowing, substrate-eating habit, have lost the parapodia characteristic of most polychaetes. Only the bundles of setae mark their positions. Thus they superficially resemble oligochaete earthworms in form and habit. There is even a clitellum-like glandular region marking the posterior border of the "thoracic" region. Its function is unknown; it is certainly not similar to that of the clitellum of oligochaetes, since these worms do not form egg cocoons. There are also small paired fingerlike respiratory "gills" on most of the trunk segments. Bloodworms continually pass wet sand through their gut. Experiments in which sand stained with methylene blue is used alternately with unstained sand indicate that they can replace the sand in their gut in as short a time as fifteen minutes, and that a thriving colony of *Thoracophelia* may cycle, every five years, an amount of sand equivalent to the entire beach in which they live.

Like other intertidal animals bloodworms are good burrowers. Their method of burrowing is unique. Unlike the coelom of most annelids, the coelom of the bloodworm is continuous throughout most of the body length, each intersegmental septum being represented only by a strand of tissue running from the dorsal body wall to the gut, throughout most of the body. There is, however, a double septum separating the coelomic cavity of the head from that of the rest of the body. This septum is reflected back on the dorsal side as a large muscular saclike injector organ communicating with the head coelom by a pair of small pores in the anterior face of the septum. When bloodworms are burrowing through the wet sand, they do so by first withdrawing the coelomic fluid from the head so that it enters the injector organ. They then constrict the circular muscles of the head and anterior part of the body, pushing the head forward into the sand. The muscles of the wall of the injector organ contract, forcing the fluid into the head and blowing it up like a small bulb by which the anterior end of the worm is anchored. The circular muscles are then relaxed and the body of the worm is pulled forward by contraction of the longitudinal muscles of the body wall. In soft wet sand bloodworms can burrow with surprising rapidity.

Another peculiarity of bloodworms, which explains a characteristic sign of their presence on a beach, is their specialization for anal respiration. The last few segments of the body are enlarged and bear setae directed obliquely upward. The dorsal wall of the rectum bears two tracts of strong cilia, which beat in an anterior direction. At the time the falling tide uncovers a bed of bloodworms, the worms are active, burrowing about close beneath the surface of the wet sand. As a worm comes to just below the surface and turns down again, the enlarged anal segments with their bristles dislodge a few grains of sand that fall into the burrow behind the worm, producing a minute hole in the sand. Thus the area occupied by the bloodworm colony becomes marked with thousands of pinpoint-size holes, which in the aggregate may darken the area and which differ from holes excavated by other burrowing forms such as beach hoppers in being smaller and having no sand piled at the entrance. Perhaps this serves a ventilating function for the colony, allowing more air into the burrows than would otherwise come through the close-packed sand. At the moment the hole is formed and the posterior portion of the worm passes beyond, one can sometimes observe an expansion of the rectum and a rippling effect produced by the two ciliary tracts, as if the worm were taking a breath. In any event, the holes do not indicate that the worms have come out on the surface for a time during the night, as was formerly thought.

In its basic features the life cycle is typically that of a polychaete. The sexes are separate, and the eggs develop into typical planktonic trochophore larvae that spend about four or five days in the plankton before developing into metatrochophores and seeking the bottom.

Bloodworm colonies are sharply delimited at both their upper and lower margins, probably by moisture requirements. At the upper margin of the colony, between tides, the worms may be somewhat deeper in the sand. Apparently they cannot thrive where the sand is almost completely saturated all the time. They also burrow more deeply as the tide comes in or if seawater is poured over the area of part of the colony. They are thus restricted to the belt where the sands close to the surface remain moist but not saturated during most of the low tides.

Scavengers

Sand hoppers, or beach fleas

In addition to the plankton and the adsorbed organic matter and detritus in the

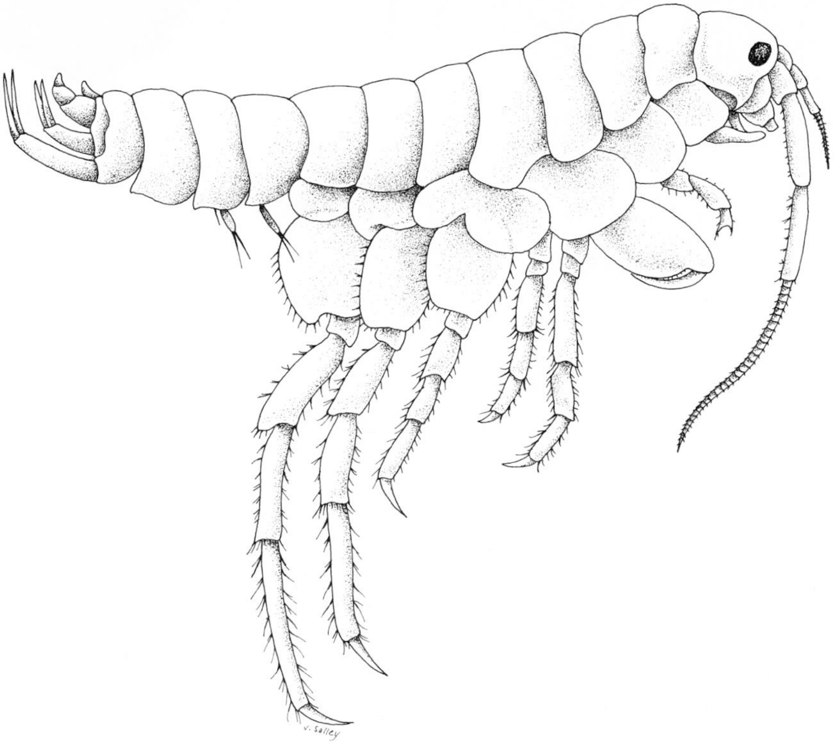

FIG. 13-9. *Orchestia selkirki,* one of the beach hoppers. (After Stebbing, T. R. R. 1888. Challenger reports. Vol. 29.)

sand, a third abundant source of food on the beaches is the flotsam left by the waves and tides. Animals from the surface waters that have drifted too close to shore and been thrown up on the beach, seaweeds and sessile animals torn from nearby rocky areas by the turbulence of spring tides or storms, and plant and animal debris washed down from the land by rivers provide food for many organisms. The most ubiquitous and successful exploiters of this food source are amphipod crustaceans of the family **Orchestiidae** or **Talitridae,** genera such as *Orchestia* (Fig. 13-9), *Orchestoidea,* and *Talitrus.*

Like certain species of periwinkles from the rocky areas, these amphipods seem to be at a midpoint in an evolutionary emergence from sea to land. They no longer swim, respire, or reproduce in the ocean waters as do the vast majority of amphipods. In fact, they are poor swimmers, and drown if submerged too long. Yet they seem to require the salty dampness of the sea and are bound to its margin. Their life and activities are tuned to both the diurnal and the tidal cycles. By day and during high tides they remain out of sight in burrows excavated several centimeters into the sand of the upper beach or under piles of damp seaweed left on the upper beach by previous high tides. But when there is a falling tide during the hours of darkness, they emerge from their burrows by the thousands, roaming out over the intertidal sands as the sands are exposed and searching for and consuming all manner of organic remnants, thus cleaning the beach. Shortly before dawn, or at any time when the tide has turned and is again beginning

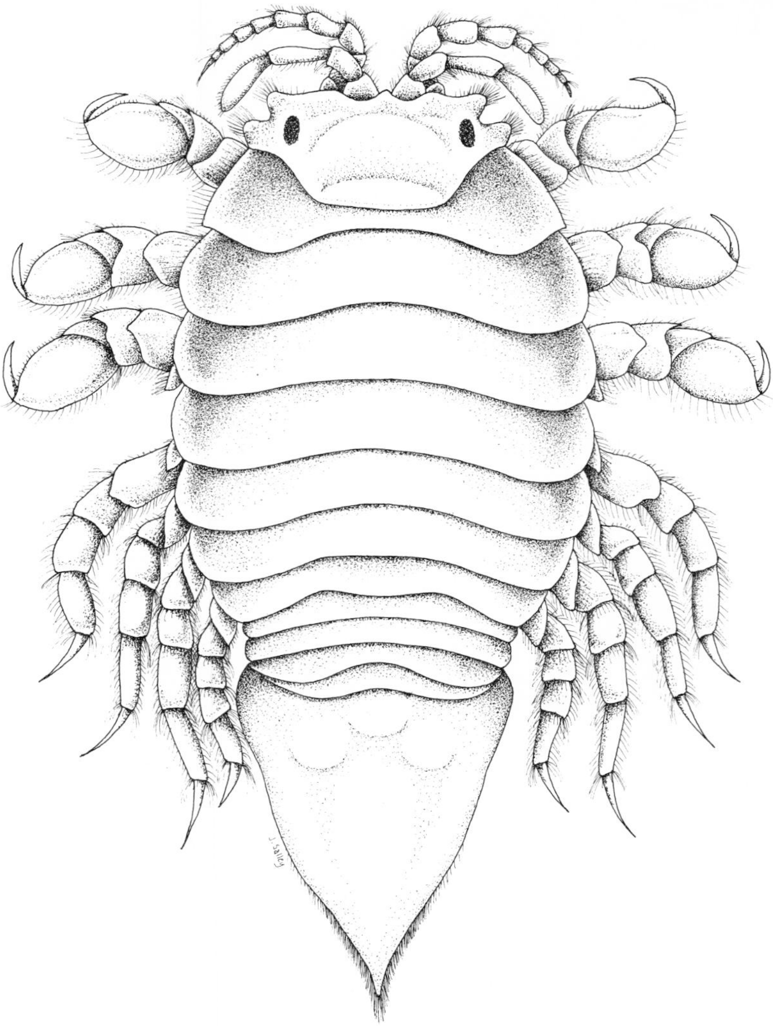

FIG. 13-10. *Chiridotea coeca* (Isopoda; Valvifera). (After Harger, O. 1880. Report on the marine isopods of New England. Report U.S. Fish Comm. for 1878.)

to flood the intertidal sands, they begin to move up the beach near the high-water line. There each sand flea furiously excavates a new burrow into which it will retreat from daylight and from high water. Before finally retiring, it plugs the burrow with sand, and withdraws to the chamber it has made at the lower end. There it will await, and somehow sense, the next occurrence of a falling tide during the hours of darkness.

Isopods

Isopods such as *Sphaeroma* (Fig. 13-11), *Exosphaeroma,* or *Cirolana* are often abundant in sand or sandy mud in the intertidal area and are active scavengers. They can be utilized for cleaning skeletons under water in the same manner that dermestid beetles are often used in dry situations.

Ghost crabs

Another group of nocturnal scavengers of the beaches, particularly in warm and

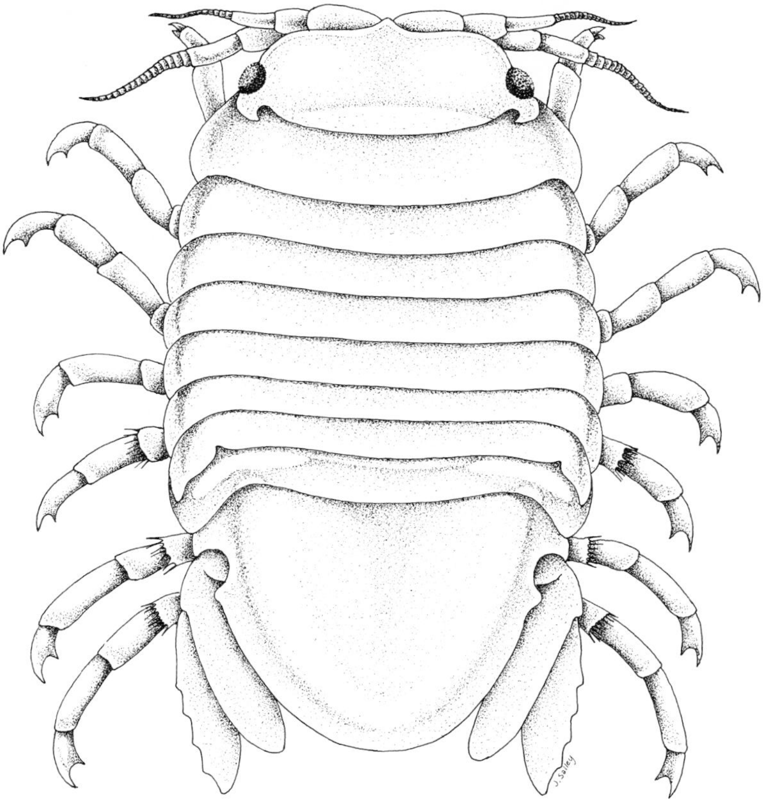

FIG. 13-11. *Sphaeroma quadridentatum* (Isopoda; Flabellifera). (After Harger, O. 1880. Report on the marine isopods of New England. Report U.S. Fish Comm. for 1878.)

tropical regions, are the ghost crabs, members of the family **Ocypodidae.** Ghost crabs are true crabs living still higher on the beach than the sand hoppers. They build their burrows near the top of the beach or even at the beginning of dunes at the back of the beach. They are still, however, bound to the sea for both respiration and reproduction. In the branchial chambers they carry a little seawater, keeping the gills wet. From time to time this water must be renewed. At such times the ghost crab approaches the waterline, waiting with one side turned toward the ocean until one wave, larger than the rest, washes over it, renewing the water in the gill chambers, and then the animal retreats again to the upper beach. The females also return to the water to spawn. Other than these two instances and occasional dashes into the water for protection from a predator on the beach, adult ghost crabs live and act like terrestrial animals.

At night they emerge from their burrows in great numbers to forage on the beach near the waterline, sometimes taking some morsels back to their burrows. They can run faster than any other crustacean and almost as fast as man. As their pale forms move rapidly over the night sands they look strangely insubstantial, like wisps of some small spirit being swept across the beach by a wind.

Like most other crabs, the early larval stages of the ghost crab are peculiar spined planktonic forms called zoeae. After the zoeal stages there is a transformation to a more crablike megalops. In the case of ghost crabs, whose megalops must make a beach landing in the turbulence of the surf, the megalops is adapted much like the mole crab, which spends all its adult life in the surf. The body is oval and protected by a hard, smooth carapace. The limbs and abdomen are tucked in closely, neatly fitting, so that the megalops can tumble in the surf without injury. Having made a safe landing, the megalops digs a small burrow in which it undergoes another molt, transforming to the adult shape. The young crabs make their burrows in the wet sands covered by high tide. As they grow larger the burrows are made farther up the beach until in the fully adult stage they are at the uppermost edge of the beach.

Where the winters are cold, as along the shores of the eastern United States, ghost crabs appear to hibernate in deep burrows throughout the winter, reappearing on the beach in the spring.

Mole crabs, coquina cams, bloodworms, ghost crabs, and sand hoppers, each in their own different way, have become adapted in structure, physiology, life cycle, and behavior patterns for life in the zone of open beaches. In these areas there are pounding waves, ever-shifting sands, fluctuating tides, and exposure to wind, rain, sunlight, heat, and cold. The environmental stresses are so great, so diverse, and so variable that most organisms are prevented from establishing themselves there. Yet sources of abundant food are there—the plankton, the organic film and microscopic animals and plants to be found on and among the sand grains, and the flotsam continually thrown up on the beaches from the sea. Those few animals which have become adapted and are able to succeed in this difficult environment and exploit these food sources find themselves virtually without serious competition and can dominate their environment in overwhelming numbers.

Thienemann (1920) stated that it is a fundamental principle of biocoenotics that the more specialized or extreme the habitat (biotope), the poorer in species but the richer in individuals will be the biocoenosis.

Terrestrial forms on sandy beaches

Scavengers

Animals from the sea are not the only ones to have become adapted to live on or regularly visit the beaches. On many beaches, where windrows of seaweeds are frequently thrown up, swarms of anthomyiid flies—the kelp flies (Fig. 13-12), genus *Fucilia,* and others—are as ubiquitous as the sand hoppers. Their maggots live in the decomposing seaweeds, and the small

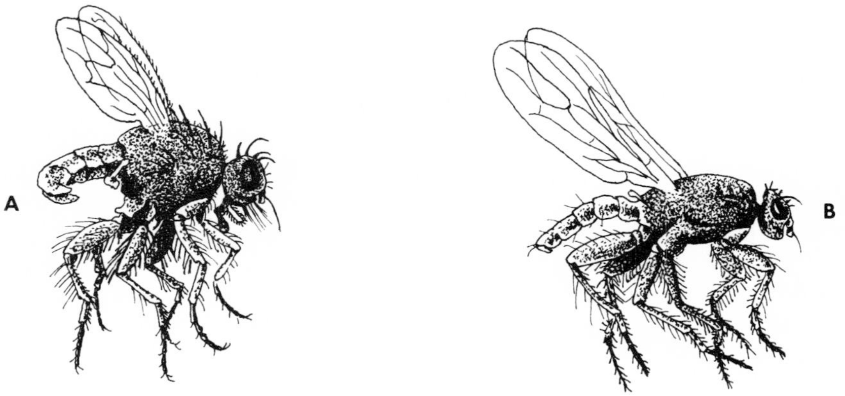

FIG. 13-12. A, *Fucellia costalis* and, **B,** *Coelopa frigida,* kelp flies.

brown sculptured barrel-shaped pupae can often be found there. Small elongate rove beetles with truncated elytra, family **Staphylinidae,** especially genus *Bledius,* are also regularly found there, as well as numbers of other beetles of the families **Nitidulidae, Tenebrionidae, Scarbaeidae,** and others.

Predators

Since much of the life of the beaches lives burrowed into the sand most of the time, several of the most notable predators are adapted for probing the sands. Such are the long-billed shore birds—sandpipers, godwits, curlews, and others often seen in numbers probing the sands for bloodworms and beach hoppers. The small sanderlings race up and down the beach in tight little groups just at the margin of the waves, probing the sand furiously for a few moments, at the lowest points they can reach as each wave recedes, for young *Crago,* small shrimps, amphipods or isopods, and small mole crabs often found there and then race back as the next waves drives them up the beach.

The seagulls are ubiquitous predator-scavengers patrolling all coastal areas—rocky, sandy, muddy, protected—and open waters for small fish and for any animal or animal remains they can pick up at the water surface or on the coast. Many of them have developed special habits, such as dropping shelled animals from a height onto rocks or hard sand to break them open. On the East Coast the ring-billed gulls tread the soft sand at the waves' edge to bring up the small coquina clams. When low tides occur during daylight hours, the gulls are always busy, searching the intertidal areas for any animal unfortunate enough to have been left stranded or exposed.

A smaller number of predators are adapted to prey on small animals on the sand. Cliff swallows nesting on the coastal bluffs are often seen sweeping low over the beach, capturing kelp flies or other insects flying over the sands. Swift-running, big-jawed tiger beetles (**Cicindelidae**) are common at times on the open sand, as well as along the margins of estuaries and other bodies of water. Fast-running hunting spiders such as *Trochosa* and *Tarantula* also occur on beaches.

At night still other animals forage on the beaches. *Peromyscus* (the deer mouse) and other rodents living among the bunch grasses or in the woods above the beach venture out onto the upper beach in search of insects or morsels of food left in the flotsam. Skunks and raccoons are also frequent visitors in many areas.

Still other predators—various predaceous annelids and molluscs, such as olive shells and *Polinices*—burrow in the substrate. Annelids of the family Glyceridae (Fig. 13-13) are among the largest, most active of these burrowing predators in sandy beaches, whereas the huge nereid *Neanthes brandti,* which may attain a length of more than 2 feet and a breadth of an inch or more, is found in sandy mud along the upper reaches of marine-dominated estuaries.

Periodic visitors from the sea

A few animals that are truly marine, driven perhaps by predation pressure on the eggs or young farther offshore, find protection for these stages by depositing them at a higher level of the subtidal or in the intertidal. There has been some speculation that the emergence of the early ancestors of the amphibians onto the land was more or less forced in this manner. In any event, several species of fish, a few invertebrates, and the sea turtles regularly leave the sea long enough to bury their eggs in the sand. In the case of the sea turtles, which descended from land-inhabiting reptiles, the reproductive tie to the land probably represents, as does their air breathing, an incompleteness to their adaptation to marine life.

The evolutionary changes involved in changing from fully aquatic to fully terrestrial modes of respiration and reproduction or vice versa seem to be the most difficult to make. Thus a number of marine

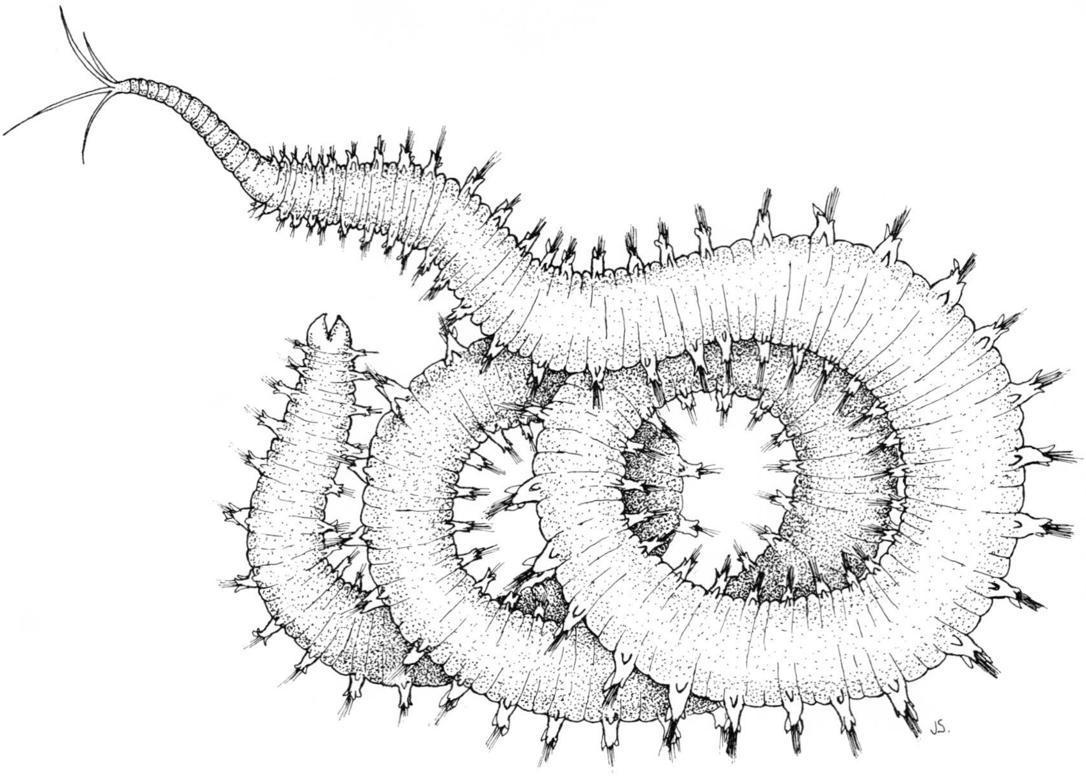

FIG. 13-13. *Glycera lapidum*, an active predator found in lower parts of sandy beaches. (After McIntosh, W. C. 1915. British marine annelids. Vol. II. Ray Society, London.)

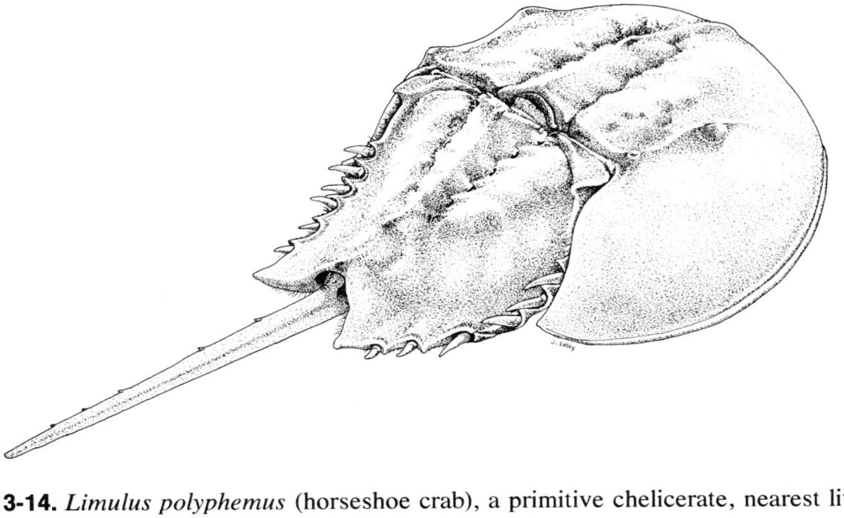

FIG. 13-14. *Limulus polyphemus* (horseshoe crab), a primitive chelicerate, nearest living relative of the trilobites.

forms, such as ghost crabs, land-dwelling hermit crabs, some mudskippers and periwinkles, and others that seem now partially emergent from the sea, must still provide for gill respiration, return to the water to spawn, and in many cases spend part of their life cycle in the plankton. On the other hand, marine mammals and reptiles all still breathe air, and, except for the cetaceans and sea snakes of the subfamily **Hydrophinae,** go onto the land to have their young or lay their eggs. Even the cetaceans in many cases seek out relatively shallow lagoons in which to bear their young.

Horseshoe crabs

Along the eastern coast of North America from Maine to Mexico, and along the eastern coast of Asia and parts of the East Indies, lives the peculiar primitive "living fossil" *Limulus,* the closest living relative of the trilobites, which dominated ancient Paleozoic seas. *Limulus* (Fig. 13-14) usually lives well offshore, but in the breeding season it comes into shallow waters, sometimes in great numbers. The female is about 0.5 meter in length, and the male is considerably smaller. Early in the summer they pair. The female comes ashore with one or more of the small males, each hanging to the tail of the one ahead. She excavates a hole in the sand in which the eggs are deposited, fertilized by the males, and left for the waves to cover. About midsummer the young hatch and make their way into the water, to disappear until they return fully grown the following year. At hatching time a beach may be swarming with thousands of the young. The young, lacking the long caudal spine of their parents, look even more like trilobites than do the adults.

Grunion

Of all the fish that spawn high in the intertidal, the most famous and in many ways remarkable is the grunion, a small atherinid fish, *Leuresthes tenuis.* From Lower California to San Francisco, from the end of March to about August, every second week beginning the second night after a full moon and just after the spring tides have reached their highest, the moonlit beaches may be alive with silvery fish about 15 to 20 cm. long. Each wave brings thousands of fish high up onto the beach. The females quickly burrow into the wet sand, tail first, until only the head down to about the pectoral fin is above the sand. While she deposits eggs, the male curls around her on the surface of the sand and deposits milt, which runs down into the sand and fertilizes the eggs. Then with the succeeding wave they slither back into the sea while others are brought onto the beach.

Since the eggs are laid just after the turn of the tide, succeeding waves are weaker as the tide recedes and only push more sand over the eggs rather than washing them out. Also, since the eggs are laid after the highest spring tide, on the descending series of tides, the eggs will be left in the sand for two weeks before the next series of rising spring tides washes them out. The eggs are actually ready to hatch in about ten or eleven days after being deposited in the sand, but they do not hatch until they are agitated in the seawater, which occurs at the next series of spring tides. At this time the young are washed out of the sand and carried into the surf, and a new spawning run of the adults begins. Because of the progression of the tides throughout the lunar month, alternate series of spring tides occur later at night when few people are on the beach to observe the run, giving the appearance of a four-week cycle.

Reproductive behavior of this type entails the most exquisite adjustment and timing of physiological processes, instincts, and behavior patterns to the diurnal and lunar cycles and illustrates the profound effect that such subtle influences as the lunar cycle may have on organisms. If the eggs are not laid at precisely the right part of the tidal cycle, they will be washed from the sand prematurely and will perish.

Only the high spring tides can be utilized.

In places such as the upper end of the Gulf of California, where some of these tides, during spawning season, occur during daylight hours, there is some loss to predation by birds. The numbers of grunion in a spawning run, however, are so great, and their period of vulnerability is so brief, that the local predators are quickly sated, stop feeding, and do not constitute a serious threat to the success of the spawning. The grunion are gone before an overwhelming concentration of birds from greater distances can be assembled.

How the fish, living offshore, sense the proper moment in the tidal cycle to make their run and how the developmental rate of the eggs became so precisely attuned to the lunar cycle is a source of wonder. It seems probable that a long period of selection was involved, once the habit of depositing the eggs in the beach sand had begun. Since any that were not deposited at the right times would perish, the progeny would all be derived from those fish which happened to be properly in tune with the lunar cycle, and all deviants would probably be rigidly eliminated. Continued selection and elimination of deviants keeps the pattern precise for the population as a whole.

Throughout the spawning season the female grunion carries three classes of eggs in the ovary, except when recently spawned out. They are extremely immature ova, intermediate ova, and bright orange ripe ova, which (as the spawning date comes) are shed into the lumen of the ovary, ready to be spawned. After spawning, the intermediate-stage ova ripen just in time for the next series of spring tides, and a new batch of immature ova develops.

Studies of growth and of concentric striations at the outer edge of the scales indicate that grunions live about three years, usually spawning two or occasionally three seasons.

Sea turtles

Sea turtles, family **Cheloniidae,** are distributed throughout the tropical seas. Although the adults are wide-ranging in tropical seas, they return to the coasts where they were born, to mate and lay eggs. The females come up onto the sandy beach, excavate a hole, bury their leathery-shelled eggs, and return to the ocean. When the young hatch, they scramble up out of the covering sand and make for the water. If they happen to emerge during the hours of daylight, they are usually doomed because gulls and other predaceous birds carry them off before they can make it to the water. A remarkable aspect of sea turtles' behavior is their homing ability. This ability to swim hundreds of miles through trackless seas to a particular beach on a particular small island where they happen to have been born is one of the marvels of animal navigation, rivaled only by the feats of certain birds, the Atlantic eels of Europe and America, and the salmon.

PROTECTED SANDY AREAS

As one leaves the open ocean beach and goes to quieter and more protected sandy flats, or offshore beyond the range of great wave turbulence, the substrate is far more stable. Many more kinds of animals can live here. Intertidal flats of sand or sandy mud are inhabited by many kinds of burrowing animals that require a more permanent type of burrow than can be maintained in the constantly shifting sands of the open beaches.

The majority of the burrowers are either filter feeders or animals that are able to ingest the detritus that collects at the surface of the sand or mud.

As in other habitats, the plankton feeders tend to be dominant. On the flats most of them are either bivalve molluscs, annelid worms, or burrowing crustaceans. Lugworms (*Arenicola* and *Abarenicola*), the parchment worm *(Chaetopterus),* and the spoon worm *(Urechis),* and crustaceans such as the ghost shrimp *(Callianassa),* which circulate water through their burrows, obtaining either their food or oxygen, or both, from the passing stream, have U-shaped burrows with two openings at the

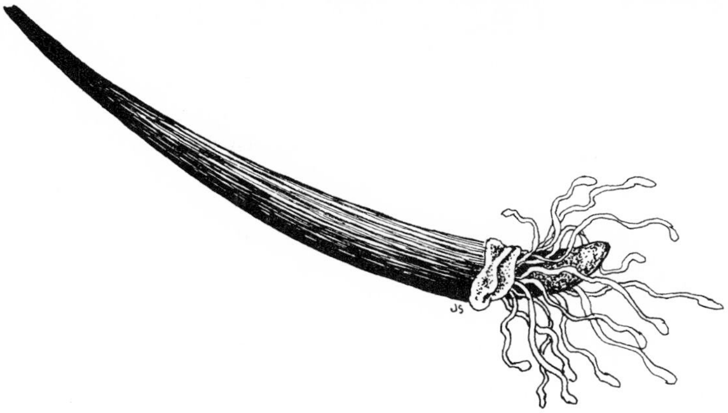

FIG. 13-15. *Dentalium dentale,* tooth shell. (After Arnold, A. [Foote]. 1901. The sea-beach at ebb-tide. Century Co.; reprinted in 1968 by Dover Publications, New York.)

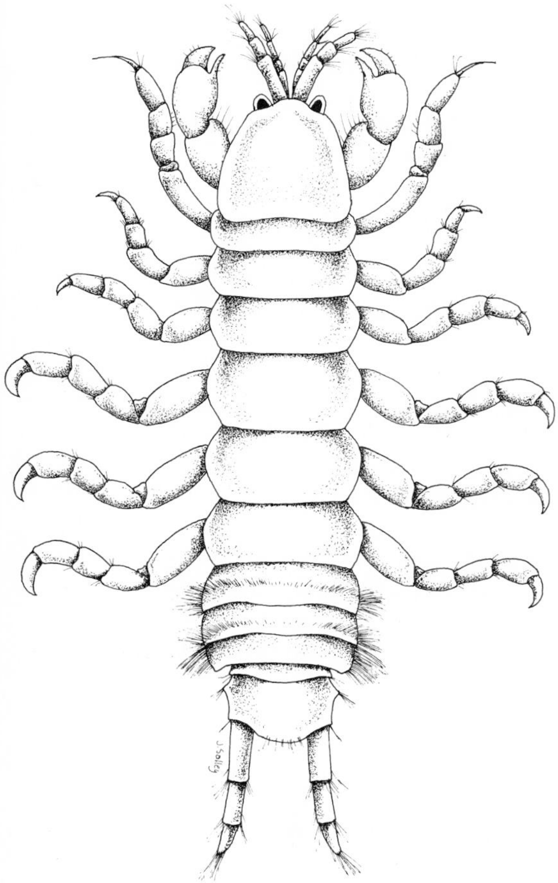

FIG. 13-16. *Tanais vittalus.* (After Harger, O. 1880. Report on the marine isopods of New England. Report U.S. Fish Comm. for 1878.)

surface. These burrows with the circulation of water through them are favorable homes for a variety of other small animals such as pea crabs, small gobies, and commensal copepods.

Bivalve molluscs

All levels of the flats contain bivalves, from the plump, heavy-shelled ribbed cockles at the surface, which can be collected with a rake, to large, deeply buried clams such as the gaper clam *Tresus* (formerly *Schizothaerus*) *nuttalli* or the geoduck, *Panope generosa,* which may be as much as 0.5 or 1 meter below the surface. In general the structure of a clam, or even merely its shell, tells much about its habits and ecology. Those clams that live well below the surface have correspondingly long siphons. The larger the siphon, the more prominent, as a rule, is the mark left on the empty shell by the line marking the mantle attachment to the shell along its edge—the indentation in the line marking the attachment of mantle muscles, known as the pallial sinus.

The soft-shell clam *Mya arenaria* and the gaper clam *Tresus nuttalli* are examples of large, deeply burrowing clams with long siphons. These clams are commonly a characteristic and dominant element in their biocenoses. The soft-shell clam is widely distributed along boreal and temperate shores in both the Atlantic and the Pacific. The gaper clam is prominent along the Pacific coast of North America, and its shells also sometimes constitute the principal fossil remains in Pleistocene marine terraces.

Large clams with big siphons commonly withdraw their siphons quickly when alarmed, as by the vibrations caused by someone walking near them or, in some cases, by a shadow passing over them. As they withdraw, they may squirt water as much as 30 or 60 cm. into the air.

Since the circulation of water is from the ventral siphon into the large ventral mantle chamber, through the gills into the dorsal mantle chamber, and out through the dorsal siphon, and since both siphons leave the clam together at the posterior end, often even bound together in a common muscular tube, clam burrows have only one opening at the surface. The same is true of such animals as phoronids, in which the digestive system is looped, bringing the anus near the mouth so that the mouth and food-gathering tentacles are together, with the anus at the mouth of the burrow.

Annelid worms

Lugworm burrows are usually in muddy sand near the upper reaches of intertidal flats. The incurrent end of the burrow is in a depression in the sand, whereas the excurrent opening is at the top of a small black mound. The darkness is caused by reduced mud, with hydrogen sulfide having been expelled there. Detritus coming in with the incurrent water together with some sand or mud is swallowed, and the sand is expelled as coiled castings around the excurrent opening of the burrow. As is the case with earthworms in terrestrial habitats, the work of a bed of lugworms tends to stir, cleanse, and renew the muddy sand in which they live. It has been calculated that where they are abundant, they may work over almost 2000 tons of mud per acre each year.

The lugworm usually lies in the bend of the U-tube burrow. When the burrow is covered with water, it stretches its head out around the entrance to feed. Around midsummer they spawn. In this case gametes are not shed into the water, but the eggs are contained in small gelatinous translucent pinkish sacs, with one end anchored in the sand by a threadlike extension. Each egg mass may contain many thousands of eggs or developing young.

Another worm abundant in some flats on both sides of the Atlantic is the trumpet worm, *Cistenides,* which constructs a conical tube of a neatly fitted mosaic of sand grains fitted and cemented onto a chitinous base. The tubes are shaped something like the shells of scaphopod molluscs and serve much the same function, being open at both ends. The small posterior end sticks out of the surface, enabling the worm to bring a respiratory current down to its gills as it forages in the soft substrate, digging with alternating vigorous shoveling motions of two groups of comblike bristles on either side of the head and sorting desired food particles with the tentacles around the mouth.

According to A. T. Watson, the young worms first secrete a membrane about themselves and then begin to collect sand grains with their tentacles and pass them to the mouth, where they are rolled about and felt. Selected grains are then fitted individually at chosen spots at the edge of the tube. As the worm grows, it continues to build the anterior end of the tube. Thus, as Watson put it, "Each tube is the life work of the tenant, and is most beautifully built with grains of sand, each grain placed in position with all the skill and accuracy of a human builder."*

Echiurida

Spoon worms, or Echiurida, are widely distributed in sandy mud bottoms and mudflats, often in areas similar to those in which ghost shrimp and *Upogebia* are found. The

*From Watson, A. T. 1928. Observations on the habits and life history of *Pectinaria (Lagis) koreni* Mgr. [edited and introduction by P. Fauvel]. Liverpool Biol. Soc., Proc. Trans. **42:**25-60.

worms are thick-bodied, and indistinctly or not segmented, with a fleshy spatulate "prostomium" in front of the mouth. Like the lugworms, they live in U-shaped tubes. The circulation of water through the tubes is maintained by peristalsis-type movements of the body. By means of glands in the prostomium and anterior end of the body, they fashion a funnel-shaped mucus or slime net in the anterior part of the burrow. The water being circulated through the burrow by the peristaltic action of the body is filtered through this slime net, which the worm swallows from time to time with its entrapped detritus and small organisms. Some species are said to sweep the ciliated ventral surface of the proboscis over the mud near the entrance of the burrow and to sweep detrital particles to the mouth along ciliated tracts.

Spoon worms show some affinity to annelids in that they possess a spacious schizocoel, trochophore larvae, and nephridia, but they are so different in many ways that they are usually regarded as constituting a separate phylum. Formerly they were sometimes grouped with the sipunculids and priapulids in a phylum called Gephyrea, usually regarded as an appendage to the Annelida.

Priapulids and sipunculids

Priapulids and sipunculids are also substrate dwellers. The priapulids live in muddy substrate. Like the Echiurida, they are thick fleshy worms, with a complete digestive tract ending in a terminal anus at the caudal end. The gut is straight and wide. In most species there are one or two large caudal appendages bearing respiratory papillae, looking somewhat like bunches of grapes. The pharyngeal region is in the form of an introvert, or eversible, proboscis, which is plump and covered with rows of small spines. Lang (1948), Hyman (1951*b*), and others regard the body cavity as a pseudocoel and have classed them with the assemblage termed Aschelminthes. This assemblage includes diverse forms distinguished primarily by a pseudocoel type of body cavity and a noncellular cuticular investment of the body, and usually by a tendency toward constancy in the number of somatic nuclei and by a syncytial epithe-

FIG. 13-17. A sipunculid worm with the introvert extended, showing tentacles in feeding position. (Oregon Institute of Marine Biology photograph.)

lium in which the nuclei are gathered into special tracts, or papillae. However, according to the researches of Shapeero (1961, 1962), the priapulids actually have a true coelom lined with peritoneum and should be classed among the eucoelomate animals, perhaps as an independent phylum. Priapulids seem to prefer cool waters and are not found in tropical or subtropical regions. Some of them occur in deep water, down to over 7000 meters.

The sipunculids, or peanut worms, are a much larger group commonly found on somewhat coarser substrates, as in the sand

A

FIG. 13-18. A, Another animal found near the surface of sand or mud in shallow water of estuaries, etc. is the cockle *Clinocardium nuttalli*. This one from South Slough, Coos Bay, Oregon, is viewed from the rear to show the short siphons. **B,** Olive snails, *Olivella biplicata,* partially buried in the sand, with the snorklelike siphons extended. **C,** *Amphitrite,* a polychaete worm that lives in a sandy mucous tube, the body buried in the substrate or in a crevice or underside of a rock. The anterior end is crowned with three pairs of much-branched red gills, and many long extensile and contractile tentacular filaments used to find detrital particles of food and transport them to the mouth. **D,** Acorn worms, balanoglossids. Burrowing substrate-ingesting worms long thought to be related to the vertebrates. Supposed chordate characteristics include a putative notochord, pharyngeal gill slits, and a segmental enterocoelom. The tornaria larvae resemble rather closely the larvae of certain echinoderms. **E,** Ghost shrimp, *Callianassa californiensis,* inhabits rather deep burrows in the mud flats. **F,** *Upogebia pugettensis,* another, somewhat larger anomuran, shrimplike crustacean living in burrows in mud flats or estuaries. (**A, C, D,** and **E,** Oregon Institute of Marine Biology photographs; **F,** M. W. Williams photograph.)

B

C

D

Continued.

FIG. 13-18, cont'd. For legend see opposite page.

FIG. 13-18, cont'd. For legend see p. 304.

among eelgrass roots, in fissures and cracks in rocky areas where sandy sediment has accumulated, or under rocks on sandy or sandy mud substrate. In the sipunculids there is also an introvert, but the mouth, when everted, is surrounded by a circle of soft, branched tentacles (Fig. 13-17) with which they obtain particles from the substrate to ingest. The intestine is unique in that the distal portion of it is wound in a rather tight spiral about itself, ending in an anus in a forward position in what might be termed the shoulder region, just posterior to the base of the introvert. There are numerous genera, and the group as a whole enjoys worldwide distribution.

OTHER SHALLOW BENTHIC COMMUNITIES

Only a few examples of the many shallow benthic communities that have been recognized will be briefly mentioned here.

Eelgrass communities: Zostera and Phyllospadix (Fig. 7-7)

Zostera occurs as a dominant feature of many types of mud and sand flats in bays, estuaries, and protected coastal situations. The roots prefer a reduced environment containing silt and free hydrogen sulfide. *Zostera* plays a role in producing and stabilizing the flats. Much of the silt entering the estuary is trapped in *Zostera* beds and prevented from washing out to sea.

In most mud flats, *Zostera* occupies the shallow channels and lower portions of the flats, covered by water most of the time and separated from the pickleweed salt marsh and *Ruppia* bordering the flats by an area of more barren mud.

During the spring and summer, *Zostera* grows rapidly, providing shelter for a great variety of crustaceans, molluscs, small fish, and other animals and commonly supporting a heavy growth of epontic diatoms, attached hydroids, small algae, bryozoans, coralline algae, protozoa, and other organisms, together with an abundance of small free forms, such as flatworms, nematodes, copepods, and others—all of which find substrate, shelter, and food there. Where *Zostera* or the surfgrass *Phyllospadix* occur in more marine situations in the North Pacific, peculiar fixed jellyfish (*Haliclystus*) are commonly found attached to them. In such places where the plants are growing from sand and detritus consolidated by the root system on rocky substrate, small sea urchins, sponges, hydroids, and Bryozoa are common on the lower parts of the plants, and many annelids, sipunculids, and long-armed brittle stars (*Amphiodia*) may be found among the roots.

On sand or mud flats of estuaries different sets of annelids are present among the roots, including long slender capitellids such as *Heteromastus* and *Notomastus,* small clams such as *Macoma* and *Tellina,* and as the *Zostera* beds expand during the summer they cover areas occupied by cockles, lugworms, *Pectinaria,* larger clams such as *Tresus* and *Saxidomus,* ghost shrimps, the hooded tubes of *Pista pacifica,* and other sand flat or mud flat dwellers. The soft-shelled clam *Myra arenaria* and the giant nereid *Neanthes brandti* are sometimes found under eelgrass but for the most part occur along the upper edges of the flats near the high tide line, where the flats give way to woods or to the salt marsh pickleweed and *Ruppia,* rather than to the eelgrass (Fig. 13-19).

During late summer or early fall the eelgrass flowers and produces seed. The blades then die back and rot, serving as nutrient for a bloom of bacteria, small flagellates, and other microorganisms providing abundant food for the larvae of bivalves such as oysters and of other animals that spawn in the warm temperatures of late summer.

Eelgrass communities are among the richest, most productive of biotic communities, important not only for the species limited to eelgrass beds but also for providing shelter and nursery for the young of many other animals, including a number of commercially important fish and crustacea.

FIG. 13-19. Salt marsh association frequently found in upper parts of estuaries. Pickleweed and marsh rosemary. Holes made by Oregon mud crab *(Hemigrapsus oregonensis)*. (M. W. Williams photograph.)

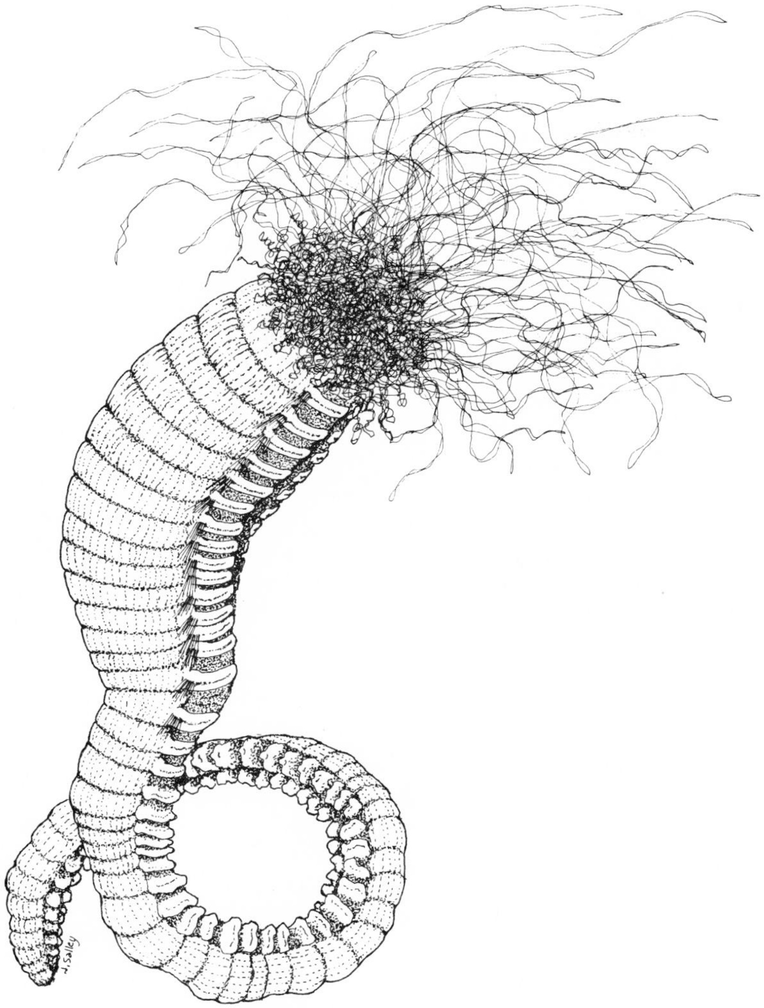

FIG. 13-20. *Thelepus crispus.*

Macoma baltica shallow-mud community

The *Macoma baltica* shallow-mud community is one of the most common of the boreal Atlantic waters. It is characterized by the clams *Macoma baltica* and *Mya arenaria* and the cockle *Cardium edule*. The lugworm *Arenicola marina,* the gastropod *Hydrobia ulvae,* and the amphipod *Corophium volutator* are also prominent members. This shallow-water community usually extends from the tidal zone to about 10 meters' depth in muddy bottoms. Where there is somewhat finer sediment with some clay or silt, the bivalve *Scrobicularia plana* may occur, whereas in places where the mud is more sandy, *Cardium* tends to be more abundant. Since *Macoma, Cardium,* and *Hydrobia* have relatively long lifespans of two or more years, the rate of turnover is rather low.

Thorson (1958) demonstrated that communities similar to this one but with different species of *Macoma* and comparable species of *Cardium, Mya, and Arenicola* occur in the Arctic, along the North American Atlantic coast, and in the Northeast Pacific.

Tellina-Donax shallow-sand community

The *Tellina-Donax* shallow-sand community is found on rather exposed sandy shores from about the lower edge of the tidal zone out to about 10 meters, where it grades into the offshore sand communities. In the Atlantic boreal to temperate regions this assemblage is dominated by the bivalves *Tellina tenuis* and *Donax vittatus*. other characteristic animals are the polychaetes *Nephtys caeca* and often *N. hombergi,* the lugworm *Arenicola marina,* and the sand star *Astropecten irregularis*. The amphipod *Bathyporeia pelagica* is also common. In slightly deeper water with finer sediment, *Tellina fabula* may largely replace *T. tenuis,* and the heart urchin *Echinocardium cordatum* becomes abundant.

Venus community

Where the sea is slightly more open and the sandy bottoms go to a depth of about 50 meters, the boreal communities are often dominated by the bivalves *Venus gallina* or species of *Spisula,* the latter tending to become dominant where the sand is rather loose. Other bivalves characteristic of this community are *Mactra* and *Psammobia*. The polychaete *Nephtys* occurs in this community as well as in the *Tellina-Donax* complex, as does also *Ophelia* and sometimes *Pectinaria*. Other polychaetes common in *Venus* communities include *Glycera* and *Lumbriconereis*. The heart urchins *Echinocardium* and/or *Spatangus purpureus* are common, as are sometimes the sand star *Astropecten* and the gastropod *Natica*. Where the sand is somewhat compacted or firmer, *Tellina* may be present as well as the brittle starfish *Ophiothrix fragilis*.

Where substrate is coarser, composed of shell gravel and coarse sand, larger, more solid bivalves such as *Venus fasciata, Glycymeris glycymeris, Ensis arcuata,* and *Tellina crassa* tend to dominate. The heart urchin *Echinocardium flavescens* occurs here, as does also the lancet *Branchiostoma (Amphioxus)*.

Syndosmya (Abra) muddy-sand community

The *Syndosmya (Abra)* muddy-sand community occurs in relatively sheltered, sometimes somewhat estuarine situations, fairly rich in organic matter. The predominant animals are small, thin-shelled bivalves, such as *Syndosmya alba, S. prismatica, Nucula turgida,* and *Cultellus pellucidus*. The brittle star *Ophiura texturata* and the heart urchin *Echinocardium cordatum* are common echinoderms in these situations. Polychaetes include species of *Glycera, Nephtys,* and *Scalibregma inflatum*. The abundance of small clams and polychaetes makes this community a good feeding ground for fish. Offshore these communities may become more dominated by brittle stars and grade into associations dominated by *Amphiura filiformis* and *Echinocardium cordatum*. The tall-spired *Truritella communis* is common in these communities and in places may at times dominate them. Other common gastropods are *Aporrhais pespelicani* and *Philine aperta*.

Farther offshore the sandy muds tend to grade into finer-grained, soft mud areas in which the brittle star *Amphiura chiajei* and the heart urchin *Brissopsis lyrifera* are conspicuous elements. The small clams *Syndosmya prismatica* and *S. nitida* and species of *Nucula* are fairly abundant, together with numerous polychaetes such as *Nephtys incisa, Glycera rouxi,* and *Lumbriconereis impatiens* and sedentary worms such as capitellids and maldanids. At the outer (deeper) portions of these communities the brittle star *Ophiura sarsi* may become prominent, and additional polychaetes not found prominently in the shallower zones, such as *Terebellides stroemi, Chaetozone setosa,* and *Melinna cristata,* are present.

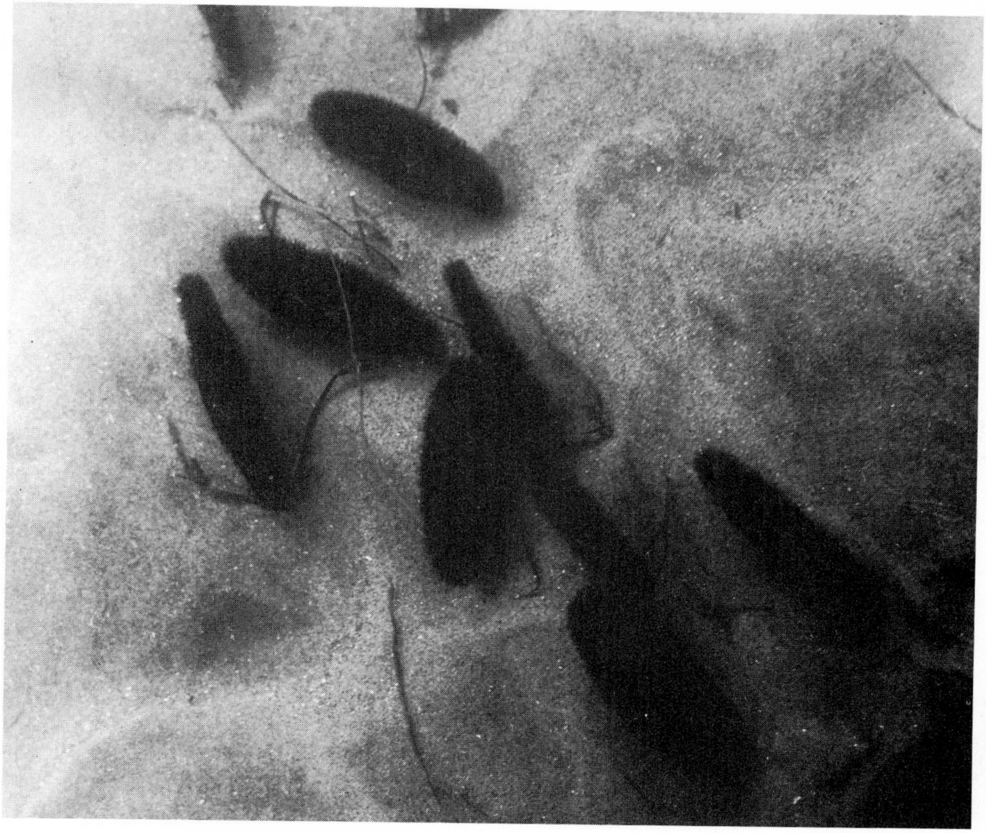

FIG. 13-21. Sand dollar bed, in which *Dendraster excentricus* lies typically almost half buried vertically or obliquely on edge in the sand. West Coast. (M. W. Williams photograph.)

Dendraster excentricus community

In somewhat protected shallow areas or in shallow sand beds beyond the intertidal zone along the western coast of North America there occur beds of the purple sand dollar, *Dendraster excentricus* (Fig. 13-21), commonly found partially buried in an oblique position in the sand, with about one half to two thirds of the disc above the sand surface. Heart urchins are also common in the offshore sandy bottoms.

Turtle grass community

In the West Indies, around Bermuda, and in the Florida Keys in the coral reef zone on coral sands, marine grasses—the turtle grass *Thalassia testudinum,* together with lesser admixtures of manatee grass and shoalgrass—form a characteristic biocoenose. Turtle grass is a broad-leaved marine grass with stout, jointed rhizomes with heavy sheaths at the base. Usually epiphytic algae such as *Fosliella, Ceramium,* and *Aegira* are to be found growing on it. It grows on mud and sand in quiet, protected waters, where it may form continuous meadows, but it also grows where there is some wave action, often forming, in these more exposed situations, patches of growth conspicuously elevated above the surrounding bottom as a result of accumulations of pebbles and shell fragments among the rhizomes. Coralline algae growing among the *Thalassia* add their skeletons to the mass and give it more thickness. *Thalassia* mats of this type represent a cal-

FIG. 13-22. Bed of turtle grass with loggerhead sponge, *Speciospongia vespera,* and tall gorgonian in foreground. (Chesher photograph.)

careous substrate resembling unconsolidated aeolian limestone, whereas mangroves, in contrast, usually fill in with peaty substrate tapering off into mud and soil. The seaward edge of such *Thalassia* patches is usually abrupt, not tapering, and the mat is about 15 cm. thick.

Characteristic of such patches of turtle grass are the large starfish *Oreaster* and the sea biscuit, the echinoid *Clypeaster subdepressus,* usually buried in the sand. Massive loggerhead sponges (Fig. 13-22) and growths of numerous alcyonarians such as gorgonians, soft corals, sea fans, and sea pens are common. Several species of large, predaceous conchs—such as the horse conch *Pleuroploca gigantea,* which occasionally grows to 60 cm. in length, the chief predator, which is the beautiful queen conch *Strombus gigas,* and the cask shell *Tonna galea*—also occur here, as do occasional octopuses, *Octopus vulgaris,* which are also found on European and Mediterranean shores. Some of the characteristic fish are the pipefish, *Syngnathus* spp., which are often difficult to see because of their body shape and posture, which mimic the turtle grass blades. The related sea horse *Hippocampus* is also characteristic, as well as the cowfish *Lactophrys trigonus* and several species of corals. Long-spined active sea urchins *Eucidaris* and *Diadema* (Fig. 13-23, *A*), the spiny lobster *Panulirus argus,* the mantis shrimp, the sea hare *Aplysia dactylomela,* and several species of coral are characteristic of this association. Sea turtles, the loggerhead turtle, green turtle, and hawksbill, sometimes visit the *Thalassia* beds to prey on sea biscuits, conchs, and other animals. Sea cucumbers are also abundant, including a variety of the circumtropical large black sea cucumber *Holothuria atra.*

As is commonly true of tropical communities, there tend to be more species present and greater diversity in this community than in comparable temperate or boreal communities, such as eelgrass communities, or in the various soft substrate communites already mentioned.

Somewhat similar communities are found in other tropical regions such as the Indo-Pacific, but many of the species are represented by ecologically more or less equivalent species. In numerous instances closely related, so-called twin species are found, one in the Indo-Pacific and the other in the Caribbean region (Chaper 16).

Mississippi Delta communities

The Mississippi Delta region offers an interesting example of an area in which a number of types of bottom communities that would normally occupy larger, rather separated areas are brought together in a small area. The Mississippi River discharges huge quantities of sediments and of

A

B

FIG. 13-23. Tropical American group of sea animals. **A,** *Diadema antillarum,* long-spined sea urchin typical of tropical waters. **B,** *Oreaster reticulata,* a large starfish, eating a dying sponge. **C,** Flamingo-tongue snail feeding on sea fan. **D,** Hermit crab feeding on plate urchin *(Clypeaster subdepressus),* photographed at night with flash camera. **E,** Scorpion fish, camouflaged against the bottom. **F,** Sea urchin, *Toxopneustes rosei,* with protective cover of debris held over it by means of the tube feet, and, **G,** with the debris removed. (Chesher photographs.)

FIG. 13-23, cont'd. For legend see p. 313.

C

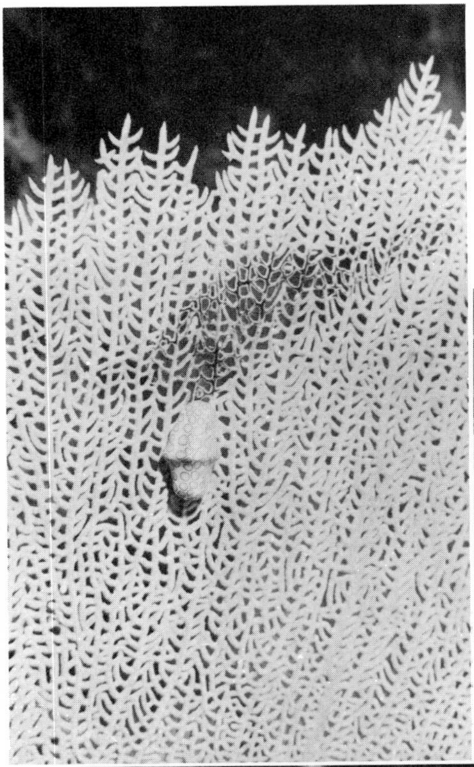

D

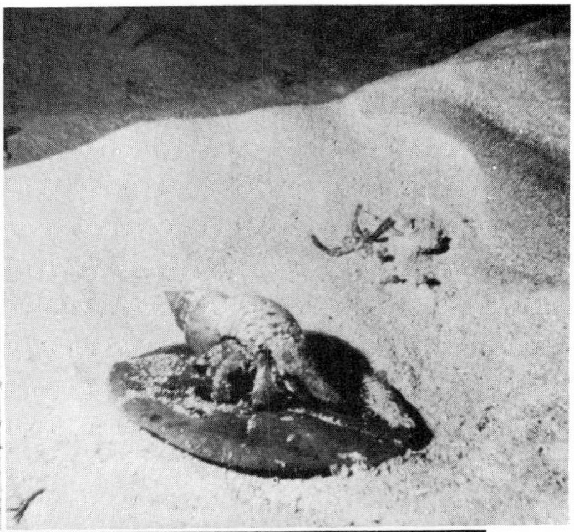

E

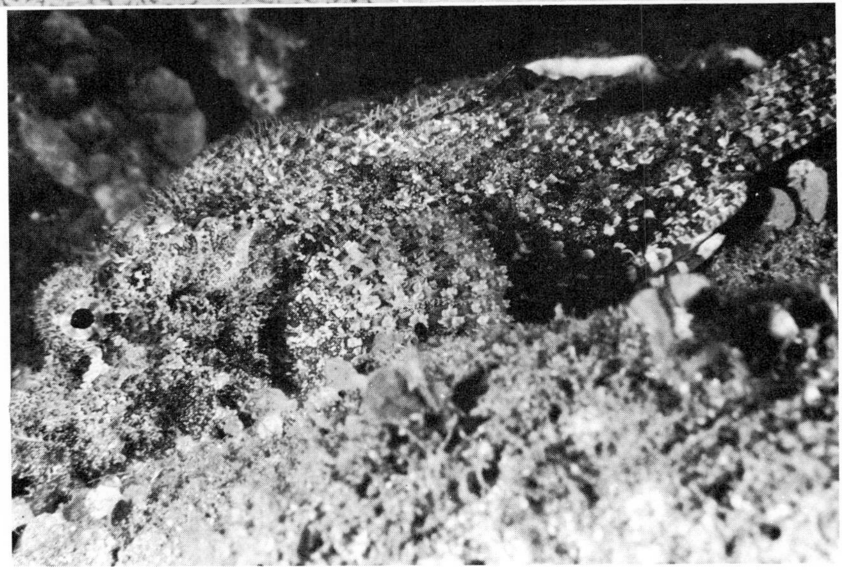

F

G

FIG. 13-23, cont'd. For legend see p. 313.

cold freshwater into the warmer, saline waters of the Gulf of Mexico. The result is an unusual degree of patchiness of the sediments and nonuniformity of salinity and of temperature of the waters. There are, then, in the area many relatively small patches of sediment resulting from localized differences in temperature, salinity, and flow of the water. They provide a large number of ecological niches in a relatively small area, almost as varied as the conditions found in rocky substrates. Furthermore, the climate varies from freezing weather in winter to subtropical temperatures during summer. Actually, then, there are several communities of benthic soft bottom animals represented in a patchy manner in a relatively small area. Areas of this sort generally show a considerably greater number of species than are found in more uniform areas of the same size, but they normally do not develop such heavy stands of any particular species.

THE DEEP-SEA BENTHOS

Deep-sea benthic communities

The formidable difficulties involved in sampling and photographing the benthos of the abyss, as well as the impossibility to date of making any but the most limited direct observations in this region, have greatly retarded our understanding of the communities of organisms found on the deep-sea bottom. Prior to 1872 only a few scattered collections had been made, notably those of Allman and Milne-Edwards, who described the profuse growth of hydroids alcyonarians, corals, worms, mussels, and bryozoans on a broken trans-Atlantic cable recovered from a depth of 2160 meters; the dredgings of O. Torell between Greenland and Spitsbergen at 2500 meters; and the work of M. Sars and G. O. Sars in Norwegian waters and that of Charles Wyville Thomson and Carpenter northwest of Scotland.

The field was really opened up by the *Challenger* Expedition under the leadership of Charles Wyville Thomson, from 1872 to 1876—probably the most important and successful expedition of deep-sea exploration ever undertaken. The fifty volumes of the '*Challenger' Reports* open up a new era in marine biology and oceanography. Subsequently there have been numerous important expeditions in various parts of the world and important technological advances, especially in the last two decades, which have enormously extended and consolidated our knowledge of the deep-sea conditions and of the life of the abyss.

Two deep-sea benthic zones are commonly recognized, the **archibenthic** and the **abyssal.** The archibenthal zone includes the continental slopes beyond the edge of the continental shelf, down to approximately 1000 meters; and the abyssal zone, the floor of the sea below this level. The boundary between the two is not always sharp. Probably the best distinction is the limit of the hemipelagic sediments, which were derived in part from terrigenous sediments swept out over the continental slope and containing more organic matter than typical eupelagic sediments of the abyssal zone proper. Temperatures in these depths are low and constant, varying from a little below 0° C. in polar regions to highs of

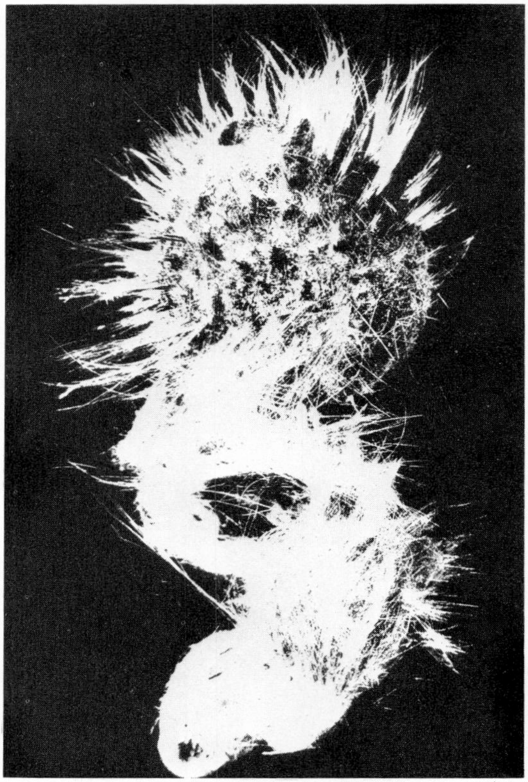

FIG. 13-24. *Pheronema giganteum* F. E. Sch. (actual size about 26 cm. long), one of the glass sponges. (From Ijima, I. 1926. *In* Reports of the Siboga Expedition, 1899-1900. Vol. 6. The Hexactinellida of the Siboga Expedition. E. J. Brill, Leiden, Netherlands.)

TABLE 13-1. Numbers of species and of individual animals taken per dredge or trawl haul in representative types of marine sediments*

	Species	Specimens
Dredging		
Hemipelagic deposits	25	57
Globigerina ooze	5	6
Red clay	1.7	4.2
Trawling		
Hemipelagic deposits	32	90
Globigerina ooze	15	39
Red clay	9	30

*From Murray, J. 1895. Summary of the scientific results obtained at the sounding, dredging, and trawling stations of H.M.S. *Challenger*. Challenger reports. Vol. 2, pp. 797-1608.

TABLE 13-2. Some eurybathic animals and the depths from which they have been taken*

Porifera	
Thenea muricata	30 to 3440 meters
Stylocordyla borealis	2 to 3000 meters
Tentorium semisuberites	26 to 2970 meters
Polychaeta	
Lumbriconereis impatiens	At least to 3000 meters
Glycera rouxi	At least to 3000 meters
Notomastus latericeus	At least to 3000 meters
Hydroides norvegica	At least to 3000 meters
Pomatoceros triqueter	5 to 3000 meters
Amphicteis gunneri	20 to 5000 meters
Cirripedia	
Verruca stroemia	Littoral to 3000 meters
Cumacea	
Diastylis laevis	9 to 2980 meters
Eudorella truncatula	9 to 2820 meters
Isopoda	
Antarcturus furcatus	10 to 3010 meters
Lamellibranchiata	
Limopsis aurita	38 to 3175 meters
Astarte sulcata	10 to 2000 meters
Scrobicularia longicallus	36 to 4400 meters
Gastropoda	
Neptunea curta	8 to 2580 meters
Neptunea islandica	30 to 3000 meters
Puncturella noachina	8 to 2000 meters
Scissurella crispata	12 to 2300 meters
Natica groenlandica	3 to 2300 meters
Nitica affinis	0 to 2600 meters
Scaphander punctostriatus	35 to 2800 meters
Asteroidea	
Pseudarchaster pareli	15 to 2500 meters
Henricia sanguinolenta	0 to 2450 meters
Ophiuroidea	
Ophiacantha bidentata	5 to 4400 meters
Ophiopholis aculeata	0 to 2040 meters
Ophiura sarsi	10 to 3000 meters
Ophiocten sericeum	5 to 4500 meters
Echinoidea	
Echinocardium australe	0 to 4900 meters
Holothuroidea	
Mesothuria intestinalis	20 to 2000 meters

*From Ekman, S. 1953. Zoogeography of the seas. Sidgwick & Jackson, Ltd., London.

about 4° C in the abyssal zone and about 9° C. in the archibenthic zone a little north of the equator. Hydrostatic pressures here reach levels at which the rates and nature of certain enzymatic and chemical reactions may be affected, and the physiology of abyssal forms may differ in certain characteristic ways from that of shallow-water species.

The nutritional value, or organic content, of the sediments decreases as one goes farther from shore. Murray (1895), reporting on the dredging and trawling stations of the *Challenger* Expedition, compared hemipelagic and eupelagic sediments with regard to the number of species and number of specimens of macrobenthos taken, omitting the South Sea stations, which were richer in both the hemipelagic and the eupelagic regions (Table 13-1).

Ekman (1953) divides the deep-sea fauna into two components: the **eurybathic** species and the **endemic deep-sea** species.

The eurybathic species are forms occupying a wide range of depths and commonly found on the continental shelf as well as in the greater depths. Table 13-2 lists a number of characteristic eurybathic species, with the range of depths from which each is taken.

The endemic deep-sea species and usually the genera to which they belong are mostly limited to the deep sea and are not found on the shelf. Some of the most characteristic follow.

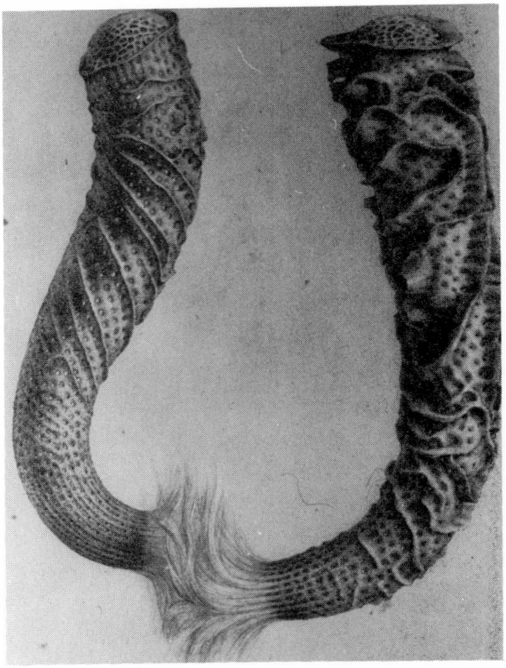

FIG. 13-25. *Euplectella aspergillum,* glass sponge characteristic of deep-water soft substrates.

Porifera

- **Glass sponges, class Hexactinellida:** Entire group, about 375 species comprising some 80 genera grouped into 15 families and 2 subclasses, is characteristically found in deep water. The few found on upper shelf are mostly antarctic, although bottom of Sagami Bay, Japan, famous for its *Euplectella,* is only 150 to 300 meters deep.

  Some representative genera are *Aphrocallistes, Caulophacus, Euplectella* (about 15 species) (Fig. 13-25), *Farrea* (11 species), *Hyalonema* (about 90 species), *Monorphaphis,* and *Rossella* (mainly antarctic). Among Demospongiae, *Thenea* and *Cladorhiza* largely deep-sea animals.

Coelenterata

- **Hydroids:** Most are shelf forms, but a few, including huge (up to 2 m. high) *Branchiocerianthus imperator,* exclusively deep-sea animals.
- **Alcyonaria:** Gorgonins and pennatularians well represented in deep water. Among gorgonians, *Caligorgia, Chrysogorgia* (about 30 species), *Narella,* and *Thouarella* almost wholly archibenthic. Twelve of 14 families of pennatularians, or sea pens, have archibenthic or abyssal species. Families Anthoptilidae, Funiculinidae, Kophobelenmonidae, Protoptilidae, and Umbellulidae comprise almost wholly archibenthic and abyssal forms. Chunellidae purely abyssal.
- **Corals:** Most deep-sea corals solitary and found in greatest numbers in Sulu Sea of Indo-Pacific. *Lophohelia* and *Amphihelia* are representative.
- **Zoantharians:** *Epizoanthus* contains both shallow-shelf and deep-sea species. *Izozoanthus* a truly deep-sea genus.
- **Antipatharians:** All but 2 genera represented in deep-sea fauna, and 5 exclusively so.

Ctenophora: Group chiefly pelagic. Peculiar platyctenids, somewhat resembling polyclad flatworms superficially, are benthic, sometimes found in fairly deep water. *Tjalfiellidea* found at depth of 475 to 575 meters off western Greenland, attached to sea pen *Umbellula lindahli.*

Annelida: Most deep-sea polychaetes belong to genera also represented in shelf faunas.

Pogonophora: This peculiar group, tubes of which for some time thought to be those of deep-sea annelids, seems to be limited to and characteristic of deep-sea benthos.

Arthropods

- **Decapod crustacea**
  - **Brachyura:** Brachyurans, or true crabs, rather poorly represented in deep-sea fauna, considering size of group and its

great prominence in shelf fauna. Two families, Homolodromiidae and Homolidae, and subfamily Tymolinae, with much-reduced eyes, are deep-sea forms. Single genera from other families, such as *Platymaja* and *Geryon,* and some species from genera represented chiefly in shelf fauna, are also deep-sea forms. Some calm-water forms with long appendages.

Anomura: In contrast with true crabs, Anomura contain many deep-sea forms. Group Galatheidea contains truly deep-sea genera *Uroptychus, Galacantha,* and *Munidiopsis,* modified for deep-sea life with reduced eyes and greatly lengthened antennae. Over 100 species of *Munidiopsis,* most of them not occurring above a depth of 400 meters, and half not above 800 meters. Large genus *Munida* preponderantly archibenthic. Among Paguridea, the family Pomatochelidae, often living in *Dentalium* shells, are mostly deep-sea forms, and, except in cold regions, Lithodidae are mostly deep-water forms. Over half of family Axiidae are deep-water forms.

Macrura: There are numerous deep-sea Reptantia—most of the family Eryonidae, with *Polycheles,* and *Willemoesia* usually abyssal, and showing reduced eyes. Among the Astacura are *Phoberus, Thaumastocheles,* and *Nephropsis.* Natantia, or shrimplike forms, have deep-sea families Holophoridae and Glyphocrangonidae, as well as bathybenthic genera belonging to other groups, such as *Benthecicymus, Benthonectes,* and *Amalopenaeus.*

Amphipoda and **Isopoda:** Rather poorly represented are Amphipoda and Isopoda, most deep-sea genera containing only a few species. Isopod *Munnopsis,* from medium depths, shows great elongation of some appendages, representing an adaptation to living in loose ooze. *Nannoniscus* and amphipod *Trischozoztoma* confined to deep waters and contain a few species each.

Cumacea: The 2 families of deep-sea Cumacea are both small families—Platysympodidae, with 2 monotypic genera, and Procampylaspidae, with 1 genus containing only a few species. Other families generally contain some deep-sea species, and *Macrocylindrus* and *Bathycuma* are purely abyssal.

Barnacles: *Megalasma, Hexalasma,* and *Verruca* are purely deep-sea forms, and *Scalpellum* includes a number of archibenthic and abyssal species.

Pycnogonids: Most genera of Pycnogonida tend to be eurybathic. *Pipetta* contains a few purely abyssal species from Indo-Malayan and Antarctic regions.

Molluscs: A number of cephalopods, both squid and octopods, are characteristically demersal deep-water forms. Among decapods, families Cirroteuthidae, Opisthoteuthidae, and Stauroteuthidae may be mentioned, comprising about 25 species. Of the octopods, *Bathypolypus* and *Benthoctopus* contain about 20 species. Deep-sea species of most other mollusc groups known, but most are members of genera or families better represented in shelf fauna.

Echinoderms: Two of the classes echinoderms, Crinoidea and Holothuroidea, are among the most characteristic and distinctively deep-sea faunal elements. Well over half the known species of crinoids are purely archibenthic or abyssal, and several other species found on shelf occur also in deeper water. Mobile unstalked comatulids mostly shelf forms, but even in this group, families Atelecrinidae and Thalassometridae, and *Trichometra* and *Thaumatometra* of Antedontidae, are deep-water forms. Majority of typical stalked crinoids, including all families Pentacrinidae, Apiocrinidae, Hyocrinidae, Phyrnocrinidae, and Bathycrinidae, are deep-water forms. Last-mentioned family contains members found more than 5000 meters deep.

Two of the 5 orders of Holothuria, the Elasipoda and the Molpadonia, are restricted to deep, mostly abyssal areas. Among the other 3 orders, most of the family **Synallactidae,** order Aspidochirota, with about 16 genera, are abyssal. The genera *Staurocucumis* (Dendrochirota) and *Acanthotrochus* (Apoda) also are deep-sea forms. The holothurians are the most important group of deep-sea substrate ingesters. In some regions they may be responsible for a considerable reworking and redistribution vertically of marine sediments. For example, in certain areas where there has been a deposition of volcanic ash at a given period, instead of forming a clean-cut layer of sediment marking the time of the deposition, the ash particles are found distributed throughout a considerable vertical accumulation of the sediments. This is thought to be the work of sea cucumbers. There is some evidence that great herds of sea cucumbers may move along the sea floor, working over an area and moving on to new areas somewhat in the manner of a herd of sheep.

The asteroids, echinoids, and ophiuroids also all have characteristically deep-sea groups, although in most cases they are not as distinctive or as well marked off from the shelf-inhabiting forms as is the case with the crinoids and sea cucumbers. Among the starfish, the long-armed **Brisingidae,** and the **Porcellanasteridae,** characterized by high marginal plates, are mostly abyssal. The species *Albatrossaster richardi* of the latter family occurs at 6000 meters or more. Some species such as *Ctenodiscus crispatus* that occur in shallow arctic waters are archibenthic or abyssal in other latitudes. Two other families containing mostly deep-sea species are the **Benthopectinidae** and **Zoroasteridae.**

The ophiuroid family **Asteronychidae** is a deep-sea family, and most other ophiuroid families contain one or more deep-sea genera.

Among the Echinoidea the suborder Meridosternata (families **Pourtalesiidae, Urechinidae,** and **Calymnidae**) are purely deep-water forms in distribution. The Pourtalesiidae are remarkable in having an elongate body, with the mouth sunk into a deep furrow in the anterior part. Other deep-sea families of echinoids include **Aspidodiadematidae, Echinothuriidae, Pedinidae,** and **Saleniidae.** Some of the echinothuriids are remarkable for their large size and their soft bodies, which collapse if lifted from the water.

Tunicates: Most tunicates are either part of shelf fauna or eurybathic. Only about one fifth of the genera—and most of these have only a few species—are typically archibenthic or abyssal. Deep-sea families are Pterygascidiidae, Hypobythiidae, and Hexacrobylidae, all rather small families.

Fish: Although most deep-sea fish are pelagic, there are a considerable number of deep-water demersal or benthic forms. Only a few groups mentioned here. Deep-water elasmobranchs include *Centrophorus* and *Centroscyllium*. Whole group of Holocephali preponderantly deep-water forms. *Chimaera* archibenthic and abyssal, and *Harriotta* is purely abyssal.

Among teleosts, whole family Macruridae (approximately 140 species) is archibenthic or abyssal. Early writers included most of these fish in *Macrurus,* which has subsequently been divided into a number of genera.

The family **Liparididae** (about 125 species) likewise is comprised mostly of archibenthic and abyssal forms. Large genera *Careproctus* and *Paraliparis* occur in shelf fauna in cold regions but submerge to deeper water in other regions. Most genera of **Zoarcidae** are archibenthic or abyssal. *Bassozetus* and *Porogadus* are purely abyssal, and *Lycodes, Lycenchelys, Neobythites,* and *Dicrolene* are archibenthic to abyssal. Also about 15 monotypic abyssal genera in this group.

Halosauridae and **Notanthidae** (approximately 35 species) almost wholly abyssal. Among Pediculati, family Ogcocephalidae contains 3 deep-sea genera, *Dibranchus, Halieutaea,* and *Malthopsis,* in addition to shelf forms. Several genera of codlike fish, Gadiformes, preponderantly archibenthic and abyssal.

Distributions of the deep-sea benthos

The extent to which the deep-sea benthos displays regional differences of zoogeographical significance is difficult to assess accurately because of the formidable problems of securing adequate samples. The fact that large areas are often of relatively uniform character somewhat offsets the difficulty. Nevertheless, the amount of data and of direct and indirect observation at the floor of the deep seas is so meager compared to that related to shelf biota that generalizations must, of necessity, be broader and less precise.

Since, as a rule, temperature decreases and salinity slightly increases with depth, the deep oceanic basins tend to be somewhat isolated from each other by temperature barriers along the rises between them and sometimes by temperature differences of their contained water. When the waters of such basins also differ in other respects such as oxygen content or biological "conditioning," the differences become reinforced, making exchanges of the more sensitive and stenothermal elements of their benthos unlikely. For benthic organisms the ridges and rises between adjacent basins may present barriers much as do mountains on land.

A surprising discovery of thriving deep-sea communities of benthic invertebrates and fishes near areas heated by upwelling of volcanic lavas at about 3000 meters' depth or more near the Galápagos Islands was recently made by divers from Oregon State University and the Scripps Institution of Oceanography using the famous research submarine *Alvin*. Since there is no photosynthesis at these depths, the nutritional basis for such communities was unclear. It has been hypothesized that perhaps nutrition is based on energy obtained by chemoautotrophic bacteria in the neighborhood of these vents.

In the deep sea, **eurybathic** species ranging over wide limits of depth, like cosmopolitan species in the superficial layers, are common but contribute less to zoogeographical characterization of a given region than do **stenobathic** forms. Archibenthic species such as most Pogonofera, found chiefly on the continental slopes, have wider geographical distributions as a rule than do strictly shelf-dwelling species but not as wide as do the abyssal forms. The East Pacific Barrier, for example, is effective for archibenthic as well as shelf-dwelling species, but it is of little significance for truly abyssal ones.

Early impressions that the deep-sea fauna of the world is relatively uniform were based on insufficient material and on failure to distinguish between bathypelagic species, which show a much greater tendency to be cosmopolitan, and bathybenthic species. It is now clear that the Atlantic abyssal fauna is distinct from that of the Pacific and Indian oceans, especially at the species level. The polar seas also have somewhat isolated and distinctive abyssobenthic faunas. The upper temperature limit for the strictly polar abyssal faunas seems to lie at about 0° C. The deep polar basins containing water at or below this temperature are relatively isolated from other abyssal areas, especially in the north, although Baffin Bay, which is separated by shallow areas from the major arctic basins, does have an arctic abyssal fauna.

The finding of some archaic types of animals known previously only as fossils or most closely related to animals known as fossils, during early deep-sea explorations, led for a time to the concept of the abyss as a sort of refuge for primitive, outmoded types of animals that had long since disappeared from other areas because of competition with more recent species or because of the greater environmental changes to which the shallow waters are sometimes subjected. It was hoped that the study of abyssal faunas would go far to fill in gaps of our understanding of the evolution and relationships of various groups. To some extent it has done so. Many stalked crinoids have fossil records going back as far as the Eocine, and some genera even to the Jurassic. The discovery of *Neopilina,* a representative of the primitive molluscan group Monoplacophora, thought to have been long extinct, is another case in point.

However, by no means are all the "living fossils" of the sea abyssal, as shown by the spectacular discovery of the coelacanth fish *Latimeria* in the Indian Ocean off eastern Africa and the Comoro Islands in water not more than about 600 meters deep. Furthermore, archaic types do not dominate the abyssal fauna to a greater extent than they do the shelf fauna. Such familiar archaic types as *Lingula, Limulus, Heliopora, Nautilus, Branchiostroma,* and *Cestracion* are all shelf-dwelling forms, or they live in the upper strata of water. The abyss also contains many types that are more specialized and have more recently evolved than their nearest relatives, now living on the shelf.

Summary

With respect to the benthic fauna of the deep sea as a whole, it can be said that it is characterized by a large number of forms that are almost completely lacking in the shelf fauna. Among these forms are several fairly high taxa, such as whole families, suborders, or orders of organisms found only in the deep sea. The faunas are poor as compared with the shelf faunas, but they contain a relatively high proportion of endemic forms, restricted to the deep sea. This is particularly true of the abyssal fauna on eupelagic bottoms. The deeper and farther from land one goes, the fewer are the eurybathic species found, and the fauna become more abyssal in character.

This finding is in line with Thienemann's rule that the more specialized and rigorous the environment, the poorer in species it becomes. The second part of this generalization, that although the number of species decreases there is usually a great increase in the number of individuals of those species that do inhabit such an environment, does not apply to the deep-sea fauna, since the relatively small amount of food in the abyssal regions imposes a low absolute limit on the development of the benthos.

Recent surveys using more adequate deep-water benthos-sampling equipment, such as the anchor dredge and epibenthic sled of Sanders and Hessler (1969), have demonstrated that the benthic fauna of deep waters display a high degree of diversity, comparable to that of the fauna found in tropical shallow waters—a striking confirmation of the general rule that long-continued stability of the environment is a major factor leading to the diversity of its biota.

The shelf faunas of the world seem to have originated in the tropics, from which they spread into northern and southern cooler regions, becoming progressively adapted to living in colder waters. Most of the archibenthic and abyssal faunas, even of warm regions, appear to have been derived from the northern or southern shelf faunas after their adaptation to cool or cold temperatures. Elements of these cold-adapted shelf faunas subsequently descended into deeper water, both directly in the regions in which they found themselves and by submergence along cool isotherms beneath the warmer subtropical and tropical waters. This general pattern of dispersal is especially true for the benthos and to a lesser extent for bathypelagic and abyssal nekton.

FIG. 14-1. Comparison in wing shapes of the frigate bird (right) and red-tailed hawk (left), showing adaptations for oceanic flight in high-speed winds (frigate bird) and slow, soaring flight over land (hawk). Slow flight requires extra spaces in the wing to prevent stalling. As the air is forced through these spaces, it increases in velocity over the upper surface of the wing, preventing stall-inducing eddies. This type of wing shape will not work at high speeds. The bird must alter its shape by partially closing the wing for faster flight. A seabird, on the other hand, must have a drag-resistant, streamlined wing shape for sustained flight in high-speed winds.

CHAPTER

14 FROM LAND TO SEA

adaptations of terrestrial vertebrates to the marine environment

ANATOMICAL ADAPTATIONS FOR FEEDING AND PROPULSION IN WATER

Most of the reptiles and mammals that became aquatic fish-eaters show elongation of the face, with the development of long, slender jaws armed with sharp teeth. The nostrils are commonly placed well back, sometimes dorsal, rather than at the tip of the snout, enabling them to breathe more easily while floating near the surface. Those which subsist on larger aquatic vegetation, such as the Sirenia or the Galápagos sea lizards, tend to have blunt muzzles and teeth modified for chewing and grinding. A third category, those which feed on the benthic, often hard-shelled animals such as molluscs, crustaceans, and echinoderms, are likewise often rather short-muzzled forms, with the teeth, or the hard horny surfaces substituting for teeth (as in the turtles), fitted for crushing their prey. The walrus has special tusks used as an aid in hauling out on ice, and by males for fighting and sexual display. Apparently these tusks are not used as clam-digging tools, as formerly thought. The cylindrical mouth and large tongue are used as a cylinder and piston, pressed against the substrate to suck the clams from their burrows.

The plankton feeders, best exemplified by the whalebone whales, exhibit specialized adaptations for straining the seawater and retaining the macroplankton. To a much lesser extent, the crab-eating seal is modified in a parallel manner, straining the water between its lateral teeth.

Propulsion in water is a different problem from propulsion on land. Land animals returning to the sea have solved the problem in two ways, each of which entails a host of secondary modifications. Some have developed the limbs as the primary organs of propulsion, whereas others depend on sinuous movements of the body or tail to move through the water in a more efficient, fishlike manner.

The animals that depend on the limbs for their movement in the water are usually relatively short-bodied, short-tailed animals with well-developed, and in some cases much elongated, necks and relatively small heads. The hind limbs are retained, and one or both pairs become either webbed or paddlelike in form. The upper arm and thigh bones are elongated but rather massive. In all forms in which the limbs develop as paddles, lower bones—the radius and ulna of the fore limbs and the tibia and fibula of hind limbs—become greatly shortened, whereas the digits are much elongated and often contain a larger number of skeletal elements than those in other animals. The ends of the long bones, the bones of the wrist and hand, and the bones of the ankle and foot tend to be cartilaginous rather than well ossified, making the appendage more flexible. The dolichosaurs, nothosaurs, and plesiosaurs all possessed relatively short, nonpropelling tails and long, mobile necks. The turtles all have well-developed necks.

Those animals that propel themselves with their tails generally have more torpedo-like bodies, with short necks or none at all, and often show shortening and coalescence of cervical skeletal elements. If the sinuous movements are made horizontally and the swimming is largely level, the tail tends to be laterally flattened and, in the best adapted forms, to develop as a fishlike vertical fin. Those which have adopted vertical undulatory movements, used largely for ascending or descending rapidly, develop horizontal flukes, as in the Cetacea and Sirenia. The limbs, no longer used for propulsion, are primarily used as balancing organs. The fore limbs become paddlelike, but in most instances the hind limbs are reduced or wholly lost. The body is often elongate, and in fish-eating forms the jaws are usually long and narrow. The bones of the upper arm and thigh are greatly shortened. The spinal column becomes more flexible.

Many groups tend to lose external structures such as hair or scales and to develop a smooth skin. The bones tend to become lighter and spongier, except in some of the bottom-feeding types such as Sirenia, in which the massive bones apparently help them stay submerged in shallow waters while feeding.

In both the reptiles and mammals, some of the most completely adapted tail-propelled groups have developed dorsal fins that serve as a type of stabilizer in rapid swimming. The ichthyosaurs among the reptiles, and the dolphins among the mammals, probably represent culminations of this line of evolution and show remarkable convergence in general form and presumably in their habits. Truly marine reptiles all show a ring of sclerotic bones in the eye, enabling the eye to compensate for pressure in deep water.

Although some successful groups in both the reptilian and mammalian lines, such as the sea turtles and the pinnipeds, have never cut their reproductive tie to the land, the most completely adapted marine forms in both groups, the ichthyosaurs, Hydrophinae, and cetaceans, all bear their young alive in the water and never return to the land for any purpose.

The inability to breathe under water remains as the last great barrier preventing these animals from adopting a way of life wholly similar to that of the fish.

Osmotic relations. Marine vertebrates, of course, all must solve the problem of osmotic relationships. The reptiles and mammals lack special salt-excreting glands such as are found in teleost fish and some birds. They apparently meet the problem through restricting the intake of seawater as much as possible and obtaining much of their water requirements from the body fluids of their prey. Since the kidneys do a much more thorough job of resorption of water in the kidney tubules, the urine tends to be minimal in volume and much more concentrated than in terrestrial animals. It is possible that the fat, which serves as a thick insulating layer in homoiothermic forms, also plays a role in the storage of water, as in some desert animals.

Heat loss. For homoiothermic animals living in cool to cold waters, minimization of heat loss is of great importance. Adaptations involve the reduction of the surface-to-volume ratio through large size and compact shape, insulation through thick, fatty layers beneath the skin, and, in some, water-repellent underfur or feathers that prevent direct contact of water with the skin over most of the body. Other adaptations include reduced peripheral circulation, especially in the extremities, and high metabolic rate coupled with the ability to eat and rapidly digest large qualities of food.

Diving. Special adaptations of mammals such as cetaceans and pinnipeds, which perform prolonged deep dives, will be mentioned in the discussion of marine mammals (p. 339). Presumably the ichthyosaurs, mosasaurs, and perhaps some other marine reptiles had somewhat similar or parallel adaptations.

Diving birds are restricted to water relatively near the surface. There are basically four different methods used by diving fishing birds, each entailing different structural and behavioral adaptations.

Birds that fly above the water, watching for fish at or near the surface, in some cases swoop down and grasp the fish in their talons in the manner of ospreys. Others, such as some pelicans and terns, plunge headlong into the water in a steep dive, taking the fish in their beak. In both of these groups there is a premium on keen vision, speed and accuracy in diving, and modification of either the feet or the beak for seizing and holding fish.

Another category of diving birds swim beneath the water surface at times in pursuit of fish or invertebrate prey. The majority of swimming birds of this type, such as cormorants, grebes, and loons, have broad webbed feet set well back on the body, muscular legs, and sharp daggerlike beaks. They propel themselves beneath the water with their feet, holding the wings against the body for streamlining. Penguins, on the other hand, use the wings as paddles, and are sometimes described as ''flying'' through the water.

CONTEMPORARY MARINE REPTILES

In contrast to the marine mammals and birds, which appeared in the Tertiary and which, at least until the advent of man, were still undergoing evolutionary diversification and development, the marine reptiles as well as reptiles on land experienced their great flowering during the long Mesozoic era. Those found today are mere remnants of a once much greater and more diverse assemblage.

Sea turtles. The best known and most widespread contemporary marine reptiles are the sea turtles, family **Cheloniidae,** comprising 5 widely distributed tropical or subtropical species. Four species—the green turtle, or edible turtle, *Chelonia mydas;* the hawksbill turtle, *Eretmochelys;* the loggerhead turtle, *Caretta caretta;* and the Ridley or Golfina *Lepidochelys kempi*—are familiar along the coasts of the southeastern United States and the Gulf states. All the sea turtles are large species with nonretractile heads. The limbs are modified into flippers for swimming. The forelimbs are especially well developed and are the chief organs of propulsion. The carapace and plastron are completely covered with horny shields, the tortoiseshell of commerce. Sea turtles seem to have a strong homing instinct, sometimes performing extended migrations between their feeding grounds and the beaches on which they spawn. The females come out of the water, dig pits in the sand of the upper beach, and deposit a clutch of eggs (sometimes up to 200) about 5 cm. in diameter, with a leathery shell.

The largest of the marine turtles—in fact, the largest of all living turtles—is the Atlantic leatherback, *Dermochelys coriacea,*

FIG. 14-2. Green sea turtle. (Photograph by M. W. Williams, taken at Steinhart Aquarium in San Francisco.)

the only member of the family **Dermochelidae.** The leatherback sometimes attains a weight of three fourths of a ton and measures up to 3.5 meters. They were exceeded in size only by the extinct *Archelon ischyros* from the Upper Cretaceous of North America, thought to have attained a weight of over 3 tons.

Instead of the usual carapace and plastron, the leatherback has a mosaic of small dermal bones buried in its leathery skin, completely free from the internal skeleton. The larger of these skin bones form seven rows dorsally, which appear externally as longitudinal keels running the full length of the trunk. Ventrally there are five rows of smaller bones, beneath which are vestiges of bones representing the plastron of other turtles.

In other anatomical characters *Demoschelys* resembles such Cretaceous forms as *Protostega* or *Archelon,* some of which had lost nearly all the costal plates and had only reduced neurals and marginals. It is argued that perhaps these large, soft-bodied turtles, because of changing conditions or habits over a long period of time and possibly exposure to new, more dangerous enemies such as the zueglodonts, again found a survival premium in greater protection. Having already lost their ancestral shells, they were unable to retrace the evolutionary steps leading to the shells of other Chelonia but met the need through a secondary proliferation of the marginals and other epithecal elements. If this view is correct, *Dermochelys* represents a highly specialized, rather than a primitive, line of marine turtles.

The turtles of the Caribbean and tropical eastern Pacific regions have been studied extensively by Carr (1965) and others.

The green sea turtle is herbivorous, feeding on the extensive meadows of turtle grass *(Thalassia)* of shallow tropical regions. The other three species are all carnivorous, preying on conchs, crabs, echinoderms, and other large invertebrates. Because of this turtle's low position in the trophic hierarchy, the fact that it is practically without competition as a grazer of turtle grass, and the fact that prior to the coming of Europeans they were not subject to excessive predation, the green turtle built up dense populations vastly exceeding in numbers the populations of the other three species combined. The green turtle feeds in the *Thalassia* beds all day, then clambers up onto exposed rocks to sleep.

The green turtle formed the principal food for explorers and settlers as the West Indies–Caribbean and Gulf of Mexico regions were opened up for development and was the principal export from many of these island areas. As this exploitation and settlement proceeded, the famous rookeries were destroyed, beginning with those on Bermuda and progressing to those of the Greater Antilles, the Bahamas, and the coasts of Florida.

The major turtle fishery still in operation is at the feeding grounds around the Cayman Islands. However, the greatest threat to the turtles is not posed by the Cayman Island hunters, since they take the turtles at their feeding grounds, where they are most resilient. The disappearance of wild, inaccessible beaches is far more dangerous. Modern motor schooners, helicopters, jeeps, etc. make even the remotest areas, such as the breeding grounds at Mosquito Bank off Nicaragua, readily accessible. Female turtles are extremely vulnerable as they come ashore, concentrated together in limited areas. They are turned over on their backs as they come in and later collected for shipment.

Turtles present an unusual opportunity for conservation if it can be effected in time. There is still a skeleton breeding stock left, and some of the best breeding beaches are among those least altered by man as yet. The extensive meadows of *Thalassia* have not been much damaged. If the remaining rookeries can be adequately protected from development and from turtle hunters, this valuable renewable resource need not be destroyed as so many have been. At present the protective laws and their enforcement are still inadequate, and the outlook for the future of the green turtle is not good.

Marine turtles commonly perform extended migrations between their feeding grounds and rookeries (Chapter 8).

The other regions of the world where turtle densities are still high in places are in the southwest Pacific. Bustard and Tognetti (1969), in a four-year study of one of the nesting beaches on the Great Barrier Reef, Australia, showed that where population density of turtles is high and feeding areas are adequate, the limiting factor determining the size of the population attained becomes the area of suitable rookeries. Turtles concentrate in great numbers on these beaches during the breeding season. When the beaches are too crowded the late-comers, as they excavate nesting sites, destroy many of the egg clutches left by earlier comers. This strictly density-dependent phenomenon, when it takes places on crowded rookeries, reduces the next generation's breeding population each year until equilibrium is reached. The effect is negligible at low population densities. The rookeries in the Great Barrier Reef may be the last where populations are still sufficiently high that space in the rookeries becomes the limiting factor.

Bustard and Tognetti also showed that the same individual turtles did not return to the rookery annually but at four-year intervals; the turtles crowding the rookery on successive years are not the same individuals.

Sea snakes. Snakes are probably the most recently evolved group of reptiles, originating from lizards sometime during the Cretaceous period. There is no known geological history of distinctively aquatic groups of snakes. Most snakes swim well, with rapid, undulatory movements of the body. Poisonous snakes, with special development of the poison glands and fangs for the injection of the poison into their victims, are known only from relatively recent geological times. It is probable that venomousness is the last important new specialization to have occurred among reptiles.

One family of contemporary snakes, the **Hydrophidae,** has become adapted to wholly aquatic marine life. They comprise a group of about 15 genera and some 50 species of highly venomous snakes related to cobras, all of them living in warm seas of the Indian Ocean and the western Pacific. The adaptations fitting sea snakes for marine life are not as striking as those found in other groups of terrestrial animals that have returned to the sea. The body and tail have become flattened from side to side, facilitating swimming. The broad transverse scales characteristic of the underside of most snakes are reduced to a few vestiges or are lacking. The snakes are viviparous, bearing their young at sea and never coming on land for any purpose whatever, except for the Laticaudinae, which lay their eggs on land above the high-water mark. The Hydrophinae bear living young at sea and are helpless on land. Fish comprise the major part of their diet.

Marine lizards. The only contemporary marine lizard is the Galápagos sea lizard, or marine iguana, *Amblyrhynchus cristatus,* which lives along the rocky sea beaches of the Galápagos Islands and never ventures far inland from the shore. They are large, heavy lizards a meter or more in length and attain a weight of about 9 kg. They are dark colored and rather sluggish in movement on land. The tail is laterally flattened, and the feet are slightly webbed. In the water they are rapid and graceful swimmers, propelling themselves with serpentine movements of the body and tail, the legs being held motionless close to the sides. Their food consists of seaweeds.

These lizards are of interest in indicating one of the ways in which terrestrial reptiles and other animals have become aquatic. Tempted by an abundance of food in shallow waters, they ventured farther and farther to obtain it. A selective premium on adaptation to the aquatic environment has led to flattening and muscularization of the tail as a swiming organ and to the ability to remain submerged for longer periods than ordinary lizards. Darwin reported that one of the sailors of the *Beagle* attempted to drown one by submerging it attached to a weighted line but that it was still active when pulled up an hour later.

Although it has sometimes been stated that these lizards also seek protection from terrestrial enemies by swimming, the opposite situation seems to have prevailed, at least until recently. In fact, it may be the prevalence of sharks and the total lack of terrestrial enemies on the Galápagos Islands that have been responsible for the lizards' retaining as much of a terrestrial habit as they have. Darwin reported that he was unable to scare them into the sea. It was easy, he said, to drive them down to any little rocky point overhanging the sea, but he was unable to frighten them into the water. On being picked up and thrown in several times, the lizard invariably returned directly to the same place. They seem to prefer to spend most of their time on the rocks along the shore, venturing into the water only to feed. Reproductively, they are also still tied to the shore, depositing their eggs in tunnels dug 30 to 45 cm. deep into the sandy slopes above the high-tide line.

The mode of swimming seems to indicate that long-tailed aquatic reptiles did not use their limbs as primary organs of propulsion in the water.

MARINE BIRDS

Marine birds are reliable indicators of the productivity of the surface waters of the oceans. Wherever they are found in great concentrations, there the waters will be teeming with life. Several different groups of birds have independently become adapted, each in its own distinctive and peculiar ways, to obtaining their sustenance from the seas. Some of them have become so completely adapted to oceanic life that they practically never come to the land except to breed. Most seabirds breed either on offshore islands or on steep cliffs facing the sea, where they are protected from most predators. These breeding grounds are often overwhelmingly crowded by the nesting populations, with thousands, hundreds of thousands, or even millions of birds crowded together in relatively small areas, using every available site. In particularly rich areas, such as around the offshore islands of Peru, the breeding islands

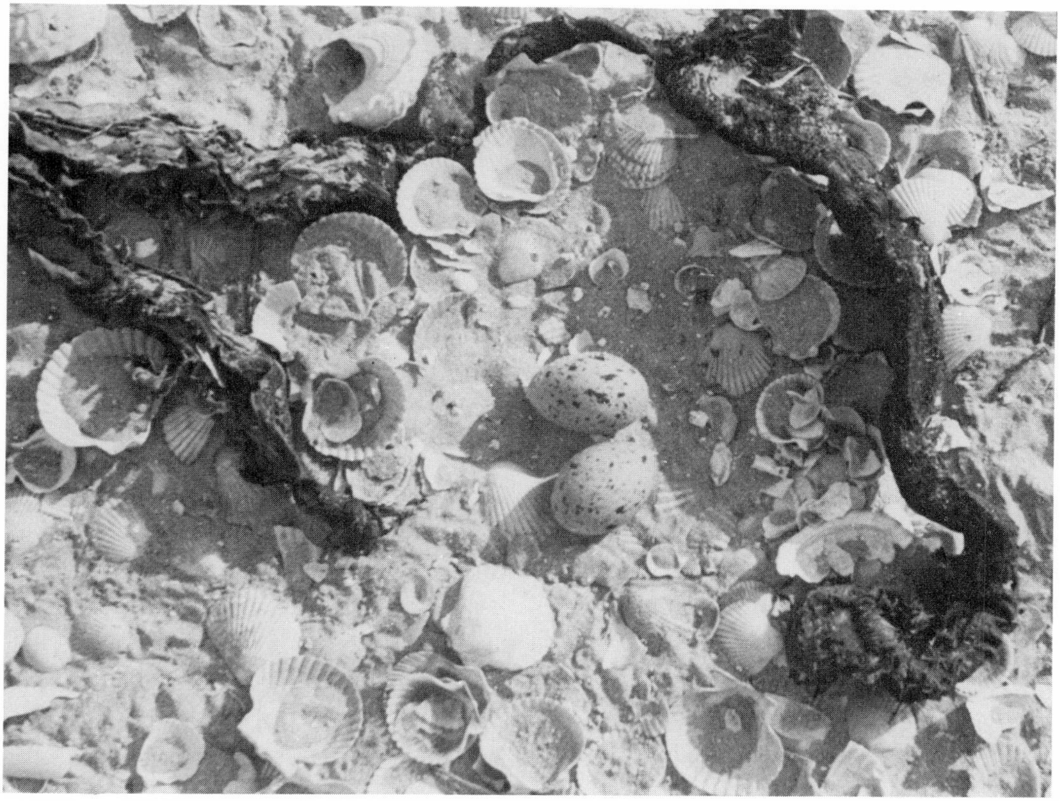

FIG. 14-3. Nest of the oyster catcher, Scammon's Lagoon, Lower California. Shells of pectins, cockles, wavy-top shells, and *Polinices* are seen in the sand, and the nest is framed with turtle gut from a nearby Mexican fishing camp. (M. W. Williams photograph.)

become so heavily covered with the bird droppings, or guano, that they form the basis for a profitable fertilizer industry.

Because of their great activity and high metabolic rate, birds are voracious feeders. A few examples give an idea of the scale of their feeding activity and the richness of marine life required to support great numbers of seabirds.

The colony of gulls, *Rissa tridactyla*, nesting at Spitsbergen during the early summer—something over 2000 birds—have been estimated to eat approximately 6 million euphausiids daily.

The 30 to 40 million guano birds off Peru are known to consume some 4 million tons of anchovies annually.

On the mud flats of southern England, the 30,000 or so overwintering oyster catchers account for some 640 million cockles through the winter.

The sandpiper, *Calidris alpina*, may eat approximately 450 nereid worms per bird daily.

During the nonbreeding portions of the year some oceanic birds fly many hundreds of miles, or even several thousand miles, over the open ocean, far from land. That they manage to find their way unerringly back to their nesting site, often only a small island, across hundreds or thousands of miles of trackless ocean is one of the great marvels of animal navigation. Certain species have been transported experimentally all the way across the North Atlantic from the British Isles to North America in covered cages so that there was no chance of their seeing anything en route.

After release they were in some cases back in the same nest within a two-week period. Others have proved their capacity to fly back to Midway, two small islands in the Central Pacific, after having been similarly transported to such distant places and different directions as Washington, Alaska, and Japan.

The means by which this navigation is effected is still not clear. A promising suggestion is that possibly the birds are sensitive to polarized light and that the set or angles of polarization by reflection from sky and sea, which differ in each part of the world, depending on one's relation to the sun, may in some way guide them back. A fine sense of time would be required as well, since the angles of light change throughout the day and are largely reversed between morning and evening, as well as undergoing some seasonal changes.

Birds that spend long periods at sea also must be adapted to meeting their water requirements from the physiologically dry, salty seawater, something that their terrestrial ancestors and relatives are unable to do. Like the teleost fish, they meet this requirement in two ways—through the evolution of special salt excreting mechanisms and through reduction and concentration of the urine so that only the minimum required for elimination of their nitrogenous wastes is formed, most of the water being absorbed in the kidney tubules.

Major groups of marine birds

Order Procellariiformes—albatross, petrels, fulmars, shearwaters

Order Procellariiformes, sometimes called the tube-nosed swimmers because of the elongation of the nasal bones covering the nasal aperture for a distance (Fig. 14-4), comprises the most truly oceanic group of birds. They are long-winged, webfooted birds ranging in size from the huge

FIG. 14-4. This fulmar illustrates the tubular nostrils and horny plates covering the bill, characteristic of the order Procellariiformes.

FIG. 14-5. Two sooty shearwaters, a Leach's petrel, and a Laysan albatross. Order Procellariiformes, families Diomedeidae and Procellariidae.

wandering albatross with a wingspread of up to 3.5 meters, the greatest of any living bird, to the smallest seabirds, certain petrels no larger than most of our familiar songbirds (Fig. 14-5). Most of them nest on remote oceanic islands but spend months or even years at sea, sometimes several thousand miles from their nesting site, mostly out of sight of land. They can drink seawater with impunity, excreting the excess salt from enlarged nasal glands near the base of the beak. When captured or seriously disturbed, they emit a highly unpleasant smelly, oily musky fluid from the nose and mouth. It has been suggested that this fluid, in addition to its defensive use, serves as a means of eliminating excess vitamin A, accumulated because the diet is rich in vitamin A.

Some of these birds perform elaborately stylized courtship and recognition rituals. Incubation is shared by the parents and in the larger species may last up to seventy or eighty days. The young are fed regurgitated food by both parents, and in some of the largest species remain in the nest as long as eight months or more, leaving only when finally deserted by their parents.

FAMILY DIOMEDEIDAE—ALBATROSS. There are 13 species of albatross. Nine of them are confined to the cold waters of the southern oceans and three to the North Pacific. One species is equatorial but feeds in the cool waters of the Humboldt Current in the eastern Pacific. Most species breed only every other year. The nest is either a hollow in the ground or a low mound of moss and twigs. Albatross are masters of gliding flight and can glide for hours on seemingly motionless wings. They alight on the water to feed, eating mostly fish and squid. Some of them follow ships to feed on the refuse. When taking off from the water, they must run over the surface with outstretched, flapping wings for a considerable distance to attain enough speed to become airborne. If approached by a boat while on the water, they often try to escape by swimming rather than flying because of the difficulty of takeoff.

Perhaps the most famous species is the great wandering albatross, *Diomedea exulans,* of the southern oceans. When mature, the male is a huge white bird with black wingtips, a black trailing edge of the wing, and mottled tail and wing coverts. The young are brown, becoming whiter at each molt. They spend as much as four or five years at sea before returning to their nesting site and are thought to sometimes circumnavigate the globe in the Southern Hemisphere during their wanderings.

The Laysan albatross, or gooney (Fig. 14-5), became well known during World War II in Midway and on other Pacific islands for its persistence in occupying its nesting sites in spite of the conversion of these sites into airstrips by the military.

The only albatross of equatorial waters is the Galápagos, or waved albatross, which has its rookery on Hood Island of the Galápagos group and fishes in the cool waters of the Humboldt Current.

The short-tailed albatross, *D. albatrus,* formerly the most abundant species to reach the offshore waters of western North America, was virtually exterminated by professional plume hunters, who raided their nesting sites, and by a volcanic eruption in its home island. By 1933 it was thought to be extinct, but by 1957 some twenty had returned to breed, and occasional sightings have been made of this species off the west coast of North America from British Columbia to Oregon.

FAMILY PROCELLARIIDAE—PETRELS, FULMARS, SHEARWATERS. Members of the family Procellariidae are also long-winged pelagic birds that spend most of their time far out at sea. They range in size from small petrels no larger than our ordinary song birds to large fulmars and shearwaters from about the size of a pigeon to that of a goose. They feed at the surface of the water on small surface animals or floating flesh.

Fulmarus glacialis, the fulmar of the North Atlantic, North Pacific, and arctic waters, is about the size and color of a herring gull. It breeds on cliffs, eats fish and offal, and often follows whaling boats, feeding on the blubber. The eggs, oil, and feathers are valued. Related species occur in the North Pacific and in the Antarctic.

Ossifraga gigantea, the giant fulmar, sometimes called Mother Carey's goose, is almost as large as an albatross. It occurs in the southern seas and ranges northward along the western coast of North America to the United States.

Priocella antarctica, the silver-gray fulmar, is one of the most abundant birds on the pack ice throughout the circumpolar antarctic waters.

Several species of shearwaters, genus *Puffinus,* ranging in size from about that of a pigeon to that of a large gull, are widely distributed in both the Atlantic and the Pacific. They usually skim close to the waves in flight. The Manx shearwater, *P. puffinus,* is common along the eastern North Atlantic, and the muttonbird, *P. tenuirostris,* is a common Pacific species. The latter was destroyed in great numbers for oil and feathers.

Several genera of usually smaller seabirds are known as stormy petrels, or Mother Carey's chickens. Most of them are dark colored with a patch of white at the rump. The largest is probably the antarctic giant petrel, *Macronectes giganteus,* about the size of a large gull.

Wilson's petrel, *Oceanites oceanicus,* to which the names stormy petrel and Mother Carey's chickens are often applied, is a small species somewhat resembling swallows in flight. It flies with rapid wingbeat and shows the distinctive white patch at the base of the tail. In rough weather the birds skim along in the troughs of the waves. They breed on barren islands in the Antarctic ocean but fly as much as 16,000 km. to spend the summer along the Gulf Stream, going as far north as Labrador, and then return to the Southern Hemisphere in winter for the southern summer. This remarkable migration of up to 32,000 km. each year, mostly out of sight of land, enables them to utilize the warm months of the ocean in both hemispheres. They eat small squid and crustaceans found in the surface waters. Feeding flocks look as though they are dancing on the water as they face into the wind just above the water, holding their wings out stiffly and paddling with their webbed feet. This petrel is the smallest of the oceanic birds and the most abundant species of bird on earth.

The genus *Prion* includes the whalebirds of the southern oceans. These birds are characterized by their peculiar broad lamellate bill. They are gregarious and often follow whaling vessels to feed on the blubber, oil, and offal.

Another rather atypical group of petrels are the diving petrels of the Southern Hemisphere, genus *Pelecanoides.* They are rather short-winged, short-tailed forms somewhat resembling auks in appearance and habits and are expert divers.

Order Sphenisciformes—penguins

Penguins are stout-bodied, short-legged, flightless seabirds of the Southern Hemisphere, ranging from about the size of a duck to a length of over a meter and a weight of 36 kg. On land they walk clumsily with erect posture and are well known for their tuxedo-clad appearance and clownish behavior. All but one of the 16 living species are confined to the Southern Hemisphere, most of them with rookeries in New Zealand, Australia, southern South America, and South Africa and on numerous islands in the Southern Ocean. Two species occupy the fringes of Antarctica, and one is found on the Galápagos Islands.

Penguins are gregarious, nesting in crowded rookeries and commonly swimming in flocks. They swim low in the water with little more than the head and neck exposed, diving and surfacing like porpoises. Unlike most swimming birds, they use the wings for propulsion in the water. The wings are flipperlike with only rudimentary scalelike quills, incapable of flexure but moved with a rotary motion by specially developed muscles. In swimming, the feet are used only for steering. Penguins are expert divers and swimmers and subsist on shellfish, fish, krill, small squid, and other animals they can take in the surface waters or on the shallow bottom.

Most species make nests of stones or bits of vegetation, in the open, and incubate their one to three eggs in the usual fashion, both parents participating in the chore. The emperor penguin, *Aptenodytes fosteri,* of Antarctica and the king penguin, *A. patagonica,* of the Falkland Islands, Kerguelen, etc. are the largest living species. They lay a single egg, incubate it by carrying it on their feet and covering it with a loose fold of skin. The emperor penguin's rookeries are usually on sea ice adjacent to the shoreline of Antarctica. The male stands on the ice through much of the antarctic winter, incubating the egg on his foot, while the female wanders off to the edge of the pack ice and fishes. She returns in the spring when the egg is ready to hatch, and both share in the care and feeding of the young. During the summer all of them go out to the edge of the pack ice to spend the summer fishing in the rich waters and fatten up for the rigors of the coming winter.

The other antarctic species is the well-known Adélie penguin (Fig. 14-6), a smaller species abundant around Antarctica. It nests on offshore islands and the edge of the mainland during the summer and winters out on the pack ice. Each pair has a fixed nesting site to which it returns each year.

The smallest penguin is the blue penguin, *Eudyptula minor,* abundant around Wellington Harbor and other shoreline

FIG. 14-6. Adélie penguin, order Sphenisciformes.

areas in New Zealand and said to extend along the coast of eastern Australia. It is only about 30 cm.

A number of fossil species of penguins are known. One of them, *Palaeeudyptes antarcticus* from the Eocene of New Zealand, reached a length of more than 2 meters.

Order Steganopodes—pelicans, cormorants, boobies, frigate birds, tropic birds

The members of order Steganopodes are large fishing birds with big beaks, usually hooked at the tip, and totipalmate feet—all four toes united by a broad web. Most of them are strong fliers and constitute a characteristic element of the coastal fauna in most parts of the world. Some are often seen far out at sea as well.

Pelicans, family **Pelicanidae,** are noted for their particularly large beaks, from the lower half of which is suspended a large elastic gular pouch used in catching fish. Some pelicans fish in groups, herding small schools of fish into shallow water, where they can be scooped up in their large pouches. The brown pelican fishes more like a giant kingfisher, soaring about rather high above the surface and plunging into the water in a spectacular dive. Brown pelicans breed in great numbers off the west coast of South America, contributing, along with other birds, to the rich guano deposits there. The white pelican nests inland along lakes of the Great Basin from Great Slave Lake in Canada south to Texas and winters along the seacoast. Flocks of white pelicans are often seen soaring at great heights. Pelicans are awkward on land but strong, graceful flyers. They are commonly seen flying in groups, in a line, alternating a few slow strokes of their large wings with an approximately equal distance of gliding. They are voiceless or nearly so.

Cormorants, family **Phalacrocoracidae,** are large black long-bodied birds with slender snakelike necks and moderately long bills hooked at the tip (Fig. 14-7). They eat fish, which they chase and catch under the water, both the feet and the wings being

FIG. 14-7. Pelagic cormorant: range, Pacific; order Steganopodes, family Phalacrocoracidae.

used in swimming. They occur in the Pacific from Alaska to New Zealand, being most abundant in the tropics. In the Atlantic they occur along both the eastern and western coasts. Cormorants usually fly low over the water, with neck and bill extended forward and the broad wings beating faster than the wingbeat of any gull.

In Japan cormorants have been used to catch fish, a ring being placed around their throat to prevent them from swallowing the fish.

Lack (1945) found that where two species, the shag *Phalacrocorax aristotelis* and the cormorant *P. carbo,* were living in the same area, they were not really in direct competition with each other. The shag ate mostly cluepeoid fish and sand eels, whereas the cormorant subsisted chiefly on shrimps and prawns, flatfish, gobies, and other fish.

On Albermarle Island of the Galápagos, the Hancock Expedition of 1933 reported that there was still a colony of several hundred flightless cormorants but that their continued existence was in danger because of man, of whom they showed no fear.

Boobies and gannets, family **Sulaidae,** are large seabirds, somewhat smaller than pelicans but more powerful in flight. They occur in temperate and tropical seas, sometimes great distances from land. They feed on fish. They are great divers, often diving from heights of 60 to 100 meters, clasping their wings to their bodies shortly before entering the water, plunging in at great speed, and going 30 to 40 meters through the water before rising. In tropical waters boobies are also expert at catching flying fish.

In tropical and subtropical seas within easy flying distance from land, wherever there are boobies there will usually be some frigate birds, family **Fregatidae,** also called man-of-war birds. Frigate birds are somewhat larger than boobies but are more expert flyers. Unlike other birds of this order, the frigate birds do not actually enter the water. They plunge toward the sea with a hissing noise and seize fish or other animals in the surface water with their beaks. They are also habitual robbers of other fishing birds, especially boobies and gannets. When a booby rises from the water heavily laden with fish it has captured and swallowed during its dive under water, a frigate bird may swoop down on the booby, causing it to regurgitate the fish, which the frigate bird seizes in midair. Male frigate birds are marked by a large bright red inflatable gular pouch used in courtship and territorial display during the breeding season. The female has a plain white breast.

Frigate birds lack the oil on their feathers that makes most aquatic birds impervious to wetting. They must therefore remain dry, since if they go into the water, it will soak them so that they are unable to rise. For this reason frigate birds must stay within easy flying distance of the land, to which they return at night. On land they are awkward and must alight in trees or bushes or near an overhang or cliff from which they can launch themselves, rather than on flat ground. They have relatively small, light bodies and short legs, but large wings with a wingspread of up to 2.4 meters. They are probably the most expert flyers and aerial acrobats of all large predatory birds.

Order Charadriiformes—shorebirds, gulls, auks, terns, puffins, guillemots

Order Charadriiformes is a large assemblage of birds. There are about 314 species, which are usually grouped into 124 genera, 16 families, and 3 suborders. Although a smaller group than the Passeriformes, with its 65 families, the Charadriiformes are probably more diverse. The various anatomical characteristics on which the order is based are not constant throughout the group. Three large families, the **Laridae, Scolopacidae,** and **Charadriidae,** contain 72% of the total number of species.

Most members of the order are various shades of brown or gray or are black and white. In several species the bill and feet are brightly colored. There is usually little sexual dimorphism.

Differences in feeding habits are strongly reflected in the structure of the bill, which varies from the short thick finchlike bill in the seed-eating **Thinocoridae** to the long slender bills of some of the **Scolopacidae,** used for probing deeply in soft sand or mud.

The eggs are laid on the ground with or without nesting material. The young are already down-covered on hatching and in some groups can actively get about and forage. Most of the seabirds belonging to this group nest in colonies. Species living in the tropics are nonmigratory, but those nesting in the temperate, boreal, or arctic zones tend to be strongly migratory in habit.

The great majority of members of the order are associated with large bodies of water. They usually stay near the shore, but some of the seagoing members of the group are pelagic.

Suborder Lari

FAMILY LARIDAE—GULLS AND TERNS. The sea gulls, genus *Larus,* more than any other bird come to mind when one mentions marine birds. They have practically worldwide distribution, occurring along ocean coastlines except in the South Pacific between south America and Australia. Some of them are also found around inland waters. They are large, heavyset birds, usually larger than pigeons, with long wings webbed feet rather strong, heavy bills, and short, unforked tails. They are strong flyers. Some tend to congregate around fishing docks, wharves, etc. or to follow boats in near coastal waters to feed on the refuse. They are both predators and scavengers, eating fish, shellfish, crustaceans, and refuse. Some of them often prey on the eggs or young of other birds. When excited or feeding they are noisy, large groups of them keeping up a continual and characteristic din with their short screams. Most of them are white, or white with some black or gray. The young tend to be brown or mottled. A few of the better-known species are discussed below.

The herring gull, *Larus argentatus,* is widely distributed in the Atlantic along both European and American shores. It migrates north in the Spring to breed, in North America breeding from Maine and the Great Lakes northward. Tinbergen (1953) has published fascinating studies of the behavior of this species and others, which are among the classics in the early studies in ethology.

The mew, or sea mew, *Larus canus,* is the common European gull. Along the American Atlantic shores the ring-billed gull, *L. delawarensis,* is a rather small, graceful species often seen along sandy beaches, treading the sand at the waterline and searching for coquina clams. Other northern species are the glaucous gull, *L. hyperboreus,* and the great blackbacked gull or cob, *L. marinus,* in which the back and upper part of the wings is slate-colored. The smallest gull, *L. minutus,* is the black-headed "little gull" of Europe, only about 28 cm. long. The western gull, *L. occidentalis,* is one of the several common species along the Pacific coast of North America. The adult is just over 0.6 meter in length, with dark mantle and black wing tips but otherwise mostly white. The feet are a pink, pale flesh color, and the stout yellow bill has a red spot near the tip of the mandible. The California gull, *L. californicus,* is often seen in large flocks, inland as well as along the coast. These birds may serve as strong checks on certain insect and rodent populations. In Salt Lake City, Utah, a monument has been erected to this bird, credited with saving the crops of the early Mormon settlers from the ravages of the Mormon cricket.

Some of the other genera of gulls include *Bruchigavia,* of which the Australian silver gull, with its coral-colored bill, feet, and legs, is one of the most beautiful species of gulls. *Pagophila* includes the beautiful ivory gull, or snowbird, a northern circumpolar species in which the adult is pure white with black feet. Another northern genus is *Rhodostethia,* which includes Ross's gull, a rather small, rare species with a wedge-shaped tail and the lower parts of rosy color. Genus *Rissa* includes a few species known as kittiwakes, in which the hind toe is short or rudimentary. The only true gulls with forked tails are the Arctic fork-tailed gull, genus *Xema,* and the Galápagos swallow-tailed gull, *Creagrus*.

The terns, subfamily Sterninae, sometimes called sea swallows, are usually smaller than the gulls and have more slender bills and weaker feet. The tail is commonly forked. They are more graceful and dashing in their flight. They fly at some height above the water, with the bill pointed downward, and plunge into the water to capture small fish. Most terns nest in colonies. Typical terns, genus *Sterna,* are mostly white with a black cap and a bluish-gray mantle. A few of the better-known species are discussed below.

The Caspian tern, *S. caspia,* is a widely distributed form occurring inland, as in the Great Lakes Region, as well as along seacoasts. It is the largest of the terns. The sooty tern, *S. fuliginosa,* is another widely distributed form preferring the warmer waters of the tropical coasts and West Indies. The royal tern, *S. maxima,* occurs along the southern United States and southward and is rather strikingly beautiful, with its black crest and bright orange bill. The fairy bird, or little tern, *S. minuta,* of Europe, Asia, and Africa is one of the smallest terns. The arctic tern, *S. paradisaea* (Fig. 14-8), performs one of the most remarkable migrations (Chapter 8) of any bird. Its breeding grounds are in the Arctic. In the winter it migrates south to an area south of the Antarctic Circle, a distance of 35,000 km. round trip.

FIG. 14-8. Arctic tern, order Charadriiformes, suborder Lari, family Laridae.

A few of the other genera of terns include the noddies, genus *Anous,* dark brownish, short-tailed terns found in tropical regions, said to be stupid and tame. The fairy tern, genus *Gygis,* of the South Pacific, pure white except for its black beak, is one of the most beautiful species. It is remarkable in its nesting habit. A single egg is laid in a tree crotch, or in a depression or knothole on the bare branch of a tree. It balances there through the two to three weeks' incubation despite the comings and goings of the parent and the blowing of the winds and hatches without falling. The young bird remains there until able to fly.

FAMILY STERCORARIIDAE—JAEGERS AND SKUAS. Jaegers and skuas are large strong-flying rapacious marine gull-like birds, usually dark blackish or sooty brown above and lighter below, with a hooked and cered bill. The female tends to be larger than the male. The birds are largely polar and boreal in distribution and are noted for harassing weaker birds, somewhat in the manner of frigate birds, until they drop or disgorge their prey. They also prey on the eggs and young of other birds. In the Northern Hemisphere some of the species are the jaeger, or pomarine, *Stercorarius pomarinus,* which nests in the Arctic. It is larger and darker than the parasitic jaeger, *S. parasiticus,* which also occurs in the northern seas. In both, the middle tail feathers are longer than the rest, extending well beyond the rest of the tail, those of the pomarine being rather obtuse, whereas in the parasitic jaeger they are pointed. The great skua, or sea crow, *Megalestris skua,* is a dark brown species of the North Atlantic, especially along European coasts. In the Antarctic the south polar skuas, *Catharacta skua* and *C. maccormicki,* are the boldest and most rapacious birds of prey.

FAMILY RHYNCHOPIDAE—SKIMMERS. One of the most oddly modified Lari in bill structure and feeding habits are the skimmers of America, southern Asia, and Africa. In these birds, related to terns, the lower mandible is strongly compressed and extends out in front beyond the upper. The pupil of the eye is vertical. The birds are partly nocturnal or crepuscular in habit and feed by flying rapidly just above the water surface, with the end of the mandible in the water, skimming off small animals. The American black skimmer, *Rhynchops nigra,* is about 0.5 meter long, mostly black above and white below, with the base of the bill and the feet red and the end of the bill black. It occurs along the southern coasts of the United States and southward in warm regions.

Suborder Alcae—auks, puffins, guillemots, auklets, murres. These heavy-bodied, short-winged, short-tailed, short-legged, web-footed diving birds with thick, compact plumage belong in the family **Alcidae.** The legs are set far back, giving them an erect posture on land reminiscent of that of penguins (Fig. 14-9). They are diving birds, eating fish, crustaceans, squid, and some larger plankton organisms. There are about 22 species, all limited to the northern Atlantic, northern Pacific, and Arctic regions. Most of them fly rapidly with rapid wings strokes and are expert swimmers and divers, using the webbed feet in swimming. Most of them nest in rookeries on cliffs and islands largely protected from terrestrial predators. The eggs are strongly tapered, reducing their tendency to roll off the precarious nesting sites.

FIG. 14-9. Pair of common murres shares an island perch with a pigeon guillemot while another murre approaches for a landing. Order Charadriiformes, suborder Alcae, family Alcidae.

One of the most unusual members of this group, the great auk, *Pinguinus impennus,* was a large flightless bird about 0.8 meter high, sometimes called the arctic penguin. The great auk was formerly abundant on both sides of the North Atlantic as far south as Massachusetts on the American side and the northern British Isles on the European side. It was ruthlessly slaughtered by the millions for its feathers, and the eggs were also collected. The last known living specimens were killed near Iceland in 1844.

The puffins, often called sea parrots, are among the oddest-appearing seabirds. *Fratercula arctica* of the North Atlantic is black above and around the neck, sharply contrasting with the white underparts of the large white cheek patches, which include the eyes. The bill is high, laterally compressed, deeply grooved, and short; the triangular basal portion is black, bordered with yellow, and the distal portion is bright orange. The legs and feet are also bright orange, with black claws. The tufted puffin, *Lunda cirrhata* (Fig. 14-10), of the Pacific is dark blackish brown all over except for the white cheek patches, the orange and yellow bill, and a large yellow plume over each eye extending back over the neck to the back. Its legs and feet are also bright orange. The horned puffin, *Fratercula corniculata,* is a Pacific species similar to the North Atlantic form but with a small fleshy, hornlike appendage on the eyelid.

The razor-billed auk, or scout, *Alca torda,* found on both coasts of the North Atlantic and on the northern islands, is about 40 cm. long, black above and white below. The bill is laterally compressed, sharp-edged, black, and crossed by a white band. A smaller, short-billed auk, the rotche or dovekie, *Alle alle,* is common on the northern ice flows. It breeds along the coasts of the Arctic Ocean and ranges southward in the winter.

FIG. 14-10. Tufted puffin: range, North Pacific; order Charadriiformes, suborder Alcae, family Alcidae.

Guillemots, sometimes called sea pigeons or sea doves, are dark-plumaged birds of this family, inhabiting rocky cliffs along the seacoast. *Cepphus grylle* is the common black guillemot of the North Atlantic, and *C. columba* is the pigeon guillemot common along the North Pacific coasts.

The rhinoceros auklet, *Cerorhinca monocerata,* of the North Pacific is curious in having a deciduous horn on the bill.

Murrelets, genus *Synthliboramphus,* are rather small members of this group mostly inhabiting islands of the North Pacific but ranging south to Mexico and Japan. The species *S. wumizusume* is the crested murrelet of Japan.

Genus *Uria* includes the true murres, sometimes also called guillemots, of which there are several species.

Other seabirds

Numerous other birds are found along the beaches, in near-shore waters, and in the waters of bays and estuaries. They include numerous shorebirds, such as sandpipers, godwits, willets, curlews, sanderlings (Fig. 14-11), and others of the Charadriiformes; many anserine birds, ducks and their relatives, some loons, grebes, and other diving birds; and a number of large wading birds such as herons, cranes, egrets, and others. Some of the Falconiformes such as ospreys also may fish in marine water. Most of these birds

FIG. 14-11. Sanderlings—common bird of the sandy beaches where it forms large flocks. Order Charadriiformes, family Charadriidae.

belong to groups that might better be termed aquatic rather than marine, since they also occur around inland waters and are not so distinctively modified for or adapted to specifically marine conditions as the groups already discussed.

MARINE MAMMALS

There are three major groups of marine mammals, each of which has arisen independently from different mammalian stocks and become adapted for life in the ocean. Life in the ocean is so different from that on land that it has impressed certain general tendencies or convergencies in the evolution of all three groups. All three have diverged sufficiently from the original stocks that it is no longer readily apparent from casual examination of presently existing species to which groups of terrestrial mammals they are most closely allied.

Having become basically adapted for life in the oceans, all three groups have undergone a kind of evolutionary radiation, that is, have undergone diversification and adapted themselves to exploiting different ecological niches—different possibilities offered by the marine environments.

Common features found in all three groups are (1) modified limbs and body shape, (2) thermal regulation, (3) modified respiratory system, and (4) osmotic adaptations.

Changes in limbs and body shape enable the animal to adapt to a swimming life. These changes involved conversion of the limbs into paddlelike flippers and in some cases the loss of the hind limbs and the development of the posterior end of the body into flukes somewhat resembling a fish's tail but horizontally placed.

Thermal regulation involved the provision for heat conservation in an environment generally considerably colder than the animal's body temperature. It was achieved by a variety of means, including thick skin with insulating layers of fur and/or fatty, oily tissue beneath, the reduction of peripheral circulation in the skin and appendages, a metabolism producing more internal heat, and an increase in size. The larger volume-to-surface area ratio in larger animals enables them to conserve heat more efficiently than is the case with the majority of mammals.

Modifications of the respiratory system enable marine mammals to breathe efficiently without taking saltwater into the lungs and to hold their breath much longer than most mammals so that they are able to make prolonged, deep dives.

"Holding their breath" in the case of deep-diving marine mammals is a different process from that in humans and other terrestrial animals.

The dives are made with the lungs empty, thus avoiding the twin problems of having to use excessive energy to overcome buoyancy while making the dive and of having nitrogen forced into the blood under pressure at considerable depths, which may form bubbles in the blood as the pressure is diminished during ascent, giving the animal the bends.

The blood of deep-diving mammals is rich in hemoglobin, and there is a further store of respiratory pigment, myoglobin, in the muscles, with even stronger affinity for oxygen—giving the muscles a dark red almost blackish aspect. Oxygen transport from the lungs to the blood and muscles is rapid and more efficient than in terrestrial animals, the volume of air inspired is greater, and the shape and movements of the animals are such that they swim under water with great efficiency—getting a maximum of swimming motion for the energy used.

During a deep dive a considerable economy in oxygen use is achieved by sphincters in the arteries leading to the viscera and the integument so that circulation to these parts of the body is cut off and all the blood goes where it is most needed—to the skeletal muscles and the brain. The heartbeat also slows down, and the muscles themselves are capable of tolerating a greater oxygen debt than is the case with terrestrial animals—even as great as functioning anaerobically for a time if need be. The neural control centers that eventually force inspiration are insensitive to higher concentrations of carbon dioxide and lactic acid in the blood than is the case with terrestrial forms.

On rising to the surface from a dive, the animal promptly takes a great breath, effects a rapid efficient carbon dioxide–oxygen exchange between the blood and lungs, and exhales as it begins another dive or period of submergence.

Osmotic adaptations prevent the intake of excessive amounts of saltwater, enabling marine mammals to get along without freshwater. They do not possess special salt-excreting glands, as do the teleosts and the procellariiform seabirds. They seem to have accomplished the adjustment by limiting the intake of water as much as possible—by having an impervious skin and obtaining most of their water from the fish diet, which is more dilute than seawater. The resorption of water in the kidneys is efficient, and it seems possible that the extensive fatty insulating layers beneath the skin may also play a role in storage of metabolic water, as has been found to be the case in some desert animals.

These changes have, of course, resulted in these animals' becoming either cumbersome and awkward on land or wholly unable to leave the water.

It is remarkable that in as short a time, geologically speaking, as 50 million years or so, the seagoing mammals have achieved such radical changes and become so successful in their environment. The whales have become completely adapted to marine life and never come onto land, even to bear young. Some of them have evolved into the largest animals that have ever lived, considerably exceeding in bulk the largest of the great dinosaurs. The seals have produced the greatest populations of large carnivorous mammals on earth. Both of these groups seemed to hold secure and dominating positions in the world's oceans until the spectacular explosion of population and technology on the part of man in the last few hundred years.

Order Cetacea—whales and porpoises

Cetaceans are all large to very large marine mammals in which there has been a complete loss or extreme reduction of hair. The forelimbs are paddlelike, are used as stabilizers in swimming, and show a secondary increase in the number of bones in the digits over that found in other mammals. The hind limbs have been lost, the pelvic bones being reduced to internal vestiges. The body is insulated by a layer of oily blubber beneath the skin, which also aids in providing bouyancy and perhaps in the storage of metabolic water. The nostrils of all modern species are located on top of the head rather than at the fore-end, as in other mammals, and have special flaps, using water pressure to close the blowhole while the animal is submerged. The brain is large, with convoluted cerebral hemispheres. The stomach of some of them is complex, with up to four or five chambers. Females have one pair of mammary glands, inguinal in position. The posterior end of the body is strongly muscular and shaped into a pair of horizontal flukes resembling a fish's tail, the semirotary motion of which gives thrust somewhat in the same manner of a ship's propeller. It has been estimated that a 27-meter whale may generate thrust equivalent to that provided by a 520-horsepower motor. The largest whales can make bursts of speed at 20 knots and can swim all day at 10 knots or better.

Suborder Archaeoceti. The cetaceans are believed to have been derived from the earliest line of carnivorous mammals, the creodonts of the Eocene period, or possibly from certain insectivores. Whales of this early period were long-bodied species. In contrast to modern forms, they possessed differentiated, or heterodont, teeth with three incisors, one canine, four premolars, and two or three molars on either side of each jaw. Also, the position of the nostrils and the relations of the maxillae are intermediate in character between those of primitive land mammals and modern cetaceans. They are grouped together in the suborder Archaeoceti, or Zeuglodontia.

They achieved a worldwide distribution, fossils having been found in Europe, America, and Australia.

From this early group the modern whales presumably differentiated during the late Oligocene or early Pliocene.

Suborder Odontoceti—toothed whales. In this group at least some teeth are present throughout life, and there is no baleen. The nostril, or blowhole, is single. The skull is specialized. It is asymmetrical and has reduced nasals, and the flattened maxillae are above the supraorbital processes of the frontals. The sternum is composed of several elements that fuse into one piece only in the adult. Several pairs of ribs join the sternum. Although whales possess no vocal cords, they produce a variety of sounds. The Odontoceti all produce very high frequency clicks, have acute hearing, and presumably use echo location to find objects and prey in the water and perhaps to communicate as well. They also produce some squealing or whistlelike sounds within human auditory range, probably made in the larynx. This group is diverse and comprises 7 distinct families.

The beaked whales, or bottlenose dolphins, family **Ziphiidae,** have the functional teeth reduced to one or two pairs in the lower jaw. Additional vestigial teeth are also commonly present. There is a pair of longitudinal grooves in the lateroventral throat region. A dorsal fin is present at about the posterior third of the body. The tail's hind margin is entire, not notched. The largest existing species of this family is *Berardius bairdi,* a North Pacific species reaching a length of 12 meters. Another large species, *B. arnuxi,* about 9 meters long, is common in antarctic waters and occurs off New Zealand and Argentina.

Another member of this group is the bottle-nosed whale, *Hyperoodon ampullatus.* The male is larger than the female, about 9 and 7 meters, respectively, and the teeth remain buried for most of the lifespan, usually only one pair erupting in old males. A peculiarity of this species is a large bony mass on each maxilla, which increases in size disproportionately with age, producing a large, gibbous swelling of the forehead in older individuals.

The dolphins, family **Delphinidae,** are a large group that includes some forms commonly called porpoises and others usually termed whales. The skull (Fig. 14-12) is

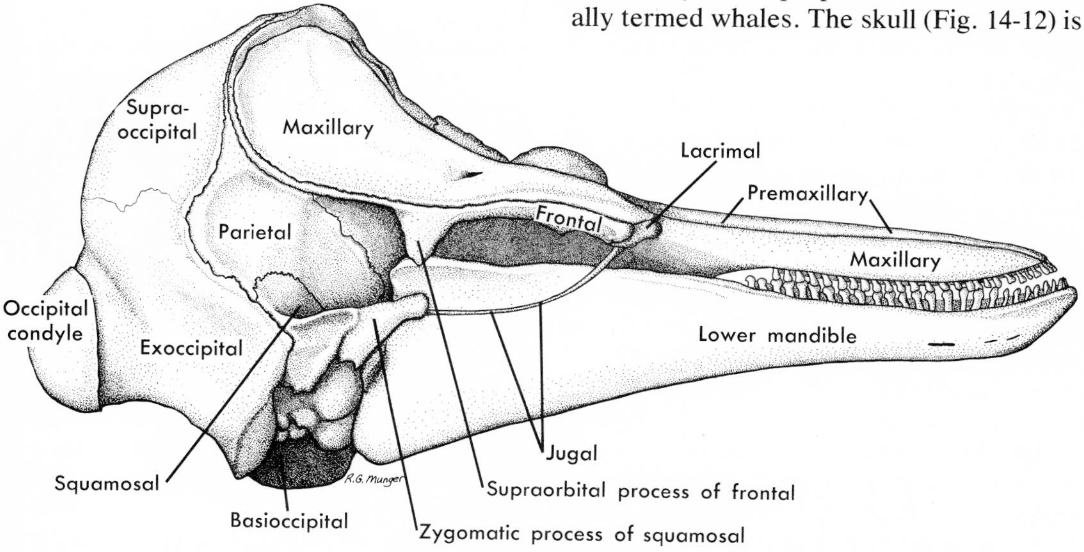

FIG. 14-12. Skull of a small-toothed whale, the harbor porpoise *(Phocoena phocoena),* showing the peculiar foreshortening effected by overlapping of the frontal and maxillary bones.

FIG. 14-13. Bottlenose dolphins leap in unison at Sea World, San Diego, California. (M. W. Williams photograph.)

asymmetrical, with the structural midline displaced to the left and the right bony nares larger than the left. As in all the Odontoceti, there is only one external blowhole, located on the top of the head. The snout in many species is produced into a "beak." The tail is notched in the middle. Many, but not all, have a prominent dorsal fin. There is a large mass of blubber between the blowhole and the end of the snout, known as the melon. Melon oil has been used as a lubricant.

The subfamily Delphininae contains many familiar dolphins of the short-beaked genus *Lagenorhynchus,* species of which are found in most oceans. The bottlenose dolphin, *Tursiops truncatus* (Fig. 14-13), has a well-defined beak 5 to 8 cm. long, has twenty to twenty-two teeth in each row in both jaws, and is dark brownish above and white below. It is a favorite in large oceanariums because of its intelligence and playfulness, the ease with which it can be trained to do a variety of tricks, and its built-in smile.

The cosmopolitan common dolphin, *Delphinus delphis,* is mostly black above and white below, with wavy streaks of white, brown, yellow, or gray on the sides and has forty to fifty teeth in each row. This is another familiar form, being common in the Atlantic and Mediterranean and one of the commonest species around the British Isles.

The subfamily Orcinae comprises the killer whales and related forms. In this group there is no well-defined beak. There is a prominent dorsal fin. The teeth are generally massive and conical, ranging from eight to nineteen in each row in the various species.

Orcinus orca, the cosmopolitan killer

whale or grampus, is the best known. The male is almost twice as long as the female, approximately 9 and 5 meters, respectively. The dorsal fin is high and falcate in the female and young, becoming triangular and up to 1.5 meters in height in the older males. Killer whales are distinctively colored, with sharply defined black and white areas, black above and white below. They have a conspicuous white patch above and behind the eye and a grayish saddle behind the dorsal fin. The flippers are rounded, and in older males the flippers, dorsal fin, and tail flukes increase disproportionately in size. They are voracious predators, eating fish, marine birds, and mammals, and even attack larger whales but are intelligent, docile, and popular performers in marine aquaria.

Porpoises of the genera *Phocaena* and *Neomeris,* having one triangular dorsal fin or none, the tail notched posteriorly, and the teeth with expanded, spade-shaped crowns, are sometimes grouped into a separate family, the **Phocaenidae.** *Neomeris* includes the finless black porpoise of Asiatic coasts from India to China; it occurs also in rivers, going up the Yangtze more than 600 miles to beyond Tung-T'ing Lake.

Another group of mostly tropical dolphins, genera *Steno, Sousa,* and *Sotalia,* comprise the **Stenidae**—long-snouted forms also occurring in estuaries and rivers as well as in the ocean.

The typical river dolphins, family **Platanistidae,** are long-snouted forms with the skull less compressed in the front-to-back direction than in the Delphinidae and with all the neck vertebrae free. They occur both in South American rivers such as the Amazon and along Asiatic coasts and rivers from India to China.

The blackfish, caaing whale, or pilot whale, *Globicephala melas* (so named because of its high, globose forehead), is another common form, the oil of which is a fine lubricant. The animals are gregarious, sometimes occurring in large schools. Schools of blackfish have on occasion become stranded on beaches after strong tides. On one occasion 147 of them were stranded on an East Lothian beach near Edinburgh, creating temporarily a major disposal problem.

Two typically arctic and circumpolar, or northern, North Atlantic forms are the narwhal and the beluga. The narwhal, *Monodon monoceros,* is peculiar for its unique dentition. It is reduced to a single pair of teeth, which only exceptionally erupts in the female. In the male the left tooth grows out as a straight, spirally marked tusk sometimes more than about 3 meters long. Occasional specimens have both teeth developed into tusks. The beluga, or white whale, *Delphinapterus leucas,* is dark gray when young and becomes pure white as an adult. This gregarious form is found in the arctic and Hudson Bay areas, sometimes as far south as the Gulf of St. Lawrence, and in the eastern North Atlantic, on rare occasions as far south as the northern part of the British Isles. These two forms lack a dorsal fin, the neck vertebrae are free, and the flippers are broad and bluntly rounded. They constitute the family **Monodontidae.**

The largest of the Odontoceti are the great sperm whales, family **Physeteridae.** The sperm whale, or cachalot, *Physeter catodon,* may attain a length of more than 18 meters, the massive head occupying about one third of the total length and extending forward well beyond the narrow lower jaw and mouth. The lower jaw bears twenty to twenty-six large teeth up to 20 cm. long on each side. The upper teeth are more or less vestigial and variable in number. The dorsal fin is represented by a series of low humps beginning back of the middle of the body and diminishing in size posteriorly. The remarkable melon, or spermaceti organ, in front of the nares is developed as a fibrous case containing up to a ton or more of clear, colorless oil, giving this animal by far the most enormous nose on record. The blubber also yields a high-grade oil. Small amounts of spermaceti are also found in the back.

Sperm whales occur in the warm parts of all oceans. Old males sometimes wander as far north or south as the polar seas. They eat chiefly large squid, including the giant squid. Large sucker marks and scars on the head of the whales sometimes attest to the dramatic battles that must take place in the depths between the sperm whales and the giant squid.

Sperm whales can dive to greater depths than any other air-breathing animal. A cable repair ship once recovered a drowned specimen entangled in a transoceanic cable from a depth of 1134 meters, more than half a mile. Since the spermaceti, oil, and blubber can absorb about six times as much nitrogen as an equal weight of blood, it is thought that in addition to providing insulation and buoyancy, they may help the whales in making deep dives followed by relatively rapid ascents without having "the bends," a serious difficulty from which human divers suffer if the pressures under which they are working in deep water are too suddenly released by rapid ascent. The pressure forces nitrogen from the air in their lungs into the blood. When this pressure is released faster than the nitrogen equilibrium can be restored by way of the lungs, bubbles of nitrogen form in the blood vessels, with painful, sometimes fatal results.

The hard, parrotlike beaks of the large squid sometimes fail to pass through the digestive tract, remaining to form gradually a pathological foul-smelling dark sticky or waxy material known as ambergris. It may be spewed up by the whales and subsequently found floating on the water or cast up on a beach, or it may be taken by whalers when they cut up the whale. For a long time ambergris, added in minute amounts to fine perfumes, was the best substance known for making perfumes hold their scent. In recent years other perfume fixatives have partially replaced it but ambergris still brings up to $10 an ounce.

Suborder Mysticeti—whalebone, or baleen, whales. The origin of the Mysticeti is not clear, although presumably they arose from some early group of Odontoceti, since although the teeth are lacking in the adult, numerous vestigial teeth are present during the late embryological stages, more than are characteristic for any primitive group of land mammals or zeuglodonts. In the adults they are replaced by an elastic horny substance, not true bone, termed baleen. Baleen occurs in the form of a series of triangular plates anchored in the roof of the mouth on each side, separated by brief intervals. The plane surfaces of these plates are at right angles to the long axis of the head, and the inner, lingual edges of the plates are frayed out into a fringe of bristles that combine to form a matted sieve, or strainer, for the collection of macroplankton. Baleen is more related to hair than to bone and like hair has its origin in the skin. Baleen plates are longest, up to 3.6 meters, in the right whales and considerably shorter, broader, and less flexible in the rorquals. A pad of softer, hornlike substance, the intermediate substance, occurs between adjacent plates. There may also be transverse rows of small subsidiary plates on the inner edges of the larger ones.

In contrast to the Odontoceti, the skull is symmetrical, and the blowhole has double, posteriorly divergent slits. The lower jaw is large, the mandibles consisting of two very large bones, bowed outward and loosely united in front. Only the first pair of ribs join the sternum, which consists of only one piece.

Baleen whales feed by swimming through clouds of zooplankton with open mouths, like gigantic, animated plankton nets. The water is forced out at the sides of the mouth between the baleen plates, and the macroplankton organisms are caught in the fringe of the bristles along the inner edges of the plates, to be licked off by the tongue and swallowed.

Pequegnat (1958) gives some figures on the energy requirements and the amounts of food needed to meet them for the antarctic blue whale, the largest animal known. According to him, an adult blue whale weighing about 90 tons, swimming at 10 to 12 knots most of the day, and using an assumed 20% efficiency of the muscles in propelling it, requires 780,000 calories per day just for propulsion. Another 230,000 calories are required for its metabolism and maintenance of body heat—a total of more than a million calories daily. Since, according to him, the blue whale feeds only during the antarctic summer, it must take in double this amount or more to provide for the stored energy used the rest of the year—stored in the form of blubber. Thus, during the feeding season an adult blue whale requires approximately 3 million calories daily. The krill, mostly *Euphausia superba,* on which this whale largely feeds, represent a little over 1000 cal./kg. Therefore the blue whale requires approximately 3 tons of krill daily during its feeding season to meet its needs. Prior to the extensive antarctic whaling of this century, it is estimated that there were at least half a million blue whales, which must have meant that the blue whales alone consumed some 270 million tons of krill each summer. Since they probably do not harvest more than 20% of the crop, there must be a minimum of 1350 million tons of krill produced annually in the antarctic waters. The actual figure is probably well in excess of this one, since several other species of whales also feed there in the antarctic summer, as well as numerous other eaters of krill, such as seabirds, fish, and squid.

Because of the short, efficient food chain for baleen whales, consisting only of phytoplankton and krill, a great deal more whale flesh can be produced than would be possible for a predator such as the killer whale, which is at the apex of a longer food chain involving at least two or three intermediary predators such as fish, squid, and seals. At each step in any food chain, there is approximately a 90% drop or greater in production because of the low efficiency of conversion of organic matter from one organism to another. Thus, even though blue whales are much larger, it is probable that there were many more blue whales than killer whales before the former were decimated by man.

According to Pequegnat's estimates, the antarctic feeding grounds must produce about half a ton or more of krill per acre during the antarctic summer.

The Mysticeti are usually divided into three families. The largest family, **Balaenopteridae,** includes the rorquals and the humpback whale. They have a relatively broad, flat rostrum and short, rigid baleen plates, and the anteroventral half of the body is marked by parallel longitudinal ridges and grooves. A small dorsal fin is present about two thirds of the way back along the body or slightly posterior to this area. The giant blue whale and several other species belong to the genus *Balaenoptera.* They range in size from the lesser rorqual, *B. acutorostrata,* found in temperate to polar latitudes of both hemispheres and not exceeding 10 meters in length, to the gigantic blue whale, *B. musculus,* also known as Sibbald's rorqual, which has been known to reach 30.5 meters. The humpback, *Megaptera novaeangliae* attains a length up to 15 meters. It is cosmopolitan, commonly migrating along inshore waters between higher and lower latitudes. The humpback is noted for its unusually long flippers, nearly one third as long as its body, as well as for its generally dumpier form and more arched back.

The right whales, family **Balaenidae**, were so named by early whalers because they float after being killed, whereas the other whales, the "wrong whales," usually sank and were lost. In these whales the skull is strongly arched, the baleen plates are long, narrow, and flexible, and the body is not marked by the parallel grooves in the anterior region. The neck vertebrae are fused. In genus *Balaena* there is no dorsal fin. The flippers are broad, with five digits in each flipper. The head is approximately one fourth to one third of the total length. The genus *Caperea,* which includes the pygmy right whale, has a less arched head, constituting only about one fifth of the total length. A dorsal fin is present, and there are only four digits in each flipper.

Family **Eschrichtidae** includes the California or Pacific gray whale. This animal is about 14 meters long. It has no dorsal fin, the flippers are four fingered, and there are scattered hairs over the entire head and lower jaw. There are only two or three longitudinal ventral grooves on each side in the throat region. The neck vertebrae are not fused.

The gray whale is a North Pacific species, migrating south in the winter to breed in shallow lagoons off Lower California in the eastern Pacific and off Japan and Korea in the western Pacific. It spends the summer in high northern waters of the Arctic and the Bering Sea. It seems to prefer relatively shallow waters and is frequently observed during its migrations swimming close to shore, sometimes even going into the zone of breakers. At one time it was nearly exterminated by whaling but has responded well to protection. A greater danger at present is that the breeding grounds may be altered by man.

Order Carnivora

Suborder Pinnipedia—seals, sea lions, walrus, elephant seals. All three families comprising this group appeared during the Miocene, but their immediate ancestry is uncertain. All are large aquatic carnivores and are gregarious and polygamous. They haul out on shore or on the ice to have their pups. The bulls are larger than the cows and in most cases establish harems defended from encroachment by other bulls. They are awkward on land but are expert swimmers and divers. The limbs are webbed flippers, with short, upper limb segments but long, broad-webbed manus and pedes. The teeth are simpler than those of land mammals, the premolars and molars being of similar character. Most species have been extensively hunted for hides, oil, and meat. Some have been nearly exterminated. Others have been placed under international protection and are harvested now on a more or less sustained-yield basis.

The three families of the group are Otariidae, Phocidae, and Odobenidae.

The eared seals, sea lions, and fur seals, family **Otariidae,** have small but well-formed ears. The hind limbs are independent, mobile, turned forward when on land, and used to facilitate movement on land. The nostrils are at the tip of the snout. The pelt bears a soft underfur. The neck is rather long.

Perhaps the most famous of this group is the northern fur seal, *Callorhinus ursinus,* of the North Pacific, which hauls out on the remote islands of the Pribilof group in great numbers each year to mate and bear young. They spend the winter roaming through the Pacific, sometimes as far as south of the equator. The Pribilof Islands are the only place at which they come to land, except for, recently, San Miguel Island off the California coast. The bulls haul out first, in June, and establish their territories, which they maintain and defend throughout the summer. Each successful bull acquires a harem as the females move in, presumably drawn to suitable areas for having their pups rather than being attracted to particular bulls. During the summer the females with pups come and go on fishing trips, which are necessary so that they may feed the young, while the bulls remain at their territorial posts, not feeding until sometime in August, by which time they have become emaciated. The migration patterns of the males and females during the rest of the year seem to differ, the females apparently roaming the ocean to greater distances as do also some of the younger males, whereas the older bulls seem to stay closer to home. Formerly, older bulls used to come down the coast of North America at least as far as Washington and Oregon or northern California, but since the time of the last major pelagic sealing operations off the West Coast of the United States, about 1900, they remain farther north. How these far-ranging pelagic seals manage to migrate back to and find their hauling grounds in the Pribilof Islands is one of the remarkable mysteries of animal navigation.

In 1968 a small colony consisting of one bull, a harem, and about forty pups established itself on San Miguel Island off the California coast. All the pups were tagged to facilitate study of this group. The wisdom of this action is questionable, since the type of tag used is one that causes considerable tissue irritation and chafing as the young animals clamber over the rocks, resulting in infections and some losses. A better, rivet type of tag, with which it is hoped this difficulty will be overcome, has subsequently been developed.

The southern fur seals, genus *Arctocephalus,* comprise a few species found in southern Australia, New Zealand, southern South America, and southern Africa. They spread northward along the western shores of South America and Africa in the cool Humboldt and Benguela currents. One species was formerly abundant at Guadalupe Island in the Galápagos but was ruthlessly slaughtered and for a time believed to be extinct. The Hancock Expeditions of 1932 and 1933 found some animals and reported that they seem to be making a comeback.

Two species of sea lions familiar along the West Coast of the United States are the Steller's sea lion, *Eumetopias jubata* (= *E. stelleri*), and the smaller, darker California sea lion, *Zalophus californianus* (Fig. 14-14). The Steller's sea lion is more northern in its distribution, being found in the North Pacific from about the Santa Barbara Islands northward, whereas the California sea lion ranges from Vancouver Island southward to Cape San Lucas,

FIG. 14-14. California sea lions haul out at Point Reyes, California. (M. W. Williams photograph.)

Lower California, and the Gulf of California. Its occurrence in the northern portion of its range is sporadic and irregular. Where sea lions are not protected, considerable numbers are shot by fishermen under the impression that they compete seriously for the fish and damage gear. Analyses of stomach contents, however, indicate that they feed mostly on fish species other than those currently intensively fished and are not really in serious competition with fishermen. Their presence in large numbers in an area is a good indication of its general productivity in fish, squid, etc., rather than a threat to fishermen, and their elimination for this reason would seem wholly unjustified. The California sea lion is the species most commonly trained to do tricks in circuses.

The hair seals, family **Phocidae,** have no external ears, the hair is short and coarse without a soft underfur, the nostrils are somewhat dorsal in position, and the hind limbs are partly fused together with the tail and are of little use on land, being dragged along posteriorly. In swimming the body is propelled by strong side-to-side strokes of the paddlelike hind flippers, each acting alternately in the power stroke. The smallest seal is a freshwater species of Lake Baikal in Siberia, *Pusa sibirica,* which attains a length of only about 1 meter. The remaining 17 species are all larger marine forms, ranging in size from about 1.5 meters in small species to 5 to 6 meters in the huge elephant seal. There are more species, and they occur in greater numbers, in the high northern and southern latitudes. Normally only one pup is born to a cow during a given season. It is nourished by the mother for a relatively short time, during which it grows rapidly. The rich seal milk contains about 50% fat. At birth most seals have soft, silky fur, which is later shed and replaced by coarser hairs. In a few species the molt is precocious, being partly or wholly completed before birth. The cows are impregnated soon after the birth of the pups, but implantation of the embryo may be delayed so that in some species at least, and perhaps in all, only about seven of the eleven intervening months between copulation and birth of the young represent gestation. Seals possess a thick coat of blubber from which a high grade of oil is obtainable. Seals commonly live about thirty years or more, but the average life-span is much shorter because of the vulnerability of the young to accident and predation.

The cosmopolitan harbor seal, *Phoca vitulina,* is perhaps the most familiar species to shoreline residents of Europe and America. It frequents coastal waters and bays, sometimes even ascending rivers for a short distance. It tends to occur in relatively small groups, never abundant enough in any one area to be commercially exploited or to warrant efforts to reduce the herds.

Two related northern species, however, the harp seal, or saddleback, *Phoca (Pagophilus) groenlandica,* and the ringed seal, *Phoca (Pusa) hispida,* are the objects of extensive commercial sealing. In the Southern Hemisphere the leopard seal, *Hydrurga leptonyx,* the antarctic Weddell seal, *Leptonychotes weddelli,* and the elephant seal, *Mirounga leonina,* have been the principal objects of exploitation. The northern elephant seal, *M. angustirostris,* which occurs along the western coast of North America, was slaughtered until, for a time, it was thought to be extinct, but in recent years it seems to be making a comeback. There is a sizable rookery at the island of Año Nuevo, north of Monterey Bay, California, and individuals have recently been seen as far north as Cape Arago, Oregon.

Coastal species such as the harbor seal are commonly rather sedentary, but some of the pelagic species, such as the harp seal, the Weddell seal, and the crabeater seal, regularly make rather extended migrations, live most of the time in the open sea, and only come out on floating ice. The crabeater seal, *Lobodon carcinophagus,* is unusual and somewhat whalelike in its manner of feeding, living on planktonic crustaceans that it strains from the seawater by passing the water through a sievelike area formed

FIG. 14-15. Walrus herd on ice floes at edge of the ice pack, Cape Lisburne, Alaska. (M. W. Williams photograph.)

by the serrated cusps of its lateral teeth.

The third family, **Odobenidae,** includes the walrus. These animals are very large and are earless like the hair seals but have hind limbs like those of the sea lions. Large males may weigh more than a ton and are thick and heavy in the neck and shoulder region. Walrus are unique in having the upper canine teeth developed into long protruding tusks, those of the bulls considerably larger and thicker than those of the cows (Fig. 14-15). The other teeth are simple, more or less peglike, and are used to crush shellfish. Both the larger Bering Sea walrus, *Odobenus obesus,* and the closely related Atlantic form, *O. rosmarus,* have been extensively hunted for their valuable skins, for the oil that can be obtained from the blubber, and for the ivory tusks, especially the large tusks of the male. They have become very rare now except in the far northern portions of their former ranges.

Suborder Fissipedia—sea otter. The sea otter, *Enhydra lutris,* family **Mustelidae,** formerly occurred in considerable numbers along the Pacific coast of North America. It attains a length of about 1.2 meters, possesses a typical slender otterlike body with relatively short legs and large webbed hind feet. The tail is cylindrical and blunt. Sea otters are active, intelligent, alert, and playful. They swim singly or in small groups, usually near shore, along the margins of the giant kelp beds. They dive readily and eat a variety of marine animals including fish and invertebrates taken along the shallow bottom. Frequently, when they have taken a sea urchin, shellfish, or crab from the bottom, they also bring up a stone. On surfacing they will float on their backs, hold the stone on their chest, and pound the prey against it to break the shell.

The fur of the sea otters is one of the most valuable on the market. The dark brown underfur is thick and rich, and the outer coat has somewhat coarser, gray-tipped hairs. They were therefore hunted nearly to extinction before stringent protection, together with their own scarcity, rendered hunting unprofitable. Since the late 1950's or early 1960's they seem to have been making a comeback in the Monterey region of California and to be spreading northward from there. There are also some along parts of the Canadian and Alaskan coasts. Prior to the last Amchitka nuclear test, groups were brought from that area and liberated near Port Orford and Cape Arago, Oregon, where they seem to be establishing themselves successfully.

Another marine member of this order, the sea mink *(Mustela macrodon)*, has been extinct since about 1700.

Order Sirenia—dugongs, manatees, sea cows

The Sirenia are large bulky aquatic mammals superficially resembling whales in the general torpedo-like shape of the body, tapering to expanded horizontal flukes posteriorly, with the hind limbs lost and the forelimbs modified into paddles. But as far back as 1816, De Blainville recognized the basic differences between this group and the Cetacea and correctly grouped them with the ungulates related to the proboscidean stem. They are sluggish animals, frequenting shallow waters and estuaries and feeding on water plants and seaweeds. The head is proportionately smaller than in the whales, and the muzzle is broad, with mobile, transversely expanded lips and coarse, heavy vibrissae, or bristles. The eyes are small, the nostrils rather dorsal in position, and the auditory openings without pinnae. The head merges indistinctly into the massive forepart of the trunk. The skin is thick and is naked or has sparse hairs. Females have one pair of mammary glands, pectoral in position.

The bones of the skull and thoracic region are exceptionally massive. There is no clavicle and no pelvic bones except a small pair of vestigial bones suspended at some distance from the vertebral column. The brain is much less convoluted and developed than that of the cetaceans.

Sirenians generally show little fear of man and are readily approached. They bear one offspring at a time, for which they show considerable affection and concern.

Although some fossil sirenian teeth seem to bear a resemblance to the teeth of early Eocene artiodactyls, and others to those of the Eocene rodent *Ischyromys*, there is much more evidence pointing to the Proboscidea as the stem group. The earliest fairly complete skeleton, *Eotheroides*, from the upper Eocene of Egypt, shows, according to Andrews (1906), a number of resemblances to that of the early proboscidean *Moeritherium* found in the same formations. He also cited a number of anatomical details in which modern sirenians and elephants show a resemblance in spite of the great differences in external appearance and mode of life.

The middle and late Tertiary sirenians, such as *Halitherium* and *Halianassa*, were larger than modern manatees and dugongs and had a pair of large vertical tusks in the upper jaw. These tusks are present in dugongs but have been lost by the manatees.

There are three groups of recent sirenia, the dugongs, the manatees, and the Steller's sea cows. Several species of dugongs and manatees have been named, but it is doubtful that they are really distinct species rather than races or varieties.

The dugongs, genus *Dugong* (= *Halicore*), occur in coastal waters of the Red Sea, on the east coast of Africa, in the Indian Ocean, the East Indies, southern Asia, and Australia. In them the tail is bilobate, the molar teeth number five or six, and the upper incisors of the male are tusklike. The upper lip is only shallowly cleft. They are more marine in habit than are the manatees and eat mostly the larger seaweeds. It has sometimes been stated that they attain a length of 6 meters, but authentic specimens are much shorter, adults usually not more than 3 meters. Dugong oil is unusually clear and free from bad taste. About 45 to 55 liters can be obtained from a full-grown specimen. It contains no iodine but is said to have the same therapeutic properties as cod-liver oil. The meat is delicious, resembling beef, and is highly prized in some areas. The dugongs have therefore been extensively hunted and practically eliminated from large areas of their former range, especially along the coasts of Asia.

Although the features of sirenians, ob-

FIG. 14-16. Manatees.

served closely, are a far cry from the type of beauty usually attributed to mermaids, the sight of dugongs in a semiupright position in the water, holding a calf with one flipper and suckling it much in the manner of a human mother, was the foundation of the mermaid myth started by early Greek sailors.

Manatees (Fig. 14-16), genus *Trichechus* (= *Manatus),* have a broad, rounded tail, rather than a bilobed one. Eleven pairs of molar teeth are formed in each jaw, but since the anterior ones are lost before the hind ones erupt, there are seldom more than six at any one time. The upper lip is deeply cleft. Manatees *(Trichechus manatus)* occur in the Caribbean region, in shallow waters of the West Indies, Florida, and northern South America and along the west coast of Africa. They ascend rivers, feeding on aquatic plants in fresh and brackish water to a greater extent than the dugongs. Adults attain a length of 2.4 to 4.5 meters and weigh about 680 kg. They are sluggish and are usually found singly or in small groups. Females bear one calf at a time. It is thought that the calf stays with the mother for more than a year. Occasionally a female is seen with two calves of different ages. The manatee occurring along the west coast of Africa displays some skeletal differences from the Caribbean variety and has been named *T. senegalensis.*

The Steller's sea cow, *Hydrodamalis gigas* (= *Rytina stelleri)* was discovered in 1741 by Bering and Steller near Kamchatka in the extreme North Pacific around two islands now known as Bering and Copper islands. Steller was shipwrecked there, spending several months on the islands before being rescued. On his return to Germany he published an account of the sea cows' habits and anatomy. They were abundant and were the largest species of sirenians, attaining a length of 7.6 meters. He reported that they were sluggish, unsuspicious, and affectionate. They differed from other sirenians in having no teeth, the masticating surfaces being merely dense hard bony pads of the jaws. They ate the large seaweeds growing abundantly around these two islands. After the existence of this herd of magnificent large sea cows became generally known, commercial hunters quickly descended upon them, and they were exterminated by 1768, only twenty-seven years after their discovery, in spite of their remote and secluded habitat.

Other mammals

A few other mammals, such as the crab-eating raccoon *(Procyon cancrivorus)* and certain mice that are found only around estuaries, while not marine in the sense of the mammals discussed above, are restricted to habitats along the edges of the oceans or of embayments.

FIG. 15-1. Professor Claude E. ZoBell, of the Scripps Institution of Oceanography, La Jolla, California—one of the leaders in the study of marine bacteria—shown with pressure chambers used in culturing marine bacteria under high pressure, simulating conditions in deep water where hydrostatic pressures may be so intense that they change the rates or character of certain enzymatic processes of biological importance. (Reproduced with the permission of Dr. ZoBell.)

CHAPTER

15 THE REDUCERS: MARINE BACTERIOLOGY

The microbiology of the oceans long remained a practically unexplored territory. There are, in the first place, formidable difficulties and expense involved in securing adequate samples from various depths and regions in the sea. Furthermore, it is necessary to carry out cultural procedures on shipboard immediately after the samples are obtained if their interpretation is to be clear. In addition, first, medical bacteriology, then, somewhat later and to a lesser extent, soil and agricultural bacteriology, and finally microbial genetics and physiology have occupied the attention of the overwhelming majority of bacteriologists.

The open ocean presents special problems not only for the microbiologist but also for any minute nonphotosynthetic organisms that live in it and are unable to ingest particulate food. Such organisms must assimilate dissolved organic molecules that can be transported through their cell membranes. They obtain these molecules by secreting exoenzymes into their environment, which break down larger molecules and aggregations of organic materials. These enzymes are specific in their action, each kind of enzyme acting only on particular chemical groups and thus only on a particular substrate or a group of closely related substrates that contain the molecular configuration in question. Dissolved organic matter in the sea is too dilute to directly provide efficient feeding of this kind for the bacterial populations there. Enzymes secreted by a bacterium suspended alone in the seawater would quickly be washed away and diluted beyond useful concentration. Any molecules of suitable type formed by the action of such enzymes on particles of organic matter in the seawater would be further diluted. The return to the bacterium of usable food molecules would be practically nil. Under these conditions bacteria could not grow and reproduce.

However, all surfaces tend to attract and hold a film of molecular and colloidal particles. Thus the living and dead bodies of all marine organisms, their shells, tests, molted exuviae, feces, and even fine clay or silica particles suspended in the sea constitute sites of concentration of available food for bacteria. Many bacteria are able to utilize directly the smaller transportable organic molecules such as glucose, even at the low concentrations at which they are present in seawater, but cannot utilize large organic molecules without first breaking them down enzymatically into smaller units. The larger macromolecular organic residues have a much greater tendency to adsorb to and be held at the surface layer of particles than do the small ones. In addition, many of these particles themselves are composed of organic matter that can be utilized.

For these reasons, mobility plus a sessile habit are important for bacteria in the open ocean. They must be able to find and attach themselves to surfaces of larger particles or objects in the water. With the increased food supply available on such a surface and the increased efficiency of digestive enzymes when they are secreted against a surface that tends to hold them and their substrates together and to catalyze the reactions between them, the bacteria are able to grow and multiply. An additional advantage of adhering to surfaces is that several kinds of bacteria may be growing together, producing a greater variety of digestive enzymes, so that the variety of organic materials that can be attacked is increased. Some species that cannot directly attack the original substrate of the particle can break down some of the intermediate products resulting from the action of enzymes of other species. Decomposition then proceeds faster, in many steps, with different species of bacteria participating at different levels of the process. Many marine bacteria, on adhering to a surface, secrete a slimy material that holds them there and in which they multiply. Bacteria themselves are also in the size range at which surface-active forces are important and tend to be adsorbed to the surfaces of larger particles and objects.

In addition to the living microorganisms, it seems probable that free enzymes secreted by some of them, or released on autolysis, may play a role in some of the transformations of organic matter in the sea. Since some of their enzymes can continue to be active after death of the cells, an estimate of viable bacterial populations is not always a reliable index of the level of biochemical activity of the milieu. *Thiobacillus thiooxidans* can produce negative pH values in some media but could not be cultured in seawater media below pH 2.5, at which pH they seemed to lyse. Some strains have been reported as growing below this value, on the basis of continued enzymatic activity, but since viability was not determined, these could be instances in which the enzymes outlived the organism.

One must also be cautious in extrapolating, from the mere presence of organisms with certain biochemical capacities, conclusions regarding the actual chemical processes occurring in the body of water from which they were taken. The physiological activities of a microorganism may vary considerably, depending on the environmental conditions. For example, the end products of sulfur oxidation by purple sulfur bacteria depend on the oxidation-reduction potential of the milieu. Agar-digesting bacteria will not digest agar if a more readily available carbohydrate is present. Nitrogen-fixing bacteria often fail to fix, or fix much less, free nitrogen if nitrogen sources such as ammonia or nitrates are present in their environment.

As Kriss and colleagues (1964) have emphasized, those organic substrates most suitable for bacteria will be readily utilized, some of them *in statu nascendi,* that is, they will be utilized as rapidly as they are produced. Thus chemical analysis of the water may fail to disclose their presence in significant amounts, even though they are being continuously produced and used on a fairly large scale. For this reason, an analysis of the populations and types of microorganisms present in a water mass may be much more revealing of the extent and nature of the biochemical processes taking place there than a chemical analysis of the water. Only the more resistant organic residues would tend to accumulate, as a type of organic marine humus. Chemical analysis alone may, by directing attention to those compounds playing the least active roles, actually be misleading.

Of great interest in this connection is the phenomenon in which, in any sample from a large body of water held for a few days in a bottle or flask, a sudden rapid increase occurs in the population of heterotrophic bacteria. The quantity of heterotrophs in isolated water samples increases several hundred or even several thousand times. This increase is the result of a marked increase in only that part of the microbial biocenosis which uses labile organic matter as a substrate. Thus the number of bacteria capable of multiplying in peptone media increases hundreds or thousands of times, whereas the total number of bacteria present, as revealed by direct microscopic count, may be only a few times greater than at the beginning.

This invalidates the claims of bacteriologists such as Kuznetsov (1955) and Sorokin (1955, 1957), who tried to measure the multiplication of chemosynthetic bacteria by measuring the incorporation of radioactive carbon-14 added to isolated samples of water or mud. That the fixation of the carbon takes place mainly by bacteria multiplying in the flasks through heterotrophic assimilation of CO_2 is also indicated by the fact that in samples to which readily assimilable organic substrates such as glucose are added, the amount of radioactive carbon absorbed by bacteria is several times higher.

The vigorous multiplication of heterotrophs in sample flasks is due to a double effect of the glass–water surface interface. In the first place, molecules of labile organic compounds, which may be present in the water in amounts too small for the bacteria to utilize, are concentrated at such an interface, making them more available. Even more important, however, is a denaturation of nonlabile organic matter at such a boundary, the denatured proteins being more readily hydrolyzed by bacterial enzymes. The sharp increase in labile organic matter, reflected by a marked rise in the population of heterotrophic bacteria, takes place at the expense of nonlabile organic compounds already present in the water but unavailable in that form to these bacteria.

The ocean depths are rich in such aquatic humus. Experiments with seawater stored in flasks indicate that its transformation into forms assimilable by microorganisms invariably occurs at contact boundaries between suspended particles and the water. This phenomenon greatly increases the significance of the sessile habit of marine bacteria, since each particle suspended in the seawater or settling through it creates a microzone where accumulation and transformation of aquatic humus occur. The development of heterotrophs on the particle is aided, bringing about its partial or complete mineralization. This also explains the dramatic rise in the number of heterotrophs found at the surface of the sea floor, above the number found a short distance either above or below this surface.

These facts emphasize the importance of making bacteriological analyses of water samples immediately after their collection. They also have important implications for the interpretation of such conventional tests as the biochemical oxygen demand (BOD) test of waters. This method consists in agitating a water sample to saturate it with atmospheric oxygen and then pouring aliquots into two or more oxygen flasks. The amount of oxygen present at the beginning is compared with the amount present in the flasks containing water kept for five, ten, twenty, or more days, and the oxygen loss due to oxidation of organic matter in the flasks, the BOD, is thereby determined in milligrams of oxygen per liter.

The BOD data merely show that the biochemical utilization of oxygen in the flasks continues at the expense of stable organic matter transformed at the solid-liquid interfaces of the flask into forms that can be broken down by the enzymes of heterotrophic bacteria.

Since no such rise in the number of heterotrophs occurs in the natural body of water from which the sample was taken, as occurs in the flasks, the results of the conventional BOD test cannot possibly be used to estimate the value or rate of oxygen demand in the sea itself, as Skopintsev had already pointed out in 1960.

Since almost all bacteria must utilize organic matter, their numbers will bear a direct relationship to the amount of organic matter present in the water. In the ocean the production of organic matter occurs in the euphotic zone through the activities of phytoplankton organisms. The distribution of bacteria in the oceans tends to parallel that of plankton, with the peaks of numbers of marine bacteria coming a little later in time and a little lower in the water than the phytoplankton peaks.

Recently it has been found that the phytoplankton of some regions, such as the Antarctic and the Arctic, exerts a powerful antibacterial effect in the bodies of animals. Organisms such as euphausiid shrimps that concentrate the phytoplankton and predators that prey on the euphausiids, or even secondarily on such predators, show marked antibiotic properties in their gut contents and tissue fluids. Thus penguins feeding on the euphausiids have been reported to be bacteriologically sterile, or nearly so. Young nestling penguins fed on regurgitated matter from their parents, and even skua gulls feeding on the young penguins, also show this condition. Their tissue fluids and gut contents have marked antibacterial properties. It is possible that this antibiotic property of the phytoplankton may play an important role in the ecology of these regions.

As mentioned in the introductory paragraphs, the overall result of bacterial activity is to break down the accumulations of organic matter represented in the bodies, exuviae, secretions, feces, skeletons, tests, and other remains of plants and animals, releasing their chemical constituents in the form of simple soluble inorganic ions and gases that can be utilized again by the plants. This process is termed the **mineralization of organic matter.** It is of the utmost importance in the cycling of available chemicals necessary for life and thus for the continuation of life on earth. In the sea the principal places for mineralization of organic matter are (1) in the upper waters, where all the organic matter is produced, slightly below the level at which the greatest amount of phytoplankton is found because of the tendency of dead and decomposing particles to slowly settle, and (2) at the bottom of the sea at the surface of the marine mud, where organic detrital particles and larger remains that have not been completely disintegrated during settling tend to accumulate and where the surfaces of particles making up the sediment serve as sites for accumulation and denaturation of dissolved "marine humus."

CHARACTERISTICS OF MARINE BACTERIA

Marine bacteria tend to differ characteristically from terrestrial and freshwater bacteria in a number of ways. Many of them require seawater or seawater media, at least for primary isolation and good growth, whereas most terrestrial forms do not do as well in such media as they do in media made with freshwater. Not only the salt content of the water but also, apparently, trace quantities of other elements or compounds of biological origin are important, since, as a rule, synthetic seawater prepared from purified chemicals and distilled water, although nutritionally adequate, osmotically similar, and chemically similar insofar as all major ions are concerned, is not usually as satisfactory for cultivation of marine bacteria as is natural seawater.

Since phytoplankton cells store little reserve carbohydrate and produce much less cellulose than do terrestrial plants and practically no lignin, and since the carbohydrates formed are used almost as fast as they are formed in the metabolism of the plant cells, the proportion of saccharolytic (carbohydrate-utilizing) to proteolytic (protein-utilizing) bacteria in the oceans is much less than on land. In the sea, one of the organic residues formed abundantly is **chitin,** the hardened cuticle of the exoskeleton of arthropods and some other animals. This residue differs from cellulose in being a nitrogenous material rather than a carbohydrate, but its formation and tendency to accumulate in the seas may be compared to that of cellulose on land. In both cases there is an abundantly formed organic residue, resistant to digestion by animals and to attack by the majority of species of bacteria. The chitinolytic, or chitin-attacking, bacteria of the sea play a role somewhat comparable to that of the cellulolytic, or cellulose-attacking, bacteria on land.*

*The Kjeldahl method of analysis for nitrogen does not detect the secondary amino group of glucosamine; thus an analysis of animals with appreciable concentrations of chitin, by this method, will give results that are too low.

The production of quantities of chitin may result in a pelagic reservoir of not readily available nitrogen, or even a permanent leak in the marine nitrogen cycle. Chitinolytic bacteria are the principal digesters.

In general, populations of marine bacteria do not survive unusual environmental stresses such as heat or changes in their chemical environment as well as do populations of terrestrial bacteria. Their marine environment normally varies within narrower limits, and resistant stages are usually absent.

Because most marine bacteria normally grow under cool or cold conditions, there is a greater proportion of types that prefer a cool environment. Preservation of such marine products as fish or shrimp by refrigeration is therefore less effective than is the case with meat products of terrestrial origin, since many of the bacteria associated with these marine products actually prefer to grow at about ordinary refrigeration temperatures. In the sea the proportions of spore-forming bacteria and of cocci in the total bacterial population are much less than on land. There are more gram-negative, polar flagellated, nonsporing rods and spirals.

It should not be thought that, because the marine humus is not readily attacked by the typical heterotrophic forms that grow so well in peptone media, there are no organisms at all that attack it directly. Direct microscopic techniques, such as studying the growth that occurs on slides or other surfaces suspended at various depths in the water, have revealed a number of types of microbial cells of unusual morphology, most of which have not been successfully grown on laboratory media. Some appear to be obligate periphyton, since they appear only on overgrowth slides. The lists of species isolated from marine sources do not yet fully reflect, even regarding morphological types, the full diversity of forms present in the sea. Only with further development of physiological-ecological methods and the use of carefully designed enrichment media will marine biologists begin to approach a detailed picture of the microbiological biocenoses of the oceans.

The problem of the taxonomy of marine bacteria also presents grave difficulties. Microbial taxonomy is beset with unusual difficulty in any case because of the impossibility of knowing the totality of characteristics revealing the biological potentialities of the forms under study and of including even the most common range of mutational variation in their descriptions. In the case of bacteria found in the sea, there is the added problem of knowing whether or not they are closely related to terrestrial species that have already been described (often inadequately). Even though the majority of terrestrial bacteria washed into the sea in vast numbers daily do not long survive, a number of species may be capable of living in either environment. Whether or not a number of forms now known only from the sea also occur in terrestrial environments is uncertain because no one has made much of an effort to look for them there. To describe all the bacteria found living in the sea as new species simply because they occur in the oceans is no real solution to the problem. In the present state of our knowledge, it is perhaps just as well to use species names in a rather inclusive sense. If a marine bacterium is not clearly different from a previously described terrestrial form in morphological or physiological characteristics, it may be regarded as a variant, race, or ecotype of the named species.

The large group of filamentous, branching Actinomycetales, so abundant and prominent terrestrially, seems to be lacking in the oceans except where washed in from adjacent land. On the other hand, the peculiar group of filamentous, nonbranching, botryose organisms known variously as the Krasil'nikoviae or the Calamobotryocidales are known only from the oceans, in which they are apparently cosmopolitan.

BACTERIA ON THE FLOOR OF THE OCEAN

At the substrate-water interface on the floor of the ocean a large microbial population is supported by the denaturation of dissolved organic humus on the surfaces of particles of sediment and by the accumulation of organic detritus settling down from above. Here the final stages of unfinished mineralization of organic residues take place. Below the surface the bacteria quickly use up all the free oxygen, and their metabolism of organic matter results in an accumulation of hydrogen sulfide so that the subsurface and deeper layers are anaerobic, black, and smell of hydrogen sulfide. The number of bacteria declines sharply downward from the surface, and the proportion of anaerobic bacteria increases. It is a curious fact, however, that even from the deeper layers of anaerobic mud, a considerable number of aerobic bacteria can be cultured. It may be that they have remained in a state of suspended animation ever since they became buried below the surface of the marine mud. Since the rate of accumulation of marine sediment is on the order of 1 mm. per century, these deeply buried aerobic bacteria must have existed there is this state for hundreds of thousands of years in some cases.

Oddly enough, the aerobic bacteria buried deeply in anaerobic sea bottom sediments for many years in a state of suspended animation, without oxygen or food, mostly grow equally well on seawater media or on freshwater media when cultured, whereas those from the top of the sediment, where they are still actively metabolizing and multiplying, for the most part will grow only on seawater media.

ZoBell and Michener (1938) also found that old stock cultures of marine bacteria that had been held for months or years without transfer, or transferred only infrequently, grew better on freshwater media than did bacteria from active cultures that had been transferred daily to slightly more hypotonic media in an effort to acclimatize them to freshwater media! Old cultures also grew over a wider range of temperatures than did those which had been maintained in an active state by frequent transfer.

It appears that physiologically young bacteria are more sensitive to adverse conditions than are senescent cells. Perhaps during senescence they synthesize and accumulate enough of certain growth factors to carry them along for a time on even unfavorable media and conditions if they can obtain oxygen and ordinary nutrients.

At the surface of marine sediments, where bacteria occur in the greatest numbers, they exercise their strongest effects on pH and thus on the processes of calcification and lithification. Their growth on the particles of sediment, often accompanied by secretion of cementing materials, may cause a cementing together of these particles and thus bring about a change in the character of the ooze and hasten its consolidation and lithification.

Table 15-1, from ZoBell and Anderson (1936), shows in round figures the numbers of aerobic and anaerobic bacteria cultured from a sample of marine sediment core taken at 2230 meters (over a mile) below the surface of the water.

The activity of the microorganisms in the marine mud also gradually alters the proportions of chemical elements of organic origin in the sediment, making them more petroleum-like, as shown in Table 15-2. The relative proportions of oxygen, nitrogen, and phosphorus are reduced, and with the loss of these elements, the proportion of carbon and hydrogen increases.

It is thought that in shallow seas, where vast amounts of organic matter may collect, the activities of marine bacteria in the sediments bring about the initial changes leading to the formation of petroleum. After these altered marine deposits become deeply buried in sediments and subjected to great pressures and perhaps high temperatures, they finally become converted into petroleum deposits.

TABLE 15-1. Aerobic and anaerobic bacteria per gram of sediment*

Depth in ocean floor (cm.)	Aerobic bacteria per gram	Anaerobic bacteria per gram
0-10	62,000,000	8,900,000
40-50	91,000	23,000
240-250	2,000	900
500-510	580	26

*From ZoBell, C. E., and D. Q. Anderson. 1936. Vertical distribution of bacteria in marine sediments. AAPG Bulletin **20**(3):258-269. Reprinted by permission of the American Association of Petroleum Geologists.

OTHER EFFECTS OF MARINE BACTERIA

Marine bacteria are important in numerous other ways in addition to their primary role in the mineralization of organic matter.

Many are ingested along with minute detrital particles by filter feeders such as bivalve molluscs and copepods, and are also eaten by small animals such as protozoans, rotifers, and nematodes, thus serving directly as food for many animals.

Where numerous, they may affect the pH of the water. A change in pH values in turn affects such processes as the deposition or solution of calcium carbonate and the rate of sedimentation of fine particles in the water, such as silt and clay carried out to sea from river mouths. Their tendency to adhere to and multiply on the surfaces of particles in the water may directly increase the tendency of such particles to agglomerate and settle. Marine bacteria may thus be of some geological importance.

In localized areas where the organic content of the water is above average and the weather calm, they sometimes deplete the

TABLE 15-2. Percentage of elements of organic origin in marine sediment*

	Carbon	Hydrogen	Oxygen	Nitrogen	Phosphorus
Marine sapropel	52	6	30	11	0.8
Recent sediment	58	7	24	9	0.6
Ancient sediment	73	9	14	3	0.3
Petroleum	85	13	0.5	0.4	0.1

*From ZoBell, C. E. 1946. Marine microbiology. By permission of The Ronald Press Co., New York.

oxygen in the surface waters, causing the death of many animals there.

The settlement and growth of marine bacteria on all submerged surfaces constitute a first step in the colonization of these surfaces by marine organisms, making them more favorable for the settlement and growth of larvae of barnacles, mussels, hydroids, sponges, bryozoans, and algae. Thus they play a role in the settlement of the so-called "fouling organisms" of ship bottoms, underwater pipes, etc. They do so by rendering such surfaces less smooth and shiny, by changing the pH at the surface-water interface, by themselves providing food for some of the larvae, and perhaps by affording some protection from any toxic elements that may be put in paints for the protection of such surfaces.

Marine bacteria may also contribute some of the biological compounds, traces of which seem to be important in "conditioning" the seawater for the growth of other forms. The dissolved metabolites may also be liberated from larger aggregates of organic matter during their decomposition by bacteria.

Marine bacteria are directly important to man in causing spoilage of marine food products and deterioration of nets, ropes, and cables and in initiating the fouling of all submerged surfaces such as boat bottoms.

Marine bacteria thus play a primary role in influencing the physical, chemical, biological, and geological processes in the oceans and ocean floors.

DISTRIBUTION OF HETEROTROPHIC BACTERIA IN THE OCEANS

In recent years extended expeditions, particularly those of Kriss and associates under the sponsorship of the Soviet Union, have done much to extend and clarify our understanding of the distribution, number, kinds, and activities of bacteria in the oceans, on a worldwide basis. Kriss and associates (1964, 1967) described some 186 kinds of heterotrophic bacteria, 15 yeasts, 12 mycobacteria, and 2 actinomycetes taken at all depths and in all major ocean regions. They were able to distinguish characteristic tropical species and associations of bacteria and other species and groups characteristic of cold polar waters, to trace the movements of major water masses and currents by analysis of their contained bacteria, and to arrive at some idea of the turnover of organic matter in various types of ocean waters through quantitative study of the populations of heterotrophic bacteria they contained. Unlike most larger organisms, many bacteria are not seriously limited by differences in hydrostatic pressure and can be found in any depths where nutrient conditions and temperature permit their growth.

Tropical waters were, in general, found to contain a much larger quantity of bacteria than did temperate, boreal, and polar waters, although not necessarily more kinds. It was also found that as a rule the bacteria typical of cold waters possessed a wider range of biochemical abilities (that is, they could utilize a greater variety of organic substrates) than did those characteristic of tropical waters. No ready explanation for this difference was apparent, although it was speculated that perhaps the comparatively high concentration of readily assimilable allochthonous organic matter in tropical water causes the high density of heterotrophs in these regions, obviating the necessity for enzymatic reactions that would ensure total utilization of organic materials. In the high latitudes where the water is poorer in food reserves, adaptations may have occurred in the heterotrophs, allowing them to use the organic matter more economically by breaking it down and converting it more thoroughly.

In all regions the proportion of heterotrophic bacteria able to metabolize inorganic nitrogen sources (that is, growing well on media containing mineral sources of nitrogen) was strikingly high—86% to 99% of the cultures.

The percentage of actively proteolytic species and of species fermenting carbohydrates varied in different regions, being highest in the North Atlantic and arctic

TABLE 15-3. Most frequently taken species of heterotrophic marine bacteria and other microorganisms*

Number	Species or variety	Stations	Group	Community	Association
1	*Bacterium agile* Jensen, var. B	120	1	1	Warm
2	*Bacterium agile* Jensen, var. A	89	1	1	Warm
3	*Pseudomonas sinuosa* Wright, var. A	61	2	2	Cool
4	*Bacterium agile* Jensen, var. D	52	1	1	Warm
5	*Bacterium candicans* (Frankland), var. B	52	3	1	Warm
6	*Bacterium candicans* (Frankland), var. A	46	1	1	Warm
7	*Bacterium agile* Jensen, var. C	32	4	3	Warm
8	*Pseudomonas sinuosa* Wright, var. B	30	5	1	Warm
9	*Bacterium parvulum* Conn, var. B	27	6	4	Cool
10	*Bacterium agile* Jensen, var. F	25	7	5	Intermediate
11	*Pseudomonas sinuosa* Wright, var. C	24	8	6	Cool
12	*Micrococcus albicans* Trevisan, var. E	24	9	7	Cool
13	*Bacterium candicans* (Frankland), var. D	21	10	3	Warm
14	*Bacterium agile* Jensen, var. E	15	11	8	Warm
15	*Pseudobacterium biforme* Eggerth	14	12	9	Cool
16	*Pseudomonas sinuosa* Wright, var. E	14	13	10	Cool
17	*Pseudomonas sinuosa* Wright, var. F	12	14	11	Intermediate
18	*Bacterium agile* Jensen, var. H	11	15	12	Warm
19	*Bacterium agile* Jensen, var. K	11	16	13	Intermediate
20	*Bacillus catenula* Migula, var. D	11	17	14	Cool
21	*Micrococcus radiatus* Flügge, var. B	10	18	15	Cool
22	*Bacterium candicans* (Frankland) var. C	9	19	16	Cool
23	*Bacillus idosus* Burchard, var. A	9	20	14	Cool
24	*Rhodotorula mucilaginosa* (Jörg)	9	21	16	Cool
25	*Debaryomyces rosei* (Kudryavtsev)	9	22	17	Cool
26	*Pseudomonas sinuosa* Wright, var. G	8	23	18	Cool
27	*Micrococcus radiatus* Flügge, var. H	8	24	19	Intermediate
28	*Pseudomonas liquida* Frankl, var. B	8	25	3	Warm
29	*Rhodotorula glutinis* (Fres)	8	26	17	Cool
30	*Bacterium parvulum* Conn, var. E	7	27	1	Warm
31	*Pseudomonas sinuosa* Wright, var. D	7	28	15	Cool
32	*Pseudobacterium variabilis* (Distaso), var. B	7	29	20	Cool
33	*Sarcina subflava* Ravenel, var. A	7	30	21	Cool
34	*Micrococcus albicans* Trevisan, var. D	7	30	21	Cool
35	*Bacillus catenula* Migula, var. I	7	31	22	Cool

*Modified from Kriss, A. E., I. E. Mishustina, N. Mitskevich, and E. V. Zemtsova. 1964. The microbial population of oceans and seas. [Transl., 1967, by K. Syers; edited by G. E. Fogg.] Edward Arnold (Publishers) Ltd., London.

regions and lowest in the equatorial and South Pacific regions. In the Indian Ocean the percentage of active cultures was intermediate, being higher in the antarctic regions and lower in the equatorial portions.

In order to bring out more clearly certain features of the distributions of bacteria, the data of Kriss and co-workers were analyzed according to the method of McConnaughey (1964) as modified by McConnaughey, Moestafa, and Lestari (unpublished). Treated in this manner, the 215 kinds fall into 115 groups, which in turn associate as 72 communities. These 72 communities, in turn, form two clearly distinct major associations, one containing 24 species and varieties, centered in the tropics, and the other being a looser association composed of two vaguely separated portions and containing 182 kinds, found primarily in cooler waters and higher latitudes. Nine species, most of them taken in only one haul each, were not associated with any other groups.

The circulation of waters in the oceans, of

course, carries tropical bacteria into high latitudes, and vice versa, in certain regions and at certain times and depths. When a water mass carries its microbial population into regions far removed from their normal distribution, it remains identifiable by its microbial content, as well as by other, more often studied characteristics, as long as the bacteria continue to live, which is probably about as long as the water mass itself maintains its distinctive identity.

The grouping of the first 35 species and varieties, all of them taken at seven or more different stations, is illustrated in Table 15-3. It can be rearranged and simplified by grouping the species as follows:

Species centered in warm waters	No. of stations
Bacterium agile var. B.	120
Bacterium agile var. A	89
Bacterium agile var. D	52
Bacterium candicans var. B	52
Bacterium candicans var. A	46
Bacterium agile var. C	32
Pseudomonas sinuosa var. B	30
Bacterium candicans var. D	21
Bacterium agile var. E	15
Bacterium agile var. H	11
Pseudomonas liquida var. B	8
Bacterium parvulum var. E.	7
Intermediate species*	
Bacterium agile var. F.	25
Pseudomonas sinuosa var. F.	12
Bacterium agile var. K.	11
Micrococcus radiatus var. H.	8

*The intermediate species include forms that were taken about equally with warmwater and cool-water types, and in all latitudes, or that clearly centered in intermediate latitudes.

Species centered in cool waters	No. of stations
Pseudomonas sinuosa var. A	61
Bacterium parvulum var. B	27
Pseudomonas sinuosa var. C	24
Micrococcus albicans var. E	24
Pseudobacterium biforme	14
Pseudomonas sinuosa var. E	14
Bacillus catenula var. D	11
Micrococcus radiatus var. B	9
Bacterium candicans var. C	9
Bacillus idosus var. A	9
Rhodotorula mucilaginosa	9
Debaryomyces rosei	9
Pseudomonas sinuosa var. G	8
Rhodotorula glutinis	8
Pseudomonas sinuosa var. D	7
Pseudobacterium variabilis var. B	7
Sarcina subflava var. A	7
Micrococcus albicans var. D	7
Bacillus catenula var. I	7

It should be noted that the names and rankings used by Kriss and colleagues to designate their isolates are not universally accepted. Microbial taxonomy is beset with unusual difficulties, and when there is a large series of fresh isolates, it is impossible to know, for example, whether morphologically similar forms that differ in some cultural or biochemical characteristics are different species or are mutant strains of a variable species. Many of these questions could not be readily approached without a far greater array of tests than would be practical or possible on shipboard, or even in most land-based laboratories.

FIG. 16-1. Captain Cook's ship, H.M.S. *Endeavour*.

CHAPTER

16 BIOGEOGRAPHY OF THE SEAS

Biological provinces, or biogeographical regions, are characterized most importantly by the endemic organisms they contain. Both the number and proportion of endemic species and the taxonomic rank of endemic groups are important. In general, any higher taxonomic group represents a longer evolution than most of its present species or subgroups. If an entire genus is endemic to a given region, this fact has more weight than the endemism of an equal number of species of a more widely distributed genus. Similarly, an endemic family or order carries more significance than a species or genus.

Endemism may result either from failure to spread beyond the region of origin or from restriction of a formerly greater range. A high degree of endemism in a given region generally denotes the relative isolation of that region for a prolonged period of time, the degree of endemism attained being roughly proportional to the degree and duration of the isolation. Obviously, given isolating factors do not affect all groups of organisms to the same extent. Something that may completely block the spread of certain types of animals or plants may have no significance at all with respect to the spread of other groups.

Biogeographical research results in the delimitation of a hierarchy of regional units and subunits analogous to the hierarchy of taxonomic units resulting from systematic research. However, in both cases the final aim is not the creation of a hierarchy of units in itself but the understanding of the evolutionary and geographical history that these systems reflect.

In an assessment of the fauna within a biogeographical system, it is advantageous, when possible, to use mathematical values for the various elements, combining them into more comprehensive figures by summation to arrive at values for the degrees of endemism or nonendemism of various groups and for degrees of affinity or independence of the region in comparison with other regions, to arrive at conclusions as objectively as possible. Ekman (1940) laid the basis for such a procedure. To use methods of this type satisfactorily, it is, of course, first necessary to have adequate faunistic and floristic information, which can be provided only by detailed surveys.

The distribution of organisms is the result of interaction between the physiological properties of the organisms, the quality or characteristics of the environment, and the history of both these components of the ecosystem. Throughout time the faunas and floras have been continuously changing as the result of infinitely complex interactions between animate and inanimate nature. The biogeographical conditions that have emerged at any particular time are the result of all that has gone before and cannot be well understood out of their historical context.

The distributions of terrestrial organisms are controlled by climate and by water barriers, especially broad seas, and to a lesser extent by terrestrial barriers such as mountain ranges. For marine organisms the most important barriers are land, temperature changes, and wide, deep regions such as those of the eastern Pacific, which is too wide for most planktonic larval forms of benthic organisms to survive while drifting.

Historically there have been three principal ways by which terrestrial organisms moved from one land mass to another, or to offshore islands: (1) corridors—at periods when the land masses were continuous there would be free movement from one to the other; (2) filter bridges—connections that allow only certain types free transit (for example, land connections in harsh climates); (3) sweepstakes routes—accidental, fortuitous transport, rare and very irregular in occurrence, such as rafting on floating logs, or dispersal to great distances by wind.

Rafting exerts a double type of selection. Large organisms are much less likely to be rafted in the first place, and small animals are more likely to starve to death en route if the distance is great.

Wind dispersal is highly selective—favoring microorganisms, plants with wind-borne spores or seeds, and flying animals such as insects or birds.

Islands tend to become populated from such rare fortuitous arrivals. The farther offshore an island, the less representative will its flora and fauna be of the biotic communities of the mainland from which they were derived. The isolated and small populations contribute to genetic differentiation from the mainland populations. Islands populated in the distant past sometimes serve as refuges for archaic types, eliminated from the mainland by competition with more recently evolved forms that never reached the islands.

For aquatic organisms, lakes, ponds, or separated watersheds are the nearest analogy to islands. Deep marine basins also tend to be isolated and to develop faunistic idiosyncrasies.

The biota of separated regions often evolve ecologically equivalent forms from diverse origins—sometimes showing striking morphological and physiological convergence. Examples are legion, such as the ant bear of South America and the aardvark of Africa. Similar convergence may be evident between forms occupying the same niche at different periods of geological history—isolation in time, rather than space—as in the case of the ichthyosaurs (reptiles) and porpoises (mammals).

With the breakup of Pangea after the Mesozoic era, the land masses were split apart. The resultant genetic isolation of their populations led to the evolution of morphological and physiological divergence among formerly more or less homogeneous biota.

Those groups which were in early stages of their evolution at the time continued to evolve separately on the various continents, giving rise on each to about as many distinctive subgroups and species as they would have done on the entire land mass had the continents remained continuous. Thus recent groups such as birds, mammals, insects, and flowering plants have evolved much wider diversity in a relatively short time than did the amphibians, reptiles, and archaic plants of Pangea during a much longer time. The presence of these new, better-adapted forms, together with climatic changes, contributed to restrict the further evolution of the older, already highly specialized types and led to the extinction of most of them.

The climatic changes attendant on the changed positions and greater emergence of the continental masses (Chapter 19) at the same time contributed to the diversity attained by the newly evolving forms by providing a greater variety of ecological niches to be colonized.

In the seas, the distribution of organisms is complicated by the dimension of depth. The epipelagic organisms and the faunas and floras of the continental shelves show different patterns of distribution than do the animals of bathypelagic or abyssal depths.

In certain tropical and subtropical regions of the open ocean, extensive "deserts" occur in which both the number of species and also of individual organisms is low—for example, those described by Wood (1965) in the Indian Ocean between 10 and 30 degrees latitude south, and in the west central South Pacific between 25 and 35 degrees latitude south. The primary reason is the stabilization of the warm surface waters and lack of adequate vertical circulation in these regions. Epipelagic faunas and floras are commonly much richer and more diverse in neritic waters over the continental shelves, where they are directly influenced by land and where turbulence, upwelling, and general circulation of the water are greater. Furthermore, production is greater over the continental shelves because here the benthos as well as the plankton and nekton is largely in the euphotic zone.

The most significant transition between what is termed the shelf fauna and the deep-water archibenthic or abyssal fauna lies at the "mud line," the upper limit of loose mud sediment, as was recognized by John Murray in his reports on collections made by the *Challenger* Expedition. This line may or may not coincide with the edge of the continental shelf in different regions, but in a general way it seems to do so.

ORGANISMS OF THE CONTINENTAL SHELVES AND NERITIC WATERS

The temperature of surface waters is of crucial importance in determining the distribution of epipelagic and shelf organisms. Most species are living much closer to the upper limits of their thermal tolerance than to the lower, since most biochemical reactions are faster and more efficient in this part of the range. An organism is usually more quickly incapacitated by moving from its normal environmental temperature into warmer surroundings than by moving into correspondingly cooler ones. Furthermore, where warm and cool waters meet, the warmer, less dense water tends to remain at the surface and the cooler water to sink below it. The cool-water organisms, instead of invading the warm surface waters, are more likely to be carried down with or to swim down to water more to their liking, showing the phenomenon of submergence, or else to be killed by the warmer water.

The result is that warm tropical waters are relatively free from incursions of organisms from the colder waters to the north and south. Life itself and perhaps the majority of principal groups of organisms appear to have originated in warm tropical waters, whereas the colder waters to the north and south and in the depths have been populated as a result of movement of organisms from warmer waters into them, together with the organisms' adaptation to life in cooler waters. Once adapted to life in cold waters, organisms have less tendency to recolonize the warm waters, both because of the greater physiological risk in moving from cooler to warmer waters and because of the competition from the forms already filling the ecological niches there.

Furthermore, the colder-water regions are subject to progressively greater ranges of environmental stress and seasonal changes in temperature and illumination. The colder and darker periods bar species adapted only to warmer waters, and warmer periods bar species adapted to still colder waters. The growing season becomes progressively more abruptly seasonal, and the annual fluctuations in conditions tend to favor those species best able to take advantage of the growing seasons through rapid growth. The surface waters of cold regions tend to be more fertile in the spring because of the annual winter turnover and mixing of waters from different depths.

In the tropics, on the other hand, conditions are far more stable. There is far greater opportunity for ecological succession to set in. Instead of the great seasonal blooms involving a relatively small number of favored species, there is a more steady state of equilibrium characterized by far greater diversity but much less tendency to produce an immense biomass of particular species. Because of its relative isolation and greater diversity, the tropical zone of the world as a whole displays a high degree of endemism. Many genera, families, and even higher systematic groups are wholly, or almost wholly, confined to the tropics. Some are circumtropical, appearing in all major divisions of the zone. Others are limited to one or two of the main divisions. The circumtropical species form only a small fraction of the total enormous number of species found in the tropics.

Tropical and subtropical regions

Two great biocenoses, or associations of organisms, characterize and are limited to shallow tropical waters around the world. They are the coral reef and the mangrove swamp. They are of the utmost biogeographical importance because they occupy dominant positions in many parts of the tropics and because they provide the habitat for special faunas of great variety. No other faunal groupings can compare in variety and beauty with those of the coral reefs.

The subtropical waters to the north and south of the tropical zone proper contain a more or less thinned-out tropical fauna, lacking the more stenothermal warmwater forms but having most of their constituents in common with the tropics, and a far smaller number of species in common with the temperate fauna on the other side. Although the subtropical waters do contain some endemic elements, these elements are fewer in number and lower in taxonomic rank than the endemic constituents of tropical waters.

The separation of tropical shallow marine fauna into three major regions has been brought about by the continents that lie across the equator in a north-south direction and by the vast deepwater stretch, the East Pacific barrier, which lies uninterrupted between outer Polynesia and Hawaii on the west and the American continents on the east. These barriers divide the tropical belt into three major components: the **Indo–West Pacific,** the **tropical Atlantic,** and the **tropical East Pacific.**

Indo–West Pacific

The Indo–West Pacific region, centered in the great Malay Archipelago off southeast Asia, extends west across the northern part of the Indian Ocean to the eastern coast of tropical Africa and the Red Sea, and east through Polynesia in the South Pacific to Hawaii, the Marquesas Islands, and the Tuamotu Archipelago. North and south it extends from the northern coasts of Australia to the Korea Strait and Japan.

The homogeneity of this vast region is best attested to by those groups of animals that are most readily transported great distances by currents and that have therefore attained distribution throughout, but limited to, this region. The fish, more than any other group, fit this category. Even strongly coastal species can be transported by currents over wide stretches of pelagic regions in either larval or adult condition without harm. Thus chance plays a larger part in the distribution of fish than is the case for most other elements of the shelf fauna. Günther, as long ago as 1868, noted that the number of species of tropical fish ranging from the east coast of Africa through the East Indies to Polynesia was large but that few of these species appeared in the tropical East Pacific along the coasts of the Americas. The fish fauna of the Indo–West Pacific is much richer than that of the other tropical regions, and contains all the families and the majority of genera of the tropical marine fish fauna of the world (except for 4 small families).

Another group of animals characteristic of the region are the sea snakes, family Hydrophidae, with 15 genera and more than 60 species ranging over the whole of this region, but only 1 species occurring outside the region, along western Africa. The dugongs, genus *Dugong (= Halicore)* are also limited to this region.

Among the invertebrates, the swimming crab *Neptunus pelagicus* has achieved a similar distribution, as has also the coconut crab, *Birgus latro,* distributed during the pelagic zoeal stage. Several stomatopods with long-lived pelagic larval stages have also achieved wide distribution in the region, as has the spiny lobster *Palinurus japanicus*. Of the crabs of the Red Sea, no less than 30% are also found in Hawaii.

All the families and subfamilies of crinoids, except one monotypic West Indian family, are represented in the Indo–West Pacific. Of the 14 families and approximately 70 genera of comatulids, 6 families and 4 species are exclusively Indo–West Pacific. Thirty-eight species of other endemic echinoderms inhabit the entire extent of the Indo–West Pacific region, and several others are limited to parts of the region.

The faunistic center of the region is the Malay Archipelago, the world's greatest archipelago, which contains extensive areas with less than 200 meters' depth. It is to be expected that such a region would be a great center for the development of shallow-water fauna and flora. The Malay Archipelago and surrounding seas are extraordinarily rich in species. The faunas of the islands of the Central Pacific are, for the most part, derived from this rich fauna but show with increasing distance a progressive diminution in the number of the species characteristic of the faunal center and some increase in endemic species of their own. The Hawaiian Islands on the extreme northern and eastern boundaries of the region, although still showing strongest Indo-Pacific affinities, lack a rather large number of the groups characteristic of the faunal center and have a fairly large proportion of endemic species of their own, with a small contingent of forms derived from the western American fauna.

A similar diminution of characteristic Malayan species and groups and an increase in the number of local endemic species can be noted in a western direction to the Red Sea and eastern Africa.

The northern Australian coastal fauna is poor in endemic echinoderms and closely related to the Indo-Malayan fauna. The Great Barrier Reef region off the northeastern coast of Queensland is more independent than northwestern Australia, presumably because until the late Quaternary the area of the shallow Torres Strait was above the surface, constituting a barrier isolating this area from the heart of the Indo-Malayan area, so that it remained more closely related to the eastern coast of New Guinea.

The Red Sea is of special interest because it is known to have been formed only about a million years ago at the end of the Pliocene or the beginning of the Quaternary period. It is a sea of exceptionally high salinity because of the intense evaporation between the African and Arabian deserts. Its temperature is relatively high, as much as 21° to 25° C. to as deep as 200 meters. Of about 430 species of decapod crustaceans known from the Red Sea, some 31% of the Macrura and 33% of the Brachyura are endemic. For crinoids the high figure of 70% endemic species has been given, but this may be partly because of imperfect knowledge of the fauna of the tropical Indian Ocean floor. In any event, the case of the Red Sea demonstrates that the rather short geological period of 1 million years is sufficient for a fairly large number of species to undergo evolutionary changes sufficient to differentiate them as new species.

Tropical Atlantic

The West Indies, the world's second largest archipelago, occupy a faunistic relation to the Atlantic somewhat akin to that of the Malay Archipelago of the western Pacific. In each case, warm tropical currents crossing the ocean over vast distances from the east bathe the region in warm waters. Although not possessing as rich a fauna as the Malay Archipelago, the West Indies do possess a rich and varied shelf fauna, especially as compared with other Atlantic regions. There are extensive coral reefs, but they are not as rich in coral species and genera as those of the Indo-Pacific. The endemic fauna is richer in species and genera than is that of any other part of the Atlantic. Of the stenothermal warmwater crabs, approximately 70% of the species and 17% of the genera of the area are endemic. All 26 of the Atlantic genera of reef corals are present in the West Indies. There is a rather large number of endemic fish genera, most of them, however, with few species. Some of the world's finest mangrove swamps have developed in the region of the Florida Keys and along the southwest coast of Florida.

In contrast, the tropical eastern Atlantic fauna is rather poor, comprising a markedly smaller number of species and genera. The communication between the tropical faunas of the eastern and western Atlantic has been restricted in recent times, as shown by the contrast in the percentage of amphi-Atlantic species in comparison with the number of genera. Of the crabs and echinoderms, which are the best-known groups, considering only the stenothermal warm-water forms, there is a total of about 300 crabs representing 115 genera, and 160 species of echinoderms representing 62 genera. Of these, only about 8% of the species of crabs and 15% of the species of echinoderms are amphi-Atlantic, whereas the corresponding figures for the genera are 22% and 42%. Of the reef corals, only 5 of the Atlantic genera occur along the western African coast.

Curiously enough, the West Indian fauna shows a closer relationship with the fauna of the Indo–West Pacific than it does with that of the eastern Atlantic, which is in the same ocean with it. If only present-day geography is considered, this phenomenon would be almost inexplicable, considering the complete barriers to communication between these tropical faunas, on both the east and the west. The relationship with the Indo-Pacific is particularly evident at the generic level. For example, in the case of the crabs, 36.5% of the genera are also represented in the Indo-Pacific. For the echinoderms, approximately 65% are common. Of the 26 genera of reef corals, 21 are also found in the Indo–West Pacific. The number of examples of other groups showing similar relationships is considerable. With respect to species, however, the relationship is slightly closer with the eastern Atlantic fauna. This discrepancy between the relationships of genera and of species means that in recent times, as long as the present species have been in existence, communication across the Atlantic has been easier than between the Atlantic and the Indo-Pacific, but that at an earlier date, within the lifetime of present-day genera but not within the time of the majority of present Atlantic species, there was open communication between the Atlantic and the Indo-Pacific. This communication is further indicated by the presence along the eastern Atlantic coasts of a number of additional fossil species and genera of Indo–West Pacific origin, and by evidence that there was formerly a much closer relationship between the tropical faunas of the eastern and western Atlantic. In more recent times the tropical fauna of the eastern Atlantic was largely destroyed during the ice ages and replaced by a more temperate fauna, the tropical fauna being largely restricted to the West Indies region.

Recent geological discoveries show that prior to and during the Mesozoic era, not only was there open communication in an east-west direction completely around the world in the tropical regions, dividing the land masses into northern and southern groups, but the major continental platforms themselves were in a different relationship to each other. They formed two great land regions—a northern mass termed Laurasia, comprising parts of what are now North America, Greenland, and northern Eurasia, and a southern mass termed Gondwanaland, comprising what are now South America, Africa, India, Australia, and Antarctica, with the Tethys Sea communicating from east to west between them. This open communication presumably existed for some time after continental drift had begun to tear apart these great land masses and to raft them into their present positions during the Cretaceous and Tertiary. During the Cretaceous and early Tertiary periods the Atlantic coastal waters of northern Africa and Europe contained a rich tropical fauna continuous at first with that of the Indo-Pacific. This fauna has now largely disappeared from these regions and to a considerable degree from the entire Atlantic. The present tropical Atlantic shallow-water fauna, chiefly centered in the West Indies, can be regarded as a relic from this ancient, more extensive tropical Atlantic fauna, which at a still earlier date was an extension of, and was continuous with, the fauna of the Indo-Pacific.

Tropical East Pacific

The Isthmus of Panama forms an insurmountable barrier to the spread of marine animals between the tropical Atlantic coast and the tropical American Pacific coast. The faunas of these areas are somewhat less closely related than are those of the West Indies region and the eastern Atlantic. Nevertheless, the tropical East Pacific fauna shows a closer relationship to that of the West Indies than it does to that of the Indo-Pacific. Again, like the relationship between the West Indies fauna and the Indo-Pacific fauna, it is most evident at the generic level, indicating that the present barrier has been in place long enough for considerable speciation to have taken place, but not so long that it has excluded present-day genera.

Among the crabs endemic to tropical American waters, for example, about 2% of the species but 18% of the genrea are common to both sides. For the echinoderms the figures are about 0.3% and 10%, respectively. However, the Pacific side has about 38% of the total number of endemic American crabs and 41% of the echinoderms, in both cases more than are found on the Atlantic side. Those species which are common to both sides appear to be ancient species that have not changed in morphology since the time the barrier was formed. Among other groups as well, amphi-American genera are more numerous than amphi-American species. Another indication of the close relationship between the two faunas is the occurrence of a number of so-called twin species, species that are very much alike, belonging to the same genus, one on each side of the barrier. Günther (1868) and others, on the basis of biogeographical resemblance between the two fauna, postulated a direct connection between the two in the past, which was later confirmed by geologists, who demonstrated an ancient channel across Central America. The relatively small number of amphi-American species shows that this communication ceased to exist before most of the species reached their present state of development. The age of the present barrier, then, is greater than the average age of the species, but the existence of several pairs of twin species shows that the connection must have been of comparatively recent date in geological terms.

These conclusions agree well with geological findings that the last direct connection existed during the Lower Pliocene. During the Eocene, Oligocene, and Miocene, the two oceans had direct connections for considerable periods. The western American tropical fauna is, then, in the main, of Atlantic origin. The deep, broad East Pacific barrier has evidently been an effective barrier for a longer time than has the Isthmus of Panama.

The relationship of the tropical western American fauna to the tropical West Indies fauna is considerably stronger than its relationship with the temperate western American faunas on either side of it.

Summary

During the Mesozoic and Cretaceous periods there was a broad sea continuous around the tropical belt of the world, dividing the land masses into northern and southern components. Only the eastern Pacific deep-water barrier interrupted the general continuity of the tropical fauna During the Miocene and Pliocene periods, the geographical situation was being drastically altered by the gradual continental drift, by the formation of the Central American isthmus, and by the land bridge between Asia and Africa. The ancient Tethys Sea was divided into the present Indo-Pacific, Mediterranean, Atlantic, and American Pacific components, largely isolated from each other. Climatic changes occurring at about the same time brought about a decided cooling, especially in the Atlantic, which decimated the former tropical fauna, particularly in the eastern Atlantic.

The relationships between the present tropical faunas of the Indo-Pacific, the Atlantic, and the eastern Pacific reflect the ancient Tethys period, whereas their differences represent the effects of isolation and of different climatic regimes in more recent times. The comparative poverty of the Atlantic tropical fauna today as compared with that of the Indo-Pacific does not necessarily mean that the Indo-Pacific was always the greatest and the only important center of origin and distribution of tropical faunas, nor does it mean that the Atlantic is poorer in tropical fauna for the same reason Hawaii is—that it is on the periphery of the area of influence. The former Atlantic tropical fauna seems to have been as rich as that of the Indo-Pacific, but it has been decimated by climatic changes.

Temperate to cold northern regions

Mediterranean-Atlantic area

The western Mediterranean and neighboring portions of the Atlantic Ocean from about Cape Blanco on the African coast to the English Channel form a faunistic region with rather ill-defined boundaries or transitional zones. The Strait of Gibraltar is not an important biogeographical boundary. The southern portion of this Atlantic area is known as the Mauretarian region, and the northern, as the Lusitanian. The Lusitanian region extends northward, merging into the boreal region.

There are a number of endemic elements, some restricted to one or another portion of the region, and others found throughout. The well-known precious coral *Corallium rubrum* occurs in the Mediterranean and in the Mauretanian region southward as far as the Cape Verde Islands, as does the sea pen *Pennatula rubra*. Ten of the 18 squid species of the Sepiolidae are endemic here, most of them exclusively in the Mediterranean. About 40% of the echinoderms collected on the shelf in this region are endemic, perhaps 15% to 16% of them restricted to the Mediterranean. Among the endemic pelagic fauna are the sardine *Sardinia pilchardus* and the anchovy *Engraulis encrasicholus*. Only about 6% of the genera of shelf fauna are endemic. A considerable similarity has been noted by various biogeographers between the Mediterranean and Japanese faunas, with a number of species being either the same or closely related twin species between the two regions. This resemblance was explained through the discovery of the extent of the ancient Tethys Sea, in much the same manner as were the similarities between the West Indian and Indo-Pacific faunas.

The water of the Mediterranean is somewhat more saline than most ocean waters because of the high rate of evaporation in the region and the fact that it does not receive a major freshwater inflow from large river systems except the Nile. Surface waters tend to flow into the Mediterranean from the Atlantic, and somewhat deeper strata flow out from the Strait of Gibraltar, bearing warm, highly saline water that strongly influences the temperature and salinity of neighboring Atlantic areas and that bears members of the Mediterranean-Lusitanian fauna, coming to the surface at certain times of the year near the entrance of the English Channel or along the seaward sides of the British Isles. This warm saline deep water from the Mediterranean makes the temperature-depth relationships of this part of the Atlantic somewhat abnormal, there often being no considerable fall in temperature until a considerable depth is reached.

During the Tertiary the main connection between the Mediterranean and the Indo-Pacific region was north of the present Red Sea area between India and southern Asia. It was many times broader, and lasted longer, than the brief Miocene union of the eastern Mediterranean and northern Red Sea region, which was not at that time open to the Indian Ocean. In the Middle Pliocene there was a connection with the Mediterranean, but it was not important in faunal exchange, since the Nile disgorged great masses of freshwater into the gulf, forming a freshwater barrier, and the connection was rather brief. Since then there had been no direct connection between the Mediterranean and the Red Sea until the completion of the Suez Canal in 1869. The faunas of the two areas, the eastern Mediterranean and the northern Red Sea, were distinct with respect to species. Only a few circumtropical species and perhaps 4 or 5 others were common to the two areas. In 1924 a study was made of the canal by a group from Cambridge, and these studies were later continued by others. It was shown that most of the species of animals in the canal came from the Red Sea because strong tidal currents from that end carried them up the canal as far as the Bitter Lakes, whereas the northern and central parts of the canal lacked such currents. Fish, because of their greater ability to swim, entered the canal from both ends. For many years the high salinity of the Bitter Lakes, which the canal traverses, proved to be a barrier to effective migrations through the canal. The salinity of the lakes varied from 46‰ to 50‰ at the surface and was even greater in deeper water. In recent years, however, continued dilution has rendered this barrier less effective. A number of Red Sea species have entered the eastern Mediterranean, and there has been some exchange of fish in both directions.

Sarmatic Inland Sea

During the Eocene and Oligocene the Tethys Sea included large areas of central and southeastern Europe, as well as the Mediterranean and parts of northern Africa and western Asia. During the Upper Miocene this sea diminished, the western part becoming shallow and the eastern part more and more brackish. Stenohaline marine forms such as corals and echinoderms, brachiopods and cephalopods, and dogfish died out and were replaced by an extensive brackish-water fauna, probably one of the most extensive brackish-water faunas ever developed. This great Eastern European–Western Asian portion of the former Tethys Sea became separated from the Mediterranean and is known as the Sarmatic Inland Sea. At its greatest extent it covered much of the Balkan Peninsula, parts of southern Russia, most of Hungary, and the areas of the present Black Sea and Caspian Sea. It flooded the valley of the Volga as far north as Kazan and large parts of the lower Kama and Ural river valleys. The water continued to become more brackish. The western end became connected with the Mediterranean during the Lower Quaternary; in the eastern part the Caspian became separated from the Black Sea during the Middle Pliocene but for a long time made connection with Lake Aral. During the second interglacial period the water of this Caspian-Aral Sea was about 80 meters higher than the level of the present Caspian Sea. The fauna of this Sarmatic Caspian-Aral basin was largely brackish and mixed to a considerable extent with freshwater forms. At present only remnants of this formerly richest-known brackish-water fauna are left in the Sea of Azov, in various Black Sea estuaries and lagoons, and in the Caspian Sea. The Caspian Sea fauna is almost purely of Sarmatic origin. The fauna of the Black Sea, however, now is composed largely of a limited number of species of Mediterranean origin that can stand the low salinities of the Black Sea and a smaller Sarmatic component mostly limited to more brackish lagoons or estuaries. In both the Black Sea and the Caspian, inadequate circulation has resulted in stagnation of the deep water and the accumulation of hydrogen sulfide; thus the benthos is limited to shallow marginal zones. In the Black Sea the benthos extends down to about 100 to 150 meters, occupying approximately 23% of the Black Sea floor. In the Caspian Sea the hydrogen sulfide does not come as near the surface, and the benthos extends to a depth of 400 meters. The Caspian Sea has been isolated from the world's oceans for about 30 million years and has thus preserved its highly individual character.

Boreal North Atlantic

Although the North Atlantic is one of the most thoroughly studied of all ocean regions, it is difficult to define the borders of this region in a precise manner. The reasons for this difficulty are (1) the lack of clear faunistic boundaries resulting from the gradual change in water temperatures in a north-south direction; (2) the lack of effective physical barriers; and (3) the fact that the wide seasonal fluctuations in temperature allow vegetatively eurythermal but reproductively stenothermal organisms, both from colder and warmer areas, to live in the region, and to reproduce at different times when the water temperatures suit them best.

The region is best defined in terms of endemic boreal species, their distributions, and centers of abundance, rather than by negative characteristics such as the northern or southern limits of occurrence of a selected number of species.

The northern European boreal region is centered in the North Sea and extends northward off the coast of Norway. It is dominated by the relatively warm North Atlantic Current, a mixture of waters mainly derived from the Gulf Stream but with an admixture of colder water from eddies off the cold East Greenland Current and a cold outflow from the Baltic through the Skagerrak. This current continues northward and branches, one branch continuing into the southern part of the Barents Sea, where it bars southward extension of truly arctic species, and the other branch forming an extensive eddy south of Spitsbergen in the Greenland Sea. Arctic waters approaching the European coasts are, then, always of a mixed character.

On the American side, the coast south of Cape Cod has a peculiar mixed character because of the warm offshore Gulf Stream but colder inshore waters derived in part from the Labrador Current, especially in winter. The waters of the New England coast, Newfoundland, and the Gulf of Saint Lawrence are essentially boreal in character but with a strong arctic admixture.

Many edemic species occur on both sides of the Atlantic, as do many additional nonendemic species also found in these waters, so that a manual of the organisms of the boreal region of either side of the Atlantic serves fairly well on both sides. The current patterns are such, however, that shallow-water American forms are more readily transported toward Europe than the reverse. The result is that more of the American boreal species are amphi-Atlantic, whereas the European boreal contains a somewhat larger number of species peculiar to the European side. Both hydrographically and faunistically, the European boreal is more clearly definable.

The boreal waters support some of the world's most productive fisheries. A few of the outstanding examples are the herring *Clupea harengus,* the cod *Gadus morrhua,* the haddock *Melanogrammus aeglefinus,* and the plaice *Pleuronectes platessa.*

Some of the more familiar invertebrates are the soft-shell clam *Mya arenaria,* the mussel *Mytilus edulis,* the barnacle *Balanus balanoides,* the lugworm *Arenicola marina,* the periwinkle *Littorina littorea,* the starfish *Asterias forbesi* and *Solaster endeca,* the sea cucumber *Cucumaria frondosa,* and many others.

One of the outstanding features of pelagic boreal waters is the great seasonal swarms of the red copepod *Calanus finmarchicus,* or herring feed. Although this species is not confined to the boreal Atlantic, undergoing submergence and appearing in other latitudes and in the North Pacific, the immense aggregations appearing in or near surface waters are characteristic of the Atlantic boreal and subarctic waters.

Baltic Sea

Extending inland from the North Sea between the Scandinavian countries and the rest of Europe, the Baltic, with a surface area of 422,000 km.^{2}, constitutes the world's largest brackish-water basin at the present time. The salinity varies both geographically and seasonally according to the circulation of marine waters and the influx of freshwater. In general it decreases with the distance inland, reaching minima in the gulfs of Bothnia and Finland. Near the entrance to the Baltic, the so-called Belt Sea, salinities vary around 15‰, plus or minus 5‰. At the upper end, in the gulfs of Bothnia and Finland, they drop to about 2‰.

Stenohaline marine forms are, of course, excluded; the number of species of marine animals and plants decreases toward the interior, and the number of brackish-water and freshwater forms increases. Parts of the gulfs of Bothnia and Finland are frozen over for five or six months or more each winter. The decrease in the number of marine species from the mouth of the Kattegat to the upper part of the Baltic proper between Darsser Ort—Gedser and Bornholm is shown in Table 16-1. Few of them extend into the gulfs of Bothnia and Finland and then only occasionally into the lower parts of these bodies.

A number of North Sea fish are found at times in the Baltic as far as the mouths of the gulfs. The most important is a brackish race of herring, *Clupea harengus membras,* which is of importance economically for the whole region. The widespread euryhaline jellyfish *Aurelia,* the clam *Mya arenaria,* a recent immigrant, and a few other marine forms are distributed fairly well throughout the extent of the Baltic proper.

Euryhaline freshwater and brackish-water species show reversed gradients of numbers of species, being most numerous in the gulfs of Bothnia and Finland, with some of them extending their ranges some distance into the upper parts of the Baltic proper. Such species, for example, are the rotifers *Keratella cochlearis, Notholoca longispina,* and *Asplanchna priodonta;* some of the planktonic freshwater and brackish-water crustaceans such as *Bosmina;* sludge worms such as *Tubifex;* some pond snails such as, *Lymnaea ovata* and *Theodoxus fluviatilis,* and a number of other animals including such fish as the stickleback *Gasterosteus,* the whitefish *Coregonus,* the bream *Abramis,* and the perch *Perca fluviatilis.* The most euryhaline and widely distributed fish in the Baltic are the Atlantic salmon *Salmo salar* and the eel *Anguilla anguilla.*

The Baltic has not always been in communication with the North Sea. During the

TABLE 16-1. Decrease in numbers of species in the Baltic Sea, from the Kattegat to the gulfs of Bothnia and Finland*

	Kattegat	Belt Sea	Upper Baltic	Gulfs of Bothnia and Finland
Sponges	27	16	0	0
Sea anemones	18	4	2	0
Polychaetes	160*	70	15	4
Amphipods	132	36	13	5
Bivalves	87	34	24	5
Echinoderms	35	8	2	0

*Modified from Ekman, S. 1953. Zoogeography of the seas. Sidgwick & Jackson, Ltd., London.

ice ages it was sometimes glaciated, and at times it was an inland lake or was in communication with the high Arctic. A number of relict high arctic species such as the circumpolar *Cottus quadricornis,* the mysid *Mysis oculata,* the amphipod *Pontoporeia affinis,* and a few others occur in the upper reaches of the Baltic; fossil beds of *Portlandia arctica* and other high arctic bivalves, together with remains of the Greenland seal *Phoca groenlandica,* attest to former high arctic fauna of the region.

Arctic Ocean

The Arctic Ocean is a great, nearly landlocked basin. It communicates with the North Atlantic over relatively shallow sills, the Nansen's Rise, between 1000 and 2000 meters deep, east of Greenland, and a similar ridge across the Davis Strait, west of Greenland. Its communication with the North Pacific through the Bering Strait is much more restricted.

The biogeographical bounds of the marine Arctic do not parallel the lines latitude, however, since in both the North Atlantic and the North Pacific cold currents flowing southward out of the Arctic Ocean along the western shores and warmer water flowing northward along the eastern shores cause a marked north-south skewing of temperature relationships. Thus, arctic conditions extend southward along the Atlantic coasts of Greenland and northern Canada because of the eastern Greenland and Canadian currents and along the Asian coast of the Pacific because of the Oyashio Current. Boreal conditions extend northward along the Norwegian coast to Barents Sea and along the Alaskan coast some distance beyond the Bering Strait.

The Arctic Ocean itself is divided into two major zones, each more than 4000 meters deep, by the Lomonosov Ridge, which runs from just north of northwestern Greenland, through the area of the North Pole, to the New Siberian Islands off central Siberia. Both the Eurasian Basin and the Canadian Basin are, in turn, divided into two parts by secondary ridges paralleling the Lomonosov Ridge. In the Eurasian Basin this ridge is a continuation of the Mid-Atlantic Ridge and rift system; in the Canadian Basin it is simply a rise variously known as the Alpha Rise, Marvin Ridge, or Fletcher's Rise. The continental shelves along most of the Eurasian side of the Arctic Ocean are broad, extending out as much as 700 km. or more, forming a series of relatively shallow seas. Beginning just north of Norway and going east, these are the Barents Sea, Kara Sea, Laptev Sea, East Siberian Sea, Chukchi Sea, and the Beaufort Sea north of Alaska and northwest Canada.

The coastal waters and the surface layer of the Arctic Ocean are strongly influenced by drainage from the surrounding land, the melting of ice, and the influx of North Pacific water of somewhat lower salinity than that of the Atlantic. The result is a layer of water about 100 to 300 meters deep, with lowered salinity (below $34^0/_{00}$) and, for the most part, with temperatures below the freezing point. Beneath is a somewhat thicker layer of Atlantic water coming from between Greenland and Spitsbergen. It is slightly warmer, commonly 0.5° to 1.5° C., and more saline, about $34^0/_{00}$. It forms a stratum extending down to approximately 700 meters. In the depths of the basins the temperatures are again negative, and the salinity is nearly $35^0/_{00}$. These layers are not everywhere as distinct as was formerly thought. The patterns of inflow and outflow from the Arctic Ocean are fairly stable at present.

The Arctic Ocean is at least partially covered by ice at all times, the extent of ice coverage varying from about one tenth of the surface to complete freezing, depending on the season and the climatic cycle. One-year-old ice is seldom more than 2 or 3 meters thick, but the ice of large ice islands rafted from the shelf ice along northeastern Canada or Ellesmere Island may be more than 12 meters thick.

Most of the extensive open-sea ice flows are from 2 to 4 meters thick. In summer the top portion melts and runs off into the sea through cracks or holes in the ice. In winter the underlying water freezes, restoring the thickness of the flow. Some addition also occurs at the top through snowfall.

Where very shallow shelf areas, within the depth to which freezing extends, underlie the ice in summer, the winter freezing incorporates surface irregularities such as rocks and pebbles, and surface mud or sand. In summer, as the ice becomes thinner by melting from the top, it rises taking with it the frozen mud, pebbles, and stones. It may also drift out of the area into deeper water. The next winter's freezing, in deeper water, thus incorporates this material at an intermediate level in the ice. In about three years, due to successive freezing at the bottom and thawing at the top, much of the mud and stones may be exposed at the surface of the ice. Here the larger, heavier, objects tend to sink somewhat into the ice and remain embedded. The mud on the ice surface absorbs heat and accelerates the summer melting; in places, holes are thawed all the way through, and the mud is washed off and sinks through them. After two or three more years, the ice may again appear clean except for the larger rocks it carries. Thus small materials—muds and clays—tend to be carried shorter distances and deposited nearer their source than larger objects, which may be carried until the ice finally reaches the edge of the ice pack and melts completely.

This phenomenon explains why ice off Eastern Siberia is often brownish and dirty looking, whereas that carried south along the east coast of Greenland is white but commonly carries larger rocks in it. The coarse materials may be carried great distances from their source before being deposited again.

The salt leaches out of sea ice one season, so that when it melts, the ice dilutes the surface water. Where winds and storms sweeping across the ocean push the ice against the shore, extensive areas of pack ice may be compressed, broken up, and refrozen as a rough, irregular surface that is difficult to negotiate. The scouring by ice also keeps the intertidal rocks of many areas bare.

Arctic Ocean fauna. The fauna of the Arctic Ocean includes a rather high proportion of endemic species, most of which are circumpolar in distribution. About 80 species of fish belonging to 45 genera occur in the Arctic Ocean. Of these, 48 species, or 60%, are known only from arctic waters or in a few cases also as relict species present also in such places as the upper Baltic. Of the 45 genera, 18, or 40%, are endemic.

Numerous endemic invertebrates are also present, including the amphipod *Onismus,* with 8 purely arctic species; *Pseudalibrotus,* with 4 arctic species, only 1 of which occurs also in northern boreal waters; and *Acanthostepheia,* with 3 arctic species. The isopod *Mesidothea* is purely arctic, although *M. entomen* still occurs as an arctic relict in the Caspian and the Baltic. The bivalves *Portlandia arctica* and *Yoldia hyperborea* are common in the polar sea.

A group of species living only in the coldest waters can be distinguished as "high-arctic species." Others preferring slightly warmer water near the subarctic waters are known as "low-arctic species." Some species occur throughout the Arctic in a "pan-arctic" distribution.

Several low-arctic and some pan-arctic species that occur on the shelf areas in the Arctic show the phenomenon of submergence and occur also in deeper water in boreal latitudes.

The arctic fauna is, in general, poorer in species than the faunas of lower latitudes, but those which do live there sometimes occur in great numbers in accordance with the general rule that extreme environments, where adequate nutrition is available, will support small numbers of species but immense numbers of individuals of those species which can exploit them.

The short arctic summer is a season of intense growth and reproduction. Planktonic animals such as *Calanus,* which may have two or even three broods a year in more southern latitudes generally have only one in the Arctic. There is also a tend-

TABLE 16-2. Biological seasons in the Barents Sea*

Property	Winter Dec.-March	Spring April-May	Spring June	Summer July-Sept.	Autumn Oct.-Nov.
Surface temperatures of water	Minimum	Low rising	Rising	Maximum	Decrease
Salinity of surface water	Maximum	Decrease	Decrease	Minimum	Increase
PO_4 and NO_3 in surface waters	Increase to maximum	Rapid fall	Poor	Decrease to minimum	Enrichment begins
Light	Minimum	Increase	24 hours	Decrease	Decrease to minimum
Zooplankton hatching	None	Intense	Decrease	Some	Dying, cessation
Phytoplankton	Minimum	Sharp bloom to maximum	Decrease	Local second maxima	Rapid decrease
Zooplankton abundance	Decrease to minimum	Low rising	Increase to maximum	Intermittent second and local increases	Decrease
Net production or consumption of zooplankton	Consumption	Production	Production	Nearly balanced	Consumption
Changes in character of zooplankton	Dying of summer species Winter *Calanus*, etc.	Bloom of Protozoa; reproduction of *Calanus* and *Thysanoessa*	Strong development of *Calanus* and *Thysanoessa*	Flowering of warmwater Atlantic spp. *Calanus* descends to deeper water	Die-off of summer plankton Reappearance of *Calanus* and halophilic eurythermal species

*Modified from Manteufel, B. P. 1941. Plankton and herring in the Barents Sea. Trans. Knipovitch Pol. Sci. Inst. Sea Fish., Oceanogr. 7 (Russian).

ency for them to grow to slightly larger size as individuals. Table 16-2, from Manteufel's (1941) studies of the Barents Sea, summarizes the seasonal changes in the surface waters characteristic of arctic regions.

North Pacific

The continental shelves form a great uninterrupted arc around the North Pacific from Lower California to Japan, since the Bering Strait is in shallow water. That portion of the shelf on the Asiatic side of the Bering Sea north of the Aleutian Islands to the Kamchatka Peninsula, Kurile Islands, and Sea of Okhotsk is arctic or subarctic in character because of the cold Oyashio and Liman currents. South of the Aleutians however, the north-flowing Japan Current, which is the Pacific equivalent of the Gulf Stream, turns eastward and crosses the North Pacific as the North Pacific Current, bringing its warm waters to the American shores at about the latitude of Oregon and Washington. Here it diverges, sending one part, the Alaska Current, as a great eddy northward to the Gulf of Alaska, where it is deflected to the west. The southern branch flows parallel to the coast as the California Current to about Point Conception, where the eastward trend of the shore leaves the current farther out to sea, still trending in a southern direction and turning west off Lower California as the North Equatorial Current.

On the American side, conditions similar to the boreal of Europe extend along the entire shelf region from Alaska to Lower California. Upwellings of deep water near shore along the coasts of California and

Lower California keep the shelf waters cooler than they would otherwise be, whereas the waters from the North Pacific Current moderate the water temperatures of the Pacific Northwest, the Canadian coast, and the Gulf of Alaska. Some of this warm eddy in the Gulf of Alaska also passes north between the easternmost Aleutian Islands along the coast of Alaska to the Bering Strait, so that the American side is noticeably warmer than the Siberian. The overall result is that surface temperatures over the shelf all the way from the Gulf of Alaska to the southern part of Lower California are remarkably uniform, usually varying between 10° and 20° C. For most of this stretch the temperatures usually range between 13° and 18° C. Many of the more eurythermal species of the shelf fauna are distributed the whole distance, or nearly so. In the northern part of the North Pacific these conditions do not follow the main shelf but rather are found only along the narrow southern edge of the Aleutian chain. At the western end of the Aleutians lies a deep-water region, forming a physical and faunistic boundary between the American and Asiatic shelf fauna.

Just as the cold Labrador Current (derived largely from the arctic Canadian Current) brings cold water south along the western coast of the Atlantic while the North Atlantic Current (derived largely from the Gulf Stream) moderates the temperatures along most of the eastern North Atlantic coast, so do the cold Oyashio and Liman currents flowing south along the western Pacific to northern Japan, while the North Pacific Current moderates the coastal temperatures of most of North America. On the western sides of both oceans, then, there is a much more abrupt transition between warm temperate conditions and cold high boreal or subarctic conditions, but on the eastern sides there are long stretches of relatively mild, cool temperate or boreal conditions.

There is a high degree of endemism along the North American Pacific coast, roughly half the species of marine organisms being endemic to this region, in contrast to the European coasts of the Atlantic, to which not more than a fourth of the resident species are endemic. There are also more endemic genera and families.

The American North Pacific is the world's leading area with respect to the number and variety of starfish, exceeding even the richest areas of the East Indies. Of the approximately 92 species, about 60% are endemic. Approximately 31% are also found in arctic or subarctic regions, and only about 8% are common to more southern regions. Most of the thirteen endemic genera of starfish are monotypic, but *Pisaster* contains 3 species and *Mediaster, 2. Pisaster ochraceous,* the most abundant and conspicuous intertidal species, comes in several strikingly different color phases, all often found together, and ranges along the whole Pacific coast of North America from the Gulf of Alaska to Lower California. One of the most striking of the monotypic genera is *Pycnopodia,* the large, soft-bodied, "twenty-rayed starfish," *P. helianthoides,* being a characteristic element of the lowest intertidal and subtidal rocky areas. In addition to the endemic genera, several more widely distributed genera such as *Henricia, Solaster, Pteraster,* and *Dipsacaster* have more species in this region than anywhere else.

Among the molluscs, octopuses, chitons, and buccinid gastropods are best developed in the American North Pacific.

The crab family Lithodidae has 13 genera and 26 species in this region, none in the subtropical waters to the south, and 2 on the antiboreal South American coast. Other crustacean genera especially well represented in the North Pacific are *Pandalus* with about 10 species, *Spirontocaris* with more than 40, and *Crangon* with about 17.

Among the fish 18 of the 20 genera of surf fish, Embiotocidae, are endemic to the American North Pacific, the other 2 being found in Japan.

The giant kelps, Laminariales, consist of some 30 genera, mostly endemic to this region, although a few genera such as *Laminaria* and *Alaria* are much more widely distributed.

The entire Pacific coast Alaska to Lower California is not, however, a wholly uniform region from a faunistic standpoint. Subjective analyses by early astute biologists, such as Dana (1853), Dall (1921), Bartsch (1921), and others, and later statistical analyses by Schenck and Keen (1936) and, still later, by Newell (1948) have demonstrated that there are faunal provinces or subregions. The most clear-cut faunal break, or transitional zone, occurs at Point Conception, California, where the California Current leaves the coast. Another is at about Forrester Island north of Vancouver Island.

Newell, on the basis of statistical analysis of the molluscan distributions, divides the coast into three broad provinces north of the warm subtropical **Panamic Province:** the **California Province,** extending from Cape San Lucas along the outer coast of Lower California north to Point Conception in southern California; the **Oregonian Province,** from Point Conception north to the Forrester Island; and the **Bering Province,** north from Forrester Island to the Bering Strait.

Newell also made interesting contributions to the methodology of analysis and emphasized a point frequently overlooked by others concerned with the problem of analysis of faunal provinces, that is, the discrepancy between the distributions of short-range species (rare species) and the distributions of long-range, abundant species. The short-range, rare species tend to cluster in ecotonal situations at the borders of biotic communities or provinces. With respect to the numbers of species present in various parts of it, a biotic community or province, then, is typically structured more like a doughnut than a discus. The greater part of its extent is characterized by a relatively small number of species, some of them having an extended range and a tendency to occur in dense stands wherever local ecological factors permit, whereas the border areas, or transitional zones, are characterized by larger numbers of species, some of them having a limited range.

Thus, on the local level at least, a high proportion of endemic species may indicate a transitional area rather than an ancient center of the regional fauna. The endemism of species that are abundant and dominant in a region is probably of more critical importance from a biogeographical standpoint than the endemism of an equal number of rare species, although the latter may well indicate the borders of such regions.

Epipelagic ecosystems of the oceanic Pacific

The major oceanic areas of the Pacific represent very old, long-stabilized ecosystems relatively untouched by man. Distribution patterns of plankton and other organisms indicate that there are eight major biotic provinces, arranged almost symmetrically with respect to latitude. They are the subarctic and subantarctic, at the extremes; the boreal and antiboreal, or transitional systems; the northern and southern central provinces, occupying the central regions of the great gyres; and, in the tropics, an equatorial and an eastern tropical Pacific system.

In the subarctic and subantarctic systems the waters are relatively rich in plant nutrients because of the high degree of thermal mixing, and primary production is light limited rather than nutrient limited. There are short, intense growing seasons, a relatively shallow euphotic zone, and weak seasonal temporary thermoclines.

The transitional systems are less well understood than the others. Upward mixing occurs here and, in the north at least, some east-to-west water movements. Neritic influences are much less apparent in the northern transitional zone, and primary production seems to be nitrogen limited.

The two central systems are regions of general downwelling, highly oligotrophic, both nitrogen and phosphorus limited, and far removed from neritic influence.

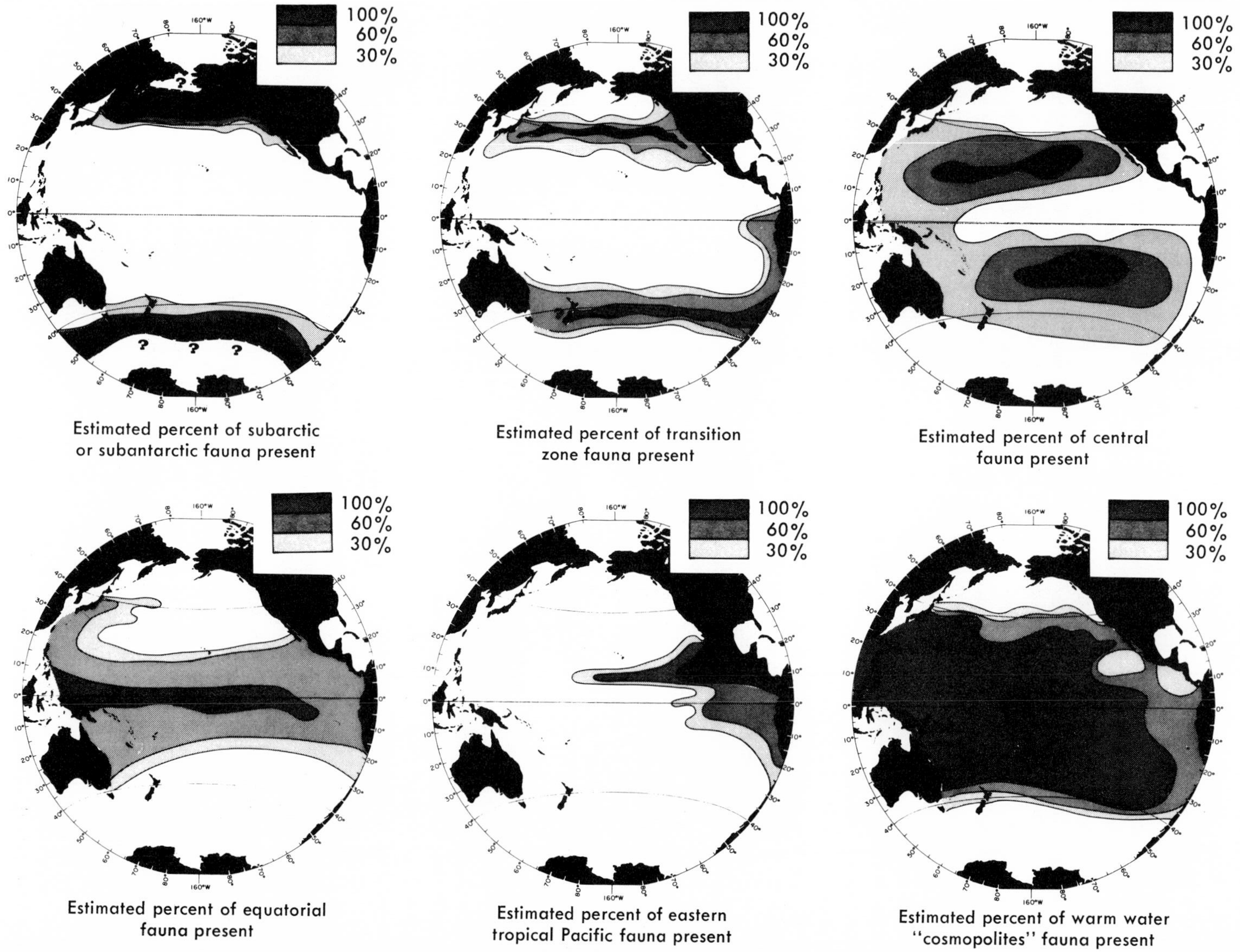
100%
60%
30%
160°W
Estimated percent of subarctic or subantarctic fauna present
Estimated percent of transition zone fauna present
Estimated percent of central fauna present
Estimated percent of equatorial fauna present
Estimated percent of eastern tropical Pacific fauna present
Estimated percent of warm water "cosmopolites" fauna present

FIG. 16-2 For legend see opposite page.

The two tropical systems are complex. Some upwelling occurs on either side of the equator between the narrow Equatorial Countercurrent and the edges of the westward-flowing Northern and Southern Equatorial currents. Also in parts of these areas the euphotic zone extends below the thermocline. The thermocline itself is a stable, permanent feature. In the eastern Pacific system there is the classical trade wind–caused upwelling, input from the edges of the California and Humboldt currents, and mixing caused by the Equatorial Countercurrent. There is also a well-developed oxygen minimum layer at relatively shallow depths.

Each of these eight provinces is characterized by a group of species that co-occur there, and all of which occur in the central, or "core," areas of the provinces. Corresponding northern and southern provinces and, of course, the two tropical provinces also share a number of species. In addition, there are a number of generally more eurythermal species, especially in warmer waters, that are not limited to any particular province.

These relationships are illustrated in Figs. 16-2 and 16-3.

The ecotonal regions bordering and lying between these biotic provinces are quite different in character from the "core areas" of the provinces. Large-scale advective water movements and horizontal transport of individuals and species into and out of these regions are much more marked.

FIG. 16-3. The patterns of the basic (100% "core" regions) biotic provinces of the oceanic Pacific. (From McGowan, J. A.: The nature of oceanic ecosystems. In C. B. Miller [ed.] 1974. The biology of the oceanic Pacific. Oregon State University Press, Corvallis. Used with permission.)

FIG. 16-2. The 100% level includes all subarctic species, but 60% have a somewhat broader range, particularly in the California Current, and 30% have an even broader range. Question marks indicate a lack of adequate, quantitative sampling. (From McGowan, J. A. The nature of oceanic ecosystems. In C. B. Miller [ed.] 1974. The biology of the oceanic Pacific. Oregon State University Press, Corvallis. Used with permission.)

Faunal mixtures, containing some, but not all, of the species from adjacent provinces, plus some species characteristic of ecotones, occur here. The state of the ecotonal regions results as much from large-scale horizontal water movements as it does from interactions and feedback loops between the components of the system, and is subject to much greater fluctuation.

Summary

When the Northern Hemisphere is considered as a whole, it becomes apparent that the major faunistic break is between the warm tropical-subtropical regions and the cool-to-cold temperate-boreal–arctic. All the families of organisms and between 50% and 75% of the genera found in arctic regions are also present in the boreal and temperate areas, whereas only 38% of the families and about 8% of the genera characteristic of temperate regions are also found in the tropics and subtropics.

Temperate to cold southern regions

Whereas the shelf regions of the Northern Hemisphere are almost continuous around their northern portions, those of the Southern Hemisphere are more limited in extent and are separated from each other by extensive tracts of deep sea. The temperate regions of southern and eastern Africa and Australia are rather closely linked with tropical regions already considered, but those of the western coasts of Africa, South America, and Antarctica are more isolated. The relations of the antiboreal and antarctic regions are totally different from those of the boreal and arctic regions. Instead of a nearly landlocked polar sea, there is an ocean-surrounded polar continent, still in the grip of a prolonged ice age. The Southern Ocean is broadly and directly contiguous with all the major oceans—Pacific, Indian, and Atlantic—and is continuous without interruption around the world.

Surface circulation of waters is characterized by the **West Wind Drift,** a great circumpolar current that washes the shores of all temperate oceanic islands of the Southern Ocean and the southern parts of South America, Africa, and Australia. Where partially intercepted by the southern tips of South America and Africa, it gives off north-flowing branches of cold current along the western coasts of these continents—the **Humboldt,** or **Peru, Current** along South America and the **Benguela Current** along Africa. The shelf areas on the western sides of these continents are for the most part rather narrow, and the extensive upwelling along the coasts contributes to the great productivity of the waters of these currents and strongly influences the coastal climate.

Immediately adjacent to the antarctic continent there is an eddy, the **East Wind Drift.** The antarctic waters of the southern part of the West Wind Drift are moving in a northeastern rather than a strictly eastern direction because they are continually being replaced near Antarctica by upwelling from intermediate depths. These Antarctic waters meet and submerge below warmer, less dense antiboreal water (so-called subantarctic water) at the **Antarctic Convergence.** Surface temperatures in the Antarctic Convergence are about 3.5° to 4.5° C. in summer and about 1° to 2° C. in winter. To the south they become colder, and to the north, warmer. The Antarctic Convergence lies, in general, between the latitudes of 50 and 60 degrees south. Its position is not absolutely fixed. It forms the most important faunal boundary of the surface waters of the Southern Ocean.

Approximately 10 degrees to the north of the Antarctic Convergence is another convergence, the **Antiboreal Convergence** (often termed Subtropical Convergence), lying roughly along the 14° to 15° C. summer isotherm or the 10° to 12° C. winter isotherm. The Antiboreal Convergence does not touch the continental coasts, however, and so does not signficantly influence their shelf faunas. It is of importance for the epipelagic oceanic faunas. Because of the general eastern direction of flow of southern temperate and antiboreal surface waters, groups of epipelagic, or shelf-dwelling, organisms having their centers of origin or dispersal in this zone in general distribute themselves to the east of such centers, but because of the extreme distances over open ocean between shallow areas, only shelf organisms with exceptionally long-lived pelagic larval stages or in which the adults are able to endure such long journeys would be able to make it to the next available habitats. Thus in these regions, faunas originating in southern Australia may be expected to be well represented in New Zealand but have only a few outlying members near the southern end of South America and practically none in southern Africa. A rich southern African fauna, however, may well have some outlying members in Australia. Similarly, a southern South American shelf fauna may be expected to be sparsely represented in the islands of the South Atlantic and perhaps in southern Africa but not in Australia. In the Australian region the Antiboreal Convergence is rather ill defined and diffuse, spread over about 5° of latitude.

Australasia

The biogeography of the Australasian region is complicated by its position extending from warm regions in the north to cold in the south, by the rather complicated pattern of currents, some of which display seasonal reversals, and by the relative isolation of the region from other biogeographical areas for a long period.

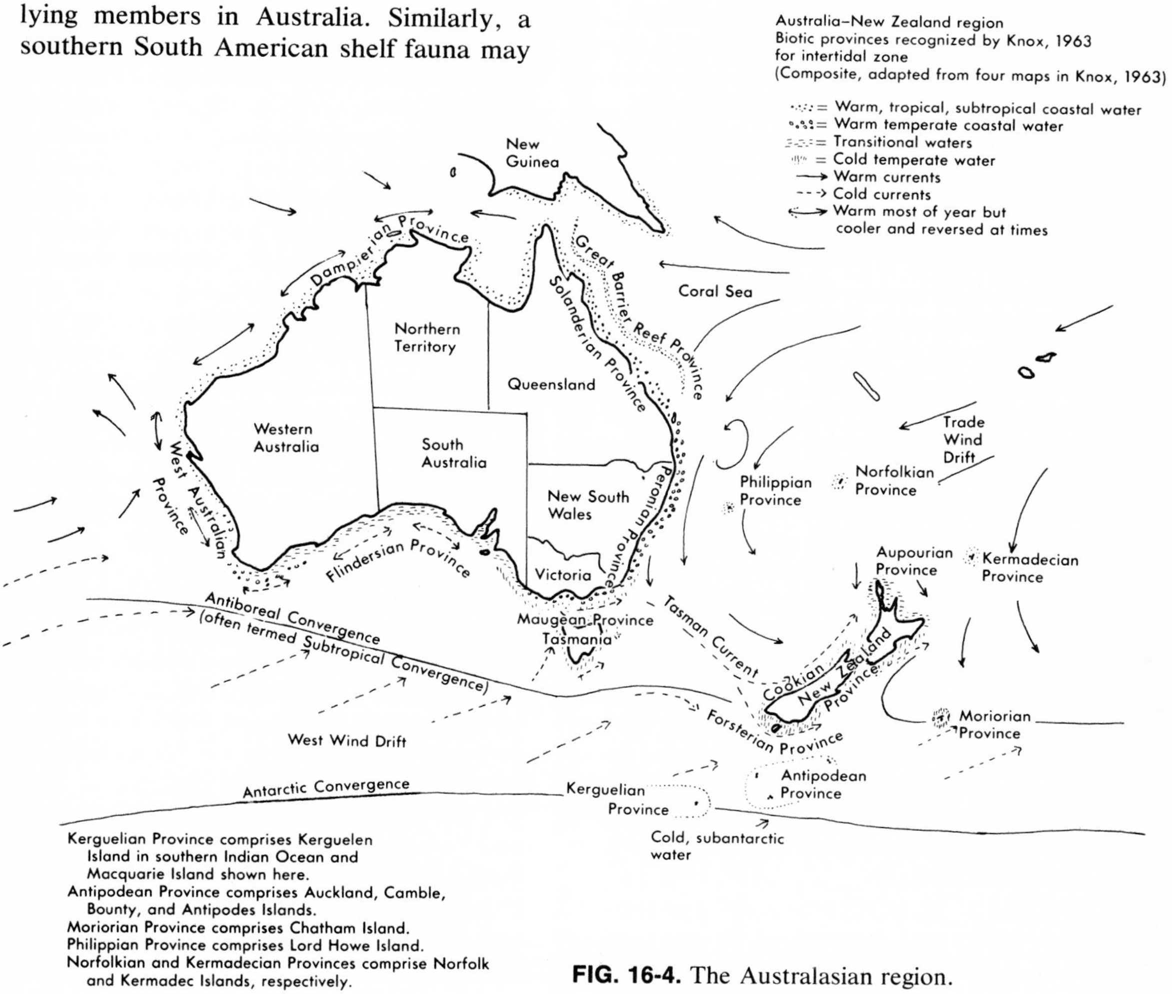

FIG. 16-4. The Australasian region.

Australia has derived most of its intertidal fauna from the tropical Indo-Pacific. To the south, as one encounters cooler waters, there is a considerable thinning out of the number of species able to survive. Because of its isolation. from other land masses, there is not much recruitment of a compensating cool-water intertidal biota, especially since the current systems in the area tend to favor the spread of warm-water species into the cooler areas rather than the reverse.

A summary of some of the important features of the region is shown in Fig. 16-4. Knox (1963) recognizes no less than sixteen biological provinces for the shallow-water marine areas—seven along the shores of Australia proper, three in New Zealand, and six small provinces for various islands or isolated groups in the region whose shore biota differs recognizably from those of any of the other provinces and from each other. These provinces are briefly outlined below.

Australian provinces. The **Dampierian Province** extends along the northern shore of Australia west of Torres Strait to the Houtman Abrolhos Islands off western Australia. This tropical-subtropical region has variable east-west trending currents along the northern coast and north-south currents in the western portion. The waters come in part from the north through the Malay Archipelago, in part from the east from the Coral Sea, and in part from the northeast portion of the Indian Ocean. The coastal waters are warm, ranging from about 24° to 29° C. along much of the northern coast and from 22° to 27° C. in the western portion. Much of the biota is pantropical in character, a smaller portion derived from species endemic to the Malay Archipelago and northern Australia. It is still questionable whether Torres Strait marks a sufficiently distinct faunal boundary between the Dampierian Province and the Solanderian Province along the northern coast of Queensland to merit their designation as separate biotic provinces. It seems probable that for some time in the Pleistocence, the Torres Strait was closed, so that these areas were temporarily separated from each other. Because the northern coasts are less well known than any other regions of the Australian shelf, conclusions about its distinctness are still doubtful. There do seem to be a number of endemic Dampierian species, especially toward the northwestern portion of the province, at least among the echinoderms, which have been better studied than most groups of the area.

The **Solanderian Province** of the rocky mainland coast of northern Queensland is a second tropical-subtropical area closely related to the northern Dampierian and overlapping the northern end of the temperate Peronian Province between Double Island Point and Point Vernon, between latitudes 25 and 27 degrees south. The coastal waters are warm, about 21° to 28° C. The zonation, or sequence of dominant intertidal organisms, differs in characteristic ways from that of the more southern Peronian Province. The upper littorinid zone is dominated by *Nodilittorina pyramidalis,* and the lower littorinid *(Melarhaphe)* zone, by *Melarhaphe melanacme*. Below this the small barnacle *Chthamalus malayensis* is dominant down to about the high-watermark of the neap tide or slightly below it. A conspicuous oyster zone is formed below the barnacles by *Crassostrea amasa,* extending downward to approximately the mean low level of spring tides. From the lower edge of this oyster zone a larger barnacle, *Tetraclita squamosa,* extends to the algal zone beginning slightly below the lower level of low neap tides. The common chiton of this province is *Liolophura giamardi*. The common limpets are *Patelloida saccharina, Cellana conciliata,* and *Chizacmea heteromorpha*.

The **Great Barrier Reef Province** parallels the Solanderian, offshore, and is also bathed with warm waters. It is dominated by reef-building corals and shows an affinity with eastern New Guinea and the western Pacific more strongly than with the Indonesian region. There are, of course, many faunal elements in common with the nearby mainland coast, but there are also characteristic differences. Endean (1957) concluded that the echinoderm faunas of the Great Barrier Reef and the Queensland coast were distinct entities; the mainland species had more elements in common with the East Indies, and the Barrier Reef complex showed more relation to New Guinea and the western Pacific. The intertidal zonation of Heron Island, a typical Barrier Reef island, differed in several respects from that on the mainland. The littorinids were much less abundant and lower in the intertidal. There was no upper barnacle *(Chthamalus)* zone, and the oyster zone was less well developed. The large barnacle was represented by a different species, *Tetraclita vitiata,* and the algal zone of the mainland was largely replaced by a lithothamnion-zoanthid-coral zone. The commonest chiton is *Acanthozostera gemmata.* The only limpet is *Penepatella inquisitor.*

The **Peronian Province** of New South Wales overlaps the Solanderian Province to the north in southern Queensland and the coldwater Maugean Province to the south along the coast of Victoria. The water temperatures run from about 17° to 20° C. There are several noticeable differences in the intertidal zonation. The lower littorinid *(Melarhaphe)* zone is dominated by *Melarhaphe unifaciata.* The upper barancle zone is dominated by *Chthamalus antennatus.* Two of the most conspicuous differences are that the zone occupied by oysters *(Crassostrea)* in the north is here dominated by the barnacle *Tetraclita rosea* and that the zone occupied by the barnacle *T. squamosa* in the north is here occupied by calcareous tubes of the serpulid worm *Galeolaria caespitosa.* The algal zone contains many more tunicates, becoming, in effect, an algal-ascidian zone. Typical echinoderms include *Asterina inopinata, Halopneustes pycnotilus, Centrostephanus rodgersii,* and *Heliocidaris tuberculata. Heliocidaris erythrogramma* and *Patirella gunii* occur both here and in the Flindersian Province of South Australia and southern West Australia.

The **South Australian,** or **Flindersian, Province** extends from western Victoria, along the shores of South Australia and most of the southern shoreline of Western Australia, where in the western part it intergrades with the West Australian Province. The exposed capes are of Paleozoic rock. Coastal waters range from about 13° to 19° C. The intertidal zonation is marked by a supralittoral band of the littorinid *Melarhaphe unifaciata,* the lower part of which overlaps a prominent black band of *Calothrix fasciculata.* Most of the littoral zone is dominated by barnacles—*Chamaesipho columna* in the upper part, followed by the surf barnacle *Catophragmus polymerus* on exposed slopes, and *Balanus nigrescens,* together with coralline algae, *Galeolaria caespitosa,* and limpets. The algal zone is dominated by species of the brown alga *Cystophora,* with some *Sargassum bracteolosum* and *Ecklonia radiata.* The Peronian barnacle *Tetraclita rosea* is absent.

Around the shores of Tasmania and adjacent shores of Victoria is a province having still cooler water, the **Maugean Province,** which is rather closely related to both the Flindersian and the Peronian provinces. The water temperatures are mostly between 11° to 15° C. One conspicuous difference is the dominance of the giant fucoid seaweed *Durvillea potatorum* in the seaweed zone. Among the mat of corallines in the lower littoral, large chitons, genus *Peronoplax,* are dominant, and a large ascidian, *Pyura stolonifera,* is abundant. In slightly deeper water the kelp *Macrocystis angustifolia* becomes dominant.

The **West Australian Province** rounds the southwestern corner of the continent, intergrading with the western ends of the Dampierian Province on the north and the Flindersian Province on the east. The water temperatures here are mostly about 17° to 20° C. The supralittoral zone commonly has an upper band of blue-green algae, below which are the littorinid *Melarhaphe unifaciata* or *Tectarius rugosus.* In the midlittoral zone the large barnacle *Balanus nigrescens* tends to dominate, along with the limpets *Notacmea onychitis* and *Patellanax peroni* and the chiton *Clavarizona hirtosa*. In the algal zone, *Ecklonia radiata* and species of *Cystophora* are abundant, and many of the rocks are covered with a lithothamnion coralline algal encrustation. The abalone *Haliotis roei* is dominant here. In general the West Australian Province differs from the Flindersian Province in lacking a number of the characteristically cooler-water forms, and from the Peronian in lacking a number of the characteristic species of that similarly situated province of the eastern coast, whose places are taken by endemic western species. The long stretches of cooler waters along the South Australian coast and of warmer water to the north have apparently been rather effective barriers to active interchanges between the biotas of these two Australian temperate areas.

New Zealand provinces. As shown in Fig. 16-4, New Zealand is situated in such a way relative to Australia, and to the oceanic currents in the region, that it would be expected to recruit biota rather strongly from Australian centers of distribution, but that exchange in the reverse direction is likely to be minimal. Its position, lying obliquely across the latitude of the Antiboreal Convergence, results in rather different water temperature conditions in the northern and southern areas.

Knox (1963) recognized three marine biotic provinces—a coldwater Forsterian Province, of the southern part of South Island and Stewart Island; an intermediate Cookian Province, embracing the northern two thirds of South Island and approximately the southern half of North Island; and finally the warmer-water Aupourian Province, comprising the northern portion of North Island.

The **Forsterian Province** is a mixed subantarctic and cold temperate region differing in several respects from the Cookian Province with which it intergrades. It possesses a strong endemic element, especially among the algae, and numerous more widespread subantarctic species are also well represented. The endemic species include *Hymenena semicostata, Brogniartella australis,* and *Apophloea lyallii*. Among the prominent subarctic species, *Pachymenia lusoria, Delesseria crassinerva, Myriogramme crispata,* and *Codium dimorphum,* together with such circumpolar antarctic species as *Adenocystis utricularis, Ballia scoparia, Chaetomorpha darwinii, Schizoseris davisii,* and others, are well represented. The echinoderm fauna strongly intergrades with that of the Cookian. Of the more than 60 species of shelf echinoderms known from the region, only about 10% are regarded as typical Forsterian species. Subantarctic species include *Trachythyore amokurae* and *Oncus brevidentis*. Characteristic Forsterian molluscs include the limpet *Cellana strigilis* var. *redimiculum* and the siphonariid *Kerguelenella stewartiana*. The supralittoral zone is characterized by the blue-green alga *Verrucaria* and an encrusting red alga, *Hildenbranchia,* in the most southern areas. The littorinids *Melarhaphe cincta* and *M. oliveri* are abundant. Throughout the midlittoral zone, bands of algae are superimposed on the barnacles and mussels. *Chamaesipho columna* is the dominant barnacle. *Bostrychia arbuscula* extends through the barnacle zone, and below this is a band of *Apophloea lyallii*. The leafy red alga *Pachmenia lusoria* forms a band in the lower midlittoral, and patches of the mussels *Modiolus neozelandicus* and *Mytilus edulis* var. *aoteanus* also occur. In some areas, algae grow so densely that they largely exclude barnacles in the midlittoral zone.

The **Cookian Province** comprises mixed cold temperate waters of the central regions of New Zealand, with some gradation in temperatures and differences in biota along the north-south extent of the province, as well as some characteristic differences between the biota of the east coast and that of the west coast. On the east coast the intertidal comprises a supralittoral zone with an upper band of lichens, below which the littorinid snails *Melarhaphe cincta* and *M. oliveri* are dominant. The upper littoral is marked by a barnacle zone comprised of *Chamaesipho columna* and, in the northern portion, *C. brunnea*. In areas of strong wave action, *Elminius plicatus* is found with *Chamaesipho*. A zone of mussels, species *Modiolus neozelandicus,* may mix with the lower part of the barnacle zone, and below these occur *Mytilus edulis* var. *aoteanus*. Tubes of the serpulid worm *Pomatoceros cariniferus* may form dense encrustations in localized protected areas. Limpets are abundant throughout the midlittoral zone, sometimes extending into the upper littoral, especially species of *Cellana, Patelloida, Sypharochiton,* and *Siphonaria*.

Along exposed coasts the lower littoral is dominated by large fucoids, *Durvillea antarctica* and *D. willana,* with a variable undergrowth of corallines and *Lithothamnion*. In more protected places species of *Cystophora, Carpophyllum,* and *Xiphophora* tend to replace the *Durvillea*.

One of the major differences between the east coast and the west coast in the Cookian Province is the absence of the large brown algae from the lower littoral and upper sublittoral zones of the west coast. The horse mussel, *Modiolus,* is found in more prolific growth, sometimes covering entire upper and midlittoral zones and extending into the lower littoral.

The **Aupourian Province** of northern New Zealand is an area of transitional warm temperate waters. The supralittoral zone is characterized by an upper band of yellow and grey lichens, below which the littorinids *Melarhaphe cincta* and *M. oliveri* dominate the rocks, with black patches of blue-green algae and the lichen *Lichina pygmaea*. *Chamaesipho brunnea* is the main barnacle of the upper littoral, giving way to *C. columna* and *Elminius plicatus* in the midlittoral zone, along with the limpets *Cellana ornata* and *C. radians,* and the whelks *Lepsiella scobina* and *Lepsia haustrum*. Bands or patches of the algae *Apophloea sinclairii* and *Bostrychia arbuscula* occur among the barnacles. In sheltered positions the oyster *Saxostrea glomerata* dominates the lower midlittoral zone. In other places the serpulid *Pomatoceros cariniferus* occurs at this level. The brown algae *Xiphophora chondrophylla* and *Carpophyllum maschalocarpum* dominate the lower littoral, with undergrowth of corallines and various smaller red algae. Below these the large brown algae *Lessonia variegata, Ecklonia radiata,* and species of *Carpophyllum* and *Sargassum,* and the red algae *Pterocladia lucida, Melanthalia abscissa,* and *Vidalia colensoi* occupy the upper sublittoral zone.

Island provinces. The smaller provinces designated by Knox in the region are three island groups north of New Zealand in relatively warm waters and three to the south in cold water.

The northern islands are Lord Howe Island, off New South Wales **(Philippian Province);** Norfolk Island, directly north of New Zealand **(Norfolkian Province);** and the Kermadec Islands, northeast of New Zealand **(Kermadecian Province).** Lord Howe Island marks the southermost occurrence of coral reefs.

The southern islands are Macquarie Island, regarded as an extreme eastern outpost of the **Kerguelian Province;** three island groups and an island south of New Zealand—the Auckland Islands, Campbell Island, the Bounty Islands, and the Antipodes Islands—grouped together as the **Antipodean Province;** and finally the Chatham Islands, southeast of new Zealand, constituting the **Moriorian Province.** One of the most striking endemics of these southern islands is the monotypic penguin, genus *Megadyptes*.

Southern South America

Along the west coast of South America the north-flowing Humboldt Current, plus extensive upwelling, combine to keep the surface temperatures of coastal waters low along almost the whole extent of the continent and to make the waters off Peru and northern Chile extraordinarily productive.

The fact that there is an uninterrupted continental shelf extending from the North Pacific to southern South America has also strongly influenced the coastal fauna. A number of groups from the north show submergence in the tropical regions and reappear in the southern fauna, either as the same or as closely related species. Others, which made the crossing at some time in the past, are not now found in the tropics but have representatives both north and south. The phenomenon of north-south **bipolarity** is probably better shown along the west coasts of the Americas than anywhere else.

Southern South America and a number of oceanic islands in the region have a colder-water antiboreal biota distinct from that of the more temperate regions to the north. The antiboreal region extends north to about the level of Chiloe Island on the west, but its extent along the southeastern coast of South America is less clear because this region is perhaps the least well known of any major coastal region.

In the antiboreal region, winter surface temperatures (August) range from 4.5° to 9.5° C. and summer temperatures (February) from about 8° to 16° C. The temperate regions show winter temperatures of 10° to 16° C., rising as high as 19° to 20° C. in summer. There is a rather abrupt transition in the north to subtropical and tropical waters where the Humboldt Current turns west to become the South Equatorial Current.

Approximately every seven years a disastrous change in current patterns brings warm water of the Equatorial Countercurrent farther south than usual in the winter along the coast of Ecuador, converging with and overlying part of the Humboldt Current. As these waters mix, the exceptionally rich plankton and other animals in the northern portion of the Humboldt Current are destroyed wholesale. Large amounts of organic matter decomposing in the water and sinking to the bottom produce sufficient hydrogen sulfide to poison the water and blacken the paint of ships passing through the area. Many of the guano-producing birds die, and others leave the area. The coastal weather is adversely affected, producing heavy rains that cause serious erosion of the relatively unprotected soils. This phenomenon is known as "El Niño."

The antiboreal fauna of South America has a high degree of endemism, about 52% of the echinoderms and 73% of the fish being endemic. The proportion of endemic genera is much less—about 8% for echinoderms. The largest proportion of nonendemic elements are antarctic or subantarctic elements, the largest contingent being 20 species, or 16% of the total shared with southern Australia, southern New Zealand, and the oceanic islands to the south. This figure is relatively high, considering the vast distances separating South America from these regions.

A few of the most abundant or characteristic elements of the South American antiboreal fauna include the starfish *Anasterias antarctica* and *A. pedicellaris,* endemic to the area, and the antarctic *Cycethra verrucosa*. The antarctic ophiuroids *Astrotoma agassizii, Gorgonocephalus chilensis, Ophiomyxa vivipara, Ophiomitrella falklandica, Ophiactis asperula,* and *Ophiocten amitinum,* the latter two being especially abundant, are also characteristic. Sea urchins include 6 endemic monotypic genera, one of them, *Tetrapygus,* extending northward along the west coast as far as Peru. The endemic genus *Austrocidaris* comprises 3 species.

Among the fish the connection with the Antarctic seems to be weaker, only about 3% being common to both regions, compared with 38% among echinoderms. The fish fauna of antiboreal South America seems to occupy a more independent position. Some of the characteristic endemics of the region are 2 genera—*Cottoperca,* with 3 species, and *Notothenia,* with 8 species—and 17 other species of the Nototheniiformes. Six genera of Zoarcidae, including *Austrolycus,* with 4 species, and

Crossostomus, with 3, are also endemic.

Some examples of forms showing bipolarity include, among the fish, the families Zoarcidae, Liparididae, and Scorpaenidae. Among the invertebrates several genera of the crustacean family Lithodidae, such as *Lithodes* and *Paralomis,* and numerous echinoderm genera, including *Florometra, Antelaster, Ctenodiscus, Diplopteraster, Echinus, Solaster, Porania, Bathybiaster,* and *Ophiocten,* show this phenomenon. The genera *Henrica* and *Pteraster,* common in both Europe and North America, show submergence in the tropical regions.

Southern Africa

The biotic zonation around the southern extremity of Africa is complicated by the meeting here of warm southwest-flowing currents from the Indian Ocean with the cold waters of the West Wind Drift. The Mozambique Current, called the Natal Current farther south, flows southward along the eastern coast of Africa, keeping the shore temperatures essentially subtropical, about 20° C., to about Algoa Bay. South of Cape Agulhas, the southernmost tip of Africa, where this warm current is also known as the Agulhas Current, it meets and is split up by the cold West Wind Drift. Some of the cold water from the West Wind Drift wedges itself between the Agulhas Current and the mainland, flowing close to the southern coast as far as Algoa Bay. At Cape Agulhas the immediate coastal waters may be several degrees warmer than the offshore waters, which farther out give way to waters of subtropical temperatures. The patterns vary somewhat with the seasons and with vagaries of the local weather. Here, more than in most areas, it becomes clear that in the biogeography of the oceans, the concern is with certain kinds of water rather than with fixed areas, boundaries and locations.

The western coast is bathed by cold, north-flowing waters of the Benguela Current and further cooled by upwelling, a situation analogous in many respects to that of western South America. Water temperatures here are in the neighborhood of 14° to 15° C. in the south, increasing northward and becoming subtropical beyond the region of Cape Frio. The southwest African temperate region from Cape Frio to St. Helena Bay is usually termed the **Namaqua Biotic Province,** following Michaelsen (1915), who contributed much to our early understanding of the biology of this region. Being a region of high illumination, with cool coastal currents fertilized by upwelling, the waters here are rich in phytoplankton and are productive. Dinoflagellates in particular become seasonally enormously abundant. ''Red tides'' are a regular recurrent feature between Walvis Bay and Lüderitz Bay in the spring, often causing high mortality among the fish and invertebrates of the area. They occur sporadically also in other areas along the southwest African coast. Since the organic matter resulting from these mass mortalities accumulates on the sea floor, a nearly azoic benthic region smelling strongly of hydrogen sulfide and containing the bodies and skeletons of animals killed during the red tides has formed. It is approximately 30 km. wide and extends down to depths of 140 to 150 meters. During the rest of the year pelagic organisms and fish repopulate the upper strata of water, but the benthos is almost extinguished.

The region is generally unfavorable to a high development of the benthos because of a sandy sea floor and lack of protection from the waves. In the calmer, deeper offshore water, decomposition of algae and other organisms contributes to the stagnation of the water and the smothering of benthic organisms. Along the immediate shoreline and in the intertidal zone, on the other hand, the fauna is rich. The water movements prevent the deposition of sapropelic mud, and the organisms have the advantage of nutrient-rich waters with abundant plankton. The fauna in general shows affinity with that of the rest of the

Atlantic Ocean rather than with that of the Indian Ocean, and numerous species appearing here also occur along the European coasts. Examples of such species are the brittle star *Ophiothrix fragilis,* the tunicate *Leptoclinides capensis,* the molluscs *Venus verrucosa* and *Cerithium vulgatum,* and the polychaetes *Nephthys hombergi* and *Sabella pavonina.* Other characteristic species of the intertidal fauna are the molluscs *Mytilus meridionalis, Cominella delalandei,* and *Patella granatina.*

A number of echinoderms characteristic of the North Atlantic are also represented here by twin species or varieties. Some of these are *Astropecten irregularis* var. *pontoporaeus, Hippasteria phrygiana* var. *capensis, Marthasterias glacialis* var. *rarispina* and var. *africana, Brissopsis lyrifera* var. *capensis,* and *Echinocardium flavescens* var. *capense.* Similar examples could be cited for other groups.

Darwin postulated that temperature decreases during the ice ages may have made possible a continuity of distributions of the northern and southern regions that has been interrupted by the rewarming of tropical warm waters since the last glacial epoch. This cooling effect seems to have been especially strong in the eastern Atlantic.

Another group of species of southern and soutwestern Africa seems to have been derived from distant regions of South America or the southern Atlantic islands by way of the West Wind Drift. Still others have a circumpolar distribution, being found in the colder parts of the Australasian region, in southern South America, and in southern and southwestern Africa. A few of these widely distributed forms, carried around the world by the West Wind Drift, include the lugworm *Arenicola assimilis,* the crustaceans *Squilla armata* and *Jasus lalandii,* and the ascidian *Corella eumvola.* Fish of the genus *Callorhynchus,* described from southern Africa, South America, and southern Australia and New Zealand, may be merely widely separated varieties of a circumpolar species.

Southern, or Antarctic, Ocean

The Southern Ocean is more than the confluence of three great oceans, the Pacific, Atlantic, and Indian. It has an identity of its own, even though its northern boundaries cannot be precisely defined. Perhaps the most logical way to determine the boundary, if there must be a boundary, would be to include all the waters involved directly in the West Wind Drift, which would place the northern limit at the Antiboreal Convergence. Biologically, however, the Antarctic Convergence is the sharpest line of demarcation between typically antarctic surface waters and those of other oceans. The strongest flow in the West Wind Drift coincides with the Antarctic Convergence.

Unlike the other oceans, the Southern Ocean is continuous around the world, the narrowest part being the 1000 km. separating the southern tip of South America from the Antarctic Peninsula (formerly the Palmer Peninsula). Elsewhere it is more than 2000 km. from the coast of Antarctica to any continental landmass.

Although the Southern ocean occupies more than 75 million km., or 22% of all the ocean area of the world, its heat content is only 10% of the oceanic total. The heat exchange between Antarctica and the rest of the world profoundly influences the nature and patterns of the world's climate, and the temperatures and current patterns of the world's oceans.

Near the shores of Antarctica, especially in the embayments forming the Ross, Amundsen, Belingshausen, and Weddell seas, large areas are covered by ice much or all of the year. Cold water from the coastal areas sinks to the bottom, forming deepwater abyssal slow-moving currents trending northward along the low contours of the ocean floors, detectable in the Atlantic even north of the equator. The sinking water is replaced by upwelling from deep waters just above the north-trending bottom water, moving south as a countercurrent and upwelling near Antarctica as the **Antarctic Divergence.** On reaching the surface layers, most of this water is caught up in the West Wind Drift and begins to move in a northeast direction until it reaches the Antarctic Convergence, where it sinks beneath the warmer antiboreal water, forming a layer of antarctic intermediate-depth water moving down and to the north between the Antarctic Convergence and the Antiboreal Convergence. The surface antiboreal waters between the two major convergences are, in general, moving in a southeast direction, comprising the northern portion of the West Wind Drift.

Along much of the shoreline of Antarctica, especially in the seas, there are large eddies, so that the coastal water is moving west rather than east.

Fossils discovered in Antarctica ever since the first explorers ventured ashore near the turn of the century demonstrate clearly that this continent has not always been covered with ice. In fact, the succession of life there has been shown to be much the same as in the other major Southern Hemisphere continents until well beyond the time of the appearance of seed plants relatively recently in the geological history of the world. Proof can be seen in fossil trees with with well-defined rings, indicating rapid growth in a temperate climate. These trees are another facet in the growing body of evidence that Antarctica, rather than having always been an isolated landmass, may, until about the end of the Mesozoic era, have been at the heart of a great southern continent comprising South America, Africa, Antarctica, Australia, and perhaps part of southern Asia. The splitting up of this former supercontinent, Gondwanaland, and the rafting of Antarctica to the geographical pole of the planet, making this pole a thermally isolated region rather than part of the world ocean, seems to have been the precipitating cause of the glaciation that still holds the continent in the grips of a prolonged ice age. The former closer association of the lands now comprising the continents of the Southern Hemisphere would also go far to explain the apparent worldwide distributions of some of the ancient groups of animals and plants without the necessity of hypothesizing numerous and improbably long land bridges, as was frequently done by earlier paleogeographers.

Antarctic Ocean fauna. The high fertility of the surface waters of the Antarctic resulting from continuous large-scale upwelling, the extreme nature of the environment, and the short growing season through the antarctic summer combine to produce the world's most dramatic outbursts of phytoplankton growth, dominated by rapidly reproducing species of diatoms. The zooplankton, too, tends to be dominated by a relatively small number of species of rapidly reproducing forms.

One species in particular, the red euphausiid shrimp *Euphausia superba,* commonly called krill, tends to dominate large areas of surface waters during the antarctic summer. Krill is the major link between the phytoplankton and the larger animal life in the Antarctic. One of the larger members of the macroplankton, it is a little over 5 cm. long, large enough to be preyed on directly by many kinds of fish, squid, birds, and even mammals. It thus provides for these animals an unusually efficient one-step food chain enabling a much larger biomass of large animals to be supported than would be the case if most of them preyed on animals of intermediate size. The world's largest animals, the great whalebone whales—the blue whale, finback whale, humpback whale, and others—congregate here during the antarctic summer. Whalebone whales are basically immense animated plankton nets having the proper mesh to capture macroplankton organisms. It has been estimated (Pequegnat, 1958) that a mature blue whale must eat about 3 tons of krill daily during its feeding season. It is estimated that, prior to recent intensive whaling in the region, at least half a million whales congregated in the Antarctic each summer, which would mean that at a conservative estimate at least 1350 million tons of krill, about 1000 pounds per acre, were produced to support them, since the whales certainly did not consume more than 20% of the total annual production. Actually, the figure must be much higher because of the hordes of other predators that also feed upon the krill.

Krill is an ideal food for predators such as the whales. Since the euphausiids congregate in immense shoals in the surface waters, only a minimum of energy is required to fish them in great quantities. Furthermore, they are a nutritious, high-protein food.

Hardy and Gunther (1935), studying the distributions of phytoplankton, krill, whales, and dissolved phosphate (a plant nutrient) in the region of South Georgia Island, made the interesting discovery that these distributions are closely correlated and that from a knowledge of any one of them, the distributions of the other three could be accurately derived. Clouds of phytoplankton are inversely correlated with high phosphate content of the water (because most of the phosphate has been used up in producing the phytoplankton). High phosphate content of the water, shoals of krill, and the presence of blue whales were positively and significantly correlated. Hardy was even able to predict, solely from records of the distribution of whales, during a year in which their distributions in the region were unusual, what the content of as-yet-unopened bottles of plankton from a plankton survey made the same year would be.

The life histories of the krill and some of the other members of the macroplankton are nicely timed to the circulation of waters in the Antarctic Ocean. During the summer the actively reproducing adults are drifting in a generally northeast direction with the surface waters, until they and the eggs and young come to the Antarctic Convergence. Here they are carried below the surface. The larval stages sink into the intermediate depths in which the water is drifting south. The waters at the depths in which they are drifting (being a thicker layer of water moving the same distance and delivering the same volume of water in a given time) are moving more slowly than the surface waters, so that the young euphausiids begin coming to the surface as newly transformed adults the second summer after their submergence. Thus, by changing their level in the water, populations of drifting planktonic organisms are able to maintain themselves in a favorable region in spite of the continual movement of the surfaces waters out of the region. The West Wind Drift has caused a general circumpolar distribution of most pelagic antarctic organisms, but even so, there are areas where localized conditions result in greater-than-average concentrations.

The krill are fed on directly by whalebone whales, the crabeater seal, Adélie penguins, hordes of flying birds such as petrels, fulmars, and terns, and hordes of small fish and squid.

The small fish and squid form the principal food of larger fish, emperor penguins, fishing birds such as the shag, and most seals, such as the Weddell seal, Ross's seal, and the leopard seal. The skua feeds on young penguins, steals fish from other fishing birds, and takes some small fish and squid. Sperm whales feed principally on squid. Leopard seals prey on penguins, fish, and other seals. The voracious killer whales attack any large animal, including other whales larger than themselves.

The benthos is supported, as everywhere, by mixed plankton in areas shallow enough so that they are in depths occupied by phytoplankton and by thinned-out zooplankton and organic detritus from above in deeper water. The unusual richness of the Antarctic Ocean enables it to support a luxuriant benthos, some groups such as sponges being better developed here than anywhere else. At the shoreline the ice keeps the rocky intertidal and subtidal free of attached animals and large algae so that the margin of the sea is barren. In many areas this barrenness extends downward approximately 10 meters below the level of the floating ice, to the lowest level at which ice crystals will form on a submerged object; the formation of ice crystals and their detachment as they grow perhaps prevent the establishment of small, early stages of attached benthos. Below this level there is an abrupt colonization of the rock with a rather rich benthos.

Although feeding is richly provided for in the Antarctic, nesting sites for birds are at a premium. The rookeries tend to be densely populated with thousands of birds, which use every available site during their nesting seasons.

There is a high level of endemism in the fauna of the Antarctic Ocean. According to Ekman (1953), about 73% of the 256 species of echinoderms known from the region are endemic to the Antarctic Ocean, and approximately 27% of the genera, several of them monotypic, are endemic. Among the fishes, 90% of some 78 species and 65% of the 37 genera are listed as endemic. The family Notothenidae is an especially noteworthy group of endemic antarctic fish.

One of the most extreme and curious adaptations to the cold, oxygen-rich antarctic waters occurs in certain fish of the family Chaenichthyidae, related to the notothenids. These so-called **icefish** attain a length of 60 cm. or more in some species and are large-headed, thin-bodied, rather sluggish fish with no red blood cells and no hemoglobin. About 16 species are known, all of them living in cold, stable, food-rich antarctic waters, where they do not have to expend great amounts of energy obtaining food. The oxygen dissolving into their colorless blood plasma from the surrounding water is sufficient to maintain their sluggish metabolism. Examination of kidney and spleen tissues, where blood is formed in other fish, shows no evidence of the formation of blood cells. Thus their anemia is probably total, at least in some species. The blood volume is, however, unusually large for fish of their size and weight, enabling them to transport more dissolved oxygen than would otherwise be the case. Only in stable, cold, oxygen-rich, food-rich waters, such as those of various antarctic areas, does it seem that such an anemic race of fish could even begin to evolve.

Summary

The oceans of the Southern Hemisphere are all broadly and directly continuous with the Southern Ocean, which girdles the globe north of Antarctica. The great circumpolar east-flowing West Wind Drift carries surface waters around the world, giving rise to north-flowing cold currents along the western shores of South America and Africa, where it is partially interrupted by the southern tips of these continents. Because of this drift, organisms with centers of dispersal in this part of the world tend to disperse toward the east from such centers or to the north along western Africa and South America, but not to the west.

In comparison with the Arctic Ocean, the Antarctic Ocean is far richer in its fauna, particularly the benthic elements, and displays a much greater degree of endemism, probably because its formation took place long before that of the north polar sea. Its relationships to other oceans create the conditions of massive interchange of waters, with extensive upwelling of water from intermediate depths, that make the surface waters exceptionally productive. The greater productivity of the surface waters is, in turn, reflected in higher productivity at all depths than would otherwise be the case.

Although the Arctic Ocean has a vastly greater extent of shallow-water shelf areas, which are directly continuous with richly populated continental shelves of both sides of the boreal Atlantic and also with those of the northern Pacific, and although for the most part the shores of Antarctica slope off into deep water much more abruptly and the Antarctic is isolated by great distances, by the direction of the surface current systems, and by great differences in ecological conditions from other regions with shallow-water shelf faunas, nonetheless the shelf fauna of the Antarctic is much richer in species as well as in biomass than is that of the Arctic. Clearly Antarctica has been a center for the development of a marine fauna for a long time and has had an uninterrupted cold climate ever since at least the early Tertiary.

MESOPELAGIC ZONE

Depths ranging from about 200 to 1000 meters are below the zone where there is sufficient light to support active photosynthesis but not entirely beyond the reach of light. These depths are also of more uniform, usually colder, temperature than the overlying waters and of slightly higher salinity. There are no functional producers here, although some chlorophyll-bearing diatoms, flagellates, coccolithophores, etc. are regularly found in these depths, apparently living heterotrophically.

The cooler temperature imposes a slower metabolism and hence a lower food demand; thus animals living in these depths live on less food than those living near the surface. They are still near the food-producing euphotic zone and are the first to benefit from the continuous settling and transport of organic nutrients from above. Since the increased density and viscosity of the water also slows down the sinking of organisms and particles from the upper layers, this zone tends to become a storehouse for sinking organic matter.

Many of the animals in the upper portion of this zone undertake diurnal migrations into the food-rich layers above them at night, returning to the twilight zone during the day. Some animals from still deeper waters migrate into the twilight zone during the day, when it contains a maximum of animals, to feed.

The mesopelagic zone has a rich animal life, composed mostly of predators, scavengers, and some detritus feeders. Many of the fish are small, with large eyes and mouths, with patterns of luminescent organs and, in some, long tactile organs. Among the most typical are species of the genus *Cyclothone* (Fig. 11-19) and members of the family Myctophidae. Other characteristic members include species of *Argyropelecus, Stomias,* hatchetfish, and Uranoscopidae.

In addition to the fish, there are great numbers of squid, swarms of euphausiids, copepods, and other crustacea. Certain coelenterates, such as siphonophores and jellyfish of the order Coronatae, are also often abundant here.

There is a tendency for the animals to be black or dark reddish, with luminescent organs usually ventral or lateral in position. Many of the animals are extremely sensitive to light, have upward-directed eyes, and show marked preference for light in rather narrow ranges of luminosity or intensity, usually dim. The daily vertical movement of isolumes in the water is probably the principal stimulus for the diurnal vertical migrations of many of them.

Because these depths tend to be of more uniform character than the surface layers, but yet suitable for support of a rather abundant community of animals freer from geographical barriers than those of any other zones, they contain the most extended animal communities to be found anywhere in the world. A relatively uniform community of mesopelagic animals commonly underlies several well-marked communities of epipelagic organisms separated from each other by temperature and other surface water differences.

ISOLATED SEAS

Mediterranean Sea

The Mediterranean is a special case in many respects. The Gibraltar Ridge prevents the exchange of all but the superficial layers of water with the Atlantic, so that the cool, deep, well-oxygenated Atlantic water is prevented from entering. There is a flow of surface water into the Mediterranean and of subsurface, more saline water out over this sill. Strong evaporation increases the salinity sufficiently to cause the warm surface water to become denser and to sink; thus the deep water in the Mediterranean is warmer than correspondingly deep water in the adjacent Atlantic. The temperature difference, the difference in salinity, the outflowing current of subsurface waters, and the height of the sill of the Gibraltar Ridge all combine to prevent the entry of Atlantic deep-water species into the Mediterranean. Thus there is no stenothermal coldwater abyssal fauna in the Mediterranean; there is only an archibenthic-abyssal fauna, which diminishes in richness toward the eastern part of the sea. Because of the unusual temperature distribution, littoral and sublittoral species extend their ranges deeper than they do in most areas, and the archibenthos extends all the way to the bottom.

Another factor that greatly affects the character of the present biota of the Mediterranean is the fact that as recently as about 8 million years ago, the Mediterranean became completely separated from the Atlantic and transformed into a great inland sea. This sea gradually evaporated until the Mediterranean basin became a hot dry below-sea level desert area. At first it contained a series of great salt lakes, which gradually shrank and became a more and more concentrated brine, until finally about 6 million years ago the basin was a hot dry wasteland comparable to Death Valley. During this time the rivers flowing into the basin cut deep gorges that subsequently became largely filled with sediments after the Mediterranean became refilled during the Pliocene, when connection with the Atlantic, via Gibraltar, was reestablished.

The refilling of the Mediterranean basin must have been an awesome spectacle, with the world's greatest waterfall, estimated to have been approximately a thousand times bigger than Niagara Falls, cascading about 10,000 cubic miles of water per year into the basin for a century or more until it was refilled.

The drying of the Mediterranean of course completely eliminated its marine biota. The present marine biota consists wholly of forms that have come in over the Gibraltar rise from the Atlantic subsequent to the reestablishment of the Atlantic connection, plus a few species in the eastern end of the Mediterranean that have recently entered from the Red Sea by way of the Suez Canal, and a few species accidentally or deliberately introduced by man. There has never been any direct connection of the present Mediterranean waters with extremely deep bathypelagic or abyssal waters of any ocean.

Red Sea

The Red Sea, of more recent origin than the Mediterranean, displays similar conditions even more intensified. It is only 2800 meters deep in its greatest depth and is separated from the Indian Ocean by a sill no more than about 100 meters in depth at its lowest point. Thus the deep water of the Indian Ocean is excluded. Evaporation here is even more intense than over the Mediterranean, and since the sea is shallower, the temperatures and salinities along the bottom are even higher than those in the Mediterranean. Bottom temperatures are about 21.5° C., and the salinity is about 40.5‰. Many groups of stenohaline animals, such as sponges, echinoderms, and ascidians, found in the shallower waters disappear before reaching the bottom. The deeper parts of the bottom are colonized chiefly by gastropods and decapod crustaceans of types that in other regions live on the shelf or are, at most, archibenthic. For the Red Sea as a whole, at the species level there is a rather high degree of endemism, indicating that approximately a million years of relative isolation is sufficient to produce numerous changes at the species level. The level of endemism of higher taxa is low, however, the fauna being clearly related to that of the Indo-Pacific.

Sea of Japan

The Sea of Japan, at present, has a depth of some 400 meters, but it is separated from the Pacific by sills no deeper than 165 meters. During comparatively recent periods of the ice ages, it is thought to have undergone a history of elevations during the glacial periods and subsidence during the interglacial periods, which at times converted it into a great freshwater lake. This relatively recent history and its present isolation from the deep waters of the Pacific account for the remarkable fact that in spite of its considerable depth and its nearness to regions with well-developed abyssal faunas, the Sea of Japan lacks a specifically abyssal fauna, the deeper parts of the benthos being colonized only by a number of eurybathic species, some of which have descended from the shelf to about 3500 meters.

Other seas

The Black Sea, Caspian Sea, and Sea of Azov are remnants of the former Sarmatic Inland Sea and have been briefly discussed on p. 370.

PART THREE

TIME AND MARINE LIFE

CHAPTER

17 THOUGHTS ON THE ROLE OF THE SEAS IN THE ORIGIN AND EARLY EVOLUTION OF LIFE

In considering origins our approach must of necessity be more indirect and speculative rather than observational and descriptive. The origins of life on earth have receded so far into the remote past that they wholly preclude any direct attack on the problem. Furthermore, it was not until living things had already become widespread, highly organized, and diversified that they began to leave recognizable traces. Most of the traces left by the life of past ages are still locked up in ancient rocks or have been altogether destroyed.

The best we can do is to study the hierarchies of living things as we know them today from the simplest viruses to the most complex animals, study the properties of living matter as such, and study the progression through time of living things as we find it recorded in fossil-bearing rocks. By extrapolating backward from the various trends observed in the evolution of living matter and extrapolating forward from what we can deduce to date with respect to abiotic chemical and organochemical evolution, we can try to arrive at some kind of conceptual meeting ground with respect to the conditions under which such an amazing system could have arisen and the period of time involved.

More and more in recent decades the conviction has grown that life is not something set apart from and, as it were, foreign to, the material universe, merely using some of the materials in a mysterious way for its purposes. Rather, life is just as much an inherent property of the universe as is any other manifestation of the matter-energy-time complex. According to this view, given suitable conditions for a sufficiently prolonged period of time, matter will inevitably manifest the properties of life. Suitable conditions are, so far as we know, exceedingly rare in the universe. The universe is, however, inconceivably vast, and it seems probable that there are millions of planets supporting some forms of life. The complexity of living things is so extraordinarily great and their evolution involves such long periods of time and is so chancy and irreversible that it is unlikely that any of the same species of animals or plants would evolve independently on two different planets. In this sense, it is probable that man stands alone. Yet any planet on which life arose and flourished for long periods would, of necessity, evolve groups of organisms ecologically and possibly structurally comparable to those found on earth.

KINDS OF ORGANISMS

Living things as we know them today can be arranged into six major groups, or kingdoms, based on different orders of complexity or basically different modes of nutrition: the **noncellular organisms,** the **procaryotic cellular organisms** or **Monera,** and four groups of **eucaryotic organisms:** the **Protista,** the **Metaphyta,** the **Fungi,** and the **Metazoa.** Recently such basic differences

FIG. 17-1. Animal "gardens" 18 meters below the surface of the Caribbean Sea near St. John, Virgin Islands. (National Park Service, M. W. Williams photograph.)

between the "methane bacteria" and other Monera have been brought to light that some biologists propose separating them as another kingdom.

Noncellular organisms

The noncellular organisms are the **viruses, episomes,** and **plasmids** as we know them today. In its active, so-called vegetative stage, a virus consists of a single strand or circle of nucleic acid immersed in the chemically rich, highly organized environment of a cell of some higher organism. There it takes advantage of the host's metabolic machinery, replicates itself at the expense of chemicals present in the host cell, and finally directs the synthesis of specific protein, which eventually coats the strands of virus nucleic acids individually forming virions that are able to survive outside the host cell and to infect new cells of the same type. Some viruses, notably the bacteriophages, may vary this cycle by joining the nucleic acid ring of the host cell, replicating synchronously with the replication of the host nucleic acid, and becoming to all intents and purposes genes, or functional units, of the host nucleic acid.

Another category of entities, even simpler than the viruses, are small rings of nucleic acid known as episomes and plasmids, which exist independently of the host nucleic acid in the cytoplasm and replicate there. Episomes and plasmids must be passed from infected cells directly to their daughter cells or to non-infected cells, usually by conjugation or direct contact of the cells, since they do not direct the formation of specific proteins and become complete virions. Like the virus nucleic acid, episomes in some cases show a definite relationship to the genome of the host cell and may behave as if they were part of the cell's own genetic mechanism. Plasmids are similar to episomes but do not integrate at any time with the host cell nucleic acid. The same particle may act as an episome in one strain of bacteria and as a plasmid in another. Both viruses and episomes may, on rare occasions, exhibit the phenomenon of crossover with host-cell genetic material, exchanging one or two genes of their own for one or two genes of the host cell's nucleic acid.

Some biologists prefer not to regard precellular stages in the evolution of life, or the viruses, episomes, and plasmids as organisms—reserving the term for those forms that display the cellular level of organization. It is to be noted that both episomes and viruses exhibit properties of life only when immersed in the extremely rich, highly organized, biochemical environment of a host cell. They show a definite relationship to the nucleic acid of the host cell, are specific regarding what kinds of cells they can infect, and consist only of nucleic acid or of nucleic acid and a specific protein coat, or at most a small number of specific proteins Another major difference between the viruses and cellular organisms is that in viruses only the nucleic acid replicates directly, whereas the protein is synthesized through redirected activities of the host cell's metabolic machinery, after which the complete virion is assembled. The cells of cellular organisms, on the other hand, duplicate themselves as complete units.

It seems probable that viruses, episomes, and plasmids as we know them are not holdovers from the primitive, precellular, ancestral forms of life but are, rather, detached bits of genetic material of cellular organisms continuing to function and evolve independently of the genome of the cells in which they live. Nonetheless, they may, by analogy, give some insight into the probable order of structural complexity and environmental needs of ancient precellular entities.

There is a big gap between the postulated precellular entities, presumably little more than gigantic macromolecules of nucleic acid replicating in a medium rich in the organic chemicals necessary for building their structure, to even the simplest of cellular organisms, with its hundreds of proteins and enzymes, its carbohydrates, lipids, and high-energy phosphate compounds, its intricate membrane system, and its digestive, respiratory, and excretory processes, all of which are organized into a functioning whole that reproduces as a complete organized unit, controlled in some way by the intimate internal ordering of coded units in the nucleic acid that constitutes its genome. Once the cellular level of organization was achieved, it was enormously successful and has continued to grow, proliferate, differentiate, and cover the earth with a living mantle that is largely responsible for the physical and chemical character of the earth's surface, its atmosphere, and its waters.

Procaryotic cellular organisms: Monera

The simplest of the cellular organisms do not have a nucleus with a definite nuclear membrane, nor do they have a number of chromosomes. The nucleic acid consists apparently of a single long strand of nucleic acid that is structurally a closed circle, although it may be complexly folded and not give the appearance of a circle in electron micrographs. The procaryotic cellular organisms comprise the bacteria, actinomycetes, and blue-green algae. The latter differ from the first two groups in having developed a mechanism of photosynthesis. Certain specialized groups of sulfur bacteria are also photosynthetic but by a different mechanism that does not involve the production of free oxygen as a by-product.

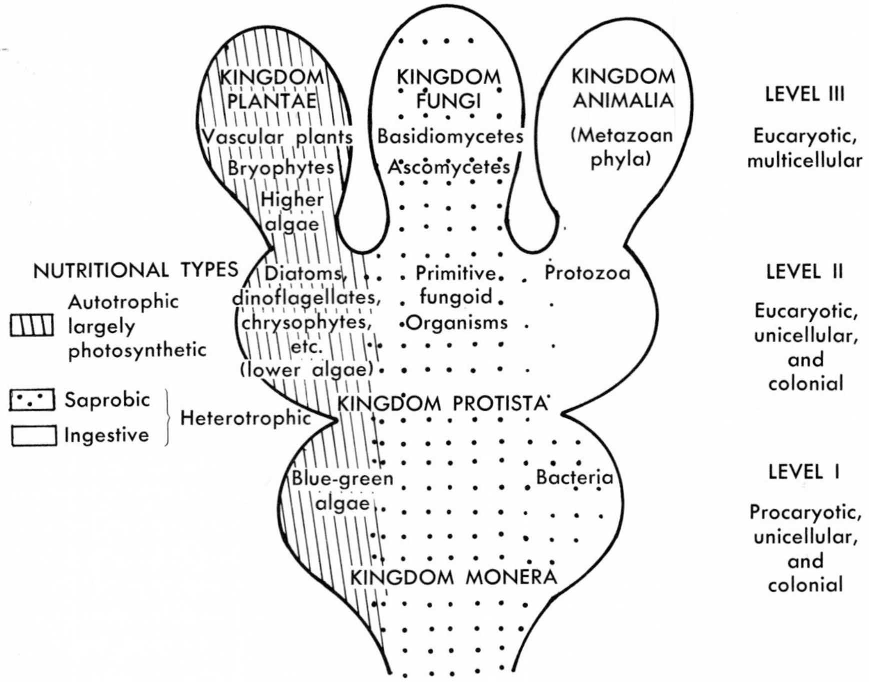

FIG. 17-2. Diagram of the five-kingdom arrangement of cellular organisms.

Eucaryotic organisms

Another significant evolutionary jump, although less difficult to visualize than the transition from noncellular to cellular organisms, occurs between the procaryotic cellular organism and the eucaryotic, or nucleated, types. In the nucleated types the cells have well-organized nuclei surrounded by nuclear membrane, and cell division is by a process of mitosis in which discrete units of DNA are condensed as a definite number of chromosomes.

The fact that membrane-bound cell organelles, such as chloroplasts and mitochondria, have their own DNA and replicate themselves, and the fact that their membranes are different from the membranes of the cells in which they occur, have led to an interesting hypothesis regarding the origin of eucaryotic cells. According to this idea the invasion of a procaryotic cell by one or more other kinds of procaryotes that became established as intracellular parasites or symbionts may have led to a regrouping of the host cell's genetic material into a membrane-bound nucleus for protection, and the relationship between the host cell and its guests became so intimate that the latter evolved into cell organelles, thus giving rise to the eucaryotic cell.

The organisms in which the cells are nucleated comprise four groups: the Protista, the Metaphyta, the fungi, and the Metazoa.

The Protista

The Protista are organisms with nucleated cells that have remained at the cellular grade of construction. There is one phylum, Protozoa, an immense assortment of species commonly divided into two subphyla—the Plasmodroma, with monomorphic nuclei similar in character to those of the cells of other animals and plants, and the Ciliophora, with peculiar dimorphic nuclei, a micronucleus and a macronucleus, which sets them rather sharply apart from other groups. Among the Plasmodroma, the class Flagellata contains both plantlike photosynthetic forms, often classed with the algae, and holozoic animal-like forms. There is no sharp break between the photosynthetic flagellates and the higher plants, such as occurs between the holozoic protozoans and the higher animals, probably because the mode of obtaining food is much the same in all plants. But the shift from ingestion at the cellular level to ingestion into a communal stomach by way of a mouth posed a more difficult evolutionary obstacle, with fewer possible transition stages.

The Metaphyta

The plant kingdom, or the metaphyta, comprises all higher plants, the multicellular photosynthetic forms.

The Fungi

The fungi comprise a diverse series of nonphotosynthetic saprobic plants, the familiar molds, mildews, toadstools, mushrooms, etc. In most fungi the vegetative body, or thallus, consists of elongated filamentous branching cells or series of cells, the mycelium. Several types of spores, both asexual and sexual, are formed. In higher fungi, the mycelium develops complex fruiting bodies, which in turn bear the spores.

The Metazoa

The animal kingdom, or the metazoa, comprises all the multicellular animals.

ORIGIN AND EARLY EVOLUTION OF LIFE

The basic common thread running through the entire series of living things is the presence of gigantic, self-replicating macromolecules of nucleic acid. We must then look to the formation of nucleic acid as the transitional step between chemical and biological evolution. So far as can be determined, such molecules need a fluid or semifluid highly organized environment extremely rich in complex organic compounds to replicate themselves. A major stumbling block for early theorists in connection with the origin of life was the need to account for such a pool of complicated organic compounds in a world without life, when the only sources known for such compounds were living organisms. For some time only an extremely long period of slow, fortuitous chemical evolution could be postulated, with any organic molecules formed accumulating over the ages because there were no bacteria or other organisms to destroy them. These molecules were visualized as accumulating in shallow seas warmed by the sun and gradually polymerizing into more complex units until the conditions for the formation and functioning of self-replicating molecules had been met. The theory was not satisfying and could be formulated only in the broadest and vaguest terms.

Several developments in the last few decades have done much to clarify possible mechanisms of chemical evolution and to demonstrate that it probably proceeded much faster than had formerly been thought possible. Speculation about the origin of life received new impetus in the 1920's from two biochemists—Oparin in the Soviet Union and Haldane in Britain. Since an oxygen-rich atmosphere such as now exists would not favor the development of organic matter from inorganic matter, they postulated an original reducing atmosphere consisting mainly of water, methane, and ammonia. From this, by a series of changes complicated organic molecules were thought to have evolved, and were able to accumulate in the oxygen-free, life-free environment. Urey further developed the idea of chemical evolution, drawing on geological, geochemical, and astronomical evidence.

Finally, the demonstration by Urey's student, Miller (1953), that an anaerobic mixture of gases similar to that postulated for the primitive atmosphere would, in the presence of ionizing radiation, quickly generate organic compounds by condensation of carbon radicals, and their polymerization greatly shortened the time period that had been thought necessary for biochemical evolution. Miller, starting with such a mixture of gases circulating over a vessel of warm water and exposed to electrical discharges, came up with several types of organic compounds, including amino acids and polymerized groups of amino acids, in a week's time.

From this demonstration it was postulated that the primitive anoxic earth's atmosphere, bombarded by intense ultraviolet light not damped by an ozone layer as at present, was the site for the formation of a variety of organic molecules. Some of these molecules would find protection from destruction by this same ultraviolet light by entrapment in spaces between clay particles on shores of estuaries; thus a reserve could be built up on which could be based further steps in the biochemical evolution. The shorelines, moreover, may have been a place of greatly increased organic materials concentration due to wind skimming from hundreds of miles of surface, with surfactant molecules in the foam and surface film.

Perhaps an even more likely site for the accumulation of such molecules, protected from the ultraviolet light but still within the sphere of influence of much of the visible light, would be along the thermoclines in the primitive oceans. Tropical thermoclines would represent the most extensive, relatively stable regions where such molecules could collect, prevented from sinking by the increased density of the water below, gently stirred and mixed by internal waves at the interface between the two layers of water, and also many of them perhaps carried up in convection currents such as those of the Langmuir circulation, concentrated in the zones of convergence, and polymerized into larger aggregates.

The shoreline would seem a less likely place for successful biochemical evolution to proceed far because of the time factor. Shorelines change much faster. Storms, depositions of new sediment, drying, burning by ultraviolet light, changes in sea level or local changes in the level of the land, etc. would continually be undoing the work of perhaps many centuries of biochemical evolution in such areas, but along the thermoclines in the open ocean, the process could go on uninterrupted for thousands of years, during which the concentration of organic materials would be increasing rapidly because of their formation in the atmosphere and entrapment in the sea. Thus the upper part of the oceans, particularly at the level of the thermocline, would become an increasingly concentrated broth of organic compounds, and the polymerization of these compounds into ever more complex units would be favored.

For details regarding how such polymerizing organic compounds would build themselves into increasingly complex and individualized units (blobs of polymer jelly, termed coacervates, with the evolution of enzymes in the interstices between them or proteinoid microspheres), the evolution of increasingly efficient mechanisms of synthesis of their component molecules (the walling off of functional units by single or double monomolecular lipid films forming eobionts, eventually with internal as well as external membranes, leading to compartmentalization and diversity of function within the single proto-organism), and the formation of ribonucleic acids, of deoxyribonucleic acid, and eventually of cells, the references indicated under "Origin and Early Evolution of Life" in the Annotations to the Bibliography should be consulted.

Even the so-called prebiological, or chemical, stages in the evolution of life probably did not represent simply chance reactions but were, rather, to a large degree self-guiding and directional processes partaking in some sense in a hereditary process. Simons' principle that hierarchic systems will evolve more rapidly than nonhierarchic systems of comparable size and complexity, and the principle that a hereditary organization occurs with a higher probability than a nonhereditary organization resulting in the same degree of complexity, probably apply here, as well as to systems recognized as living.

Once the stage had been reached where organisms capable of reproducing themselves had been formed, their increase in the relatively rich, predator-free primitive seas must have been rapid until the time came that so much of the organic soup of the primitive sea was incorporated into their bodies that it became too thin to support further rapid increase of the biomass. As this critical point was approached, the rapid increase in the biomass would have slowed down, perhaps rather abruptly, and diversity would have begun to increase. Mutants that, during the phase of rapid growth of the population, never attained an appreciable proportion of the population might continue to increase, while the formerly rapidly growing parent populations became stabilized in numbers or even began to decrease. Ecological succession would set in. Any forms that had evolved the process of photosynthesis, building up new organic matter from inorganic substrates by using absorption of energy from the visible spectra, would have been especially and selectively favored and would have increased rapidly. They may well have been in the form of bacteria-like blue-green algal cells. Likewise, any cells that had developed the capacity to ingest other cells as a source of organic nourishment would now be especially favored. The dichotomy between holophytic and holozoic modes of nutrition would become accentuated, whereas the relative importance of saprozoic nutrition would have diminished.

Whether life originated in shallow nearshore waters such as estuaries and lagoons or along the tropical thermoclines in the ocean, the latter would appear to be the earliest places where the free oxygen evolved as a by-product of photosynthesis could have accumulated sufficiently to make possible the change from anaerobic fermentative metabolism to the much faster, more vigorous oxidative metabolism characteristic of most organisms today. Near the shore the oxidation of the great amounts of mineral substances still in a reduced state would use up the oxygen as fast as it was produced for a much longer period of time than in the upper strata of the ocean along the thermocline.

The early oceans at this stage, then, probably had a layer of photosynthetic blue-green algalike cells concentrated along the thermocline, with perhaps also coatings of these or similar cells growing on the substrate in shallow areas where they could obtain light. The chief Pre-Cambrian fossils are stromatolites, aggregates formed by the growth of microscopic blue-green algal cells and trapped sediment particles, which built mounds. The largest stromatolites towered as much as 50 feet above the sea floor and had a 30-foot base. The origin and early evolution of holozoic forms using oxidative metabolism probably took place along the thermoclines, as Weyl (1968) has pointed out. Such forms might have been able to utilize the mat of growth on the substrate in shallow areas at times and places when more oxygenated seawater swept over them or to penetrate marginal areas for short periods to feed on the presumably lush protected growth in shelf areas if they could swim back into more oxygenated water when necessary. Evolution of both plant and animal forms could then have proceeded for a long time and attained a high degree of diversification before the animals were able to colonize shallow-water shelf areas, marginal lagoons and estuaries, etc. to any great extent. Only when enough oxygen had accumulated so that most of the exposed reduced mineral surfaces and suspended particles in the water had been oxidized, could colonization occur.

In the evolution of early animals along the thermocline, density would have been a critical factor. The development of forms with dense, compact muscular systems or heavy skeletons or shells would have been precluded. Within the bounds of this limitation, however, a considerable diversity could have been achieved well before extensive colonization of the shallow regions became possible. According to Weyl, the period of this colonization was probably during late Pre-Cambrian or early Cambrian times. The fact that considerable evolution and diversification of invertebrates had already occurred prior to this time helps account for the relatively sudden appearance of a great variety of shallow-water types. Once in the shallow water, where increased density is no longer a liability but often an advantage, the evolution of all types of skeletal and muscular systems, without regard to density, would have occurred. Indeed, the study of Cambrian fossils does show a progression from early Cambrian forms with lighter, more fragile shells or exoskeletons to types in the late Cambrian with heavier, better developed ones. It was only with the evolution of forms having heavier bodies and hard parts that fossils began to be preserved on a sufficiently large scale that we are likely to find any of them today.

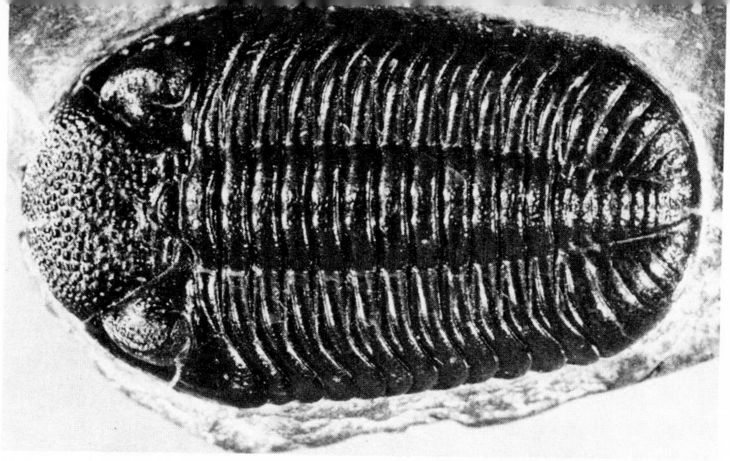

FIG. 18-1. Fossil of trilobite, *Phacops rana* Green, one of the most abundant and widely distributed species of the Paleozoic era.

CHAPTER

18 THE PROGRESSION OF LIFE THROUGH THE AGES

To present in any detail a summary of the fossil record would be far beyond the scope of this book. This chapter will merely recapitulate some of the general trends and consider in summary form one type of marine biological community, the reef community, which has the twin virtues, from the point of view of the student of paleontology, of leaving massive fossil remains and of developing only near the surface of marine waters under conditions of prolonged environmental stability relatively free from stress. This sensitivity to environmental change makes the reef community particularly valuable as an indicator of paleoecological conditions and changes on a worldwide basis.

Fig. 18-2 presents the principal reef-forming organisms against the background of time, showing in which eras and periods each type existed and during which portions of this time they contributed most significantly to reef formation (thickened portions of the lines). Fig. 18-3 is a similar time background with some of the other most familiar animal and plant types similarly positioned. A few major geological developments are indicated in the legend to provide points of reference.

Following are some of the major trends revealed by the fossil record:

1. There has been a tendency toward expansion—an overall increase through time in the number and diversity of living species and in the total biomass produced. This expansion has not been at a constant rate. There have been periods of relatively rapid expansion—occupation of new habitats and increasing amount and diversity of life in old habitats. There have also been periods of ecosystem collapse and decreased amount and diversity of life.

2. The ecosystem is never a static equilibrium but is in a continual state of change, with expansions and contractions of the various populations of animals and plants

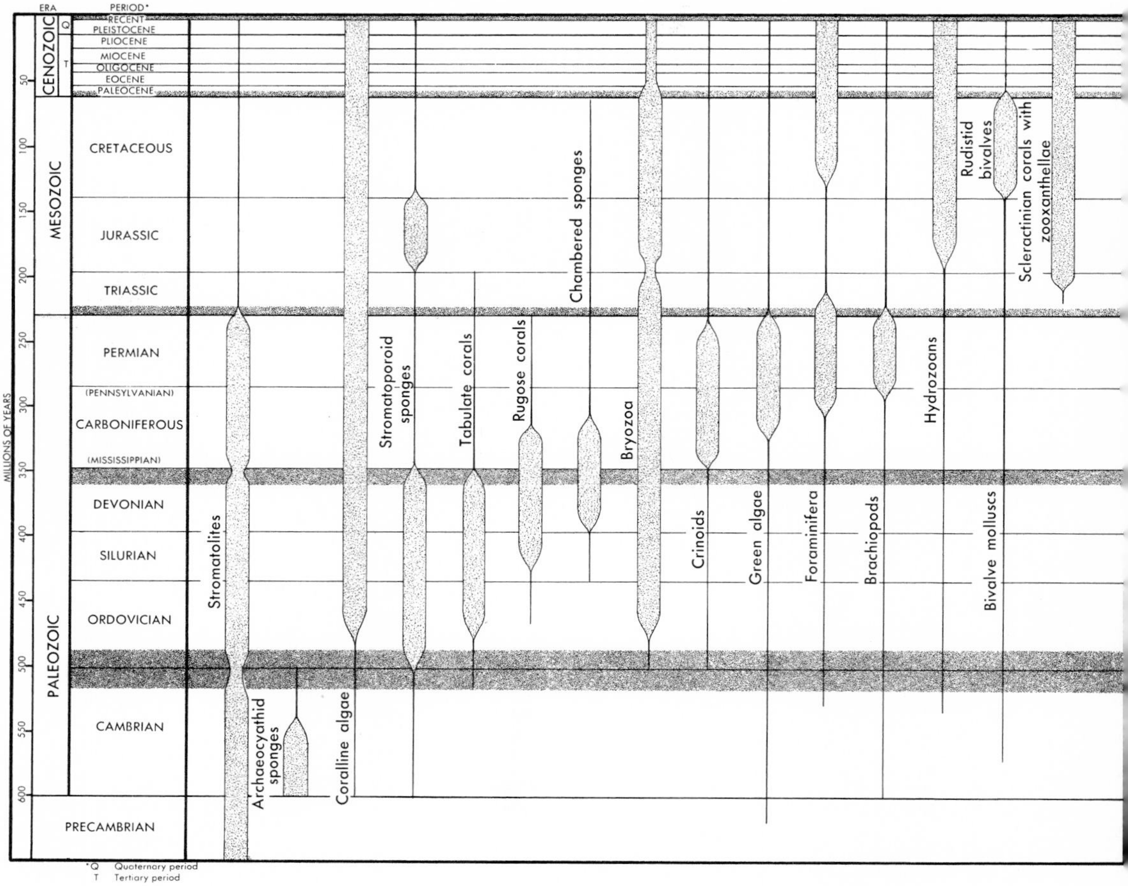

FIG. 18-2. History of reef-forming organisms correlated with the sequence of major periods of the earth's history since the Paleozoic, the earliest era represented by an adequate fossil record. The thickened vertical lines represent the times during which a particular group contributed most significantly to the formation of reefs. There has been no attempt to make the widths proportional to the contribution being made by each group at any given time or at different times. The horizontal shaded bars represent times of worldwide ecosystem collapse, during which much of the biota of the world was eliminated.

and replacement of some groups by others.

3. Although evolution is not necessarily progressive, increasing complication and improvement in organic structure and function have occurred. These progressive changes involve increasingly precise adaptations to particular ecologic niches, specialized modes of energy capture, increasingly efficient utilization of energy, marked structural changes making new ways of life possible, and increased levels of awareness, perception, and consciousness.

4. In addition to darwinian replacement of given species by others, which affects species or groups more or less individually and goes on at all times, there have been hundreds of episodes of mass extinctions in which many animals and plants of diverse groups, living in different habitats, have become extinct, whereas others living with them have persisted or have given rise to new groups that replaced them. Some of these episodes have been fairly localized, but many involve large areas—even whole continents or seas—and a few

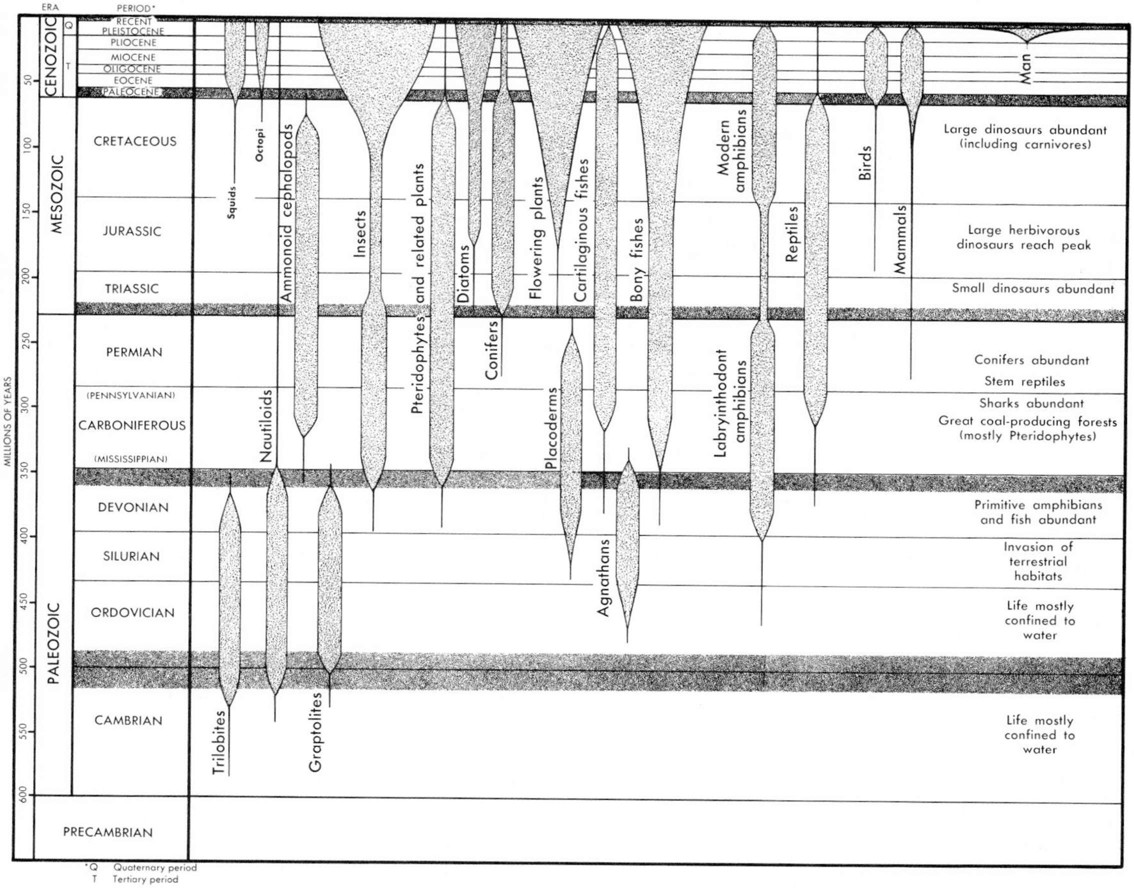

FIG. 18-3. History of familiar animal and plant groups. See Fig. 18-2 for explanation of chart. Predominant organisms are noted on right side of chart. Some major geologic developments are indicated below to provide points of reference. These events often had significant effects on climates and ecosystems. *Late Paleozoic to early Mesozoic:* continental uplift; continental plates gathered together as one supercontinent (Pangea); ice sheets formed at southern end of continent; draining of shallow seas; worldwide ecosystem collapse. *Late Mesozoic to early Cenozoic:* continental plates separated by sea floor spreading; widespread continental uplift; draining of shallow seas; worldwide ecosystem collapse. *Middle Cenozoic to Pleistocene:* Mediterranean drying; thermal isolation of poles; great ice ages (Fig. 18-7).

have been worldwide. The four great worldwide periods of ecosystem collapse of the past, and the one currently in progress, are indicated by horizontal shading across Figs. 18-2 and 18-3.

An ecosystem collapse does not necessarily entail a tremendous reduction in total biomass, although it may do so. It is, rather, a selective elimination of many species and groups of organisms. Generally eliminated are large long-lived organisms and highly specialized forms adapted to narrow and particular ecological niches. Subsequently, if complex mature ecosystems are again evolved, their places are taken by new and different forms evolved from populations of smaller and less specialized organisms that survived the collapse or were able to adapt to changing conditions during the period of collapse.

Although these periods of mass extinction appear to be sudden on a geological time chart because of the extreme compression of the time scale, in actuality they spanned periods of thousands or even millions of years and would not have been readily apparent to anyone living at the time. During any one collapse the extinctions were by no means all simultaneous, either for different species or for members of the same species in different regions.

PALEOZOIC ERA
Cambrian period

During the Cambrian and Ordovician periods, life was almost wholly confined to water. Continental areas were barren, harsh, soilless regions, perhaps more reminiscent of the present surface of the moon than of the earth's land surfaces today. Pre-Cambrian limestones laid down by living organisms consist of peculiar laminated limestone mounds or columns called stromatolites, with aprons of limestone debris. These are thought to have been laid down by blue-green algal cells and consist

FIG. 18-4. Reconstruction of an early Paleozoic (Middle Cambrian) sea floor in western North America, showing free-swimming trilobites, the dominant form of early Paleozoic life. (From Burris, T. L., and H. J. Spiegel. 1976. Earth in crisis: an introduction to the earth sciences. The C. V. Mosby Co., St. Louis; courtesy Field Museum of Natural History, Chicago, Ill.)

of calcium carbonate secreted by them, together with trapped sand grains and other particles.

The first true reef communities, in the sense of reefs built up by more than one kind of organism contributing significantly to the formation of the reef, are found in Cambrian rocks and consist of stromatolite growths mixed with low thickets of archaeocyathid cup sponges. For reasons not altogether understood, the archaeocyathids largely vanished about mid-Cambrian (Fig. 18-4), leaving only the primitive algae for the next 60 million years or so as builders of stromatolite reefs. This was not due to competition with another, more efficient reef builder, since none was present during the remainder of the Cambrian.

The end of the Cambrian period was marked by a prolonged period of massive extinctions—the first great worldwide ecosystem collapse—and a cessation of reef building.

Ordovician period

The Ordovician period saw a renewal of reef building, with several new groups participating in reef formation. Stromatolites were still prominent but were commonly joined by coralline algae; stony stromatoporoid sponges, some platelike, others bushy; and bryozoans and the first stony coelenterates, the tabulate corals.

Silurian period

During the Silurian period the groups mentioned above, for the Ordovician, continued as the chief reef formers. Rugose corals also became a significant part of reef communities. This period also saw the invasion of terrestrial habitats by various primitive plants and animals and the rise of the first vetebrates—primitive agnath fishes.

Devonian period

Most of the Devonian period was a time of tremendous expansion and differentiation of life types. Chambered sponges were added to the flourishing reef communities. Toward the end of the Devonian period, however, there were apparently worldwide climatic changes leading to a second great period of extinctions. Among the reef-building elements, stromatoporoid sponges and tabulate corals were all but exterminated. Reef building ceased or greatly diminished everywhere.

Carboniferous period

During the early Carboniferous period the reef communities were rather impoverished and consisted mostly of stromatolites, coralline algae, chambered sponges, rugose corals, and bryozoans. Crinoids began to be a prominent element in them. By the late Carboniferous the chambered sponges and rugose corals had been eliminated, whereas crinoids, green algae, and foraminiferans had become prominent elements. The placoderm fish dominated the Carboniferous period, and the Chondrichthyes and Osteichthyes appeared. On land there developed large forests and low-lying swamps dominated by primitive fernlike plants; they were the basis for the eventual formation of extensive coal beds. Among this biota were found early labyrinthodont amphibians and a variety of terrestrial arthropods, including primitive insects.

Permian period

The Permian period was marked by continuation of the expansion of life forms that took place in the Carboniferous. Reef communities flourished, comprised of stromatolites, green algae, foraminiferans, bryozoans, coralline algae, and a strong blooming of brachiopods with thousands of species. Crinoids continued prominent. By the end of the Permian, the world's land masses were grouped together in a great supercontinent, pangea, and the remarkable development leading toward the ruling reptiles was well under way. Toward the end of the Permian, ice sheets developed on the southern end of the continent, and there was a period of cooling and worldwide collapse of ecosystems. Half the known families of animals and plants both terrestrial and marine, were exterminated. The Permian reef associations died out, and no reefs were formed anywhere for the next 10 million years. Stromatolites largely disappeared.

Widespread continental emergence with draining of shallow seas as water was withdrawn to form the ice sheets may have been responsible for the generally cooler, harsher climates during this period.

MESOZOIC ERA

Triassic period

It was not until about the middle of the Triassic period that renewed warming trends had developed to the point that reef building began again. The new reefs differed considerably from previous reef communities, consisting mostly of coralline algae, bryozoans, and the appearance of a new group, the scleractinian, or stony, corals so prominent in today's reefs.

FIG. 18-5. Restoration of middle Mesozoic (Jurassic) sea floor, showing free-swimming belemnoids and oysterlike pelecypods. The marine reptile *Ichthyosaurus* is shown in the background. (From Burris, T. L., and H. J. Spiegel. 1976. Earth in crisis: an introduction to the earth sciences. The C. V. Mosby Co., St. Louis; courtesy Field Museum of Natural History, Chicago, Ill.)

Foraminiferans were much reduced as a reef-building element.

Jurassic period

During the Jurassic period (Fig. 18-5) stony corals became more important as reef builders, hydrocorals appear as a significant element in the reefs, and green algae, bryozoans, and resurgent stromatoporoid sponges are also important. The latter, however, did not persist beyond the Jurassic. The continental plates making up the supercontinent of Pangea had begun to separate. On land dinosaurs were the dominant vertebrates, primitive mammals continued a modest expansion, and the first birds appeared. Reptiles had also invaded the aquatic habitats, producing completely aquatic forms—ichthyosaurs, the reptilian equivalent of our present-day porpoises.

FIG. 18-6. Restoration of late Mesozoic (Cretaceous) sea floor in central United States, showing many long, straight ammonoids and one coiled variety, as well as pelecypods and gastropods. (From Burris, T. L., and H. J. Spiegel. 1976. Earth in crisis: an introduction to the earth sciences. The C. V. Mosby Co., St. Louis; courtesy Field Museum of Natural History, Chicago, Ill.)

Cretaceous period

The first 20 million years or so of the Cretaceous period (Fig. 18-6) were rather unfavorable for reef formation but were followed by a long period of generally warm climates. The continental masses had separated into northern and southern masses, leaving a broad shallow sea, the Tethys Sea, connecting the Indo-Pacific with the eastern Atlantic. North and South America were also separated, leaving a continuous belt of tropical seas around the world. Large areas of the present continental land masses were also covered by shallow seas. The sea bottoms were about 14° C. warmer than at present, and only about 18% of the earth's surface was above water compared with almost 30% today. It was primarily a tropical warmwater world.

There was a great expansion of living forms. Extensive reef formation occurred, including regions well beyond the present tropical belt, especially to the north. The reefs show major increases in stony corals, hydrocorals, bryozoans, and coralline algae, as well as a resurgence of foraminiferans and the appearance of a new group, the rudistid bivalves, as an important reef element.

On land the ruling reptiles continued to dominate, but the expansion of mammals and birds also had begun, and seed plants, mostly conifers but also some angiosperms, had appeared and had begun to invade habitats not favorable for the primitive fernlike groups. Reptiles attained the greatest diversity of their history, producing additional aquatic groups such as plesiosaurs, and even flying forms—the pterodactyls.

The end of the Cretaceous was marked by the greatest period of extinctions to date. About one third of all the families of animals present in the late Cretaceous died out. The rudistid bivalves vanished, and the ammonites disappeared from the seas. On land the dinosaurs and numerous other groups of specialized reptiles disappeared, and many of the primitive plants were exterminated. The separation of the continental plates with the development of a deepening Atlantic basin drained off much of the shallow seas, resulting in general cooling fatal to the generally tropical biota. No new reefs were formed during the 10 million years of the Paleocene.

Marine reptiles of the past

The Mesozoic era is especially well known as the era of the ruling reptiles—a time when the reptiles reached a peak in their evolutionary history with the production of a remarkable diversity of forms, including marine and flying reptiles as well as some of the largest terrestrial animals that have ever existed. Some of the marine forms are briefly discussed here.

Sauropterygians

Plesiosaurs. In number and variety of species and genera and in worldwide distribution the plesiosaurs constituted, by a rather wide margin, the largest, most successful group of marine reptiles that has ever existed. They are remarkable for the great relative elongation of the neck in many species and for the extreme variation in its structure. Elongation was effected both by increase in the number of cervical vertebrae over that found in other reptiles and by increase in the length of some of the elements, especially those near the base of the neck. The number of cervical vertebrae in various plesiosaurs ranged from thirteen in relatively short-necked forms to twenty-six. In the long-necked species the neck was relatively thick at the base, tapering toward the head, and the head was relatively small. In *Elasmosaurus platyurus,* the head was 0.6 meter long, the neck 7 meters, the body 2.7 meters, and the tail about 2 meters—a total of 12.5 meters. In the short-necked *Brachauchenius lucasi,* on the other hand, the head was 0.8 meter long, the neck was less than 0.6 meter, and the body about 0.5 meter. The length of the tail is uncertain, but it was not greater than 0.5 meter. The vertebrae were biconcave but with shallow saucerlike or nearly flat ends, much different from the conical fishlike cavities of the strongly amphicoelous ichthyosaurian vertebrae.

The trunk was not much elongated, having only twenty-five to thirty vertebrae, and the tail was always shorter than the trunk, tapering rapidly to the end. The ribs in the cervical region were short but locked together posteriorly in a way permitting little lateral motion. Those of the trunk were attached high on the arch to the end of the stout transverse processes by a single head and ended freely below, with no attachment to a breastbone or other ventral bony structure. Their shape and position indicate that in life the body of plesiosaurs tended to be broad and somewhat flattened dorsoventrally.

The limbs were modified into broad paddles, and unlike other aquatic animals, the hind limbs were not reduced and were probably as effective as the forelimbs in swimming. The humerus and femur were broad, massive, and elongate, as in the sea turtles. Their strong, muscular rugosities suggest that the plesiosaurs possessed powerful swimming muscles. The paddles were undoubtedly more powerful than those of any other aquatic air-breathing animals. All plesiosaurs had five digits on each hand or foot but often an excessive increase in phalanges, some digits of certain species containing as many as twenty-four bones.

Some species were relatively broad-headed, with long, slender, recurved, pointed teeth in deep sockets. Others were slender-jawed and gavial-like, with smaller, more numerous teeth. The larger posterior surface of the skull for attachment of the masseter muscles is evidence of powerful muscles for biting and seizing prey.

Small plesiosaurs were about 3 meters long when adult, but some of the largest species attained lengths of about 15 meters, the neck comprising about one half the total length.

The rigid nature of the jaws, united in front and incapable of lateral movement behind, indicates that they could not swallow large prey whole. The flat-headed plesiosaurs probably tore their food apart much in the manner of crocodiles, with quick, powerful jerks. The narrow-jawed species probably ate fish only, which their small, sharp, more numerous teeth fitted them to catch. Small fish were swallowed whole, and this group of plesiosaurs probably were not dangerous to larger animals.

The swimming must have been more turtlelike than fishlike, with the tail used as a rudder rather than as an organ of propulsion. They could hardly have been really swift swimmers like the ichthyosaurs and apparently seized their prey by swift downward lunges of the neck rather than by quick swimming. Many of them, like some modern crocodiles, apparently swallowed fairly large numbers of pebbles to be used as gastroliths in a muscular stomach for grinding food.

Unlike the ichthyosaurs, they appear to have been of more or less solitary habits, and some of them, at least, made journeys of considerable distance at sea. They have been found in deposits believed to be of deep-water origin and hundreds of miles from the nearest sources of the gastrolith pebbles found with their skeletons. Most species, however, probably inhabited coastal waters.

Like the other great reptiles—the mosasaurs, pterodactyls, dinosaurs, and others—they became extinct at the close of the Mesozoic era from causes not yet fully understood. As far as can be determined, it was a considerable time before their place was taken by any other creatures.

Ichthyosaurs. Like the whales among mammals, the ichthyosaurs among reptiles were the most perfectly adapted marine animals of terrestrial descent of their day. Their fossils are abundant in ancient Jurassic rocks, and we know them as a flourishing, fully developed group. But like the whales, they occupied a somewhat isolated position. So perfectly were they adapted to marine conditions that it is difficult to relate them to the earlier terrestrial forms from which they arose. Paleontologists of the nineteenth century regarded them as descended from fishes, so marked is the convergence shown.

Externally, ichthyosaurs were porpoise-like in general appearance. The body was naked, and the head was produced into a long, slender snout and closely joined to the body, with no indication of a neck. The body was somewhat torpedo-shaped—cylindrical and expanded in front, with a large thorax and abdomen, then tapering off rapidly and ending in a large, strong fishlike tail. The front paddles were close behind the head. The hind paddles were reduced in size, their function being replaced by that of the broad tail.

The vertebrae were amphicoelous and deeply biconcave, like those of fish. However, at the time when the ichthyosaurs must have arisen, probably from the earliest Cotylosauria stem, all reptiles had biconcave vertebrae. In the ichthyosaurs, this type of vertebra persisted, as it has in the fish, because in the water such vertebrae are best suited for the quick, pliant movements of the spinal column required of most aquatic animals, although other reptile lines developed firmer backbones. In modern porpoises, which closely resemble these ancient reptiles in size, shape, and probably in habits, the small, flat-ended vertebrae are widely separated by discs of flexible cartilage.

There was no sacrum, that is, no united vertebrae in the hip region for attachment and support of the pelvis, since no such support was needed. This same adaptation is also found in cetaceans and sirenians. The long, slender ribs resemble those of fish-eating porpoises, but their articulation to the vertebrae by two attachments to the centrum and none to the arch differs from that of all other animals. The hind legs were small, and the pelvis was weak and suspended below the spinal column in the fleshy, abdominal wall. It is probable that if the ichthyosaurs had persisted to the present time, they would by now have lost the hind legs altogether, as have the cetaceans. The earlier ichthyosaurs of the Triassic (often classed as a separate family, the Mixosauridae) had a larger pelvis, longer paddles that were more leglike than in the later forms, a less well developed tail, and somewhat longer, less fishlike vertebrae. For propulsion, they apparently used their paddles more and their tail less than did their later Jurassic descendants.

The ichthyosaurs were apparently gregarious and inhabited the open ocean. Remains are often found in deep-water sediments, complete and in a natural position, showing that they were not disturbed after the death of the animals as those of shallow-water coastal creatures usually are.

Ichthyosaurs were viviparous. Some fossils show as many as seven embryonic skeletons associated with an adult. The remains of food found in some of the fossils indicate that they fed mostly on fish, squid, belemnites, and various invertebrates. Their mouth and teeth were adapted for seizing and holding slippery prey but not for tearing or chewing their food or for swallowing large prey. The jaws were closely united, permitting none of the expansion characteristic of snakes.

The ichthyosaurs attained a nearly worldwide distribution but died out before other contemporaneous groups of marine reptiles, such as the plesiosaurs and mosasaurs.

Mosasaurs. Mosasaurs were large aquatic lizards distantly related to the monitor lizard, *Varanus*. They appeared during the Cretaceous; enjoyed a rather rapid adaptive radiation, soon reaching its culmination in size, numbers, and distribution; and disappeared before the close of the Cretaceous. Their skeletons have been found in Australia and New Zealand, Europe, and North America and are especially abundant and well developed in the Kansas chalk deposits.

The skull of mosasaurs was large in proportion to the total size, nearly one sixth of the total length. Both jaws were armed with numerous strong, sharp, conical, recurved teeth, and there were also additional teeth on the pterygoid bones in the back of the mouth. Unlike the teeth of other lizards, they were inserted on large tumid bony bases rather loosely attached in shallow pits. They were relatively easily dislodged and were readily replaced by new ones throughout life, so that adult mosasaurs had teeth of all sizes at the same time. The jaws had considerable freedom of movement both laterally and vertically and were rather loosely joined in front. Furthermore, they possessed a unique joint just back of the teeth, permitting movement between the front and back portions of the jaw. The large skull and heavy armament of teeth and mobile jaws indicate extremely predaceous and pugnacious habits.

As in other aquatic reptiles, the eyes were protected by a ring of bony plates, and the ears had a thick, cartilaginous drum in place of a simple membrane. The neck was short and strong. The body was elongate, and the tail was long and strong, and laterally flattened, with the beginning of expansion toward the free end into a caudal fin. The vertebrae were procoelous like those of terrestrial lizards and snakes, giving greater strength to the spinal column but less flexibility than in other aquatic reptiles, a condition doubtless carried over from terrestrial lizard ancestors.

The limbs were adapted for swimming, the hind limbs reduced in size as in other forms that propel themselves by means of the tail. The bones of the upper arm and thigh were much shortened and broadened, and in the most specialized types such as *Tylosaurus*, the wrist and ankle bones were mostly cartilaginous. This tendency of the ends of the long bones and the bones of the wrists and ankles to become less well ossified in purely aquatic animals is so strong that aquatic habits are suspected at once in any reptile showing them, even though the skeleton may be imperfectly known.

The body was covered with small, overlapping scales, somewhat smaller and smoother in comparison to the size of the animal than those found on modern monitor lizards. The top of the skull was covered by horny plates, as in most lizards.

Mosasaurs ranged from about 4.5 meters to over 12 meters in length and were doubtless one of the fiercest, most voracious groups of predators that have existed. They were able to take rather large prey, which they probably swallowed whole, head first, in the manner of snakes. Their jaws and teeth were not fitted for tearing off bite-size chunks from larger animals, nor for chewing, but rather for seizing and holding prey. It is a curious thing that in a form of which the fossils are so abundant, no embryonic or young individuals have been found. This would seem to indicate that mosasaurs were oviparous, in which case they probably laid their eggs on shores and beaches as do sea turtles and crocodiles. The waters in which mosasaurs lived contained many kinds of predaceous fish, together with other mosasaurs; thus it would seem perilous for the young, perhaps a foot or so in length, to venture into them. It is possible that the females entered estuaries and rivers to find protected places and that the young remained in such places until they were large enough to cope with conditions in the sea. Much less is known of the fauna of lakes and rivers of this period.

Thalattosaurs. The thalattosaurs were a group of small marine reptiles, only 1 to 2 meters in length, first studied by Merriam from Upper Triassic rocks in northern California. They are known only from purely marine deposits containing little or no material of terrestrial origin.

The skull displays adaptation to the aquatic environment in the elongation of the snout, with the nares well back from the end and dorsal in position. The jaws were armed with conical pointed teeth in the front, the teeth becoming rounded, rugose, and obtuse farther back. The palate was largely covered by similar teeth, which must have been used for crushing crustacean or mollusc shells—perhaps the light shells of ammonites, which were abundant in these seas at that time. The conical pointed teeth of the forepart of the jaw indicate that they probably also seized swifter-swimming prey such as fish.

The eyes show the ring of ossifications surrounding the eyeball, a characteristic of truly aquatic reptiles.

The limbs were structurally much like those of the mosasaurs. The ribs had a single articulation to the centrum, a characteristic shown by lizards, snakes, and mosasaurs and unlike those of other reptiles. The vertebrae were amphicoelous, as were the vertebrae of all primitive reptiles existing at the time of the thalattosaurs. The ball-and-socket joints of the vertebrae of present reptiles did not arise until long after the disappearance of the thalattosaurs.

The thalattosaurs apparently never achieved a worldwide distribution, nor did they enjoy a long existence in geological terms. Apparently they were eliminated in competition with the forerunners of the plesiosaurs, ichthyosaurs, and mosasaurs or perhaps by other marine predators such as some of the fish.

Order Thalattosuchia

The thalattosuchians represent an early marine offshoot from the great crocodilian stem during the Upper Jurassic period. In general form and appearance they resembled what we might imagine present-day gavials would become if they should become marine in habit. They were 3 to 6 meters long, with relatively small heads drawn out into a long, slender snout, as in present gavials. The eyes had a ring of sclerotic bones as in other truly marine reptiles but which is found in no other crocodiles. Their eyes were directed laterally, rather than more or less upward. The tail was long and strong, making up about half the total length and developed into a fleshy terminal fin. There was no sternum, a structure found in all other crocodiles. The forelimbs were paddlelike and attached well forward, serving as balancing organs like the pectoral fins of most fish, rather than as organs of propulsion. In contrast to the legs of other marine forms propelled chiefly by the tail, the hind legs were larger than the forelegs and not much different in structure from those of gavials. The pelvis was well developed and firmly attached to the sacrum. The presence of well-formed hind legs and of a ventral armor of bony ribs seems to indicate that the marine crocodiles visited the shores periodically, probably to lay eggs. Perhaps failure to become viviparous was a limitation that may have contributed to their early extinction in competition with the truly viviparous aquatic carnivores.

Order Pterosauria

The flying reptiles, or pterodactyls, of Jurassic through Cretaceous times seem to have been the reptilian equivalent of the seabirds of today. The wings were supported by an enormous extension of the fifth, or "little," finger, which stretched out a long but rather narrow skin plane. The thumb and first three fingers were small and clawed.

Pterodactyls ranged in size from that of moderate-sized birds such as terns to large species somewhat larger than any contemporary bird. They did not have powerful wing muscles, and thus probably launched themselves from cliffs near the ocean and soared over the water, perhaps snatching fish from the surface, as frigate birds sometimes do. Having relatively weak wings, the larger species, at least, may have been unable to take off from the water surface readily or from flat places on land.

The brain was large for a reptile, with an especially well-developed birdlike cerebellum.

All remains of pterodactyls have been found in marine sediments.

CENOZOIC ERA

In the late Eocene a new radiation of scleractinian corals occurred, but before the end of the Eocene and throughout the Oligocene, climates become generally cooler and more seasonal, and reef building declined except in the Caribbean and Indo-Pacific. By the Miocene, sea floor spreading had rafted Australia to its present tropical position, where the coral-coralline algae complex began to build up what was to become the Great Barrier Reef system—today's most extensive reef complex.

Even more significantly, Antarctica was rafted toward its present polar position, which it reached during the Pliocene, isolating the south polar area from thermal interchange with the world's oceans and initiating formation of the present Antarctic ice cap, which further lowered sea levels and augmented the worldwide cooling trend both on land and in the oceans.

During the Pliocene the movement of the North American and Eurasian continental masses also resulted in the nearly complete thermal isolation of the Arctic Ocean as well. The Mediterranean basin became separated from the Atlantic and dried out, becoming a hot, dry, below–sea level desert marked by salt basins and alkaline flats, and the Arctic Ocean froze over, further contributing to the development of a strongly zoned cooler world climate, with the tropics limited to a belt near the equator. Restoration of the Gibraltar connection with the Atlantic during the late Pliocene resulted in a refilling of the Mediterranean basin from Atlantic surface waters, but there has been no connection with deep waters. Finally, the emergence of the Panama connection between North and South America formed a complete bar-

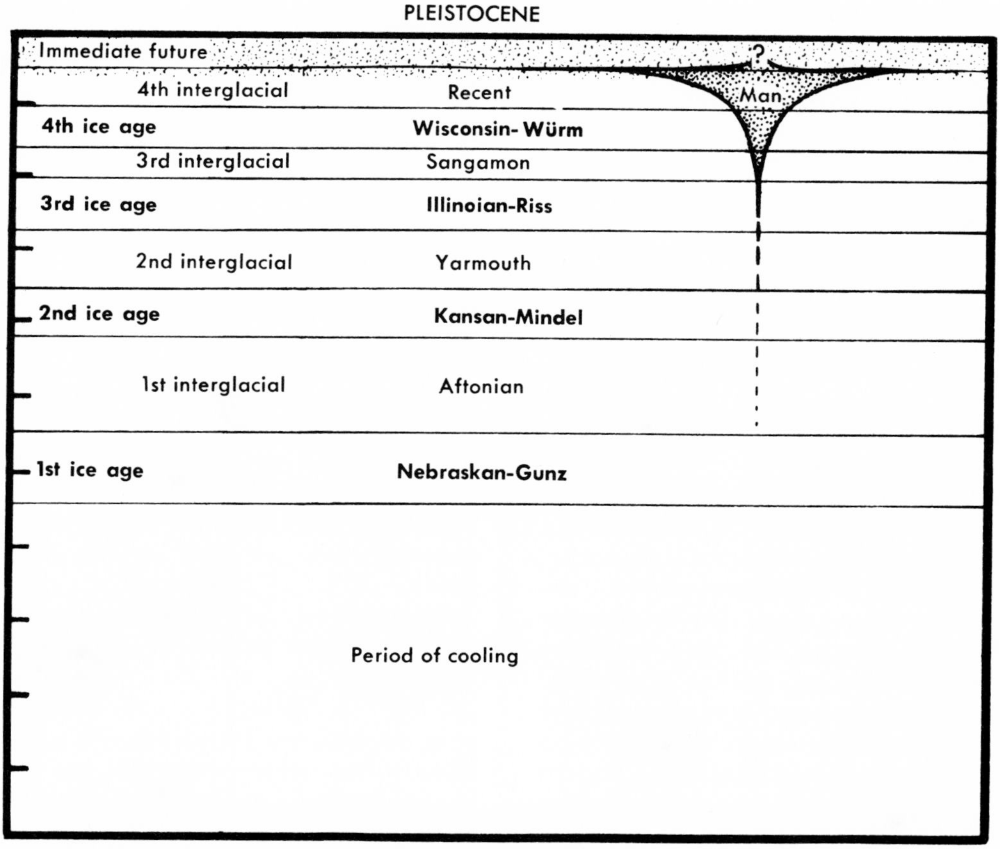

FIG. 18-7. Late Pleistocene, showing the history of man in relation to the great ice ages. The names of the various periods are given. Each ice age has two names (hyphenated on illustration). Left, the North American ice sheet; right, the ice sheet covering Europe and Siberia. The horizontal gray band at the top indicates the fifth worldwide ecosystem collapse currently in progress. Each division on the left margin represents 100,000 years.

rier between the Atlantic and Pacific oceans except in the far north by way of the Arctic Ocean and in the far south, south of the tips of South America and Africa.

The Cenozoic era is commonly divided into two principal subdivisions, the **Tertiary,** which includes the Eocene, Oligocene, Miocene, and Pliocene epochs, and the **Quaternary,** including the Pleistocene and recent times. The Recent epoch is really a part of the late Pleistocene, marked by the sudden rise and dominance of man.

During the Pleistocene thermal interchanges between the Atlantic and Arctic oceans seem to have been largely responsible for initiating a dramatic series of climatic changes in the Northern Hemisphere—a series of ice ages and interglacial periods (Fig. 18-7). We are presently enjoying the relatively mild climates of the fourth interglacial period. These climatic changes are further discussed in Chapter 19.

During the Tertiary, marine faunas of different regions of the world became increasingly isolated from each other. World climate became more strongly zoned and the tropics more and more restricted as the Tertiary drew to a close. Evolutionary changes in marine biota took place more or less independently of each other in different regions of the world, and at different rates, so that exact correlation of the time of deposition of marine deposits in different parts of the world through use of indicator species of fossils is more difficult and less reliable for late Tertiary and Quaternary deposits than for most others.

The Cenozoic can be characterized as the era in which diatoms became the principal phytoplankton of the seas, especially in cool-to-cold waters, and teleost fish the principal aquatic vertebrates, exceeding all other vertebrate groups in number of species and diversity of form and habits. Reef communities were largely dominated by scleractinian or stony corals containing symbiotic zooxanthellae and associated with coralline algae.

On land there was a great development of mammals, birds, flowering plants, and insects. Some of the early Tertiary mammalian groups reinvaded marine habitats, giving rise to two great groups of marine mammals, the Cetacea and the Pinnipedia, plus a few minor groups such as manatees.

The fifth worldwide ecosystem collapse

The most striking development of the late Pleistocene has been the sudden rise of man. Being the most omnivorous of all large animals and being able through his intelligence to grow and store food, accumulate and transmit information, and open up and make use of sources and levels of energy hitherto unavailable to any organisms, man has undergone the most explosive, dramatic population expansion, by several orders of magnitude, of any large animal species. This, plus his sudden development of new technologies for exploiting the world's resources, exerts on the world's ecosystems as a whole effects vastly more profound and abrupt than anything caused by any other species throughout the entire history of the world.

These effects appear to be too sudden, too novel, and too massive for present world ecosystems to absorb. We are now witnessing the fifth great worldwide ecosystem collapse. This one is unique in being the first major collapse caused by, or at least greatly augmented and speeded up by, the growth and activities of an animal species rather than by geological and climatic changes totally beyond the control or understanding of any of the organisms affected. It is probably also unique in the abruptness of the collapse, which may well make it the most devastating yet in the history of life, since living populations have less time to respond and adjust to the changes and some of the changes are themselves of a novel character to which biotic communities have not previously been subjected.

CHAPTER

19 SEA FLOOR SPREADING, CONTINENTAL DRIFT, ICE AGES, AND WORLD CLIMATE

SEA FLOOR SPREADING AND CONTINENTAL DRIFT

The theory of continental drift was first espoused by Wegener (1912) and others on rather circumstantial evidence such as the coincidences in shape of the major continents, which are such that one can visualize them fitted rather closely together like the pieces of a jigsaw puzzle. To this evidence were added curious coincidences between the faunas and floras of such continents as would have been joined together in this manner at the time of origin of the major groups of animals and plants of today. Geological evidence in the form of similarities and continuities of geological formations that would fit together in a logical manner in such an arrangement of the continents lent further plausibility to the idea, as did similarities in the fossil faunas and floras of Mesozoic age or older.

Many scientists believed, however, that the major ocean basins and continental platforms had probably held much the same general relationships to each other throughout the known history of the world, and isostatic changes in the level of land or of areas of the sea floor, together with changes in sea level and the formation of land bridges during certain periods between some of the continents, could account for the coincidences noted by the proponents of the continental drift theory more easily than the postulated migrations of continents. The difficulties of imagining the physical causes and mechanisms that would account for the supposed movements of whole continents seemed to pose a greater problem than the finding of alternative explanations for the coincidences noted. In other words, the theory seemed to raise more serious and difficult problems than it solved.

During the past three decades, however, an impressive array of new evidence from several different disciplines has again brought the theory of continental drift into prominence and seems to have demonstrated its essential correctness.

In the first place, a great worldwide system of ridges and rifts showing tensional features, as if the earth were gradually being pulled apart along this system, has been discovered. The Mid-Atlantic Ridge extends the full length of the Atlantic Ocean, passes through Iceland, and continues north into the Arctic Ocean, where it bisects the Eurasian Basin. In the Pacific the East Pacific Rise continues the system from the antarctic region to the Gulf of California, from the head of which it continues northward as the San Andreas Fault. Another branch extends north and west from the Antarctic portion of the Pacific system into the Indian Ocean, where it runs north into the Red Sea, finally turning back as the Great Rift Valley in Africa. In many parts of the world this rift system has a longitudinal central valley or trench, with the mountains or rises arranged more or

FIG. 19-1. Floor of Indian Ocean, with midocean ridges, fracture zones, major basins, continental shelves, and extensive mud aprons deposited by great rivers.

less symmetrically along either side of it. It is also characterized by the presence of major fracture zones running at right angles to the main rift system, along which horizontal displacements, sometimes of many miles, have taken place, such as might be expected if lateral movements of great blocks of the earth's crust were taking place, not all at exactly the same rate, along the course of the rifts (Fig. 19-1).

In other parts of the world there are mountain systems running somewhat parallel courses that show compressional features as if they had been pushed up by horizontal pressures in the earth's crust. Along the western rim of the Pacific, for example, is a great system of offshore deep trenches, followed on their western sides by island chains or continental margins showing compressional characteristics.

It has also been determined that these rift and ridge systems are the principal sites of tectonic activity of the world. The majority of all recorded earthquakes are centered along them, as is most volcanism.

Evidence that the volcanic rock found near the axis of the mid-ocean ridges is younger than that found at some distance to either side and that the rocks on either side are progressively older the farther they are from the central rift has come from a study of magnetic anomalies. When rock or sediments are formed, magnetic elements are oriented in a definite manner with respect to the magnetic poles of the earth, forming as it were, permanent compasses oriented according to the position of the magnetic poles at the time the rock was formed. From the study of terrestrial formations, it has become clear that at certain times in the past the magnetic field of the earth collapsed and then was reconstituted. It appears that the magnetic poles may be reconstituted either in their original positions or in a reversed position. Several such reversals have been discovered and dated by the known ages of the strata in which they occur. Similar correlated magnetic reversals have also been found in some deep-sea sediment cores.

Along either side of the mid-ocean ridges, symmetrically placed patterns of magnetic reversal have been found corresponding to the magnetic reversals already known from their vertical locations in strata and cores from other regions. Along the sides of the mid-ocean ridges, however, these patterns of magnetic reversal are laid out as horizontally disposed bands paralleling the axis of the ridge systems. This arrangement would seem to indicate that new rock has been forming along the axis of the ridges and has been transported laterally in both directions. The distances between successive magnetic reversals, together with the known dating of the reversals with which they have been correlated, give us a means of estimating the rate of this lateral spreading of the ocean floor. The rate, for the most part, seems to be about 1 to 4 cm. per year. Where the spreading is relatively fast, as along the East Pacific Rise, the resulting submarine topography is one of gently rising and undulating hills. Along the Mid-Atlantic Ridge, where the rates of spreading are somewhat slower, steeper, more rugged mountain ranges have been formed, paralleling the axis of the rift on both sides, apparently because much more new lava is poured out and piled up along the axis of the rift before it is carried away by the lateral movements of the earth's crust.

Other lines of evidence pointing to the same conclusions lie in the lack of, or thinness of, eupelagic sediments near the axis of the ridge systems and their increasing thickness laterally. The thinner sediments contain only the recent top layers of the marine sediment, but the thicker sediments at greater distances from the axis contain progressively older sediments as well.

Even the history of continental faunas seems to bear out the idea. The fact that, in a much shorter geological period of time, mammals and birds have each diversified into about twice as many orders as did the reptiles over a much longer time span has been attributed to the fact that during the period of the origin and early diversification of the reptiles, there were only one or two great land masses. Hence there was less isolation of different populations of the more primitive reptiles. In the case of the mammals and birds, however, the continents drifted apart shortly after the origin of the groups, isolating different populations of their more primitive members, each of which proceeded to undergo adaptive radiation to about the same extent that the whole group would have done had the populations not been isolated.

The crustal movements seem to be caused by gigantic convection currents in the mantle beneath the earth's crust. The material being brought to the surface at the midocean ridge systems should not be thought of as pushing the rest of the sea floor away laterally. Rather, widespread movements of great segments of the earth's crust, which floats on the more plastic material of the mantle, are pulling the crust apart along these lines, and the new crustal material originating there can be thought of as a type of compensatory upwelling. In some regions, land masses are included with the parts of the crust being rafted. Along the western Pacific the corresponding downwelling occurs just to the east of the main continental mass, producing the deep trenches where the mantle turns downward and pulls the crustal material with it back into the mantle. The pressures of this down-turning current, which perhaps in part is deflected downward by the continental platform and which perhaps meets another, similar current of the mantle flowing in the opposite direction beneath the continental platform, produce the island chains and mountain ridges to the west of the trenches, which show such marked compressional characteristics.

There is some evidence that after great trenches such as those along the eastern sides of the Pacific island chains are formed, they eventually become filled with sediments and, for reasons not yet clearly understood, are later raised, forming new island chains. New deep trenches are formed to the seaward. The shallow seas between the earlier island chains and the mainland eventually also fill with sediments, thus extending the continental border.

Be that as it may, it is now abundantly clear that the earth's crust is not as stable as had formerly been supposed. Present continents have not always occupied their present relative positions. There has been, at least since the end of the Mesozoic era, an extensive exchange of materials between the earth's crust and the mantle, involving for the most part crustal material of the floors of the oceans where the crust is thinnest. The major continental platforms, although they have undergone changes in position, and, of course, erosional and tectonic changes of their surfaces and edges and even perhaps been split apart, have nonetheless existed as recognizable entities for a far longer time than have the present floors of the oceans. The Atlantic Ocean is a relatively new ocean, formed in the gap as the American continents and the great Eurasian-African land masses were rafted apart.

This explains why deep-sea sediments have not formed layers approaching the thickness calculated for them on the basis of known rates of sedimentation and assuming relatively stable oceanic basins since at least Cambrian times. In a way, this is rather disappointing to paleontologists. It had formerly been believed that, since the sea floors are relatively quiet regions protected from most of the rapid erosional and other changes so noticeable on and near land, they must have, locked in their deeper sediments, a far more complete and far better record of the past world history than could be found on land. Now, just as technology is developing to the point at which we can look forward to obtaining good samples from these ancient oceanic sediments, we find that they do not exist—that the present floors of the oceans, for the most part, date only from about the end of the Mesozoic era and that all their priceless, more ancient sediments have slipped away forever beneath the earth's crust, returning to the mantle whence they came and from which, even now, new ocean floor materials are rising and spreading out along the midocean ridges.

For extremely ancient fossils and geological history of the earth and its oceans, we must, then, continue to rely on sediments formed where ancient seas flooded areas of continental platforms, laying down sediments that have been outside this great circulation of crustal and mantle materials but that have been subjected to the vicissitudes of terrestrial environments. (See pp. 417 and 423 for possible relation of continental drift to changes in world climate and the onset of the ice ages.)

ICE AGES AND WORLD CLIMATE

Curiously enough, it now appears that the Arctic ocean was relatively open and ice free during early stages of the great Pleistocene glacial ice ages, but was largely frozen over during the late glacial and most of the interglacial periods, such as the present. This phenomenon, according to the theory of Ewing and Donn (1956), is the result of a type of cyclic exchange of great volumes of water between the Atlantic Ocean, the Arctic Ocean, and continental masses of ice. According to this theory, during interglacial periods the glaciers melt and retreat, and their water is added to the oceans, gradually raising the sea level. The increased exchange of water between the Arctic Ocean and the North Atlantic would have the effect of cooling the Atlantic water and of warming the Arctic Ocean, finally causing it to become ice free. A relatively warm, ice-free polar sea whipped by arctic winds would contribute great amounts of water vapor to the air, resulting in blizzards and storms, which would deposit, over thousands of years, a blanket of ice over the northern portions of nearby continents. Eventually the withdrawal of water from the oceans, forming these ice sheets, would reach a critical point at which the exchange of water over the shallow sills between the North Atlantic and the Arctic Ocean would no longer be sufficient to warm the polar sea. The Arctic Ocean would freeze over, cutting off the supply of water vapor and putting an end to the massive deposition of continental ice. The reduced exchange between the Atlantic and the polar sea would result in a warming of the Atlantic waters, and a gradual recession of the ice sheets would begin, again contributing to a rising of sea levels. This theory accounts nicely for known changes in sea levels as shown by marine terraces, etc. and has subsequently been further bolstered by archeological findings along the shores of the Arctic Ocean, which seem to indicate that it was, in fact, ice free during early parts of the last ice age.

Since their elaboration of this theory of ice ages in 1956, Ewing and Donn have modified it in certain important aspects, especially with respect to the later stages of the growth of the ice sheets and the termination of ice ages. In its new form the theory may be restated as follows:

The migration of the geographical poles to regions of thermal isolation (or perhaps the thermal isolation of the geographical poles by migration of the continents) in the Tertiary brought about the development of a strongly zoned world climate with high latitude temperatures at the threshold for glaciation. Continental glaciation probably began in Antarctica, where it has persisted to the present, contributing to a general cooling of both hemispheres.

Northern Hemisphere glaciation was initiated in the arctic latitudes by increased precipitation caused by an ice-free Arctic Ocean. However, after the glaciers reached a certain critical size, their refrigerating effect was sufficient to cause them to become self-perpetuating in North America and Europe, where they could draw on abundant moisture provided by the North Atlantic and, in western North America, the North Pacific. In Siberia the central desert limited glacial growth because of lack of moisture. The freezing of the Arctic Ocean marked the cessation of glacial growth in that region. The size of the Siberian ice sheet probably represents the approximate size the glaciers would have attained elsewhere if their growth had been fed chiefly, or only, by moisture provided by the Arctic Ocean.

As the glaciers grew in extent, they finally reached the point at which warming from the south was in dynamic equilibrium with their own refrigerating effect. By this time, or sometime thereafter, the continued cooling of the oceans lowered their vapor pressure, by as much as 50% over much of the North Atlantic and to a smaller but significant extent elsewhere. Thus the atmospheric moisture available for building glaciers was reduced beyond the equilibrium value, and a retreat began. Retreating glaciers leave wet soil in the regions of their margins, which greatly increases the absorption of radiation, the albedo of wet soil being half that of dry soil. This warming effect would be enhanced through heat absorption by dirty, stagnant, marginal ice. Once the retreat began, the warming effects continued, and the general level of atmospheric moisture remained low because of a significant lag between the beginning of the retreat and the warming of the surface waters of the oceans. The oceans were still being fed great quantities of nearly ice-cold freshwater, which tended to remain at the surface because of its lower density.

By the time the temperatures of the ocean surfaces had increased sufficiently to provide a significant increase in moisture, the ice sheets had been reduced beyond the critical size at which, by their refrigerating effect, they could benefit from this increase and had retreated to regions dependent chiefly on the Arctic Ocean for moisture. Thus they were unable to build themselves back. The retreat continued at an accelerated pace. Only on Greenland, which was surrounded by water and had sufficient moisture to keep its ice pack built up, was there no retreat.

The critical point is that both the initiation and the termination of ice ages were caused by changes in precipitation rather than by initial changes of temperature. The growth and decline of the glaciers were responsible for, rather than caused by, changes in world temperature. The lag between the onset of glacial retreat and the warming of the oceans allowed the retreat to proceed beyond the point of no return, until such time as the Arctic Ocean itself is again ice free, providing moisture to initiate a new cycle of glaciation.

Probably the reason the ice ages began when they did and were not a chronic feature of the world's history throughout earlier geological ages lies in the different relationships of the continental masses and oceans during the earlier periods. It was not until continental drift rafted the continents into their present spatial relationships that the poles were in positions of thermal isolation and the thermostatic effect of the exchange between the Arctic Ocean and the North Atlantic could cause these tremendous climatic oscillations.

PART FOUR

THE IMPACT OF MAN

This earth is sacred

The following letter, written in 1855, was sent to President Franklin Pierce by Chief Sealth of the Duwamish Tribe of the state of Washington. It concerns the proposed purchase of the tribe's land. Seattle, a corruption of the chief's name, is built in the heart of the Duwamish land.

The Great Chief in Washington sends word that he wishes to buy our land. The Great Chief also sends us words of friendship and good will. This is kind of him, since we know he has little need of our friendship in return. But we will consider your offer, for we know if we do not so, the white man may come with guns and take our land. What Chief Sealth says, the Great Chief in Washington can count on as truly as our white brothers can count on the return of the seasons. My words are like stars—they do not set.

How can you buy or sell the sky—the warmth of the land? The idea is strange to us. Yet we do not own the freshness of the air or the sparkle of the water. How can you buy them from us? We will decide in our time. Every part of this earth is sacred to my people. Every shining pine needle, every sandy shore, every mist in the dark woods, every clearing and humming insect is holy in the memory and experience of my people.

We know that the white man does not understand our ways. One portion of the land is the same to him as the next, for he is a stranger who comes in the night and takes from the land whatever he needs. The earth is not his brother, but his enemy, and when he has conquered it, he moves on. He leaves his fathers' graves, and his children's birthright is forgotten. The sight of your cities pains the eyes of the redman. But perhaps it is because the redman is a savage and does not understand. . . .

There is no quiet place in the white man's cities. No place to hear the leaves of spring or the rustle of insect's wings. But perhaps because I am a savage and do not understand—the clatter only seems to insult the ears. And what is there to life if a man cannot hear the lovely cry of a whippoorwill or the arguments of the frogs around a pond at night? The Indian prefers the soft sound of the wind darting over the face of the pond, and the smell of the wind itself cleansed by a mid-day rain, or scented with a piñon pine. The air is precious to the redman. For all things share the same breath—the beasts, the trees, the man. The white man does not seem to notice the air he breathes. Like a man dying for many days, he is numb to the stench.

If I decide to accept, I will make one condition. The white man must treat the beasts of this land as his brothers. I am a savage and I do not understand any other way. I have seen a thousand rotting buffaloes on the prairies left by the white man who shot them from a passing train. I am a savage and I do not understand how the smoking iron horse can be more important than the buffalo that we kill only to stay alive. What is man without the beasts? If all the beasts were gone, men would die from great loneliness of spirit, for whatever happens to the beast also happens to man. All things are connected. Whatever befalls the earth befalls the sons of earth.

Our children have seen their fathers humbled in defeat. Our warriors have felt shame. And after defeat, they turn their days in idleness and contaminate their bodies with sweet food and strong drink. It matters little where we pass the rest of our days—they are not many. A few more hours, a few more winters, and none of the children of the great tribes that once lived on this earth, or that roamed in small bands in the woods, will be left to mourn the graves of a people once as powerful and hopeful as yours.

One thing we know which the white man may one day discover. Our God is the same God. You may think now that you own him as you wish to own our land. But you cannot. He is the Body of man. And his compassion is equal for the red man and the white. This earth is precious to him. And to harm the earth is to heap contempt on its creator. The whites, too shall pass—perhaps sooner than other tribes. Continue to contaminate your bed, and you will one night suffocate in your own waste. When the buffalo are all slaughtered, the wild horses all tamed, the secret corners of the forest heavy with the scent of many men, and the view of the ripe hills blotted by talking wires, where is the thicket? Gone. Where is the eagle? Gone. And what is it to say good-by to the swift and the hunt, the end of living and the beginning of survival.

We might understand if we knew what it was that the white man dreams, what hopes he describes to his children on long winter nights, what visions he burns into their minds so they will wish for tomorrow. But we are savages. The white man's dreams are hidden from us. And because they are hidden, we will go our own way. If we agree, it will be to secure the reservation you have promised. There perhaps we may live out our brief days as we wish. When the last redman has vanished from the earth, and the memory is only the shadow of a cloud moving across the prairie, these shores and forest will still hold the spirits of my people, for they love this earth as the newborn loves its mother's heartbeat. If we sell you our land, love it as we've loved it. Care for it as we've cared for it. Hold in your mind the memory of the land, as it is when you take it. And with all your strength, with all your might, and with all your heart—preserve it for your children, and love it as God loves us all. One thing we know—our God is the same. This earth is precious to him. Even the white man cannot be exempt from the common destiny.

NOTE: Certain historical discrepancies in this letter have been called to my attention since the pages of this book were set in type. For example, it seems unlikely that Chief Sealth could have had direct experience with trains, talking wires, and massive pollution prior to 1855. On looking into the matter, I find that these do not appear in early versions of his *Message to the White Chief* (see Lorch, F. W., et al. [eds.] 1946. Of time and truth. The Dryden Press, Inc., New York). At some point along the line, the truth was apparently not as sacred as the earth to someone. It is a real shame, since there is so much truth both in the powerful and moving statement that Chief Sealth did make and in the substance of the additions that have been made to it, if not in their attributions to Chief Sealth and the historical timing. The fudging blemishes a powerful and beautifully written statement. The basic message should remain undiminished.

FIG. 20-1. One-meter plankton net being launched from deck of M.S. *Horizon,* a research ship of the Scripps Institution of Oceanography. (M. W. Williams photograph.)

CHAPTER

20 THE GROWTH OF THE MARINE SCIENCES

SCIENTIFIC GROWTH

The history of science is a history of great bursts and surges, now in one direction, now another, as new insights, new methods, and new techniques open up unexpected directions for inquiry or provide rationale and instrumentation that make possible inquiry in areas formerly amenable only to speculation.

In previous centuries these basic new insights and the development of new, greatly improved methods of scientific inquiry were relatively scattered, isolated events both in time and space. Scientists were few in number, communication was slow, and the spread and use of new ideas were slow and erratic. The social milieu was often unfavorable for dissemination of new ideas. Some of the basic early discoveries and techniques were developed independently in several different civilizations.

In the last two centuries, particularly in the twentieth century, the situation has changed radically. All the sciences have suddenly come to fruition with quantum leaps in technology, in sophistication of theoretical and practical methodology, in the numbers of gifted people devoting full time to scientific and technical inquiry, and in the rapidity with which new developments find diverse applications. All the sciences have become intimately interlinked. Instrumentation and conceptual approaches developed in one field quickly find application throughout the whole body of science. Each branch of science has expanded so rapidly in so many directions at once that no brief treatment can give any adequate idea of the development and ramifications of even a very restricted branch of science—let alone of a great conglomerate such as the marine sciences, in which nearly all branches of science are involved.

The irregular spurts and surges of science still characterize its growth; however, these advances have become so numerous, find applications so quickly, and ramify so quickly throughout the whole body of science that they merge into one great overall explosive growth wholly unprecedented in the history of mankind.

Marine sciences have always involved an element of excitement, adventure, risk, and romance to a greater degree than have most, from the early days of explorer adventures to modern underwater researchers with their scuba gear, submersibles, and underwater habitats.

EARLY DEVELOPMENTS IN SHIPPING AND NAVIGATION

The Egyptians and Phoenicians

Prerequisite to any real understanding of the seas was the ability to travel across them and to navigate. The arts of shipbuilding and navigation were developed independently by several early maritime civilizations. The earliest boats and ships we know of were built in Egypt and date back to 4000 B.C. They had no keel, sternpost, or internal framework, but some did use the sail. As far back as the sixth century B.C. the Phoenicians had developed a flourishing maritime civilization in the region now known as Lebanon, and not only had developed trade routes along the whole length of the Mediterranean but had circumnavigated Africa and ventured north as far as Britain and perhaps west as far as the Azores. Their seagoing ships were manned by two or three banks of oarsmen and were divided into three types: (1) the "longships," or *bylbos*—galleys built for speed and maneuverability and armed with a sharp ram at the bow; (2) "round ships," known as *gōlah,* with broader beam and greater cargo capacity for trade; and (3) large barges known as *hippos,* because the prow was a camel's or horse's head, used in calm waters. Most of the Phoenicians' voyages and explorations were carried on in coastal waters within sight of land and thus were more in the nature of piloting—using landmarks for orientation—rather than true navigation. But the Phoenicians are also credited with discovery and use of the polestar. They learned a great deal about winds and currents along their trade routes.

The Greeks

The Greek civilization of the third century B.C. developed shipping patterned somewhat after that of the Phoenicians. Perhaps their greatest contribution to exploration and navigation was the great voyage of Pytheas in 325 B.C. He sailed out through the Strait of Gibraltar, eluding the Phoenicians, who held it at that time, turned north along the coasts of Spain and France, crossed the channel to Britain, and sailed north to the Orkney Islands. There is some evidence that he may have gone north as far as Norway and perhaps west to Iceland and beyond. Unfortunately only secondhand accounts of his voyage have survived, and the exact methods he used for navigating are not known. It is known that the Greeks had learned to determine latitude by day lengths at various times of the year, but the determination of longitude remained a very vexing problem. Pytheas also first reported the Atlantic tides and their correlation with the phases of the moon.

At about the same time that Pytheas was preparing his expedition, Aristotle produced two pioneering studies during his second stay in Athens. The *Meteorologica* is a compilation of all that was known about physical aspects of the world, together with Aristotle's own observations and conjectures. His *Historia Animalium* is the pioneer work in marine biology. He described 180 different species of the Aegean fauna and made excellent observations of their natural history.

The Europeans

From the time of the Phoenicians, Pytheas, and Aristotle, very little in the way of significantly new developments in exploration, navigation, or marine sciences occurred in Europe for several centuries until the latter part of the middle ages. In the middle of the first century A.D. Pliny introduced a new era into the history of science, in which original exploration and discovery gave way to encyclopedism—the synthesis and rearrangement of what was already known. His *Historia Naturalis* was a sort of culmination of ancient science, summarizing the knowledge of his time—a work of compilation and synthesis, rather than one of original discovery. This trend characterized most of the work done during the middle ages in Europe. Scholars turned to authority, to the wisdom of the past, rather than to direct observation and empirical methods for their enlightenment. The art of shipbuilding did, however, progress, and by the time of the Renaissance the major European nations had fleets of large sailing vessels for trade and warfare in European, Mediterranean, and North African waters.

The Polynesians

The Polynesians independently developed a remarkable maritime culture in the South Pacific islands, culminating in the great voyages of Hawaii Loa from Raiatéa, west of Tahiti, to the Hawaiian Islands in about 450 A.D., and 800 years later a second voyage by the priest Paao in 1275 A.D. to Hilo. These voyages were made in huge double 50-foot canoes built especially for the purpose from planks of the giant koa tree. They were accurately hewn to shape, lashed together with coconut sennet, and caulked with a mixture of coconut fiber and breadfruit gum. Then they were provided with a mast and triangular sail and in calm waters propelled by paddles. The Polynesians developed an intimate knowledge of seamanship and a skill in navigation that were far beyond anything in Europe at the time. Hawaii Loa is reputed to have known

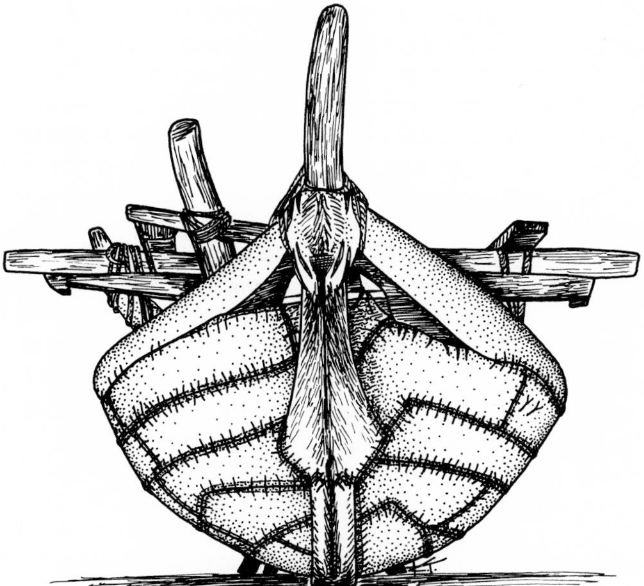

FIG. 20-2. Large Polynesian canoe.

150 stars by name, and just when certain stars would be directly above various islands. His voyage to the Hawaiian Islands involved five months at sea, skillful navigation by the sun, moon, and stars, use of cues by migrating plovers and curlews, the use of frigate birds to spot land, and an intimate familiarity with cloud forms and the shape of wave trains as affected by distant land.

The navigational achievements of the Polynesians have received little attention in the west because their civilization was unknown to Europeans until the nineteenth century A.D. and their culture was largely destroyed by the Westerners, who never took the trouble to really try to understand it. Another factor is that, although very conscious of their history, the Polynesians did not preserve it as a written record, but rather in the form of chants, or meles, and rituals, the significance of which was not generally recognized by Westerners. These chants and rituals were much more than myths or legends and reveal a detailed knowledge of natural history, navigation, and the history of the people, with genealogies going back over centuries.

THE REAWAKENING

The outstanding figure in Western Europe in science during the mid-medieval times was "the Venerable Bede" (A.D. 673-735). His earlier work, *De Natura Rerum Liber,* written about 703, was encyclopedic in the tradition of Pliny, but in *De Temporum Ratione,* about 730 A.D., he gives an excellent discussion of the retardation tides and their relation to the moon, the difference between spring and neap tides, the effects of wind in altering the time and height of high or low water, and the fact that high flood tides are followed by correspondingly low ebb tides. He illustrated his work from observations on the tides in Britain, especially in the region of Northumbria. He also was aware of the change in the time of high tide with latitude along the British coast.

Bede's writings were widely circulated in Europe during the ensuing centuries and are important not only for their scientific contributions but also because they are one of the few examples of new scientific work to emerge in Western Europe during the long interval between the decline of Greek science and the revival of the late middle ages.

In the fourteenth and fifteenth centuries A.D. a new spirit of empiricism began to manifest itself. Leading scholars came to realize that personal observation is an indispensable corrective to dogmas handed down from the past.

Leonardo da Vinci (1452-1519), one of the greatest, most versatile geniuses of all time, among other things, collected a large body of data about water, its movements, and the behavior of waves. He also studied marine fossils from Italian mountains and correctly deduced that Italy must have lain below the sea at the time these deposits were formed.

The voyage of Columbus to the new World in 1492 marks the first great new exploratory voyage by Europeans since before Christ. Although the Norsemen (Vikings) had actually preceded his crossing of the Atlantic and had reached North America by the tenth century, the voyage of Columbus was more important because of its impact on the centers of Western civilization, the resulting intellectual excitement and ferment, and the riches that soon began to pour into Europe from newly established overseas colonies.

In the ensuing thirty years North and South America had been fairly well delineated, the Pacific Ocean discovered by Balboa, and the world circumnavigated by Magellan. William Bourne, Bernard Palissy, William Gilbert, and other sixteenth-century scholars continued to develop the empirical method, depending on personal observation and experiment, which found its chief defender and exponent in Francis Bacon (1561-1626).

It was also during the sixteenth and early seventeenth centuries that Copernicus, Galileo, and Kepler revolutionized mankind's concepts of the universe and of the place of the world in the scheme of things.

Early in the seventeenth century the mathematician William Oughtred, in *The Circles of Proportion,* 1633, called for organized efforts to collect information about the oceans in a way that curiously foreshadowed the work and methods of Matthew Fontaine Maury two centuries later, which have proved so fruitful, and of the International Council for the Study of the Sea. Oughtred wrote:

> If the masters of ships and pilots will take the painses in the journalls of their voyages to dilligently and faithfully set downe in several columnes, not only the Rumbe they goe on, and the measure of the ships way in degrees, and the observation of latitude and the variation of their compasse, but also their conjectures and reasons of the connection they make of the aberrations they shall find, and the quality or condition of their ship, and the diversities and seasons of the windes, and the secret mothions or agitations of the seas, when they beginne, and how long they continue, how farre they extend, and with what inequallity; and what else they shall observe at sea worthy of consideration, and will be pleased to freely communicate the same with artists, such as are indeed skilfull in the mathematicks and lovers and inquirers of the truth: I doubt not but there shall in convenient time be brought to light many necessary praecepts, which may tend to the perfecting of navigation, and the health and saftie of such whose vocations doe enforce them to commit their lives and estates in the vast and wide ocean to the providence of god.

This suggestion went unheeded until two centuries later.

Seventeenth century

Next to Newton, the greatest British scientist of the seventeenth century was Robert Boyle (1627-1691), a man of exceptionally wide-ranging interests. He inspired many others to perform experiments, in addition to those carried out by himself. His six essays on the seas were very influential. He did much to clarify problems of the relation of temperature and salinity to depth, and clarified the concepts about hydrostatic pressure.

The beginnings of microscopy and the opening up of the microbial world to the consciousness of man by Anton van Leeuwenhoek must be ranked as another major scientific breakthrough of the seventeenth century.

Toward the end of the seventeenth century the enthusiasm of mid-century scientists for marine science declined. There is probably no single reason for this decline, but two factors may have contributed to it:

1. The publication of Newton's *Principia* accelerated a trend away from chemistry, meteorology, and oceanography and similar practical sciences in favor of physics, mathematics, and astronomy. But Newton's achievement was so outstanding that few were able to add significantly to what he had done; the net result was a period of semistagnation.
2. Marine sciences of the time were faced with nearly insuperable difficulties. In an age when technology was rudimentary at best, the need for instrumentation for precise measurements in all branches of oceanography, although clearly appreciated, could not be met. Although many types of apparatus were designed, they were not satisfactory without modern materials and techniques. Testing them was usually very laborious, and the results were discouraging.

Many people also believed that the problems of marine sciences had been adequately solved. Newton had explained the tides. Halley had established the causes of trade winds and explained to most people's satisfaction the cause of the current in the Strait of Gibraltar. Boyle had dealt with the problems of temperature and salinity. Only Boyle, Oldenberg, Hooke, and a few others appreciated the value and need for more accurate measurements of such things as small differences in salinity, temperature, and chemistry of various parts of the sea. Toward the end of the century even Hooke became discouraged, believing that he was alone in an intellectual environment of indifference. War had made marine research more dangerous, but indifference made it nearly impossible.

Eighteenth century

Captain Cook's voyages (1768-1771, 1772-1775, 1778-1779). The next great strides in marine science were made by Captain James Cook, who was not only one of the great explorers and adventurers of all time, but also a scientist of high caliber in his own right. He made important contributions to navigation, seamanship, mapping, and geography. He and the scientists he included on his voyages made important surveys of the geography, geology, faunas, and floras of far-flung Pacific areas from the Antarctic Ocean to Australia, the Pacific islands, and the Pacific Northwest of North America. His voyages inaugurated the great "collecting phase" of eighteenth-century natural sciences.

Using the new chronometer invented by John Harrison in 1761, Cook solved the vexing problem of exact determination of longitude. He also was the first to systematically employ antiscorbutic diets for his crews and demonstrate their great value, reducing losses to scurvy from the usual 33% on long voyages to zero on his second voyage.

Cook's first voyage on H.M.S. *Endeavour* left England in 1768. He succeeded in mapping Eastern Australia, Java, Sumatra, New Zealand, and many Pacific islands. Everywhere they landed he and the scientists with him made collections and observations of the flora, fauna, geology, customs, and tools of the natives. The artist Parkinson made remarkable drawings of everything.

His third and last voyage resulted in the discovery of many Pacific islands, including the Hawaiian Islands. From there they sailed north and explored the west coast of North America and finally back to Hawaii, where Cook was killed during an unfortunate brawl between some of his crew and the natives.

Alexander von Humboldt's voyage (1779-1804). The voyages of Captain Cook were followed at the turn of the century by that of Alexander von Humboldt to South and Central America. Although not as extensive as Cook's voyages, Humboldt's voyage was a very ambitious and significant achievement, and included wide-ranging studies in biology, geology, meteorology, astronomy, and social life of the peoples encountered. Its reports comprise thirty volumes and are a milestone of scientific achievement.

Nineteenth century

Edward Forbes. The first in-depth studies of sublittoral marine benthic faunas were made by Edward Forbes (1815-1854), a brilliant young naturalist who at 12 years of age had completed a manuscript entitled "A Manual of British Natural History in All Its Departments." In 1839 he induced the British Association for the Advancement of Science to set up a committee to investigate the marine zoology of Britain. In the late 1830's and early 1840's he and John Gordon carried out a series of dredgings in the Shetland Sea and adjacent waters, the results of which were presented to the British Academy for the Advancement of Science. He authored *A History of British Starfishes* and made studies of the zonation of living organisms as related to depth. He pioneered in the study of population dynamics, and in interpreting present distributions and conditions in terms their geological history. Equipment available at that time limited dredgings to relatively shallow shelf waters.

His studies of the amount of living organisms as related to depth, together with theoretical considerations respecting the extreme conditions—no light, extreme hydrostatic pressure, etc.—at great depths led him to conclude that there could be no life below about 300 fathoms. Forbes was also a brilliant and inspiring teacher who exerted a strong influence on Charles Wyville Thompson and others who were to become, in turn, outstanding pioneers in marine sciences.

Although directly responsible for overturning some of Forbes's conclusions on important points, Thompson fully appreciated the value and originality of Forbes's contributions. His tribute to Forbes is a model of generosity and understanding appreciation:

> Although we now look somewhat differently on cer-

tain fundamental points—to Forbes is due the credit for having been the first to treat these questions in a broad philosophical sense and to point out that the only means of acquiring a true knowledge of the rationale of the distribution of our present fauna is to make ourselves acquainted with its history; to connect the present with the past. Every year adds enormously to our stock of data and every new fact indicates more clearly the brilliant results which are to be obtained by following his methods, and by emulating his enthusiasm and indefatigable industry.

Tragically Forbes died at the early age of 39 years.

Charles Darwin. About the same time that Forbes was beginning his studies, another young British naturalist, Charles Darwin, was offered, in 1831, the post of naturalist on the *Beagle,* a ship commissioned to do mapping and exploration around South America and in the Pacific. Everywhere they went Darwin made careful observations and collections of the fauna, flora, and geological specimens. He was particularly impressed by the unique conditions found on many oceanic islands, especially the Galápagos Islands and various coral atolls, and by the strange fossils discovered in South America and elsewhere. He was the right man at the right time.

Had they gone much earlier, the state of knowledge and theory would have been such that the significance of islands would probably have been missed altogether. Somewhat later—much of the story would have been largely erased. Today on a worldwide scale the unique island populations are vanishing—many never having been seriously investigated. Man brought himself, his cultivated plants, cats, pigs, goats, rats, roaches, weeds, etc. to islands the world over—and the rare, antique, strange, fragile, beautiful faunas and floras are disappearing without a trace.

Darwin spent the next twenty years after the return of the *Beagle* in semiseclusion, studying, compiling mountains of data, and meditating on the theory of evolution that was shaping up in his mind. Two events—the publication of Thomas Malthus' essay on population, which provided the key concept of population pressure and of the ensuing struggle for survival and survival of the fittest, and the independent, intuitive leap to the same conclusions by the explorer Alfred Russel Wallace in Java—finally galvanized Darwin into putting much of his material together and publishing his monumental *Origin of Species* in 1859.

Hewett Watson—a contemporary of Darwin—wrote to him after its publication: "It has the characteristics of all great natural truths, clarifying what was obscure, simplifying what was intricate, adding greatly to previous knowledge." One might add to this a fourth characteristic—providing a focus and stimulus for further highly significant and fruitful inquiries.

One of the great unsolved mysteries of the seas in the mid-1800's was still the problem of whether life could exist at great depths. Forbes had concluded that it could not.

Charles Wyville Thompson. In the summers of 1868 to 1870 Charles Wyville Thompson persuaded the British admiralty to put the small vessel the *Lightning* and subsequently the *Porcupine* at his disposal for deepwater dredging and sampling. Although the ships were not well suited for the purpose and were plagued by rough weather, they did achieve highly significant results, bringing up a variety of living animals from greater depths than ever before, and proving that life exists at least to considerably greater depths than Forbes had thought possible. They also made the first deepwater temperature measurements, and some of their depth-temperature profiles were so good that the thermocline is clearly recognizable in them—a great achievement considering the state of the technology for such work at the time. Their bottom mud samples also confirmed the similarity of the globigerina ooze, earlier noted by Brooks in 1854, to the chalk cliffs so prominent along parts of the English channel.

Thompson wrote the first general reference on oceanography, *The Depths of the Sea,* in 1873. His book, although outdated in many ways, is still useful.

Certain peculiar more or less amorphous membranous or pellicle-like formations found in preserved bottom samples about this time were studied by Huxley and Haeckel and regarded by them as a primitive living material. Huxley named it *Bathybius haecklii*. They proposed a theory quite the opposite of that of Forbes—that the sea bottom might be largely covered by a primitive living slime, from which other forms of life may have arisen.

Occasional animals brought up on sounding lines from deep water, and in particular a group of fifteen species found living encrusted or attached to a defective communications cable that had been hauled up from more than 1000 fathoms in the Mediterranean near Sardinia after the cable had been there for three years, indicated that animals might occur at much greater depths than had been thought possible.

The results of the dredgings from the *Lightning,* the *Porcupine,* and the *Shearwater* in 1868 to 1871 from depths as great as 2,435 fathoms were so intriguing, and the excitement generated by Darwin's new theory of evolution, together with the diverse and opposing views about the possibility of life at great depths, aroused so much interest that Charles Wyville Thompson with the help of the Royal Society persuaded the British Admiralty to outfit a ship and sponsor an expedition for worldwide study of the ocean floor.

The Challenger Expedition. In December, 1872, the *Challenger,* under Captain George Nares, with a scientific staff headed by Charles Wyville Thompson and John Murray, sailed from Portsmouth as the first and one of the greatest deepwater and ocean-floor scientific expeditions ever mounted. They were gone three and one-half years and covered about 68,890 miles. They took deepwater soundings and dredgings at 360 deepwater stations scattered over 140 million square miles of ocean floor, and to depths as great as about 5½ miles.

The achievements of the *Challenger* Expedition were tremendous. For the first time soundings, dredgings, and bottom samples were taken from the ocean floor in all parts of the world. For the first time the general contours and depths of the ocean floors of much of the world were roughed out. They made systematic well-preserved collections of marine animals and plants from all over the world, including 4417 species and 715 genera new to science. They made the first worldwide systematic plots of ocean currents and temperatures. They demonstrated that animals were to be found even at the greatest depths—but no trace of the primordial *Bathybius* was found anywhere. Their chemist, J. Y. Buchanan, finally determined that *Bathybius* was not a living organism but only a precipitate of sulfate of lime, formed when certain types of bottom mud samples were preserved in alcohol.

After the return of the *Challenger* to Britain a commission was set up at the University of Edinburgh to organize and evaluate the results. Thompson did not live to see the completion of this work, but after his death in 1882 it was ably carried on by John Murray. The results were eventually embodied in 50 large-folio volumes. Most of these are monographs of various groups of marine organisms by some of the outstanding biologists of the time, including Alexander Agassiz, H. N. Moseley, and Ernst Haeckel, whose three-volume monograph of the Radiolaria is one of the most famous and handsomely illustrated biological monographs ever produced. He also made outstanding studies and illustrations of various other groups. John Murray's studies of marine muds and other sediments laid a firm foundation for our understanding of the nature of sediment. The myth of the lost continent of Atlantis was laid to rest. Many important contributions to hydrography, meteorology, chemistry of seawater, geology, and petrology were also made.

The *Challenger* Expedition can be said to have opened up the exploration of the marine world and paved the way for the development of modern oceanography in much the same sense that Darwin's theory of evolution brought our understanding of biology into a new focus and provided direction and stimulus for the great advances in all branches of biology that have taken place in the last 120 years.

After the *Challenger* Expedition most of the marine exploratory work for the remainder of the nineteenth century centered around the discovery and description of new forms of life in the oceans. Only in the twentieth century did analytical oceanography develop into a great science in its own right.

Early American marine science

In America marine science in the mid-nineteenth century was carried out especially by Matthew Fontaine Maury, Alexander D. Bache, the self-taught mathematician-physicist William Ferrel, and the biologist Louis Agassiz.

Maury was director of the United States Navy's Depot of Charts and Instruments from 1847 to 1861. He greatly expanded its activities and also began the systematic collection and study of data from ships all over the Atlantic regarding winds, currents, depths, etc. In this endeavor he was uniquely successful. In return for data, his office supplied shipmasters with continuously updated charts, showing how best to take advantage of winds and currents to shorten sailing time. In 1854 he published the first contour map of the bottom of the Atlantic Ocean. The following year he published his famous *The Physical Geography of the Sea,* which was an immediate success, going through six editions in four years and being translated into several languages. Although some of the theoretical aspects were oversimplified and contradictory, Maury's book did serve to expand many people's knowledge of the seas and to stir up interest. It also spurred to greater effort those persons intent on proving him wrong on various points.

William Ferrel was a self-taught mathematician and physist with an interest in marine problems. In 1856 he published an "Essay on Winds and Currents of the Sea" in the *Nashville Journal of Medicine*. This essay contains the first good general discussion of the effect of the earth's rotation on winds and ocean currents.

It also disproved some of Maury's assumptions. Unfortunately it was not widely read or understood—and no corrections were made in later editions of Maury's book on the points at issue. Alexander D. Bache

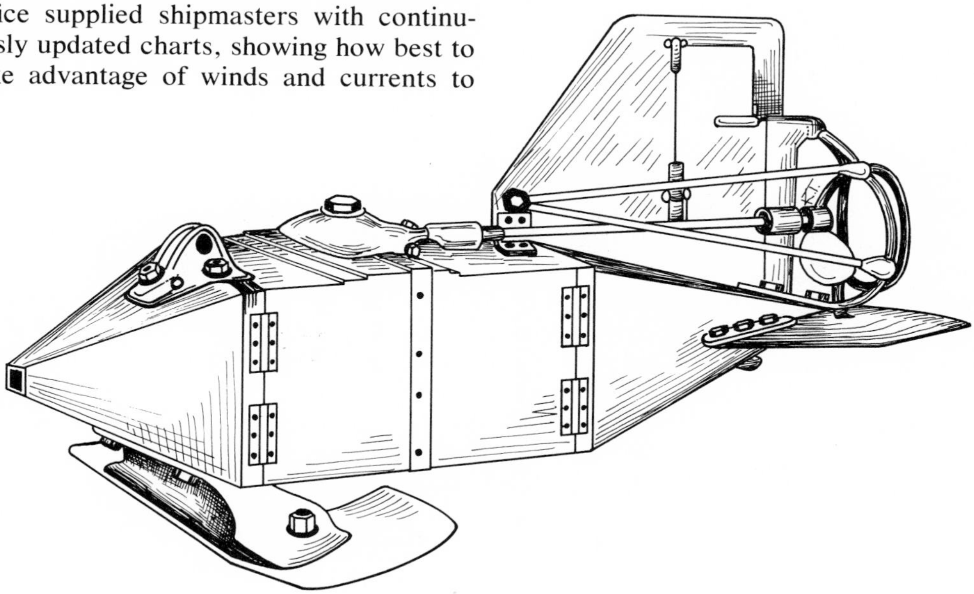

FIG. 20-3. Hardy continuous plankton sampler.

in 1843 succeeded Hassler as head of the U.S. Coastal Survey. He remained for 24 years and made it the strongest scientific organization in the government. In 1863 he also became head of the newly formed National Academy of Sciences, the formation of which he had helped to promote. In both positions he exerted a strong influence on the development of marine science. By encouraging Louis Agassiz to make use of some of the survey vessels for biological as well as oceanographic investigations, he also promoted marine biology.

Louis Agassiz and his son Alexander Agassiz did more to promote marine biology than any other Americans of the nine-

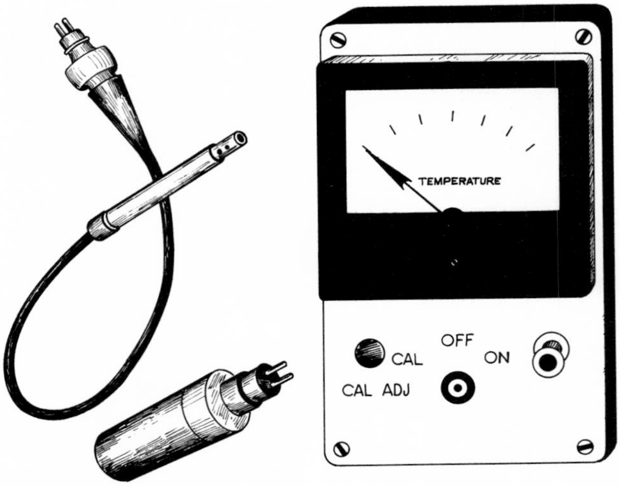

FIG. 20-4. Linearized thermistor thermometer.

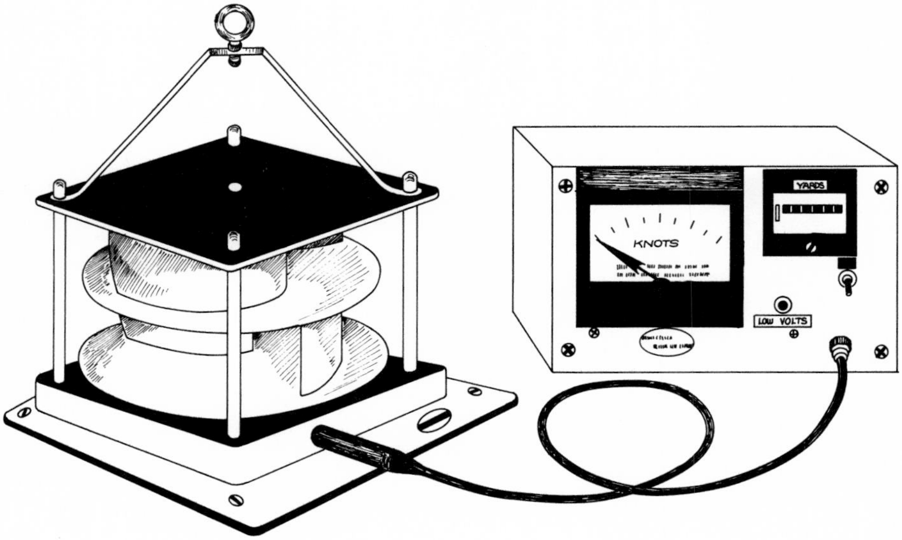

FIG. 20-5. Current measuring system.

teenth and early twentieth centuries. Louis Agassiz founded the famous Museum of Comparative Zoology at Harvard, did much marine biological work in cooperation with the U.S. Coastal Survey, and began the first marine station in America, the Anderson School. His son Alexander, after making a fortune through careful management of Michigan copper mines, devoted his fortune and his great talents to expansion of the Museum of Comparative Zoology founded by his father, and to marine exploration in both the Atlantic and Pacific. Louis Agassiz was also a great teacher and popular speaker, and did much to interest others in this work and to rally support for it.

In the period from the late 1880's to World War II, government support for sciences in the United States declined. Both the Navy Depot of Charts and Instruments, which in 1866 had been divided into two separate entities—the Naval Hydrographic Office and the Naval Observatory—and the U.S. Coastal Survey, which in 1878 became the U.S. Coast and Geodetic Survey, lapsed into a period of more routine work. Funds were difficult to

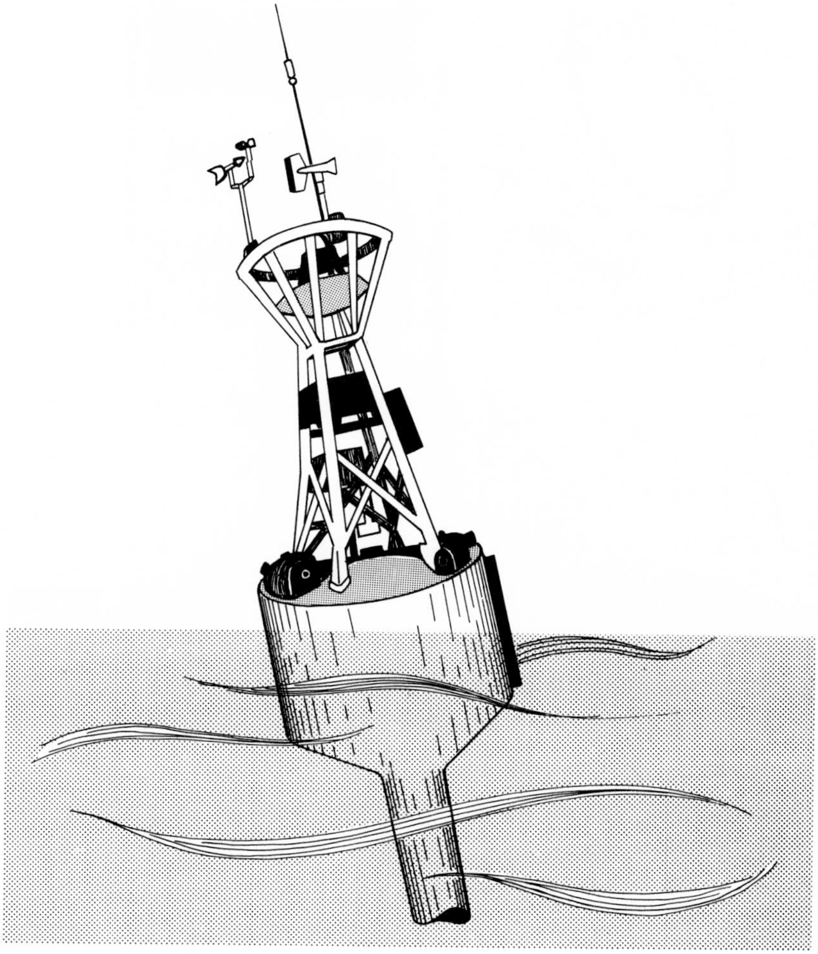

FIG. 20-6. Lockheed environmental buoy system. This modern monitoring device combines automatic recording for a variety of environmental parameters.

obtain and inspiration seemed lacking. J. S. Pillsbury's study of the Gulf Stream (1890) was the last major oceanographic work to come from the Coast and Geodetic Survey.

Whereas in Great Britain a strong tradition of mutual cooperation between the Admiralty and civilian scientists had been built up and maintained, in the United States the two groups drifted apart, perhaps in consequence of the unfortunate experiences and animosities engendered between naval officers and civilian scientists of the Wilkes Expedition (U.S. Exploring Expedition, 1838-1842).

After the Civil War the United States Congress became progressively more provincial in outlook, economy oriented, and disinclined to invest money in anything that did not promise immediate practical results. Scientific endeavor came to depend increasingly on private sources of funding.

Europe: late 1800's to World War II

Before the turn of the century a major change occurred in the European North Sea Fishery. The yawl-rigged sailing smacks gave way before the more efficient steam trawlers. With the rapidly increasing population and greater fishing intensity, it became imperative to know more about the

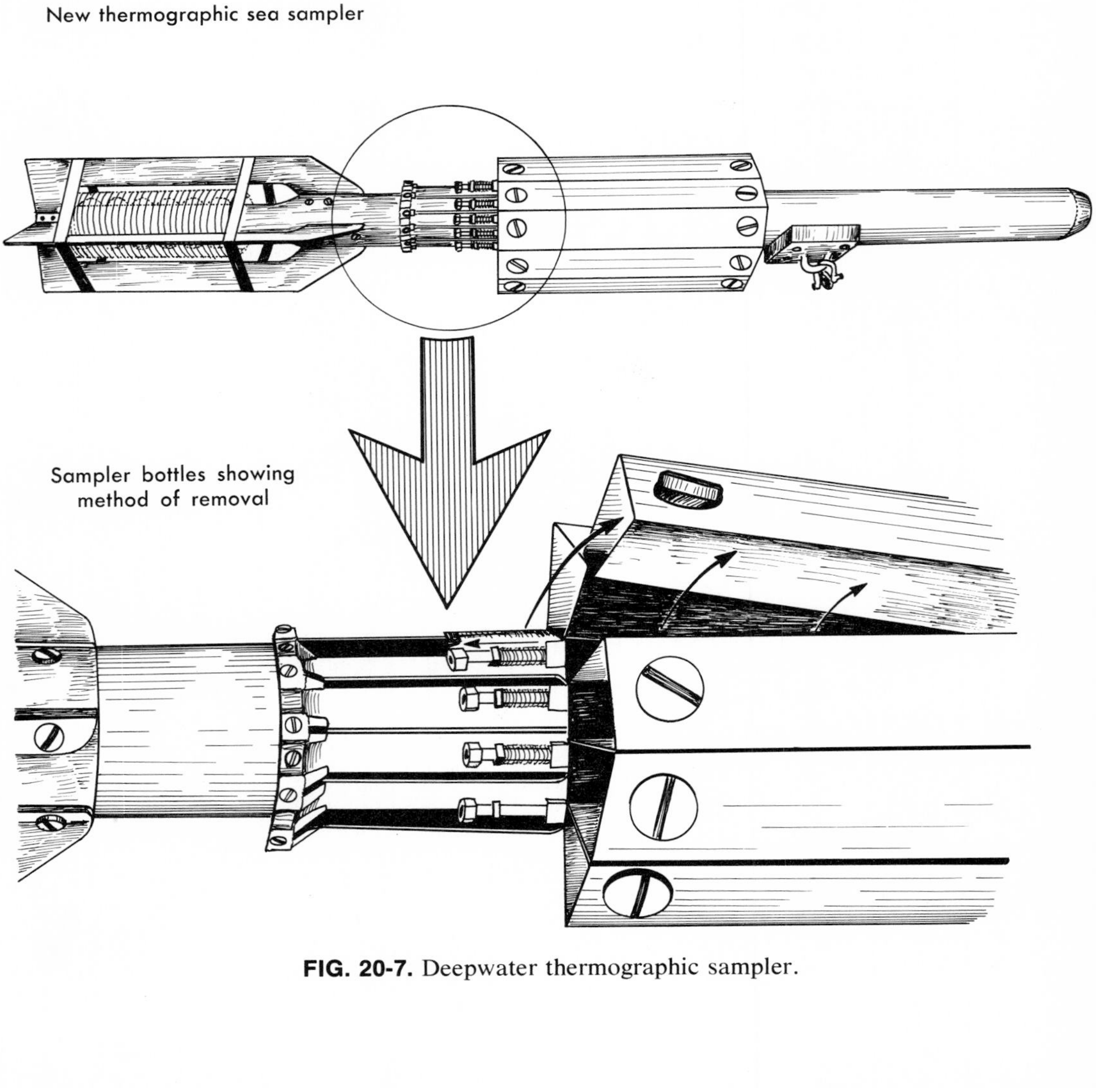

FIG. 20-7. Deepwater thermographic sampler.

biology and movements of the commercial species. Annual fluctuations in the fishing grounds were variously attributed to overfishing, to irregular migratory movements of the fish, or to movements of water masses carrying the fish with them.

In 1902 the International Council for the Exploration of the Sea (I.C.E.S.) was established in Copenhagen, Denmark. Representatives of all the countries with fishing interests in the North Sea, the Baltic, and the North Atlantic participated, including most of the prominent fisheries scientists of Europe, such as the Danish zoologist C. G. Johannes Petersen, Johann Hjort of Norway, William Garstang of England, and Johannes Schmidt of Denmark. The I.C.E.S. organized cooperative surveys in which ships from all the countries involved participated and pooled their data, and served as a focal point for cooperative research on fisheries problems, for promulgating new regulations, and for initiating marine biological research of all kinds.

By 1932 the destructive practices of the whaling industry had so reduced world stocks that it was evident to all that there would soon be no whaling at all if corrective action was not taken. The International Whaling Commission was set up to gather data on the numbers of the various species and to set quotas to protect threatened species. For many years their work was hampered by inability to adequately monitor the destruction, by the fact that compliance with their regulations was voluntary, and by the fact that some whaling nations were not members and did not subscribe to the regulations. The fact that the commission was set up and run by the whaling interests also worked against its success, since it was unduly influenced by them and the quotas set were often unrealistically high from a sustained-yield standpoint. In recent years, as the situation has continued to become more critical, some of these shortcomings have been at least partially remedied. Whaling will be discussed in more detail in Chapter 22.

Marine stations and oceanographic institutions

By the 1870's the need for shore-based laboratories and facilities for processing the collections and data gathered on expedi-

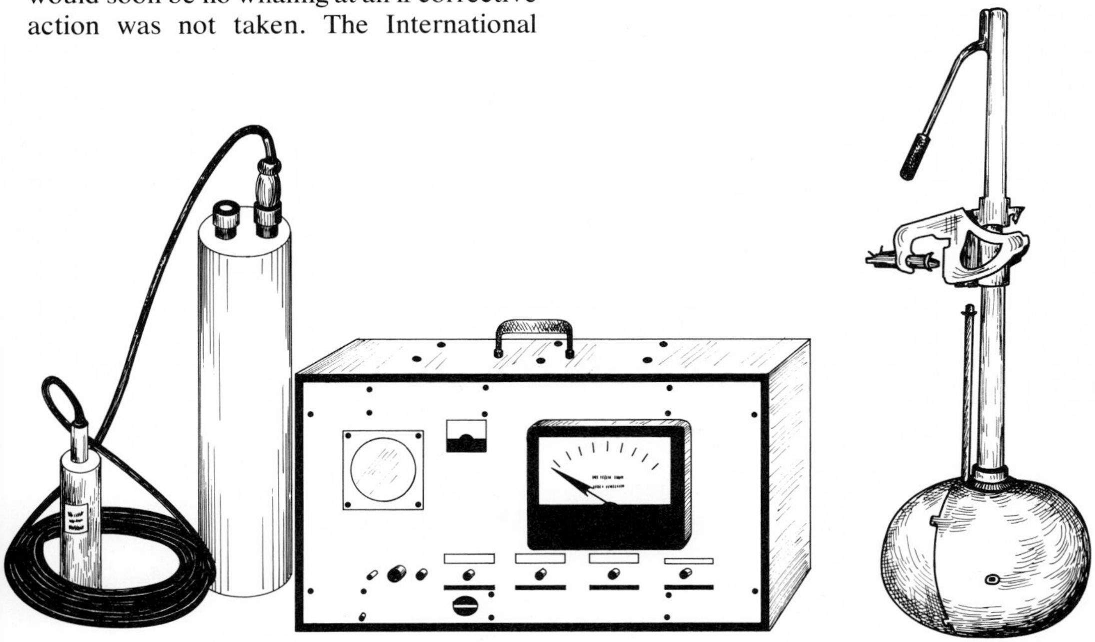

B

FIG. 20-8. A, Acoustic tracking and ranging system, and **B,** receiver antenna.

FIG. 20-9. *Flip,* a unique ship designed to be towed to station, then upended to provide a stable platform at sea from which to work.

tions, and for the study of near-shore biota and processes, was widely felt.

Lacaze-Duthiers established the zoological station at Roscoff in 1871, and Anton Dohrn founded the great international zoological station at Naples in 1872. The latter station has continued to be one of the world's leading institutions for marine biological research of all kinds.

In America the U.S. Fish Commission was established in 1871 through the efforts of Spencer Fullerton Baird, with a laboratory at Woods Hole, Massachusetts. Baird was a first-rate biologist with broad interests who, as its first director, instituted programs of basic research, as well as summer educational programs. In 1873 Louis Agassiz opened the first summer school for marine biology, the Anderson School of Natural History on Penikese Island in Buzzards Bay. This school, though short-lived (two years), was the inspiration and precedent for the establishment of the Marine Biological Laboratory at Woods Hole some years later. While serving as director of the fish commission, Baird for several years promoted the idea of establishing a marine station, patterned after that at Naples, at Woods Hole. He was not to live to see the fruition of this dream, but in 1888, one year after his death, the Marine Biological Laboratory at Woods Hole became a reality, with Dr. Charles Otis Whitman as its first director. Like its counterpart in Naples, this station, too, has flourished and become one of the world's leading institutions for biological research. Subsequently, in 1930, Woods Hole was to become the site of another important institution, the Woods Hole Oceanographic Institution, with Henry Bigelow as its first director.

On the American west coast two major marine oceanographic stations emerged in the early part of the twentieth century. The Friday Harbor Oceanographic Laboratory of the University of Washington, founded by Trevor Kincaid in 1903, and the Scripps

Institution of Oceanography at La Jolla, California, founded as the Scripps Institution for Biological Research, by William Emerson Ritter in 1911, as part of the University of California.

Prior to World War II more than a hundred marine stations or oceanographic institutions had been established in Europe and America. This number has increased since the war, as has the number of similar institutions in other parts of the world, notably in Russia, Japan, China, Australia, India, South Africa, Indonesia, and Senegal.

WORLD WAR II AND THE POSTWAR PERIOD

World War II drastically changed the picture in marine research. The development of sophisticated submarines, aircraft, and finally nuclear weapons so changed the character of warfare that navies the world over realized the need to revamp many of their basic concepts and methods. A vast amount of research therefore became necessary. The U.S. Navy soon realized its need for the expertise of civilian scientists.

The National Defense Research Committee (N.D.R.C.) and, a year later, the more comprehensive Office of Scientific Research and Development (O.S.R.D.) were organized to focus scientific effort on problems of weapons research. They operated largely by contracting for research to be done by universities or private institutions. The problems to be researched were determined largely by the armed services, but the manner in which it was to be done was largely left up to the scientists. This policy provided a fairly flexible framework for relating civilian scientists to government-sponsored research and proved to be very fruitful. By 1941, when the United States entered the war, about 450 contracts, involving about 2000 scientists, had been negotiated. In addition there was a great growth of laboratories and research groups within the armed services themselves.

Prior to the war the great problem for oceanographers had been to trim their research ambitions to fit inadequate available budgets. This problem suddenly vanished. The sudden influx of men and money had both positive and negative effects. Security procedures and compartmentalization of knowledge and information are basically incompatible with science and were a source of irritations, delays, and frustrations. Many of the new personnel were not adequately trained, and some highly trained, competent persons could not obtain security clearance.

Most important, the fact that research problems were selected by the armed forces led to a radical redistribution of interests and activities within the field of oceanography. In some cases, such as the urgent problems of submarine and antisubmarine warfare, this pointed up basic fields that were not well understood, such as the properties of sound transmission in the upper layers of water, where marked temperature changes occur.

The development of the Ewing-Vine bathythermograph, which enabled continuous temperature/depth profiles to be taken from moving ships or submarines, and of highly sophisticated sonar instrumentation and recording devices led to rapid advances in such areas, and to the discovery of unexpected phenomena such as the deep scattering layer, and the amount and diversity of underwater sounds of biological origin.

In any event, oceanography emerged from the war a much larger, more sophisticated enterprise, irrevocably tied to large-scale governmental financing, and hence to government policies, and with more emphasis on the physical aspects and less on the biological, than before.

The great advances in instrumentation that occurred during and since the war made possible effective research in many non–weapons-related projects and areas that had been discontinued, or held in abeyance, during the war, or that had been impossible previously.

Institutions such as the Woods Hole

Oceanographic Institution and the Scripps Institution of Oceanography were greatly expanded, and numerous new marine stations and oceanographic institutions were founded. One of the most significant is the Lamont Geological Observatory, with Maurice Ewing as its director.

The continued cold war rivalry between the United States and Russia has given an air of almost wartime urgency to the situation, and governmental support for oceanographic research has continued at a high level.

In the past two decades the urgency of the ecological situation has become more apparent. Biological research has again become a very prominent part of the total endeavor, with the emphasis shifted from the discovery and description of new forms, to study of basic biological processes and their interrelations with physicochemical factors of the environment, and of the effects of environmental changes both manmade and otherwise.

Government support for sea grant colleges, analogous to the earlier land grant colleges, which have played such an important role in agricultural research and development, has given further impetus to this work.

INTERNATIONAL IMPLICATIONS: THE LAW OF THE SEA

The growing worldwide exploitation of fisheries, the increasing interest in seabed minerals, oil, and other resources, and possible new military uses of the seas have become matters of international concern. Efforts are now in progress to develop, through the United Nations, a comprehensive law of the sea, and to avoid the chaos, confusion, and strife that might well develop if these matters are simply left to the competitive struggle between nations.

The doctrine of freedom of the seas, as stated by the Dutch jurist Hugo Grotius in 1609, was the accepted law of nations until recent years when problems of overfishing, pollution, increased commercial and military uses, and discovery of offshore mineral resources made this simple concept no longer practical. The basic concept of freedom of the seas simply divided the sea into the territoral sea, a narrow strip controlled by each coastal state it bordered, and the high seas, where complete freedom reigned. The territorial sea of 3 miles was originally determined by the distance a cannon could shoot. It 1945 the United States in the Truman Proclamation extended United States jurisdiction over natural resources to its continental shelf. Other nations quickly followed with similar claims, and the United Nations has engaged in a number of conferences to codify these changes.

The appeal for international control of the oceans and their vast resources was made by Arvid Padro, Maltese delegate to the United Nations in 1967. He proposed that the United Nations declare the seabed and ocean floor beyond the limits of national jurisdiction and to be the "common heritage of mankind," not subject to appropriation by any nation for its sole use. He urged the creation of a new kind of international agency to assume jurisdiction, as a trustee for all countries, over the seabed and to supervise the exploitation of its resources, with the profits to be used to promote the development of poor countries, thus excluding from the arms race an area comprising nearly three fourths of the globe.

A seabed committee was appointed in 1968, a declaration of principles adopted in 1970, and a Seabed Disarmament Treaty signed in 1971. Since then, three United Nations Law of the Sea Conferences have been held in an effort to create the necessary agreements and institutional framework.

This conference is the most far-reaching, significant, complex, and difficult international lawmaking conference ever held. Involved are six major areas of conflicting interests and concerns: fisheries, navigation (on, above, and beneath the surface of the seas), oil resources, mineral resources, pollution, and scientific research. The contestants include rival great powers, developed nations versus undeveloped nations, and coastal states versus landlocked, or "geographically deprived," nations.

Resources at stake include 80% of the world's animal life, three fifths of the petroleum resources, and a trillion and a half tons of manganese nodules rich in nickle, copper, cobalt, and manganese. Without conflict-resolving international institutions of law, the frantic scramble by the technologically advanced to claim these resources will result in worldwide disputes, conflicts, and possible environmental calamity. Measures are now before the Congress to allow the United States to unilaterally claim stretches of deep seabed for United States companies to mine. If these measures become law, they will probably dash all hope for a comprehensive law of the sea treaty in the foreseeable future.

If the Law of the Sea Conference in its coming sessions manages to overcome the remaining hurdles, involving deep seabed mining and the establishment of an international seabed authority, it will have set the stage for peaceful resolution of these problems, and may serve as a model for new laws of the world, including a new international economic order and disarmament.

The new Law of the Sea Treaty, as the negotiating text now stands, will divide the seas into six complicated, sometimes overlapping areas:

1. Internal Waters. These are areas such as bays, inside the baselines from which the other areas are determined. Each coastal state has full national sovereignty.
2. Territorial Sea. Each state has the right to establish the breadth of its territorial sea up to 12 nautical miles from a baseline established according to the convention.
3. Contiguous Zone. The contiguous zone may extend beyond this to 24 nautical miles beyond the baseline. Within these zones the coastal state has the right to "prevent infringement of its customs, fiscal, immigration and sanitary standards.
4. Exclusive Economic Zone. This may extend up to 200 miles. In this area each coastal state would have exclusive or extensive rights to all resources, living and nonliving, and the responsibility for setting up and controlling appropriate conservation measures. It would have jurisdiction over waters, as well as the seabed and its subsoil, but for maritime purposes must treat the area essentially as high seas.
5. Continental Shelf Extending Beyond 200 miles. "Broad margin" countries have a shelf that extends in some cases several hundred miles beyond the 200-mile Exclusive Economic Zone. In this area they have jurisdiction over the seabed and its subsoil, but not over the waters above.
6. International Area. This is the area left after all preceding claims. Here are most of the famed manganese nodules, which are to be mined under the aegis of International Seabed Authority.*

This proposal represents a considerable reduction from Arvid Pardo's original concept. Unfortunately for developing countries, which would have benefited from revenues from these resources, 95% of the harvestable living resources, most of the oil and gas, and a significant part of the mineral resources lie within the 200-mile economic zones.

By a method of consensus delegates working in three committees have reached wide acceptance on a number of points:

1. A 12-mile territorial sea (with another 12-mile contiguous zone)
2. A 200-mile national economic zone that will cede exclusive resource control to coastal nations
3. Revenue sharing from oil drilling beyond the 200-mile zone with the international community
4. The principle of a governing inter-

*Hudson, R. 1977. The international struggle for a law of the sea. Global Report no. 1, Oct. Center for War/Peace Studies.

national authority for the deep seabed
5. Unimpeded transit in the high seas and through present international straits
6. Principles and responsibilities with respect to pollution
7. Access to the sea for landlocked countries
8. Need for creation of a Law of the Sea Tribunal for binding settlement of disputes
9. Scientific research (to be defined)
10. Transfer of technology to developing nations
11. Obligation to preserve the marine environment

Success of the conference depends on resolution of deep seabed mining and establishment of an international authority. Accommodation must be reached between the industrialized countries and the developing countries over the composition and power of the Authority's international mining enterprise. Industrialized countries want the actual mining to be done by their corporations, which have made heavy investments. The developing countries, still hopeful for revenues, want an enterprise that will have real powers to carry out the mining. For a time, protection from competition for the land-based mineral-producing developing countries will be needed.

The issues facing the conference in its future meetings encompass all the problems of the international community: the arms race, the gap between the rich and poor, the multinational corporations, environmental degradation, food management, energy, communication, resource management, the impact of technology on institutions, and relations between nations. The Law of the Sea Treaty is therefore a vital part of world order. Recent Presidential appointments to the United States delegation give hope that with the goodwill of all nations, developed and developing, there will finally be a comprehensive treaty adopted in coming sessions of the Law of the Seas Conference.

FIG. 21-1. Manner of fishing with a flying lure at Banda. (From Weber, M. 1902. *In* Reports of the Siboga Expedition, 1899-1900. Vol. 1. Introduction and description of the expedition. E. J. Brill, Leiden, Netherlands.)

CHAPTER

21 UTILIZATION OF MARINE RESOURCES

Ever since the dawn of man's history the seas have served as a source of foods and other products for coastal populations, supplementing the use of terrestrial resources that have always been the principal sustenance for our species. Ancient Indian kitchen middens along the Pleistocene marine terraces of the Pacific coast attest to the use of *Cryptochiton, Kathrina,* clams, fish, and in some places marine mammals.

As man's numbers and technical proficiency increased, so too did the utilization of marine resources. However, our utilization of ocean resources has remained relatively primitive, corresponding to the preagricultural gathering and hunting phases of man's development on land. Although the total amount of photosynthesis in the oceans is comparable to, or perhaps a little greater than, that on land, phytoplankton is difficult to harvest in significant amounts, the effort expended exceeding the return. In the seas we have confined ourselves, with a few minor exceptions, to skimming off the animals at the higher trophic levels rather than managing them on an efficient basis, which might ensure continued returns. Meanwhile, our methods of gathering and hunting have greatly improved in both scale of operations and efficiency. The typical response to decreased returns due to overfishing has been to increase the fishing effort and the efficiency of fishing methods.

During the past century mankind has entered into the terminal phases of the logarithmic population expansion that characterizes any population of organisms confronted with an open environment—that is, an environment in which the carrying capacity for that species vastly exceeds the population of the species. This phase of logarithmic population growth is invariably followed either by a period of equilibrium, in which the numbers become more or less stabilized, or by a period of decline, which may be more or less sharp, depending on the extent to which the increase in the species destroys vital features of its own environment. In the case of man the situation has been greatly exacerbated by rapid technological developments just as he was coming into this phase of his biological history, which, in effect, opened up the environment repeatedly and also suddenly lowered death rates, resulting in a currently much greater than normal terminal population increase. The discovery of and expansion into the New World continents, together with the rapid technological developments that have enabled man to exploit levels of his environment hitherto unavailable to any animal, have enabled him to continue to expand his population and exploit his environment to a degree never before even remotely approximated by any large animal throughout the world's history.

The rapidity with which basic changes in his way of living have taken place has made man also the most bewildered and confused species of animal ever to exist on the earth. It has endowed many individuals who have been in a position to benefit from this rapid expansion with the naïve Chamber of Commerce attitude that such rapid expansion is a normal and desirable state of affairs and that our technology can be depended on to continue to open up more and more new facets of our environment for exploitation as we deplete the old ones.

Suddenly now, we begin to realize that this is not the case—that we have now reached a major ecological crossroads, that a high standard of living, or perhaps any living at all, will depend on our ability to bring the world population of mankind into a balanced equilibrium at a much lower numerical level than the present 4 billion. Such a change will, of course, entail complete reorganization and change in our economics, politics, religion, and philosophy, which, in the Western world at least, are geared to a continuous state of expansion.

In such a grave crisis, it is natural that most people first turn to the old approach that served us so well in the past and begin to look about for new facets of the environment to open up and exploit, for new technological developments that will, like a magician, pull more rabbits out of the hat for us to eat.

Thus all the great powers and major economic interests are now turning eagerly to the oceans as the last great reservoirs of untapped resources and are also busy with plans to decimate the few remaining terrestrial wild areas, such as the Amazon Valley and the jungles of western New Guinea, equatorial Africa, and Borneo, to squeeze what can be squeezed out of them.

At this stage of our social and biological evolution, when populations are on a runaway collision course with disaster and there are no effective plans in sight to limit them, other than by widespread starvation or nuclear war, and in the present state of international economic and political anarchy, when the major effort of all great powers is devoted to sterile, wasteful military posturings and confrontation, the discovery now of any further really large new, untapped resources could be a tragic development for mankind rather than a blessing.

Exploitation of new ocean resources can provide constructive outlets for the energies of a rather large number of people, but the overall effect could be to continue to divert man's attention from the directions in which possible solutions to the human dilemma actually lie. The resources discovered will be seized by the presently powerful economic and military interests to increase their power, increasing the gap between the haves and the have-nots. The bulk of any such resources will be wasted by the military instead of being conserved or used in rational ways. It could simply mean that if the time ever comes when man does bring his populations into planned and viable equilibrium with his environment, that environment will have been already much more thoroughly impoverished, and man will have to exist at a far lower level than would otherwise have been the case. If he doesn't bring his populations into a viable equilibrium, such discoveries will simply act as temporary palliatives, enabling the present disastrous increases to continue a little further and making the scale of starvation or of killing by war or pestilence just that much greater but not much farther off in the future.

The only certain element in this clouded picture of man's immediate future is that, barring nuclear war, populations will continue to increase rapidly for a few more years, after which they will be greatly reduced or exterminated.

COMMERCIAL FISHERIES

The first question that arises in connection with increased utilization of marine resources during the terminal agonies of population increase is: To what extent can present ocean fisheries be expanded to provide additional food?

It should be borne in mind that although the oceans occupy about 71% of the world's surface, this does not at all represent a vast, largely untapped potential fishing grounds. Most of the fish occur in coastal areas, where more than 90% of all fishing is done. Coastal waters are only 7% of the ocean area. The oceans provide only about 2% of the food currently consumed by mankind. Practically all the very good fisheries of the world have already been discovered and are being utilized. Prior to the 1950's, new fishing grounds were usually discovered by venturesome fishermen. Since that time, fishery oceanographers, through systematic exploration and study of oceanic conditions bearing on fisheries, have been responsible for locating new fishing areas and also within already fished areas provide fishermen with information useful in their scouting and catching operations—forecasts of seasonal and long-term changes in abundance and catchability of fish populations.

As a result of these efforts and of vastly improved fishing technology, especially by Russia and Japan, including the building of big new fishing fleets with superior gear, factory ships, and transport of finished goods to and from the factory ships and fleets so that they do not have to return to land frequently, the fisheries' harvest has more than tripled since World War II, and all the world's major fishing grounds are now heavily utilized.

Although the news media constantly reiterate that the harvest from the oceans will largely solve the problem of feeding the world's hungry (2.5 billion of the world's 4 billion people), this is not, in fact, the case. Most of the catch is being channeled to the already well fed, the only people who can afford to buy it. The hungry are relegated to the sidelines.

Of the total 52 million metric tons taken in 1967, only 8 million went to feed hungry people—the rest went to the well fed. Much of the 1967 catch, 45%, or 20 million metric tons, was channeled into reduction plants to make fish meal and fish oil, but 83% of the fish meal and oil goes to the rich nations, mostly to make high-protein animal food for broilers, egg layers, hogs, etc.

The oceans have become the reserve of the well fed. We mobilize the great fisheries not to feed the hungry or the populations living nearest to them—but to deplete their resources to feed the rich nations. Salted and dried fish, long a staple item in the diet of many underdeveloped regions, are being produced in smaller quantities than formerly, whereas more goes into frozen fillets and into animal food for meat and egg production of the rich nations.

If the total food fish catch of the world were allotted to the United States, it would not provide much more than our current protein intake through meat.

Table 21-1, modified from one supplied by Dr. Vishniac and updated to 1964, should be studied in conjunction with Tables 4-2 to 4-4.

From Table 21-1 it can be seen that feeding our expanding populations by a severalfold expansion of ocean fisheries is not within the realm of possibility. At most, the fisheries can be only not quite doubled before reaching the limit imposed by the total annual productivity of the oceans.

TABLE 21-1. Ocean cropping*

Commercial fish catch for 1964 (including invertebrate fisheries, but not including sport fishing and unrecorded fishing in many areas of the world)	~5.2 × 10^7 tons fresh weight
Commercial fish catch in terms of carbon	~3.9 × 10^6 tons
Annual carbon productivity of the oceans	~5.3 × 10^{10} tons
Maximum possible carbon productivity	~1.6 × 10^{11} tons

*Assuming a three-link food chain with 5% efficiency of transfer, an annual cropping of 3.9 × 10^6 tons of carbon as fish requires 3.2 × 10^{10} tons of carbon as living organisms as an ecological basis. Conclusions about possible expansions of ocean fisheries must be based on estimated annual carbon fixation. The figure for maximum possible productivity is merely a hypothetical figure based on an assumed 100% utilization of incident light, with no limitations imposed by amounts and rates of turnover of phosphorus and nitrogen.

Actually, the potential level of further expansion of ocean fisheries is less than that indicated by the tables because a considerable fraction of ocean production goes into species not readily exploited on a commercial basis and because of the fairly extensive unrecorded fishing that makes the actual present cropping somewhat higher than that shown. Furthermore, limitations imposed by amounts and rates of turnover of phosphorus, nitrogen, and other elements are real.

In other words, we are already cropping the ocean fisheries dangerously close to the upper limits of biological possibility. Only wise and careful management practices will enable us to maintain even the present levels of cropping without destroying more species most useful to us, as has already happened to some.

Some authorities are more optimistic, estimating that the harvest may be increased as much as fourfold.

Present international conventions making the oceans into regions of free-for-all competitive exploitation by peoples of all countries, in whatever manner their whims and those of their governments dictate, are not conducive to good management practices. Limited international cooperation in conservation has been obtained on a voluntary basis in connection with a few fisheries clearly threatened with almost immediate destruction. The work of the International Whaling Commission is an example. The Food and Agriculture Organization of the United Nations has also done much good work, especially in bringing together and making available comprehensive studies of the biology of important species. These efforts need to be greatly expanded, the conservation regulations made more enforceable, and the regulatory agencies put under the charge of competent biologists whose primary concern is the preservation of the ecosystem rather than the extraction of the maximum annual profits from it, and who in their employment are not directly dependent on the industries they are supposedly regulating.

More efficient use of so-called trash fish and parts of fish commonly wasted, for the manufacture of high-protein fish flour, offers the possibility of a significant augmentation of our benefit from fisheries without a correspondingly large increase in the present level of fishing, as does also increased use of canned, frozen, and other forms of preserved fish.

The most significant impact that ocean fisheries could make on the problem of hunger does not lie in the area of further increasing the world fishery harvest, but rather in turning the flow of fish and fish products away from the rich countries, making ocean fisheries the common property of all mankind, subject to joint planning and allocation to meet the real needs of humanity.

Bivalve molluscs

Oysters, clams, mussels, and pectens have been esteemed food items since antiquity and have formed the basis for

FIG. 21-2. Oyster farm, Tomales Bay, California. Eastern oysters spread out on the flats to fatten. After several months they are gathered in baskets and marketed. (M. W. Williams photograph.)

flourishing coastal fisheries in many regions. In the last few decades, methods for culturing and growing oysters and mussels have greatly increased the potential yield and made of their production an industry more like agriculture than merely a gathering and processing operation.

The United States alone produces over 40,000 tons of oysters (Fig. 21-2) annually (meats—not including the shells), about 17,000 tons of meats of the surf clam *Spisula,* and about 13,000 tons of pecten meats, plus smaller amounts of numerous other clams and mussels.

In the shallow flats of the southern Baltic it is estimated that there are about 4 million tons (live weight) of the mussel *Mytilus edulis.*

Bivalves, especially mussels, can probably produce more food for human consumption, in a shorter time, with less expense and effort on the part of the farmer than any other animal. They are a "key industry" animal, converting plant material into high-grade protein. Their powers of reproduction are colossal and their growth rapid.

Squid

Squid are sufficiently abundant in many areas to constitute an important adjunct to the ocean fisheries. Off Monterey, California, for example, more than 5000 tons are taken annually.

Shrimp

Shrimp fisheries have attained worldwide importance over the past several decades. The shrimp harvest along the Atlantic and Gulf coasts of the United States is valued at more than $82 million in a good year, about 22% higher than the famous Pacific salmon fishery.

Shrimp culture, pioneered especially in Japan, has further increased yields and is an important coastal industry in Japan and numerous other places. In Japan, *Penaeus* are mass cultured in laboratory tanks until they reach a length of 4 to 5 cm. They are then transferred to big saltwater ponds at about 10° C., where in a year they mature to a length of about 12 cm. and a weight of 25 gm. and are ready to market.

In experimental ponds near Galveston Bay, Texas, yields of 27 to 28 tons/km.^{2} of pond area have been attained.

United States exports of canned shrimp, valued at slightly more than a dollar per pound, have ranged between 2 and 4 million pounds a year.

FARMING THE OCEANS

The avenues open for obtaining greatly augmented supplies of food from the oceans lie in large-scale "farming" of certain species and in the harvesting of organisms low in the trophic hierarchy. One such organism, for example, is the antarctic krill. The size and shoaling habits of this immensely abundant, nutritionally rich crustacean should make it a favorable object for commercial exploitation. Now that we have largely eliminated the great baleen whales, perhaps man will come to occupy their ecological niche so far as the krill are concerned. Commercial farming of bivalve molluscs, which feed directly on the plankton, can probably also be considerably expanded, as well as the farming of certain crustaceans and edible seaweeds.

OTHER CONSIDERATIONS

Farming and fishing the seas for food will increasingly have to compete with other aspects of exploitation of the seas that may not be compatible, such as underwater mining, oil drilling, and perhaps military uses, as well as to cope with the generally increased destruction through pollution and alteration. Most of the best fisheries are in shelf areas, relatively near land, where they are subjected to a maximum of such influences. In still other cases, even though the adult fish may occur well offshore, the eggs or young are dependent on near-shore environments highly susceptible to destruction.

In localized areas with sandy or muddy substrate, the production of fish can sometimes be much increased by artificial reefs, providing habitats for rock-dwelling species of invertebrates and seaweeds and attracting desirable species of fish. The creation of suspended artificial reefs from systems of floating platforms or buoys farther offshore may also offer interesting possibilities. This method may be combined with a system of fish traps that would make harvesting the fish easy and perhaps serve as sites for large-scale farming of useful bivalves such as mussels or races of oyster adapted to the open-ocean salinities.

The I.C.L.A.R.M.

One of the more hopeful kinds of development in recent times is represented by such groups as the I.C.L.A.R.M.—International Center for Living Aquatic Resources Management. This relatively small-scale, unpretentious, apolitical organization was formed because of the realization that present large-scale fisheries operations are self-destructive and in no way geared to meet the real needs of humanity.

This group concentrates on three areas of activity that show great potential and that are not disruptive of the societies and ecosystems with which they deal, as are so many of the large-scale projects undertaken by governments or large institutions of the developed countries. These areas are (1) aquaculture, (2) funding of small-scale, imaginative research-and-development projects unlikely to attract funding from more conventional sources, and (3) working with fishermen in poor, distressed areas and using slightly better gear, or methods that are within the means of the local fishermen to develop or acquire.

Their activities are all geared to local conditions and are of such a nature that the benefits go to the local populations, where they are needed.

Recently this group has moved its headquarters from Honolulu to the Philippines because of a policy decision that it should be in one of the developing countries.

Freshwater from the sea

Another way in which the oceans may make a greater contribution to man may be through furnishing freshwater on a large scale in certain areas. A variety of schemes for desalinization have been proposed, most of which suffer from the fact that the energy required to accomplish the desalinization, plus the disposal problem created by excessive quantities of concentrated brine or salt (produced as a by-product), make them impracticable except in localized situations where freshwater is at a high premium. It has even been suggested that huge icebergs may be towed to the vicinity of metropolitan areas such as Los Angeles to be used as sources of fresh water and to provide a cooling effect.

One of the more interesting proposals that may be feasible in warm areas with moist trade winds most of the year was made by Gerard and Worzel (1967). They propose to use the power of the wind, by means of big windmills, to pump seawater up through large-diameter pipes from below the thermocline, at about 9° C. This cold water would be run through very large condensers located onshore, where they would intercept the humid wind. According to their calculations, a condenser 200 meters long by 10 meters high should deliver about a million gallons of freshwater per day. There would be no harmful brine or salt encrustations, as in desalinization schemes. The deep marine water used could be discharged into a nearby lagoon, where it would increase the fertility of the water, raising its productivity and supporting various schemes of aquaculture. The big condenser would act as an air conditioner for any settlement to the lee of it, supplying more pleasant, cooler, dryer breezes than those which were customary in the region. It may even be possible to recover some power by using the water descending from the condenser to turn a turbine or perhaps by taking advantage of the temperature difference between the deep water at 9° C. and the surface waters at 20° C. to drive a turbine by boiling propane at surface-water temperatures and then cooling and recovering the propane by using the cooler deep water. The cycling of the propane would obviate any need for a continual supply of more fuel. There has even been a suggestion that if such an installation were built on one of the larger Pacific coral atolls and the used deep water discharged into the lagoon, the lagoon may become fertile enough to support the farming of whalebone whales or in any case to support a greatly augmented fishery.

Electrical power from the sea

Despite perennial interest in the possibility of harnessing the power of the tides for the generation of electricity, only a modest beginning has been made in that direction.

There are only a few places where the difference in level between high and low tides gives sufficient head to make generation practical, and the required dams across bays or estuaries would have environmental impact that would require careful study in each case. The Bay of Fundy on the east coast of Canada, with its 50-foot head, is probably the most favorable site.

At present the Soviet Union has a tidal plant that generates about 400 kilowatts, and there is one in northern France with a 240,000-kilowatt capacity.

Two other suggestions have been made that would appear at first sight to have little potential for adverse environmental impact. One is to utilize the thermal difference between surface and deep waters in warm regions, utilizing warm surface waters in a heat exchanger to boil ammonia or some other highly volatile fluid, the vapor from which would drive a turbine and then be condensed with colder water and recycled. The cold, nutrient-rich, deeper water used in the cooling system would enrich local surface waters where discharged, augmenting the fishery, or perhaps be utilized in aquiculture projects. Such systems would be most feasible in tropical regions, where there is the greatest thermal difference between surface and underlying water.

The other suggestion is to tap the immense energy of ocean currents. The Gulf Stream alone is said to carry more than fifty times the flow of all freshwater rivers combined. If "underwater windmills" could be installed at places where the currents are fairly strong and steady, a pollution-free energy source of minimal environmental impact could perhaps be acquired. However, no such installation has yet been designed or tested, and it may be that the low revolutions per minute obtainable would make this economically impracticable.

FIG. 22-1. Crews near Santa Barbara, California, trying in vain to save the beaches from oil pollution by casting great quantities of hay into the water to soak up the oil. Later burning of some of this oil-soaked hay contributed to air pollution. (Wide World Photos, Inc.)

CHAPTER

22 MAN'S IMPACT ON MARINE ENVIRONMENTS

Since man is a terrestrial animal, the effects of his extraordinary proliferation and recent developments in technology have been most marked on land. Here we are involved in the greatest, most rapid redistribution and change in the faunas and floras of the world that has ever occurred. Very large numbers of the dominant elements of the flora and fauna of vast regions have been suddenly eliminated and replaced by domesticated types of crop plants and ornamentals and by domestic animals molded and shaped according to the whims and desires of man and bearing little resemblance to any "naturally occurring species." Plants and animals have been introduced on a large scale into regions where they did not formerly occur, with concomitant upsets and repercussions for the entire ecosystems of those regions. Dry regions have been irrigated, wet regions drained, water-table levels changed, and great areas sprayed and dusted with long-lasting poisons that have entered the geochemical cycle of the entire world, producing sometimes devastating, sometimes subtle effects that we can hardly begin to assess as yet. Erosion and pollution have so changed the waters that the rivers and streams of entire continents have taken on the character of open sewers, carrying off so much in the form of plant nutrients, soil, sewage, and industrial wastes that they can only be regarded as hemorrhages of the land rather than part of its normal circulation.

Fossil fuels are being burned at such a rate that the character of the atmosphere is being radically changed in a manner that threatens to get entirely out of hand and perhaps eliminate the present world biota altogether. Radioactive materials are being created in an altogether new order of magnitude. The level of background radioactivity has already been appreciably raised throughout the world. The proliferation of nuclear weapons and of nuclear industrial plants and laboratories is such that it is no longer just a high probability, but a virtual certainty, that either by accident or by design, large regions will be rendered uninhabitable and the levels of radioactivity throughout the world raised to such heights that they will exert profound effects on whatever life remains.

The changes that have taken place in the oceans to date are not so dramatic or obvious as those on land, but they are important, and since our technology is now making possible large-scale industrial utilization of marine resources, they will increase rapidly in scope and effects in the immediate future. Many of these changes are secondary effects of changes taking place on land. Others are the result of the direct intervention of man in the oceans. Only a few of the changes are considered briefly, as examples.

OVEREXPLOITATION

Depending on the particular circumstance and the nature of the resource, the results vary all the way from a slow decline in the abundance and availability of the resource to its abrupt and permanent elimination or extinction. Examples are legion. Steller's sea cow was completely wiped out only twenty-seven years after its discovery. The great auk, or arctic penguin, was likewise exterminated. The Caribbean monk seal is gone. The blue whale, elephant seal, sea otter, white albatross, and many other once-abundant sea animals have been brought to the verge of extinction. Several species of seals have been eliminated over the greater part of their former ranges. The great fish, such as swordfish, sailfish, and tarpon, are threatened.

In fisheries the usual signs of overfishing are a decline in the average size of the fish of a given species that are taken and an increase in the intensity or efficiency of the fishing effort necessary to take the same catch. The normal response to a declining fishery is just such increases in fishing effort, together with improvements in the efficiency of the gear or fishing methods and expansion of the area fished, if possible. Extensive trawling for demersal fish in limited areas such as the Grand Banks or the banks of the North Sea, which have long been celebrated fishing grounds, can be especially destructive because the big trawls are destructive to the entire benthic community on which the fish depend. When World War I forced a four-year halt in fishing operations in the North Sea, the fish were both larger and more numerous at the resumption of fishing than they had been just before the war, demonstrating that the fishery was already on the decline at that time.

Whaling

The history of the whaling industry is typical and may serve as an example. Serious whaling began in the North Atlantic in the sixteenth century. The Atlantic right whale was the first species to come under heavy attack. This was an abundant, large, black, oil-rich, 60-foot whale, termed the "right whale" because when harpooned and killed, the carcass would float, whereas certain other whales, "wrong whales," would sink and be lost. By the end of the seventeenth century the Atlantic right whale had practically disappeared. Whalers turned their attention to the Greenland right whale, a species with abundant and valuable baleen as well as oil and meat. Soon these, too, were nearly exterminated. About 1862 a Norwegian whaler invented the harpoon gun with grenade, with which whalers could kill even the gigantic blue whale. Power-driven catcher boats with pumps for pumping air into the carcasses immediately after the kill came into use, making all species of whales easy prey. Whaling continued for a time west of Greenland, around Baffin Bay, but by the early part of the twentieth century the industry was dead in the North Atlantic. The large whales had all but vanished from the Northern Hemisphere. Whalers, with much improved equipment, now turned to the great antarctic summer feeding grounds. Modern equipment includes huge factory ships that serve as mother ships for fleets of smaller whaling boats. Helicopters and radar devices are used to spot the whales. Each whaling boat quickly slaughters whale after whale, pumps air into the carcasses so that they float, marks them with flags or even mounts radio transmitters on them, and then, at the end of the day's work, comes back, affixes lines to them, and tows the whole group to the factory ship. During the first half of the twentieth century, whalers slaughtered more than five times as many whales as had been killed by all the famous nineteenth-century whaling fleets combined.

By 1932 it was obvious to all that the whaling industry was in serious trouble and would soon be out of business if this continued. The International Whaling Commission was set up with the functions of obtaining as reliable data as possible on the number of different kinds of whales and setting quotas for each species every year, with a view to conserving threatened species. Because of economic and political pressures, the quotas set are often much too high from a sustained-yield standpoint. The commission is handicapped by inability to enforce its quotas or even to assign quotas for particular nations. It simply sets an overall quota for each species, after which representatives of the countries concerned negotiate the division of the quota among their whaling fleets. It seems to be increasingly difficult for them to reach agreement. Furthermore, not all the nations subscribe to the authority of the commission. Some whalers who abide by the quotas during the summer in the Antarctic also take whales at other times during their migration or at their breeding grounds.

In the past decade, as the whaling situation became more critical, some of these shortcomings have been at least partially remedied. The quotas are more realistic, and the most threatened species, the blue, humpback, right, and bowhead whales, together with the California gray whale, are now "fully protected." The gray whale has made a rather good comeback, but development in Lower California along some of its most important breeding grounds, such as Scammon's Lagoon, may pose a new threat to it.

The Federal Marine Mammal Protection Act of 1972 is a progressive step in the right direction.

The gigantic blue whale, the largest animal that ever lived, is now on the verge of extinction. It is now supposedly under complete protection, but this action was probably taken too late. The blue whale ranges over the open sea, from the Antarctic to the equator. It appears that the population has been reduced to such a low level that too few of them are able to find mates and reproduce often enough to maintain the species.

At the present writing a special problem of the slaughter of porpoises in connection with the tuna fishing is under intense debate and research. It seems that schools of porpoises habitually swim above large schools of tuna, perhaps sharing with them the mesopelagic prey fish resource, trapped between the tuna below and the porpoises above. In any event, when the fishermen set their nets around the porpoise school in order to obtain the tuna, great numbers of porpoises are incidentally killed in the nets. Efforts to provide escapes for them, and to induce them to use these devices, have been only partially successful. In the case of United States boats, the Marine Mammal Protection Act makes it illegal to use or sell any of the porpoises so killed—resulting in a total waste of this resource. The United States fishing industry maintains that it cannot compete with fishing fleets of other nations not subject to this act if not permitted to fish with the same methods.

Pacific sardine

One of the most dramatic cases of the sudden failure of a great fishery occurred in the Pacific sardine fishery centered at Monterey Bay, California. During its heyday, that fishery supported a large fleet of fishing boats and a number of large canneries located along the edge of the bay near Pacific Grove, known as Cannery Row. During the mid-1930's over half a million tons were taken annually. In the late 1930's and early 1940's there was a sharp decline, and the catches became more irregular. By 1947 the fishery was clearly threatened. Abundant money suddenly became available from both private and governmental sources for research programs on the biology of the sardine (Fig. 22-2). The California Cooperative Sardine Research Program was organized in 1949, eventually involving most of the interested institutions and agencies along the Pacific coast. By 1952 the canneries had begun to close down, and soon the industry was dead. Most of the equipment was sold to firms operating in South America, and part of Cannery Row was converted into a West Coast branch of Ward's Biological Supply House.

Data developed through the efforts of the Cooperative Sardine Research Program seem to indicate that in this instance the causes of the catastrophic decline were complex, involving extensive hydrographical changes in the North Pacific as well as the activities of man. During the years in which the sharpest decline occurred, average water temperatures of the region were from 1° to 3° cooler than during the good years of the fishery. The sardines shifted their spawning southward, off Lower California, and also decreased sharply in abundance.

The hatching time of sardine eggs is retarded from 54 hours at 17° C., which is about optimum for them, to 73 hours at 15° C. This increases significantly the period during which the eggs are subject to predation. The cooling also slows the growth of the young fry, increasing their period of greatest vulnerability. Murphy (1961) estimated that the prolonged period of vulnerability caused by a 3° decrease in temperature could decrease survival by a factor of ten.

Furthermore, zooplankton was much more abundant in the water during the cool years. Since many zooplankters are predators of the fish eggs and young fry, preda-

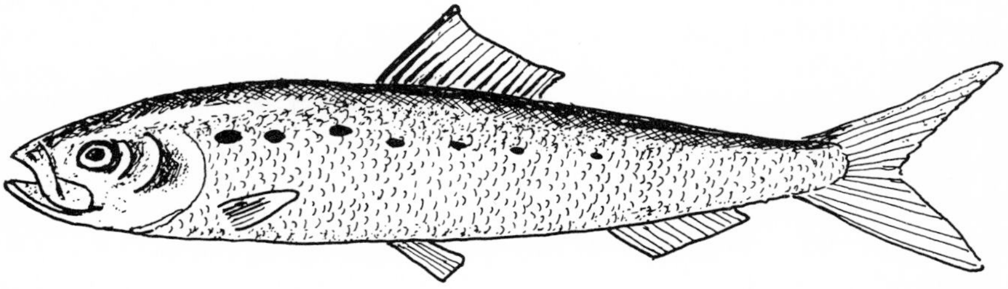

FIG. 22-2. Pacific sardine, which supported a once-great fishery at Monterey, California. Actually, the Pacific "sardine" is a species of pilchard, *Sardinops caerulea*.

tion pressure on these vulnerable stages actually increased at the very time their period of vulnerability was being prolonged due to the slowing of the rates of development.

Intensified fishing effort during the period when the decline was beginning also diminished the adult reproductive stock at the same time that their replenishment was falling off because of the greatly decreased survival through the early stages.

In 1957 oceanic temperatures in the North Pacific again began to rise, rather abruptly. Southern species of fish and other organisms began to appear farther north than before. Episodes of attack by sharks farther north were reported. The coastal sea level rose approximately 5 inches, and there was a slight upsurge in sardine catch. The abruptness of this general rise in ocean temperatures and level shows that mechanisms of interaction between ocean and atmosphere can produce large-scale effects more rapidly than had been supposed and in ways not yet well understood.

Changes in "ocean climate," even of only a few degrees, may have marked effects on the distribution and abundance of pelagic organisms. Fishing pressure by man may, for heavily fished species, greatly increase the probability that slight ecological changes will have catastrophic effects for the species. Clearly, problems of conservation and management in oceanic fisheries are not simple, nor are the results always predictable.

There is not always a direct relation between conservation efforts to limit overexploitation, and future crop of a commercial species. It is not always true that saving more reproductive adults will result in more young and a bigger population in the next generation.

In California, fisheries biologists, concerned about possible overexploitation of squid, discovered that in the breeding season the squid gathered in such huge numbers to mate and lay eggs and that such a thick mat of eggs was produced at the breeding sites that only the top 10% of the mass got enough oxygen to support development. A considerably greater exploitation of the parent stock would therefore be possible without reducing the numbers of the next generation.

Many fish lay so many eggs that, under favorable conditions, a small number of females could replenish the stock. As a rule, protection afforded by man is minor compared to the natural impact of predators, small changes in temperature, etc.

This doesn't mean conservation efforts are fruitless. It means, rather, that the biology of each species must be well understood to properly assess the probable effects of, and best timing for, protective measures.

Thinning and culling a fish population may have beneficial effects, as in gardening and forestry. Old fish, for example, grow slowly, utilize food inefficiently, and may even eat young of the same species. A well-managed fishery is more productive because the old are eliminated.

Conservation—too little and too late

The following is the usual progression of events when natural resources are threatened by overexploitation: While a resource is abundant, those persons engaged in exploiting it resent and resist the "meddling" of anyone who wants to study it carefully, to recommend less wasteful ways of exploiting it, or to suggest regulations for conserving it. Only when the industry is clearly and imminently threatened do they call on and support the activities of such persons, often too late. At such times the larger operators, who can afford to do so and who have the greatest investment to lose if the industry fails, become very public spirited and conservation conscious, eager to set the industry up on a sustained-yield basis. Smaller operators who cannot stay in business under suddenly more stringent conservation regulations are forced to sell out at a loss. In the event that enough is salvaged to enable the industry to continue its operations, the large operators, now on a sustained-yield basis or more nearly so, in all sincerity claim and are credited with far-sighted, public-spirited vision for having adopted or submitted to the regulations that alone enable them to continue operation—regulations that should have been put into effect when the industry was first opened up and which these same operators may have fought tooth and nail as long as it was profitable for them to do so. Because of their far-sightedness, public spirit, and experience, they go on to maintain that it is they, not some outside regulatory body, who should have the decisive voice with respect to planning and regulating further resource exploitation in their areas of operation.

Since it is now becoming rapidly clear that most major resources are becoming threatened, a continuing upsurge of interest in conservation and in social regulation of major industries can be expected. However, the lack of any international authority capable of enforcement, the distrust and rivalry between countries, the newness of many of the enterprises, and the fact that much of the damage they do will be under water, where few will be aware of it, will make the proper planning and regulation of ocean industries exceedingly difficult to achieve. (See, however, United Nations efforts in discussion of law of the sea, pp. 443 and 444.)

CHANGES IN ABUNDANCE AND DISTRIBUTION OF ORGANISMS

The removal of any important species from the scene or its relegation from a position of dominance to one of scarcity is never an isolated phenomenon. Every such change also has far-reaching effects throughout the entire ecosystem. Changes in the relative abundance of organisms closely linked to it in the food cycle follow immediately, and from these, still further changes. Some species may be eliminated, and others may be caused to proliferate as never before. Parasites whose life cycles involve some of these species may also be greatly affected. So complex and intertwined are the relationships within the ecosystem that any changes within it or changes in the pressures on it have repercussions throughout the entire system that are difficult to predict.

An example of the far-reaching effects of changes in the abundance of species is the recent explosive growth in the populations of the formerly rare coral-eating *Acanthaster planci,* the well-named "crown of thorns" (Fig. 12-1). This starfish has already destroyed 100 square miles of Australia's Great Barrier Reef and most of the coral along a 24-mile stretch of the coast of Guam and is decimating corals at places as widely scattered as Borneo, Fiji, Palau, Saipan, Wake, and Midway.

If this devastation continues unchecked, it could destroy the entire system of marine life on which millions of islanders depend for their protein food, subject the islands to erosion from which the reefs now protect them, and remove one of the world's most far-flung, gigantic biocenoses, unsurpassed for variety and beauty and for biogeographical and economic importance. The effects would be worldwide—a staggering loss from both the economic and aesthetic standpoints.

Although the reasons for the great proliferation of *Acanthaster* have not been precisely pinpointed as yet, most biologists concerned about the problem believe that the most probable explanation lies in the widespread dredging, channel blasting, and other projects that have been disrupting the coral habitats on a constantly increasing scale ever since World War II and providing abundant new surface areas of killed coral on which the larvae of *Acanthaster* can safely settle. The increased devastation of corals in the area by enlarged populations of *Acanthaster* then permits this process to become self-augmenting, leading to rapid destruction of the reef. Elimination of some planktonic microfeeding species that formerly fed on the larvae is also a possibility. This could have been caused by DDT or by radioactive residues from atomic tests. The reduction or elimination of triton, a large predatory gastropod that preys on adult *Acanthaster*, by shell collectors has also been mentioned as a contributing factor in some areas.

Reefs that have been destroyed have never been found to come to life again.

Introduction of new species by man

Through both deliberate and inadvertent introduction of organisms into areas where they did not previously exist, an accelerating process because the mobility of man has increased so sharply, we are now involved in one of the fastest, most extensive distributional revolutions in the history of the earth. Like the elimination of key species, the successful introduction of new species into an area also has far-reaching, often unpredictable effects.

Thorson (1971) relates the effects of the inadvertent introduction of the carnivorous snail, *Rapana thomasiana*, from Japan to the Black Sea. The snail may have come as egg capsules on the shells of oysters imported from Japan about 1947. Once introduced, it spread, becoming extremely abundant, and totally destroying a number of flourishing oyster and mussel beds. The result was the diminution in the number of bottom-feeding fish that had been dependent on them.

Next, the sand eel, *Ammodytes cicerellus*, which came in through the Bosporus from the Mediterranean, adopted the snail larvae as its chief item of diet and flourished massively. The fish *Sargus*, which had been eating worms and crustacea, switched to eating sand eels.

The small race of the hermit crab, *Clibanarius*, characteristic of the Black Sea, has now grown to the same impressive size as that of the Mediterranean race, since it now has plenty of large *Rapana* shells—its previous small size being due to having only small shells available to inhabit.

It is still too early to tell what final new ecological balance may emerge as a result of the introduction of *Rapana*.

So-called fouling organisms—the communities of barnacles, mussels, algae, bryozoans, hydroids, tube worms, tunicates, and other, smaller associated organisms that grow on ship bottoms, along with boring organisms such as shipworms and gribbles—have been carried all over the world. As a result, many of them have established themselves in new areas, sometimes displacing elements of the indigenous biota. Among those most noticed by man have been the wood-boring species. It is probable that man is chiefly responsible for both the abundance of these forms and their worldwide distribution. Prior to the advent of man, the marine environment contained little wood—only the occasional trees washed into the sea by rivers or toppling into the margin of the sea from forested sea cliffs. Its occurrence in most ocean areas was so rare and so fortuitous that it could hardly have supported widespread thriving populations of specifically wood-boring marine animals. Their occurrence must have been limited.

Man, with his extensive lumbering, his wooden-hulled ships, his pilings and other wooden structures built along the margins of the seas, has for the past two or three thousand years provided a vastly augmented, steadily increasing supply of wood in the marine environment, together with free transportation for wood borers. What were previously minor species of limited distribution suddenly were presented with an open environment and have become important pests on a worldwide scale. Shipworms have been important pests for all early commercial and naval vessels with wooden hulls from before the time of the Roman galleys. In the early 1700's they damaged the dikes of Holland so seriously that they gravely threatened large areas of the Netherlands with inundation. The first careful study of the group was the treatise of Sellius (1733), written at that time, in which he proved that shipworms were molluscs. He did basic anatomical work on them and reviewed all that was known about them. Another, more recent spectacular case of shipworm damage occurred after introduction of the worms into San Francisco Bay sometime before World War I. The worms spread unnoticed and rapidly among the thousands of pilings of piers, wharves, railroad trestles, marine warehouses, and other structures along the bay front. Many of them began to collapse by the 1920's, causing millions of dollars worth of damage.

The Oyster Creek nuclear power plant in New Jersey, which obtains its cooling water by reversing the flow of Forked River between the plant and the sea and discharges the warm salty water into Oyster Creek, so altered conditions in Oyster Creek that yacht basins and wharves, heretofore free from this pest, soon became heavily infested.

In 1957 a worldwide testing program was instituted. Studies are being made from samples collected on a worldwide basis, and methods of protecting wooden structures are being systematically evaluated.

The construction of canals has provided another means by which various species have become redistributed. The Suez Canal from the Mediterranean to the Red Sea, completed in 1869, was not for a long time an effective avenue for interchange of faunas because of the high salinity of the Bitter Lakes, through which it passes. Over the years this salinity has been decreased through interchange with waters by way of the canal and has been reduced from about 68‰ to 45‰, so that in recent years, passage of several species has become possible. Most of the exchange has been from the Red Sea into the Mediterranean because tidal movements from the Red Sea carry organisms far up the canal from that end but not in the reverse direction at the Mediterranean end. Over 150 species of Red Sea animals now occur in the eastern Mediterranean, 9 of them commercially exploited.

The Welland Canal permitted the Atlantic sea lamprey to invade the Great Lakes. About 100 years after its contruction the lampreys underwent an explosive increase in the Great Lakes that destroyed the valuable whitefish and trout fisheries. The governments of the United States and Canada have spent about $16 million on control, and of course the financial loss to the fishing industry was far greater.

At present there is under consideration by the United States a plan to construct a new sea-level canal through the Isthmus of Panama, possibly using nuclear explosives to assist with the earth moving. This plan is viewed with alarm by many biologists who believe that it may set off a chain reaction of major ecological adjustments extending across the oceans. It may also change the physical environment, especially along the Pacific side, by allowing warm Caribbean surface water to spread out over a large area now cooled by upwelling produced by offshore winds, producing effects comparable to El Niño along the coasts of Ecuador and Peru. If, on the other hand, water movement through the canal should be in the other direction, the cooler Pacific water could be destructive to Gulf and Caribbean tropical communities.

It is believed that in any case, a major intervention of this kind in the world ecosystem is too grave a responsibility for unilateral decision by one country and that an adequately financed international commission of marine biologists and ecologists should be set up first to survey the region carefully, to assess the probable and possible impacts of the canal on the ecosystems of both sides of the isthmus, and to monitor the changes that occur if and when the canal is constructed. Such a commission should be multidisciplinary, should be independent of any single government or governmental agency, university, or private concern, and should be well funded. It should be in operation for several years before the final decision to build such a canal is made.

The commission itself should make the decision—not simply be advisory—and if it decides to build the canal, it should also have general supervision of the actual construction, including such decisions as whether or not to use nuclear explosives. Military considerations should be allowed no role whatsoever in making the decision or in determining the characteristics of the canal. **The only acceptable reasons for deliberate large-scale interventions in the world ecosystem are great potential benefits to mankind as a whole together with general world-wide scientific and popular concurrence that the particular intervention in question is desirable.** Adequately financed, completely independent international commissions of competent scientists, empowered to make the decisions in question and to supervise their implementation, offer a more rational approach to such problems than do unilateral decisions made by the competitive heads of state and military leaders.

It would be far better to put up with some delay and inconvenience in shipping for a few years than to blunder into a large-scale ecological disaster. Even if the ecological results of such a canal should be beneficial rather than disastrous, it would be better to have biological and oceanographical studies made so that a maximum return of scientific knowledge and understanding could be gained from the venture. Furthermore, this procedure would set a first example in deliberate cooperative international planning and decision-making based on the best scientific knowledge obtainable. This kind of attitude and cooperation is sadly needed and could in the long run be of far greater benefit to mankind than the canal itself.

The possible use of nuclear explosives during the construction of the canal is another facet of the problem with grave implications that should not be decided hastily or unilaterally. Whether or not to use such methods should be decided not by physicists and engineers or politicians interested in making the tests for their own sake but by biologists whose primary concern is environmental quality, after careful study in each instance.

POLLUTION

There are many types of pollution of major importance in connection with marine environments, some of which are (1) changes in estuaries, bays, lagoons, and coastal habitats through pollution from the land and through filling, dredging, construction, shipping, and other activities; (2) worldwide dissemination of long-lived pesticides and other chemicals; (3) oil pollution; (4) worldwide contamination by radioactive materials; (5) changes in the earth's atmosphere that may affect the carbon dioxide–oxygen–carbonate balance in a manner destructive to the entire world biota; and (6) thermal pollution. Each of these classes of change will be briefly considered simply to point up something of the nature and scale of the changes being made.

Estuaries, bays, lagoons, and coastal habitats

Even ordinary pollution with only local municipal sewage and wastes may have marked effects on the coastal biota. For example, Table 22-1, compiled from data presented by Leighton and Boolootian (1963) in connection with an entirely different type of study, compares the growths of marine

TABLE 22-1. Effects of municipal pollution in Santa Monica Bay

Point Dume	Flat Rock
Very abundant species	
Gigartina canaliculata	Scattered occurrence
Rhodoglossum affine	Rare
Common but with discontinuous distribution	
Pelvetia fastigiata	Rare
Egregia laevigata	Rare
Gigartina leptorhynchos	None
Gigartina spinosa	None
Polysiphonia pacifica	Rare
Phyllospadix scouleri	Scattered occurrence
Of scattered occurrence	
Eisenia arborea	None
Macrocystis pyrifera	None
Dictyota flabellata	Rare
Cystoseira osmundacea	None
Gelidium purpurascens	Rare
Pterocladia pyramidale	Rare
Zanardinula cornea	None
Bossea sp.	Rare
Enteromorpha sp.	Rare
Ulva sp.	None
Rare at Point Dume	
Coralline algae	Very abundant
Codium fragile	None

algae and eelgrass at two comparable sites on opposite sides of Santa Monica Bay, California, both of which formerly supported flourishing growths of these plants. Point Dume has remained relatively free from local pollution, whereas the Flat Rock area has been subjected to municipal pollution.

From Table 22-1 it can be seen that every species of marine plant noted underwent noticeable diminution in amount or was wholly eliminated, with the sole exception of the formerly rare coralline alga, which came to dominance in the polluted area. Any such marked change in the local flora would, of course, entail great changes in the abundance and species composition of the fauna, in addition to the direct effects of the pollution on the animals.

Bays, estuaries, and coastal lagoons not only support important fisheries in their own right, close to the land and readily available, but also serve as nurseries for the eggs and young stages of numerous other fish that are usually found offshore as adults. The young stages of the fish and also of invertebrates such as crabs, bivalve molluscs, and other valuable animals are often particularly sensitive to alterations in their environment.

Extensive deforestation and agricultural activity, even far inland, results in erosion that greatly increases the sediment load and adds materials dissolved from fertilizers and insecticides to the water of most rivers and streams. Sewage and industrial wastes from towns and factories or mills add a variety of metallic ions, detergents, organic matter, acids, and other poisons to the water, as well as reduce its oxygen content. When the waters from such overloaded, partially sterilized rivers come to their estuaries, the excessive sediment in the water tends to smother gill-breathing animals and to choke nonselective sediment feeders, and the metallic ions, acids, or insecticides, sometimes even in only trace quantities, may wipe out sensitive embryonic and larval stages of many animals. The deposition of excessive sediment may smother much of the infauna.

Impoundment of water by huge inland dams for flood control, by irrigation projects, and by hydroelectric developments also radically alters the stream flow and the character of the estuaries into which the streams discharge, as well as interferes with the spawning migrations of anadromous fish such as salmon. Dams and irrigation projects also cause rivers to drop most of their heavier sediments far inland, so that beaches that had formerly received most of their sand from these rivers may be stripped of sand for miles by alongshore transport (p. 30) and may receive too little replacement sand. Added to this situation is the fact that most bays and estuaries are the sites for development of towns and ports. The patterns of circulation are changed by land fills, dredging, and the building of roads, causeways, bridges, boat basins, etc. Not infrequently, large parts of an estuary are filled in to make more land for real estate developments. Of the twenty-three estuaries along the coast of California, all but three have already been biologically ruined, and these remaining three are on the way out. Even estuaries remote from actual direct operations of man suffer.

In connection with proposed alterations of estuaries, it should always be borne in mind that the area of estuaries is truly minute compared with that of adjacent land masses, and that the environmental values peculiar to estuaries are found nowhere else.

In particular, the filling of estuarine areas to create cheap land or for easy disposal of dredgings should be discouraged everywhere, since it completely and permanently eliminates a large percentage of one set of values to replace them with a minute percentage of another set of values for which there is usually ample space already on land.

Dredging and filling and the discharge of harmful wastes into estuaries should be held to an absolute minimum and undertaken only as a last resort where clear and great harm may come to a region if they are not done. They should never be allowed for such purposes as "attracting industry" or enabling private groups to enrich themselves through abuse of the estuary.

Chemical pollution

Woodwell and associates (1967), in a study of East Coast estuaries, state that long-term, detailed observations show substantial reduction over the preceding decade in the populations of shrimp, amphipods, summer flounder, blue crab, spring peeper, Fowler's toad, woodcock, and other DDT-sensitive species even in estuaries and marshes remote from direct human disturbance. They also state that the analysis of the water means little in assessing the threat, since water has the least DDT of the major components of the ecosystem. Often, although the water may show only traces of DDT, or even when it is undetectable, the animals may contain nearly lethal concentrations. It is concentrated progressively in animals higher in the trophic scale. Table 22-2 is extracted from a larger table showing progressive increase in DDT levels in animals at the Carmans River estuary in the vicinity of Long Island, New York.

It has also been found in the past few years that DDT influences the calcium metabolism in some subtle way. Birds of prey in particular, which accumulate the highest levels, have eggs with thinner, more fragile shells, have more losses from breakage, and are declining rapidly in numbers.

TABLE 22-2. Accumulation of DDT in trophic levels*

Source	DDT in ppm
Water of estuary	0.00005
Zooplankton	0.040
Shrimp	0.16
Flying insect (Diptera)	0.30
Fundulus	1.24
Tern	2.8 to 5.17
Cormorant (immature)	26.4
Ring-necked gull (immature)†	75.5

*Based on data from Woodwell, Wurster, and Isaacson (1967).

†In the ring-necked gull, the levels approach those found in animals dying of DDT poisoning.

The plankton of estuaries is also extremely sensitive to some commercial herbicides, especially substituted urea compounds such as Monuron, Diuron, Neburon, etc., and in some cases cannot tolerate them in concentrations as low as 0.5 ppb.

Estuaries and coastal waters bear the brunt of the load of insecticides and poisons washed out of fields and woods into the streams. For example, it is estimated that the Mississippi now carries 10,000 kg. of pesticides into the Gulf of Mexico annually. The San Joaquin River in California discharges 1900 kg. per year into San Francisco Bay.

Tests on Barbados Island indicate that the easterly trade winds carry about 600 kg. of DDT per year into that distant part of the tropical Atlantic (Risebrough and associates, 1968). PCB, a toxic compound widely used in the manufacture of plastics, paints, and other products, is also found now in marine fish and birds along the coast of southern California and also appears to be progressively concentrated at higher trophic levels. Radioactive substances commonly behave in a similar way.

The joint Group of Experts on the Scientific Aspects of Marine Pollution (GESAMP) has reported that annually 2×10^5 tons of lead from the combustion of motor fuels are rained into the seas, and that annually 8000 to 10,000 tons of mercury reach the seas from industrial waste and from the burning of fossil fuels.

People who became upset about the addition of a little fluorine to the drinking water, which at least has been shown to have some beneficial effects, should become very much disturbed over the wholesale addition of all these known poisons, insecticides, herbicides, heavy metals, industrial sewage products, and radioactive compounds, to the world's atmosphere and waters, so that now not only every man, woman, and child alive but every organism on earth is forced to involuntarily partake of them. What the long-term effects will be no one can say, but they will undoubtedly be profound. Most of these substances are an entirely new addition to the world's ecosystem, developed only within the past three or four decades. They have been found in animals all over the world, even from such remote areas as northern Alaska, Antarctica, the tip of Lower California, and Barbados Island.

Oil pollution

A recurrent problem of ever-growing magnitude—pollution by oil—is a grave threat to coastal ecology everywhere. Because of its striking and devastating effects along resort beaches and in boat basins, as well as along less conspicuous stretches of the coast, this situation has received much publicity. It deserves much more. Oil and crude oil, spread out as a surface layer over the water, wash up along the shores, coating everything with a sticky tarry or oily film, fouling the gills and respiratory passages of intertidal animals so that they choke to death, and killing the surface plankton. Seabirds whose feathers are fouled by the oil are unable to fly and either drown or die slow, miserable deaths along the beaches. The fur and respiratory passage of marine mammals are fouled, and, of course, an oil-fouled coastline is ruined for a long time for fishing or recreation.

Since the more obvious effects of some oil spills seem to be taken care of in a few weeks with a combination of evaporation and biological degradation, some scientists supposed that the overall effect on marine life was negligible. However, more prolonged studies of subtidal organisms that cannot escape now indicate that residues of the oil persist far longer than had been thought possible—they continue to spread through the ocean floor sediments for months, and the toxic effect on a wide variety of marine organisms lasts for years.

At Woods Hole Oceanographic Institution, for example, a study has been made of a small oil spill that occurred in Buzzards Bay, Massachusetts, in 1969. The immediate kill was 95% of the fish, crabs, lobsters, and clams in the bay. Control stations were set up outside the immediate area. These have demonstrated a spreading contamination throughout the sediments, eventually covering an area of 22 $km.^2$ of offshore water, tidal rivers, and marsh. Substantial residual components of fuel oil remained in affected oysters kept for six months in clean running seawater after their exposure.

Golf ball–sized pieces of tar, formed from the heavier components of oil, are now found in massive amounts on the surface of the Atlantic Ocean. Similarly, tiny spherules of plastic, known as suspension beads, have been found throughout the Atlantic and Caribbean areas.

Oil pollution comes from many sources—some of it from oily wastes from oil fields or refineries near the coast, producing local problems in their areas; some from tankers and ships; and some from offshore oil wells.

The tonnage of crude oil carried in tankers is steadily growing, and ships of unprecedented capacity are being put into use. When one of these has an accident or is sunk off a coastal region, major ecological catastrophe ensues.

Some companies have caused considerable local pollution by cleaning their tankers near the coasts and rinsing them with seawater. Such practices, together with occasional accidents through negligence, became such a nuisance, especially along the Atlantic coast of the United States, that in 1924 Congress passed the Oil Pollution Act, authorizing the government to make efforts to remove spilled oil from the water or the shorelines and to sue the responsible companies for costs and damages. Under this act approximately a hundred prosecutions a year were filed, resulting in a considerable improvement in the practices of oil firms, although by no means wholly solving the problem. In 1966 some amendments to this act were included in the Clean Water Restoration Act of that year, originally intended to strengthen the government's hand in such cases. However, because of pressuring from oil interests, during the House-Senate conference on the bill the definition of "discharge" was changed to "any grossly negligent, or willful spilling." Proving "gross negligence" or "willful spilling" is almost an impossibility, and no one has been prosecuted under the new law.

During World War II many tankers were sunk, especially in the Atlantic. The Coast Guard estimates that over 200 million gallons went down in the United States flagships alone. Since then, whenever an oil slick appears offshore in the ocean, it has been customary to blame it on leakage from the sunken tankers. This is doubtless true in some cases, although it will be a decreasing source as times goes on. Recent developments in chemical analysis of oils will probably enable oil slicks to be precisely identified in the future and compared with samples obtained from tankers operating in the region.

A few examples will serve to indicate the variety of types of oil accidents occurring today.

In the early summer of 1965 one of the United States warships stationed in the Mediterranean not far from the French Riviera accidentally discharged several thousand gallons of oil into the water. A sailor assigned to flush out and clean the water tanks opened the wrong valves. This embarrassing accident caused great damage to the expensive and crowded resort beaches along the French Riviera and hundreds of thousands of dollar's loss to local resort businesses. It was especially distressing to the Navy because the French were at the time trying to get rid of American forces stationed in France, and the accident added much to the "Yankee go home" sentiments already being expressed both officially and unofficially and in signs along the highways. The fact that the navy assigned all available personnel to help try to clean up the beaches and even shipped up boatloads of fresh, clean sand from as far away as Naples, at United States taxpayers' expense, did little to mitigate the irritation aroused by this *faux pas*. Fortunately, this accident caused little real ecological damage, since the French Riviera is a particularly barren strip of gravelly beaches already rendered almost biologically sterile by vast overdevelopment that has occurred there. It would have been hard to choose a better place for such an accident, since it engendered maximum publicity and awareness of the problem without doing much harm.

A far more serious accident was the *Torrey Canyon* disaster in March, 1967. The *Torrey Canyon* was a huge tanker filled with crude oil from Kuwait, which went aground on rocks about 15 miles off the Cornwall coast of England. This accident resulted in the spillage of some 117,000 tons of crude oil. The problem was complicated by the massive attempt to save the beaches by dispersing the crude oil slick with detergents. More than 20,000 tons of at least twelve kinds of detergent were used. This action proved to be a great mistake because the detergents were water soluble and toxic, ensuring the death of almost the entire biota beneath the oil slick, as well as that on the surface. They also failed to keep the oil from the beaches. One of the most used, BP 1002, contained a highly toxic surface-active organic-solvent stabilizer responsible for massive mortalities. The combination of crude oil and detergents wiped out the intertidal plants and animals for miles and killed all larval forms in the offshore waters. This destruction, together with the persistent gummy coating of rocks and beaches, will impede recruitment of new biota for a long time. The detergents further delayed recovery by killing off the bacteria capable of degrading the oil. Another effect, which may not be altogether undesirable, has been that the extensive killing of intertidal gastropods has resulted in unexpected extension of algal meadows in several areas.

Persons interested in this problem should read the report on this disaster by scientists from the Plymouth Laboratory, edited by Smith (1968).

The Santa Barbara oil spill of 1969 from a Union Oil Company offshore drilling rig was also damaging, and received a great deal of publicity because it was in a famous resort area largely dependent on tourism. At its height the slick covered more than 800 square miles of ocean and damaged about 50 miles of coastline. It was particularly damaging to seabirds because of the numerous channel island rookeries in the area.

Other oil spills, both before and since, such as the wreck of the tanker *Tampico Maru* in Lower California in 1957 and the numerous blowouts and fires from offshore rigs in the Gulf of Mexico as well as off Australia, have perhaps been even more damaging in some cases but received less attention. The *Tampico Maru* spill occurred in a remote area of Lower California. No efforts to clean it up or mitigate its effects were made. Continuing studies made by North and others (1964) and by Mitchel and others (1969) showed that although there was considerable recovery of some species, many biological dislocations were still evident six years after the spill, and changes in the biota were still visible after twelve years.

The year 1976 and the first half of 1977 were especially alarming with respect to oil spills. There were several serious tanker accidents, then the blowout of the ''Bravo'' drilling rig in the North Sea in June, 1977, which was capped after spewing out more than a million gallons per day for five days. This accident carried great potential for damage, since it came just at the time that billions of fish eggs were floating near the surface. Stocks of plaice, sole, haddock, whiting, and herring were threatened. Furthermore, it was at this time that billions of birds were gathering in the region to breed. Just as this book goes to press, the American supertanker *Amoco Cadiz* is breaking up off the coast of Brittany in one of the worst spills yet, spewing some 70 million gallons of oil into the sea, more than twice as much as in the *Torrey Canyon* disaster.

The damage caused by a particular spill depends on the nature and extent of the spill, whether crude oil or refined products are involved, and the character of the ecosystems affected by the spill. Also, as the *Torrey Canyon* disaster illustrates, it depends on the nature of the measures taken to try to mitigate the effects of the spill.

The present oil rush in northern Alaska, together with the pipeline, poses a grave new threat to arctic ecosystems. Recovery of damaged arctic ecosystems will be slower than that elsewhere, since the cold temperatures will slow down both evaporation and biological degradation of the oil, as well as impede human efforts to mitigate the effects.

The present thrust to gain offshore oil concessions off southeast Asia and Indonesia also pose grave threats to these regions and their peoples.

Like the expansion of fisheries, such projects are promoted as great strides in the development of these regions and their peoples and in freeing them from poverty and degradation. In reality, the benefits are siphoned off to the rich countries, plus the ruling cliques of the underdeveloped areas, who are made richer and more powerful. The poor are relegated to the sidelines and further impoverished by finding their coastlines and fisheries damaged or threatened. The oil reserves of these areas are depleted to feed the military-industrial complexes of the more developed nations largely responsible for their poverty in the first place.

Radioactive contamination

Radioactive substances arise from military testing of nuclear bombs, from industrial and research nuclear reactors, and to a limited extent from natural causes such as bombardment of the earth by cosmic rays. These substances behave in many ways like the insecticides and other poisons in distributing themselves throughout the biosphere and in that they are unevenly concentrated from the environment by different organisms, with a tendency to concentrate in animals of the higher trophic levels.

All the elements can be made radioactive, some of them in several isotopic forms. In general, the atoms of an element in the radioactive state behave chemically in the same manner as do normal atoms of that element. They are therefore incorporated into the intimate molecular structure of all organisms in amounts that are proportional to the ratios of radioactive to normal atoms of the element in question available in the environment and dependent on the degree to which the organism in question utilizes or concentrates that element. Each is further characterized by a half-life decay period, representing the time during which half the atoms present will decay. These half-life periods range from as short a time as 10^{-7} seconds to 10^{6} years. Elements that have moderate to long half-life periods and that are either required in the bodies of organisms or are so closely related chemically to other required elements that they are incorporated along with them, are especially dangerous.

The most serious effects produced are alterations in the genetic material of the cells, causing heritable mutations, most of which are deleterious. For this reason, the effects are most serious in actively proliferating tissues such as epithelium, bone marrow, or germ cells and least serious in tissues that are not proliferating such as adult nerve cells. For the same reasons, growing embryos are more likely to suffer serious damage than are adult organisms.

In proliferating somatic tissues, the incorporation of radioactive elements may lead to somatic mutations, producing in localized areas of the body the cells that differ in some respect from normal, or it may lead to the escape of the cells from regulatory control by the body and their proliferation as cancer or leukemia. On the other hand, it may injure the cells in such a way that their normal proliferation is prevented, as in the case of white cells arising in the bone marrow, leading to leukopenia—a serious diminution of white cells that makes a person or animal much more susceptible to infections.

In germ cells, mutations are, of course, expressed as heritable changes in the offspring or in extreme cases as failure of the zygote to develop into a viable offspring at all. In human populations deleterious mutations accumulate to a greater extent than in most animals because the individuals bearing them are sedulously nursed along with all modern medical skill.

There is no such thing as a safe threshold of radioactivity below which there will be no damage. Any dosage of ionizing radiation, no matter how small, can induce a mutation if it happens to be on target. The chance for damage and the degree of damage, of course, increase proportionally to the dosage. However, low levels of exposure spread over a large population will produce just as many mutations as high levels spread over a smaller population.

Since a large proportion of nuclear weapons testing has been done in remote areas of the seas, and since the oceans have also been used to some extent as disposal sites for excess dangerous radioactive materials produced as by-products of nuclear reactors both on land and on ships and submarines, and because most of the radioactive materials in particulate form in the atmosphere eventually fall into the oceans or are washed into the oceans from the land, the oceans of the world have received a disproportionate amount of radioactive materials. Many marine organisms concentrate some of these isotopes to an astonishing extent and pass them along to other animals in the food chain, including man.

Various organisms concentrate certain trace elements from their environment anywhere from one hundred- to ten thousand-fold. When these organisms are eaten by others, the concentration is further increased.

Metals present in the water in very low concentrations are retained on chelating or complexing macromolecules that provide both appropriate chemical groups and spatial geometry for such uptake. The study of such agents in organisms has led to synthesis of macromolecular chelating agents possessing unusually high specificities. Some of these agents may eventually provide feasible methods for extraction of rare metals from seawater. They also point up the importance of biological processes in the formation of mineral deposits on the sea floor.

The speed with which radioisotopes are distributed throughout the biosphere and concentrated in organisms is illustrated by the fact that only a few months after the Russian nuclear tests, Osterberg (1962) was able to demonstrate high levels of zirconium-95, niobium-95, and cesium-141 in euphausiids off Astoria, Oregon, and somewhat lower levels in euphausiids off Newport, Coos Bay, and northern California. The euphausiids also contained high levels of zinc-65, but this had already been present as a result of pollution from the Hanford Plant, the amount of zinc-65 present being related to distance and direction from the mouth of the Columbia River. The niobium, zirconium, and cesium were definitely linked to the fallout from the Russian nuclear test. The amounts of fallout isotopes present in different areas depend on the patterns of fallout that produce hot spots in some areas and lower concentrations in others.

For a summary and analysis of this aspect of environmental pollution, the reader is referred to Polikarpov (1964), *Radioecology of Aquatic Organisms,* and to publication 651 of the National Academy of Sciences–National Research Council (1957), entitled *Effects of Atomic Radiation on Oceanography and Fisheries*. The general conclusion from these studies and others is that further additions of radioactive substances of any kind to the world's ocean are inadmissible.

Atmospheric pollution

According to a report by Eshleman (1969) on recent studies of the atmospheres of other planets, the earth may be in grave danger from extensive atmospheric pollution and burning of fossil fuels. During the past 100 years the burning of fossil fuels has increased the percentage of atmospheric carbon dioxide by 15% and is now using up 15% of the biological production of oxygen annually. Smog, jet contrails, deforestation, and pollution of the earth's waters, together with the increased carbon dioxide in the atmosphere, tend to raise the temperature of the atmosphere. A critical level may be reached (it is not as yet known what this critical level is) where the increased temperature would drive more water vapor and carbon dioxide into the atmosphere from the oceans, producing in turn an enhanced greenhouse effect that would end only when the earth's atmosphere had become a heavy, hot, suffocating, carbon dioxide–rich mixture unfit for most of the present biota of the world and leading to the greatest mass extinction since the beginning of life on earth. As Eshleman says, "It would seem that our relative ignorance of the possible global effects of small changes in the composition and temperature of the earth's atmosphere should give us serious and immediate concern about man's use of the atmosphere as a garbage dump."*

Thermal pollution

The great increase in the number and size of electrical generating plants over the past few decades, and especially the advent of large thermonuclear plants, has made the problem of thermal pollution acute in many areas.

These power plants use immense quantities of water to cool their condensers or, in the case of nuclear plants, to control the temperature of the reactors. The warmed water, often also containing some toxic metallic ions or radioactive elements, is discharged into rivers, lakes, estuaries, or coastal waters where they may produce drastic biological effects (the effects of the Oyster Creek nuclear power plant have been mentioned, p. 460).

Most organisms are already living close to the upper limits of their thermal tolerance because this is the range in which chemical processes in their bodies are most efficient. An upward change of only a few degrees can induce far-reaching biological

*From Eshleman, R. V. 1969. The atmosphere of Mars and Venus. Sci. Amer. **220**(3):78-88.

effects such as fish kills, plankton die-off, or marked changes in the species composition of the affected waters.

Even nonlethal temperatures may disrupt biological systems fundamentally by producing changes in timing of biological events such as production of gametes, periods of growth, or emergence from dormant stages. A mistiming of seasonal temperature changes can set in motion a chain of events that could seriously disrupt normal biological interactions in a region.

It has been thought that the oceans are so vast that they are immune to the effects of heat pollution and that generating plants could therefore be safely located along the coast. However, this solution may not be correct, since marine life is usually most concentrated in coastal areas.

In addition to the local effects of each power plant, there are generalized global effects, resulting from rapid widespread industrialization and increased burning of fossil fuels. It has been estimated that at the current rate of heat addition to the biosphere, plus the blanketing effect of carbon dioxide and particulate matter added to the atmosphere, the average global surface temperature may be raised as much as 1° or 2° C. in a few decades. This temperature increase could have profound biological effects on a worldwide basis. It may also hasten the melting of the arctic ice, adding enough quantities of water to the oceans to raise their level sufficiently to flood some coastal cities. Furthermore, it may provide enough moisture in high northern latitudes to initiate another cycle of glaciation, bringing about the next ice age somewhat ahead of schedule.

Persons interested in more detailed discussion of temperature effects may well begin with a perusal of *Temperature and Aquatic Life,* by Holdaway and others (1967). Much valuable information is also contained in *Thermal Pollution—1968.**

*Thermal pollution—1968. Hearings before the Subcommittee on Air and Water Pollution of the Committee on Public Works, U.S. Senate, 90th Congress, 2nd Session. U.S. Government Printing Office, Washington, D.C.

SELECTIVE ACTION OF CONTAMINANTS

It has been noted in connection with pesticides, radioactive nuclides, mercury, and other poisons that there is a marked tendency for them to become more concentrated in animals of the higher trophic levels—that is, they are progressively concentrated as they are passed along biological food chains. Furthermore, they are concentrated with time as well. Long-lived animals tend to ingest and retain more than short-lived, equally voracious animals. The probability of damage to a long-lived animal is also enhanced by the prolonged exposure to the poison in its body. There is also often an augmentation, or amplification, of the effects of a given deleterious agent in the presence of others or in organisms already suffering from some other cause.

Another facet of the problem is that populations of smaller short-lived, rapidly reproducing organisms can often adjust to specific poisons through the survival of those rare mutant individuals in the population which are for some reason less sensitive to the agent in question and from which populations of insensitive individuals can be reconstituted. Populations of larger, long-lived animals are in general too few in numbers and too slow in rate of replacement to respond in this manner.

Thus the environmental changes now coming to a head, with great quantities and a great variety of new chemicals and poisons introduced into the biosphere, all mitigate selectively against large, long-lived, slowly reproducing animals.

Man is, of course, one of the largest, longest lived, most slowly reproducing animals still prevalent on the face of the earth. He is already weakened by widespread malnutrition and by deliberate preservation of genetic weaknesses through his medical skills. He is also exposed to the greatest variety of unfamiliar chemicals. In addition to the pollutants already discussed, to which all animals and plants of the world are exposed, man is exposed directly to a great many more in the form of drugs, medicines, food additives and preservatives, cosmetics, urban smog, and industrial chemicals. Man is, then, one of the species most vulnerable to the particular type of ecosystem collapse he is bringing about.

ROLE OF THE MILITARY

No honest review of the world ecological crisis or of the plight of man today can overlook the role of the military.

The industrial and resource requirements of modern military forces are so vast and the nature of their weapons and of the byproducts from the manufacture and testing of these weapons is so deadly, so long-lived, and so accident-prone that it is rapidly becoming evident that the continued maintenance of such forces by any countries is incompatible with long-continued human life on earth—even if these forces are never used in war.

The military and their parasitic supportive industries are the world's number one wasters of our most valuable nonrenewable mineral, fuel, and energy resources. They are also the greatest, most dangerous polluters and destroyers of the world's environments. They constitute the gravest, most immediate ecological problem facing mankind today anywhere on earth. Getting rid of all significant military forces will not of itself resolve the grave ecological crisis facing our species, but is the greatest single step that could be taken and is prerequisite to any meaningful approach to dealing with these problems.

CONCLUDING THOUGHTS

The overall result of the rise of man has been to contribute greatly, and at an ever-accelerating pace, to the instability of the ecosystem. It is safe to say that for the next few years, at any rate, the uncontrolled population explosion will continue to contribute to this instability. We will continue to destroy the few remaining wild regions, to introduce deadly poisons and radioactive substances into the biosphere in increasing amounts and variety, and to burn fossil fuels. Every such change in the ecosystem entails the extermination of numerous species of animals and plants. We are in the midst of one of the great periods of mass extinction of species, a marked contraction in the numbers of kinds of living things on earth. Every increase in the instability of the ecosystem has the general effect of reducing the diversity of the living things inhabiting it.

The same thing is happening to man himself and to his cultures. The unprecedented increase in the mobility of man and in his technology has led to a general weakening and destruction of nontechnical cultures all over the world and to the spread of Western gadgets and the ways of life that they entail. Cities the world over are becoming full of cars, gasoline stations, department stores and supermarkets, smog, and decaying ghettos in the center but expanding suburbs for the well-to-do. Radio and television, telephones, and airplanes link all peoples so closely that what is done anywhere in the world has immediate repercussions throughout all of society. Whether we like it or not, human society is rapidly becoming one entity and exhibiting greater and greater uniformities throughout. The evolution of diverse cultures independently of each other has become a thing of the past. Because of the uncontrolled population increases and the inequities of our social and economic arrangements for the concentration and distribution of the world's wealth, the general level of human life will be low and poverty-stricken. It will continue this way longer, and recover to a lesser degree, insofar as we discover and use up the remaining uptapped resources of the world at this stage of our development.

We have already passed a point of no return where the ratio between the needs of our expanding population and the resources necessary to support it are such that simply to exist we are forced to destroy our environmental capital faster than it can be renewed. There is no hope in our time or in that of our children of avoiding an accelerated general environmental deterioration and a further lowering of the quality of life for our species. These facts are difficult for many of us in the United States or Western Europe to believe, since through our rapid exploitation of a virgin continent, our new technology, and our worldwide economic imperialism, we are, though only about 12% of the world's population, utilizing well over half the total economic production of the world and have thereby produced momentarily the most affluent society that has ever existed. While doing so, we have brought about an ever-accelerating destruction of the ecological base on which all humanity depends for its existence.

It cannot be emphasized too strongly that the real problem facing humanity is not inadequate potential for production of food and other necessities, or even population growth per se, but one of social organization and goals.

As long as industry, agriculture, and fisheries are dominated by large corporations, large land holders, absentee investors, and governmental groups intent on working them primarily for profit, only those who control such enterprises and who are wealthy enough to buy their products will benefit.

As long as most significant enterprises are not under the control of, and not primarily benefiting, the people actually laboring in them, they will serve only the well-to-do, and the general condition of mankind will continue to deteriorate.

It has been repeatedly demonstrated that, contrary to current mythology, large corporate agricultural enterprises and huge mechanized farms are less efficient, more wasteful, and more destructive to the land than are farms owned and operated by the people who work them. Furthermore, large investor-oriented enterprises tend to use most of the best land for the most profitable cash crops regardless of the needs of the local populations. Local populations then become wholly dependent on imports for most of their foods and other necessities—a far more precarious situation than one in which all regions are more diversified and more self-sufficient in their production. Similar considerations also apply to fisheries and to industry.

A world lacking basic diversity in its cultures, with reduced diversity in its faunas and floras, and crowded with poverty-stricken people will be a less interesting and a less viable world. Unless and until mankind overcomes its pathological fear of man, disarms, corrects present gross social and economic inequities, and finds and utilizes acceptable ways of bringing the human population into equilibrium with the ecosystem at a much lower numerical level than the present world population, he is condemned to inhabit a social and biological slum.

GLOSSARY

For clarification of terms not listed, consult the Index.

abioseston Nonliving inorganic particles suspended in the water. *See* bioseston.

aboral Away from the mouth; at the opposite pole of the body from the mouth.

abyssal Extremely deep; the deeper regions of the sea; depths greater than 4000 meters.

abyssobenthic On or near the bottom in the abyssal zone.

abyssopelagic In the water well above the bottom in the abyssal zone.

acaryotic *See* akaryotic.

acclimate; acclimatize To accustom or become accustomed to a new climate or environment; adapt.

aciculum In polychaete annelids, a large seta, larger than and distinct from the ordinary setae, and almost completely buried in the fleshy parapodium.

acraspedote Without a velum; the condition of "true" jellyfish, or Scyphozoa, as contrasted with hydrozoan medusae.

actinotroch larva A larval form in Phoronida, somewhat related to trochophore larvae but bearing ciliated arms.

actinula A larval form in certain medusae in which the planula larva develops arms but does not settle down as a fixed polypoid form before becoming a medusa; and intermediate growth stage in coelenterates without a polypoid generation.

adductor muscle A muscle that pulls together or closes the valves or shells of an organism—as the adductor muscles of barnacles or bivalve molluscs.

adenosine triphosphate (ATP) End product of stepwise oxidation of hexose sugars during respiration—a high-energy phosphate compound resulting from oxidation of glucose and used as energy source in biosynthetic pathways.

adradial In the tetramerous radial symmetry of most medusae, the sector midway between the main radial (perradial) canals; adradial canals are additional canals occupying this position in some medusae.

aesthetask Sense organs on the antennae of certain copepods.

agar-agar Gelatinous solidifying agent obtained from certain red algae such as *Gelidium,* widely used for solidifying bacteriological media and for other purposes.

agglomerate Objects or structures closely crowded into a cluster.

Agulhas Current A warm current, continuation of the south-flowing Mozambique Current from the eastern side of Africa, around the southern end of the continent to Cape Agulhas.

air bladder In many teleost fishes, a gas-filled sac originating as an outgrowth from the pharynx and serving as a hydrostatic organ.

akaryotic Cells without a definite nucleus with a nuclear membrane, and without mitotic apparatus or chromosomes.

akineton Plankton without any power of self-movement, such as many spores or planktonic eggs.

Alaska Current A north-flowing, then west-flowing eddy from the North Pacific Current in the Gulf of Alaska.

algal film The growth of small to microscopic algae, mixed with other small organisms, which forms on all surfaces exposed to marine waters in the euphotic zone. A source of food for many animals, especially chitons and many gastropods that graze on it, using a filelike lingual ribbon, the radula, to scrape it off rock surfaces into the mouth.

alima larva A second larval stage of certain Stomatopoda, after the pseudozoeal stage.

allochthonous Acquired from elsewhere—nonnative.

allogenetic plankton Carried into the region by currents, wind, etc. but normally living and breeding elsewhere.

allogenous detritus Detritus carried into a region from some other area.

allopatric With separate, mutually exclusive areas of geographical distribution.

alongshore current (commonly termed **longshore current**) The movement of sand and other loose particles in a given direction along the coast because of the prevailing angle of the wave trains to the coastline.

Alpha Rise A rise in the floor of the Arctic Ocean, in the Canadian Basin, roughly paralleling the Lomonosov Ridge.

alveolus A pit or cavity in a surface, as in the lorica of various tintinnids. Differs from a pore in not passing all the way through.

ambergris A foul-smelling pathological waxy material

formed largely from squid beaks in the digestive tract of certain toothed whales such as the sperm whale. It has been valued as a stabilizer added in minute quantities to fine perfumes.

ammocoetes larva The larval stage of lampreys, remarkable for its resemblance to the Cephalochordata.

amphiblastula larva A larval stage developing from the fertilized egg cell in certain sponges, comprised of small flagellated cells in one hemisphere, which eventually give rise to the choanocytes, and bulkier nonflagellated cells in the other hemisphere, which give rise to the other cells in the sponge body.

amphicoelous Concave at both ends; descriptive of the vertebrae of fish.

anabolism The constructive processes in metabolism; the building up of the compounds and structural elements that comprise the tissue or substance of an organism. *See* catabolism.

anadromous Migrating from the sea into freshwater to spawn. *See* catadromous.

anal fin In fish, a midventral fin behind the anus.

ancestrula In Bryozoa, the first zooecium of a colony; the zooecium formed by the larva on settling.

angstrom unit A minute unit of distance useful in describing such things as light waves; one one-hundred millionth of a centimeter.

animal exclusion theory The theory that the patchiness and the inverse relationship in surface distributions of phytoplankton and of zooplankton are caused in part by failure of the zooplankton to make it to the surface waters when impeded in its diurnal vertical migration by dense clouds of phytoplankton.

anisogamus With some visible differentiation between gametes. Generally used where the gametes are not greatly different but are not so much alike that they are difficult or impossible to distinguish, as in the isogamous condition.

annulus The transverse groove bearing the horizontal flagellum in dinoflagellates.

antapex The end opposite the apex; the base.

Antarctic Convergence An important biogeographical and hydrographical feature in the Southern Hemisphere where northeast-drifting surface waters from the Antarctic meet and sink beneath southeast-drifting warmer waters. It lies, for the most part, in or near the 50- to 60-degree latitude band south.

Antarctic Divergence The region near the shores of Antarctica where water of intermediate depths rises to the surface in extensive upwelling, replacing the surface waters that are drifting in a northern and eastern direction in the West Wind Drift.

antenna In arthropods, a preoral sensory appendage of the head. Antennae are paired. In Crustacea there are two pairs, the first pair commonly called antennules, whereas the term antennae is reserved for the second pair. Feelers.

antennal scale The short, flattened exopodite, or outer branch, of the second antennae in mysids and various decapod crustaceans.

antennules In Crustacea, the first pair of feelers.

anteriad Toward the anterior end.

antherozoid A swimming male gamete in certain algae and other primitive plants.

Antiboreal Convergence The second greatest zone of convergence in the Southern Hemisphere, located approximately 10 degrees north of the Antarctic Convergence. Often called the Subtropical Convergence.

antizoeal stage The first larval stage of various Stomatopoda.

aperture An opening. In shelled organisms such as Foraminifera, the main or largest opening in the test.

apophysis In Radiolaria, a horizontal outgrowth at right angles to the longitudinal axis of the radial spines.

appendix interna In Euphausiacea, a median process uniting the two pleopods of a given somite; also termed the stylambis.

archibenthic On the continental slope; between the relatively shallow continental shelf and the main ocean bottom, or abyssal plain.

arcuate Arched or curved.

areolation The crowding of a surface with small alveoli, or pits, so that it forms a type of network, or meshwork.

arthrobranch In decapod Crustacea, a gill arising from the arthrodial membrane between the coxopodite and the thoracic wall at the base of a thoracic appendage.

article In Arthropoda, one of the segments of a jointed appendage.

aspinose Without spines, as in some Radiolaria. In aspinose pores, the opening of the pore is not armed with a spine.

atoll An isolated ring of coral islands and reefs surrounding a large central lagoon.

ATP *See* adenosine triphosphate.

atrium An enclosed space in an animal body, lined with ectodermal epithelium and morphologically outside the animal and surrounding the pharyngeal region, into which water from the pharynx flows and into which the excretory and reproductive organs usually empty; opens to exterior by the atrial opening, or atriopore. Found in tunicates, cephalochords, etc. The term is also used for the thin-walled chamber(s) of the heart, which receives blood from the respiratory organs or from the body, and from which blood goes into the ventricle.

auricularia larva The free-swimming, ciliated larva of certain holothurians. It possesses earlike, blunt extensions of the body rather than long, slender larval arms.

autecology Study of distributions of individual organisms or species within a given environment and their relationship to that environment. *See* synecology.

autochthonous Native, indigenous.

autogenetic plankton Plankton organisms that live and breed in the region in which they are found.

autopelagic plankton Surface-living plankton.

autotrophic Literally "self-feeding." Said of organisms, such as plants, that synthesize needed organic foodstuff from inorganic substrates. *See* heterotrophic.

auxiliary cell In certain red algae, a cell to which the zygote extends a tubular growth, the ooblast, through which it migrates to the auxiliary cell. This cell participates in the formation of the fruit of the sexual plant.

auxospore In diatoms, an enlarged cell (in some cases, at least, resulting from union of gametes), that restores size to that line of cells in which the size was diminished at each cell division during the last few preceding divisions.

auxotrophic Not requiring organic carbon to form protoplasm, but requiring organic constituents such as auxins or vitamins.

avicularia Beaklike zooids in some Bryozoa colonies, such as *Bugula,* thought to function in keeping the colony free from debris or from larvae of other organisms that might settle on it.

axenic Said of a culture of some organism that contains no other organisms.

Axopodium In Protozoa, a pseudopodium containing an axial filament.

baleen The horny material comprising the plates in the mouth of the whalebone whale, through which water is strained in feeding, used to capture macroplankton. Also termed whalebone.

bank Shallow marine area surrounded by deeper waters; often excellent fisheries.

barrier reef A coral reef parallel to, but at some distance from the shore.

basipodite In biramous crustacean appendages, the basal portion of the appendage before the branching.

bathypelagic Living in deep water below the euphotic zone but above the abyssal zone, from about 1000 meters to about 4000 meters; free in the water, not living on the bottom.

bends A painful and dangerous condition created in divers or in any air-breathing animal in deep water by too rapid a release of the hydrostatic pressure on rising, causing formation of bubbles of nitrogen gas in the blood.

Benguela Current The major north-flowing cold current along the western coast of Africa.

benthic Living in or on the bottom.

benthos The organisms living in or on the bottom.

bifurcate Forked; divided into two branches.

bilateral symmetry Organization of the body so that its longitudinal axis divides two equivalent sides, right and left, and two nonequivalent sides, dorsal and ventral.

binary fission A form of asexual reproduction in which an organism divides into two more or less equivalent parts, each of which continues to live as a whole organism.

biocoenosis A community of organisms found in a given area or region or habitat.

biological clock Inherent physiological rhythm in organisms that usually manifests itself in changes in behavior or physiology timed to certain recurrent changes in the normal environment of the organism and that may continue for a time even when the environmental stimulus to which the organism is timed has been removed.

bioluminescence The production of light by living organisms.

biomass The weight of the standing crop of living organisms, or of selected kinds of living organisms, in a given unit of the environment at a particular time or over a given time period.

bioseston Living particulate matter in the water, and organic particles derived directly from living organisms. *See* abioseston.

biosphere That portion of the surface of the world, its waters, and its atmosphere permeated with living organisms or their products.

biotic community The group of species of organisms living in, and characteristic of, a given habitat.

biotope A specific habitat.

bipinnaria larva A larval stage of various starfish in which the ciliated bands are drawn out into arms.

bipolarity The geographical distribution of species, or closely related forms, in both the Northern and Southern hemispheres but not in the intervening equatorial belt.

biramous Two-branched, as various crustacean appendages.

black zone A band of growth of microscopic blue-green algal cells and other minute microplants at the extreme upper limit of the intertidal zone, in the splash zone. The black zone represents the transition from essentially terrestrial to essentially marine habitats.

bonification Estimation of fertility and standing crop.

boreal Cool or cold temperate regions of the Northern Hemisphere. The corresponding region of the Southern Hemisphere is sometimes termed antiboreal.

botryose Having the form of a cluster of grapes.

brachiolarian larva The last stage in starfish larvae just prior to metamorphosis into the starfish form. Brachiolarian larvae develop additional larval arms, not involving the ciliated band, in connection with fixation to the substrate.

bract In Siphonophora, thick gelatinous structures of various shapes that aid in swimming and in the protection of the colony. Bracts have been regarded as modified medusoids or, more recently, as merely modified tentacles.

buccal mass In molluscs and various other organisms, the mouth and pharynx when heavily muscularized and used as an organ of prehension and mastication.

budding Asexual reproduction in which a small portion of an organism grows out from and eventually develops into another individual or another equivalent part of a colony.

buffer Salts or other substances that are to some degree amphoteric, able to remove either hydrogen or hydroxyl ions from solution, thus retarding changes in pH of the solution when either acids or bases are added and making for a greater constancy in the pH.

cadophore In Thaliacea of the family Doliolidae, a posterior dorsal extension of the body that carries the buds in the aggregate phase of the life cycle.

Cambrian The earliest of the geological ages from which a good fossil record has been preserved, beginning approximately 500 million years ago and lasting approximately 80 million years.

Canadian Current A south-flowing cold current from the Arctic Ocean along the northern coast of Canada.

capacity adaptation Compensatory changes that enable an organism to carry on normal physiological activities over the usual range of fluctuation in such factors as temperature and illumination, occurring in its habitat. *See* resistance adaptation.

captacula The feeding tentacles of tooth shells—*Dentalium.*

carapace In Crustacea, a fold of skin and exoskeleton of the dorsal body wall, covering the head, part or all of the thorax, and in decapods forming lateral gill chambers.

carinate Having a keel or ridge.

carotenoid Yellow or orange pigments chemically related to carotene. Carotenoid pigments are lipid soluble and are an important coloring matter in many marine organisms.

carpogonium In red algae, the female unicellular sex organ, bearing a terminal filament on which the drifting spermatium from the male lodges, giving rise to the carpospores when fertilized.

carposporangium In red algae, a spore-bearing organ on the gonimoblast filament, derived from the zygote.

carpospore A spore resulting from division of the zygote in red algae.

catabolism The breakdown of organic molecules during metabolism. *See* anabolism.

catadromous Migrating from freshwater into the sea to spawn. *See* anadromous.

caudal Posterior, or toward the tail.

central capsule In Radiolaria, the inner portion of the protoplast, bounded by a membrane and containing the nucleus or nuclei.

centric In diatoms, having the valves arranged radially and symmetrically around a central point. A member of the order Centrales.

cephalic Pertaining to the head, or cephalon. Toward the head.

cephalosome In copepods, the apparent head when viewed dorsally, when this consists of the head plus one or more fused thoracic somites.

cephalothorax In Crustacea, the head and thorax when united under a single carapace.

cercaria A larval stage of digenetic trematodes. The cercaria leaves the mollusc host and either directly, or indirectly by way of a metacercarial stage, invades the vertebrate host, where it matures into the sexual adult worm.

chela In Crustacea and some other arthropods, a pincer formed when the terminal article of a limb is opposable to a process from the penultimate article, as in crabs. An appendage ending in this manner is termed a chelate appendage.

chemoautotrophic Deriving metabolic energy from the oxidation of inorganic chemicals occurring outside the organism, a condition found in a few minor groups of bacteria—the nonphotosynthetic autotrophic forms. Chemotrophic; chemolithotrophic.

chitin A polymer of glucosamine that forms an impervious stable substance synthesized by many animals, such as arthropods, as an external covering. Impregnated with calcium salts, it may form a firm, rigid exoskeleton.

chlorinity Grams of chlorine and bromine ions per thousand grams of seawater. Determined by simple titration with silver nitrate. Used in calculating salinity.

chloroplast In plants, a cytoplasmic structure, plastid, containing chlorophyll. The site of photosynthesis.

chromatophore In plant cells, plastids containing yellow or brown pigments in addition to chlorophyll, giving them a color other than clear green.

cicatrix In fossil Nautiloidea, a scar on the initial shell chamber, thought to represent the point of attachment of an embryonal shell.

cinereous Ash gray, the color of wood ashes.

cingulum In dinoflagellates, the horizontal groove around the cell.

circular canal In coelenterate medusae, a tube of the gastrovascular cavity, lying at the periphery of the bell, encircling it, and connected to the stomach by the radial canals.

cirrus A slender, tentacle-like process. In ciliate protozoans, a fused group of cilia.

claspers In various animals, appendages or organs of a male, used to hold the female and facilitate copulation.

clavate Club-shaped.

cleaning symbiosis A relationship between different species of animals in which one, usually a smaller species, picks parasites, bits of infected tissue, etc. from the body surface, mouth, or gills or a larger animal.

clone Genetically similar individuals derived by asexual reproduction from one individual or, in Ciliata, from a pair of exconjugants.

coccolith The tiny calcareous concretion, or plates, of the coccolithophores, often forming a significant part of calcareous marine sediments.

coelom A body cavity lined with mesodermal epithelium, the peritoneum.

coenenchyme The fleshy cellularized mesogloea of alcyonarians, usually traversed by gastral tubes, solenia, connecting the gastral cavities of the polyps.

coenocyte A tissue in which the cytoplasm is continuous rather than divided into individualized cells around each nucleus.

coenosarc The living tissue in hydroid colonies, joining all the polyps.
collarette In Chaetognatha, a thickened ectodermal area in the neck region of the body.
colloblast Adhesive cell on the tentacles of Ctenophora of the class Tentaculata. Used in the capture of prey.
columella In gastropod shells, the central axis of shell around which the whorls are spiraled.
comb plates The flagellated platelets in Ctenophora responsible for swimming.
commensalism An association between two species in which one of the pair is benefited, whereas the other seems to be neither benefited nor harmed.
community *See* biotic community.
compensatorium In certain Bryozoa, a water-filled invagination of ectodermal epithelium from which the water can be expelled as the animal retracts into its zooecium, making room for the retracted body.
conceptacle In brown algae of the subclass Cyclosporeae (the Fucales), a round cavity borne in the swollen tip of the branched thallus and containing the reproductive organs.
concrescent Growing together, coalesced.
conjugation In the Ciliata, a sexual process in which there is a temporary cytoplasmic connection between two individuals. While the connection lasts, there is an exchange of haploid nuclear elements, each of which fuses with a corresponding nucleus of the other individual, reconstituting a diploid condition. Then the individuals separate. The descendants of a pair of exconjugants are all genetically alike and constitute a clone.
constant plankton Perennial holoplankton in a given region.
consumers Organisms that feed on already formed organic matter. All "animals," and a few other forms.
convergence A zone where surface waters meet and are carried down beneath the surface. *See* divergence.
copepod An entomostracan crustacean of the order Copepoda. Copepods are ubiquitous in both marine and fresh water and constitute one of the most important basal groups of aquatic food chains.
copepodid An immature copepod, the stage occurring between the last naupliar stage and the mature adult. Copepodids look much like adult copepods but are not sexually mature.
coracidium A larval stage in certain pseudophyllidean tapeworms in which the egg is surrounded by a ciliated or flagellated embryophore and swims freely in the water.
corallite A cup-shaped structure in coral skeleton in which the polyp sits.
cordate Heart-shaped.
cormidium In Siphonophora, a group or colony of individuals borne together on the coenosarc of the stalk. In some species the cormidia are easily detached and are usually found separately in plankton hauls.
coronal constriction In jellyfish of the order Coronatae, a circular constriction of the bell.
coronal pores In some Radiolaria, small pores surrounding the parmal pores. *See* parmal pores.
cortical shell In some Radiolaria, the portion of the skeleton forming an outer framework.
costae Ridges of a shell.
coxa In Arthropoda, the basal segment of one of the jointed appendages; in biramous appendages, the first segment or article of the basipodite.
craspedote Having a velum, a small shelf of tissue forming an inward-directed ring from the edge of the bell in hydromedusae.
crenulate An edge having a series of small indentations, giving it a scalloped appearance.
critical tide levels The average levels attained by each of the tides of the daily tide cycles during the year.
cruciate; cruciform In the form of a cross.
crystalline style In bivalve molluscs and a few other filter-feeding molluscs, a rodlike gelatinous structure in the first part of the intestine, or in diverticulum thereof, which is rotated by ciliary action and gradually worn down anteriorly by rubbing on the gastric shield in the stomach. It functions to stir the stomach contents and slowly release digestive enzymes incorporated into it.
ctenidia The typical gills of most aquatic molluscs.
cyst A resistant wall or shell secreted around an organism, or around certain stages in the life cycle of an organism, to protect it during periods unfavorable to ordinary life activities.

dactylozooid In coelenterates, a modified, mouthless, often rodlike polyp, heavily armed with nematocysts, which serves to touch and catch prey or to defend the colony.
decussate Crossed or intersected.
deep-scattering layer An intermediate-depth zone, above the bottom, that reflects sound from echo-sounding equipment, giving a record of a "false bottom." It is formed by the aggregation of certain small animals at particular depths during the day. At night, as the animals usually swim upward into water nearer the surface, the deep-scattering layer tends to rise and become more diffuse, and individual layers within it lose their identity.
delta An extension of low land formed by deposition of sediments in the mouth of a river. Deltas are usually fan-shaped, with the river forming numerous channels through them.
demersal Said of a swimming organism that prefers to spend most of its time on or near the bottom—as the flatfish, octupuses, and many shrimps.
density The weight per unit volume.
detritus Particulate matter, especially that of organic origin, floating in the water or settling to the bottom.
dextral To the right. In gastropods, having the aperture of the shell on the right side when the shell apex is upward and the aperture is facing the observer (on the observer's right). *See* sinistral.
diadromous Said of fish; species that do a spawning migration through an estuary either to freshwater or the open sea. *See* anadromous and catadromous.

diameter In diatoms, the greatest width.

dichotomous Branching into two parts at each division.

digenetic trematodes Flukes, trematode worms in which the sexual adult stage is found in vertebrates and the asexual larval stages in molluscs, usually gastropods.

digitiform Finger-shaped.

dilatancy Tendency of fine sand and mud, under certain conditions of low water content (less than 22%), to lose interstitial water when pressured, making them harder and more resistant to shear. Opposed to thixotropy.

diploblastic Formed of two layers of cells; an outer epithelium and an inner gastrothelium; the grade of construction of simpler coelenterates such as *Hydra*.

distal Pertaining to the part of an appendage or other structure farthest from the point of attachment. Away from the point of origin.

divergence A zone where water from below is upwelling and spreading out over the surface. *See* convergence.

diversity In biogeography, an expression that is a function of both the number of species present in a given region or habitat and also the relative proportions of their numbers. For example, given two faunas, each with 500 species—if, in one of them, two or three species were overwhelmingly dominant and the others relatively rare, whereas in the second many of the species appeared in considerable numbers, the second fauna would exhibit greater diversity. Diversity tends to be correlated with stability of ecological conditions, other things being equal.

doliolarian larva A short-lived free-swimming nonfeeding larva of certain shallow-water crinoids, such as *Antedon*.

drill Carnivorous gastropod, such as oyster drill and whelk, that uses its radula to drill a hole through the shell of its prey.

drogue Device for measuring current.

ecdysis The molting of a cuticle or exoskeleton, as in the Arthropoda and the Nematoda.

echo-sounding equipment Acoustical equipment designed for determining depth of water by bouncing sound waves off the bottom and recording the time required for the echo to reach a recording instrument. Fathometer.

ecosystem The entire system of organisms and the physical and chemical aspects of their environment with which they interact.

ectocrine External secretions or hormones that affect other organisms.

ectoderm The embryonic cell layer of the Eumetazoa forming the outer epithelium, lining of the mouth, and various other structures.

endemic Found only in a given region.

endite In Crustacea, an inner lobe or cuticular extension from certain basal segments of the appendages —as the endites of amphipod and isopod thoracic limbs that form the brood pouch in the females, or the gnathobases of appendages of trilobites and of *Limulus*.

endoderm The embryonic cell layer from which the stomach and its diverticula and glands are formed in Eumetazoa.

endogenous detritus Detritus originating in the locality where it is found.

endogenous plankton Plankton originating in the region where it is found.

endopodite In Crustacea, the inner ramus of a biramus appendage. Also endopod.

endopsammon The microscopic fauna of sand and mud.

endoskeleton A skeleton produced within the flesh of an animal and typically remaining embedded there.

endostyle A glandular ciliated groove along the ventral side of the pharynx in Urochorda, Cephalochorda, and lamprey larvae.

Entomostraca A collective term for all the lower subclasses of Crustacea, most of which are of small size and hatch from the egg as nauplius larvae.

ephyra The earliest free-swimming stage of scyphozoan medusae that have just been released from the polyp, or scyphistoma. Also ephyrula.

epicone In dinoflagellates, the portion of the cell above or anterior to the horizontal groove, or cingulum.

epifauna Animals living on the surface of firm substrates. *See* infauna.

epipelagic Living in the upper zone of the ocean, where there is light enough for photosynthesis.

epiphyte A plant growing on another plant; sometimes extended to all organisms growing as sessile forms attached to plants.

epiplankton Plankton occurring near the surface, in the upper 200 meters.

epipodite A laterally directed extension from one of the basal segments of an arthropod limb.

episome A small bit of nucleic acid in the cytoplasm of a cell, replicating independently of the cell's own genome and transferable to another cell of the same or closely related type by direct contact.

epitheca The upper or larger half of the frustule of a diatom, or the plates above the cingulum in peridinian dinoflagellates.

epizoon An organism growing attached to the external surface of an animal.

epontic Periphyton attached to any type of surface.

eretmocaris stage Postlarval stage in certain decapod Crustacea, characterized by possession of long eyestalks.

erichthus larva The second larval stage in Stomatopoda.

estuary The brackish-water areas influenced by tides, where the mouths of rivers come to the sea.

euhalabous Plankton, especially phytoplankton living in water of 30‰ to 40‰ salinity.

euphotic zone The upper strata of water that receive sufficient light to support photosynthesis.

euplankton Organisms that spend most or all of their lives as plankton.

eurybathic Able to tolerate a considerable range of depths.

euryhaline Able to tolerate a wide range of salinities.

eurythermal Able to tolerate a wide range of temperatures.

eutrophic plankton Plankton in areas of especially high productivity.

exogenous plankton Plankton originating in some other locality than that in which it was found.

exopodite In Crustacea, the outer branch, or ramus, of a biramus appendage. Also exopod.

exoskeleton In Arthropoda, the hardened cuticular covering of the body, serving for protection of soft parts and as sites for muscle attachments.

extracapsular cytoplasm In Radiolaria, the cytoplasm peripheral to the central capsule.

exumbrella The top, or convex outer surface, of medusae. The aboral surface.

false bottom The deep-scattering layer.

fascicle A bundle or cluster of threadlike, rodlike, or bristlelike structures bound together or arising from the same place, as a fascicle of setae.

fenestra A small opening, as in a shell.

ferruginous The color of iron rust.

filament A threadlike structure.

filiform In the shape of a thread.

fiord *See* fjord.

fjord (also spelled **fiord**) Long narrow arm of the sea bordered by steep cliffs; usually formed by glacial erosion.

flagellum A whiplike structure. In Protozoa of the Flagellata, the characteristic locomotor organ.

fluke A parasitic flatworm belonging to the digenetic trematodes.

foot In ordinary usage, the terminal portion of a walking appendage that contacts the substrate. In molluscs, the ventral surface of the body, variously modified in different groups for creeping, digging, or serving as prehensile arms.

fossette A small hollow or pit.

fracture zone In geology, a zone along which there has been displacement. In the oceans, major fracture zones are found at right angles to the Midocean Ridge system.

frustule The siliceous shell of a diatom.

fucoxanthin A brownish xanthin pigment found in seaweeds.

furcilia larva The last larval stages in euphausiids.

fuscous Brown-colored and tinged with gray; swarthy.

fusiform Tapering toward both ends, as in many pennate diatoms.

gametangium In algae, a special structure in which gametes are borne.

gamete One of the germ cells, a sperm or an egg.

gametophyte In algae, the generation giving rise to gametes.

gastrozooid In Siphonophora and in Doliolidae, individuals of the colony specialized for feeding.

geniculate Elbowed, with a kneelike bend, as in the antennae of various isopods and other Crustacea.

genital somite; genital segment In copepods, the somite or fused pair of somites bearing the genital openings, usually the first segment of the urosome.

gill A general term for special structures evolved for underwater respiration.

gill arch The tissues of the pharynx, which supports the gills; it is usually supported by cartilage or bone, in fish, between the gill slits.

gill rakers Comblike extensions from the inner side of one or more gill arches in plankton-feeding fish that retain the plankton as water passes through the gill slits.

girdle The connecting bands of the two halves of diatom frustules.

glabrous Having a smooth surface.

globigerina ooze A calcareous sediment on the floor of extensive areas of the ocean, containing, among other things, great numbers of foraminiferan shells of the genus *Globigerina*. Because of the solubility of the shells under great pressure, this ooze is not found in the deepest parts of the ocean.

gnotobiotic Literally "known biology," said of cultures grown in a chemically defined medium without other organisms or with only known organisms present in the culture.

gonangium In hydroids, the modified polyp from which medusae buds are developed.

Gondwanaland The southern land mass during the Mesozoic era, composed of the present South America, Africa, Australia, India, and Antarctica. *See* Laurasia.

gonimoblast filament A carpospore-bearing filament developed directly from the zygote in some red algae.

gonodendron In Siphonophora, processes in the gonophores bearing the sex cells.

gonophore In Siphonophora, modified medusoid individuals or reproductive zooids bearing the sex organs.

gonozooid In colonial Coelenterata and tunicates, individuals bearing gonads.

greenhouse effect Possible self-sustaining increase in temperature, water vapor, and carbon dioxide in the atmosphere; heat absorption resulting from increased temperature and carbon dioxide from burning of fossil fuel and pollution.

groin A small jetty or sea wall extended out from the shoreline, usually for the purpose of holding beach sand in place.

gross production The total of the organic matter being produced in a given unit of time in the system studied, without any correction for respiration or catabolism. *See* net production.

hadal zone The bottom in the deepest (below 8000 meters) parts of the oceans having few life forms (only 370 out of 157,000 known benthic species).

haliplankton Marine or inland saltwater plankton.

haptic Clinging tightly, as the Gnathostomulida, which are strongly haptic, clinging to the sand grains.

hekistoplankton Flagellated nanoplankton.

helix A coiled spiral, as in gastropod shells.

hemiplankton Organisms spending only part of their lives in the plankton; meroplankton.

heterogamous Having differentiated gametes, sperm, and egg cells.

heterotrophic The mode of nutrition of organisms that cannot synthesize new organic matter from inorganic substrates. Requiring already formed organic matter. This includes all animals and most bacteria. *See* autotrophic.

hinge joint In Copepoda, the major articulation between the prosoma and the urosome.

holdfast The rootlike or disclike, anchoring portion of the thallus of a seaweed.

holopelagic Pelagic throughout the entire life cycle.

holophytic Having plantlike nutrition, synthesizing new organic matter from inorganic substrates. Photosynthetic.

holoplankton Organisms that are planktonic throughout their entire life cycle.

holozoic Having animal-like nutrition. Fed by ingestion of organisms of other organic matter.

homoiothermal Having a body temperature that is relatively constant and mostly independent of the temperature of the environment; warmblooded. Opposed to poikilothermal.

Humboldt Current (also called **Peru Current**) The great north-flowing coldwater current along the west side of South America.

hyaline Transparent or semitransparent, with a glassy or gelatinous consistency.

hydrocoele An embryonic coelomic vesicle in echinoderms and in tornaria larvae of Hemichordata, communicating to the outside by a water pore.

hydroecium In Siphonophora, the extended part of the swimming bell, covering the proximal part of the stem of the colony.

hydrotheca In Hydrozoa, the cuplike extension of the perisarc around the feeding polyp, into which it can withdraw. Present in members of the Calyptoblastea.

hyphalomyraplankton Plankton of brackish water.

hypocone In Dinoflagellata the lower posterior portion of the cell back of the cingulus.

hypodermis In Arthropoda, the epithelial cells beneath the cuticular exoskeleton.

hypoplankton Demersal plankton; plankton taken close to the bottom.

hypotheca In diatoms, the smaller or lower valve of the shell; in armored dinoflagellates the lower plates below or back of the cingulum.

infauna Animals living in and on soft bottom.

injector organ In certain burrowing polychaetes such as Opheliidae, a muscular sac back of the prostomium that withdraws or injects coelomic fluid of the head coelom.

interradial In medusae, the portion midway between the primary radial (perradial) canals of the gastrovascular cavity.

isogamous Having gametes that show no obvious differentiation into distinctive male and female gametes. Gametes that look alike.

isolume A theoretical warped plane or line in the water connecting points with equal light intensity.

isotherm A theoretical warped plane or line in the water connecting points of equal temperature.

Japan Current The north- and northeast-flowing current offshore from Japan, also called the Kuroshio Current.

karyotic Having nucleated cells.

knephoplankton Plankton found between 30 and 500 meters' depth.

krill Shoals of euphausiids, which serve as whale food; in Antarctic waters especially, the red *Euphausia superba*.

Kuroshio Current The major warm northeasterly current off the shores of Japan, sometimes termed Japan Current.

Labrador Current A cold south-flowing current along the coast of eastern Canada.

lactate dehydrogenase (LDH) Enzyme catalyzing reduction of pyruvic acid to lactic acid in anaerobic respiration.

lambert A unit of illumination; the luminance produced by a uniform diffuser with a total flux of one lumen per square centimeter.

lamelliform Flat and leaf-shaped.

lamina A leaflike structure or object; a thin layer.

laminate Formed of thin layers.

Langmuir Circulation The system of cells of convergence and divergence that forms in open water at right angles to the direction of prevailing wind. If the water is flowing at a different angle, the small rips, or zones of convergence, formed may not be directly crosswind in direction.

Laurasia The northern member of the two supercontinents supposed to have existed prior to Tertiary times, composed of what is now North America, Greenland, and northern Eurasia. *See* Gondwanaland.

laver The green alga, *Ulva,* dried and sold for food.

LDH *See* lactate dehydrogenase.

lecithotrophic Said of larvae that are nourished from nutrients in the egg rather than by feeding on plankton.

leptopel Large organic molecules and aggregates of molecules of approximately colloidal size dissolved or suspended in the water.

leuconoid Descriptive term for sponges in which a complex series of incurrent and excurrent channels lead to and from internal chambers lined with choanocytes. Most Demospongiae are leuconoid.

light saturation The amount of light at which a given kind of plant cell will carry out photosynthesis most actively.

Liman Current A cold south-flowing current in the western North Pacific.

linear Forming a line, as in an end-to-end arrangement of objects. In graphic representations, usually taken to mean forming a straight line on the graph rather than a curve.

lingual ribbon The radula in molluscs.

liter, litre A measure of volume, 1000 cubic centimeters, about equal to one quart.

lithocyst A cystlike hydrostatic organ in the margin of the disc in coelenterate medusae.

Lithostyle A sensory club or tentaculocyst on the free margin of the disc of a jellyfish, usually containing a calcareous concretion with a cushion of sensory cilia. Lithocysts are a modified form of lithostyle.

littoral zone The intertidal zone. Some writers include the entire shelf area.

longshore current *See* alongshore current.

lophophore The horseshoe- or double-spiral–shaped ridge of tissue, and its supporting structures, surrounding the mouth and bearing ciliated tentacles in the Bryozoa, Brachiopoda, and Phoronida.

lorica A secreted noncellular protective case, as in Tintinnidae.

lumen (1) An opening and (2) a unit of illumination equivalent to one meter candle or lux.

lunate Crescent-shaped.

lunule An impressed area just below the beak in the shell of a bivalve mollusc.

macrobenthos The larger organisms of the benthos, more than 1 mm. in length; usually do not pass through a 1 mm. mesh sieve.

macronucleus The larger vegetative nucleus in ciliate Protozoa, which does not undergo mitosis during division nor participate in nuclear exchange during conjugation.

macroplankton Plankton organisms ranging in size from about 1 mm. to 1 cm. in length.

magnetic anomaly A reversal in the direction of magnetic orientation in rocks or sediments, indicating a reversal of polarity of the earth's magnetic field.

major articulation In copepods, the articulation between the prosoma and the urosome.

mantle In molluscs, the dorsal body wall, covering the visceral hump and secreting the shell in shelled forms.

mantle cavity The space between the mantle and the rest of the body, in molluscs, in which the mantle has grown down around the body at any point. The mantle cavity commonly contains the ctenidia and the openings of the digestive, reproductive, and excretory systems.

manubrium In jellyfish, an ectodermally lined protrusion from the surface of the subumbrella, bearing the mouth.

marine humus The less digestible organic residues that accumulate in solution or suspension in the seas.

marine terrace A flat, platformlike area showing the position of a former beach, or area cut back by the sea.

maxillae In Crustacea, the two pairs of appendages immediately back of the jaws, or mandibles. The first pair are often termed maxillulae, the term maxillae being used for the second pair.

maxillipeds In Crustacea, the first two or three pairs of thoracic appendages back of the maxillae, when turned forward and modified to aid in the handling of food.

medullary shell In Radiolaria, the part of the shell forming an inner framework.

medusa The sexual stage of Hydrozoa and Scyphozoa, usually in the form of a free-swimming "jellyfish."

megaplankton Plankton organisms larger than 1 cm. in length or diameter.

megascleres In sponges, large spicules forming the chief supporting framework.

meiobenthos Small benthic organisms that will pass through a 1 mm. mesh sieve but be retained by a 0.1 mm. mesh; mostly 0.5 to 2 mm. long; same as meiofauna.

melon In toothed whales, a mass of oily blubber between the blowhole and the end of the snout.

membranelle In ciliates, a group of cilia fused together to form a vibratile plate or membrane.

meroplankton Organisms occurring in the plankton only periodically, or only for a portion of their life cycle.

merus In Crustacea, the third segment of the leg in the thoracic legs.

meshalobous Phytoplankton, or plankton in general, living in brackish water with salinity ranges of 5‰ to 20‰.

mesogloea The gelatinous noncellular or secondarily cellular material between the outer epithelium and the gastrothelium in many coelenterates. It is particularly well developed in the medusae of Hydrozoa and Scyphozoa and the coenenchyma of Alcyonaria.

mesopelagic Intermediate depths below the euphotic zone but above the completely aphotic zone. The dysphotic depths. Organisms found in these depths.

mesoplankton Planktonic organisms between 0.5 and 1 mm. in size.

metabolite A chemical produced as a result of the metabolism of organisms.

metagenesis The alternation of sexual and asexual generations, as in many hydroids.

metanauplius In Euphausiacea, a postnaupliar larval stage in which appendages in addition to the original three pairs of naupliar limbs are present but before metamorphosis to the calyptopis stage has occurred.

metapleural folds In Cephalochordata, ventrolateral fin folds in the pharyngeal region of the body. Probably metapleural folds of early Chordata gave rise to the paired fins of fish.

metasoma In Copepoda, the thoracic somites behind the cephalosome and in front of the major articulation.

microbenthos Benthic organisms small enough to pass through a sieve with a 0.1 mm. mesh.

micrometer (also **micron**) One one-thousandth of a millimeter.

micron *See* micrometer.

micronucleus In Ciliata, the small nucleus (or nuclei), which is comparable to the nucleus of other cells. It undergoes mitosis during cell division and undergoes reduction divisions to the haploid state prior to conjugation.

microplankton Plankton organisms in the size range from 0.5 mm. down to 60 μm.

microscleres In sponges, small spicules scattered throughout the mesenchyme.

Midocean Ridge A system of rifts and parallel mountain ranges or hills found in all major oceans, thought to be sites of the upwelling of new ocean floor material from the earth's mantle, from which the ocean floors are gradually spreading out laterally.

millilambert One one-thousandth of a lambert. A very small unit of illumination used in expressing bioluminescence of some plankters.

miracidium The larval stage that emerges from the egg in digenetic trematodes and carries infection to the molluscan host.

mitraria larva A modified trochophore larva of some annelid tube worms, having two bundles of long setae extending below the larva.

mixotrophic Able to exist by either autotrophic or heterotrophic nutrition, as conditions necessitate.

monaxons In sponges, spicules formed by growth along a single axis.

monomyarian In bivalve molluscs, having only one adductor muscle.

monoton plankton Heavily dominated by one species.

monoxenous Said of a parasite limited to a single host species.

mutualism A relationship between two species in which both benefit, in many cases becoming completely dependent on the relationship, as in the case of some cellulose-eating forms such as termites and the microbial fauna of their intestines.

myxotrophic Able to live autotrophically but also under certain conditions able to live heterotrophically or phagotrophically.

naked dinoflagellates Members of Gymnodinioidea that do not have protective plates on the cell surface.

nannoplankton *See* nanoplankton.

nanoplankton Very small plankton organisms in the size range from 5 to 60 μm. Plankton organisms too small to be retained by a plankton net. Most writers on oceanography and plankton spell this nannoplankton. The root nano- comes from the Latin *nanus* or Greek *nanos,* meaning dwarf or small, and is spelled with the single *n* in most other combinations and scientific writings (nanosecond, etc.). *See also* ultraplankton.

natatory Having the function of swimming.

nauplius A larval stage in many of the lower groups of Crustacea, having only three pairs of appendages and a single median eye.

nectocalyx In Siphonophora, one of the members of the colony modified as a swimming organ.

nectophore In Siphonophora, a nectocalyx, or swimming bell.

nekton Aquatic, free-swimming organisms whose swimming activity largely determines the direction and speed of their movement, including fish, cephalopods, some larger crustaceans, etc.

nematocyst The typical stinging cell of coelenterates.

neritic Relating to the waters over the continental shelves.

net production Gross production minus the amount lost in the same time by catabolic processes.

neuston Extremely small organisms living in or near the surface film at the top of the water.

nodule In pennate diatoms, part of the complex system, including the raphe and nodules, that enables them to move.

North Atlantic Current The eastward continuation of the Gulf Stream across the North Atlantic.

North Equatorial Current The west-flowing major current of the North Pacific.

notochord In Chordata, a stiff rod of turgid vacuolated cells extending longitudinally dorsal to the digestive tract, between it and the dorsal nerve cord, and serving as an axial skeletal support. In adults of most vertebrates, it is replaced by the centrum of the vertebral column.

nucleic acid Macromolecules characteristic of all living things and containing the coded information by which their genetic characteristics are determined.

nullipore Lime-secreting algae. Coralline algae.

oceanic plankton Plankton characteristic of the open ocean beyond the continental shelves.

ocellus In Arthropoda, a simple eye. Also used for light-sensitive pigmented structures in other groups, as in medusae.

oecia In colonial Bryozoa, special reproductive zooids containing egg cells.

offlap In geology, the situation where terrestrial deposits are laid down over marine deposits, as when the sea level becomes lower relative to the land in a given region.

oikoplast In Larvacea such as *Oikopleura,* the cell or cells that secrete the gelatinous house.

oligohalabous plankton Plankton found in water with salinity of less than 5‰.

oligotrophic plankton Plankton of seas with a low level of nutrients, where light penetrates deeply into clear, relatively barren water.

oligoxenous Said of a parasite with a small range of host species.

onlap In geology, the condition in which marine sediments are deposited over terrestrial sediments, as when sea level becomes higher than before, relative to the land in the region.

oostegite The thin cuticular endites from the thoracic legs in female isopods and amphipods, forming the brood pouch.

ooze Soft deposits on the floor of the ocean.

otocyst A balancing organ comprised of a small vessel

filled with fluid and containing a solid particle or particles that impinge on sensory hairs or cilia.

ovisac In copepods, the egg sac(s) secreted by the ovaries in which the eggs are retained prior to hatching.

Oyashio Current A cold current from the Arctic, in the western North Pacific.

pallial line In bivalve molluscs, a line on the shell showing the attachment of mantle muscles to the shell.

pallial sinus In bivalve molluscs that have a well-developed siphon, an emargination in the pallial line, in the posterior end of the shell where the siphon musculature is attached.

palp In arthropods, the exopodite of mouth-part appendages, sensory in function.

palpon In Siphonophora, a simple mouthless zooid serving as an accessory digestive zooid.

pantomictic plankton Plankton not dominated by any particular species.

parameter One of the constants by which a population or environment is defined.

parapyle In Radiolaria of the order Phaeodaria, the two lateral pore fields in the central capsule.

parmal pores In some Radiolaria, the primary pores through the shell, surrounding each radial spine.

parthenogenesis Development of an egg without fertilization by a sperm.

pectin A complex carbohydrate produced in plants.

pedalium In certain jellyfish, an expanded gelatinous peduncle bearing one or more marginal tentacles or sense organs.

peduncle In some jellyfish, an extension of the central part of the subumbrella, bearing the manubrium at its end.

pelagic In the water—open-sea species above the benthos (includes around 3000 species of animals).

pennate A diatom with the valves bilaterally or asymmetrically arranged on a longitudinal axis. Members of the order Pennales.

pereion In Crustacea, the thorax.

pereiopods Thoracic appendages, in Crustacea.

perforate Provided with holes or pores.

periodic plankton Organisms that regularly appear in the plankton only at certain times, as at dusk.

periphyton Microorganisms attached to something.

perradius In jellyfish, the axes of the disc in which the primary radial canals lie.

person In colonial coelenterates, urochordates, etc., one of the individuals of the colony.

Peru Current Northern part of the Humboldt Current.

pervalar axis In diatoms, the axis through the center of the valves or through the length of the diatom cell.

phaeoplankton Plankton of the upper 30 meters.

phagotrophic Ingest particulate organic matter.

phenology Study of the quality and quantity of seasonal changes in plankton.

Phi scale A scale used in expressing the relative sizes of particles.

photogenic Capable of generating light, as photogenic tissue in luminescent organs.

photokinesis Aggregation in a particular region of the spectrum; or movement with speed that varies with variation in the light intensity.

photophore A luminescent organ found in certain fishes and crustaceans.

photoreceptor A light-sensitive sense organ.

photosynthesis The synthesis of organic matter from inorganic substrates, using light as a source of energy.

phototaxis Response to simulus by light; movement toward or away from a light source.

phototropism Positive phototaxis; attraction toward light.

phragmocone In extinct cephalopods of the group Belemnoidea, a chambered portion of the shell.

phyllosoma larva The flattened, transparent planktonic larva of spiny lobsters.

phytoplankton Photosynthetic planktonic organisms such as diatoms and dinoflagellates.

pilidium larva A helmet-shaped planktonic larva of some nemerteans.

plankter A planktonic organism.

planktobiont An organism living solely in the plankton. A holoplanktonic organism.

plankton The floating and drifting organisms in the water whose movements from place to place are due to the movements of the water rather than to their swimming.

plankton spectrum The qualitative and quantitative composition of the plankton population at a given time and place.

planktophile Living mainly as plankton, although able to live otherwise.

planktoxene Only occasionally in the plankton.

pleon In Crustacea, the abdomen.

pleopod In Crustacea, an abdominal appendage, especially those anterior to the uropods.

pleurobranch In decapod Crustacea, a gill inserting on the pleura of the thoracic wall above the insertion of a thoracic appendage.

pleustron Animals at the surface of the sea extending into the air.

pluteus larva The ciliated planktonic larva of sea urchins and other Echinoidea.

pneumatophore In Siphonophora, the float.

podobranch In Crustacea, a gill attached to the basal segment of one of the thoracic legs.

poikilothermal Having a body temperature that fluctuates with the environment; cold-blooded. Opposed to homoiothermal.

polymictic plankton With several species present in great numbers.

polyp In Coelenterata, the stage commonly developing from a planula larva. In Hydrozoa and Scyphozoa, the polyp is usually a sessile form multiplying asexually, whereas the sexual form is commonly a free-swimming medusa or an attached reduced medusoid. In Anthozoa, there is no medusa stage, and the gonads are borne on the mesenteries in the polyp.

polythalamous Many-chambered, as in the shells of some Foraminifera.

polytroch Having several ciliated bands, as in some trochophore larvae.

polyxenous Parasites with broad host ranges.

pore canal In echinoderm and hemichordate larvae, a canal from the hydrocele to the exterior.

pore field In Radiolaria, an area on the central capsule-bearing pores.

porosity An expression relating the size and number of pores in a given porous structure to the area. It is a function relating to total area represented by pores to that of the pore-bearing structure. (In ordinary usage in laboratories, the porosity of something, such as a filter, refers only to the size of the pores.)

postlarva In Crustacea, a young individual of essentially adult morphology but not yet sexually mature.

prevalent plankton A species that constitutes more than half the plankton in a given place.

primary consumers Organisms that eat plants. Herbivores.

primary production The creation of new organic matter from inorganic substrates.

producers Organisms that carry on primary production. Plants.

proloculum In Foraminifera, the initial, or first, chamber of the test.

proostracum In extinct cephalopods of the group Belemnoida, a dorsal platelike portion of the shell.

prosoma In Copepoda, the forepart of the body in front of the major articulation.

proteolytic Capable of degrading proteins. Said of enzymes that digest proteins.

protoplankton Planktonic protophyton and Protozoa and sometimes also the bacteria.

protoplast The protoplasm within a cell wall; the protoplasm of a cell.

prototroch In trochophore larvae, the ciliated band in a preoral position.

proximal Toward the origin of an appendage.

pseudochitinous Superficially like chitin but chemically different.

pseudocoelom A body space, as in the Aschelminthes, between the gut and the body wall, not lined by a peritoneum and derived from the blastocoel rather than from an internal split in the mesoderm.

pseudoplankton Organisms not normally planktonic, carried into the plankton by turbulence.

pusule A vacuole-like structure in dinoflagellates.

pycocline Density gradient.

pyrenoid A structure in the plastids of plants concerned with the synthesis of starch.

Q 10 Expression denoting the factor by which the rate of a reaction or process is changed over a temperature interval of 10° C.

$$Q\,10 = \left(\frac{R1}{R2}\right)^{\frac{10}{t1 - t2}}$$

where R1 and R2 are rates of the reaction or process and t1 and t2 are the temperatures.

radial canals In coelenterate medusae, tubes of the gastrovascular system passing from the centrally located stomach to the periphery of the disc.

radioles In many annelid tube worms, the feathery processes extending from the anterior end, used for food-gathering and also as sensory structures.

radula The tonguelike lingual ribbon of most molluscs except bivalves.

ramus A branch; in Crustacea, one of the branches of a biramous appendage.

raphe The longitudinal slit in the side of pennate diatoms.

raptorial Pertaining to structures such as claws or pincers used to seize prey or enemies. The term is also used to describe organisms that actively seize their prey, as raptorial birds.

redox potential discontinuity zone (RPD zone) In sediments the transitional zone between surface layers, in which some free oxygen is present, and deeper anaerobic layers. It is often gray in color, rather thin, and contains the greatest concentration and diversity of microorganisms.

reducers Organisms such as bacteria whose life activities result in the mineralization of organic matter.

resistance adaptation Adaptation that enables an organism to survive the occasional environmental extremes that may occur in its habitat. *See* capacity adaptation.

reticulopod Pseudopodia forming an anastomosing network, as in Foraminifera.

reticulum A network.

rhopalia In Scyphozoa, marginal sense organs that are reduced and modified tentacles containing an endodermal statolith.

rhopaloneme A type of nematocyst found only in Siphonophora.

rostrum In Crustacea, a forward prolongation of the carapace beyond the head.

RPD zone *See* redox potential discontinuity zone.

salinity The equivalent amount of sodium chloride, usually expressed as parts per thousand, which would give to the water the same osmotic properties as those which result from the total amount of dissolved salts in the water; loosely, the saltiness of the water.

salt pump A physiological mechanism that concentrates an ion against the osmotic gradient.

saprophytic A mode of nutrition found in nongreen plants and many bacteria that require organic food absorbed through the body surface. If the substrate furnishing this food cannot be absorbed directly, it is partially broken down or digested outside the organism by exoenzymes before absorption.

saprozoic A somewhat similar mode of nutrition in some animals such as tapeworms, differing in that the subsequent metabolism of the organic food is of a different type from that found in plants.

scapulet In jellyfish of the group Rhizostomeae, outgrowths of the oral arms near the bell, bearing supplementary mouth openings. Scapulets are not found in all rhizostomes, but only in certain families and genera.

scotoplankton Plankton occurring between the bottom and 500-meter depth. Deepwater plankton.

sculpture Elevated or depressed marks on a surface, as on a shell.

scutum One of the paired movable calcareous plates of a barnacle.

scyphistoma The sessile polypoid stage in the life of some Scyphozoa.

secondary production The sum of the organic matter produced as a result of predation on primary producers and their products. Net secondary production would be this figure less the loss due to catabolism.

sessile With respect to organisms, those which are fixed to the substrate. With respect to structures that are sometimes stalked or raised, the condition in which they are not so stalked or raised.

seston Particulate matter in the water.

setae Bristlelike stiff cuticular structures, as in the parapodia of polychaetes, on the head of chaetognaths, or on the appendages of various crustaceans, etc.

sibling species Species that are much alike but that occur in separated areas or separate ecological niches. Twin species.

single-species blooms Plankton blooms very much dominated by one species.

sinistral Left-turning, with shell aperture on the left. *See* dextral.

siphonoglyph In most Anthozoa except corals, a ciliated canal at one or both ends of the gullet serving as a means of passing a current of water in and out of the gastrovascular cavity.

somatocyst In many siphonophores, the portion of the gastrovascular canal lying in front of the proximal part of the stem.

sound A long passage of water connecting two larger bodies of water and generally wider than a strait.

spermaceti In sperm whales, the clear high-grade oil found in the melcon, or spermaceti organ, in the rostrum.

spermatangium Unicellular male sex organs in red algae.

spermatium A nonflagellated gamete as in red algae.

spermatophore A packet of sperm within a complicated case secreted by the male, as in copepods and cephalopods.

spicules Needles or variously shaped structures of calcium carbonate or siliceous material found in the flesh or tissues of certain organisms, as in sponges.

spongin The fibrous horny flexible material making up the skeleton of various Demospongiae. In the Keratosa, of which commercial sponges are representative, it comprises the entire skeletal framework of the sponge. Other groups also have spicules or may have only spicules.

statocyst In decapod Crustacea, a small chitinous walled sac in the basal segment of the antennule. In the base is a ridge bearing three sets of hairs, each innervated with a separate neuron. After each molt the decapod puts into it a number of sand grains that serve as statoliths, and enable the statocyst to serve as an organ of equilibrium. If iron filings are substituted, the animals will orient to a magnet.

statolith In jellyfish, a small calcareous concretion within a lithocyst.

stenobathic Having a limited vertical range, or range of depth tolerance.

stenohaline Having a narrow range of salinity tolerance.

stenothermal Having a narrow range of temperature tolerance.

stolon A tubular or rootlike growth over the substrate from which individuals or colonies of sessile organisms such as hydroids, endoproct Bryozoa, etc. grow.

strait A relatively narrow passage of water between the larger bodies of water, generally narrower than a sound.

stratosphere In oceanography, nearly uniform masses of cold water in the high latitudes and of cold bottom water in middle and low latitudes.

styliform Having the form of a dagger, or stiletto.

subchelate A chela in which the dactyl simply presses back upon or opposes a somewhat enlarged penultimate segment of the appendage, rather than opposing a corresponding process of that segment.

subumbrella In jellyfish, the lower or inner side of the bell, usually the concave side, bearing the mouth or manubrium.

swimmeret In higher Crustacea, the biramous appendages of the abdominal segments; pleopods.

symbiosis Literally a "living together," a term used by some biologists to include strictly parasitic, commensal, and mutualistic relationships between organisms but by others restricted to mutualistic relationships.

sympatric With overlapping geographical distribution.

syncytium Multinucleate; protoplasm containing several nuclei without corresponding division into cells.

synecology The study of the grouping and association of species within an environment and the relationships of such groups or communities to each other and to the environment. *See* autecology.

tambak Impoundment of marine water near shore used as fish culture ponds in Java and other tropical Pacific areas, especially for culture of milkfish *(Chanos)*.

tapetic Felt-forming; descriptive of growth of certain blue-green algae, diatoms, etc.

telotroch In trochophore larvae, a circlet of cilia in front of the anus.

telson In higher Crustacea, the caudal end of the body behind the last true somite. The telson contains the anal opening, and together with the uropods comprises the tail fan.

tentaculocyst A lithostyle.

tentilla Lateral branches of the tentacle of gastrozooids in Siphonophora.

tergum In Arthropoda, the skeletal plate covering the dorsal part of each free segment.

tetraxon In sponges, spicules with four rays, each pointed in a different direction.

theca A case, lorica, or shell surrounding an animal.

thermal dome An area in which upwelling of water in the centers of cyclonic gyres brings cooler water to the surface. More or less permanent thermal domes are caused by gyres, for example, one in the Pacific off Central America (Costa Rican thermal dome) and one in the Atlantic off Angola. Thermal domes and upwellings along open ocean divergences have marked effects on weather and climate, cloud cover, and rainfall. They also enrich the surface waters and result in increased productivity.

thixotropic The property of becoming more liquid when stirred or shaken—characteristic of certain gels and of some muds.

torsion In Gastropoda, the twisting of the visceral hump during larval life, bringing the mantle cavity and gills to a forward position.

transapical The width in a diatom shell.

triaxon Sponge spicules with three axes crossing at right angles, resulting in six rays, or points. Characteristic of Hexactinellida.

trophosphere In oceanography, the upper layer of oceanic waters in middle and low latitudes where there are strong currents and warm temperatures.

tubulus In some Radiolaria, a wide radial cylindrical extension of the skeleton.

tychoplankton Not regularly part of the plankton; brought up from the bottom by turbulence.

ultraplankton Planktonic organisms less than 5 μm in length or diameter.

uropod In higher Crustacea, the last pair of abdominal appendages, flattened and, together with the telson, making up the tail fan.

zoospore A flagellated plant spore capable of swimming.

zygote A fertilized egg cell.

BIBLIOGRAPHY

This list comprises only a representative sampling of the literature of marine biology—it is by no means a complete listing.

Abbott, I. A., and M. Kurogi. 1971. Contributions to the systematics of benthic marine algae of the North Pacific. Japan-U.S. cooperative Science program. Sapporo, Japan. Also Suppl. by G. Hollenberg and I. A. Abbott.

Abbott, I. A., and G. J. Hollenberg. 1976. Marine algae of California. Stanford University Press, Stanford, Calif. 827 p.

Abbott, R. T. 1954. American seashells. D. Van Nostrand Co., Inc., Princeton, N.J.

Adams, M. 1976. Ecology of eelgrass (*Zostera marina* [L]). Fish communities. I, II. J. Exp. Mar. Biol. Ecol. **22**(3):269-312.

Adolph, E. F., and P. E. Adolph. 1925. Regulation of body volume in fresh-water organisms. J. Exp. Zool. **43**:105-149.

Agassiz, A. 1906. Reports on scientific results of the expedition to the Eastern Tropical Pacific, in charge of Alexander Agassiz, by the U.S. Fish Commission Steamer "Albatross" from Oct., 1904, to March, 1905. Part V. General report of the expedition. Mem. Mus. Comp. Zool. Harv. **33**:xiii, 1-75.

Agassiz, E. C., and A. Agassiz. 1865. Seaside studies in natural history: marine animals of Massachusetts Bay. Ticknor & Fields, Boston.

Ahlstrom, E. H. 1952. Oceanographic instrumentation. Nat. Acad. Sci. Pub. **309**:36-50.

Aix, W. W. von. 1962. Introduction to physical oceanography. Addison-Wesley Publishing Co., Inc., Reading, Mass.

Alcock, A. 1895-1900. Materials for a carcenological fauna of India. Historical Naturalis Classica no. 64. Reprinted in 1968 by Stechert-Hafner, Inc., New York. 456 p.

Alexander, R. McN. 1966. Physical aspects of swim bladder function. Biol. Rev. **41**:141-176.

Aleyev, Y. G. 1977. Nekton. Dr. W. Junk BV, Publishers, The Hague. 435 p.

Allee, W. C. 1931. Animal aggregations. The University of Chicago Press, Chicago.

Allee, W. C., A. E. Emerson, O. Park, T. Park, and K. P. Schmidt. 1949. Principles of animal ecology. W. B. Saunders Co., Philadelphia. 837 p.

Allen, E. J. 1919. A contribution to the quantitative study of plankton. J. Mar. Biol. Ass. U.K. **12**:1-8.

Allen, E. J., and E. W. Nelson. 1910. On the artificial culture of marine plankton organisms. J. Mar. Biol. Ass. U.K. **8**:421-474.

Allen, W. E. 1920*a*. Quantitative and statistical study of plankton of the San Joaquin River and its tributaries in and near Stockton, Calif., in 1913. Univ. California Pub. Zool. **22**:1-292.

Allen, W. E. 1920*b*. Some work on marine phytoplankton in 1919. Trans. Amer. Microbiol. Soc. **40**:177-181.

Allen, W. E. 1921. Problems of floral dominance in the open sea. Ecology **2**:26-31.

Allen, W. E. 1922*a*. Quantitative studies on marine phytoplankton at La Jolla in 1919. Univ. California Pub. Zool. **22**:329-347.

Allen, W. E. 1922*b*. Quantitative studies on inshore marine diatoms and dinoflagellates of Southern California in 1920. Univ. California Pub. Zool. **22**: 369-378.

Allen, W. E. 1924. Surface catches of marine diatoms and dinoflagellates made in 1923 by U.S.S. "Pioneer" between San Diego and Seattle. Univ. California Pub. Zool. **26**:243-248.

Allen, W. E. 1930. Methods in quantitative research in marine microplankton. Bull. Scripps Inst. Oceanogr. (Tech. Ser.) **2**:319-329.

Allen, W. E. 1938. "Redwater" along the west coast of the United States in 1938. Science **88**(2272):55-56.

Allen, W. E. 1939. Cutthroat competition in the sea. Sci. Monthly **49**:111-119.

Allen, W. E. 1946. "Redwater" in La Jolla Bay in 1945. Trans. Amer. Microbiol. Soc. **65**:149-153.

Amos, W. H. 1952. Marine and estuarine fauna of Delaware. Univ. Delaware Mar. Lab. Annu. Rep. **1**:14-20.

Amos, W. H. 1954. Biological survey of the Delaware River estuary. Univ. Delaware Mar. Lab. Annu. Rep. **2**:21-31.

Amos, W. H. 1955. Guide to aquatic animals of Delaware and to the waters in which they live. Thesis. University of Delaware. 140 p.

Amos, W. H. 1956. Selective bibliography concerning the hydrography and biology of the Delaware River estuary. Univ. Delaware Mar. Lab. Ref. 56-57. 12 p.

Amos, W. H. 1959. The living wreck. Estuarine Bull. **4**:8-10.

Amos, W. H. 1960. Life in bays. National Audubon Society, Nature Program. Doubleday & Co., Inc., Garden City, N.Y. 64 p.

Amos, W. H. 1966. The life of the seashore. McGraw-Hill Book Co., New York. 231 p.
Andel, T. H. van. 1968. Deep-sea drilling for scientific purposes: a decade of dreams. Science **160:** 1419-1424.
Andersen, H. T. [ed.]. 1969. Biology of marine mammals. Academic Press, Inc., New York.
Anderson, A. 1973. Chaos at sea. The rape of the seabed. Saturday Review–World, Nov. 6, pp. 16-19.
Anderson, A. E., E. C. Jonas, and H. Odum. 1958. Alteration of clay minerals by digestive processes of marine organisms. Science **127:**190-191.
Anderson, D. L. 1962. The plastic layer of the earth's mantle. Sci. Amer. **207**(1):52-59.
Anderson, L. G. 1977. The economics of fisheries management. The Johns Hopkins University Press, Baltimore.
Anderson, P. K. 1971. Omega: murder of the ecosystem and suicide of man. William C. Brown Co., Dubuque, Iowa.
Andree, K. 1920. Geologie des Meeresbodens. Borntraeger. Leipzig.
Andrews, C. W. 1906. A descriptive catalogue of the Tertiary vertebrata of the Fayûm, Egypt. Sirenia Rep. Brit. Mus. (N.H.), pp. 197-218.
Annandale, N. 1922. The marine element in the fauna of the Ganges. Bijdr. Dierk. Amsterdam (Feest-Num. W. Weber):143-154.
Apstein, C. 1907. Das Plankton im Colombo See auf Ceylon. Zool. Jahrb. Abt. Syst. **25:**201-244.
Apstein, C. 1910. Das Plankton des Gregory-Sees auf Ceylon. Zool. Jahrb. Abt. Syst. **29:**661-680.
Arakawa, K. V. 1970. Scatological studies of the Bivalvia (Mollusca). *In* F. S. Russell [ed.]. Advances in marine biology. Vol. 7. Academic Press, Inc., New York.
Arber, A. 1920. Water plants: a study of aquatic angiosperms. Cambridge University Press, New York. Reprinted, 1963, by J. Cramer. Hafner Publishing Co., Inc., New York.
Armstrong, F. A. J. 1954. Phosphorus and silicon in seawater off Plymouth during the years 1950-1953. J. Mar. Biol. Ass. U.K. **33:**381-382.
Armstrong, F. A. J. 1957. Iron content of seawater. J. Mar. Biol. Ass. U.K. **36:**509-517.
Armstrong, F. A. J. 1968. Inorganic suspended matter in seawater. J. Mar. Res. (Thompson Anniversary vol.) **17:**23-34.
Arnold, A. F. 1901. The sea beach at ebb tide. D. Appleton-Century Co., New York.
Arrhenius, G. 1961. Geological record on the ocean floor. *In* M. Sears [ed.]. Oceanography. American Association for the Advancement of Science pub. no. 67. Washington, D.C.
Ashmole, N. P. 1971. Sea bird ecology and the marine environment. Vol. I. *In* D. S. Farner and J. R. King [ed.]. Avian biology. Academic Press, Inc., New York.
Atema, J., J. H. Todd, and J. E. Bardadi. 1969. Olfaction and behavioral sophistication in fish. *In* Cirfaffmann [ed.] Olfaction and taste. Proceedings of the Third International Symposium. Rockefeller University Press, New York.
Atkins, W. R. G. 1922. Influence upon algal cells of an alteration in the H ion concentration of seawater. J. Mar. Biol. Ass. U.K. **12:**789-791.
Atkins, W. R. G. 1926. The phosphate content of seawater in relation to the growth of algal plankton. J. Mar. Biol. Ass. U.K. **14:**447-467.
Atkins, W. R. G. 1945. Autotrophic flagellates as a major constituent of the oceanic phytoplankton. Nature (London) **156:**446-447.
Atkins, W. R. G. 1953. Seasonal variation in copper content of seawater. J. Mar. Biol. Ass. U.K. **31:**493-494.
Atkins, W. R. G., and H. H. Poole. 1958. Cube photometer measurements of angular distribution of submarine daylight and total submarine illumination. J. Cons. Perm. Int. Explor. Mer **23:**327-336.
Atwood, W. E., and A. A. Johnson. 1924. Marine structures: their deterioration and preservation. National Research Council, Washington, D.C. 534 p.
Augener, H. 1906. West indischen polychaeten. Cambridge Museum. 108 p.
Augener, H. 1927. Polychaeten von sudöst and Süd-Australian. Videns. Medd. Kjöbenhavn. **83:**71-279.
Aurich, H. J. 1949. Die Verbreitung des Nannoplanktons in Oberflächenfasser von der Nordfriesischen Küste. Ber. Deut. Komm. Meeresforsch. **11:** 403-405.
Austin, T. S. 1957. Summary of oceanographic and fishery data, Marquesas Islands area, Aug.-Sept., 1956 (Equapac). Special report: Fisheries no. 217. Fish and Wildlife Service, U.S. Department of the Interior, Washington, D.C. 186 p.
Austin, T. S., and V. E. Brock. 1959. Meridional variations in some oceanographic and marine biological factors in the Central Pacific. International Oceanographic Congress Preprints. American Association for the Advancement of Science, Washington, D.C.
Baas-Becking, L. G. M., I. R. Kaplan, and D. Moore. 1960. Limits of the natural environment in terms of pH and oxidation-reduction potentials. J. Geol. **68:**243-284.
Baer, J. G. 1951. Ecology of animal parasites. University of Illinois Press, Urbana.
Bainbridge, R. 1949. Movements of zooplankton in diatom gradients. Nature (London) **163:**910-912.
Bainbridge, R. 1953. Studies on the interrelationships of zooplankton and phytoplankton. J. Mar. Biol. Ass. U.K. **32:**375-445.
Bainbridge, R. 1957. Size, shape, and density of marine phytoplankton concentrations. Biol. Rev. **32:**91-115.
Bakus, G. J. 1964. Effects of fish-grazing on invertebrate evolution in shallow tropical waters. Occasional Papers, Allan Hancock Found. Pub. **27:**1-29.
Bakus, G. J. 1966. Some relationships of fishes to benthic organisms (algae, invertebrates) on coral reefs. Nature (London) **210:**280-284.

Bakus, G. J. 1969. Energetics and feeding in shallow marine waters. Int. Rev. Gen. Exp. Zool. **4:**275-369.

Bal, D. V., and L. B. Pradhan. 1952. Records of zooplankton in Bombay waters during 1944-1947. J. Univ. Bombay N. S. [B] **20:**75-80.

Ballantine, D. 1961. A biologically defined exposure scale for the comparative description of rocky shores. Field studies. Part I. 19 p.

Banning, G. H. 1933. Hancock Expedition to the Galápagos Islands, 1933 (general report). Bull. Zool. Soc. San Diego, no. 10.

Banse, K. 1955 Über das Verhalten von meroplanktischen Larven im geschichteten Wasser. Kieler Meeresforsch. **11:**188-200.

Banse, K. 1956. Über den Transport von meroplanktischen Larven aus dem Kattegat in die Kieler Bucht. Ber. Deut. Komm. Meeresforsch. **14:** 147-164.

Barbour, T. 1912. A contribution to the zoogeography of the East Indian Islands. Mem. Mus. Comp. Zool. Harv. **44:**1-203.

Barghoorn, E. S. 1971. The oldest fossils. Sci. Amer. **224**(5):30-53.

Barghoorn, E. S., and J. W. Schopf. 1966. Microorganisms three billion years old from the Precambrian of South Africa. Science **152**(3723):758-763.

Barghoorn, E. S., and S. A. Tyler. 1965. Microorganisms from the Gunflint chert. Science **147**(3658): 563-577.

Barlow, J. P. 1955. Physical and biological processes determining the distribution of zooplankton in a tidal estuary. Biol. Bull. Mar. Biol. Lab. Woods Hole **109:**211-225.

Barnard, J., R. J. Menzies, and M. C. Băcescu. 1962. Abyssal crustacea. Columbia University Press, New York. ix + 222 p.

Barnard, K. 1914-1932. Contributions to the crustacean fauna of South Africa. 11 parts. Capetown Museum. 879 p.

Barnes, C. A., and T. G. Thompson. 1938. Physical and chemical investigations in the Bering Sea and portions of the North Pacific Ocean. Univ. Washington Pub. Oceanogr. **3**(2):35-79.

Barnes, H. 1959*a*. Apparatus and methods of oceanography. Vol. I. Chemical apparatus and methods. John Wiley & Sons, Inc., New York. 341 p.

Barnes, H. 1959*b*. Oceanography and marine biology: a book of techniques. The Macmillan Co., New York.

Barnes, R. D. 1968. Invertebrate zoology. 2nd ed. W. B. Saunders Co., Philadelphia.

Barnet, L., and the editors of *Life*. 1960. The wonders of life on earth. Time, Inc., New York.

Bartsch, P. 1921. Ocean currents, the prime factor in the distribution of marine molluscs on the west coast of America. Special Publications, Berenice P. Bishop Mus. no. 7, pp. 505-526.

Bary, B. McK. 1959*a*. Biogeographic boundaries: the use of temperature-salinity-plankton diagrams. International Oceanographic Congress Preprints. American Association for the Advancement of Science, Washington, D.C.

Bary, B. McK. 1959*b*. Species of zooplankton as a means of identifying different surface waters and demonstrating their movements and mixing. Pacific Sci. **13:**14-54.

Bass Becking, L. G. M., and E. J. F. Wood. 1955. Biological processes in the estuarine environment. Parts I and II. Kon. Ned. Akad. Wetensch. Proc. **B58:**160-181.

Bates, M. 1960. The forest and the sea. Random House, Inc., New York.

Bayer, F. M., and B. O. Harding. 1969. The free-living lower invertebrates. The Macmillan Co., New York.

Bayliss-Smith, T. P., and R. G. A. Feachem [eds.] 1977. Subsistence and survival: rural ecology in the Pacific. Academic Press, Inc. xiv + 428 p.

Bé, A. W. H. 1968. Shell porosity of recent planktonic Foraminifera as a climatic index. Science **161:** 881-884.

Beanland, F. L. 1940. Sand and mud communities in the Dover Estuary. J. Mar. Biol. Ass. U.K. **24:** 589-611.

Beebe, W. 1928. Beneath tropic seas. G. P. Putnam's Sons, New York.

Beebe, W. 1935. Half a mile down. John Lane, London. 344 p.

Beklemishev, K. V. 1957. Spatial interrelationships of marine zooplankton and phytoplankton. Trudÿ Inst. Okeanol. **20:**253-278.

Belser, W. L. 1959. Bioassay of organic materials in seawater. International Oceanographic Congress Preprints. American Association for the Advancement of Science, Washington, D.C.

Benson, A. A., and R. F. Lee. 1975. The role of wax in oceanic food chains. Sci. Amer. **232**(3):76-89.

Bentley, J. A. 1960. Plant hormones in marine phytoplankton: zooplankton and seawater. J. Mar. Biol. Ass. U.K. **39:**433-444.

Berger, W. H. 1967. Foraminiferal ooze: solution at depths. Science **156:**383-385.

Berkeley, E., and C. Berkeley. 1948. Annelida Polychaeta Errantia. Canadian Pacific Fauna no. 9b 1: 1-100; 1952. Annelida Polychaeta Sedentaria. 9b(2): 1-139. Fisheries Res. Board of Canada. University of Toronto Press, Toronto.

Bernal, J. D. 1967. The origin of life. The World Publishing Co., Cleveland.

Bernard, F. 1939. Étude sur les variations de la fertilité des eaux méditerranéennes. J. Cons. Perm. Int. Explor. Mer **14:**228-241.

Bernard, F. 1948. Recherches sur le cycle de *Coccolithus fragilis* Lohm. flagellé dominant des mers chaudes. J. Cons. Perm. Int. Explor. Mer **15:** 177-188.

Bernard, F. 1955*a*. Densité du plancton vu au large de Toulon depuis la bathyscaphe F.N.R.S. III. Bull. Inst. Oceanogr. Monaco **1063:**1-16.

Bernard, F. 1955*b*. Étude préliminaire quantitative de la repartition saisonnière du zooplancton de la baie d'Alger. Bull. Inst. Oceanogr. Monaco **1065:**1-28.

Bernard, F. 1958. Données récentes sur la fertilité élémentaire en la Mediterranée. Rapp. P.-v. Réum. Cons. Perm. Int. Explor. Mer **144:**103-108.

Bernstein, R. 1973. Poisoning of the seas. Saturday Review–World, Nov. 20, pp. 14-16.

Berrill, N. J. 1966. The life of the ocean. McGraw-Hill Book Co., New York.

Berrill, N. J., and J. Berrill. 1951. The living tide. Dodd, Mead & Co., New York.

Berrill, N. J., and J. Berrill. 1959. 1001 questions answered about the seashore. Dodd, Mead & Co., New York.

Bert, P. 1833. Sur la cause de la mort des animaux d'eau douce que l'on plonge dans l'eau de mer, et réciproquement. C. R. Acad. Sci. (Paris) **97:**133-136.

Bigelow, H. B. 1914*a*. Explorations in the Gulf of Maine, July-Aug., 1912, by the U.S. Fisheries schooner "Grampus." (Oceanography and notes on plankton.) Mem. Mus. Comp. Zool. Harv. **58:** 393-419.

Bigelow, H. B. 1914*b*. Oceanography and plankton of Massachusetts Bay and adjacent water, Nov., 1912, to May, 1913. Mem. Mus. Comp. Zool. Harv. **58:**393-419.

Bigelow, H. B. 1915. Exploration of the coastal water between Nova Scotia and Chesapeake Bay, July-Aug., 1913. Mem. Mus. Comp. Zool. Harv. **59:** 149-360.

Bigelow, H. B. 1917. Exploration of the coastal water between Cape Cod and Halifax in 1914-1915 by the U.S. Fisheries schooner "Grampus." (Oceanography and notes on plankton.) Mem. Mus. Comp. Zool. Harv. **61:**365-442.

Bigelow, H. B. 1922. Exploration of the coastal water off the Northeast United States in 1916 by U.S. Fisheries schooner "Grampus." Mem. Mus. Comp. Zool. Harv. **65:**85-188.

Bigelow, H. B. 1926*a*. Plankton of the offshore waters of the Gulf of Maine. Bull. Bur. Fish. (Washington) **40**(2):1-509.

Bigelow, H. B. 1926*b*. Physical oceanography of the Gulf of Maine. Bull. Bur. Fish. (Washington) **40**(2):511-1027.

Bigelow, H. B. 1938. Plankton of the Bermuda oceanographic expeditions. VIII. Medusae taken during the years 1929-1930. Zoologica **23:**99-189.

Bigelow, H. B., L. C. Lillick, and M. Sears. 1940. Phytoplankton and planktonic protozoa of offshore waters of the Gulf of Maine. I. Numerical distribution. Trans. Amer. Phil. Soc. **31:**149-191.

Bigelow, H. B., and M. Sears. 1939. Studies of the waters of the continental shelf, Cape Cod to Chesapeake Bay. III. Study of zooplankton. Mem. Mus. Comp. Zool. Harv. **54:**183-378.

Bird, E. C. F. 1969. Coasts: an introduction to coastal morphology. M. I. T. Press. Vol. I, 398 p.; vol. II, 320 p.

Blaxter, J. H. S., and R. I. Currie. 1967. Effect of artificial lights on acoustic scattering layers in the ocean. *In* N. B. Marshall [ed.]. Aspects of marine zoology. Academic Press, Inc., New York.

Blegvad, H. 1914. Food and conditions of nourishment among communities of invertebrate animals on or in the sea bottom in Danish waters. Rep. Dan. Biol. Sta. **22:**41-78.

Blegvad, H. 1916. On the food of fish in Danish waters. Rep. Dan. Biol. Sta. **24:**17-72.

Blegvad, H. 1925. Quantity of fish food in the sea bottom. Rep. Dan. Biol. Sta. **31:**27-56.

Blegvad, H. 1929. Mortality among animals of the littoral region in icy winters. Rep. Dan. Biol. Sta. **35:**49-62.

Boden, B. P., and E. N. Kampa. 1953. Winter cascading from an oceanic island and its biological implications. Nature (London) **171:**426-427.

Boden, B. P., and E. N. Kampa. 1967. Influence of natural light on vertical migrations of an animal community in the sea. *In* N. B. Marshall [ed.]. Aspects of marine zoology. Academic Press, Inc., New York.

Bogorov, B. G. 1934. Investigation of nutrition of plankton-consuming fishes. Bull. Inst. Mar. Fish. Oceanogr. U.S.S.R. **1:**1-18.

Bogorov, B. G. 1939. The particularities of seasonal phenomena in the plankton of Arctic seas and their significance for ice forecastings. Zool. J. U.S.S.R. **18:**735-747.

Bogorov, B. G. 1941. Biological seasons in the plankton of different seas. C. R. Acad. Sci. (U.S.S.R.) **31:**404-407.

Bogorov, B. G. 1946. Peculiarities of diurnal vertical migration of zooplankton in polar seas. J. Mar. Res. **6:**25-32.

Bogorov, B. G. 1956. Unification of plankton research. Coll. Int. Biol. Mar. Sta. Biol. Roscoff. Ann. Biol. **33:**299-315.

Bogorov, B. G. 1957. Standardization of marine plankton investigations. Trudȳ Inst. Okeanol. **24** (R).

Bogorov, B. G. 1958*a*. Biogeographical regions of the plankton of the northwestern Pacific Ocean and their influences on the deep sea. Deep-Sea. Res. **5**(2): 149-161.

Bogorov, B. G. 1958*b*. Estimates of primary production in biogeographical regionalization of the ocean. Rapp. P.-v. Reun. Cons. Perm. Int. Explor. Mer **144:**117-121.

Bogorov, B. G. 1958*c*. Perspectives in the study of seasonal changes of plankton and of the number of generations at different latitudes. *In* A. A. Buzzati-Traverso [ed.]. Perspectives in marine biology. University of California Press, Richmond.

Bogorov, B. G., and C. W. Beklemishev. 1955. On the phytoplankton production in the western North Pacific [in Russian]. C. R. Acad. Sci. (U.S.S.R.) **104:**141-143.

Bogorov, B. G., and E. Kreps. 1958. On the possibility of burying radioactive wastes in the abysses of the ocean. Priroda **9:**45.

Bogorov, B. G., and M. E. Vinogradov. 1955. Some

essential features of zooplankton distribution in the northwestern Pacific [in Russian]. Trudȳ Inst. Okeanogr. **23:**113-123.

Boltovskoy, E. 1963. The relationship between Foraminifera and their environment. Mus. Argent. Cienc. Nat., Rev. Hidrobiol. **1**(2):22-107.

Boltovskoy, E. 1965. Los Foraminiferos recientes. Buenos Aires, Librart S.R.L. 510 p.

Boolootian, R. A. [ed.]. 1966. Physiology of Echinodermata. John Wiley & Sons, New York. 822 p.

Boney, A. D. 1965. Biology of seaweeds of economic importance. Advances in marine biology. Vol. 3. Academic Press, Inc.

Boney, A. D. 1966. Biology of marine algae. Hutchinson's Publishing Group, Ltd., London. Hutchinson. 216 p.

Borgese, E. M. 1968. The ocean regime: suggested statute for the peaceful uses of the high seas and the sea-bed beyond the limits of national jurisdiction. Center for the Study of Democratic Institutions, Santa Barbara, Calif. 40 p.

Borgese, E. M. [ed.] 1978. Ocean Year Book. Vol. 1. University of Chicago Press, Chicago.

Borgstrom, G. 1961. Atlantic Ocean fisheries: catching, processing, marketing. Fishing News Books, London. 350 p.

Borgstrom, G. 1970. The harvest of the seas: how fruitful and for whom? *In* H. W. Helfrich [ed.]. The environmental crisis. Yale University Press, New Haven, Conn. pp. 65-85.

Borgstrom, G. 1972. The hungry planet. The Macmillan Co., New York. 552 p.

Bossanyi, J. 1958. Preliminary survey of small natant fauna in the vicinity of the sea floor off Blyth, Northumberland. J. Anim. Ecol. **26:**353-368.

Bottemanne, C. J. 1959. Principles of fisheries development. North Holland Publishing Co., Amsterdam. xvi + 677 p.

Boughey, A. S. 1968. Ecology of populations. The Macmillan Co., New York.

Bowen, E. J. [ed.]. 1965. Recent progress in photobiology. Academic Press, Inc., New York.

Bowman, R. I. [ed.]. 1964. Proceedings of the Galápagos International Scientific Project of 1964. University of California Press, Richmond.

Boyko, H. [ed.]. 1965. Salinity and aridity. W. Junk, Publishers, The Hague, The Netherlands. 300 p.

Braarud, T. 1935. The Ost Expedition to the Denmark Strait, 1929. II. Phytoplankton and its conditions of growth. Hvalråd Skr. **10:**1-173.

Braarud, T. 1961. Cultivation of marine organisms as a means of understanding environmental influences on populations. *In* M. Sears [ed.]. Oceanography. Pub. no. 67. American Association for the Advancement of Science, Washington, D.C.

Braarud, T. 1962. Species distribution in marine phytoplankton. J. Oceanogr. Soc. Jap. **20:**628-649.

Braarud, T., K. R. Garden, and J. Grontoed. 1953. The phytoplankton of the North Sea and adjacent waters in May, 1948. Rapp. P.-v. Réun. Cons. Perm. Int. Explor. Mer **133:**1-89.

Braarud, T., and A. Klem. 1931. Hydrographical and chemical investigations in the coastal waters off More and in the Romsdalsfjord. Hvalråd Skr. **1:**1-88.

Brady, G. S. 1880. Report on the Ostracoda, dredged during the years 1873-1876. Challenger reports (Zoology). Vol. 1.

Brady, G. S. 1883. Report on the Copepoda obtained by H.M.S. *Challenger* during the years 1873-1876. Challenger reports (Zoology). Vol. 8.

Brahmachary, R. L. 1967. Physiological clocks. Int. Rev. Cytol. **21:**65-89.

Brand, T. von. 1934. Methods for determination of nitrogen and carbon in small amounts of plankton. Biol. Bull. Mar. Biol. Lab. Woods Hole **69:**221-232.

Brand, T. von, and N. W. Rakestraw. 1941. The determination of dissolved organic nitrogen in seawater. J. Mar. Res. **4:**76-89.

Brand, T. von, N. W. Rakestraw, and C. E. Renn. 1937-1947. Decomposition and regeneration of nitrogenous organic matter in seawater. I-VI. Biol. Bull. Mar. Biol. Lab. Woods Hole **72:**165-175; **77:**285-296; **79:**231-236; **81:**63-69; **83:**273-282; **92:** 110-114.

Brandhorst, W. 1958. Nitrate accumulation in the northeast tropical Pacific. Nature (London) **182:**679.

Brandhorst, W. 1959. Nitrification and denitrification in the eastern tropical North Pacific. J. Cons. Perm. Int. Explor. Mer **25:**3-20.

Brandt, K. 1898. Beitrage zur Kenntnis der chemischen Zusammensetzung des Planktons. Wiss. Meeresuntersuch. Abt. Kiel **3:**43-90.

Brandt, K., and C. Apstein [ed.]. 1901-1942. Nordisches Plankton. (Zoology, 7 vols.; botany, 1 vol.) Lipsius & Tischer, Publishers, Kiel and Leipzig. Reprint. A. Asher & Co., Amsterdam.

Bray, A. W. 1923. Investigation into the fouling of ships' bottoms by marine growths. Report of Bureau of Construction and Repair, U.S. Navy Depot, Washington, D.C. 40 p.

Breeder, C. M. 1948. Field book of marine fishes of the Atlantic coast. G. P. Putnam's Sons, New York.

Briggs, J. C. 1974. Marine zoogeography. McGraw-Hill Book Co., New York. x + 475 p.

Briggs, P. 1968. Men in the sea. New York. Simon & Schuster, Inc., New York. 128 p.

Brongersma-Sanders, M. 1957. Mass mortality in the sea. *In* J. W. Hedgpeth [ed.]. Treatise on marine ecology and paleoecology. Vol. I. Ecology. Geological Society of America, New York.

Brotskaja, V. A., and L. A. Zenkevich. 1939. Quantitative evaluation of the bottom fauna of the Barents Sea. Trudȳ Vses. Nauchno-issled. Inst. Morsk. Rȳb. Khoz. Okeanogr. **4:**99-126.

Brouardel, J., and L. Fage. 1953. Variation, en mer, de la teneur en oxygène dissous au proche voisinage des sediments. C. R. Acad. Sci. (Paris) **237:** 1605-1607.

Brown, E. T. 1896. On British hydroids and medusae. Proc. Zool. Soc. Lond. 459-500.

Brown, E. T. 1903. Report on medusae from Norway and Spitzbergen. Bergens Mus. Arb. **4:**1-36.

Brown, E. T. 1906. Biscayan plankton. IX. The medusae. Trans. Linn. Soc. Lond. Ser. 2. Zool. **X**(6): 163-187.

Brown, E. T. 1908. Medusae of the Scottish National Antarctic Expedition. Trans. Roy. Soc. Edinburgh **XLIV**(1909):233-251.

Brown, E. T. 1926. Report on the medusae. Zoological results of the Cambridge Expedition to the Suez Canal, 1924. Trans. Zool. Soc. Lond, **22:**105-115.

Brown, F. A. 1958. Studies of the timing mechanisms of daily, tidal, and lunar periodicities of organisms. *In* A. A. Buzzati-Traverso [ed.]. Perspectives in marine biology. University of California Press, Richmond. (Brown and his co-workers have also published several papers on endogenous rhythms of fiddler crabs and other organisms, reference to which can be found in this paper or that of Pittendrigh.)

Brown, F. A., and J. W. Hastings. 1970. *In* J. D. Palmer [ed.]. The biological clock: two views. Academic Press, Inc., New York. 94 p.

Brown, M. E. [ed.]. 1956-1957. Physiology of fishes. Academic Press, Inc., New York. 2 vol.

Bruce, H. E., and D. W. Hood. 1969. Diurnal inorganic phosphate variations in Texas bays. Pub. Inst. Mar. Sci. **6:**133-165.

Bruce, J. R. 1928. Physical factors on the sandy beach: tidal, climatic, and edaphic. J. Mar. Biol. Ass. U.K. **15:**535-552.

Brunelli, G. 1941. Note sul plancton della Laguna Veneta. Archo. Oceanogr. Limnol. **1:**55-56.

Bruns, E. 1958. Ozeanologie. Band I. V. E. B. Verlag der Wissenschaft, Berlin. 420 p.

Bruun, A. F. 1957. Deep-sea and abyssal depths. *In* J. W. Hedgpeth [ed.]. Treatise on marine ecology and paleoecology. Vol. I. Ecology. Geological Society of America, New York.

Bruun, A. F., S. Greve, H. Mielche, and R. Sparch [ed.]. 1956. The "Galathea" deep-sea expeditions, 1950-1952. The Macmillan Co., New York.

Bsharah, L. 1957. Plankton of the Florida Current. V. Bull. Mar. Sci. Gulf Carib. **7:**201-251.

Buchanan, J. B. 1968. The bottom fauna communities across the continental shelf off Accra, Ghana (Gold Coast). Proc. Zool. Soc. (London) **130:**1-56.

Buchner, P. 1965. Endosymbiosis of animals with plant microorganisms. Interscience Publishers, Inc., John Wiley & Sons, Inc., New York. 909 p.

Buchsbaum, R. 1948. Animals without backbones. The University of Chicago Press, Chicago. 405 p.

Budker, P. 1959. Whales and whaling. The Macmillan Co., New York.

Buljan, M. 1959. Utjecaj šum na produkcija priobalnim vodama [Influence of the forest upon the production in coastal waters]. Morskov. Rȳb. Barst. **11**(9): 193-194.

Bullard, Sir Edward. 1968. Reversals of the earth's magnetic field (Bakerian Lecture of the Royal Society). Roy. Soc., London, Phil. Trans., series A. **263**(1143):481-524.

Burke, J. M., J. Prager, and J. J. A. McLaughlin. 1962. Nutritional and ecological factors which determine dominance in phytoplankton blooms. J. Protozool. (supp.) **8:**7-8. (Abstr.)

Burkenroad, M. D. 1931. Notes on the sound-producing marine fishes of Louisiana. Copeia **1:**20-28.

Burkholder, P. R. 1959. Vitamin-producing bacteria in the sea. International Oceanographic Congress Preprints. American Association for the Advancement of Science, Washington, D.C.

Burkholder, P. R., and L. M. Burkholder. 1956. Vitamin B in suspended solids and marsh muds collected along the coast of Georgia. Limnol. Oceanogr. **1:**202-208.

Burt, W. V. 1956. A light-scattering diagram. J. Mar. Res. **15**(1):76-80.

Burt, W. V., and L. D. Marriage. 1957. Computation of pollution in the Yaquina River estuary. Sewage Ind. Wastes **29**(12):1385-1389.

Burt, W. V., and W. B. McAlister. 1959. Recent studies in the hydrography of Oregon estuaries. Res. Briefs, Fish Comm. Ore. **7**(1):14-27.

Bustard, H. R., and K. P. Tognetti. 1969. Green sea turtles: a discrete simulation of density-dependent population regulation. Science **163:**939-941.

Buzas, M. A., and T. G. Gibson. 1969. Species diversity: benthic Foraminifera in the western North Atlantic. Science **163:**72-74.

Buzzati-Traverso, A. A. [ed.]. 1958. Perspectives in marine biology. University of California Press, Richmond.

Cairns, J. 1956. Effects of heat on fish. Ind. Wastes **1**(5):180-183.

Cairns, J. 1956. Effects of increased temperatures on aquatic organisms. Ind. Wastes **1**(4):150-152.

Cairns, J. 1970. New concepts for managing aquatic life systems. J. Water Pollut. Contr. Fed. **1:**77-82.

Cairns, J. 1971. Thermal pollution—a cause for concern. J. Water Pollut. Contr. Fed. **2:**55-56.

Callaway, R. J. 1957. Oceanographic and meteorological observations in the northeast and central North Pacific, July-Dec., 1956. Special science report no. 230. Fish and Wildlife Service, U.S. Department of the Interior, Washington, D.C.

Calvin, M. 1961. Chemical evolution. Condon Lecture, Oregon State System of Higher Education. University of Oregon Press, Eugene.

Carey, C. L. 1938. Occurrence and distribution of nitrifying bacteria in the sea. J. Mar. Res. **1:**291-304.

Carr, A. 1954. The passing of the fleet. Amer. Inst. Biol. Sci. Bull. **4**(5):17-19.

Carr, A. 1961. The Pacific turtle problem. Nat. Hist. **70**(8):64-71.

Carr, A. 1965. The navigation of the green turtle. Sci. Amer. **212**(5):79-86.

Carrington, R. 1957. Mermaids and mastodons. Holt, Rinehart & Winston, Inc., New York.

Carrington, R. 1960. Biography of the sea. Basic Books, Inc., New York.

Carson, R. 1951. The sea around us. Houghton Mifflin

Co., Boston. Rev. ed. 1961. Oxford University Press, New York.

Carson, R. 1955. The edge of the sea. Houghton Mifflin Co., Boston.

Carson, R. 1962. Silent spring. Houghton Mifflin Co., Boston.

Carter, L. J. 1968. Continental shelf: scramble for federal oil-lease revenues. Science **160:**1431-1432.

Carter, L. J. 1970. Galveston Bay: test case of an estuary in crisis. Science **167:**1102-1108.

Caspers, H. 1957. The Black Sea and the Sea of Azov. *In* J. W. Hedgpeth [ed.]. Treatise on marine ecology and paleoecology. Vol. I. Ecology. Geological Society of America, New York.

Castro, J. de. 1952. The geography of hunger. Little, Brown & Co., Boston. 336 p.

Cattell, M. 1936. Biological effects of pressure. Biol. Rev. **11:**411-476.

Chapin, H., and W. Smith. 1952. The ocean river. Charles Scribner's Sons, New York.

Chapman, V. J. 1963. Mangroves of the world. Interscience Publishers, Inc.; John Wiley & Sons, Inc., New York.

Cheng, C. 1941. Ecological relations between the herring and the plankton off the northeast coast of England. Hull Bull. Mar. Ecol. **1:**239-254.

Chesher, R. H. 1969. Destruction of Pacific corals by the sea star *Acanthaster planci.* Science **165:**280-288.

Chichester, C. O. [ed.]. 1965. Research in pesticides. Academic Press, Inc., New York.

Chidester, F. E. 1924. Critical examination of the evidence for physical and chemical influences on fish migration. Brit. J. Exp. Biol. **2:**79-118.

Childress, J. J. 1968. Oxygen minimum layer: vertical distribution and respiration of the mysid *Gnathophausia ingens.* Science **160:**1242-1243.

Chow, T. J., and T. G. Thompson. 1952. The determination and distribution of copper in seawater. I. J. Mar. Res. **11:**124-138.

Chow, T. J., and T. G. Thompson. 1954. Seasonal variation in the concentration of copper in the surface waters of San Juan Channel, Washington. J. Mar. Res. **13:**233-244.

Chu, G. W. T. C., and C. E. Cutress. 1954. Human dermatitis caused by marine organisms in Hawaii. Proc. Hawaii Acad. Sci. **29:**9.

Clarke, A. C. 1960. The challenge of the sea. Holt, Rinehart & Winston, Inc., New York.

Clarke, G. L. 1936. Light penetration in the western North Atlantic and its application to biological problems. Rapp. P.-v. Réun. Cons. Perm. Int. Explor. Mer **101:**3-12.

Clarke, G. L. 1939. The relation between diatoms and copepods as a factor in productivity of the sea. Quart. Rev. Biol. **14:**60-64.

Clarke, G. L. 1940. Comparative richness of zooplankton in coastal and offshore areas of the Atlantic. Biol. Bull. **78:**226-255.

Clarke, G. L. 1941. Observations on transparency in the southwestern section of the North Atlantic Ocean. J. Mar. Res. **4:**221-230.

Clarke, M. R. 1966. A review of the systematics and ecology of oceanic squids. Advances Mar. Biol. **4:** 91-300.

Clements, F. E., and V. E. Shelford. 1927. Concepts and objectives in bio-ecology. Yearbook of the Carnegie Institute, 1926.

Cleve, P. T. 1901. Plankton from the Indian Ocean and the Malay Archipelago. K. svenska Vetensk-Akad. Handl. **35**(5):4-58.

Cloud, P. E. 1952. Facies relationships of organic reef. Bull. Amer. Ass. Petrol. Geol. **36:**2125-2149.

Clowes, A. J. 1938. Phosphate and silicate in the Southern Ocean. 'Discovery' Rep. **19:**1-20.

Coe, W. R. 1932. Season of attachment and rate of growth of sedentary marine organisms at the pier of the Scripps Institution of Oceanography, La Jolla, Calif. Bull. Scripps Inst. Oceanogr. (Tech. Ser.) **3:** 37-86.

Coe, W. R., and W. E. Allen. 1937. Growth of sedentary marine organisms in experimental blocks and plates for four successive years at the S.I.O. pier. Bull. Scripps Inst. Oceanogr. (Tech. Ser.) **4:** 101-136.

Coker, R. E. 1947. This great and wide sea. University of North Carolina Press, Chapel Hill.

Cold Spring Harbor Symposia on Quantitative Biology. 1960. Vol. 25. Biological clocks. Cold Spring Harbor, New York, Biological Laboratory.

Cole, H. A. 1956. Benthos and the shellfish of commerce. *In* M. Graham [ed.]. Sea fisheries, their investigation in the United Kingdom. Edward Arnold, Publishers, Ltd., London.

Coleman, J. S. 1933. Nature of intertidal zonation of plants and animals. J. Mar. Biol. Ass. U.K. **18:** 435-476.

Coleman, J. S., and F. Segrove. 1955. Tidal plankton over Stoupe Beck Sands, Robin Hood's Bay. J. Anim. Ecol. **24:**445-462.

Collier, A. 1953. The significance of organic compounds in seawater. Trans. N. Amer. Wildlife Conf. **18:**463-472.

Collier, A. 1958. Some biochemical aspects of red tides and related oceanographic problems. Limnol. Oceanogr. **3:**33-39.

Collier, A., S. Ray, and W. B. Wilson. 1956. Some effects of specific organic compounds on marine organisms. Science **124:**220.

Collins, A. C. 1958. Foraminifera. Great Barrier Reef Expedition. Sci. Rep. **6:**339-437.

Colman, J. 1940. On the faunas inhabiting intertidal seaweeds. J. Mar. Biol. Ass. U.K. **24:**129-183.

Colwell, R. R., and R. Y. Morita. 1974. Effect of the ocean environment on microbial activities. University Park Press, Baltimore. 604 p.

Colwell, R. R., and others. 1975. Marine and estuarine microbiology laboratory manual. University Park Press, Baltimore. 120 p.

Committee to Study the Organization of Peace. 1973.

The United Nations and the oceans. The United Nations, New York.

Commoner, B. 1966. Science and survival. The Viking Press, Inc., New York. 150 p.

Compton, I. 1957. The living sea. Doubleday & Co., Inc., Garden City, N.Y.

Connell, J. H. 1961. Influence of interspecific competition and other factors on the distribution of the barnacle *Chthamalus stellatus. In* P. S. Dawson and C. E. King [eds.]. 1971. Readings in population biology. Prentice-Hall, Inc., Englewood Cliffs, N.J.

Conover, S. A. M. 1956. Oceanography of Long Island Sound, 1952-1954. IV. Phytoplankton. Bull. Bingham Oceanogr. Coll. **15:**62-112.

Conservation Foundation. 1976. Coastal ecological systems of the United States. Conservation Foundation, Washington, D.C. 1977 p.

Conway, E. J. 1942. Mean geochemical data in relation to oceanic evolution. Proc. Roy. Irish Acad. [B] **48:**119-159.

Conway, E. J. 1943. The chemical evolution of the ocean. Proc. Roy. Irish Acad. [B] **48:**162-212.

Cooper, G. A. 1945. Phylum Brachiopoda. *In* H. W. Shimer and R. R. Shrock [eds.]. Index fossils of North America. John Wiley & Sons, Inc., New York.

Cooper, L. H. N. 1933. Chemical constituents of biological importance in the English Channel, Nov., 1930, to Jan., 1932. I. Phosphate, silicate, nitrite, and ammonia. J. Mar. Biol. Ass. U.K. **18:**677-728.

Cooper, L. H. N. 1934. Variation of excess base with depth in the English Channel, with reference to the seasonal consumption of calcium by plankton. J. Mar. Biol. Ass. U.K. **19:**747-754.

Cooper, L. H. N. 1935*a*. Iron in the sea and in marine plankton. Proc. Roy. Soc. [Biol.] **118:**419-438.

Cooper, L. H. N. 1935*b*. The rate of liberation of phosphate in seawater by the breakdown of plankton organisms. J. Mar. Biol. Ass. U.K. **20:**197-200.

Cooper, L. H. N. 1937. The nitrogen cycle in the sea. J. Mar. Biol. Ass. U.K. **22:**183-204.

Cooper, L. H. N. 1938. Phosphate in the English Channel, 1933-1938, with a comparison with earlier years, 1916 and 1923-1932. J. Mar. Biol. Ass. U.K. **23:**181-195.

Cooper, L. H. N. 1948*a*. The distribution of iron in the waters of the western English Channel. J. Mar. Biol. Ass. U.K. **12:**279-313.

Cooper, L. H. N. 1948*b*. Phosphate and fisheries. J. Mar. Biol. Ass. U.K. **27:**326-336.

Cooper, L. H. N. 1951. Chemical properties of the seawater in the neighborhood of the La Badie Bank. J. Mar. Biol. Ass. U.K. **30:**21-26.

Cooper, L. H. N. 1952*a*. Processes of enrichment of surface water with nutrients due to strong winds blowing onto a continental slope. J. Mar. Biol. Ass. U.K. **30:**453-464.

Cooper, L. H. N. 1952*b*. Factors affecting the distribution of silicate in the North Atlantic Ocean and the formation of North Atlantic deep water. J. Mar. Biol. Ass. U.K. **30:**511-526.

Cooper, L. H. N. 1952*c*. The physical and chemical oceanography of the waters bathing the continental slope of the Celtic Sea. J. Mar. Biol. Ass. U.K. **30:** 465-510.

Cooper, L. H. N. 1956. On assessing the age of deep oceanic water by carbon-14. J. Mar. Biol. Ass. U.K. **35:**341-354.

Cooper, L. H. N. 1961. Vertical and horizontal movements in the ocean. *In* M. Sears [ed.]. Oceanography. Pub. no. 67. American Association for the Advancement of Science, Washington, D.C.

Cooper, L. H. N., and A. Milne. 1938-1939. Ecology of the Tamar estuary. Parts I-V. J. Mar. Biol. Ass. U.K. Vols. 22 and 23.

Cooper, L. H. N., and D. Vaux. 1949. Cascading over the continental slope of water from the Celtic Sea. J. Mar. Biol. Ass. U.K. **28:**719-750.

Coppleson, V. M. 1959. Shark attack. Angus & Robertson, Ltd., London.

Corlett, J. 1953. Net phytoplankton at ocean weather stations "I" and "J." J. Cons. Perm. Int. Explor. Mer **19:**178-190.

Corner, E. D. S. 1961. On the nutrition and metabolism of zooplankton. I. J. Mar. Biol. Ass. U.K. **41:**5-16.

Corner, E. D. S., and A. G. Davies. 1971. Plankton as a factor in the nitrogen and phosphorus cycles in the sea. *In* F. S. Russell and M. Yonge [eds.]. Advances in marine biology. Vol. 9. Academic Press, Inc., New York.

Cornwell, J. 1971. Is the Mediterranean dying? Atlas **20**(3):16-19.

Costello, D. P., M. E. Davidson, A. Eggers, M. H. Fox, and C. Henley. 1957. Methods for obtaining and handling marine eggs and embryos. Marine Biology Laboratory, Woods Hole, Mass. 247 p.

Cousteau, J. Y., and F. Dumas. 1953. The silent world. Harper & Brothers, New York.

Cowell, E. B. 1968. Some biological effects of oil pollution. *In* R. M. Chute [ed.]. 1971. Environmental insight. Harper & Row Publishers, Inc., New York.

Cowen, R. C. 1960. Frontiers of the sea. Doubleday & Co., Inc., Garden City, N.Y.

Cowey, C. B. 1956. A preliminary investigation of the variation of vitamin B in oceanic and coastal water. J. Mar. Biol. Ass. U.K. **35:**609-620.

Cowlers, A. J. 1938. Phosphate and silicate in the Southern Ocean. 'Discovery' Rep. **19:**1-120.

Cowles, R. P. 1930. A biological study of the offshore waters of Chesapeake Bay. Bull. Bur. Fish. (Washington) **46:**277-381.

Creitz, G. I., and F. A. Richards. 1955. The estimation and characterization of plankton populations by pigment analysis. III. A note on the use of "Millipore" membrane filters in the estimation of plankton pigments. J. Mar. Res. **14:**211-216.

Cromwell, T. 1953. Circulation in a meridional plane in the central equatorial Pacific. J. Mar. Res. **12:** 196-213.

Cronin, L. E. [ed.]. Begun 1975. Estuarine research (biennial). Academic Press, Inc., New York. Vol. 1.

Chemistry, biology, and the estuarine system, 750 p.; vol. 2. Geology and engineering, 602 p.

Cross, J. C., and H. B. Parks. 1937. Marine fauna and seaside flora of the Nueces River basin and the adjacent islands. Bull. Texas Coll. Arts Ind. **8:**1-14.

Cupp, E. E. 1943. Marine plankton diatoms of the West coast of North America. Bull. Scripps Inst. of Oceanogr. **5**(1):1-238.

Curl, H. C. 1960. Primary production measurements in the north coastal waters of South America. Deep-Sea Res. **7:**183-189.

Curl, H. C. 1961. The measurement of dehydrogenase activity in marine organisms. J. Mar. Res. **19**(3): 123-138.

Curl, H. C., and G. C. McLeod. 1961. Physiological ecology of a marine diatom, *Skeletonema costatum* (Grev.) Cleve. J. Mar. Res. **19**(2):70-88.

Curl, H. C., and J. Sandberg. 1962*a*. Analysis of carbon in marine plankton organisms. J. Mar. Res. **20:**181-188.

Curl, H. C., and J. Sandberg. 1962*b*. Standing crops of carbon, nitrogen, and phosphorus and transfer between trophic levels in continental shelf waters south of New York. Rapp. et Proc. Verb. I. C. O. S. **153:**193-189.

Currie, R. I. 1953. Upwelling in the Benguela Current. Nature **171:**497-500.

Currie, R. I. 1958. Some observations on organic production in the Northeast Atlantic. Rapp. P.-v. Réun. Cons. Perm. Int. Explor. Mer **144:**96-102.

Cushing, D. H. 1951. Vertical migration of planktonic Crustacea. Biol. Rev. **19:**3-22.

Cushing, D. H. 1953. Studies on plankton populations. J. Cons. Perm. Int. Explor. Mer **19:**3-22.

Cushing, D. H. 1954. Some problems in the production of oceanic plankton. Document VIII. Presented to Commonwealth Oceanographic Conference, 1954. Plymouth, Mass.

Cushing, D. H. 1955. Some experiments on the vertical migration of zooplankton. J. Anim. Ecol. **24:**137-166.

Cushing, D. H. 1958. The estimation of carbon in phytoplankton. Rapp. P.-v. Réun. Cons. Perm. Int. Explor. Mer **144:**32-33.

Cushing, D. H. 1959*a*. On the nature of production in sea. Fish. Invest. London (ser. 2) **22**(6):1-40.

Cushing, D. H. 1959*b*. The seasonal variation in oceanic production as a problem in population dynamics. J. Cons. Perm. Int. Explor. Mer **24:**455-464.

Cushing, D. H. 1971. Upwelling and the production of fish. *In* F. S. Russell and M. Yonge [eds.]. Advances in marine biology. Vol. 9. Academic Press, Inc., New York.

Cuvier, G. L. C. F. D., Baron von. 1834-1836. Recherches sur les ossemens fossiles de quadrupèdes. . . . Mit Discours préliminaire 1812. [Letzte Aufl. posthum. hrsg. von F. Cuvier.] 12 Bde.

Daisley, K. W., and L. R. Fisher. 1958. Vertical distribution of vitamin B_{12} in the sea. J. Mar. Biol. Ass. U.K. **37:**683-686.

Dakin, W. J. 1934. The plankton calendar of the continental shelf of the Pacific coast of Australia at Sydney, compared with that of the Irish Sea. (James Johnston Memorial vol.) Liverpool University Press, Liverpool, pp. 164-174.

Dakin, W. J., and A. Colefax. 1933. Marine plankton of coastal waters of New South Wales. I. The chief planktonic forms and their seasonal distribution. Proc. Linn. Soc. N.S.W. **58:**186-222.

Dakin, W. J., and A. Colefax. 1935. Observations on the seasonal changes in temperatures, salinity, phosphates and nitrates, nitrogen, and oxygen of the ocean waters on the continental shelf off New South Wales, and the relationship to plankton production. Proc. Linn. Soc. N.S.W. **60:**303-314.

Dakin, W. J., and A. N. Colefax. 1940. The plankton of the Australian coastal waters off New South Wales. I. Univ. Sydney Publ. Zool. Monograph no. 1, pp. 1-211.

Dales, K. P. 1966. Symbiosis in marine organisms. *In* S. M. Henry [ed.]. Symbiosis. Vol. I. Academic Press, Inc., New York.

Dales, R. P. 1957. Pelagic polychaetes of the Pacific Ocean. Bull. Scripps Insta-Oceanogr. **7**(2):1-70.

Dall, W. H. 1921. Summary of the marine shell-bearing molluscs of the northwest coast of America. Bull. U.S. Nat. Mus. no. 112, pp. 1-217.

Daly, R. A. 1910. Pleistocene glaciation and the coral reef problem. Amer. J. Sci. Ser. **4:**30.

Dana, J. D. 1853. Crustacea. U.S. Explorer Expedition. II. Vol. 14. Philadelphia. 1618 p.

Darwin, C. 1897. The structure and distribution of coral reefs. D. Appleton & Co., New York. 344 p.

Darwin, C. 1916. The voyage of the "Beagle." E. P. Dutton & Co., Inc., New York.

Dasmann, R. F. 1970. Environmental conservation. John Wiley & Sons, Inc., New York. 375 p.

David, P. M. 1961. The influence of vertical migration on speciation in the oceanic plankton. Syst. Zool. **10:**10-16.

David, P. M. 1967. Illustrations of oceanic neuston. *In* N. B. Marshall [ed.]. Aspects of marine zoology. Academic Press, Inc., New York.

Davis. C. C. 1948. Effect of industrial copper as pollution upon the plankton in the Baltimore Harbor area. Pub. Chesapeake Biol. Lab. **72:**1-12.

Davis, C. C. 1955. The marine and freshwater plankton. Michigan State University Press, East Lansing.

Davis, F. M. 1923*a*. Quantitative studies on the fauna of the sea bottom: Dogger Bank. Fish. Invest. London (ser. 2) **6**(2):1-54.

Davis, F. M. 1923*b*. Quantitative studies on the fauna of the sea bottom: southern North Sea. Fish. Invest. London (ser. 2) **8**(4):1-50.

Davis, J. H. 1940. Ecology and geological role of mangroves in Florida. Carnegie Inst. Washington Pub. **517:**303-412.

Dawson, E. Y. 1956. How to know the seaweeds. William C. Brown Co. Publishers, Dubuque, Iowa.

Dawson, E. Y. 1966. Marine botany. Holt, Reinhart & Winston, Inc., New York.

Dawson, E. Y., and N. Foldvik. 1964. Seaweeds of Peru. Nov. Hedw. supp. 13. Lew Heyman, Publisher. 112 p.

Dawydoff, C. 1936. Observations sur la faune pélagique des eaux indochines de la mer de Chine meridionale. Bull. Soc. Zool. Fr. **61:**461-484.

Day, J. H. 1951. Ecology of South African estuaries. I. Review of estuarine conditions in general. Trans. Roy. Soc. S. Afr. **33:**53-91.

Day, J. H. 1974. Guide to marine life on South African shores. A. A. Balkema, Rotterdam, The Netherlands. 308 p.

Day, J. H., N. A. H. Millard, and G. J. Broekhuysen. 1953. Ecology of South African estuaries. IV. The St. Lucia system. Trans. Roy. Soc. S. Afr. **34:** 129-156.

Day, J. H., N. A. H. Millard, and A. D. Harrison. 1952. Ecology of South African estuaries. III. Knysna: a clear, open estuary. Trans. Roy. Soc. S. Afr. **33**(3):367-413.

Deacon, G. E. R. 1933. General account of the hydrology of the South Atlantic Ocean. 'Discovery' Rep. **7:**173-238.

Deacon, G. E. R. 1937. Hydrology of the Southern Ocean. 'Discovery' Rep. **15:**1-124.

Deacon, M., 1971. Scientists and the sea. Academic Press, Inc., New York. xvi + 445 p.

Debenham, F. 1960. Discovery and exploration. Doubleday & Co., Inc., Garden City, N.Y.

Deevey, G. B. 1948. Zooplankton of Tisbury Great Pond. Bull. Bingham Oceanogr. Coll. **12:**1-44.

Deevey, G. B. 1952*a*. Survey of zooplankton of Block Island Sound, 1943-1946. Bull. Bingham Oceanogr. Coll. **13**(3):65-119.

Deevey, G. B. 1952*b*. Quantity and composition of the zooplankton of Block Island Sound, 1949. Bull. Bingham Oceanogr. Coll. **13**(3):120-164.

Deevey, G. B. 1956. Oceanography of Long Island Sound, 1952-1954. V. Zooplankton. Bull. Bingham Oceanogr. Coll. **15:**113-115.

Defant, A. 1958. Ebb and flow: the tides of earth, air, and water. The University of Michigan Press, Ann Arbor.

Degener, O. 1949. Naturalists' South Pacific Expedition: Fiji. Waialua, Hawaii. Published by the author. 303 p.

deLaubenfels, M. W. 1954. Sponges of the west-central Pacific. Oregon State University Press, Corvallis. 320 p.

Denman, K. L. 1976. Covariability of chlorophyll and temperature in the sea. Deep-Sea Res. **23**(6):539-550.

Denton, E. J., and T. I. Shaw. 1962. Buoyancy of gelatinous marine animals. J. Physiol. (London) **161:**14-15.

Detwyler, T. R. 1971. Man's impact on environment. McGraw-Hill Book Co., New York. 731 p.

Devanney, J. W., G. Ashe, and B. Parkhurst. 1976. Parable beach: a primer in coastal zone economics. The M.I.T. Press, Cambridge, Mass. 160 p.

Devereux, I. 1967. Temperature measurements from oxygen isotope ratios of fish otoliths. Science **155:**1684-1685.

Dewey, J. F. 1972. Plate tectonics. Sci. Amer. **226** (5):56-68.

Dietz, R. S. 1948. Deep-scattering layer in the Pacific and Antarctic oceans. J. Mar. Res. **8:**430-442.

Dietz, R. S. 1962. The sea's deep-scattering layer. Sci. Amer. **207**(1):44.

Dietz, R. S., and J. C. Holden. 1970. The breakup of Pangea. Sci. Amer. **233**(4):30-41.

Digby, P. S. 1967. Pressure sensitivity and its mechanism in the shallow marine environment. *In* N. B. Marshall [ed.]. Aspects of marine zoology. Academic Press, Inc., New York.

Dittmar, W. 1884. Report on researches into the composition of ocean water collected by H.M.S. Challenger. 'Challenger' Rep. (phys. chem.) **1:**1-251.

Dobrin, M. B. 1947. Measurement of underwater noise produced by marine life. Science **105:**19-23.

Donn, W. L., and M. Ewing. 1966. The theory of ice ages. III. The theory involving polar wandering and an open polar sea is modified and given a quantitative basis. Science **152:**1706-1712.

Doochin, H., and F. G. W. Smith. 1951. Marine boring and fouling in relation to the velocity of water currents. Bull. Mar. Sci. Gulf Carib. **1:**196-208.

Dorst, J. 1970. Before nature dies. Houghton Mifflin Co., Boston.

Doty, M. S. 1946. Critical tide factors correlated with vertical distribution of algae and other organisms along the Pacific coast. Ecology **27:**315-328.

Doty, M. S. 1956, 1959. Current status of carbon-14 method of assaying productivity of the ocean. University of Hawaii Annual Report for 1956 and 1959.

Doty, M. S., and M. Ogun. 1957. Evidence for a photosynthetic daily periodicity. Limnol. Oceanogr. **2:**37-40.

Doubilet, D., and M. E. Long. 1977. Consider the sponge. Nat. Geog. Mag. **151:**(3):392-408.

Doudoroff, P., and M. Katz. 1950. Critical review of literature on the toxicity of industrial wastes and their components to fish. I. Alkalies, acids, and inorganic gases. Sew. Ind. Wastes **22:**1432-1458.

Douglas, J. S. 1952. The story of the oceans. Dodd, Mead & Co., New York.

Droop, M. R. 1957. Auxotrophy and organic compounds in the nutrition of marine phytoplankton. J. Gen. Microbiol. **16:**286-293.

Drummond, J. C., and E. R. Gunther. 1934. Observations on fatty constituents of marine plankton. III. Vitamin A and D content of oils. J. Exp. Biol. **11:** 203-209.

Dugan, J. 1956. Man under the sea. Harper & Row, Publishers, New York.

Dunbar, M. J. 1940. On the size, distribution, and breeding cycles of four marine planktonic animals from the arctic. J. Anim. Ecol. **9:**215-226.

Dunbar, M. J. 1942. Marine macroplankton from the Canadian eastern Arctic. II. Medusae, Siphonophora, Ctenophora, Pteropoda, and Chaetognatha. Can. J. Res. **20**[D] (3):71-77.

Dunbar, M. J. 1946. On *Themisto libellula* in Baffin Island coastal waters. J. Fish Res. Bd. Canada **6:** 419-434.

Dunbar, M. J. 1970. The evolution of stability in marine environments. Natural selection at the level of the ecosystem. *In* J. W. Nybakken [ed.]. 1971. Readings in marine ecology. Harper & Row, Publishers, New York.

Duosik, D. W. 1974. Shoreline for the public; handbook of social, economic, and legal considerations. The M.I.T. Press, Cambridge, Mass. 182 p.

Duursma, E. K. 1963. The production of dissolved organic matter in the sea as related to the primary gross production of organic matter. Netherlands J. Sea Res. **2:**85-94.

Earle, S. A., and A. Giddings. 1976. Life springs from death in Truk lagoon. Nat. Geog. Mag. **149**(5):578-613.

Ebling, J., and K. C. Highnam. 1969. Chemical communication. Edward Arnold, Ltd., London.

Edmondson, C. H. 1933. Reef and shore fauna of Hawaii. Bernice P. Bishop Museum, Special publication no. 22. 295 p.

Edmondson, C. H. 1944. Incidence of fouling in Pearl Harbor. (Occasional papers.) Bernice P. Bishop Mus. **18:**1-34.

Edmondson, C. H., and W. M. Ingram. 1939. Fouling organisms in Hawaii. Bernice P. Bishop Mus. **14:** 251-300.

Edwards, C. E. 1971. Persistent pesticides in the environment. Chemical Rubber Co. Press, Cleveland.

Ehrlich, P. 1968. The population bomb. Ballantine Books, Inc., New York.

Eibl-Eibesfeldt, I. 1966. Land of a thousand atolls (Maldive and Nicobar islands). The World Publishing Co., Cleveland.

Ekman, S. 1940. Begründung einer statistischen Methode in der regionalen Tiergeographie. Nova Acta Reg. Soc. Sci. Upsaliensis (ser. 4) **12:**2.

Ekman, S. 1953. Zoogeography of the seas. Sidgwick & Jackson, Ltd., London.

Ellis, D. V. 1959. The benthos of soft sea-bottom in arctic North America. Nature **184:**79-80.

Elsner, R. 1969. Cardiovascular adjustments to diving. *In* H. T. Anderson [ed.]. Biology of marine mammals. Academic Press, Inc., New York.

Elton, C. 1948. The ecology of invasions by animals and plants. John Wiley & Sons, Inc., New York.

Emery, K. O. 1938. Rapid method of mechanical analysis of sands. J. Sed. Petrol. **8**(3):105-111.

Emery, K. O. 1956. Deep-standing internal waves in California basins. Limnol. Oceanogr. **1:**35-41.

Emery, K. O. 1969. A coastal pond studied by oceanographic methods. American Elsevier Publishing Co., New York. viii + 80 p.

Emery, K. O., and C. O'D. Iselin. 1967. Human food from ocean and land. Science **157:**1279-1281.

Emery, K. O., W. L. Orr, and S. C. Rittenberg. 1955. Nutrient budgets in the ocean. *In* Essays in honor of Captain Hancock. University of Southern California Press, Los Angeles. pp. 229-310.

Endean, R. 1957. The biogeography of Queensland's shallow-water echinoderm fauna (excluding Crinoidea), with a rearrangement of the faunistic provinces of tropical Australia. Australian J. Mar. Freshwater Res. **8**(3):233-273.

Engel, L., and editors of *Life,* 1961, 1963. The sea. (Life Nature Library.) Time, Inc., New York.

Enright, J. T., and W. M. Hammer. 1967. Vertical diurnal migration and endogenous rhythmicity. Science **157:**937-941.

Epstein, S., R. Buchsbaum, H. A. Lowenstein, and H. C. Urey. 1953. Revised carbonate-water isotopic temperature scale. Bull. Geol. Soc. Amer. **64:**1315-1326.

Eriksen, A., and D. T. Lawrence. 1940. Occurrence and cause of pollution in Gray's Harbor. Bull. no. 2. Washington State Pollution Commission.

Eshleman, R. von. 1969. The atmospheres of Mars and Venus. Sci. Amer. **220**(3):78-88.

Esterly, C. O. 1919. Reaction of various plankton animals with reference to their diurnal migrations. Univ. California Pub. Zool. **19:**1-83.

Evans, I. O. [ed.]. 1962. Sea and seashore. Frederick Warne & Co., Inc., New York.

Evans, W. E. 1969. Marine mammals communication: social and ecological factors. *In* H. T. Anderson [ed.]. Biology of marine mammals. Academic Press, Inc., New York.

Evenau, M., L. T. Evans, and A. P. Hughes. 1965. Photoenvironment. *In* E. J. Bowen [ed.]. Recent progress in photobiology. Academic Press, Inc., New York.

Ewing, M., and W. L. Donn. 1956. The theory of ice ages. Science **123:**1061.

Ewing, M., and W. L. Donn. 1958. The theory of ice ages. II. The theory that certain local terrestrial conditions caused Pleistocene glaciation is discussed further. Science **127:**1159.

Eyden, D. 1923. Specific gravity as a factor in the vertical distribution of plankton. Proc. Cambridge Phil. Soc. Biol. Sci. **1:**49-55.

Fager, E. W. 1957. Determination and analysis of recurrent groups. Ecology **38:**586-595.

Fairbridge, R. 1966. Encyclopedia of oceanography. Reinhold Publishing Corp., New York.

Fankboner, P. V. 1971. Intracellular digestion of symbiotic zooxanthellae by host amoebocytes in giant clams (Bivalvia, Tridacnidae), with a note on the nutritional role of the hypertrophied siphonal epidermis. Biol. Bull. Woods Hole **141:**222-234.

Farfante, I. P. 1969. Western Atlantic shrimps of the genus *Penaeus.* U.S. Department of the Interior, Fish and Wildlife Service, Bur. of Comm. Fisheries. Fishery Bull. **67**(3):461-591.

Farran, G. P. 1947. Vertical distribution of plankton, *Sagitta, Calanus,* and *Metridia,* off the south coast of Ireland. Proc. Roy. Irish. Acad. [B] **51:**121-136.

Fasham, M. J. R., and M. V. Angel. 1975. Relationship of zoogeographical distributions of planktonic ostracods of the North East Atlantic to water masses. J. Mar. Biol. Ass. U.K. **55**(3):739-758.

Fasham, M. J. R., and P. R. Pugh. 1976. Observations on the horizontal coherence of chlorophyll *a* and temperature. Deep-Sea Res. **23**(6):527-538.

Feder, H. M. 1955. On the methods used by the starfish *Pisaster ochraceous* in opening three types of bivalve mollusks. Ecology **36**:764-767.

Feder, H. M. 1963. Gastropod defensive responses and their effectiveness in reducing predation by starfishes. Ecology **44**(3):505-512.

Feder, H. M. 1966. Cleaning symbiosis in the marine environments. *In* S. M. Henry [ed.]. Symbiosis. Vol. I. Academic Press, Inc., New York.

Feder, H. M. 1967. Organisms responsive to predatory sea stars. Sarsia **29**:371-394.

Fenchel, T. 1968*a*. The ecology of the marine microbenthos. II. The food of marine benthic ciliates. Ophelia **5**(1):73-121.

Fenchel, T. 1968*b*. The ecology of the marine microbenthos. III. The reproductive potential of ciliates. Ophelia **5**(1):123-136.

Fenchel, T. 1969. The ecology of marine microbenthos. IV. Structure and function of the benthic ecosystem, its chemical and physical factors and the microfauna communities, with special reference to the ciliated protozoa. Ophelia **6**(1):1-182.

Fenchel, T., and B. O. Jansson. 1966. The ecology of the marine microbenthos. I. On the vertical distribution of the microfauna in the sediments of a brackish-water beach. Ophelia **3**(2):161-177.

Fischer-Piette, E. 1931. Sur la pénétration des diverses espèces marines sessiles dans les estuaries et sa limitation par l'eau douce. Ann. Inst. Oceanogr. Monaco **10**:213-243.

Fish, C. J. 1925. Seasonal distribution of the plankton of the Woods Hole region. Bull. Bur. Fish. (Washington) **41**:91-179.

Fish, C. J. 1935. Marine biology and paleoecology. J. Paleontol. **9**:92-100.

Fish, C. J. 1954. Preliminary observations on biology of boreo-arctic and subtropical oceanic zooplankton populations. Symposium on Marine and Fresh-Water Plankton in the Indo-Pacific. Bangkok 3-9.

Fish, C. J., and M. W. Johnson. 1937. The biology of the zooplankton population in the Bay of Fundy and Gulf of Maine, with special reference to production and distribution. J. Biol. Bd. Canada **3**: 189-321.

Fisher, J. 1957. Wonderful world of the sea. Doubleday & Co., Garden City, N.Y. 72 p.

Fisher, R., and R. Revelle. 1955. The trenches of the Pacific. Sci. Amer. **193**(5):36-41.

Fitch, J. E. 1953. Common marine bivalves of California. California Department of Fish and Game. Fisheries Bull. no. 90. 102 p.

Flattely, F. W., and C. L. Walton. 1922. The biology of the seashore. The Macmillan Co., New York.

Fleming, R. H. 1939. The control of diatom populations by grazing. J. Cons. Perm. Int. Explor. Mer **14**(2):210-227.

Fleming, R. H. 1957. General features of the oceans. *In* J. W. Hedgpeth [ed.]. Treatise on marine ecology and paleoecology. Vol. I. Ecology. Geological Society of America, New York.

Forbes, E., and R. Goodwin-Qustin. 1959. Natural history of the European seas. Jon Van Voorst, London.

Ford, E. 1923. Animal communities of the level sea bottom adjacent to Plymouth. J. Mar. Biol. Ass. U.K. **13**:164-224.

Forester, G. R. 1953. A new dredge for collecting burrowing animals. J. Mar. Biol. Ass. U.K. **32**:193-198.

Fossato, V. U., and W. J. Canzonier. 1976. Hydrocarbon uptake and loss by the mussel *Mytilus edulis*. Mar. Biol. **36**(3):243-250.

Fowell, R. R. 1939. Communities harbored by oar weeds *(Laminaria)*. Proc. Swansea Sci. Field Natur. Soc. **2**:92-103.

Fox, D. L. 1947. Carotenoid and indolic biochromes of animals. Ann. Rev. Biochem. **16**:443-470.

Fox, D. L. 1950. Comparative metabolism of organic detritus by inshore animals. Ecology **31**:100-108.

Fox, D. L. 1953. Animal biochromes and structural colors. John Wiley & Sons, Inc., New York.

Fox, D. L. 1957. Particulate organic detritus. *In* J. W. Hedgpeth [ed.] Treatise on marine ecology and paleoecology. Vol. I. Ecology. Geological Society of America, New York.

Fox, D. L., S. C. Crane, and B. H. McConnaughey. 1949. A biochemical study of the marine annelid worm, *Thoracophelia mucronata:* its food, biochromes, and carotenoid metabolism. J. Mar. Res. **8**:567-585.

Fox, D. L., J. D. Isaacs, and I. F. Corcora. 1952. Marine leptopel: its recovery, measurement, and distribution. J. Mar. Res. **11**:29-46.

Fox, D. L., C. H. Oppenheimer, and J. S. Kittridge. 1953. Microfiltration in oceanographic research. II. Retention of colloidal micelles by adsorption filters and by filter-feeding invertebrates; proportions of dispersed organic and dispersed inorganic matter to organic solutes. J. Mar. Res. **12**:233-243.

Fox, H. M. 1939. The activity and metabolism of poikilothermal animals in different latitudes. Proc. Zool. Soc. London [A] **109**:141-156.

Fox, S. W. [ed.]. 1965. The origins of prebiological systems and of their molecular matrices. Academic Press, Inc., New York.

Foxton, P. 1956. Standing crop of zooplankton in the Southern Ocean. 'Discovery' Rep. **28**:193-235.

Fraenkel, G. S., and D. L. Gunn. 1940. The orientation of animals. Clarendon Press, Oxford.

Frank, P. W. 1965. The biodemography of an intertidal snail population. Ecology **46**(6):831-844.

Frank, P. W. 1968. Life histories and community stability. Ecology **49**(2):355-357.

Fraser, C. M. 1942. Marine zoology in the northeast Pacific. Trans. Roy. Soc. Canada (Ser. 3) **36**:1-18.

Fraser, J. H. 1947. Fiches d'identification du zooplancton. Nos. 9, 10. Thaliacea I, II. J. Cons. Perm. Int. Explor. Mer.

Fraser, J. H. 1948. Plankton in Scottish waters. Ann. Biol. (Copenhagen) **3:**43.

Fraser, J. H. 1949. Plankton of the Faeroe-Shetland Channel and the Faeroes. June, Aug., 1947. Ann. Biol. (Copenhagen) **4:**27-28.

Fraser, J. H. 1949-1955. Plankton investigations from the Scottish research vessel. Ann. Biol. (Copenhagen) **4:**66-67; **6:**91-99; **7:**25; **7:**76-77; **8:**32-35; **8:**104-105; **9:**32-33; **10:**31.

Fraser, J. H. 1952. The Chaetognatha and other zooplankton of the Scottish area and their value as biological indicators of hydrographic conditions. Mar. Res. Scot. **2:**1-52.

Fraser, J. H. 1954. Zooplankton collections made by Scottish research vessel during 1953. Ann. Biol. (Copenhagen) **10:**99-101.

Fraser, J. H. 1955. The plankton of the waters approaching the British Isles in 1953. Mar. Res. Scot. no. 1.

Fraser, J. H. 1956, 1959. Scottish plankton investigations. Ann. Biol. (Copenhagen) **11:**26-27, 52-54; **14:**28-30, 66.

Fraser, J. H. 1962. Nature adrift: the story of plankton. Dufour Editions, Inc., Chester Springs, Pa.

Fraser, J. H., and A. Seville. 1949*a*. Macroplankton in the Faeroe Channel, 1948. Ann. Biol. (Copenhagen) **5:**61-64.

Fraser, J. H., and A. Seville. 1949*b*. Plankton distribution in Scottish and adjacent waters in 1948. Ann. Biol. (Copenhagen) **5:**61-64.

Frey, H., and S. Frey. 1961. One hundred and thirty feet down: handbook for hydronauts. Harcourt, Brace & World, Inc., New York.

Friedrich, H. 1965. Meeresbiologie: eine Einführung in die Probleme und Ergebnisse. Gebrüder Borntraeger, Berlin.

Frolander, H. F. 1962. Quantitative estimations of temporal variations in zooplankton off the coast of Washington and British Columbia. J. Fish. Res. Canada **19**(4):657-675.

Frost, N. 1937. Further plankton investigations. Report of the Fisheries Research Institute of Newfoundland, 1936-1937.

Fry, W. G. [ed.] 1969. Biology of the Porifera. Symposium 25. Zoological Society of London and Academic Press, Inc., New York. 512 p.

Gaardner, T., and H. H. Gran. 1927. Investigations of the production of plankton in the Oslo Fjord. Rapp. P.-v. Réun. Cons. Perm. Int. Explor. Mer **42:**1-48.

Gaardner, T., and R. Sparck. 1931. Biochemical and biological investigations of the west Norwegian oyster pools. Rapp. P.-v. Réun. Cons. Perm. Int. Explor. Mer **75:**47-58.

Gardiner, A. C. 1937. Phosphate production by planktonic animals. J. Cons. Perm. Int. Explor. Mer **12:**144-146.

Gauld, D. T. 1950. A fish cultivation experiment in an arm of a sea-lock. III. The plankton of Kyle Scottish. Proc. Roy. Soc. Edinb. [B] **64:**36-64.

George, P. C. 1953. The marine plankton of the coastal waters of Calicut, with observations on hydrological conditions. J. Zool. Soc. India **5:**76-107.

Gerard, R. D., and W. J. Worzel. 1967. Condensation of atmospheric moisture from tropical maritime air masses as a freshwater resource. Science **157:** 1300-1302.

Giesbrecht, W. 1892. Pelagische Copepoden. Monograph. Fauna und flora des Golfes von Neapel. Mitt. Zool. Stn. Neapel. 10.

Giesbrecht, W., and O. Schmeil. 1898. Copepoda Gymnoplea. *In* Das Tierreich. Deutschen Zoologischen Gesellschaft, Berlin. 169 p.

Giese, A. C., and J. S. Pearse [eds.]. 1974, 1975. Reproduction of marine invertebrates. 3 vols. Academic Press, Inc., New York. Vol. 1, 558 p.; vol. 2, 319 p.; vol. 3, 311 p.

Gilbert, K., and F. McNeill. 1959. The Great Barrier Reef and adjacent isles. Coral Press Pty., Ltd., Paddington, Sidney, Australia.

Gillbricht, M. 1952*b*. Untersuchungen zur Produktionsbiologie des Planktons in der Kieler Bucht. Kieler Meeresforsch. **8:**173-191.

Gilmartin, M. 1958. Some observation on the lagoon plankton of Eniwetok Atoll. Pacific Sci. **12:**313-316.

Gislén, T. 1930. Epibioses of the Gullmar Fjord. II. K. svenska Vetensk-Akad. Stockholm **4:**1-376.

Glover, R. S. 1955. Science and the herring industry. Advances Sci. **44:**426-434.

Glover, R. S. 1961. Biogeographical boundaries: the shapes of distributions. *In* M. Sears [ed.]. Oceanography. Pub. no. 67. American Association for the Advancement of Science, Washington, D.C.

Glover, R. S. 1967. The continuous plankton recorder survey of the North Atlantic. *In* N. B. Marshall [ed.]. Aspects of marine zoology. Academic Press, Inc., New York.

Goldberg, E. D. 1957. Biogeochemistry of trace metals. *In* J. W. Hedgpeth [ed.]. Treatise on marine ecology and paleoecology. Vol. I. Ecology. Geological Society of America, New York.

Goldberg, E. D., M. Baker, and D. L. Fox. 1952. Microfiltration in oceanographic research. I. Marine sampling with the molecular filter. J. Mar. Res. **11:** 194-204.

Goldman, C. R. 1962. A method of studying nutrient-limiting factors *in situ* in water columns isolated by polyethylene film. Limnol. Oceanogr. **7:**99-101.

Golubiatnikova, T. P. 1974. Soviet investigations of saithe in the North Sea, 1972-1974. Ann. Biol. **31:** 109-110.

Goodwin, R. H. 1961. Connecticut's coastal marshes, a vanishing resource. Bulletin no. 2. Connecticut Arboretum, New London. 36 p.

Gorden, M., and M. Gorden [eds.]. 1972. Environmental management: science and politics. Allyn & Bacon, Inc., Boston. 548 p.

Gordienke, P. A. 1961. The Arctic Ocean. Sci. Amer. **204**(5):88.

Gore, R. 1977. Striking it rich in the North Sea. Nat. Geog. Mag. **151**(4):519-532.

Goreau, T. F., N. I. Goreau, and C. M. Yonge. 1971. Reef corals: autotrophs or heterotrophs? Biol. Bull. Woods Hole **141:**247-260.
Gotto, R. V. 1951. Some plankton records from Strangford Lough, County Down. Irish Natur. J. Biol. **10:**162-164.
Gourley, D. R. H. 1952. Failure of P^{32} to exchange with organic phosphorous compounds. Nature (London) **169:**192.
Graham, F. 1970. Since Silent spring. Houghton Mifflin Co., New York.
Graham, H. W. 1941. Plankton production in relation to character of water in the open Pacific. J. Mar. Res. **4:**189-197.
Graham, H. W. 1943. Chlorophyll content of marine plankton. J. Mar. Res. **5:**153-160.
Graham, H. W., and H. Gay. 1945. Season of attachment and growth of sedentary marine organisms at Oakland, Calif. Ecology **26:**375-386.
Graham, M. 1956. Sea fisheries: their investigation and use in the United Kingdom. Edward Arnold Publishers, Ltd., London.
Grainger, E. H. 1959. The annual oceanographic cycle at Igloolik in the Canadian Arctic. I. The zooplankton and physical and chemical observations. J. Fish. Res. Bd. Canada **16:**453-501.
Gran, H. H. 1912. Pelagic plant life. *In* J. Murray and J. Hjort [eds.]. Depths of the ocean. The Macmillan Co., New York.
Gran, H. H. 1927. The production of plankton in the coastal waters off Bergen, Mar.-Apr., 1922. Rep. Norweg. Fish. Mar. Res. Invest. **3**(8):1-74.
Gran, H. H. 1929*a*. Quantitative investigations during expeditions with the "Michael Sars" (July-Sept., 1924). Rapp. P.-v. Réum. Cons. Perm. Int. Explor. Mer **56**(5):1-50.
Gran, H. H. 1929*b*. Investigation of the production of plankton outside the Romsdalsfjord 1926-1927. Rapp. P.-v. Réun. Cons. Perm. Int. Explor. Mer **56** (6):1-112.
Gran, H. H. 1931. On the conditions for the production of plankton in the sea. Rapp. P.-v. Réun. Cons. Perm. Int. Explor. Mer **75:**37-46.
Gran, H. H. 1932. Phytoplankton, methods and problems. J. Cons. Perm. Int. Explor. Mer **7:**343-358.
Gran, H. H., and T. Braarud. 1935. Quantitative study of phytoplankton in the Bay of Fundy and Gulf of Maine (including observations on hydrography, chemistry, and turbidity). J. Biol. Bd. Canada **1** (5):279-467.
Grassé, P. P. 1948-1955. Traité de zoologie. 17 vols. Masson et Cie, Paris.
Graves, W. 1976. The imperiled giants. Nat. Geog. Mag. **150**(6):722-751.
Graves, W. 1977. Puget Sound. Nat. Geog. Mag. **151** (1):71-97.
Graymore, C. N. 1970. Biochemistry of the eye. Academic Press, Inc., New York. 796 p.
Green, J. 1961. A biology of Crustacea. Quadrangle Books, Chicago. 180 p.
Green, J. 1968. The biology of estuarine animals. *In* R. P. Bales [ed.]. Biology Series. University of Washington Press, Seattle.
Grier, M. C. 1941. Oceanography of the North Pacific Ocean, Bering Sea, and Bering Strait [a bibliography]. Library Series. Vol. II. University of Washington Press, Seattle. 319 p.
Grøntved, J. 1951. Investigations on the phytoplankton in the southern North Sea, May, 1947. Medd. Komm. Danm. Fisk. Havund. (Ser. Plankton) **5** (5):1-49.
Grøntved, J. 1957. Sampler for underwater macrovegetation. J. Cons. Perm. Int. Explor. Mer **22:** 293-297.
Grøntved, J. 1958. Underwater macrovegetation. J. Cons. Perm. Int. Explor. Mer **24:**32-42.
Gross, F. 1937. Notes on the culture of some marine plankton organisms. J. Mar. Biol. Ass. U.K. **21:** 753-768.
Gross, F., and E. Zeuthen, 1948. The buoyancy of plankton diatoms: a problem of cell physiology. Proc. Roy. Soc. [ser. B] **135:**382-389.
Guberlet, M. L. 1936. Animals of the seashore. Metropolitan Press. Portland, Ore.
Guberlet, M. L. 1956. Seaweeds at ebb tide. University of Washington Press, Seattle.
Guilecher, A. 1958. Coastal and submarine morphology. John Wiley & Sons, Inc., New York.
Gulland, J. A. 1971. Ecological aspects of fishery research. *In* J. B. Cragg [ed.]. Advances in ecological research. Vol. 7. Academic Press, Inc., New York.
Gullion, E. A. [ed.]. 1968. Uses of the seas. The American Assembly, Columbia University. Prentice-Hall, Inc., Englewood Cliffs, N.J.
Gunter, G. 1957. Temperature. *In* J. W. Hedgpeth [ed.]. Treatise on marine ecology and paleoecology. Vol. I. Ecology. Geological Society of America, New York.
Gunter, G., and W. E. Gunter. 1947. Mass mortality of marine animals on the lower west coast of Florida, Nov., 1946, to Jan., 1947. Science **105:**256-257.
Gunter, G., and W. E. Gunter. 1956. Some relations of faunal distributions to salinity in estuarine waters. Ecology **37:**616-619.
Gunter, G., and W. E. Gunter. 1958. A study of an estuarine area with water-level control in the Louisiana marsh. Proc. Louisiana Acad. Sci. **21:**5-34.
Gunter, G., R. H. Williams, C. C. Davis, and F. G. W. Smith. 1948. Catastrophic mass mortality of marine animals and coincident phytoplankton bloom on the west coast of Florida, Nov., 1946, to Aug., 1947. Ecol. Monogr. **18:**310-324.
Günther, A. 1868. An account of the fishes of the states of Central America. Trans. Zool. Soc. (London) **6:**377-494.
Gunther, E. R. 1934. Fatty constituents of marine plankton. J. Exp. Biol. **11:**173-197.
Gunther, E. R. 1936. A report on oceanographical investigations in the Peru Coastal Current. 'Discovery' Rep. **13:**109-276.
Günther, K., and K. Deckert. 1950. Wunderwelt der

tief See. F. A. Herbig Verlagsbuchhandlung (walterkahnert), Berlin, Grunewald. [Trans. by E. W. Dickes. 1956. Creatures of the deep. George Allen & Unwin, Ltd., London.]

Gurney, R. 1939. Bibliography of the larvae of decapod Crustacea. (Available as reprints from Stechert-Hafner, 1960.) The Royal Society, London, i-viii, 123 p.

Gurney, R. 1942. Larvae of decapod Crustacea. (Available as reprints from Stechert-Hafner, 1960.) Ray Society, London. i-viii, 306 p.

Haeckel, E. 1879. Monographie der Medusen. Gustav-Fischer, Jena. 2 vols.

Haeckel, E. 1887. Report on the Radiolaria collected by H.M.S. "Challenger" during the years 1873-1876. Challenger reports (Zoology). Vol. 18, parts 1 and 2. 1803 p.

Hallam, A. 1972. Continental drift and the fossil record. Sci. Amer. **227**(5):56-66.

Halstead, B. W. 1959. Dangerous marine animals. Cornell Maritime Press, Inc., Cambridge, Md.

Hamilton, W. J. 1967. Analysis of bird navigation experiments. *In* K. E. F. Watt [ed.]. Systems analysis in ecology. Academic Press, Inc., New York.

Hand, C. H. 1961. Present state of nematocyst research. *In* H. M. Lenhoff and W. F. Loomis [eds.] Biology of Hydra. Symposium. University of Miami Press, Coral Gables, Fla., 187-202.

Hand, C. H., and L. B. Kan. 1961. Medusae of the Chukchi and Beaufort seas of the Arctic Ocean. Montreal Arctic Institute of North America, Tech. paper no. 6.

Hansen, K. V. 1951. On the diurnal migration of zooplankton in relation to the discontinuity layer. J. Cons. Perm. Int. Explor. Mer **17**:231-241.

Hansen, K. V. 1959. Investigations of the primary production and the quantitative distribution of the macroplankton in the North Atlantic. International Oceanographic Congress Preprints. American Association for the Advancement of Science, Washington, D.C.

Harding, J. P., and N. Trebble. 1963. Speciation in the sea. Systematics Association, London. 211 p.

Hardy, A. C. 1926. A new method of plankton research. Nature **118**:630.

Hardy, A. C. 1936*a*. The continuous plankton recorder, 'Discovery' Rep. **11**:457-510.

Hardy, A. C. 1936*b*. The arctic plankton collected by the "Nautilus" Expedition, 1931. I. General account. J. Linn. Soc. London **39**:391-403.

Hardy, A. C. 1936*c*. Ecological relationships between the herring and the plankton investigated with the plankton indicator. J. Mar. Biol. Ass. U.K. **21**:147-291.

Hardy, A. C. 1939. Ecological investigation with the continuous plankton recorder. Hull Bull. Mar. Ecol. **1**:1-57.

Hardy, A. C. 1941. Plankton as a source of food. Nature **147**:695-696.

Hardy, A. C. 1947. Experiments on the vertical migration of plankton animals. J. Mar. Biol. Ass. U.K. **26**:467-526.

Hardy, A. C. 1953. Essays in marine biology. Oliver & Boyd, London. 144 p.

Hardy, A. C. 1955. A further example of the patchiness of plankton distribution. (Papers in Marine Biology and Oceanography.) Deep-Sea Res. (supp.) **3**:7-11.

Hardy, A. C. 1956, 1960. The open sea: its natural history. Part 1. The world of plankton. Part 2. Fish and fisheries. Houghton Mifflin Co., Boston.

Hardy, A. C. 1958. Toward prediction in the sea. *In* A. A. Buzzati-Traverso [ed.]. Perspectives in marine biology. University of California Press, Richmond.

Hardy, A. C., and R. Bainbridge. 1954. Experimental observations on the vertical migrations of plankton animal. J. Mar. Biol. Ass. U.K. **33**:409-448.

Hardy, A. C., and E. R. Gunther. 1935. The plankton of the South Georgia whaling grounds and adjacent waters, 1926-1927. 'Discovery' Rep. **11**:1-456.

Hardy, A. C., and W. N. Paton. 1947. Experiments on vertical migration of plankton animals. J. Mar. Biol. Ass. U.K. **26**:467-526.

Harris, E. 1957. Radiophosphorus metabolism in zooplankton and microorganisms. Can J. Zool. **35**:769-782.

Harris, E. 1959. The nitrogen cycle in Long Island Sound. Bull. Bingham Oceanogr. Coll. **17**:31-65.

Harris, J. E. 1953. Physical factors involved in the vertical migration of plankton. Quart. J. Microbiol. Sci. **94**:537-550.

Harris, J. E., and U. K. Wolfe. 1955. A laboratory study of vertical migration. Proc. Roy. Soc. (Biol.) **144**:329-354.

Harrison, F. W., and R. R. Cowden [eds.] 1976. Aspects of sponge biology. Academic Press, Inc., New York. 371 p.

Harrison, R. J. [ed.]. 1972, 1974. Functional anatomy of marine mammals. 2 vols. Academic Press, Inc., New York. xviii + 452 p.; x + 366 p.

Harrison, R. J., and J. E. King. 1965. Marine mammals. Hutchinson University Library, London.

Hart, T. J. 1934. On the phytoplankton of the southwest Atlantic and the Bellingshausen Sea, 1929-1931. 'Discovery' Rep. **8**:1-268.

Hart, T. J. 1942. Phytoplankton periodicity in the Antarctic surface waters. 'Discovery' Rep. **21**:263-348.

Hart, T. J. 1953. Plankton of the Benguela Current. Nature (London) **171**:631-634.

Hart, T. J. and R. I. Currie. 1960. The Benguela Current. 'Discovery' Rep. **31**:123-298.

Hartman, O. 1952. The literature of the polychaetous annelids. Alan Hancock Foundation. University of Southern California Press, Los Angeles.

Hartman, W. D., and T. F. Goreau. 1970. Jamaican coralline sponges: their morphology, ecology, and fossil relatives. Symp. Zool. Soc. London **25**:205-243. [Proposes new class of sponges, the Sclerospongiae.]

Harvey, H. W. 1926. Nitrate in the sea. J. Mar. Biol. Ass. U.K. **14**:71-88.

Harvey, H. W. 1928. Biology, chemistry, and physics of seawater. Cambridge University Press, New York. 194 p.

Harvey, H. W. 1934*a*. Annual variation of planktonic vegetation. J. Mar. Biol. Ass. U.K. **19**:775-792.

Harvey, H. W. 1934*b*. Measurement of phytoplankton population. J. Mar. Biol. Ass. U.K. **19**:761-773.

Harvey, H. W. 1937. Note on selective feeding by *Calanus*. J. Mar. Biol. Ass. U.K. **22**:97-100.

Harvey, H. W. 1942. Production of life in the sea. Biol. Rev. **17**:221-246.

Harvey, H. W. 1945. Recent advances in the chemistry and biology of seawater. Cambridge University Press, New York.

Harvey, H. W. 1947. Manganese and growth of phytoplankton. J. Mar. Biol. Ass. U.K. **26**:562-579.

Harvey, H. W. 1949. On manganese in sea and fresh waters. J. Mar. Biol. Ass. U.K. **28**:155-164.

Harvey, H. W. 1950. Production of living matter in the sea. J. Mar. Biol. Ass. U.K. **29**:97-136.

Harvey, H. W. 1957, 1960. The chemistry and fertility of seawater. Cambridge University Press, New York.

Harvey, H. W., L. H. N. Cooper, M. V. Lebour, and F. S. Russell. 1935. Plankton production and its control. J. Mar. Biol. Ass. U.K. **20**:407-441.

Hasle, G. R. 1959. A quantitative study of phytoplankton from the equatorial Pacific. Deep-Sea Res. **6**: 38-59.

Hata, Y. 1960. Relations between the activity of marine sulphate-reducing bacteria and the oxidation-reduction potential of the culture media. J. Shimonoseki Coll. Fish. **19**:57-77.

Hazen, W. E. [ed.]. 1964. Readings in population and community ecology. W. B. Saunders Co., Philadelphia. 388 p.

Hedgpeth, J. W. 1951. Classification of estuarine and brackish waters and the hydrographic climate. *In* H. S. Ladd and others [eds.]. Report of the Committee on a Treatise on Marine Ecology and Paleoecology, 1950-1951. Geological Society of America, New York.

Hedgpeth, J. W. [ed.]. 1957. Treatise on marine ecology and paleoecology. Vol. I. Geological Society of America, New York.

Hedley, R. H. 1964. The biology of the Foraminifera. *In* W. J. Felts and R. J. Harrison [eds.]. Academic Press, Inc., New York.

Heezen, B. C. 1956. The origin of submarine canyons. Sci. Amer. **195**(2):36-41.

Heezen, B. C., and M. Tharp. 1963. The Altantic floor. *In* A. Love and D. Love [eds.]. North Atlantic biota and their history. Pergamon Press, Inc., New York.

Heinrich, A. K. 1962. Life histories plankton animals and seasonal cycles of plankton communities in the oceans. J. Cons. Perm. Int. Explor. Mer **27**:15-24.

Helfrich, H. W. [ed.]. 1970. The environmental crisis. Yale University Press, New Haven, Conn. 187 p.

Henderson, G. T. D. 1954. The young of fish and fish eggs. Hull Bull. Mar. Ecol. **3**:215-252.

Hensen, V. 1911. Das Leben im Ozean nach Zahlungen seiner Bewohner. Erg. Plankton-Expdn. der Humboldt-stiftung. Lipsius & Tischer, Leipzig. 406 p.

Hentschel, E. 1936. Allgemeine Biologie des Südatlantischen Ozeans. Wiss. Erg. Deut. Atlantischen Exped. auf Meleor, 1925-1927. Vol. XI. 344 p.

Hentschel, E., and H. Wattenberg. 1930. Plankton und Phosphat in der Oberfläshenschicht des Südatlantischen Ozeans. Ann. Hydrogr. (Berlin) **58**:273-277.

Herald, E. S. 1961. Living fishes of the world. Doubleday & Co., Inc., Garden City, N.Y.

Herring, P. J. 1967. Pigments of plankton at the sea surface. *In* N. B. Marshall [ed.]. Aspects of marine zoology. Academic Press, Inc., New York.

Hersey, J. B. [ed.]. 1967. Deep-sea photography. The Johns Hopkins Press, Baltimore.

Hesse, R., W. C. Allee, and K. Schmidt. 1951. Ecological animal geography. Part 2. Marine. 2nd ed. John Wiley & Sons, Inc., New York.

Heyderdahl, T. 1950. Kon Tiki. Rand McNally & Co., Chicago. 304 p.

Hickling, C. F. 1970. Estuarine fish farming. *In* F. S. Russell [ed.]. Advances in marine biology. Vol. 8. Academic Press, Inc., New York.

Hickman, C. P. 1973. Biology of the invertebrates. 2nd ed. The C. V. Mosby Co., St. Louis.

Hida, T. A. 1957. Chaetognaths and pteropods as biological indicators in North Pacific waters. Special Scientific Report on Fisheries, no. 215. Fish and Wildlife Service, U.S. Department of the Interior, Washington, D.C.

Hida, T. A., and J. E. King. 1965. Vertical distribution of zooplankton in the central equatorial Pacific (July-Aug., 1952). Special Scientific Report on Fisheries, no. 144. Fish and Wildlife Service, U.S. Department of the Interior, Washington, D.C. 22 p.

Hill, M. N. [ed.]. 1962-1963. The sea. Vol. I. Physical oceanography. Vol. II. Composition of seawater and descriptive oceanography. Vol. III. The earth beneath the sea: history. John Wiley & Sons, Inc., New York.

Hillson, C. J. 1976. Seaweeds: color-coded illustration guide to common marine plants of the East Coast of the United States. Pennsylvania State University Press, State College, Pa. 208 p.

Hinrichs, N. [ed.]. 1972. Population, environment and people. McGraw-Hill Book Co., New York. 226 p.

Hoagland, R. A. 1920. Polychaetous annelids collected by the United States Fisheries steamer "Albatross" during the Philippine Expedition of 1907-1909. Bull. U.S. Nat. Mus. no. 100. Vol. 1, part 9, 603-635.

Hobson, L. A., and D. E. Ketchum. 1974. Observations on subsurface distributions of chlorophyll *a* and phytoplankton carbon in the northeastern Pacific ocean. J. Fish. Res. Bd. Canada **31**(12):1919-1925.

Hock, C. W. 1941. Marine chitin-decomposing bacteria. J. Mar. Res. **4**:99-106.

Hoffman, C. 1956. Untersuchungen über die Remineralization des Phosphores im Plankton. Kieler Meeresforsch. **12**:25-36.

Holdaway, J. L., L. A. Resi, N. A. Thomas, L. R. Parrish, R. K. Stewart, and K. M. Mackenthun. 1967. Temperature and aquatic life. Technical and Advisory Branch, Technical Services Program, Federal Water Pollution Control Administration, U.S. Department of the Interior, Lab. Investigations Series no. 6. 151 p.

Holdgate, M. W. [ed.]. 1970. Antarctic ecology. 2 vols. Academic Press, Inc., New York.

Hollibaugh, J. T. 1976. Biological degradation of arginine and glutamic acid in seawater in relation to growth of phytoplankton. Mar. Biol. **36**(4):303-312.

Holmboe, J., G. E. Forsythe, and W. S. Gustin. 1945. Dynamic meteorology. John Wiley & Sons, Inc., New York.

Holme, N. A. 1949*a* Fauna of sand and mud banks near mouth of Exe estuary. J. Mar. Biol. Ass. U.K. **28**: 189-237.

Holme, N. A. 1949*b*. A new bottom sampler. J. Mar. Biol. Ass. U.K. **28**:323-332.

Holme, N. A. 1950. Bottom fauna of the Great West Bay. J. Mar. Biol. Ass. U.K. **29**:163-183.

Holmes, R. W. 1958. Surface chlorophyll "A," surface primary production and zooplankton volumes in the eastern Pacific Ocean. Rapp. P.-v. Réun. Cons. Perm. Int. Explor. Mer **144**:109-116.

Holmes, R. W. 1960. Physical, chemical, and biological observations in the eastern tropical Pacific Ocean. Scottish Expedition (April-June, 1958). Special Scientific Report on Fisheries, no. 345. Fish and Wildlife Service, U.S. Department of the Interior, Washington, D.C.

Hood, D. W. [ed.]. 1971. Impingement of man on the oceans. Interscience Publishers, Inc., John Wiley & Sons, Inc., New York. 738 p.

Hood, D. W., and K. Park. 1962. Bicarbonate utilization by marine phytoplankton in photosynthesis. Physiol. Plantarum **15**:273-282.

Hoppe, H. G. 1976. Determination and properties of actively metabolizing heterotrophic bacteria in the sea, investigated by means of microautoradiography. Mar. Biol. **36**(4):291-302.

Howard, J. K., and S. Ueyanagi. 1965. Distribution and relative abundance of billfishes (Istiophoridae) of the Pacific Ocean. Studies in tropical oceanography. No. 2. University of Miami Press, Coral Gables, Fla. 134 p.

Howell, A. B. 1930. Aquatic mammals: their adaptations to life in the water. Reprint. Dover Publications Inc., New York. 338 p.

Hoyle, G., and T. Smyth. 1963. Giant muscle fibers in a barnacle *Balanus nubilus* Darwin. Science **139**: 49-50.

Hulbert, E. M., J. H. Ryther, and R. R. C. Guillard. 1960. Phytoplankton of the Sargasso Sea, off Bermuda. J. Cons. Perm. Int. Explor. Mer **25**:115-128.

Hull, C. H. J. 1960. Oxygen balance in an estuary. J. San. Eng. Div., Proc. Amer. Soc. Civ. Eng. **86** (SA6):105-120.

Hulstedt, F. 1956. Marine littoral diatoms from Beaufort, North Carolina. Duke University Mar. Stn. Bull. **6**:1-67.

Hutchinson, G. E. 1953. The concept of pattern in ecology. Proc. Acad. Natur. Sci. (Philadelphia) **105**:1-12.

Hutner, S. H., and J. J. A. McLaughlin. 1958. Poisonous tides. Sci. Amer. **199**(2):92-98.

Hyman, L. H. 1940. The invertebrates. Vol. I. Protozoa through Ctenophora. McGraw-Hill Book Co., New York.

Hyman, L. H. 1951*a*. The invertebrates. Vol. II. Platyhelminthes and Rhynchocoela. McGraw-Hill Book Co., New York.

Hyman, L. H. 1951*b*. The invertebrates. Vol. III. Acanthocephala, Aschelminthes, Entoprocta. McGraw-Hill Book Co., New York.

Hyman, L. H. 1955. The invertebrates. Vol. IV. Echinodermata. McGraw-Hill Book Co., New York.

Hyman, L. H. 1959. The invertebrates. Vol. V. Smaller coelomate phyla. McGraw-Hill Book Co., New York.

Hynes, H. B. N. 1960. Biology of polluted waters. University Press, Liverpool.

Idyll, C. P. 1971. Abyss: the deep sea and the creatures that live in it. Rev. ed. Thomas Y. Crowell Co., New York.

Idyll, C. P. [ed.]. 1972. Exploring the ocean world: a history of oceanography. Rev. ed. Thomas Y. Crowell Co., New York.

Ikard, F. N. 1967. Oily wastes at sea. Science **157**:625.

Inagaki, T., W. Sakamoto, I. Aoki, and T. Kuroki. 1976. Studies on the schooling behavior of fish. III. Mutual relationship between speed and form in the schooling behavior. Bull. Jap. Soc. Sci. Fish. **42** (6):629-636.

Ingle, R. M., and D. P. de Sylva. 1955. The red tide. Educ. Ser. Mar. Lab. Univ. Miami **1**:30.

Inglis, C. C., and F. J. I. Kestner. 1958. Long-term effects of training walls, reclamation, and dredging on estuaries. Min. Proc. Instr. Civ. Eng. **9**:193-216.

Ingmanson, D. E., and W. J. Wallace. 1973. Oceanology: an introduction. Wadsworth Publishing Co., Belmont, Calif. 327 p.

Iselin, C. O. D. 1939. Some physical factors which may influence the productivity of New England's coastal waters. J. Mar. Res. **2**:75-85.

Ivanoff, A. 1953. Au sujet de la diffusion de la lumière par l'eau de mer. Ann. Geophys. **9**:26-27.

Ivanoff, A. 1958*a*. Au sujet de l'utilisation d'un diagramme P-β pour caractériser les masses d'eau océanique. C. R. Acad. Sci. (Paris) **246**:2636-2639.

Ivanoff, A. 1958*b*. Essai d'hydrologie optique entre Nice et la Corse. C. R. Acad. Sci. (Paris) **246**: 3492-3496.

Ivanoff, A. 1959*b*. Optical method of investigation of

the oceans: the P-β diagram. J. Opt. Soc. Amer. **49:**103-104.

Ivanoff, A. 1961. A new water sampler and a new scattering polarizing meter for optical investigations of the oceans. International Oceanographic Congress Preprints. American Association for the Advancement of Science, Washington, D.C.

Ivanoff, A., N. Jerlov, and T. H. Waterman. 1961. Comparative study of irradiance, beam transmittance, and scattering in the sea near Bermuda. Limnol. Oceanogr. **6**(2):129-148.

Ivanoff, A., and T. H. Waterman. 1958*a*. Elliptical polarization of submarine illumination. J. Mar. Res. **1:**255-282.

Ivanoff, A., and T. H. Waterman. 1958*b*. Factors, mainly depth and wavelength, affecting the degree of underwater light polarization. J. Mar. Res. **16:** 283-307.

Iversen, E. S. 1968. Farming the edge of the sea. Fishing News (Books) Ltd., London.

Jackson, S. W. [ed.]. 1971. Man and the environment. William C. Brown Co., Publishers, Dubuque, Iowa.

Jacobs, M. B., and M. E. Ewing. 1969. Suspended particulate matter: concentration in the major oceans. Science **163:**380-383.

Japanese Hydrographic Office. 1956. Plankton observation. Equapac. Expedition. Unpublished report.

Jaschnow, W. A. 1939*a*. Plankton productivity in the southwestern part of the Barents Sea. Trans. Inst. Mar. Fish. Oceanogr. U.S.S.R. **4:**201-224.

Jaschnow, W. A. 1939*b*. Reproduction and seasonal variations in distribution of different stages of *Calanus finmarchicus* of the Barents Sea. Trans. Inst. Mar. Fish. Oceanogr. U.S.S.R. **4:**225-244.

Jaschnow, W. A. 1940. On the plankton productivity in the northern seas of U.S.S.R. Société des Naturalistes de Moscou, Moscow. 84 p.

Jerlov, N. G. 1951. Optical studies of ocean waters. (Reports of the Swedish Deep-Sea Expedition.) Phys. Chem. **3:**1-59.

Jerlov, N. G. 1953. Particle distribution in the ocean. (Reports of the Swedish Deep-Sea Expedition.) Phys. Chem. **3:**73-97.

Jerlov, N. G. 1955*a*. Factors influencing the transparency of the Baltic waters. Medd. Oceanogr. Inst. Göteborg **25:**1-19.

Jerlov, N. G. 1955*b*. Particulate matter in the sea as determined by means of the Tyndall meter. Tellus **7:**218-225.

Jerlov, N. G. 1959. Maxima in the vertical distribution of particles in the sea. Deep-Sea Res. **5:**173-184.

Jerlov, N. G., and F. Koczy. 1951. Photographic measurements of daylight in deep water. (Reports of the Swedish Deep-Sea Expedition.) Phys. Chem. **3:**1-59, 61-69.

Jerlov, N. G., and J. Piccard. 1959. Bathyscaphe measurements of daylight penetration into the Mediterranean. Deep-Sea Res. **5:**201-204.

Jespersen, P. 1924. On the quantity of macroplankton in the Mediterranean and Atlantic. Int. Rev. Hydrobiol. Hydrog. **12:**102-115.

Jespersen, P. 1935. Quantitative investigations in the distribution of macroplankton in different oceanic regions. Dana. Rep. no. 7. 44 p.

Jespersen, P. 1940. Investigations on the quantity and distribution of zooplankton in Icelandic waters. Medd. Komm. Danm. Fisk. Havund. (Ser. Plankton) **3**(5):1-77.

Jespersen, P. 1944. Investigations on the food of herring and the macroplankton in the waters around the Faeroes. Medd. Komm. Danm. Fisk. Havund. (Ser. Plankton) **3**(7):1-44.

Jespersen, P., and M. F. S. Russell. 1939-1952. Fiches d'identification du zooplancton. J. Cons. Perm. Int. Explor. Mer, no. 1-49.

Jitts, H. R. 1957. The C [4] method for measuring CO uptake in marine productivity studies. C.S.I.R.O. Aust. Div. Fish. Oceanogr. Rep. **8:**1-12.

Jitts, H. R., and B. D. Scott. 1961. Determination of zero-thickness activity in Geiger counting of C [4] solutions used in marine productivity studies. Limnol. Oceanogr. **6:**116-123.

Johns, W. 1968. Estuaries: America's most vulnerable frontiers. American Wildlife Federation. Washington, D.C. 16 p.

Johnson, C. E. 1970. Ecocrisis. John Wiley & Sons, Inc., New York. 182 p.

Johnson, M. W. 1939. The correlation of water movements and the dispersal of pelagic larval stages of certain littoral animals, especially the sand crab, *Emerita*. J. Mar. Res. **2:**236-245.

Johnson, M. W. 1942. Notes on zooplankton. Records of observations. Scripps Institution of Oceanography. Vol. I. Oceanographic observations on E. W. Scripps cruises of 1938. University of California Press, Berkeley.

Johnson, M. W. 1943. Underwater sounds of biological origin. Division of Water Research, report no. U28. University of California Press, Berkeley. 20 p.

Johnson, M. W. 1948. Sound as a tool in marine ecology. J. Mar. Res. **7:**433-458.

Johnson, M. W. 1949. Zooplankton as an index of water between Bikini Lagoon and the open sea. Trans. Amer. Geophys. Union **30:**238-244.

Johnson, M. W. 1954. Plankton of the northern Marshall Islands. Geological survey. Professional paper no. 260-F. Part 2. F. 301-314. Washington, D.C.

Johnson, M. W. 1958. Observations on inshore plankton collected during summer, 1957, at Point Barrow, Alaska, J. Mar. Res. (Thompson Anniversary vol.) **17:**272-281.

Johnson, M. W., F. A. Everest, and R. W. Young. 1947. Role of snapping shrimp (*Crangon* and *Synalpheus*) in production of underwater noise in the sea. Biol. Bull. Mar. Biol. Lab. Woods Hole **93:**122-138.

Johnson, M. W., and R. C. Miller. 1935. Seasonal settlement of shipworms, barnacles, and other wharf-piling organisms at Friday Harbor, Washington. Univ. Washington Pub. Oceanogr. **2:**1-18.

Johnson, T. W. 1968. Saprobic marine fungi. *In* G. C. Ainsworth and A. S. Sussman [eds.]. The fungi. Vol. 3. Academic Press, Inc., New York.

Johnson, T. W., and F. K. Sparrow. 1962. Fungi in oceans and estuaries. Hafner Publishing Co., Inc., New York.

Johnston, J. W., D. G. Moulton, and A. Turk [eds.]. 1970. Communication by chemical signals. I. Advances in chemoreception. Appleton-Century-Crofts, New York.

Johnston, R. 1955. Biologically active compounds in the sea. J. Mar. Biol. Ass. U.K. **34**:185-195.

Johnston, R. 1959. Antimetabolites and marine algae. International Oceanographic Congress Preprints. American Association for the Advancement of Science, Washington, D.C.

Johnstone, J. 1908. Conditions of life in the sea. Cambridge University Press, New York. 332 p.

Johnstone, J., A. Scott, and H. C. Chadwick. 1924. The marine plankton, with special reference to investigations made at Port Erin, Isle of Man; a handbook for students and amateur workers. University Press of Liverpool, Ltd.; Hodder & Stoughton, Ltd., London.

Jones, G. E. 1959. Biologically active organic substances in seawater. International Oceanographic Congress Preprints. American Association for the Advancement of Science, Washington, D.C.

Jones, N. S. 1950. Marine bottom communities. Biol. Rev. **25**:283-313.

Jones, N. S. 1956. Fauna and biomass of muddy sand deposit off Port Erin. J. Anim. Ecol. **25**:217-252.

Jones, O. A., and R. Endean [eds.] 1973-1976. Biology and geology of coral reefs. 4 vols. Academic Press, Inc., New York and London. I, Geology, 1973, 400 p.; II, Biology 1, 1974, 494 p.; III, Biology 2, 1975, 486 p.; IV, Biology 3, 1976.

Jones, W. E., and A. Demetropoulos. 1968. Exposure to wave action: measurements of an important ecological parameter on rocky shores of Anglesey. J. Exp. Mar. Biol. Ecol. **2**(1):46-63.

Jordon, D. S. 1905. A guide to the study of fishes. 2 vol. Henry Holt, New York.

Jørgensen, C. B. 1966. Biology of suspension feeding. Pergamon Press. Inc., New York.

Juday, C. 1943. The utilization of aquatic resources. Science **97**:456-458.

June, F. C., and J. L. Chamberlin. 1959. Role of the estuary in the life history and biology of Atlantic menhaden. Proceedings of the Gulf and Carribbean Fisheries Institute, 11th annual session, 1958.

Kalber, F. A. 1959*a*. Hypothesis on role of tide marshes in estuarine productivity. Estuarine Bull. **4**:3, 14-15.

Kalber, F. A. 1959*b*. Where does the shoreline begin? Delaware Conserv. **3**(3):4-6.

Kalber, F. A. 1960. Pilot study of nutrients in Delaware River estuary. Project F-13-R-3. Ref. 60-7. Federal Aid to Fish and Wildlife Recovery. University of Delaware Marine Laboratory.

Kalber, F. A., and F. G. Walton-Smith [eds.]. 1974. CRC Handbook of marine science. CRC Press, Cleveland.

Kalle, K. 1957. Chemische Untersuchungen in der Irmiger See im Juni 1955. Ber. Deut. Komm. Meereforsch. **14**:313-328.

Kampa, E. M., and B. P. Boden. 1956. Light generation in an ionic scattering layer. Deep-Sea Res. **4**:73-92.

Kandler, R. I. 1950. Jahreszeitliches Vorkommen und unperiodisches Auftreten von Fischbrut, Medusen und Dekapod-Larven im Fehmernbelt in den Jahren 1934-1943. Ber. Deut. Komm. Meeresforsch. **12**: 47-85.

Kanwisher, J. W. 1971. Temperature regulation in the sea. *In* V. G. Dethier and others [eds.]. Topics in the study of life: the BIO source book. Harper & Row, Publishers, New York.

Kelley, D. W. 1966. Ecological studies of the Sacramento–San Joaquin estuary. California Fish and Game, Fisheries Bull. no. 133.

Ketchum, B. H. 1947. Biochemical relations between marine organisms and their environment. Ecol. Monogr. **17**:309-315.

Ketchum, B. H. 1957. The effects of the ecological system on the transport of elements in the sea. Pub. no. 551 (pp. 52-59). National Academy of Sciences–National Research Council.

Ketchum, B. H. 1972. The water's edge. Critical problems of the coastal zone. The M.I.T. Press, Cambridge, Mass. 393 p.

Ketchum, B. H., J. C. Ayers, and R. F. Vaccaro. 1952. Processes contributing to the decrease of coliform bacteria in a tidal estuary. Ecology **33**:247-258.

Ketchum, B. H., C. L. Carey, and M. Briggs, 1949. Preliminary studies on the viability and dispersal of coliform bacteria in the sea: limnological aspects of water supply and waste disposal. American Association for the Advancement of Science, Washington, D.C.

Ketchum, B. H., N. Corwin, and D. J. Keen. 1955. The significance of organic phosphorus determinations in ocean waters. Deep-Sea Res. **2**:172-181.

Ketchum, B. H., and D. J. Keen. 1948. Unusual phosphorus concentrations in the Florida "red tide" seawater. J. Mar. Res. **7**:17-21.

Ketchum, B. H., J. H. Ryther, C. S. Yentsch, and N. Corwin. 1958. Productivity in relation to nutrients. Rapp. P.-v. Réun. Cons. Perm. Int. Explor. Mer **144**:132-140.

Ketchum, B. H., and D. H. Shonting. 1958. Optical studies of the particulate matter in the sea. Ref. no. 58-15. Woods Hole Oceanographic Institute.

Keys, A., E. H. Christensen, and A. Krogh. 1935. The organic metabolism of seawater, with special reference to the ultimate food cycle in the sea. J. Mar. Biol. Ass. U.K. **20**:181-196.

Kielhorn, W. V. 1952. The biology of surface zone zooplankton of a boreo-arctic ocean area. J. Fish. Res. Bd. Canada **9**:223-264.

Kincaid, T. 1942. Biotic and economic relations of plankton. California Fish Game **28**:210-215.

King, J. E., T. S. Austin, and M. S. Doty. 1957. Preliminary report on Expedition East Tropic. Special Scientific Report on Fisheries, no. 201. Fish and Wildlife Service, U.S. Department of the Interior. 155 p.

King, J. E., and J. Demond. 1953. Zooplankton abundance in the Central Pacific. U.S. Fish Wild. Serv. Fish. Bull. **54:**111-114.

King, J. E., and T. S. Hida. 1957. Zooplankton abundance in Hawaiian waters, 1953-1954. Special Scientific Report on Fisheries no. 221. Fish and Wildlife Service, U.S. Department of the Interior.

Kinne, O. [ed.]. 1970-1972. Marine ecology: a comprehensive, integrated treatise on life in oceans and coastal waters. Interscience Publishers, Inc., John Wiley & Sons, Inc., New York. 5 vols. (1, Environmental factors; 2, Physiological mechanisms; 3, Cultivation; 4, Dynamics; and 5, Ocean management).

Kitching, J., T. Macan, and H. Gilson. 1934. Studies in sublittoral ecology. I. A submarine gully in Wemburg, South Devon. J. Mar. Biol. Ass. U.K. **19**(2): 667-705.

Knight-Jones, E. W., and S. Z. Quasim. 1955. Responses of some marine plankton animals to changes in hydrostatic pressure. Nature (London) **175:**941.

Knox, G. A. 1963. The biogeography and intertidal ecology of the Australian coasts. Oceanogr. Mar. Biol. Annu. Rev. **1:**341-404.

Knudsen, N. 1901. Hydrographic tables. 2nd ed. G. E. C. Gad, Copenhagen. 63 p.

Knudsen, V. O., R. S. Alford, and J. W. Emling. 1948. Underwater ambient noise. J. Mar. Res. (Sverdrup 60th Anniversary vol.) **7:**410-429.

Kofoid, C. A., and A. S. Campbell. 1929. A conspectus of the marine and fresh water ciliata belonging to the suborder Tintinnoinea, with descriptions of new species principally from the Agassiz Expedition to the Eastern Tropical Pacific. Univ. California Pub. Zool. **34:**1-403.

Kofoid, C. A., and R. C. Miller. 1927. Marine borers and their relation to marine construction on the Pacific coast. The San Francisco Bay Marine Piling Committee, San Francisco.

Kofoid, C. A., and O. Swezy. 1921. The free-living unarmored Dinoflagellata. Memoirs University of California. Vol. 5. University of California Press, Berkeley. 562 p.

Kohlmeyer, J., and E. Kohlmeyer. 1964. Synoptic plates of higher marine fungi. 2nd ed. Weinheim J. Cramer, New York.

Kohn, A. J., and P. Helfrich. 1957. Primary organic productivity of a Hawaiian coral reef. Limnol. Oceanogr. **2:**241-251.

Koltun, V. M. 1969. Sponges of the Arctic and Antarctic: a faunistic review. *In* Biology of the Porifera. Zoological Society of London and Academic Press, Inc., New York.

Komai, T. 1951. The nematocysts in the ctenophore *Euchlora rubra.* Amer. Nat. **85:**73.

Kon, S. K. 1958. Some thoughts on biochemical perspectives in marine biology. *In* A. A. Buzzati-Traverso [ed.]. Perspectives in marine biology. University of California Press, Richmond.

Korringa, P. 1976. Developments in aquaculture and fisheries science. Vol. 1. Farming marine organisms low in the food chain. xiv + 264 p.; vol. 2. Farming the cupped oysters of the genus *Crassostrea.* x + 244 p.; vol. 3. Farming the flat oysters of the genus *Ostrea.* xiv + 238 p.; vol. 4. Farming marine fishes and shrimps. xii + 208 p. Elsevier Scientific Publishing Co., Amsterdam.

Kort, V. G. 1962. The Antarctic Ocean. Sci. Amer. **207**(3):113.

Kow, Tham Ah. 1953*a*. Plankton calendar of Singapore Strait, with suggestions for a simplified methodology. Section II. Proceedings of the 4th meeting. Tudo-Pacific Fisheries Council. Quezon City, The Philippines.

Kow, Tham Ah. 1953*b*. A preliminary study of the physical, chemical, and biological characteristics of Singapore Strait. Colonial Office. Fish. Pub. **1**(4): 1-65.

Kozloff, E. N. 1973. Seashore life of Puget Sound, the Strait of Georgia, and the San Juan Archipelago. University of Washington Press, Seattle.

Kramp, P. L. 1961. Synopsis of the medusae of the world. J. Mar. Biol. Ass. U.K. Vol. 40. Cambridge University Press, London.

Krenkel, P. A., and F. L. Parker [eds.]. 1969. Biological aspects of thermal pollution. Vanderbilt University Press, Portland. 407 p.

Kreps, E. 1934. Organic catalysts or enzymes in sea water (James Johnston Memorial Volume). University of Liverpool Press, 193 p.

Kreps, E., and N. Verjbinskaya. 1930. Seasonal changes in the phosphate and nitrate content and in hydrogen ion concentration in the Barents Sea. J. Cons. Perm. Int. Explor. Mer **5:**327-346.

Krey, J. 1958*a*. Chemical determinations of net plankton, with special reference to equivalent albumin content. J. Mar. Res. (Thompson Anniversary vol.) **17:**312-324.

Krey, J. 1958*b*. Chemical methods of estimating the standing crop of phytoplankton. Rapp. P.-v. Réun. Cons. Perm. Int. Explor. Mer **144:**20-27.

Krey, J. 1961. Der Detritus im Meere. J. Cons. Perm. Int. Explor. Mer **26:**263-280.

Kriss, A. E. 1959. Microbiology and the chief problems in the Black Sea. Deep-Sea Res. **5:**193-200.

Kriss, A. E., I. E. Mishustina, N. Mitskevich, and S. S. Abyzov. 1961. Microorganisms as hydrological indicators in seas and oceans. IV. Deep-Sea Res. **7:**225-236.

Kriss, A. E., I. E. Mishustina, N. Mitskevich, and E. V. Zemstova. 1964. Microbial population of oceans and seas. 1st Eng. transl. by K. Syers; edited by G. E. Fogg. 1967. Edward Arnold (Publishers) Ltd., London.

Kriss, A. E., and E. A. Rukina. 1952. Biomass of microorganisms and their rates of reproduction in oceanic depths. Zh. Obshch. Biol. **12:**349-362.

Krizenecky, J. 1925. Untersuchungen über die Assimilation Fähigheit der Wassertiere für im Wasser gelöste Nahrstoff. Biol. Gen. **1:**79-149.

Krogh, A. 1931. Dissolved substances as food of aquatic organisms. Rapp. P.-v. Réun. Cons. Perm. Int. Explor. Mer **75:**7-26. (Also, Biol. Rev. **6:** 412-442.)

Krogh, A. 1934*a*. Conditions of life in the ocean. Ecol. Monogr. **4:**421-429.

Krogh, A. 1934*b*. Life at great depths in the ocean. Ecol. Monogr. **4:**430-439.

Krogh, A. 1939. Osmotic regulations in aquatic animals. Cambridge University Press, New York. 242 p.

Krogh, A., and R. Sparck. 1936. A new bottom sampler for investigation of the micro-fauna of the sea bottom. Kongl. Danske. Vidensk. Selsk. Biol. Medd. **13:**1-12.

Krumbein, W. C. 1936. Application of logarithmic moments to size-frequency distributions of sediments. J. Sed. Petrology **6:**35-47.

Kuenen, P. H. 1941. Geochemical calculations concerning the total mass of sediments in the earth. Amer. J. Sci. **39:**161-190.

Kuenen, P. H. 1950. Marine geology. John Wiley & Sons, Inc., New York.

Kunne, C. 1935. Die Verbreitung der grosseren Planktontiere (ausser Fischbrut). *In* W. Mielck and C. Kunne. 1935. Helgoland Wiss. Meeresunters. **19**(7):62-118.

Kunne, C. 1950. Die Nahrung der Meerstiere. II. Das Plankton. *In* Handbuch der Seefischerei Nordeuropas. Verlag Stuttgart.

Kusmorskaja, A. 1940. Changements saisonnières du plancton de la Mer d'Okhotsk. Bull. MOIP, Biol. **49:**3-4. (F.s.)

Kuznetsov, S. 1955. Use of radioactive isotopes for photosynthesis and chemosynthesis study in bodies of water. Papers presented by U.S.S.R. at International Conference on Peaceful Uses of Atomic Energy.

Lack, D. 1945. The ecology of closely related species, with special reference to the cormorant *(Phalacrocorax carbo)* and shag *(P. aristotelis)*. J. Anim. Ecol. **14**(1):12-16.

Laevastu, P. 1959. A review of marine pollution and the work of the F.A.O., Biology Branch, in this subject. Food and Agriculture Organization. Rome (mimeographed). 58 p.

Lane, F. W. 1957. Kingdom of the octopus; the life history of the Cephalopoda. Jarrolds Publishers, Ltd., London.

Lang, K. 1948. Contribution to the ecology of *Priapulus caudatus*. Lam. Ark. Zool. **41**(5):1-12.

Langmuir, I. 1938. Surface motion of water induced by wind. Science **87:**119-123.

Lanyon, W. E., and W. N. Tavolga. 1960. Animal sounds and communications. Hafner Publishing Co., Inc., New York.

Lappé, F. M., and J. Collins. 1977. Food first: beyond the myth of scarcity. Houghton Mifflin Co., Boston. 466 p.

Larson, R. L., and F. N. Spiess. 1969. East Pacific Rise: a near-bottom geophysical profile. Science **163:**68-70.

Lauff, G. H. [ed.]. 1967. Estuaries. (Symposium, Mar.-Apr., 1964.) American Association for the Advancement of Science, Washington, D.C. 733 p.

Lawson, G. W. 1954. Rocky shore zonation on the Gold Coast. J. Ecol. **44**(1):153-170.

Lawson, G. W. 1955. Rocky shore zonation in the British Cameroons. J. W. Afr. Sci. Ass. **1**(2):78-88.

Lawson, G. W. 1957*a*. Seasonal variation of intertidal zonation on the coast of Ghana in relation to tidal factors. J. Ecol. **45:**831-860.

Lawson, G. W. 1957*b*. Some features of intertidal ecology of Sierra Leone. J. W. Afr. Sci. Ass. **3**(2): 166-174.

Lea, H. E. 1955. The chaetognaths of western Canadian coastal waters. J. Fish. Res. Bd. Canada **12:** 593-617.

Leavitt, B. B. 1935. Quantitative study of vertical distribution of larger zooplankton in deep water. Biol. Bull. Mar. Biol. Lab. Woods Hole **68:**115-130.

Leavitt, B. B. 1938. Quantitative vertical distribution of macroplankton in the Atlantic Ocean Basin. Biol. Bull. Mar. Biol. Lab. Woods Hole **74:**376-394.

Le Boeuf, B. J., and R. S. Peterson. 1969. Social status and mating activity in elephant seals. Science **163:**91-93.

Lebour, M. V. 1917. Microplankton of Plymouth Sound from the region beyond the breakwater. J. Mar. Biol. Ass. U.K. **11:**133-182.

Lebour, M. V. 1922, 1923. The food of plankton organisms. J. Mar. Biol. Ass. U.K. **12:**644-677 (part I); **13:**70-92 (part II).

Lebour, M. V. 1930. The planktonic diatoms of northern seas. Ray Society, London. 244 p.

Lee, M. O. [ed.]. 1964-1965. Biology of the Antarctic seas. *In* G. A. Llano [ed.]. Vol. 2. American Geophysical Union, National Academy of Sciences–National Research Council. Pub. nos. 1190, 1297.

Legare, J. E. H. 1957. Qualitative and quantitative distribution of plankton in the Strait of Georgia in relation to certain oceanographic factors. J. Fish. Res. Bd. Canada **14:**551-552.

Legaud, M. 1957. Orsom III. Résultats biologiques de l'expédition Equapac. Institute Français d'Océanie. Rapp. Sci. no. 1 (processed report).

Leighton, D., and R. A. Boolootian. 1963. Diet and growth in the black abalone, *Haliotis cracherodii*. Ecology **44:**227-238.

Leis, J. M., and J. M. Miller. 1976. Offshore distributional patterns of Hawaiian fish larvae. Mar. Biol. **36**(4):359-368.

Lewin, J., and J. A. Hellebust. 1976. Heterotrophic nutrition of the marine pennate diatom *Nitzschia angularis* var. *affinis*. Mar. Biol. **36**(4):313-320.

Lewin, R. A. 1954. A marine *Stichococcus* sp. which

requires vitamin B_{12} (cobalamin). J. Gen. Microbiol. **10**(1):92-96.

Lewis, J. B. 1976. Experimental tests of suspension feeding in Atlantic reef corals. Mar. Biol. **36**(2): 147-150.

Lewis, J. R. 1964. The ecology of rocky shores. English Universities Press, Ltd., London.

Light, S. F. 1975. Intertidal invertebrates of the central California coast. 3d ed. (S. F. Light's laboratory and field text in invertebrate zoology, revised and edited by R. I. Smith and J. T. Carlton.) University of California Press, Berkeley. 716 p.

Lillick, L. C. 1938. Preliminary report on the phytoplankton of the Gulf of Maine. Amer. Midl. Natur. **20:**624-640.

Lillick, L. C. 1940. Phytoplankton and planktonic protozoa of the offshore waters of the Gulf of Maine. II. Qualitative composition of the planktonic flora. Trans. Amer. Phil. Soc. **31:**193-237.

Limbaugh, C. 1961. Cleaning symbiosis. [Prepared by Howard Feder from field notes of the late Conrad Limbaugh.] Sci. Amer. **205**(2):42-49.

Lindberg, G. U. 1959. A list of the fauna of sea waters of south Sakhalin and south Kurile Islands: investigation of Far Eastern seas [in Russian]. C. R. Acad. Sci. (U.S.S.R.) **6:**173-256.

Lipman, C. B. 1926. The concentration of seawater as affecting its bacterial population. J. Bacteriol. **12:**311-313.

Lippson, J. A. [ed.]. 1973. The Chesapeake Bay in Maryland: an atlas of natural resources. The Johns Hopkins University Press, Baltimore.

Loeblich, A. R., and A. R. Loeblich, Jr. 1967. Index to the genera, subgenera and sections of the Pyrrhophyta. I. Studies in tropical oceanography. University of Miami Press, Coral Gables, Fla. 94 p.

Loeblich, A. R., and A. R. Loeblich, Jr. 1968. Index to the genera, subgenera and sections of the Pyrrhophyta II. J. Paleontol. **42**(1):210-213.

Loftas, T. 1969. The last resource—man's exploitation of the sea. H. Hamilton, London.

Lohmann, H. 1908. Üntersuchungen zur Festellung des Vollständigen Gehaltes des Meeres an Plankton. Wiss. Meeresuntersuch. Abt. Kiel. **10:**129-370.

Lohmann, H. 1911. Über das Nannoplankton und die Zentrifugierung. Int. Rev. Hydrobiol. Hydrogr. **4:**1-38.

Lohmann, H. 1920. Die Bevölkerung des Ozeans mit Plankton nach den Ergebnissen der Zentrifugensfänge wahrend der Ausreise der "Deutschland," 1911. Zugleich ein Beitrag zur Biologie des Atlantischen Ozeans. Arch. Biontol. **4**(3):1-617.

Longhurst, A. R. 1959. Benthos densities off tropical West Africa. J. Cons. Perm. Int. Explor. Mer **25:**21-28.

Loosanoff, V. L., and H. C. Davis. 1951. Delayed spawning of lamellibranchs by low temperature. J. Mar. Res. **10:**197-202.

Loosanoff, V. L., H. B. Engle, and C. A. Nomejka. 1955. Differences in intensity of settling of oysters and starfish. Biol. Bull. **109:**75-81.

Lotka, A. J. 1925. Elements of physical biology. The Williams & Wilkins Co., Baltimore. 460 p. (Republished in 1956 by Dover Publications, New York, under the title: Elements of mathematical biology.)

Love, A., and D. Love [eds.]. 1963. North Atlantic biota and their history. (Symposium at the University of Iceland.) Pergamon Press, Inc., New York.

Love, R. M. 1970. The chemical biology of fishes. Academic Press, Inc., New York. 547 p.

Loye, D. P., and D. A. Proudfoot. 1946. Underwater noise due to marine life. J. Acoust. Soc. Amer. **18**(2):446-449.

Lucas, C. E. 1936. On certain interactions between phytoplankton and zooplankton under experimental conditions. J. Cons. Perm. Int. Explor. Mer **11:** 343-361.

Lucas, C. E. 1938. Some aspects of integration in plankton communities. J. Cons. Perm. Int. Explor. Mer **13:**309-321.

Lucas, C. E. 1947. Ecological effects of external metabolites. Biol. Rev. **22:**270-295.

Lucas, C. E. 1949. External metabolites and ecological adaptation. Symp. Soc. Exp. Biol. **111:**336-356.

Lucas, C. E. 1955. External metabolites in the sea. (Papers in Marine Biology and Oceanography.) Deep-Sea Res. (supp.) **3:**139-148.

Lucas, C. E. 1956. Plankton and basic production. *In* M. Graham [ed.]. Sea fisheries: their investigation in the United Kingdom. Edward Arnold Publishers, Ltd., London.

Lucas, C. E. 1961. Interrelationships between aquatic organisms mediated by external metabolites. *In* M. Sears [ed.]. Oceanography. Pub. no. 67. American Association for the Advancement of Science, Washington, D.C.

Luck, J. M. 1957. Man against his environment: the next 100 years. Science **126:**903.

Lund, E. J. 1936. Some facts relating to the occurrence of dead and dying fish on the Texas coast during June, July, and August, 1935. Annual Reports, Texas Game, Fish, and Oyster Commission, 1934-1935, pp. 47-50.

MacArthur, R. H., and J. H. Connell. 1966. The biology of populations. John Wiley & Sons, Inc., New York. 226 p.

MacGinitie, G. E. 1939. Littoral marine communities. Amer. Midl. Natur. **21:**28-55.

MacGinitie, G. E., and N. MacGinitie. 1949. Natural history of marine animals. McGraw-Hill Book Co., New York.

Mackintosh, N. A. 1934. Distribution of macroplankton in the Atlantic sector of the Antarctic. 'Discovery' Rep. **9:**67-158.

Mackintosh, N. A. 1937. Seasonal circulation of Antarctic macroplankton. 'Discovery' Rep. **16:** 367-412.

Malins, D. C. [ed.] 1977. Effects of petroleum on arctic and subarctic marine environments and organisms. Academic Press, Inc., New York. 1. Nature and fate of petroleum. 324 p. 2. Biological effects. 512 p.

Mankowski, W. 1948. Macroplankton investigations in the Gulf of Gdańsk in June and July, 1946. Ann. Biol. (Copenhagen) **3:**110.

Manning, R. B. 1969. Stomatopod crustacea of the western Atlantic. Studies in tropical oceanography. No. 8. University of Miami Press, Coral Gables, Fla. 380 p.

Manter, H. W. 1966. Parasites of fishes as biological indicators of recent and ancient conditions. *In* J. E. McCauley [ed.]. Host-parasite relationships. Oregon State University Press, Corvallis, Ore. pp. 59-72.

Manteufel, B. P. 1941. Plankton and herring in the Barents Sea. Trans. Knipovitch Pol. Sci. Inst. Sea Fish., Oceanogr. 7 (Russian).

Marcus, E., and E. Marcus. 1967. American opisthobranch mollusks. No. 6. Studies in tropical oceanography. University of Miami Press, Coral Gables, Fla.

Marcus, H. S., J. E. Short, J. C. Kuypers, and P. O. Roberts. 1976. Federal port policy in the United States. The M.I.T. Press, Cambridge, Mass. 376 p.

Mare, M. F. 1940. Plankton production of Plymouth and the mouth of the English Channel in 1935. J. Mar. Biol. Ass. U.K. **24:**461-482.

Mare, M. F. 1942. A marine benthic community, with special reference to the microorganisms. J. Mar. Biol. Ass. U.K. **25:**517-554.

Margalef, R. 1958. Temporal succession and spatial heterogeneity in phytoplankton. *In* A. A. Buzzati-Traverso [ed.]. Perspectives in marine biology. University of California Press, Richmond.

Margalef, R. 1961. Communication of structure in planktonic populations. Limnol. Oceanogr. **6:** 124-128.

Mariscal, R. N. [ed.]. 1974. Experimental marine biology. Academic Press, Inc., New York. 384 p.

Marshall, E. K., and H. W. Smith. 1930. The glomerular development of the vertebrate kidney in relation to habitat. Biol. Bull. **59:**125-153.

Marshall, N. B. 1954. Aspects of deep-sea biology Hutchinson's Scientific & Technical Publications, London.

Marshall, N. B. 1965. The life of fishes. George Weidenfield & Nicolson, Ltd., Publishers, London.

Marshall, N. B. [ed.]. 1967. Aspects of marine zoology. Symposium no. 19 of the Zoological Society of London. Academic Press, Inc., New York.

Marshall, P. T. 1958. Primary production in the Arctic. J. Cons. Perm. Int. Explor. Mer **23:**173-177.

Marshall, S. M. 1925. A survey of Clyde plankton. Proc. Roy. Soc. Edinb. [B] **45:**117-141.

Marshall, S. M. 1933. The production of microplankton in the Great Barrier Reef region. Great Barrier Reef Expedition. Sci. Rep. Brit. Mus. (N.H.) **2:** 111-158.

Marshall, S. M. 1948. Further experiments on the fertilization of a sea loch (Loch Craiglin). J. Mar. Biol. Ass. U.K. **27:**360-379.

Marshall, S. M., A. G. Nicholls, and A. P. Orr. 1933-1934. Biology of *Calanus finmarchicus*. I-IV. J. Mar. Biol. Ass. U.K. **19:**111-138; **20:**1-28.

Marshall, S. M., and A. P. Orr. 1927. Relation of plankton to some chemical and physical factors in the Clyde Sea area. J. Mar. Biol. Ass. U.K. **14:** 837-868.

Marshall, S. M., and A. P. Orr. [eds.]. 1953. Essays in marine biology. Oliver & Boyd, London.

Marshall, S. M., and A. P. Orr. 1955. The biology of a marine copepod, *Calanus finmarchicus* Gunnerus. Oliver & Boyd, Ltd., London. 188 p.

Martin, D. F. 1968. Marine chemistry: analysis of seawater. Marcel Dekker, Inc., New York.

Matsunaga, K. 1976. Estimation of variation of mercury concentration in the oceans during the last several decades. J. Oceanogr. Soc. Japan **32**(1):48-55.

Matthews, G. V. T. 1955. Bird navigation. Cambridge University Press, New York.

Mauchline, J., and L. R. Fisher. 1969. Biology of euphausiids. *In* F. S. Russell [ed.]. Advances in marine biology. Vol. 7. Academic Press, Inc., New York.

Mayer, A. G. 1910. Medusae of the World. Carnegie Institution, Washington, D.C. Vol. 1, pp. 1-230; vol. 2, pp. 231-498; vol. 3, pp. 499-735.

Mayor, A. G. 1924. Structure and ecology of Samoan reefs. Carnegie Institute papers. Dep. Mar. Biol. **19:**1-25.

McAllister, C. D., T. R. Parsons, and J. D. H. Strickland. 1960. Primary productivity and fertility at Station "P" in the northeastern Pacific Ocean. J. Cons. Perm. Int. Explor. Mer **25:**240-259.

McClane, A. J. [ed.]. 1965. McClane's Standard fishing encyclopedia and international angling guide. Holt, Rinehart & Winston, Inc., New York.

McClendon, J. F. 1917. Standardization of a new colorimetric method for determination of CO_2 tension and CO_2 and O_2 of seawater. J. Biol. Chem. **30:**256-288.

McConnaughey, B. H. 1949. Mesozoa of the family Dicyemidae from California. Univ. California Pub. Zool. **55:**1-34.

McConnaughey, B. H. 1951. The life cycle of the dicyemid mesozoa. Univ. California Pub. Zool. **55:**295-336.

McConnaughey, B. H. 1964. The determination and analysis of plankton communities. *In* G. Rahardjo [ed.]. Penelitian Laut di Indonesia. (Special number.) 40 p.

McConnaughey, B. H. 1966. The mesozoa. *In* E. L. Dougherty and others [eds.]. The lower invertebrates. University of California Press, Richmond.

McConnaughey, B. H. 1968. The Mesozoa. *In* M. Florkin and B. T. Sheer [eds.]. Chemical zoology. Vol. II. Academic Press, Inc., New York.

McConnaughey, B. H., and D. L. Fox. 1949. The anatomy and biology of the marine polychaete *Thoracophelia mucronata* (Treadwell) Opheliidae. Univ. California Pub. Zool. **47:**319-340.

McConnaughey, B. H., and E. I. McConnaughey.

1954. Strange life of the dicyemid mesozoans. Sci. Monthly **79:**277-284.

McGowan, J. A. 1974. The nature of oceanic ecosystems. *In* C. B. Miller [ed.]. The biology of the oceanic Pacific. Oregon State University Press, Corvallis.

McGowan, J. A., and T. L. Hayward. 1978. Mixing and oceanic productivity. Deep Sea Res. (In press.)

McIntyre, J. [ed.]. 1974. Mind in the waters. Charles Scribner's Sons, New York. 240 p.

McLaren, I. A. 1966. Adaptive significance of large size and long life of the chaetognath *Sagitta elegans* in the Arctic. *In* P. S. Dawson and C. E. King [eds.]. Readings in population biology. Prentice-Hall, Inc., Englewood Cliffs, N.J. 1971.

McMurrich, J. P. 1916. Notes on plankton of the British Columbia coast. Trans. Roy. Soc. Canada (ser. 3) **10:**75-85.

Meglitsch, P. A. 1967. Invertebrate zoology. Oxford University Press, New York.

Menard, H. W. 1961. The East Pacific Rise. Sci. Amer. **205**(6):52.

Menard, H. W. 1964. Marine geology of the Pacific. McGraw-Hill Book Co., New York.

Menard, H. W. 1967. Sea-floor spreading, topography, and the second layer. Science **157:**923-924.

Menon, K. S. 1931. Preliminary account of the Madras plankton. Rep. Indian Mus. **33:**489-516.

Menon, K. S. 1945. Observations on the seasonal distribution of the plankton of the Trivandrum coast. Proc. Indian Acad. Sci. **22B**(2): 31-62.

Menzel, D. W., and J. H. Ryther. 1960. Annual cycle of primary production in the Sargasso Sea off Bermuda. Deep-Sea Res. **6:**351-367.

Menzel, D. W., and J. H. Ryther. 1961*a*. Nutrients limiting the production of phytoplankton in the Sargasso Sea, with special reference to iron. Deep-Sea Res. **7:**276-281.

Menzel, D. W., and J. H. Ryther. 1961*b*. Zooplankton in the Sargasso Sea off Bermuda and its relation to organic production. J. Cons. Perm. Int. Explor. Mer **26:**250-259.

Menzies, R. J. 1965. Conditions for the existence of life on the abyssal sea floor. *In* H. Barnes [ed.]. Oceanography and marine biology. Vol. 7. Allen & Unwin, Ltd., London.

Mesolella, K. J. 1967. Zonation of uplifted Pleistocene coral reefs on Barbados, West Indies. Science **156:**638-640.

Metcalf, M. M. 1930. Salinity and size. Science **72:**526-527.

Meyers, S. P., and E. S. Reynolds. 1957. Incidence of marine fungi in relation to wood borer attack. Science **126:**969.

Miles, J. W., and G. W. Pearce. 1957. Rapid method for measurement of rate of sorption of DDT by mud surfaces. Science **126:**169.

Miller, C. B. [ed.]. 1974. The biology of the oceanic Pacific. Proc. 33d Ann. Biol. Colloquium. Oregon State University Press, Corvallis. 157 p.

Miller, R. C., W. D. Ramage, and E. L. Lazier. 1928. Physical and chemical conditions in San Francisco Bay, especially in relation to the tides. Univ. California Pub. Zool. **31:**201-267.

Miller, S. L. 1953. A production of amino acids under possible primitive earth conditions. Science **117:** 528-529.

Miller, S. L., and H. C. Urey. 1958. Organic compound formation on the primitive earth. *In* E. Hutchings [ed.]. Frontiers in science. Basic Books, Inc., Publisher, New York.

Millott, N. [ed.]. 1967. Echinoderm biology. (Symposium of the Zoological Society of London, no. 20.) Academic Press, Inc., New York.

Milne, A. 1940. The ecology of the Tamar estuary. J. Mar. Biol. Ass. U.K. **24:**69-87.

Miner, R. W. 1950. Field book of seashore life. G. P. Putnam's Sons, New York.

Mitchel, C. T., E. K. Anderson, L. G. Jones, and W. J. North. 1969. Ecological effects of oil spillage in the sea. Proc. 42nd Ann. Conf. Water Pollution Control Fed. October 5-10, 1969, Dallas, Tex.

Mitchel, C. T., C. H. Turner, and A. R. Strachan. 1969. Biology and behavior of the California spring lobster, *Panulirus interruptus* (Randall) Calif. Fish and Game **55**(2):121-131.

Mitchell, R., and H. Ducklow. 1976. Slow death of coral reefs. Nat. Hist. **85**(8):106-111.

Miyake, Y., and K. Saruhasi. 1956. On the vertical distribution of dissolved O in the ocean. Deep-Sea Res. **3:**242-247.

Moberg, E. G. 1926. Chemical composition of marine plankton. Proceedings of the Third Pan-Pacific Scientific Congress, Tokyo.

Moestafa, S. H., and B. H. McConnaughey. 1966. *Catostylus ouwensi* (Rhizostomeae, Catostylidae), a new jellyfish from Irian (New Guinea), and *Ouwensia catostyli* n. gen., n. sp., parasitic in *C. ouwensi.* Treubia **27:**1-9.

Moore, H. B. 1930, 1931. Muds of the Clyde Sea area. J. Mar. Biol. Ass. U.K. **16:**597-607; **17:**325-358.

Moore, H. B. 1937. Marine fauna of the Isle of Man. Proc. Liverpool Biol. Soc. **50:**38-57.

Moore, H. B. 1949. The zooplankton of the upper waters of the Bermuda area of the North Atlantic. Bull. Bingham Oceanogr. Coll. **12**(2):1-97.

Moore, H. B. 1958. Marine ecology. John Wiley & Sons, Inc., New York.

Moore, H. B., and E. G. Corwin. 1956. Effects of temperature illumination and pressure on vertical distribution of zooplankton. Bull. Mar. Sci. Gulf Carib. **6:**273-287.

Moore, H. F. 1899. An inquiry into the feasibility of introducing useful marine animals into the waters of Great Salt Lake. U.S. Fish. Comm. Rep., 1899.

Morris, R. W., and L. R. Kittleman. 1967. Piezoelectric property of otoliths. Science **158:**368-370.

Muller, M. M. 1969. Production of zooplankton in the oceans: present status and problems. *In* H. Barnes [ed.]. Oceanography and marine biology. Vol. 7. Allen & Unwin, Ltd., London.

Mullin, J. P., and J. B. Riley. 1955. Spectrophotometric determination of NO in natural waters, with particular reference to seawater. Anal. Chim. Acta **12**:464-480.

Munk, W. 1955. The circulation of the oceans. Sci. Amer. **193**(3):96-108.

Murdoch, W. W. 1971. Environment: a sourcebook on resources, pollution and society. Sinauer Associates, Inc., Stamford, Conn.

Murphy, G. I. 1961. Oceanography and variations in the Pacific sardine population. California Coop. Fish. Invest. Rep. **8**:55-64.

Murphy, R. C. 1938. Birds of the high seas: albatrosses, petrels, man-o'-war birds, and tropic birds. (Color paintings: Wings over the bounding main, by Allen Brooks.) Nat. Geog. Mag. **74**:226-251.

Murphy, R. C. 1962. The oceanic life of the Antarctic. Sci. Amer. **207**(3):187-210. Reprinted in Readings from Scientific American. 1971. Oceanography. W. H. Freeman & Co., Publishers, San Francisco. pp. 287-299.

Murray, J. 1895. Summary of the scientific results obtained at the sounding, dredging, and trawling stations of H.M.S. Challenger. 'Challenger' Rep. Part I: pp. xxxiii-lii, 1-176; xix, 797-1608.

Murray, J., and J. Hjort. 1912. The depths of the ocean. Hist. Natur. Class, vol. 37. Reprint, 1964. Hafner Publishing Co., Inc., New York.

Muscatine, L. 1967. Glycerol excretion by symbiotic algae from corals and *Tridacna* and its control by the host. Science **156**:516-519.

Myers, J. J., C. H. Holm, and R. F. McAllister [eds.]. 1969. Handbook of ocean and underwater engineering. McGraw-Hill Book Co., New York.

National Academy of Sciences—National Research Council. 1957. Effects of atomic radiation on oceanography and fisheries. Pub. no. 651. Washington, D.C. 137 p.

National Academy of Sciences–National Research Council. 1969. Resources and man. W. H. Freeman Co., San Francisco.

National Academy of Sciences, Ocean Science Committee of the NAS-NRC Ocean Affairs Board. 1971. Marine environmental quality. Suggested research programs for understanding man's effect on the oceans. Washington, D.C.

National Geographic Society. 1965. Wondrous world of fishes. The Society, Washington, D.C.

Needham, P. 1930. On the penetration of marine organisms into fresh water. Biol. Zentralbl. **50**: 504-509.

Nelson, P. R., and W. T. Edmondson. 1955. Limnological effects of fertilizing Bear Lake, Alaska. Fisheries Bull. no. 102. Fish and Wildlife Services, U.S. Department of the Interior, Washington, D.C.

Nelson-Smith, A. 1970. The problem of oil pollution in the sea. *In* F. S. Russell and M. Yonge [eds.]. Advances in marine biology. Vol. 8. Academic Press, Inc., New York.

Neushul, M., W. D. Clarke, and D. W. Brown. 1967. Subtidal plant and animal communities of the southern California islands. *In* N. Philbrick [ed.]. Biology of California islands. Santa Barbara Botanical Garden, Santa Barbara.

Newcombe, C. L. 1935. Certain environmental factors of a sand beach in the St. Andrew's region, New Brunswick, with a preliminary designation of the intertidal communities. J. Ecol. **23**:334-355.

Newcombe, H. B. 1957. Magnitude of biological hazard from strontium-90. Science **126**:549.

Newell, G. E., and R. C. Newell. 1963. Marine plankton: a practical guide. Hutchinson Educational; Hutchinson & Co. (Publishers) Ltd., London. 207 p.

Newell, I. M. 1948. Marine mulluscan provinces of western North America: a critique and a new analysis. Proc. Amer. Phil. Soc. **92**:155-166.

Newell, N. D. 1972. The evolution of reefs. Sci. Amer. **228**(6):54-69.

Newell, R. C. 1970. Biology of intertidal animals. American Elsevier Publishing Co., New York. 555 p.

Newman, W. A. 1970. Acanthaster: a disaster? (with reply by R. H. Chesher). Science **167**:1274-1275.

Nicholls, A. G. 1944. 1. Littoral Copepoda from South Australia. 2. Calanoida, Notodelphyoida, Monstrilloida and Caligoida. Rec. S. Austr. Mus. **8**:1-62.

Nicholls, G. D., H. Curl, and V. T. Bowen. 1959. Spectrographic analysis of marine plankton. Limnol. Oceanogr. **4**:472-478.

Nicol, J. A. C. 1967. The biology of marine animals. 2nd ed. John Wiley & Sons, Inc., New York.

Nielsen, S. E. 1937. On the relation between the quantities of phytoplankton and zooplankton in the sea. J. Cons. Perm. Int. Explor. Mer **19**:309-328.

Nikitin, V. N. 1960. Marine biology, an AIBS translation. Acad. Nauk. S.S.S.R., Washington, D.C.

Nikolsky, G. V. 1963. Ecology of fishes. (Transl. by L. Birkett.) Academic Press, Inc., New York.

Norman, J. R. 1963. A history of fishes. 2nd ed. Hill & Wang, Inc., New York.

Norman, J. R., and F. C. Fraser. 1949. Field book of giant fishes. G. P. Putnam's Sons, New York.

Norris, K. S. [ed.]. 1966. Whales, dolphins, and porpoises. Proceedings of the 1st International Symposium on Cetacean Research (Aug., 1963), Washington, D. C. University of California Press, Richmond. 785 p.

North, W. J. 1963. Ecology of the rocky nearshore environment in southern California and possible influences of discharged wastes. Int. Conf. Water Pollution **7**(6/7):721-736.

North, W. J., M. Neushul, and K. A. Clendenning. 1964. Successive biological changes observed in a marine cove exposed to a large spillage of mineral oil. Comm. Int. Explor. Sci. Mer. Medit., Symposium Pollut. Mar. par Microorgan. Prod. Petrol. Monaco. pp. 335-354.

Nybakken, J. W. [ed.]. 1971. Readings in marine ecology. Harper & Row, Publishers, New York.

Odum, E. P. 1961. Role of tidal marshes in estuarine production. N.Y. State Conserv. **15**(6):12-15.

Odum, H. T. 1953. Factors controlling marine invasion into Florida fresh waters. Bull. Mar. Sci. Gulf Carib. **3:**134-156.

Odum, H. T. 1956. Efficiencies, size of organisms, and community structure. Ecology **37:**593-597.

Odum, H. T., and E. P. Odum. 1955. Trophic structure and productivity of a windward coral reef community on Eniwetok Atoll. Ecol. Monogr. **25:**291-320.

Oliver, W. R. S. 1923. Marine littoral plant and animal communities in New Zealand. Trans. New Zealand Inst. **54:**496-545.

Olson, T. A., and F. J. Burgess [eds.]. 1967. Pollution and marine ecology. John Wiley & Sons, Inc., New York.

Oparin, A. I. 1957. The origin of life on the earth. 3rd ed. (Transl. from Russian by A. Synge.) Oliver & Boyd, Edinburgh.

Oparin, A. I. 1961. Life: its nature, origin, and development. (Transl. from Russian by A. Synge.) Academic Press, Inc., New York.

Oparin, A. I. 1961. Origin of life in the oceans: oceanography. *In* M. Sears [ed.]. American Association for the Advancement of Science pub. no. 67, Washington, D.C.

Oppenheimer, C. H. 1955. Effect of marine bacteria on development and hatching of pelagic fish eggs, and the control of such bacteria by antibiotics. Copeia No. 1, pp. 43-49.

Oppenheimer, C. H. [ed.]. 1963. Symposium on marine microbiology. Charles C Thomas, Publisher, Springfield, Ill.

Orr, A. P., and F. W. Moorhouse. 1933. Physical and chemical conditions in mangrove swamps. Great Barrier Reef Expedition. Sci. Rep. Brit. Mus. (N.H.) **2:**102-110.

Orton, H. H. 1920. Sea temperature, breeding, and distribution in marine animals. J. Mar. Biol. Ass. U.K. **12:**339-366.

Orton, J. H. 1923. Some experiments on rate of growth in a polar region (Spitzbergen) and in England. Nature (London) **3:**146-148.

Ostenfeld, C. H. [ed.]. 1916. International plankton catalog. J. Cons. Perm. Int. Explor. Mer. **170:**1-87.

Osterberg, C. 1962. Fallout radionuclides in euphausiids. Science **138**(3539):529-530.

Osterberg, C. 1962. Zn^{6} content of salps and euphausiids. Limnol. Oceanogr. **7**(4):478-479.

Osterberg, C., L. Small, and L. Hubbard. 1963. Radioactivity in large marine plankton as a function of surface area. Nature (London) **197**(4870):883-884.

Ostwald, W. 1902. Zur Theorie des Planktons. Biol. Zentralbl. **22:**596-605, 609-638.

Oudot, C., and B. Wauthy. 1976. Upwelling et dome dans le Pacifique tropical occidental: distributions physico-chemico et biomasse vegetale. Cah. O.R.S.T.M., Ser. Oceanogr. **14**(1):27-48.

Paddock, W., and P. Paddock. 1967. America's decision: Who will survive? Little, Brown & Co., Boston.

Paine, R. T. 1971. Food web complexity and species diversity. *In* R. W. Nybakken [ed.]. Readings in marine ecology. Harper & Row, Publishers, New York.

Palmisano, J. F., and J. A. Estes. 1976. Sea otters: pillars of the nearshore community. Nat. Hist. **85**(7):46-53.

Park, K., W. T. Williams, J. M. Prescott, and D. W. Hood. 1962. Amino acids in deep sea water. Science **138**(5539):531-532.

Parker, R. H. 1955. Changes in invertebrate fauna attributable to salinity changes in bays of central Texas. J. Paleontol. **29:**193-211.

Parker, R. H. 1956. Macro-invertebrate communities as indicators of sedimentary environments in east Mississippi Delta regions. Bull. Amer. Ass. Petrol. Geol. **40:**295-376.

Patil, A. M. 1951. Study of the marine fauna of the Karwar coast and neighboring islands. J. Bombay Natur. Hist. Soc. **50:**128-139.

Patten, B. C. 1959. An introduction to the cybernetics of the ecosystem: the trophic-dynamic aspect. Ecology **40:**221-231.

Pearse, A. S. 1936. Migrations of animals from sea to land. Duke University Press, Durham, N.C.

Pearse, A. S. 1950. The emigrations of animals from the sea. Sherwood Press, Dryden, N.Y.

Pearse, A. S., and G. Gunter. 1957. Salinity. *In* J. W. Hedgpeth [ed.]. Treatise on marine ecology and paleoecology. Geological Society of America, New York.

Pearse, A. S., H. J. Humm, and G. W. Wharton. 1942. Ecology of sand beaches at Beaufort, N.C. Ecol. Monogr. **12:**135-190.

Pearse, V. B., and L. Muscatine. 1971. Role of symbiotic algae (zooxanthellae) in coral calcification. Biol. Bull. Woods Hole **141:**350-353.

Pearson, C. S. 1975. International marine environmental policy—the economic dimension. The Johns Hopkins University Press, Baltimore.

Pennak, R. W. 1943. An effective method of diagramming diurnal movements of zooplankton organisms. Ecology **24:**405-407.

Pequegnat, W. E. 1958. Whales, plankton, and man. Sci. Amer. **198**(1):84-90.

Pequegnat, W. E. 1961. New world for marine biologists. In shallow offshore waters lie fertile areas for study. Nat. Hist. **70**(4):8-17.

Perrier, R. 1936. La faune de la France en tableaux synoptiques illustres. Vol. 1. Delgrave, Paris.

Petersen, C. G. 1913. The animal communities of the sea bottom and their importance for marine zoogeography. Rep. Dan. Biol. Sta. **21:**1-44.

Petersen, C. G. 1915. Investigations concerning the valuation of the sea. Rep. Dan. Biol. Sta. **23:**27-29.

Petersen, C. G. 1918. The sea bottom and its production of fish food. I. Apparatus. Rep. Dan. Biol. Sta. **25:**1-62.

Petersen, C. G. 1924. Necessity for quantitative methods in investigation of animal life on the sea bottom. Proc. Zool. Soc. London **94:**687-694.

Petersen, C. G., and P. Boysen-Jensen. 1911. Valuation of the sea. I. Animal life of the sea bottom, its food and quantity. Rep. Dan. Biol. Sta. **20**:1-79.

Pettersson, H. 1954. The ocean floor. Yale University Press, New Haven, Conn.

Pfaffmann, C. [ed.]. 1969. Olfaction and taste. Proceedings of the Third International Symposium. Rockefeller University Press, New York.

Pfeffer, G. 1889. Zur Fauna von Süd Georgien. Mitt. Naturforsch. Mus. Hamburg 6.

Pfleger, F. B. 1960. Ecology and distribution of recent foraminifera. The Johns Hopkins Press, Baltimore.

Phifer, L. D. 1934. Phytoplankton of Eastsound, Washington, Feb.-Nov., 1932. Univ. Washington Publ. Oceanogr. **1**:97-110.

Philbrick, N. [ed.]. 1967. Proceedings of the symposium on the biology of the California Islands. Santa Barbara Botanical Garden, Santa Barbara.

Phillips, A. G. 1917, 1922. Analytical search for metals in Tortuga's marine organisms. Papers of the Department of Marine Biology, nos. 11, 18. Carnegie Institute, Washington, D.C.

Piccard, A. 1956. Earth, sky, and sea. Oxford University Press, New York.

Pickford, G. E., and B. H. McConnaughey. 1949. The *Octopus bimaculatus* problem: a study in sibling species. Bull. Bingham Oceanogr. Coll. **12**(4):1-66.

Pierce, E. L. 1947. Annual cycle of the plankton and chemistry of four aquatic habitats of northern Florida. University of Florida Press, Gainesville.

Pittendrigh, C. S. 1958. Perspectives in the study of biological clocks. *In* A. A. Buzzati-Traverso [ed.]. Perspectives in marine biology. University of California Press, Richmond.

Platt, R. B., and J. F. Griffiths. 1964. Environmental measurement and interpretation. Reinhold Publishing Corp., New York.

Plunkett, M. A., and N. W. Rakestraw. 1955. Dissolved organic matter in the sea. (Papers in Marine Biology and Oceanography.) Deep-Sea Res. (supp.) **3**:12-14.

Polikarpov, G. G. 1964. Radioecology of aquatic organisms. Scripta Bechnica. Transl. by Schultz and Klement, 1966. Reinhold Publishing Corp., New York.

Pomeroy, L. R. 1960. Residence time of dissolved phosphate in natural waters. Science **131**:1731-1732.

Pomeroy, L. R., and F. M. Bush. 1959. Regeneration of phosphate by marine animals. International Oceanographic Congress Preprints. American Association for the Advancement of Science, Washington, D.C.

Ponomareva, L. A. 1957. Zooplankton of the West Kara Sea and Baidaratskaya Bay. Trans. Inst. Oceanol. Acad. Sci. U.S.S.R. **20**:228-245.

Poore, M. E. C. 1962. Successive approximation in descriptive ecology. *In* J. B. Cragg [ed.]. Advances in ecological research. Vol. I. Academic Press, Inc., New York.

Potter, Van R. 1971. Bioethics: bridge to the future. Prentice Hall, Inc., Englewood Cliffs, N.J. 224 p.

Powell, W. M., and G. L. Clarke. 1936. The reflection and absorption of daylight at the surface of the ocean. J. Opt. Soc. Amer. **26**:111-120.

Powers, P. A. 1932. *Cyclotrichium meunieri,* sp. nov. (Protozoa, Ciliata): cause of red water in the Gulf of Maine. Biol. Bull. **63**:74-80.

Prasad, R. R. 1954. Observations on the distribution and fluctuation of planktonic larvae off Mandapan. Symposium on Marine and Freshwater Plankton in the Indo-Pacific, Bangkok, 1954.

Pratt, D. M. 1949. Experiments in the fertilization of a saltwater pond. J. Mar. Res. **8**:36-59.

Pratt, D. M. 1950. Experimental study of the phosphorus cycle in fertilized salt water. J. Mar. Res. **9**:29-54.

Pritchard, D. W. 1952. Estuarine hydrography. Advances in geophysics **1**:243-280. Academic Press, Inc., New York.

Pritchard, D. W. 1955. Estuarine circulation patterns. Proc. Amer. Soc. Civil Eng. **81**(717):1-11.

Proudman, J. 1953. Dynamic oceanography. John Wiley & Sons, Inc., New York.

Putter, A. 1909. Die Ernährung der Wassertiere und der Stoffhaushalt der Gewässer. Gustav Fischer, Jena, E. Germany.

Quastler, H. 1959. Information theory of biological integration. Amer. Natural. **93**:245-254.

Rae, K. M. 1950. The continuous plankton recorder survey; the plankton around the north of the British Isles in 1950. Ann. Biol. **7**:72-76.

Rae, K. M. 1958. Parameters of the marine environment. *In* A. A. Buzzati-Traverso [ed.]. Perspectives in marine biology. University of California Press, Richmond.

Rae, K. M., and Rees, C. B. 1947. The copepoda in the North Sea. Hull Bull. Mar. Ecol. **1**:171-238.

Rafter, T. A., and G. J. Fergusson. 1957. "Atom-bomb effect"—recent increase of carbon-14 content of the atmosphere and biosphere. Science **126**:557.

Rakestraw, N. W. 1936. The occurrence and significance of nitrate in the sea. Biol. Bull. Mar. Biol. Lab. Woods Hole **71**:133-167.

Rakestraw, N. W., and D. E. Carritt. 1948. Some seasonal chemical changes in the open ocean. J. Mar. Res. **7**:362-369.

Rakusa-Suszczewski, S., and M. A. McWhinnie. 1976. Resistance to freezing by antarctic fauna: supercooling and osmoregulation. Comp. Biochem. Physiol. **54**(3A):287-290.

Ray, C., and E. Ciampi. 1956. The underwater guide to marine life. A. S. Barnes & Co., Inc., Cranbury, N.J.

Ray, D. L. [ed.]. 1958. Marine boring and fouling organisms (1st Friday Harbor symposium, 1957). University of Washington Press, Seattle.

Raymont, J. E. G. 1955. Fauna of an intertidal mud flat. (Papers in Marine Biology and Oceanography.) Deep-Sea Res. (supp.) **3**:178-203.

Raymont, J. E. G. 1963. Plankton and productivity in the oceans. A Pergamon Press Book, The Macmillan Co., New York.

Raymont, J. E. G. 1966. The production of marine plankton. *In* J. B. Cragg [ed.]. Advances in ecology research. Vol. 3. Academic Press, Inc., New York.

Raymont, J. E. G., and B. G. A. Carne. 1959. Zooplankton of South Hampton water. International Oceanographic Congress Preprints. American Association for the Advancement of Science, Washington, D.C.

Raymont, J. E. G., and R. J. Conover. 1961. Further investigations on the carbohydrate content of marine zooplankton. Limnol. Oceanogr. **6:**154-164.

Raymont, J. E. G., and S. Krishnaswamy. 1960. Carbohydrates in some marine planktonic animals. J. Mar. Biol. Ass. U.K. **39:**239-248.

Raymont, J. E. G., and R. S. Miller. 1962. Production of marine zooplankton with fertilization in an enclosed body of sea water. Int. Rev. Hydrobiol. Hydrogr. **47:**169-209.

Redfield, A. C. 1934. On the proportions of organic derivatives in seawater and their relation to the composition of the plankton. (James Johnstone memorial vol.) Liverpool University Press, Liverpool.

Redfield, A. C. 1939. The history of a population of *Limacina retroversa* during its drift across the Gulf of Maine. Biol. Bull. Mar. Biol. Lab. Woods Hole **76:**26-47.

Redfield, A. C. 1941. The effect of circulation of water on the distribution of the calanoid community in the Gulf of Maine. Biol. Bull. Mar. Biol. Lab. Woods Hole **80:**86-110.

Redfield, A. C. 1958. The biological control of chemical factors in the environment. Amer. Sci. **46:** 205-221.

Redfield, A. C., and A. B. Keys. 1938. The distribution of ammonia in the waters of the Gulf of Maine. Biol. Bull. Mar. Biol. Lab. Woods Hole **74:**83-92.

Redfield, A. C., H. P. Smith, and B. H. Ketchum. 1937. The cycle of organic phosphorus in the Gulf of Maine. Biol. Bull. Mar. Biol. Lab. Woods Hole **73:**421-443.

Rees, C. B. 1939. Plankton of the upper reaches of the Bristol Channel. J. Mar. Biol. Ass. U.K. **23:**397-415.

Rees, C. B. 1940. Preliminary study of the ecology of a mud flat. J. Mar. Biol. Ass. U.K. **24:**185-199.

Rees, W. J. [ed.]. 1966. The cnidaria and their evolution. (Symposia of the Zoological Society of London, no. 16.) Academic Press, Inc., New York.

Reid, G. K. 1961. Ecology of inland waters and estuaries. Reinhold Publishing Corp., New York.

Remane, A. 1959. Die interstitielle Fauna des Meeressands. Proceedings of the 15th International Congress of Zoology, London.

Renouf, L. P. W. 1939. Faunistic notes from the south coast of County Cork, Eire. Ann. Mag. Natur. Hist. (ser. 11) **4:**520-525.

Renz, G. W. 1976. Distribution and ecology of Radiolaria in the central Pacific. Bull. 22, Scripps Institute of Oceanography, University of California Press, Berkeley. 267 p.

Reuszer, H. W. 1933. Marine bacteria and their role in the cycle of life in the sea. III. Distribution of bacteria in the ocean waters and muds about Cape Cod. Biol. Bull. Mar. Biol. Lab. Woods Hole **65:**480-497.

Rhodina, A. G. 1971. Methods in aquatic microbiology. (Edited, translated, and revised by R. R. Colwell and M. S. Zambruski.) University Park Press, Baltimore. 352 p.

Rice, D. W., and A. A. Wolman. 1971. Life history and ecology of the gray whale *(Eschrichtius robustus)*. Amer. Soc. Mammalogists spec. pub. no. 3, Stillwater, Okla.

Rice, T. R. 1954. Biotic influences affecting population growth of planktonic algae. Fisheries Bull. no. 54. Fish and Wildlife Service, U.S. Department of Interior.

Richards, F. A. 1957*a*. Oxygen in the ocean. *In* J. W. Hedgpeth [ed.]. Treatise on marine ecology and paleoecology. Vol. I. Ecology. Geological Society of America, New York.

Richards, F. A. 1957*b*. Current aspects of chemical oceanography. *In* L. H. Ahrens, F. Press, K. Rankama, and S. K. Runcorn [ed.]. Progress in physics and chemistry of the earth. Vol. 2. Pergamon Press, Inc., New York. pp. 77-128.

Richards, F. A., and B. B. Benson. 1961. Nitrogen/argon and nitrogen isotope ratios in two anaerobic environments: the Cariaco Trench in the Caribbean Sea, and Dramsfjord, Norway. Deep-Sea Res. **7:** 254-264.

Richards, F. A., and T. G. Thompson. 1952. Estimation and characterization of plankton populations by pigment analysis. II. A spectrophotometric method for the estimation of plankton pigments. J. Mar. Res. **11:**156-172.

Richards, F. A., and R. F. Vaccaro. 1956. The Cariaco Trench, an anaerobic basin in the Caribbean Sea. Deep-Sea Res. **3:**214-228.

Ricketts, E., and J. Calvin. 1974. Between Pacific tides. 5th ed. Revised by J. Hedgpeth. Stanford University Press, Stanford, Calif.

Riedl, R. J. 1953. Quantitativ ökölogische Methoden mariner Turbellarienforschung. Österr. Zool. Zeitschr. **4:**108-145.

Riedl, R. J. 1954. Unterwasserforschung im Mittlemeer. Nchurw. Rundsch. **2:**65-71.

Riedl, R. J. M. 1969. Gnathostomulida from America. Science **163:**445-452.

Riley, G. A. 1937. Significance of the Mississippi River drainage for biological conditions in the northern Gulf of Mexico. J. Mar. Res. **1:**64-74.

Riley, G. A. 1937-1938. Plankton studies. I. A preliminary investigation of the plankton of the Tortugas region. J. Mar. Res. **1:**335-352.

Riley, G. A. 1939. Plankton studies. II. The western North Atlantic, May-June, 1939. J. Mar. Res. **2:** 145-162.

Riley, G. A. 1941. Plankton studies. V. Regional summary. J. Mar. Res. **4:**162-171.

Riley, G. A. 1942. Relation of vertical turbulence and spring diatom flowerings. J. Mar. Res. **5:**67-87.

Riley, G. A. 1946. Factors controlling phytoplankton population on Georges Bank. J. Mar. Res. **6:**54-73.

Riley, G. A. 1947*a*. Seasonal fluctuations of the phytoplankton populations in New England coastal waters. J. Mar. Res. **6**:114-125.

Riley, G. A. 1947*b*. Theoretical analysis of zooplankton population of Georges Bank. J. Mar. Res. **6**: 104-113.

Riley, G. A. 1955. Review of the oceanography of Long Island Sound. (Papers in Marine Biology and Oceanography.) Deep-Sea Res. (supp.) **3**:224-238.

Riley, G. A. 1957. Phytoplankton of the north central Sargasso Sea. Limnol. Oceanogr. **2**:252-270.

Riley, G. A. [ed.]. 1963. Marine biology. Vol. I. First International Interdisciplinary Conference on Marine Biology. Princeton, N.J., Oct. 29 to Nov. 1, 1961. Hafner Publishing Co., Inc., New York.

Riley, G. A. 1970. Particulate organic matter in the seawater. *In* F. S. Russell and M. Yonge [eds.]. Advances in marine biology. Vol. 8. Academic Press, Inc., New York.

Riley, G. A., and R. Arx. 1949. Theoretical analysis of seasonal changes in the phytoplankton of Pusan Harbor, Korea. J. Mar. Res. **8**:60-72.

Riley, G. A., and D. F. Bumpers. 1946. Phytoplankton-zooplankton relationship on Georges Bank. J. Mar. Res. **6**:33-47.

Riley, G. A., and S. A. M. Conover. 1956. Oceanography of Long Island Sound, 1952-1954. III. Chemical oceanography. Bull. Bingham Oceanogr. Coll. **15**:47-61.

Riley, G. A., A. Stommel, and D. F. Bumpers. 1949. Quantitative ecology of the plankton of the western North Atlantic. Bull. Bingham Oceanogr. Coll. **12**:1-169.

Risebrough, R. W., R. J. Huggett, J. J. Griffin, and E. D. Goldberg. 1968. Pesticides: transatlantic movements in the Northeast Trades. Science **159**:1233-1235.

Rittenberg, S. C., K. O. Emery, and W. L. Orr. 1955. Regeneration of nutrients in sediments of marine basins. Deep-Sea Res. (supp.) **3**:23-45.

Ritter, W. E., E. L. Michael, and G. P. McEwen. 1915. Hydrographic, plankton and dredging records of the Scripps Institution for Biological Research of the University of California, 1901-1912. Univ. California Pub. Zool. **15**:1-206.

Roaf, H. 1910. Contributions to the physiology of marine invertebrates. J. Physiol. (London) **39**:438-452.

Robbins, S. F., and C. M. Yentsch. 1973. The sea is all around us. Peabody Museum, Salem, Mass. vii + 162 p.

Robertson, R. J., and H. E. Wirth. 1934*a*. Free ammonia, albuminoid nitrogen, and organic nitrogen in waters of the Pacific Ocean off the coasts of Washington and Vancouver Island. J. Cons. Perm. Int. Explor. Mer **9**:187-195.

Robertson, R. J., and H. E. Wirth. 1934*b*. Report on the free ammonia, albuminoid nitrogen, and organic nitrogen in the waters of the Puget Sound area, during the summers of 1931 and 1932. J. Cons. Perm. Int. Explor. Mer **9**:15-27.

Robison, B. H. 1976. Deep-sea fishes. Nat. Hist. **85**(7):38-45.

Rona, E., and C. E. Miliana. 1969. Absolute dating of Caribbean cores P6304-8 and P6304-9. Science **163**:66-67.

Rose, M. 1925. Contributions à l'étude de la biologie du plancton. Arch. Zool. Exp. Gen. **64**:387-542.

Ross, D. A. 1976. The Red Sea: an ocean in the making. Nat. Hist. **85**(7):74-77.

Ross, D. M., and L. Sutton. 1967. Swimming sea anemones of Puget Sound: swimming of *Actinostola* new species in response to *Stomphia coccinea*. Science **155**:1419-1421.

Roughly, T. C. 1936. Wonders of the Great Barrier Reef. Angus & Robertson, Ltd., London.

Rounsefell, G. A. 1975. Ecology, utilization and management of marine fisheries. The C. V. Mosby Co., St. Louis. 516 p.

Rounsefell, G. A., and W. H. Everhart. 1953. Fishery science: its methods and application. John Wiley & Sons, Inc., New York.

Rubinoff, I. 1968. Central American sea level canal: possible biological effects. Science **161**:857-861.

Russell, F. S. 1925-1934. Vertical distribution of marine macroplankton. I-XII. J. Mar. Biol. Ass. U.K. **13** to **19**.

Russell, F. S. 1933. Seasonal distribution of macroplankton as shown by catches in 2-meter ring trawl in offshore waters off Plymouth. J. Mar. Biol. Ass. U.K. **19**:17-82.

Russell, F. S. 1934. The zooplankton. II, III, IV. British Museum Great Barrier Reef Expedition. Sci. Rep. Brit. Mus. (N.H.) **2**:159-176, 176-201, 203-276.

Russell, F. S. 1935. On the value of certain plankton animals as indicators of water movements in the English Channel and North Sea. J. Mar. Biol. Ass. U.K. **20**:309-332.

Russell, F. S. 1937. Seasonal abundance of pelagic young of teleostean fishes in the Plymouth area. IV. The year 1936, with notes on the conditions, shown by the occurrence of plankton indicators. J. Mar. Biol. Ass. U.K. **21**:679-686.

Russell, F. S. 1939. Hydrographical and biological conditions in the North Sea as indicated by plankton organisms. J. Cons. Perm. Int. Explor. Mer **14**: 171-192.

Russell, F. S. 1953. The medusae of the British Isles. 2 vols. Cambridge University Press, London.

Russell, F. S., and C. M. Yonge. 1928. The seas: our knowledge of life in the sea and how it is gained. Frederick Warne & Co., Inc., New York.

Russell, H. D. 1971. Index Nudibranchia, Delaware Museum of Natural History, Greenville, Del.

Russell, R. C. M., and D. H. McMillan. 1952. Waves and tides. Hutchinson's Scientific & Technical Publication, London.

Rustad, D. 1952. Zoological notes from the biology station, University of Bergen. Arb. 1951. Naturvid. Rekke. 1952, 1-2.

Rustad, E. 1946. Experiments on photosynthesis and

respiration at different depths in the Oslo Fjord. Nytt. Mag. Naturvid. **85:**223-229.

Rutten, M. G. 1962. The geological aspects of the origin of life on earth. Elsevier Publishing Co., Amsterdam and New York.

Ruud, J. 1926. Quantitative investigations of plankton at Lofoten, March-Apr., 1922-1924. Rep. Norweg. Fish. Mar. Res. Invest. **3**(7):1-30.

Ryther, J. H. 1954. Ecology of phytoplankton blooms in Moriches Bay and Great South Bay, Long Island, N. Y. Biol. Bull. Mar. Biol. Lab. Woods Hole **106:**198-209.

Ryther, J. H. 1956. Photosynthesis in the ocean as a function of light intensity. Limnol. Oceanogr. **1:**61-70.

Ryther, J. H. 1959. Potential productivity of the sea. Science **130:**602-608. Reprinted in W. E. Hazen [ed.]. 1970. Readings in population and community ecology. 2nd ed. W. B. Saunders Co., Philadelphia.

Ryther, J. H. 1960. Seasonal and geographic range of primary production in the western Sargasso Sea. Deep-Sea Res. **6:**235-238.

Ryther, J. H. 1963. Geographic variations in productivity in the sea. *In* M. N. Hill [ed.]. The sea; ideas and observations on progress in the study of the seas. Vol. 2. John Wiley & Sons, Inc., New York.

Ryther, J. H. 1969. Photosynthesis and fish production in the sea. Science **166:**72-76. (An important article, which gives further confirmation to the pessimistic outlook regarding further expansion of marine fisheries. "At present rate of expansion" they cannot last over a decade. Exploitation already is close to the maximum possible for sustained yield and is exceeding it in several fisheries.)

Ryther, J. H., and R. F. Vaccaro. 1954. Comparison of the oxygen and C[4] methods of measuring marine photosynthesis. J. Cons. Perm. Int. Explor. Mer **20:**25-34.

Ryther, J. H., and C. S. Yentsch. 1957. Estimation of phytoplankton production in the ocean from chlorophyll and light data. Limnol. Oceanogr. **2:**281-286.

Ryther, J. H., and C. S. Yentsch. 1958. Primary production of continental shelf waters off New York. Limnol. Oceanogr. **3:**327-335.

Sagan, S. 1971. Origin of life. *In* V. G. Dethier and others [eds.]. Topics in the study of life: the BIO source book. Harper & Row, Publishers, New York.

Sanders, H. L. 1956. Oceanography of Long Island Sound, 1952-1954. X. Biology of marine bottom communities. Bull. Bingham Oceanogr. Coll. **15:** 345-414.

Sanders, H. L. 1960. Benthic studies in Buzzards Bay. I. Animal-sediment relationships. Limnol. Oceanogr. **3:**245-258.

Sanders, H. L. 1960. Benthic studies in Buzzards Bay. II. The meiofauna. Limnol. Oceanogr. **5:** 121-137.

Sanders, H. L. 1960. Benthic studies in Buzzards Bay. III. The structure of the soft-bottom community. Limnol. Oceanogr. **5:**138-153.

Sanders, H. L., and R. R. Hessler. 1969. Ecology of the deep-sea benthos. Science **163:**1419-1424.

Sanders, M. J., and A. J. Morgan. 1976. Fishing power, fishing effort, density, fishing intensity and fishing mortality. J. du Conseil **37**(1):36-40.

Sargent, M. C., and T. S. Austin. 1954. Biologic economy of coral reefs. Bikini and nearby atolls, Marshall Islands. Prof. Pap. U.S. Geol. Surv. 260-E, pp. 293-300.

Sars, G. O. 1901-1918. An account of the Crustacea of Norway. 6 vols. Bergen Museum.

Savage, R. E. 1931. The relation between feeding herring off the coast of England and plankton in surrounding waters. Fish. Invest. London (Ser. 2) **12**(3):1-88.

Savilov, A. I. 1957. Biological aspect of bottom fauna groupings of the north Okhotsk Sea. Trudȳ Inst. Okeanol. **20:**88-170.

Say, T. 1818. An account of the Crustacea of the United States. J. Acad. Sci. **1**(2):441. Hist. Nat. Class no. 73. (Reprinted 1967 by Stechert-Hafner, Inc., New York.)

Scagel, R. F. 1957. Annotated list of marine algae of British Columbia and northern Washington, with keys to genera. Canadian Department of Northern Affairs and Natural Resources, Ottawa.

Schäfer, W. 1972. Ecology and paleontology of marine environments. [Transl. by I. Oertel.] *In* Craig, G. Y. [ed.] (Based on North Sea studies.) University of Chicago Press, Chicago. xiv + 568 p.

Scheffer, V. B. 1976. Exploring the lives of whales. Nat. Geog. Mag. **150**(6):752-767.

Schenck, H. G., and A. M. Keen. 1936. Marine molluscan provinces of western North America. Proc. Amer. Phil. Soc. **76:**921-938.

Schlee, S. 1973. The edge of an unfamiliar world: a history of oceanography. E. P. Dutton & Co., New York. 398 p.

Schlieper, C. 1968. Methoden der meeresbiologischen Forschung. Gustav Fischer, Jena.

Schlieper, C. 1972. Research methods in marine biology. 2nd ed. (Biology Series, translation.) University of Washington Press, Seattle. 300 p.

Schmalhausen, I. I. 1968. The origin of terrestrial vertebrates. (Translated from Russian by L. Kelso.) Academic Press, Inc., New York.

Schmidt, J. 1925. The breeding places of the eel. Smithsonian Institution Annual Report for 1924. Washington, D.C., pp. 279-316.

Schmidt-Nielsen, K. 1971. Marine vertebrates: problems of salt and water. *In* V. G. Dethier and others [eds.]. Topics in the study of life: the BIO source book. Harper & Row, Publishers, New York.

Schminke, H. K. 1976. The ubiquitous telson and the deceptive furca. Crustaceana **30**(3):292-300.

Schopf, J. W. 1970. Precambrian microorganisms and evolutionary events prior to the origin of vascular plants. Biol. Rev. **45**(3):319-352.

Schröder, J. H. [ed.]. 1973. Genetics and mutagenesis of fish. Springer-Verlag, New York. xiv + 356 p.

Schultz, L. P., and E. M. Stern. 1948. The ways of fishes. D. Van Nostrand Co., Inc., New York.

Sculthorpe, C. D. 1967. The biology of aquatic vascular plants. St. Martin's Press, Inc., New York.

Sears, M. 1941. Notes on the phytoplankton on Georges Bank in 1940. J. Mar. Res. **4**:247-257.

Sears, M. 1954. Notes on the Peruvian Coastal Current. I. An introduction to the ecology of Pisco Bay. Deep-Sea Res. **1**:141-169.

Sears, M. [ed.]. 1961. Oceanography. Symposium no. 67. American Association for the Advancement of Science, Washington, D.C.

Segerstråle, S. G. 1953. Increase in salinity of the inner Baltic and its influence on the fauna. Soc. Scint. Fennica **13**(15):1-7.

Segerstråle, S. G. 1957. The Baltic Sea. *In* J. W. Hedgpeth [ed.]. Treatise on marine ecology and paleoecology. Vol. I. Ecology. Geological Society of America, New York.

Sergerstråle, S. G. 1959. Brackish-water classification, a historical survey. Arch. Oceanogr. Limnol. (supp.) **11**:7-13.

Seiwell, G. E. 1935*a*. Annual organic production and nutrient phosphorus requirement in the tropical western North Atlantic. J. Cons. Perm. Int. Explor. Mer **10**:20-32.

Seiwell, G. E. 1935*b*. Note on iron analyses of Atlantic coastal waters. Ecology **16**:663-664.

Seiwell, G. E. 1939. Daily temperature variations in the western North Atlantic. J. Cons. Perm. Int. Explor. Mer **14**:357-369.

Seiwell, H. R. 1942. Analysis of vertical oscillations in the southern North Atlantic. Trans. Amer. Phil. Soc. **85**:136-158.

Sellius, G. 1733. Historia naturalis Teredinis seu xylophagi marine *Tubuloconchoidis speciatum* Belgin. Trajecti ad Rhenum.

Seshaiya, R. V. 1959. Estuarine hydrology and biology. Curr. Sci. **28**:54-56.

Shapeero, W. L. 1961. Phylogeny of Priapulida. Science **133**:879-880.

Shapeero, W. L. 1962. Distribution of *Priapulus caudatus* Lam. on the Pacific coast of North America. Amer. Midl. Nat. **68**(1):237-241.

Sharp, C. A. 1962. Polynesian navigation. *In* J. Golson [ed.]. Polynesian navigation. Polynesian Society Memoir no. 34. Polynesian Society, Inc., Wellington, N.Z. [Critique of Sharp's theory.]

Sharp, C. A. 1963. Ancient voyagers in Polynesia. University of California Press, Berkeley.

Shaw, E. 1962. The schooling of fishes. Sci. Amer. **206**(6):128-136. Reprinted in Readings from Scientific American. Oceanography, pp. 235-243. W. H. Freeman & Co., Publishers, San Francisco.

Shelford, V. E. 1930. Geographic extent and succession in Pacific North American intertidal *(Balanus)* communities. Pub. Puget Sound Biol. Sta. **7**: 216-223.

Shelford, V. E., and E. D. Towler. 1925. Animal communities of the San Juan Channel and adjacent areas. Pub. Puget Sound Biol. Sta. **5**:33-73.

Shelford, V. E., A. O. Weese, L. A. Rice, D. I. Rasmussen, and A. MacLean. 1935. Some marine biotic communities of the Pacific coast of North America. I. General survey of communities. Ecol. Monogr. **5**:251-332.

Shepard, F. 1948. Submarine geology. Harper & Brothers, New York.

Shepard, F. 1959. The earth beneath the sea. The Johns Hopkins Press, Baltimore.

Shepard, F., and O. Emery. 1946. Submarine photography of the California coast. J. Geol. **54**:306-321.

Shimada, B. M. 1958. Diurnal fluctuations in photosynthetic rate and chlorophyll "A" content of phytoplankton from eastern Pacific waters. Limnol. Oceanogr. **3**:336-339.

Shirshov, P. P. 1937. Season changes of the phytoplankton of the polar seas in connection with the ice regime. Trans. Arctic Inst. Leningrad **82**:47-113.

Shropshire, R. G. 1944. Plankton harvesting. J. Mar. Res. **5**:185-188.

Sieburth, J. McN., and P. R. Burkholder. 1959. Antibiotic activity of antarctic phytoplankton. International Oceanographic Congress Preprints. American Association for the Advancement of Science, Washington, D.C.

Sillén, L. G. 1967. The ocean as a chemical system. Science **156**:1189-1197.

Simons, E. G. 1957. An ecological survey of the Upper Laguna Madre. Publ. Inst. Mar. Sci. (Port Aransas, Texas) **4**(2):156-200.

Sindermann, C. J. 1957. Diseases of fishes of the Western Atlantic. V. Parasites as indicators of herring movements. Res. Bull. 27. Marine Department of Sea and Shore Fisheries. 30 p.

Sindermann, C. J. 1966. Diseases of marine fishes. *In* F. S. Russell [ed.]. Advances in marine biology. Academic Press, Inc., New York. 90 p.

Sindermann, C. J. [ed.] 1977. Disease diagnosis and control in North American aquaculture. Vol. 6 of Developments in Aquaculture and Fisheries Science. Elsevier Scientific Publishing Co., Amsterdam. xii + 330 p.

Singarajah, K. V. 1975. Escape reactions of zooplankton: effects of light and turbulence. J. Mar. Biol. Ass. U.K. **55**(3):627-640.

Skopintsev, B. 1960. Organic matter in natural water (aquatic humus). Trudȳ Gos. Okeanogr. **17**(29).

Skottsberg, C. [ed.]. 1920-1956. The natural history of Juan Fernandez and Easter Island. 3 vols. Almquist & Wiksell, Stockholm.

Sleggs, G. F. 1927. Marine phytoplankton in the region of La Jolla, Calif., during the summer of 1924. Bull. Scripps Inst. Oceanogr. (Tech. Ser.) **1**:93-117.

Smayda, T. J. 1958. Biogeographical studies of marine phytoplankton. Oikos **9**:158-191.

Smayda, T. J. 1963. Ectocrine substances and limiting factors as determinants of succession in natural phytoplankton communities. *In* C. H. Oppenheimer [ed.]. Symposium on marine microbiology. Charles C Thomas, Publisher, Springfield, Ill.

Smith, F. E. 1972. Spatial heterogeneity, stability and

diversity in ecosystems. *In* E. S. Deevey [ed.]. Growth by intussusception. (Ecological Essays in Honor of G. Evelyn Hutchinson.) Trans. Conn. Acad. Arts Sci. **44**:309-335.

Smith, F. G. W., and F. A. Kalber [eds.]. 1974. CRC Handbook of marine science. CRC Press, Cleveland. Vol. 1, 640 p.; vol. 2, 404 p.

Smith, F. G. W., R. H. Williams, and C. C. Davis. 1950. An ecological survey of the tropical waters adjacent to Miami. Ecology **31**:119-146.

Smith, G. M. 1944. Marine algae of the Monterey Peninsula. Stanford University Press, Stanford, Calif.

Smith, J. E. [ed.]. 1968. "Torrey Canyon" pollution and marine life. A report of the Plymouth Laboratory. Cambridge University Press, New York.

Smith, M. 1926. Monograph of the sea snakes. Reprint, 1964. Hafner Publishing Co., Inc., New York.

Sorokin, Yu. 1955. Determination of chemosynthesis value in the water of the Rybinskoye Reservoir by the use of C^{14}. Dokl. Akad. Nauk. S.S.S.R. **105**: 1343.

Sorokin, Yu. 1957. Determination of chemosynthesis efficiency in methane and hydrogen oxidation in bodies of water. Mikrobiologia **26**(13).

Sorokin, Yu. 1958. Study of chemosynthesis in mud deposits by the use of C^{14}. Mikrobiologia **27**(206).

Sorokin, Yu. 1964. Photosynthetic production of phytoplankton in the Black Sea. Izv. Akad. Nauk. S.S.S.R. Ser. Biol. **5**:749-759.

Sorokin, Yu. 1964. Role of dark bacterial assimilation of carbon dioxide in water pools. Mikrobiologia **33**(5):880-886.

Sorokin, Yu. 1969. Processes of chemical and biological oxidation of H_2S in meromictic lakes. Mikrobiologia **37**(3):523-533.

Sorokin, Yu. 1969. Technique for determining the radioactivity of C^{14} carbonate in measurements of primary reservoir production. Mikrobiologia **37**(4): 741-744.

Soto, A. R., and W. B. Deichmann. 1967. Major metabolism and acute toxicity of aldrin, dieldrin, and endrin. Envir. Res. **1**(4).

Southward, A. J. 1961, 1962. The distribution of some plankton animals in the English Channel and western approaches. J. Mar. Biol. Ass. U.K. **41**:17-35; **42**:275-375.

Spencer, C. P. 1954. Studies on the culture of a marine diatom. J. Mar. Biol. Ass. U.K. **33**:265-290.

Spencer, C. P. 1956. Bacterial oxidation of ammonia in the sea. J. Mar. Biol. Ass. U.K. **35**:621-630.

Spooner, G. M. 1933. Observations on the reactions of marine plankton to light. J. Mar. Biol. Ass. U.K. **19**:385-438.

Spooner, G. M., and H. B. Moore. 1940. Ecology of the Tamar estuary. VI. Macrofauna of intertidal muds. J. Mar. Biol. Ass. U.K. **24**:283-330.

Sproston, N. G. 1949. Preliminary survey of plankton of the Chu-san region, with a review of relevant literature. Sinensia **20**:58-161.

Sreenivasa, R. 1975. The public order of ocean resources—a critique of contemporary law of the sea. The M.I.T. Press, Cambridge, Mass. 250 p.

Stanbury, F. A. 1931. Effect of light of different intensities, reduced selectivity, and nonselectivity on rate of growth of *Nitzschia closterium*. J. Mar. Biol. Ass. U.K. **17**:633-653.

Starr, T. J., M. E. Jones, and D. Martinez. 1957. Production of vitamin B_{12}-active substances by marine bacteria. Limnol. Oceanogr. **2**:114-119.

Stebbing, T. R. R. 1893. Crustacea. London.

Stebbing, T. R. R. 1905-1922. South African Crustacea. 762 p.; 1910. General catalog of South African Crustacea. Vol. 6, part 4. Ann. S. Afr. Mus.

Steele, J. H. 1956. Plant production on the Fladen Ground. J. Mar. Biol. Ass. U.K. **35**:1-33.

Steele, J. H. 1958. Production studies in the northern North Sea. Rapp. P.-v. Réun. Cons. Perm. Int. Explor. Mer **144**:79-84.

Steele, J. H. 1959. Quantitative ecology of marine phytoplankton. Biol. Rev. **34**:129-158.

Steele, J. H. [ed.]. 1970. Marine food chains. University of California Press, Berkeley.

Steele, J. H. 1974. The structure of marine ecosystems. Harvard University Press, Cambridge, Mass.

Steele, J. H., and C. S. Yentsch. 1960. The vertical distribution of chlorophyll. J. Mar. Biol. Ass. U.K. **39**:217-226.

Steeman, E. N. 1962. Inactivation of the photochemical mechanism in photosynthesis as a means to protect cells against high light intensities. Physiol. Plantarum **15**(1):161-171.

Steemann-Nielsen, E. 1935. The production of phytoplankton at the Faeroe Isles, Iceland, E. Greenland, and in the waters around. Medd. Komm. Danm. Fisk. Havund. (Ser. Plankton) **3**(1):1-93.

Steemann-Nielsen, E. 1937*a*. The annual amount of organic matter produced by the phytoplankton in the sound off Helsingør. Medd. Komm. Danm. Fisk. Havund. (Ser. Plankton) **3**(3):1-37.

Steemann-Nielsen, E. 1937*b*. On the relation between the quantities of phytoplankton and zooplankton in the sea. J. Cons. Perm. Int. Explor. Mer **12**:147-153.

Steemann-Nielsen, E. 1940. Die Produktionsbedingungen des Phytoplanktons in Übergangsbebiet zwischen der Nord- und Ostsee. Medd. Komm. Danm. Fisk. Havund. (Ser. Plankton) **3**(4):1-55.

Steemann-Nielsen, E. 1951. The marine vegetation of the Isefjord: a study ecology and production. Medd. Komm. Danm. Fisk. Havund. (Ser. Plankton) **5**(4): 1-114.

Steemann-Nielsen, E. 1952. Use of radio-active carbon (C^{14}) for measuring organic production in the sea. J. Cons. Perm. Int. Explor. Mer **18**:117-140.

Steemann-Nielsen, E. 1954. On organic production in the oceans. J. Cons. Perm. Int. Explor. Mer **19**: 309-328.

Steeman-Nielsen, E. 1955. The production of antibiotics by plankton algae and its effect upon bacterial activities in the sea. Deep-Sea Res. **3**:181-186.

Steemann-Nielsen, E. 1958*a*. The balance between

phytoplankton and zooplankton in the sea. J. Cons. Perm. Int. Explor. Mer **23:**178-198.

Steemann-Nielsen, E. 1958*b*. Experimental methods for measuring organic production in the sea. Rapp. P.-v. Réun. Cons. Perm. Int. Explor. Mer **144:**38-46.

Steemann-Nielsen, E. 1958*c*. A survey of recent Danish measurements of the organic productivity in the sea. Rapp. P.-v. Réun. Cons. Perm. Int. Explor. Mer **144:**92-95.

Steemann-Nielsen, E., and V. K. Hansen. 1959. Light adaptation in marine phytoplankton populations and its interrelation with temperature. Physiol. Plantarum **12:**353-370.

Steemann-Nielsen, E., and E. A. Jensen. 1957. Primary oceanic production: the autotrophic production of organic matter in the oceans. 'Galathea' Rep. **1:**49-136.

Steinbeck, J., and E. F. Ricketts. 1941. Sea of Cortez. The Viking Press, Inc., New York.

Stephen, A. C. 1930. Studies on the Scottish marine fauna: additional observations on the fauna of the sandy and muddy areas of the tidal zone. Trans. Roy. Soc. Edinb. [B.] **56:**521-535.

Stephen, A. C. 1931, 1932. Notes on the biology of some lamellibranchs. J. Mar. Biol. Ass. U.K. **17:** 277-300; **18:**51-68.

Stephen, A. C. 1938. Production of large broods in certain marine lamellibranchs, with possible relation to weather conditions. J. Anim. Ecol. **7:**130-143.

Stephenson, T. A. 1947. The constitution of the intertidal fauna and flora of South Africa. Ann. Natal Mus. **10**(3):261-358, **11**(2):207-324.

Stephenson, T. A., and A. B. Stephenson. 1949. Universal features of zonation between tide marks on rocky coasts. J. Ecol. **37**(2):289-305.

Stephenson, T. A., and A. B. Stephenson. 1950. Life between tidemarks in North America: Florida Keys. Ecology **38:**354-402.

Stephenson, T. A., and A. B. Stephenson. 1952. Life between tidemarks in North America: North Florida and the Carolinas. Ecology **40:**1-49.

Stephenson, T. A., and A. B. Stephenson. 1954. Life between tidemarks in North America: Nova Scotia and Prince Edward Island. Ecology **42:**14-70.

Stephenson, T. A., and A. B. Stephenson. 1961. Life between tidemarks in North America: Vancouver Island. Ecology **49:**1-29, 227-243.

Stephenson, W. 1949. Certain effects of agitation upon the release of phosphate from mud. J. Mar. Biol. Ass. U.K. **28:**371-380.

Stepheus, G. C., and R. A. Schiuske. 1957. Uptake of amino acids from sea water by ciliary-mucoid feeding animals. Biol. Bull. Mar. Biol. Lab. Woods Hole **111:**356-357. (See also Ibid. **115:**341-342 and Limnol. Oceanogr. **6:**175-181.)

Stetson, H. C. 1955. The continental shelf. Sci. Amer. **192**(3):82-87.

Stoddart, D. R., and M. Yonge [ed.] 1971. Regional variation in Indian Ocean corals reefs. Symposium no. 28, Zoological Society of London. Academic Press, Inc., New York.

Stommel, H. A. 1955. The anatomy of the Atlantic. Sci. Amer. **192**(1):25-30.

Stommel, H. A. 1958. The Gulf Stream. University of California Press, Richmond. 202 p.

Stonehouse, B. [ed.]. 1975. The biology of penguins. University Park Press, Baltimore. 540 p.

Strahler, A. N., and A. H. Strahler. 1973. Environmental geoscience. Hamilton Publishing Co., Santa Barbara, Calif.

Strickland, J. D. H. 1960. Measuring the production of marine plankton. Fisheries Res. Bd. Canada, Bull. **122:**1-72.

Strickland, J. D. H., and T. R. Parsons. 1960. A manual of seawater analysis with special reference to the more common micronutrients and to particulate organic matter. Fisheries Research Board of Canada, Bull. no. 125. 185 p.

Strickland, J. D. H., and L. D. B. Terhune. 1961. The study of *in situ* marine photosynthesis using a large plastic bag. Limnol. Oceanogr. **6:**93-96.

Strobbe, M. A. [ed.]. 1971. Understanding environmental pollution. The C. V. Mosby Co., St. Louis. 357 p.

Sverdrup, H. U. 1953. On conditions for the vernal blooming of phytoplankton. J. Cons. Perm. Int. Explor. Mer **18:**287-295.

Sverdrup, H. U., M. W. Johnson, and R. H. Fleming. 1946. The oceans: their physics, chemistry, and general biology. Prentice-Hall, Inc., Englewood Cliffs, N.J. 1087 p.

Swallow, J. C. 1957. Some further deep-current measurements using naturally buoyant floats. Deep-Sea Res. **4:**93-104.

Swallow, J. C., and L.V. Worthington. 1961. Observations of a deep countercurrent in the western North Atlantic. Deep-Sea Res. **8:**1-19.

Sylva, D. P. de. 1970. Systematics and life history of the great barracuda, *Sphyraena barracuda* (Walbaum). Studies in tropical oceanography. No. 1. University of Miami Press, Coral Gables, Fla.

Szidat, L. 1961. Versuch einer Zoogeographie des Sud-Atlantik mit Hilfe von Leit-parasiten der Meeresfishche. *In* W. Eichler [ed.]. Parasitologische Schriftenreihe, vol. 13. Gustav-Fischer Verlag, Jena.

Talling, J. F. 1961. Photosynthesis under natural conditions, Ann. Rev. Pl. Physiol. **12:**133-154.

Tavolga, W. N. [ed.]. 1964. Marine bioacoustics. The Macmillan Co., New York.

Taylor, H. F. 1952. Survey of marine fisheries of North Carolina. University of North Carolina Press, Chapel Hill.

Taylor, W. R. 1957. Marine algae of the northeastern coast of North America. University of Michigan Press, Ann Arbor.

Taylor, W. R. 1961. Marine algae of the eastern tropical and subtropical coasts of the Americas. University of Michigan Press, Ann Arbor.

Taylor, W. R. 1962. Marine algae from the tropical Atlantic Ocean. V. Algae from the Lesser Antilles. Contr. U.S. Natl. Herbarium **36**(2):43-62.

Teal, J. M. 1962. Energy flow in the salt marsh ecosystem of Georgia. Ecology **43:**614-624.

Thalman, H. E. 1960. Index to the genera and species of Foraminifera, 1890-1950. George Vanderbilt Foundation, Stanford University Press, Stanford, Calif.

Thienemann, A. 1920. Die Grundlagen der Biocoenotik. *In* Festschrift für Zschokke. Kober, Basel.

Thienemann, A. 1926. Der Nahrungskreislauf im Wasser. Verh. d. Zool. Ges. Leipzig (Zool. Auz.) **31:**29-79.

Thomas, W. H. 1959. Evaluation of phytoplankton production measurements. Bull. Scripps Inst. Oceanogr. (Tech. Ser.) **52**(22):8-11.

Thomson, S. H. 1961. What is happening to our estuaries? Transactions of 26th American Wildlife and Natural Resources Conference, Washington, D.C.

Thorson, G. 1936. Larval development, growth, and metabolism of arctic marine invertebrates. Medd. Grønland **100:**1-155.

Thorson, G. 1946. Reproduction and larval development of Danish marine invertebrates. Medd. Komm. Danm. Fisk. Havund. (Ser. Plankton) **4:**1-523.

Thorson, G. 1950. Reproductive and larval ecology of marine bottom invertebrates. Biol. Rev. **25:**1-45.

Thorson, G. 1952. Zur jetzigen Lage der marinen Bodentier—Ökologie. Verh. Deut. Zool. Ges. Wilhelmshaven, 1951, pp. 270-327.

Thorson, G. 1955. Modern aspects of marine level-bottom animal communities. J. Mar. Res. **14:** 387-397.

Thorson, G. 1958. Parallel level-bottom communities: their temperature adaptation and food animals. *In* A. A. Buzzati-Traverso [ed.]. Perspectives in marine biology. University of California Press, Richmond.

Thorson, G. 1971. Life in the sea. McGraw-Hill Book Co., New York. 256 p.

Tinbergen, N. 1953. The herring gull's world: a study of the social behavior of birds. William Collins & Sons Co., Ltd., London. 255 p. [Rev. ed. 1961. Basic Books, Inc., Publishers, New York.]

Todd, J. H. 1971. The chemical language of fishes. Sci. Amer. **224**(5):98-108.

Tolbert, N. E., and W. Garey. 1976. Apparent total carbon dioxide equilibrium point in marine algae during photosynthesis in sea water. *In* Photorespiration in marine plants. N. E. Tolbert and C. B. Osmond [eds.] University Park Press, Baltimore. 140 p.

Tolbert, N. E., and C. B. Osmond [eds.] 1976. Photorespiration in marine plants. University Park Press, Baltimore. 140 p.

Tombes, A. S. 1970. Introduction to invertebrate endocrinology. Academic Press, Inc., New York. 217 p.

Towler, E. D. 1930. Analysis of intertidal communities of the San Juan Archipelago. Pub. Puget Sound Biol. Sta. **7:**223-240.

Treadwell, A. L. 1914. Polychaetous annelids of the Pacific Coast, University of California Publ. in Zool. **13**(8):175-238.

Treadwell, A. L. 1924. Polychaetous annelids. Barbados-Antigua Expedition. University of Iowa Studies in Natural History **10**(4):ser. 1, no. 81.

Tully, J. P., and A. J. Dodimead. 1957. Canadian oceanographic research in the North Pacific Ocean. Pacific Oceanographic Group, Nanaimo, B.C. Fisheries Research Board of Canada.

Turekian, K. K. 1968. The oceans. Prentice-Hall, Inc., Englewood Cliffs, N.J.

Turner, R. D. 1966. A survey and illustrated catalog of the Teredinidae. Museum of Comparative Zoology of Harvard University, Cambridge, Mass.

Uchida, T. 1927, 1928. Studies on Japanese Hydromedusae. I. J. Fac. Sci., Tokyo Univ. **1:**145-241; II. **2:**73-97.

Uchida, T. 1929. Studies on Stauromedusae and Cubomedusae. Jap. J. Zool. **2:**103-192.

Uchida, T. 1933. Medusae from the vicinity of Kamchatka. J. Fac. Sci. Hokkaido University. Zool. (Ser. 6.) **2:**125-133.

Uchida, T. 1934. Metamorphosis of Scyphomedusae. Proc. Imp. Acad. Jap. **10:**421-430.

Uchida, T. 1947. Medusae from the central Pacific. J. Fac. Sci., Hokkaido Univ. Zool. (Ser. 6.) **9:**297-319.

Uchida, T. 1964. Medusae from New Caledonia. Publ. Seto Mar. Biol. Lab. **12**(1):109-112.

UNESCO. 1968. Monographs on oceanographic methodology. Zooplankton sampling. UNESCO Publications Center, New York.

Urey, H. C., A. Lowenstein, S. Epstein, and C. R. McKinney. 1951. Oxygen isotope methods—temperature of calcareous structures. Bull. Geol. Soc. Amer. **62:**399.

Ussing, H. H. 1938. The biology of some important plankton animals in the fjords of East Greenland. Medd. Grønland **100:**101-108.

Utermohl, H. 1931. Neue Wege in der quantitativen Erfassung des Planktons mit besonderer Berücksichtigung des Ultraplanktons. Verh. int. Verein Theor angew. Limnol. **5:**567-596.

Utinomi, H. 1956. Colored illustrations of seashore animals of Japan. Hoikush. Osaka, Japan. 167 p.

Vaccaro, R. F., and J. H. Ryther. 1960. Marine phytoplankton and the distribution of nitrate in the sea. J. Cons. Perm. Int. Explor. Mer **25:**260-271.

Välikangas, I. 1933. Über die Biologie der Ostsee als Brackwassergebiet. Verh. int. Verein. theor. angew. Limnol. **6:**1.

Vallentyne, J. R. 1957. The molecular nature of organic matter in lakes and oceans, with lesser reference to sewage and terrestrial soils. J. Fish. Res. Bd. Canada **14:**33-82.

Vaughan, R. W. 1919. Corals and the formation of coral reefs. Smithsonian report, 1917. Smithsonian Institution, Washington, D.C.

Vaughan, T. W. 1930. The oceanographical point of view: contributions to marine biology. Stanford University Press, Stanford, Calif. pp. 40-56.

Veevers, H. G. 1951. Photography of the sea floor. J. Mar. Biol. Ass. U.K. **30:**215-221.

Veevers, H. G. 1952. A photographic survey of certain areas of sea floor near Plymouth. J. Mar. Biol. Ass. U.K. **31**:215-221.

Verjbinskaya, N. 1932. Observations on the nitrate changes in the Barents Sea. J. Cons. Perm. Int. Explor. Mer **7**:47-52.

Vernberg, F. J., A. Calabrese, F. P. Thurberg, and W. B. Vernberg [eds.]. 1977. Physiological responses of marine biota to pollutants. Academic Press, Inc., New York. 482 p.

Vernberg, F. J., and W. B. Vernberg [eds.]. 1974. Pollution and physiology of marine organisms. Academic Press, Inc., New York.

Vernberg, W. B., and F. J. Vernberg. 1972. Environmental physiology of marine animals. Springer-Verlag, New York. 346 p.

Verrill, A. E. 1880, 1882, 1885. Remarkable marine fauna occupying the outer banks. Amer. J. Sci. (ser. 3) **20, 23, 29.**

Vinogradov, A. P. 1938. Chemical composition of marine plankton. Trudy Inst. Okeanol. **2**:97-112.

Vinogradov, A. P. 1953. The elementary chemical composition of marine organisms. Memoir 2. Sears Foundation for Marine Research, New Haven, Conn. 647 p.

Vinogradova, N. G. 1959. The zoogeographical distribution of the deep-water bottom fauna in the abyssal zone of the ocean. Deep-Sea Res. **5**: 205-208.

Vishniac, W. 1968. Autotrophy; energy availability in the sea and scope of activity. *In* C. H. Oppenheimer [ed.]. Marine biology. Vol. IV. (Proceedings of Fourth International Conference on Marine Biology, 1966.) New York Academy of Sciences, New York.

Visscher, J. P. 1927. Nature and extent of fouling of ships' bottoms. Bull. Bur. Fish. (Washington) **43**:193-252.

Volterra, V. 1926. Leggi delle fluttuazioni biologiche. Atti R. Accad. Naz. Lincei R. Cl. Sci. Fis. Mat. e Nat. **5**(1):3-10 and **5**(2):61-67.

Volterra, V., and U. d'Ancona. 1935. Les associations biologiques au point de vue mathématique. Herrmann et Cie, Paris.

Wafar, M. V. M., and S. Z. Qasim. 1975. Carbon fixation and excretion in symbiotic algae (zooxanthellae) in the presence of host homogenates. Ind. J. Mar. Sci. **4**(1):43-46.

Wagner, R. H. 1971. Environment and man. W. W. Norton & Co., New York. 462 p.

Waksman, S. A. 1933. Distribution of organic matter in the sea bottom and chemical nature and origin of marine humus. Soil Sci. **36**:125-147.

Waksman, S. A., C. L. Carey, and H. W. Reuszer. 1933. Marine bacteria and their role in the cycle of life in the sea. I. Decomposition of marine plant and animal residues by bacteria. Biol. Bull. Mar. Biol. Lab. Woods Hole **65**:57-79.

Waksman, S. A., and M. Hotchkiss. 1937. Viability of bacteria in seawater. J. Bacteriol. **33**:389-400.

Waksman, S. A., and C. E. Renn. 1936. Decomposition of organic matter in seawater. III. Factors influencing the rate of decomposition. Biol. Bull. Mar. Biol. Lab. Woods Hole **70**:472-483.

Waksman, S. A., H. W. Reuszer, and C. L. Carey. 1935. Decomposition of organic matter in seawater by bacteria. J. Bacteriol. **29**:531-543.

Waksman, S. A., and U. Vartiovaara. 1938. Adsorption of bacteria by marine bottom. Biol. Bull. Mar. Biol. Lab. Woods Hole **74**:56-63.

Walford, L. A. 1958. Living resources of the sea: opportunities for research and expansion. The Ronald Press Co., New York. 321 p.

Waller, H. O., and W. Polski. 1959. Planktonic Foraminifera of the Asiatic shelf. Contr. Cushman Found. **10**:123-126.

Wangersky, P. J. 1952. Isolation of ascorbic acid and rhamnosides from seawater. Science **115**:685.

Wangersky, P. J. 1959. Dissolved carbohydrates in Long Island Sound, 1956-1958. Bull. Bingham Oceanogr. Coll. **17**:87-94.

Ward, R. 1971. The living clocks. Alfred A. Knopf, Inc., New York. ix + 385 p.

Ward, R. 1974. Into the ocean world. Alfred A. Knopf, Inc., New York. xvii + 323 p.

Warren, C. E. 1971. Biology and water pollution control. W. B. Saunders Co., Philadelphia.

Waterman, T. H. 1954. Polarization patterns in submarine illumination. Science **120**:927-932.

Waterman, T. H. 1960. Interaction of polarized light and turbidity in the orientation of *Daphnia* and *Mysidium*. Z. Vergl. Physiol. **43**:149-172.

Waterman, T. H. [ed.]. 1961. The physiology of crustacea. 2 vols. Academic Press, Inc., New York.

Waterman, T. H., R. F. Nennemacher, F. A. Chace, and G. L. Clarke. 1939. Diurnal vertical migrations of deep-water plankton. Biol. Bull. Mar. Biol. Lab. Woods Hole **76**:256-279.

Watson, A. T. 1928. Observations on the habits and life history of *Pectinaria (Lagis) koreni*. [Edited and with introduction by P. Fauvel.] Mgr. Proc. Trans. Liverpool Biol. Soc. **42**:25-60.

Watson, G. E. 1966. Seabirds of the tropical Atlantic Ocean. Smithsonian Identification Manuals Series. Smithsonian Institution, Washington, D.C.

Watson, S. W. 1953. Virus diseases of fish. Trans. Amer. Fish. Soc. **83**:331-341.

Watt, K. E. [ed.]. 1966, 1967. Systems analysis in ecology. Academic Press, Inc., New York.

Wegener, A. 1912. Die Entstehung der Kontinente. Peterm. Mitt. 185-195; 253-256; 305-309. Shorter version in Geol. Rundsch. **3**(4):276-292.

Wegener, A. 1924. The origin of continents and oceans. Transl. from 3rd German ed. by J. G. A. Skerl. Methuen & Co., Ltd., London.

Wegener, A. 1966. The origin of continents and oceans. Transl. from 4th German ed. by J. Biram. Dover Publications, Inc., New York.

Wenk, E. 1972. The politics of the ocean. University of Washington Press, Seattle. 590 p.

Wentworth, C. K. 1926. Methods of mechanical

analysis of sediments. Univ. Iowa Stud. Natur. Hist. **11**(11):3-52.

Wetmore, A. 1936. Birds of the northern seas. Nat. Geog. Mag. **69**(1):95-122.

Weyl, P. K. 1968. Precambrian environment and the development of life. Science **161:**158-160.

Wheeler, W. M. 1901. The free-swimming copepods of the Woods Hole region. Bull. U.S. Fish Comm. 19. Washington, D.C., 157-192.

Whittaker, R. H. 1969. New concepts of kingdoms of organisms. Science **163:**150-160.

Wiborg, K. F. 1954. Investigations on zooplankton in coastal and offshore waters of western and northwestern Norway. Rep. Norweg. Fish Mar. Res. Invest. **11:**1-246.

Wiborg, K. F. 1955. Zooplankton in relation to hydrography in the Norwegian Sea. Rep. Norweg. Fish Mar. Res. Invest. **11**(4):1-66.

Wiborg, K. F. 1960. Investigations on pelagic fry of cod and haddock in costal and offshore areas of northern Norway in July-August, 1957. Rep. Norweg. Fish Mar. Res. Invest. **12**(8):1-18.

Wickstead, J. H. 1958. A survey of the larger zooplankton of Singapore Strait. J. Cons. Perm. Int. Explor. Mer **23:**341-353.

Wickstead, J. H. 1959. A predatory copepod. J. Anim. Ecol. **28:**69-72.

Wickstead, J. H. 1965. An introduction to the study of tropical plankton. Hutchinson & Co. (Publishers), Ltd., London.

Wiens, H. J. 1962. Atoll environment and ecology. Yale University Press. New Haven, Conn. 532 p.

Wieser, W. 1952. Investigations on the microfauna of seaweeds inhabiting rocky coast. J. Mar. Biol. Ass. U.K. **31:**145-174.

Wieser, W. 1953. Die Beziehung zwischen Mundhöhlengestalt, Ernaehrungsgeweise und Vorkommen bei freilebenden marinen Nematoden. Ark. Zool. **2**(4):439-484.

Wieser, W. 1954. Beitrage zur Kenntnis der Nematoden submariner Höhlen. Ergebnisse der Österreichischen Tyrrhenia Expedition, 1952. Teil II. Österr. Zool. Zeitschr. **5**(½):172-230.

Wieser, W. 1957. Free-living marine nematodes. *In* Reports Lunds University Chile Expedition, 1948-1949. Vol. 4, Ecology. Acta Univ. Lund. Andra Afd. N.S. **52**(13):1-115.

Wieser, W. 1959. Free-living nematodes and other small invertebrates of the Puget Sound beaches. Univ. Washington Publ. Zool. **19:**1-179.

Wieser, W., and J. Kanwisher. 1961. Ecological and physiological studies on marine nematodes from a salt marsh near Woods Hole, Massachusetts. Limnol. and Oceanogr. **6**(3):262-270.

Wilbur, K. M., and C. M. Yonge. 1964, 1966. Physiology of mollusca. 2 vols. Academic Press, Inc., New York.

Wiles, W. W. 1960. Use of pore concentration in the tests of planktonic Foraminifera for correlation of quaternary deep sea sediments. Bull. Geol. Soc. Amer. **70:**1699.

Wiley, M. [ed.]. 1976. Estuarine processes. Academic Press, Inc., New York. Vol. 1, 588 p; vol. 2, 444 p.

Williams, G. B. 1976. Aggregation during settlement as a factor in the establishment of coelenterate colonies. Ophelia **15**(1):57-64.

Williston, S. W. 1914. Water reptiles of the past and present. The University of Chicago Press, Chicago. 251 p.

Wilson, C. B. 1932*a*. Copepod crustaceans of Chesapeake Bay. Proc. U.S. Nat. Mus. **80:**1-54.

Wilson, C. B. 1932*b*. Copepods of the Woods Hole region, Bull. 58, Mass. U.S. Natl. Mus., 635 p.

Wilson, C. B. 1942. Copepods of plankton gathered during last cruise of the "Carnegie." Publ. Carnegie Inst., Washington, D.C. Vol. 536. 237 p.

Wilson, D. P. 1951*a*. Larval metamorphosis and the substratum. Ann. Biol. **27:**491-501.

Wilson, D. P. 1951*b*. Life of the shore and shallow sea. Nicholsen & Watson, London.

Wilson, D. P. 1958. Problems in larval ecology related to localized distribution of bottom animals. *In* A. A. Buzzati-Traverso [ed.]. Perspectives in marine biology. University of California Press, Richmond.

Wilson, D. P., and F. J. Armstrong. 1958. Biological differences between seawaters: experiments in 1954 and 1955. J. Mar. Biol. Ass. U.K. **37:**331-348.

Wilson, E. O. 1970. Chemical communication within animal species. *In* Chemical ecology. Sondheimer and Simeone [eds.] Academic Press, Inc., New York.

Wilson, E. O., and W. H. Bossert. 1963. Chemical communication between animals. *In* Recent Progress in Hormone Research. Vol. 19. G. Pincus [ed.] Academic Press, Inc., New York.

Wilson, J. T. 1963. Continental drift. Sci. Amer. **208**(4):86-104. Reprinted in V. G. Dethier and others [eds.]. Topics in the study of life: the BIO source book. Harper & Row, Publishers, New York.

Wimpenny, R. S. 1936. The distribution, breeding, and feeding of some important plankton organisms of the southwest North Sea in 1934. I. *Calanus finmarchicus* (Gunn.), *Sagitta setroa* (J. Müller), and *Sagitta elegans* (Verrill). Fish. Invest. London (ser. 2) **15**(3):1-53.

Wimpenny, R. S. 1938. Diurnal variation in the feeding and breeding of zooplankton related to the numerical balance of the zoophytoplankton community. J. Cons. Perm. Int. Explor. Mer **13:**323-336.

Wimpenny, R. S. 1946. The size of diatoms. II. Further observations on *Rhizolenia styliformis* (Brightwell). J. Mar. Biol. Ass. U.K. **26:**271-284.

Wimpenny, R. S. 1958. Carbon production in the sea at the Smith's Knoll Light Vessel. Rapp. P.-v. Réun. Cons. Perm. Int. Explor. Mer **144:**70-72.

Wimpenny, R. S. 1966. The plankton of the sea. American Elsevier Publishing Co., Inc., New York.

Winberg, G. G. [ed.]. 1971. Methods for estimation of production of aquatic animals. (Transl. from Russian by A. Duncan.) Academic Press, Inc., New York. 186 p.

Wolff, T. 1960. The hadal community: an introduction. Deep-Sea Res. **6**:95-124.

Wolken, J. J. 1971. Invertebrate Photoreceptors. Academic Press, Inc., New York. 188 p.

Wood, E. J. F. 1950. Investigations of underwater fouling. I. Role of bacteria in early stages of fouling. Aust. J. Mar. Freshw. Res. **1**:85-91.

Wood, E. J. F. 1953. Heterotrophic bacteria in marine environments of eastern Australia. Aust. J. Mar. Freshw. Res. **4**:140-200.

Wood, E. J. F. 1954. Dinoflagellates of the Australian region. I. Austr. J. Mar. Freshw. Res. **5**:171-351.

Wood, E. J. F. 1956. Diatoms in the ocean deeps. Pacific Sci. **10**:377-381.

Wood, E. J. F. 1963. Dinoflagellates of the Australian region. II and III. Div. Fish Oceanogr. C.S.I.R.O. Tech. paper **14**:1-55; **17**:1-20.

Wood, E. J. F. 1963. Relative importance of groups of algae and protozoa in marine environments of the southwest Pacific and east Indian oceans. *In* C. H. Oppenheimer [ed.]. Symposium on marine microbiology. Charles C Thomas, Publisher, Springfield, Ill.

Wood, E. J. F. 1964. Studies in the microbial ecology of the Australasian region. Nova Hedwigia **79**(1-4).

Wood, E. J. F. 1965. Marine microbial ecology. Reinhold Publishing Corp., New York. 243 p.

Woodmansee, R. A. 1958. Seasonal distribution of zooplankton off Chicken Key in Biscayne Bay, Florida. Ecology **39**:247-262.

Woodwell, G. M., C. F. Wurster, and P. A. Isaacson. 1967. DDT residues in an East Coast estuary: a case of biological concentration of a persistent insecticide. Science **156**:821-824.

Worthington, L. V. 1954. Preliminary note on the time scale in North Atlantic circulation. Deep-Sea Res. **1**:244-251.

Wright, W. R. 1976. Currents of the sea. Nat. Hist. **85**(7):30-37.

Wust, G. 1930. Meridional Schichtung und Tiefenzirculation in den west halften der drei Ozeane. J. Cons. Int. Perm. Explor. Mer **5**:7-21.

Wust, G. 1954. Die zonal Verteilung von Salzgehalt, Niedershlag, Verdunstung, Temperatur, und Dichte an der Oberflache der Ozeane. Kieler Meereforsch. **10**:137-161.

Wyrtki, K. 1961. The thermohaline circulation in relation to the general circulation in the oceans. Deep-Sea Res. **8**:39-64.

Yankwich, P. E., T. H. Norris, and J. L. Juston. 1947. Correcting for the absorption of weak beta particles in thick samples. Ind. Eng. Chem. Anal. **1950**:439-441.

Yentsch, C. S., and J. H. Ryther. 1957. Short-term variations in phytoplankton chlorophyll and their significance. Limnol. Oceanogr. **26**:140-142.

Yentsch, C. S., and J. H. Ryther. 1959. Relative significance of the net phytoplankton and nannoplankton in the waters of Vineyard Sound. J. Cons. Int. Perm. Explor. Mer **24**:231-238.

Yentsch, C. S., and R. F. Scagel. 1958. Diurnal study of phytoplankton pigments: an *in situ* study in East-sound, Washington. J. Mar. Res. (Thompson Anniversary vol.) **17**:567-583.

Yonge, C. M. 1928. Feeding mechanisms in the invertebrates. Biol. Rev. **3**:21-76.

Yonge, C. M. 1958. Ecology and Physiology of reef-building corals. *In* A. A. Buzzati-Traverso [ed.]. Perspectives in marine biology. University of California Press, Richmond.

Zalokkar, M. 1942. Les associations sous-marines de la côte adriatique au-dessous de Velebit. Bull. Soc. Bot. Genève **33**:172-195.

Zeitsschel, B. [ed.]. 1973. The biology of the Indian Ocean. SCF(SEPS) (Ecological studies, vol. 3.) Springer-Verlag, New York.

Zenkevitch, L. A. 1947, 1951. The fauna and biological productivity of the sea. Moscow. Vol 1, 506 p.; vol. 2, 588 p.

Zenkevitch, L. A. 1961. Certain quantitative characteristics of the pelagic and bottom life of the ocean. *In* M. Sears [ed.]. Oceanography. Pub. no. 67. American Association for the Advancement of Science, Washington, D.C.

Zenkevitch, L. A. 1963. Biology of the seas of the U.S.S.R. Transl. by S. Botcharskaya. John Wiley & Sons, Inc., New York.

Zenkevitch, L. A., and J. A. Birstein. 1956. Studies of deep-water fauna and related problems. Deep-Sea Res. **4**:54-64.

Zenkevitch, L. A., V. Brodsky, and M. Idelson. 1928. Materials for the study of the productivity of the sea bottom in the White, Barents and Kara seas. J. Cons. Perm. Int. Explor. Mer **3**:372-379.

Zim, H. S., and L. Ingles. Seashores. Simon & Schuster, Inc., New York.

Zimmerman, A. M. 1970. High-pressure effects of cellular processes. Cell Biology Series. Academic Press, Inc., New York.

ZoBell, C. E. 1941. Studies on marine bacteria. I. Cultural requirements of the heterotrophic aeroves. J. Mar. Res. **4**:42-75.

ZoBell, C. E. 1946. Marine microbiology. Chronica Bontanica Press, Waltham, Mass.

ZoBell, C. E. 1949. Influence of hydrostatic pressure on growth and viability of terrestrial and marine bacteria. J. Bacteriol. **57**:179-189.

ZoBell, C. E. 1952. Bacterial life at the bottom of the Philippines trench. Science **115**:507-508.

ZoBell, C. E. 1954. Occurrence of bacteria in the deep sea and their significance for animal life. Publ. Int. Union Sci. **B16**:20-29.

ZoBell, C. E. 1957. Marine bacteria. *In* J. W. Hedgpeth [ed.]. Treatise on marine ecology and paleoecology. Vol. I. Ecology. Geological Society of America, New York.

ZoBell, C. E., and E. C. Allen. 1935. Significance of marine bacteria in fouling of submerged surfaces. J. Bacteriol. **29**:239-251.

ZoBell, C. E., and F. H. Johnsen. 1949. Influence of

hydrostatic pressure on growth and viability of terrestrial and marine bacteria. J. Bacteriol. **57:**179-190.

ZoBell, C. E., and H. B. Michener. 1938. A paradox in the adaptation of marine bacteria to hypotonic solutions, Science **87**(2258):328-329.

ZoBell, C. E., and C. H. Oppenheimer. 1950. Effects of hydrostatic pressure on multiplication and morphology of bacteria. J. Bacteriol. **57:**179-190.

Zweig, G. [ed.]. 1963-1967. Analytical methods for pesticides, plant growth regulators, and food additives. Vol. 1 (1963). Principles, methods, and general applications. Vol. 2 (1964). Insecticides. Vol. 3. (1964). Fungicides, nematocides, soil fumigants, rodenticides, and food additives. Vol. 4 (1964). Herbicides. Vol. 5 (1967). Additional principles, applications, and analyses for pesticides. Academic Press, Inc., New York.

ANNOTATIONS TO THE BIBLIOGRAPHY

The following are groupings of publications representative of certain types of publication or of particular areas of marine biology. The student should realize that the bibliography given in this book is by no means a complete list of relevant publications. It is merely a sampling. Most of the references cited will furnish leads to numerous other important and interesting contributions.

Publications for the general reader as well as for marine biologists

Bates, 1960; Berrill, 1951, 1959; Beebe, 1928, 1935; Carrington, 1957, 1960; Carson, 1951, 1955; Cousteau and Dumas, 1953; Cowen, 1960; Douglas, 1952; Dugan, 1956; Idyll, 1964 (deep sea); Ingmanson and Wallace, 1973; Loftas, 1969; Piccard, 1956; Steinbeck and Ricketts, 1941; Thorson, 1971; Walford, 1958.

The following are good examples of highly illustrated, colorful books for young and old natural histroy lovers: Amos, 1966; Berrill, 1966; Engel and editors of *Life*, 1961 (*Life* volumes: *The World We Live In* and *The Sea).*

Scientific American has devoted special issues and numerous articles to oceanic and marine biological studies. Good examples of regional studies are the May, 1961 (Arctic), and September, 1962 (Antarctic), iisues. Ten of these articles were reprinted in book form as: *The Ocean,* a *Scientific American* book, published by the W. H. Freeman Co., San Francisco, in 1969 and forty-one of them in 1971 under the title *Readings From the Scientific American: Oceanography*. Freeman also makes reprints from *Scientific American* available at nominal cost. The *National Geographic Magazine* contains may fine regional studies on a less technical level. Indexes to these periodicals will give leads to many rewarding, beautifully illustrated articles. *Natural History,* published by the American Museum Natural History, also contains numerous interesting well-illustrated articles in marine biology. The Aug./sept., 1976, issue is devoted to the oceans.

Some books helpful along the seashore: photographs, drawings, and discussion of organisms commonly seen and their natural history and ecology

Carson, 1955 (eastern coast, U.S.); Dawson, 1956 (marine algae); Day, 1974 (South African seashore life); Flattely and Walton, 1922 (biology of the seashore); Flora and Fairbanks, 1962 (natural history, Puget Sound region); Guberlet, 1936, 1956 (West Coast animals and seaweeds); Hillson, 1976 (marine plants, U.S. eastern coast); Iverson, 1968 (farming the edge of the sea); Kozloff, 1973 (Puget Sound area marine life); Lewis, 1964 (ecology of rocky shores); Light and others, 1954, revised by Smith and others (Keys, West Coast animals); MacGinitie and MacGinitie, 1949 (natural history, marine animals); Ricketts and Calvin, 1968 (western coast, U.S.); Robbins and Yentsch, 1973 (New England Coast); Smith, 1944 (marine algae, West Coast); Wilson, 1951*b* (life of shore and shallow waters); Zim and Ingles, 1961 (shoreline biology).

History of marine sciences

Briggs, 1968 (biographical sketches of recent oceanographers); Deacon, 1971 (history of exploration and of marine sciences); Schlee, 1973 (history of oceanography); Ward, 1974 (account of significant developments in marine science).

Some books addressed primarily to marine biologists and oceanographers

Friedrich, 1969; Hardy, 1956; Mariscal, 1974; Marshall, 1954, 1967; Murray and Hjort, 1912; Russell and Yonge, 1928; Stommel, 1958. The foregoing works are of a general nature and contain much of interest to laymen as well as to professional marine biologists. The following are intended primarily for professional marine biologists and oceanographers. Barnes, 1959; Buzzati-Traverso [ed.], 1958; Davis, 1955; Fairbridge, 1966; Giese and Pearse, 1974, 1975; Green, 1968; Hedgpeth [ed.], 1957 (2 vol. treatise); Kinné, 1970-1972 (5 vol. treatise); Kriss and others, 1964; Moore, 1958; Nicol, 1959; Nitikin [ed.], 1959; Oppenheimer [ed.], 1963; Raymont, 1963; Sverdrup, Johnson, and Fleming, 196; Wimpenny, 1966.

Some recurrent publications, journals, annual reviews, etc. of interest to marine biologists

Advances in Marine Biology (F. S. Russell [ed.]; annual; Academic Press, Inc.)

Biochemical and Biophysical Perspectives in Marine Biology. (D. C. Maline and J. R. Sargent [eds.]; Academic Press, Inc., New York)

Biological Bulletin of the Marine Biology Laboratory, Woods Hole (monthly)

Bulletin of the Bingham Oceanographic Collection (Yale) (has been discontinued)
Bulletin of Enviornmental Contamination and Toxicology
Bulletin of Marine Science of the Gulf and Caribbean (quarterly; University of Miami Press)
Current Contents in Marine Sciences (FAO monthly—a bibliographic aid)
Deep-Sea Research (Pergamon Press, Inc.). (The 1955 supplement to vol. 3, Papers in Marine Biology and Oceanography, dedicated to H. B. Bigelow, one of the leaders in development of oceanographic studies in America, contains numerous interesting contributions by various scientists and a complete bibliography of the scientific publications of H. B. Bigelow.)
Director of Oceanographers (annual; National Academy of Science)
Ecology (American Ecological Society)
Environmental Research (annual; Academic Press, Inc.)
EQS Environmental Quality and Safety (International) (semiannual; Academic Press, Inc.; also Georg Thieme Verlag, Stuttgart)
Estuarine and Coastal Marine Science (bimonthly; Academic Press, Inc.)
Estuarine Research (biennial; begun 1975; Academic Press, Inc.)
International Journal of Oceanology and Limnology (Omnipress)
International Marine Science Quarterly (UNESCO)
Internationale Revue der gesampten Hydrobiologie
Journal du Conseil (Copenhagen)
Journal of Experimental Marine Biology and Ecology
Journal of Fish Biology (Academic Press, Inc.)
Journal of the Marine Biological Association, United Kingdom
Journal of Marine Research (Sears Foundation)
Journal of Physical Oceanography (American Meteorological Society)
Limnology and Oceanography (American Society of Limnology and Oceanography)
Marine Biology (international journal begun 1967-1968)
Marine Pollution Bulletin (monthly; Pergamon Press)
Ocean Year Book. (E. M. Borgese [ed.]; annual review of major ocean issues; begun 1978; University of Chicago Press)
Oceanic Citation Journal (Oceanic Research Institute; contains abstracts as well as citations)
Oceanic Instrumentation Reporter (monthly; abstracts of world literature; Ocean Engineering Information Service)
Oceanography and Marine Biology (annual review; Allen & Unwin, London)
Pollution Abstracts (Paul Janensch, La Jolla, Calif.)
Sea-bed 1968. 6 vols; 1969. 8 vols. (M. Y. Sachs [ed.]; annual; United Nations and other international papers and documents relative to the seabed; for research libraries, etc.)

Ecology

General and marine: Adams, 1976 (eel-grass, fish communities); Allee and others, 1949; Clements and Shelford, 1927; Dunbar, 1971 (evolution of stability); Emery and others, 1955 (nutrient budgets); Fankboner, 1971 (role of zooxanthellae); Hedgpeth [ed.], 1957 (2-volume treatise on marine ecology and paleoecology); Hesse and others, 1951 (ecological animal geography); Kanwisher, 1971 (temperature regulations); Kinne [ed.], 1970-1972 (marine ecology, 5-volume treatise); Manteufel, 1941 (Barents Sea, annual cycles); Moore, 1958; Odum, 1953, 1956; Rakussa-Suszczewski and McWhinnie, 1976 (resistance to freezing in Antarctica); Rittenberg, 1955 (nutrient cycles in sediments); Schäfer, 1972 (marine ecology and paleontology); Schmidt-Nielsen, 1971 (saltwater balance, vertebrates); Smayda, 1963 (role of ectocrine substances and limiting factors); Smith, 1972 (ecosystem structure and dynamics); Steele [ed.], 1970 (marine food chains); Steemann-Nielsen, 1935-1958 (studies in production, role of antibiotics, etc.); Vernberg and Vernberg, 1972 (physiological ecology); Wilson and Bossert, 1963 (chemical communication).

Analytical methods in ecology: Fager, 1957 (analysis of recurrent groups); Gislin, 1930 (culmination of Peterson school of particulate approach to community analysis in ecology); Lucas, 1938 (integration in plankton communities); Margalef, 1958 (information theory in ecological analysis); McConnaughey, 1964 (analysis of field data to determine communities, revision and extension of Fager's method); Newell, 1948 (analysis of linearly disposed communities or provinces of organisms, revision and extension from methods proposed by Schenck and Keen); Patten, 1959 (cybernetics of ecosystem); Petersen, 1918 (statistical analysis of soft bottom benthic communities); Poore, 1962 (successive approximation in descriptive ecology); Schenck and Keen, 1936 (analysis of linearly disposed biological provinces); Wieser, 1958 (particulate and comparative approaches compared); Williams, 1965 (and numerous other articles by Williams, and by Williams and Lambert in British Journal of Ecology) (multivariate analysis and information theory in ecological analysis).

Population ecology: Boughey, 1968; Fleming, 1939 (diatom population and grazing); Hazen [ed.], 1964; MacArthur and Connel, 1966.

Biogeography and regional studies: Ekman, 1953 (best general summary of biogeography of world's oceans); Knox, 1963 (good example of regional biogeographical study—intertidal biogeography of Australian–New Zealand region); McGowan, 1974 (Pacific oceanic ecosystems); Staele, 1974 (structure of marine ecosystems); Zeitsschel, 1973 (biology of Indian Ocean).

The great multivolume reports of various scientific expeditions are also mines of information. Among the greatest of these, the *'Challenger' Reports,* some fifty volumes, resulting from the first worldwide oceanographic expedition, opened up a new era in marine biology and oceanography and remain as a classic with which all oceanographers and marine biologists should be familiar. A few of the other notable expedition reports, regional in scope, are those from the Siboga Expedition (Dutch East Indies), the Great Barrier Reef Expedition (northeast Australia), the "Galathea" Expeditions, the John Murray Expedition, the *'Discovery' Reports,* and the aprince Albert of Monaco Expeditions, the John Hancock Expeditions, and the "Albatross" Expeditions.

Plankton

Some books devoted to plankton: Davis, 1955 (keys and illustrations for identification); Fraser, 1962; Hardy, 1956; Johnstone and others, 1924; Ostenfeld, 1916 (international catalog); Raymont, 1963 (probably the best general book on plankton and marine production).

Some shorter workds of a general nature relating to plankton: Bernard, 1955, Bogorov, 1934, 1941, 1956, 1958, 1958*c* (plankton of northwest Pacific); Buljan, 1959 (influence of forest on coastal plankton); Cushing, 1953, 1954, 1959*a*; Hardy, 1956, 1960; Harvey, 1942, 1947, 1950; Harvey and others, 1935; Kincaid, 1942; Lebour, 1922, 1923; Lucas, 1956; Nielsen, 1937; Raymont, 1966; Redfield, 1934; Ryther, 1954, 1956; Steele, 1959.

Some examples of special studies of plankton organisms from particular points of view: Bentley, 1960 (influence of plant hormones on); Corner and Davies, 1971 (relation of plankton to N and P cycles); Eyden, 1923 (relation to specific gravity); Gunther, 1934 (fatty canstituents); Spooner, 1933 (reactions to light).

Some general plankton studies from various areas: Apstein, 1907, 1910 (Ceylon); Bigelow, 1914, 1917, 1922, 1926*a* (Atlantic coast off northern U. S.); Bogorov, 1939 (Arctic, in relation to ice forecasting); Brandt and Apstein, 1901-1912 (North Atlantic); Brunelli, 1941 (Laguna Veneta, Italy); Bsharah, 1957 (Florida Current); Burckhardt, 1899 (White Sea); Candeias, 1930 (Portugal); Cleve, 1901 (Indian Ocean and Malay Archipelago); Coleman and Segrove, 1955 (New England waters); Cowles, 1930 (Chesapeake Bay, offshore); Delsman, 1939 (Java Sea); Fish, 1925 (Woods Hole, seasonal); Fraser, 1948, 1949, 1954, 1955, 1956 (British Isles region); Frost, 1937 (Newfoundalnad); Gauld, 1950 (Scottish sea lakes); George, 1953 (India, coastal); Gilmartin, 1958 (Eniwetok lagoon, Central Pacific); Gotto, 1951 (Ireland); Gran, 1927 (Norway); Gunther, 1936 (Peru Current); Hardy, 1936*b* (Arctic); Hart, 1953 (Benguela Current); Hentschel and Wattenberg, 1930 (South Atlantic); Holmes, 1958 (eastern Pacific); Japanese Hydrographic Office, 1956 (equatorial Pacific); Jaschnow, 1939, 1940 (Arctic and Barents Sea); Jespersen, 1923, 1924, 1935, 1944 (macroplankton, North Atlantic and Mediterranean); Kow, 1953 (Singapore Strait); Kusmorskaja, 1940 (Sea of Okhotsk); Wood, 1963 (southwest Pacific and east Indian Ocean); Yamazi, 1958 (Japan).

Phytoplankton

Biology: Bainbridge, 1957 (size, shape, density of phytoplankton clouds); Burke and others, 1962 (dominance in phytoplankton blooms); Curl and McLeod, 1961 (physiology and ecology of *Skeletonema costatum*); Cushing, 1958 (carbon content); Doty and Ogun, 1957 (daily periodicity in production); Droop, 1957 (nutrition of); Gran, 1912, 1927, 1929, 1931 (quantitative studies and conditions for production); Harvey, 1934, 1947 (measurement of, and relation to manganese); Margalef, 1958 (succession and heterogeneity); Rice, 1954 (biotic influences on growth of); Rustad, 1946 (photosynthesis and light at various depths); Sieburth and Burkholder, 1959 (antibiotic activity of); Smayda, 1958 (biogeography of marine phytoplankton); Steemann-Nielsen, 1955 (production and ecological effects of antibiotics); Sverdrup, 1953 (conditions for vernal bloom); Vaccaro and Ryther, 1960 (relation to NO_3).

Some regional or area studies of phytoplankton: Allen, 1919-1924 (southern California); Bigelow, 1922, and Bigelow and others, 1940 (Gulf of Maine); Bogorov and Beklemishev, 1955 (western North Pacific); Corlett, 1953 (oceanic Atlantic); Gran and Braarud, 1935 (Bay of Fundy, Gulf of Maine); Grøntved, 1952 (North Sea); Hart, 1934 (Bellingshausen Sea and southwestern Atlantic); Hasle, 1959 (equatorial Pacific); Hulbert and others, 1960 (Sargasso Sea); Lillick, 1938, 1940 (Gulf of Maine and environs); Phifer, 1934 (Puget Sound); Riley, 1942, 1946, 1947 (New England region), 1957 (Sargasso Sea); Riley and Conover., 1956 (Long Island Sound); Ryther, 1954 (Long Island Sound); Sears, 1941 (Georges Bank, Gulf of Maine); Shirshov, 1937, 1938 (polar seas, relation to ice regime); Sleggs, 1927.

Primary production and productivity of marine waters

General works on production: Cushing, 1971 (upwelling and fish production); Fasham and Pugh, 1976 (chlorophyll and temperature); Raymont, 1963; Ryther, 1963 (geographical variations in productivity), 1954 (ecology of blooms), 1956 (photosynthesis as a function of light intensity), 1959 (potential productivity), 1969 (marine production and world fisheries); Steele, 1970 (food chains).

Methods of studying production: Braarud, 1961 (cultivation methods); Brand, 1934 (determination of carbon and nitrogen in small samples); Creitz and Richards, 1955 (estimation by pigment analysis); Doty, 1956, 1959 (C^{14} method); Gran, 1927, 1929, 1931 (oxygen method); Gross, 1937 (culture); Harvey, 1934, 1942, 1950 (measurement of production and review summary of production); Harvey and others, 1935 (production and its control); Hollibaugh, 1976 (amino acids and phytoplankton growth); Krey, 1958, 1961 (chemical methods); Richards and Thompson, 1952 (pigment analysis); Ryther and Yentsch, 1957 (chlorophyll and light data); Steemann-Nielsen, 1952, 1954, 1958, and Steemann-Nielsen and Jensen, 1957 (experimental methods, especially C^{14}); Strickland, 1960 (measuring poduction); Utermohl, 1931 (use of inverted microscope for plankton counting); Winberg, 1971 (production of animals).

Some studies of production in various regions and areas: Curl, 1960 (northern coastal South America); Currie, 1953 (Benguela Current); Gaardner and Gran, 1927 (Norway); Gaardner and Sparck, 1931 (Norway); Gillbricht, 1952*b* (Kieler Bucht); Graham, 1941, 1943 (open Pacific); Gran, 1927-1931 (North Atlantic and off Norway); Hansen, 1959 (North Atlantic); Holmes, 1958 (eastern Pacific); Jaschnow, 1939, 1940 (Arctic Ocean and Barents Sea); Marshall, P. T., 1958 (Arctic); McAllister and others, 1960 (northeastern Pacific); Riley, 1946 (Georges Bank), 1955 (Long Island Sound), 1957 (Sargasso Sea); Riley and Conover, 1956 (Long Island Sound); Riley and others, 1949 (North Atlantic); Rose, 1925 (Bay of Algiers); Ryther, 1963 (Sargasso Sea); Steele, 1958 (North Sea); Steemann-Nielsen, 1935, 1937, 1940, 1951 (eastern North Atlantic); Wimpenny, 1958 (Smith's Knoll Light Vessel).

Zooplankton

Identification: Jespersen and Russell, 1939-1952 (identification).

Diurnal vertical migration and vertical distribution—relation to the deep-scattering layer: Boden and Kampa, 1967 (relation to light); Bogorov, 1946 (vertical migration in polar seas); Clarke, G. L., 1936 (relation to light); Cushing, 1951, 1955 (review and experiments); David, 1961 (relation to speciation); Enright and Hammer, 1967 (massive experiments involving several species); Esterly, 1919 (experiments on reactions); Hansen, 1951 (relation to the discontinuity layer); Hardy, 1947, and Hardy and Bainbridge, 1954 (experiments on rates of movement, etc., and use of plankton wheel); Hida and King, 1955 (vertical distribution in euqatorial Pacific); Leavitt, 1935, 1938 (movements of macroplankton); Moore and Corwin, 1956 (temperature and light); Russell, 1925-1934 (vertical distribution); Russell, 1935 (behavior in relation to light, value as indicators of water masses in English Channel, general review); Waterman and others, 1939 (deep-water plankton).

Zooplankton as indicators of hydrographic conditions and biogeographic boundaries: Bary, 1959. (See also section on diurnal vertical migration and vertical distribution, above, and section on general studies in particular regions or areas, below.)

Other aspects of zooplankton behavior: Bainbridge, 1949 (movements in a diatom gradient); Bossanyi, 1958 (demersal antant zooplankton); David, 1967 (neuston); Denton and Shaw, 1962 (bouyancy of gelatinous zooplankton).

Some general studies in particular regions or areas: Bernard, 1955*b* (Bay of Algiers, seasonal); Bernstein, 1934 (Kara Sea); Bigelow, 1938 (Bermuda, medusae); Bigelow and Sears, 1939 (New England, volumes of zooplankton); Bogorov and Vinogradov, 1955 (northwestern Pacific); Clarke, G. L., 1940 (comparative richness, coastal and offshore Atlantic); Deevey, 1948, 1952, 1956 (Block Island Sound, Long Island Sound); Delsman, 1939 (Java Sea); Fraser, 1947, 1952, 1955 (off the British Isles); Frolander, 1962 (zooplankton off Washington); Jespersen, 1923, 1924, 1935, 1940, 1944 (macroplankton of North Atlantic, Mediterranean, quantity in various oceans, etc.); Johnson, 1954, 1958 (Marshall Islands, Pacific, and Point Barrow, Alaska); Mackintosh, 1934, 1937 (Antarctic, relation of circulation and depth changes to life cycle and maintenance of populations in the region); Mauchline, 1969 (biology of euphausiids); McLaren, 1966 (chaetognath size and life-span inthe Arctic); Menzel and Ryther, 1961 (Sargasso Sea); Moore, H. B., 1949 (Bermuda); Riley, 1947 (Georges Bank, Gulf of Maine); Runnstrom, 1932 (Norway, fjords); Russell, 1934 (Great Barrier Reef, Australia); Savage, 1931 (off England); Shropshire, 1944 (plankton harvesting).

Relations between zooplankton and phytoplankton: Bainbridge, 1953; Beklemishev, 1957; Clarke, G. L., 1939; Nielsen, 1937; Steemann-Nielsen, 1937.

Continuous plankton recorder studies: Glover, 1967; Hardy, 1926, 1936*a*, 1939, 1958; Henderson, 1954; Rae and Rees, 1947; Rees, 1954.

Nanoplankton: Aurich, 1949; Lohmann, 1911; Yentsch and Ryther, 1959.

Nekton: Alegev, 1977 (evolution, ecology, and biology of nekton).

Shoreline biology

Estuaries, tidal flats, river mouths, etc.: Some general treatments of estuarine biology: Cronin [ed.], 1975 (estuaries); Green, 1968 (biology of estuarine animals); Lauff [ed.], 1967 (symposium on estuaries); Marcus and others, 1976 (federal port policy); Wiley, 1976 (estuarine processes).

Other references: alexander and others, 1932 (salinity foreshore waters); Amos, 1952, 1954, 1955, 1956 (Delaware estuary); Barlow, 1955 (processes determining distribution of zooplankton in estuaries); Bass Becking and Wood, 1955 (biological processes in estuaries); Beanland, 1940 (Dover estuary); Burt and Marriage, 1957 (pollution in estuaries); Burt and McAllister, 1959 (hydrography of estuaries); Cooper and Milne, 1938-1939 (ecology of the Tamar estuary); Day, 1951, and Day and others, 1952, 1953 (ecology of South African estuaries); Fischer-Piette, 1931 (penetration of sessile marine animals into estuaries); Goodwin, 1961 (vanishing Connecticut estuaries); Gunter and Gunter, 1956, 1958 (Louisiana estuaries); Hedgpeth, 1951 (classification of brackish and estuarine environments); Hickling, 1970 (estuarine fish farming); Holme, 1949*a*, 1950 (Exe estuary and Great West Bay, Great Britain); Hull, 1960 (oxygen balance in estuaries); Johns, 1968 (vulnerability of estuaries); June and

Chamberlin, 1959 (estuaries and menhaden); Kalber, 1959*a*, 1960 (estuaries and tide marshes); Kelley, 1966 (ecology of San Joaquin and Sacramento estuaries); Ketchum, 1972 (coastal zone problems); Lippson, 1973 (Chesapeake Bay); Milne, 1940 (ecology of Tamar estuary); Odum, E. P., 1961 (tide marshes and estuarine production); Pratt, 1949, 1950 (fertilization of saltwater pond); Pritchard, 1955 (estuarine circulation patterns); Raymont, 1955 (ecology of tidal mud flat); Rees, 1940 (ecology of tidal mud flat); Reid, 1961 (inland waters and estuaries); Segerstråle, 1953, 1957, 1959 (Baltic region, classification of brackish waters); Seshaiya, 1959 (estuarine hydrology); Shelford, 1930 (Pacific Northwest); Shelford and Towler, 1925 (animal communities); Shelford and others, 1935 (marine communities); Sinha, 1970 (estuarine pollution, bibliography); Spooner and Moore, 1940 (ecology of Tamar estuary); Teal, 1962 (energy flow in a salt marsh).

Penetration of marine animals into estuaries and fresh waters: Annandale, 1922 (into the Ganges River); Fischer-Piette, 1931; Needham, 1930; Odum, 1953; Pearse, 1936, 1950 (immigrations from sea to land); Schmalhausen, 1968 (origin land vertebrates).

Open coast and continental shelf: For general treatments, especially of the intertidal, see Ricketts and Calvin, 1968; Ballantine, 1961 (biological description of rocky shores); Bird, 1969 (coastal morphology); Ketchum, 1972 (coastal zone problems); Newell and Newell, 1963 (biology of rocky intertidal); Stephenson and Stephenson, 1949 (universal features of intertidal life), and 1950, 1952, 1954, 1961 (accounts of the intertidal in representative regions of North America); Wilson, 1951*b* (life of shore and shallow sea).

Other references: Blegvad, 1914, 1916, 1925, 1929 (Danish waters and mortality due to ice); Brotskaja and Zenkevich, 1939 (Barents Sea); Bruce, 1928 (physical factors, sandy beach); Buchanan, 1968 (shelf benthos off African Gold Coast); Coe and Allen, 1937 (growth on test blocks, southern California); Cole, 1956 (commercial shellfish); Coleman, 1933 (intertidal zonation); DeVanney and others, 1976 (coastal zone economics); Drach, 1948 (rocky shore communities); Ellis, 1960 (infaunal benthos, North American Arctic); Ford, 1923 (off Plymouth); Fowell, 1939 (*Laminaria* communities); Fox, 1950 (organic detritus and inshore animals); Frank, 1965 (limpet population demography); Gilsen, 1930 (Gullmar Fjord); Grøntved, 1957, 1958 (seaweeds); Jones, 1950, 1956 (bottom communities); Kitching and others, 1934 (sublitoral ecology); Knox, 1963 (intertidal, Australia–New Zealand); Lawson, 1954-1957 (zonation, rocky shores, West Africa); Longhurst, 1959 (benthos densities of tropical West Africa); Loosanoff and Davis, 1951 (ecology, growth, and settlement of larval bivalves); MacGinitie, 1939 (littoral communities); Mare, 1942 (microorganisms of benthic community); Moore, 1930, 1931 (mud of Clude Sea); Neushul and others, 1967 (subtidal, California islands); Newcombe, 1935 (sand beach ecology); Parker, 1955, 1956 (ecology of Texas coast and Mississippi River delta); Pearse and others, 1942 (ecology of eastern coast sand beach); Petersen, 1913, 1915, 1918, 1924 (Danish benthic communities); Petersen and Boysen-Jensen, 1911 (evaluation of benthic life); Remane, 1959 (interstitial life in sand); Sanders, 1956, 1958, 1960 (benthos, eastern coast, U.S.); Savilov, 1957 (Sea of Okhotsk); Stephen, 1930 (sandy mud ecology); Stephenson, 1947 (intertidal, South Africa); de Sylva, 1970 (barracuda life cycle); Thorson, 1936, 1946, 1950, 1952, 1955, 1958 (ecology of benthos, especially with respect to reproduction and larval life and settlement, and comparison of equivalent benthic communities of different latitudes and depths); Towler, 1930 (intertidal, Puget Sound); Wieser, 1952 (microfauna of seaweeds); Wilson, 1951*a*, 1958 (larval ecology and settlement); Wilson and Armstrong, 1958 (biological differences between different seawater masses); Yonge, 1928 (feeding mechanisms).

Boring and fouling organisms: The best general summary of work with these groups is contained in Ray [ed.], 1958. A few other works dealing with these problems or the organisms involved are Coe and Allen, 1937 (growht on test blocks); Doochin and Smith, 1951 (fouling in relation to currents); Edmondson, 1944; Edmondson and Ingram, 1939 (fouling organisms in Hawaii); Graham and Gay, 1945 (at Oakland, Calif.); Johnson and Miller, 1935 (shipworms at Friday Harbor); Kofoid and Miller, 1927 (shipworms in San Francisco Bay); Turner, 1966 (survey of shipworms); Visscher, 1927 (fouling of ship bottoms); ZoBell and Allen, 1935 (role of bacteria in initiating fouling).

Deep-sea biology

Bruun, 1957 (general discussion); Bruun and others, 1956 (''Galathea'' expedition); Cattell, 1936 (effects of pressure); Idyll, 1964 (general work on the deep sea); Menzies, 1965 (life of deep-sea benthos); Rittenberg and others, 1955 (nutrient cycles in deep sediments); Sanders and Hessler, 1969 (ecology of deep-sea benthos); Vinogradova, 1959 (abussal benthos); Waksman, 1933 (chemical and organic matter of sediemnts, marine humus); Wood, 1956 (diatoms in deep sediments); Zenkevitch and Birstein, 1956 (deep-water fauna); Zimmerman, 1970 (pressure effects on cell processes); ZoBell and others, 1949, 1950 (effects of pressure on bacteria).

Other references concerning marine organisms

Mass mortalities and red tides: Allen, W. E., 1946 (at La Jolla, Calif., 1945); Blegvad, 1929 (intertidal, due to ice); Brongersma-Sanders, 1957 (review article on mass mortalities, all causes); Collier, 1958 (biochemical aspects of red tides); Galtsoff (red tides); Gunter and Gunter, 1947, and Gunter and others, 1948 (red tide, mass mortality, Florida); Ingle and de Sylva, 1955 (red tide); Ketchum and Keen, 1948 (phosphorus concentrations in Florida red tide); Lund, 1936 (red tides along the Texas coast); McLaughlin, 1958 (poison tides); Powers, 1932 (red tide, Gulf of Maine, caused by a ciliate).

Underwater sounds of biological origin: Burkenroad, 1931; Dobrin, 1947; Knudsen and others, 1948; Johnson, 1943, 1948; Johnson, Everest, and Young, 1947; Loye and Proudfoot, 1946; Tavolga [ed.], 1964.

Biological clocks: Cold Spring Harbor Symposium, 1960; Brahmachary, 1967; Brown, 1958 (Brown and co-workers have published numerous other articles on this subject, not listed in this bibliography); Pittendrigh, 1958; Brown and Hastings, 1970 (biological clocks—two contrasting views).

Behavior of ecological significance: Allee, 1931 (animal aggregations); Baer, 1951 (ecology of animal parasites); Carr, 1965 (turtle migrations); Dales, 1966 (symbiosis in marine organisms); Ebling and Highnam, 1969 (chemical communication); Feder, 1966 (cleaning symbiosis); Fraenkel and Gunn, 1940 (orientation); Gilbert, 1962 (shark behavior); Johnston and others [eds.], 1970 (chemical communication); Lack, 1945 (feeding behavior–related sympatric birds); Shaw, 1966 (schooling, fish); Thorson, 1946 (reproductive and larval behavior); Tinbergen, 1953 (birds). A new journal, begun in 1968, is entitled *Communications in Behavioral Biology* (S. A. Weinstein [ed.], Academic Press, Inc.).

Marine microbiology

Recurrent: Advances in the Microbiology of the Sea (begun 1968).

Other references: Burkholder, 1959, and Burkholder and Burkholder, 1956 (vitamins and bacteria); Carey, 1938 (nitrifying bacteria); Colwell, 1974, 1975 (effect of ocean environment on microbial activities; manual of methods in marine microbiology); Hata, 1960 (relation between microbial activity and redox potential); Hock, 1941 (chitin-digesting bacteria); Hoppe, 1976 (heterotrophic bacteria in sea); Johnson, 1968 (saprobic marine fungi); Johnson and Sparrow, 1962 (fungi in oceans and estuaries); Ketchum and others, 1949, 1952 (coliform bacteria in seawater); Krogh and Sparck, 1936 (bottom-sampler for microbiology); Kriss, 1959, Kriss and Rukina, 1952, and Kriss and others, 1961, 1964 (1967) (oceanic microbiology); Kuznetsov, 1955 (use of C^{14} to study microbial chemosynthesis; Kriss says his method is invalid); Lipman, 1926 (effect of dilution of seawater on marine bacteria); Meyers and Reynolds, 1957 (relation of marine fungi to wood borer attack); Oppenheimer, 1955, 1958, 1963 (marine microbiology); Reuszer, 1933 (bacteria in the cycle of the sea); Skopintsev, 1950 (marine humus); Sorokin, 1955, 1957, 1958 (C^{14} measurements of chemosynthesis; subject to same criticism as Kuznetsov, above); Spencer, 1956 (bacterial oxidation of ammonia in the sea); Starr and others, 1957 (production of vitamin B_{12} by marine bacteria); Waksman, 1933, and Hotchkiss, 1937 (role of bacteria in the cycle of the sea, and viability of bacteria in seawater); Wood, 1953, 1963, 1964, 1965 (marine microbial ecology); ZoBell, 1941, 1946, 1949, 1952, 1954, 1957 (marine bacteria); ZoBell and Michener, 1938 (adaptation of marine bacteria to hypotonic water).

Biogeography of the seas

Ekman, 1940, 1953 (zoogeography of the seas); Knox, 1963 (zoogeography of Australian region intertidal); Kriss and others, 1964 (microbial populations); Murphy, 1962 (antarctic life); Szidat, 1961 (South Atlantic, use of parasites as indicators).

Physical and chemical oceanography, and meteorology important for marine biology

Some books of importance: Sverdrup and others, 1946 *(The Oceans);* Sears [ed.], 1961 *(Oceanography);* Fairbridge, 1966 *(Encyclopedia of Oceanography);* Myers and others [ed.], 1969 *(Handbook of Ocean and Underwater Engineering);* Hill [ed.], 1962-1963 *(The Sea,* 3 volumes); Hewson and others, 1944 *(Meteorology);* Proudman, 1953 *(Dynamic Oceanography);* Holmboe and others, 1945 *(Dynamic Meteorology).*
the annual review volumes, *Oceanography and Marine Biology,* edited by Barnes, also contain many valuable contributions. A good presentation of physical and biological oceanography is also to be found in *Oceanography,* 1966, publication no. 1492 of the National Academy of Sciences–National Research Council, issued in 1967, and in a projection of governmental goals and possible programs in theis field in *Marine Science Affairs—a Year of Transition,* issued by the Government Printing Office the same year, and prepared by the national Council on Marine Resources and Engineering Development.

Some individual research reports: Anderson, 1962 (earth's mantle layer); Atkins, 1926 (phosphate and phytoplankton), 1953 (copper content of seawater); Atkins and Poole, 1940, 1958 (angular distribution of light in the sea); Braarud and Klem, 1931 (hydrographical and chemical coastal Scahndinavian waters); Bullard, 1967 (magnetic reversals); Callaway, 1957 (oceanographical and meteorological observations, Pacific); Clarke, G. L., 1936, 1939, 1941 (penetration and absorption of light in seawater); Currie, 1953 (upwelling, Benguela Current); Dakin and Colefax, 1935 (oceanographical observations off Australia); Deacon, 1933, 1937 (hydrology of South Atlantic and Antarctic oceans); Emery, 1956 (internal waves); Evenau and others, 1965 (photoenvironment); Fisher and Revelle, 1955 (deep Pacific trenches); Grainger, 1959 (annual oceanographical cycle, Canadian Arctic); Grier, 1941 (oceanography, North Pacific and Bering Sea); Gunter, 1957 (temperature); Gunther, 1936 (Peru Current); Heezen, 1956 (submarine canyon); Iselin, 1939 (physical factors and productivity); Evanoff, 1953, 1957, 1958, 1959, 1961 (light); Jacobs and Ewing, 1969 (suspended particulate matter in major oceans); Jerlov, 1944, 1951, 1953, 1955, 1958, 1959 (light); Kalle, 1939 (color of the sea); Kampa and Boden, 1956 (light generation in ionic scattering layer); Ketchum and Shonting, 1958 (optical studies of particulate matter in sea); Knudsen, 1901 (hydrographic tables); McEwen, 1916 (hydrographical observations at Scripps Institution); Miller and others, 1928 (physicochemical conditions in San Francisco Bay in relation to tides); Oster and Clarke, 1935 (light); Park and others, 1962 (dissolved amino acids in seawater); Polikarpov, 1964 (important book—radioecology of aquatic organisms); Riley, 1970 (particulate organic matter); Riley and Conove, 1956 (Long Island Sound); russell and McMillan, 1952 (waves and tides); Seiwell, 1942 (vertical oscillations); Stetson, 1955 (continental shelves); Stommel, 1955 (anatomy of Atlantic Ocean); Thompson and Gilson, 1937 (chemicophysical observations, Murray Expedition); Waterman, 1954, 1960 (polarized light); Wust, 1954 (zonal divisions of upper ocean waters with respect to various factors).

Some studies of currents and ocean circulation: Boden and Kampa, 1953 (cascading); Cooper, 1961 (cascading); Cooper and Vaux, 1949 (cascading); Cromwell, 1953 (circulation in meridional plane in equatorial Pacific); Hart, 1953, Hart and Currie, 1960 (Benguela Current); Munk, 1955 (ocean circulation); Sears, 1954 (Peruvian Current); Stommel, 1958 (Gulf Stream); Swallow, 1957, Swallow and Worthington, 1961 (deep currents); Worthington, 1954 (time scale in Atlantic circulation); Wust, 1930 (deep circulation and its divisions); Wyrtki, 1961 (thermohaline circulation).

Temperature and salinity: Boyko [ed.], 1965 (book, salinity and aridity); Fox, H. M., 1939 (poikilotherms in different latitudes); Metcalf, 1930 (salinitya nd size); Pearse and Gunter, 1957 (salinity, general review).

Theory of ice ages: Ewing and Donn, 1956; Donn and Ewing, 1966.

Chemistry of seawater and of marine organisms

Chemistry of seawater: Armstrong, 1954 (phosphorus and silicon), 1957 (iron), 1968 (inorganic suspended matter); Atkins, 1922 (influence of pH on algal cells), 1953 (seasonal variation in copper); Bernard, 1939, 1958 (fertility of Mediterranean water); Brandhorst, 1958 (nitrification and denitrification, tropical Pacific); Brouardel and Fage, 1953 (oxygen next to the sediment); Bruce and Hood, 1969 (phosphate in Texas bays); Chow and Thompson, 1952, 1954 (copper); Clowes, 1938 (phosphate and silicate, Southern Ocean, 1950 (hydrology, South African waters); Cooper, 1933, 1934, 1935, 1938, 1948, 1951, 1952, 1956, 1961 (chemistry and circulation of seawaters and relation to fisheries, etc.); Goldberg, 1957 (trace metals); Gourley, 1952 (phosphorus); Harris, E., 1959 (nitrogen cycle); Harvey, 1926, 1928, 1934, 1942, 1945, 1947, 1949, 1957 (important series of articles and a short book on physics and chemistry of seawater); Hoffman, 1956, 1960 (remineralization of phosphorus); Kalle, 1953, 1957 (chemistry of shallow seas); Ketchum and others, 1955, 1958 (phosphorus); Kreps and Verjbinskaya, 1930 (nutrients, Barents Sea); Martin, 1968 (marine chemistry); McClendon, 1917 (carbon dioxide and oxygen); Miyake and Saruhasi, 1956 (vertical distribution of oxygen); Mullin and Riley, 1955 (nitrate); Murphy and Riley, 1958 (phosphate); Pomeroy, 1960 (dissolved phosphate—time in water); Pomeroy and Bush, 1959 (phosphate regeneration); Rakestraw, 1936 (nitrate); Rakestraw and Carritt, 1948 (chemistry of ocean, seasonal changes); Redfield, 1958 (biological control of chemical environment); Redfield and Keys, 1938 (ammonia); Redfield and others, 1937 (organic phosphorus); Richards, 1957 (2 articles, general, and oxygen); Richards and Benson, 1961, and Richards and Vaccaro, 1956 (anaerobic environments); Rider and Mellon, 1946 (nitrate); Riley and Conover, 1956 (chemistry, Long Island Sound); Robertson and Wirth, 1934*b* (chemistry, Puget Sound); Seiwell, 1935 (2 articles, phosphorus in tropical western Atlantic and iron in Atlantic coastal waters); Spencer, 1956 (oxidation of ammonia); Steiner, 1938 (phosphorus cycle); Stephenson, 1949 (phosphate, release from mud); Thompson and Bremer, 1935 (iron); Verjbinskaya, 1932 (nitrate).

Organic matter and biologically active compounds in the sea: Brand and Rakestraw, 1937-1947 (organic nitrogenous matter in the sea); Collier, 1953, 1958 (organic compounds in the water, their effects, and biochemical aspects of red tides); Cowey, 1956 (vitamin B_{12} in coastal waters); Daisley and Fisher, 1958 (vertical distribution, vitamin B_{12}); Fox, 1957, and Fox and others, 1952, 1953 (marine leptopel and colloidal micelles); Harvey (several articles—see section on chemistry of seawater, above); Johnston, 1955, 1959 (biologically active compounds and antimetabolites); Jones, G. E., 1959 (biologically active compounds); Ketchum, 1947, 1957 (biochemical relations between organisms and ecosystem, and transport of elements in sea); Keys and others, 1935 (organic metabolism of seawater, ultimate food cycle); Kreps, 1934 (organic catalysts or enzymes in seawater); Krizenecky, 1925 (dissolved nutrients as possible source of animal food); Krogh, 1931 (same); Park and others, 1962 (dissolved amino acids); Putter, 1909 (same); Sillen, 1967 (ocean as a chemical system); Riley, 1970 (partiulate organic matter); Vallentyne, 1957 (molecular nature of organic matter in water).

Biochemistry of marine organisms: Belser, 1959 (bioassay of marine organisms); Benson and Lee, 1975 (wax in ocean food chains); Harris, 1957 (radiophosphorus in metabolism zooplankton and microorganisms); Herring, 1967 (pigments, surface plankton); Ketchum and others, 1955, 1958 (productivity in relation ot nutrients); Kon, 1958 (biochemical perspectives); Lewin, 1954 (*Stichococcus* and vitamin B_{12}); Lucas, 1947, 1949, 1955, 1961 (external metabolites); Phillips, 1917, 1922 (metals in marine organisms); Raymont and Conover, 1961 (carbohydrates in marine organisms); Todd, 1971 (chemical communication in fishes); Vinogradov, 1938, 1953 (chemical composition, marine organisms); Wangersky, 1952, 1959 (ascorbic acid, rhamnosides, and dissolved carbohydrates); Wilson, 1951, Wilson and Armstrong, 1958 (biological differences between seawaters.

Submarine geology

Books: Dietz and Holden, 1970 (breakup of Pangea); Guilecher, 1958 (coastal and submarine morphology); Hill, 1963 (Vol. III, *The Sea*); Kuenen, 1950 (marine geology); Larson and Spiess, 1969 (East Pacific Rise); Menard, 1964 (marine geology, Pacific); Pettersson, 1954 (ocean floor); Shepard, 1948, 1959 (submarine geology).

Shorter works: Andel, 1968 (deep-sea drilling); Bascom, 1961 (deep-sea drilling); Boden and Kampa, 1953 (geological effects, cascading and water movements); Hallam, 1972 (continental drift and fossil record); Kuenen, 1941 (calculations, total mass of sediments in the earth); Menard, 1953, 1955, 1961, 1967 (sea floor spreading, fractures, etc.); Rona and Miliani, 1969 (dating Caribbean cores); Shepard and Emery, 1946 (submarine photography, California coast); Veevers, 1951, 1952 (photography, sea floor); Wilson, 1971 (continental drift).

Origin and early evolution of life

Barghoorn, 1971 (Precambrain fossils); Bernal, 1967; Calvin, 1961 (chemical evolution); Fox, 1965 (chemical evolution); Miller, 1953 (production, organic chemicals in artificial primitive anaerobic atmosphere); Miller and Urey, 1958 (origin of organic compounds in the primitive earth); Oparin, 1957, 1961 (origin of life); Rutten, 1962 (geological aspects of origin of life); Schopf, 1970 (early evolution); Weyl, 1968 (role of thermocline in Precambrian evolution).

Some references on the classification and biology of particular groups of organisms

Marine plants

Diatoms and dinoflagellates: Cupp, 1943 (planktonic diatoms); Hustedt, 1956 (littoral diatoms); Kofoid and Swezy, 1921 (unarmored dinoflagellates); Lebour, 1925 (dinoflagellates, northern seas); Lebour, 1930 (diatoms, northern seas); Patrick and Reimer (diatoms of U.S.); Wood, 1954 (dinoflagellates of Australian region).

Seaweeds and other benthic algae: Abbott and Kurogi, 1971 (north Pacific benthic algae); Boney, 1966 (biology of marine algae); Dawson, 1956 (how to know the seaweeds), 1964 (seaweeds of Peru); Guberlet, 1956 (intertidal seaweeds); Hillson, 1976 (guide to common seaweeds, East Coast of U.S.); Smith, 1944 (marine algae of Monterey Peninsula, Calif.); Taylor, 1957 (marine algae of northeastern coast, North America), 1961 (marine algae of eastern tropical American coasts), 1962 (marine algae of tropical Atlantic).

Mangrove: Chapman, 1963 (mangroves); Davis, 1940 (ecology and geological role of mangroves in Florida); Orr and Moorhouse, 1933 (physical and chemical conditions in mangrove swamps).

Marine animals

General works: Friedrich, 1969 (marine biology); Light's manual, 1976 edition (keys to California intertidal forms); MacGinitie and MacGinitie, 1949 (natural history of marine animals); Nikitin, 1960 (marine biology); Ricketts and Calvin, 1974 (intertidal U.S. West Coast forms); Robbins and Yentsch, 1975 (U.S. East Coast forms).

Invertebrate zoology

The literature in this area is too vast to even begin to summarize here. Some good American books are the 5-volume treatise of Hyman (1940-1959), and the recent texts of Barnes (1968), HIckman (1973), and Meglitsch (1967). Ricketts and Calvin (as revised by Hedgpeth), 1968, contains a useful annotated bibliography arranged by groups. Tombes, 1970 (invertebrate endocrinology).

Protozoa—Foraminifera: Boltovskoy (numerous works, South American Foraminifera); Brady, 1884 (*Challenger* Expedition Foraminifera, vol. 9, *'Challenger Reports'*); Collins, 1958 (Foraminifera of the Great Barrier Reef Expedition, Australia); Cushman (many papers and monographs on Foraminifera from 1926 to 1948); Heron and Earland, 1932-1936 (Foraminifera from the *Discovery* Expeditions, 4 parts); Pfleger, 1960 (ecology, distribution of recent Foraminifera); Thalman, 1960 (index to genera and spp. from 1890 to 1950).

Protozoa—Radiolaria: Haeckel, 1887 (*Challenger Reports,* vol. 18); Renz, 1976 (distribution and ecology in central Pacific).

Sponges: Bowerbank, 1864-1882 (British sponges, 4 vols.), 1872-1876 (general treatment of sponges); Fry, 1969 (biology of sponges); Harrison and Cowden [eds.], 1976 (sponges); Koltrum, 1969 (arctic and antarctic sponges, faunistic review); deLaubenfels, 1954 (sponges of west central Pacific).

Coelenterates: Bigelow, 1912-1940 (numerous papers on medusae); Browne, 1896-1926 (several papers on medusae and hydroids); Haeckel, 1879 (classification of medusae); Hand, 1950's to present (numerous papers on coelenterates); Kramp, 1961 (synopsis, medusae of the world); Mayer, 1910 (medusae of world); Rees, 1966 (Cnidaria and their evolution); Russell, 1953 (medusae of British Isles); Uchida, 1920's to 1960's (numerous papers on medusae of western Pacific).

Corals: One of the most extensive studies of coral reefs was made by an expedition sponsored by the British Museum of Natural History. Their reports are embodied in a series of volumes: *Scientific Reports of the Great Barrier Reef Expedition, 1928-1929, British Museum of Natural History.* One of the earliest and greatest scientific studies of coral reefs is that of Darwin, 1897. This was recently reissued as a paperback by the University of California. More recent books are those of Gilbert and McNeill, 1959, and Wiens, 1962 (book, atoll ecology). The most recent extensive review of all phases of coral reef biology and geology is the 4 vol. treatise edited by Jones and Endean 1973-1976, published by Academic Press, Inc.

Some other references on corals: Bakus, 1964, 1966, 1969 (effects of fish-grazing on organisms of coral reefs); Daly, 1910 (Pleistocene glaciation and coral reefs); Goreau and others, 1971 (productivity of coral reef); Ladd and Tracy, 1949 (general discussion of coral reefs); Mayor, 1924 (structure and ecology of Samoan reefs); Mesolella, 1967 (Pleistocene reefs on Barbados); Muscatine, 1967 (role of zooxanthellae); Pearse and Muscatine, 1971 (zooxanthellae and calcification); Roughly, 1936 (Great Barrier Reef); Sargent and Austin, 1954 (biological production and economy of coral reef); Stoddart and Yonge, 1971 (Indian Ocean corals); Yonge, 1958 (ecology and physiology of reef corals).

Annelids: Augener, 1906 (West Indies polychaetes); Berkeley, 1948, 1952 (Canadian-Pacific polychaetes); Dales, 1957 (pelagic Pacific polychaetes); Gravely, 1909 (polychaete larvae); Hartman, 1952 (literature of polychaetes, also both before and since, numerous papers on various groups and collections of polychaetes); Hoagland, 1920 (Philippine polychaetes); Moore, 1903-1923 (papers on U.S. and Alaskan western coast polychaetes); Treadwell, 1926-1943 (papers on Philippine and U.S. West Coast polychaetes).

Arthropods

Crustacea, general: Alcock, A., 1895-1900 (Crustacea of India, reprinted, Stechert-Hafner 1968); Barnard, Menzies, and Bacescu, 1962 (abyssal Crustacea); Green, 1961 (biology of Crustacea); Sars, 1901-1918 (Crustacea of Norway); Say, 1817-1818 (Crustacea of U.S.); Stebbing, 1910 (Crustacea of South Africa); Waterman, 1961 (physiology of Crustacea).

Crustacea, copepods: Brady, 1883 (copepods of *Challenger* Expedition, vol. 8, *'Challenger' Reports*); Giesbrecht, 1892, 1895 (copepods of Gulf of Naples and pelagic copepods); Giesbrecht and Schmeil, 1898 (copepods, in Das Tierreich); Nicholls, 1944 (copepods, Woods Hole region); Wilson, 1932 (copepods of Woods Hole region and of Chesapeake Bay).

Vertebrates

General: Slchmidt-Nielsen, 1971 (physiology).

Fishes: *Journal of Fish Biology* (quarterly; Academic Press, Inc.). *Fish Physiology* (a 6-volume treatise, Hoar and Randall [ed.], Academic Press, Inc.).
Other references: Brown [ed.], 1956, 1957 (physiology); Love, 1970 (chemical biology); Morris-Kittleman, 1967 (otoliths and pressure detection); Shaw, 1962 (schooling); Schmidt, 1925 (migration of eels).

Reptiles: Carr, 1954, 1961, 1965 (turtles); Williston, 1914 (marine reptiles past and present).

Birds: Ashmole, 1971 (ecology); Hamilton, 1967 (navigation); Lack, 1945 (suympatric reltaed species); Tinbergen, 1953 (behavior); Watson, 1966 (tropical seabirds); Wetmore, 1936 (birds of northern seas).

Mammals: Andersoñ [ed.], 1969 (biology of marine mammals); Budker, 1959 (whales); Cuvier, 1834-1836 (fossils); Elsner, 1969 (adaptations to diving); Evans, 1969 (communication); harrison and King, 1965 (general work on marine mammals); Howell, 1930 (adaptations to aquatic life).

Methods

Books: Barnes, 1959*a* (apparatus and methods, oceanography, 2 volumes); Costello and others, 1957 (handling marine eggs and embryos); Frey and Frey, 1961 (handbook for hydronauts); Meyers and others, 1969 (handbook, ocean and underwater engineering); Schlieper, 1968, 1972 (methods of marine biological research); Watt [ed.], 1966, 1967 (systems analysis, electronic and computer methods); Zweig [ed.], 1963-1967 (analytical methods for pesticides, 5 volumes).

See also recent annual instrumentation issues of *Science*.

Other references: Ahlstrom, 1952 (instrumentation); Allen and Nelson, 1910 (culture of plankton organisms); Belser, 1959 (bioassay, organic substances); Braarud, 1961 (use of culture methods for analysis of environmental influences); Brand, 1934, Brand and others, 1937-1947 (chemical methods and bioassay methods); Creitz and Richards, 1955 (estimation of plankton by pigment analysis and millipore filtration); Forester, 1953 (dredge for burrowing animals); Goldberg and others, 1952 (microfiltration); Goldman, 1962 (method of studying nutrient-limiting factors in the water column); Grontved, 1957 (macrovegetation sampler); Hardenberg (plankton sampling different depths); Hardy, 1936*a* (plankton indicator); Holme, 1949*b* (bottom-sampler); Knudsen, 1927 (bottom-sampler for hard bottoms); Kow, 1953*a* (making plankton calendar); Morgan, 1956 (diving safely); Mullin and Riley, 1955 (nitrate, spectrophotometric determination); Murphy and Riley, 1958 (phosphate determination); Petersen, 1918 (methods and analysis for benthic communities on soft substrate); Richards and Thompson, 1952 (spectrophotometric method, plankton analysis by pigments); Rider and Mellon, 1946 (nitrate); Riedl, 1953 (methods for study of meiobenthos).

Methods using radioactive isotopes: Calvin and others, 1949 (C^{14} methods); Devereux, 1967 (oxygen isotope ratios for temperature determination at time of deposition in calcareous structures); Doty, 1956, 1959 (C^{14} method); Epstein and others, 1951, 1953 (oxygen isotope ratio method); Jitts, 1957, and Jitts and Scott, 1961 (C^{14} method); Steemann-Nielsen, 1952, 1958 (C^{14} method); Yankwich and others, 1947 (correction for C^{14} method).

Marine biological resources and their use

Books: Borgstrom, 1972 (world food situation); Bottemanne, 1959 (fisheries development); Graham, 1956 (sea fisheries); Huberty and Flock, 1959 (natural resources—good chapter on marine resources); Korringa, P., 1976 (farming marine organisms. 4 vols.); Loftas, 1969 (exploitation of marine resources); Rounsefell, 1975 (fisheries); Sindermann, 1977 (diseases and their control in marine aquaculture); Sreenivasa, R., 1975 (ocean resources and law of the sea); Taylor, 1952 (fisheries); Walford, 1958 (living resources from the sea).

Some shorter references: Borgstrom, 1970 (amounts and distribution of marine resources); Carter, 1968 (scramble for oil leases); Emery and Iselin, 1967 (food from oceans and land); Gerard and Lamar, 1967 (freshwater from sea wind); Glover, 1955 (herring); Hardy, 1936*b*, 1941 (herring and plankton, plankton for food); Juday, 1943 (utilization of fisheries); Paulik, 1971 (ecology of Peru current); Pequegnat, 1958 (whales and krill); Raymont and Miller, 1962 (production in enclosed fertilized marine water body); Shropshire, 1944 (plankton harvesting).

Impact and plight of man

Books: Anderson, 1971 (destruction of ecosystem); Bayliss-Smith and Feachem, 1977 (rural ecology in the Pacific); Carson, R., 1962 (*Silent Spring,* effects of pesticides and other environmental pollution); Castro, 1952 *(Geography of Hunger);* Commoner, 1966 *(Science and Survival);* Detwiler, 1971 (man's impact); Dorst, 1970 *(Before Nature Dies);* Duosik, 1974 (social, economic, and legal); Ehrlich, 1968 *(The Population Bomb);* Graham, 1970 *(Since Silent Spring);* Hinrichs, 1972 (population); Hynes, 1960 *(Biology of Polluted Waters);* Jackson, 1971 *(Man and Environment);* Laevastu, 1959 *(A Review of Marine Pollution);* Lappé and Collins, 1977 (*Food First,* one of the best analyses of plight of man); Malins, 1977 (oil pollution, arctic and subarctic); Murdoch, 1971 (source book on resources, pollution and environment); Olson and Burgess [ed.], 1967 *(Pollution and Marine Ecology);* Paddock and Paddock, 1967 (coming famine and our choices); Polikarpov, 1964 (1966) (radioactive pollution); Potter, 1971 (bioethics); Smith, J. E. [ed.], 1968 ("Torrey Canyon" disaster); Vernberg and others [eds.], 1974, 1977 (effects of pollution); Warren, 1971 (water pollution). In addition to the above books, the report of the National Academy of Sciences–National Research Council, 1957 (effects of atomic radiation on oceanography and fisheries) should be read. Sinha (1970) has assembled an annotated bibliography, and *Pollution Abstracts* gives monthly notices and summaries of current literature relating to pollution.

Shorter articles: Bernstein, 1973 (marine pollution); Bogorov and Kreps, 1958 (radioactive waste disposal); Cairns, 1956, 1970, 1971 (thermal pollution); Carter, 1968 (oil leases on continental shelf), 1970 (estuary in crisis); Chesher, 1969 (destruction of Pacific corals by sea star); Chichester, 1965 (pesticides); Davis, 1948 (industrial copper); Eriksen and Lawrence, 1940 (pollution in Gray's Harbor, Wash.); Goodwin, 1961 (disappearing estuaries); Ikard, 1967 (oil pollution); Inglis and Kestner, 1958 (alteration of estuaries); Johns, 1968 (vulnerability of estuaries); Luck, 1957 (man against his environment); Matsunaga, 1976 (mercury pollution); Miles and Pearce, 1957 (pesticides); Newcombe, 1957 (biological hazard of strontium-90); Osterberg, 1962 (radioactive fallout); Rafter and Fergusson, 1957 (increase of atmospheric C^{14} because of testing, etc.); Risebrough and others, 1968 (pesticides); Ruvinoff, 1968 (sea level canal); Soto and Deichmann, 1967 (pesticides); Strahler and Strahler, 1973 (man-made geological changes); Tauk, 1973 (environmental geology); Thompson, S. H., 1961 (estuaries); Woodwell, Wurster, and Isaacson, 1967 (insecticides).

International concerns: policy, economics, law of the sea

Borgese, 1968 (The Ocean Regième—statutes for peaceful uses of the high seas); Burnell and von Simson, 1970 (prospects and hazards of impending exploitation of the seas); Gullion, 1968 (uses of the seas); Pearson, 1975 (international marine policy).

TAXONOMIC APPENDIX

The following brief outline will enable the student to obtain an idea of where various organisms fit into the general classification. It is, of course, in no sense a complete rundown, even of strictly marine organisms, and it also emphasizes, perhaps unduly, certain groups that happen for one reason or another to come more prominently to people's attention.

Following Whittaker (1969), living things are now commonly classified into five primary categories, or kingdoms, rather than the old, familiar animal and plant kingdoms. These kingdoms in turn are grouped into three levels of structural organization. At each level, differentiation into autotrophic, mostly photosynthetic, and heterotrophic modes of nutrition has occurred. At the eucaryotic levels further differentiation of the heterotrophic forms into those which obtain nutrients by absorption or transport through the cell membranes, and those which obtain them by ingestion, has taken place. None of these distinctions is absolute. Some photosynthetic organisms can also exist heterotrophically, and many organisms capable of ingesting food can also obtain part or all of their nourishment through absorption under certain conditions.

Fig. 17-2 is a simplified diagram showing these relationships.

See the index for additional figures.

KINGDOM MONERA Procaryotic cells lacking nuclear membranes, membrane-bound organelles such as mitochondria and plastids, endoplasmic reticulum, and possessing only one linkage group, or chromosome. Flagella, when present, are simple (not the 9+2-strand type). Saprobic or photosynthetic. Flagellated forms and related nonmotile types comprise the Mastigomonera, including the phyla Eubacteriae, or true bacteria; Actinomycota, or mycelial bacteria; and Spirochaetae, or the spirochaetes. Nonflagellated gliding forms and their relatives comprise the Myxomonera, including the phyla Myxobacteriae, or gliding bacteria, and the Cyanophyta, or blue-green algae.

PHYLUM Eubacteriae True bacteria.

ORDER Eubacteriales A rather heterogeneous assemblage of rod-shaped, coccoid, fusiform, or pleomorphic bacteria; peritrichous when motile.

Representative families

FAMILY Bacillaceae Spore formers. Under appropriate conditions the bacterial cell may form a single endospore, resistant to adverse conditions, and germinating to form one bacterium under favorable conditions. All other families of eubacteria are nonspore-forming. Spore formers are scarce in strictly marine environments.

Bacillus Aeorbic or facultatively anaerobic.

Clostridium Strictly anaerobic.

FAMILY Bacteriaceae Since neither the type genus, *Bacterium,* nor the type species were described in an identifiable way, this family is now used as a repository for nonspore-forming rods not clearly belonging to one of the other groups.

FAMILY Enterobacteriaceae Gram-negative nonsporing rods, peritrichous when motile. Some are characteristic inhabitants of the large intestines of higher vertebrates, including man, and hence much used as indicators of sewage pollution. Others are well-known animal or plant pathogens. Not typical of marine habitats except where found in sewage-contaminated coastal or estuarine situations. *Escherichia, Enterobacter (=Aerobacter), Erwinia, Proteus, Salmonella, Shigella,* etc.

FAMILY Micrococcaceae Rounded coccoid cells, nonmotile, occurring in various groupings, from single cocci, pairs, tetrads, regularly arranged cubical packets, or irregular clusters. A large and physiologically varied group, mostly terrestrial, and many associated with skin, of animals. Some marine. *Micrococcus, Sarcina, Halobacterium* (salt-loving or halophilic types).

FAMILY Streptococcaceae or Lactobacillaceae Gram-positive cocci in chains, or nonmotile gram-positive rods. Fermentative metabolism. Anaerobic or facultatively aerobic, usually forming minute colonies on media exposed to air. *Streptococcus, Lactobacillus.*

FAMILY Azotobacteriaceae Rounded or oval forms, peritrichous when motile, best known for their ability to fix atmospheric nitrogen. *Azotobacter, Beijerinckia.*

FAMILY Corynebacteriaceae Mostly gram-positive, nonmotile, rods or irregularly shaped coccoid bacteria, aerobic, often pleomorphic and irregularly staining. Some show "snapping division," resulting in side-by-side or palisade arrangements of individuals. Forms resembling soil corynebacteria commonly present in sediments, especially of estuaries, and on the skin of sharks. *Corynebacterium.*

ORDER Pseudomonadales Gram-negative rods, rigid, straight, curved or spiral, usually motile, with one or more polar flagella at one or both ends; some photosynthetic. A large, very diverse group with many marine forms. Photosynthetic families sometimes grouped together as suborder Rhodobacteriineae, nonphotosynthetic ones as suborder Pseudomonadineae.

Suborder Pseudomonadineae Nonphotosynthetic bacteria.

FAMILY Pseudomonadaceae Mostly aerobic straight or curved rods with polar flagella, single or in tufts; mostly grow well and rapidly on ordinary media. Straight or slightly curved forms include *Pseudomonas, Alteromonas, Vibrio, Desulfovibrio* (anaerobic), *Photobacterium* (bioluminescent), and many others. Spiral forms include *Spirillum* and others.

FAMILY Nitrobacteriaceae Autotrophic forms utilizing CO_2, carbonate, or bicarbonate ion as carbon source and obtaining energy by oxidation of ammonia, nitrite, hydrogen, sulfur, or thiosulfate. Some can also utilize certain organic substrates, whereas others fail to grow in the presence of appreciable amounts of organic matter. *Nitrobacter, Nitrosomonas, Thiobacillus, Thiobacterium, Thiospira, Hydrogenomonas,* etc.

FAMILY Caulobacteriaceae Nonfilamentous rods, often attached to substrate by simple or branched stalks; polar flagellate in free state. Some of the "iron bacteria" belong here, characterized by deposition of ferric hydroxide. *Caulobacter, Gallionella.* Other "iron bacteria" belong to the Siderocapsaceae, which deposit iron or maganese compounds in their mucilaginous capsules.

Suborder Rhodobacteriineae Photosynthetic bacteria.*

FAMILY Athiorhodaceae Purple sulfur bacteria utilizing organic substrates for growth; those able to grow in presence of air can also grow in the dark in aerobic cultures, metabolizing organic compounds. Mostly microaerophilic. In illuminated cultures with extraneous H-donors present, they can grow under strictly anaerobic conditions. Masses of cells appear yellowish brown, olive brown, dark brown, or various shades of red due to bacteriochlorophyll and carotenoid pigments. The cells do not contain deposits of sulfur. *Rhodopseudomonas, Rhodospirillum.*

FAMILY Thiorhodaceae Purple sulfur bacteria requiring light; anaerobic to microaerophilic, growing in presence of H_2S, which is oxidized to elemental sulfur, visible as droplets within the cells, or to sulfate. Develop in inorganic mineral media with H_2S. Often develop in estuaries and mud flats after the peak of the bloom of green algae such as *Ectocarpus* or *Enteromorpha* or under mats of tapetic blue-green algae, utilizing wavelengths of light complimentary to those utilized by the algae

---

*The bacteriochlorophylls found in photosynthetic bacteria differ from the chlorophylls of blue-green algae or green plants, and no free oxygen is liberated as a result of their photosynthesis.

above them. *Thiosarcina, Thiopedia, Thiocapsa, Thiodictyon, Rhodothece, Chromatium,* and others.

FAMILY Chlorobacteriaceae Green sulfur bacteria. Strictly anaerobic in environments high in H_2S and exposed to light. Sulfur usually deposited outside the cells rather than within them. Common in estuaries and in anaerobic situations on the continental shelf where there is H_2S and sufficient light. *Chlorobium, Chlorobacterium, Chlorochromatium, Pelodictyon,* etc.

PHYLUM Cyanophyta Blue-green algae. Procaryotic cells with chlorophyll and other pigments dispersed in the cytoplasm, not in plastids; motile forms with gliding movement, no flagella; no sexual reproduction known. The cells are microscopic but may be aggregated into macroscopic colonies of various forms or long filaments. Approximately 150 genera and 1400 species known. Unicellular to colonial nonmotile forms are grouped in subphylum Coccogoneae, of which the Coccogonales and Chroococcales are representative. Filamentous, often motile forms constitute the subphylum Hormogoneae, of which the Hormogonales, or Nostocales (in which divisions within the filament are always transverse), and the Stigonematales (in which longitudinal divisions also occur, resulting in distinctive changes in the organization of the filaments and individuals), are representative. The Beggiatoales are nonphotosynthetic counterparts of the Hormogonales. Genera with marine species mentioned in this book: *Anabaena, Haliarachne, Katagnymene, Lyngbya, Microcoleus, Oscillatoria, Phormidium, Richelia, Rivularia, Stigonema, Trichodesmium, Verrucaria.* There are, of course, many others.

KINGDOM Protista Eucaryotic cellular organisms, single celled or colonial.

PHYLUM Euglenophyta With one to three flagella, surface of cell differentiated as a pellicle, which may be plastic, allowing for "euglenoid" or "metabolic" movement, or which may be sufficiently rigid to give the cell a definite permanent shape. Photosynthetic members with grass-green chloroplasts, and usually a red stigma. Saprobic forms without these. Food reserves paramylum. Many freshwater and marine species.

FAMILY Euglenaceae Typical photosynthetic members, usually with chloroplasts, stigma, pyrenoids, and paramylum bodies. *Euglena, Eutreptia.*

FAMILY Astasiaceae Similar to euglenids but without chloroplasts or stigma. Commonly swim with flagellum held straight out in front, with only the tip moving. *Astasia, Menoidium, Distigma, Spheciomonas.*

FAMILY Paranemaceae Mostly small freshwater, often creeping, sometimes more or less bilateral forms. Some marine species. *Paranema, Anisonema, Heteronema, Urceolus.*

PHYLUM Chrysophyta Approximately seventy-five genera of small flagellates with xanthophyll pigments in addition to chlorophylls. Some nonmotile and coccoid. Reserve products leusocin (Chrysolaminarin) fats, and oils. Most are also phagotrophic as well as photosynthetic.

CLASS Chrysophyceae Golden-brown algae. Chromatophores golden brown. Many marine forms.

FAMILY Chromulinaceae *Chromulina.*

FAMILY Coccolithophoraceae *Discusphaera, Pontosphaera, Syracosphaera, Coccosphaera, Petalosphaera, Acanthoica, Halopappus.*

FAMILY Ochromonadaceae *Ochromonas.*

FAMILY Silicoflagellaceae *Distephanus.*

FAMILY Phaeocystaceae *Phaeocystis.*

CLASS Xanthophyceae Yellow-green algae. Mostly freshwater forms, a few marine. Chromatophores yellow-green.

CLASS Bacillariophyceae Diatoms. Characterized by a two-part silicious frustule. Chloroplasts brownish green, with the pigment diatomin somewhat masking the green chlorophylls in most forms. The silicious frustule is usually beautifully and symmetrically ornamented. Diatoms constitute one of the most important and dominant elements of the phytoplankton.

ORDER Centrales or Centricae Diatom usually circular, disc-shaped, cylindrical, or triangular. Valves radially symmetrical, with radially disposed striae or concentric markings; sometimes with spines, horns, or other protrusions; no raphe; cells non-motile; cytoplasm with numerous small chromatophores; resting spores and auxospores commonly formed; gametes, where known, anisogramous with motile flagellated male gametes. Mostly marine, planktonic. About 100 genera and 2400 species.

Cells of centric diatoms produce only a single auxospore. Although they were long thought to be asexual, it has been found that for some the process is sexual, and it seems probable that this may be true for all. In auxospore formation the protoplast rounds up to form a single egg cell, pushing the valves apart in the process. In other cells of the same species the protoplast divides repeatedly, forming 8, 16, 32, 64, or 128 motile flagellated antherozoids (called microspores before their role as gametes was understood). An antherozoid swims to and fertilizes an egg. The zygote thus formed becomes an auxospore.

Suborder Biddulphioideae Box-shaped; prevalvular axis not much longer than, or shorter than, valvular axis; valves usually oval, sometimes polygonal, circular, or semicircular; unipolar, bipolar, or multipolar—each pole with an angle, horn, or spine or with both angles and horns.

FAMILY Biddulphiaceae *Bacteriastrum, Bellerochia, Biddulphia, Cerataulina, Ditylum, Hemiaulus, Lithodesmium, Terpsinoe.*

FAMILY Chaetoceraceae *Chaetoceros.*

FAMILY Eucampiaceae *Climacodium, Eucampia, Streptotheca.*

Suborder Discoideae Disc-shaped or cylindrical, valve circular, flat, or convex or sometimes hemispherical; zonal diameter shorter than valvular diameter; sculpture radial or concentric, related to a central pole; spines or horns frequent.

FAMILY Coscinodiscaceae *Arachnodiscus, Asteromphales, Coscinodiscus, Cyclotella, Planktoniella, Aulacodiscus.*

FAMILY Melosiraceae *Melosira, Peralia*

FAMILY Skeletonemaceae *Skeletonema, Stephanopyxis*

FAMILY Thalassiosiraceae *Coscinosira, Lauderia, Schröderella, Thalassiosira*

Suborder Solonoideae or Soloniineae Valves oral or circular; cells long, cylindrical, with numerous intercalary bands; no internal septa; united in chains by their valves.

FAMILY Corethronaceae *Corethron*

FAMILY Soleniaceae *Atthenya, Dactyliosolen, Guinardia, Leptocylindrus, Rhizosolenia*

ORDER Pennales or Pennatae Elongated, oblong, or feather- or boat-shaped, the valves with two planes of symmetry, usually elongated, with the striae more or less transverse to the apical axis; no spines; a longitudinal groove, the raphe, present in the majority; those possessing a true raphe are motile, with a gliding movement; chromatophores commonly large, platelike, lobed, or somewhat rolled up into a partial cylinder, usually only one or two present; no resting spores formed; gametes, where known, isogamous, not flagellated. About 70 genera and 2900 species, somewhat more than half marine.

This group includes all the soil diatoms, the majority of freshwater species, and a great many marine diatoms, especially the benthic species. In auxospore formation two small cells produced by meiotic division become surrounded by a gelatinous envelope. The valves are pushed apart and the two protoplasts, each acting as a gamete, fuse. The zygote forms an auxospore. In some species each cell first divides into gametes, and two auxospores are formed.

Although there has long been a general impression that centric diatoms dominate the plankton while pennate diatoms are mostly benthic or epontic, several pennate species are at times dominant in oceanic phytoplankton—especially species of *Asterionella, Fragilaria,* and *Fragilariopsis.* Furthermore, numerous centric diatoms such as *Cyclotella, Melosira,* and some species of *Coscinodiscus, Biddulphia, Triceratium,* and others are benthic or epontic.

Suborder Araphidineae No raphe or pseudoraphe; nonmotile.

FAMILY Fragilariaceae *Asterionella, Fragilaria, Thalassiothrix*
FAMILY Tabellariaceae *Striatella*
Suborder Biraphidineae Well-developed raphes on each valve; commonly motile, with gliding motion.
FAMILY Cymbellaceae *Amphora, Cymbella*
FAMILY Naviculaceae *Amphiphora, Navicula*
FAMILY Nitzschiaceae *Bacillaria, Hantzschia, Nitzschia, Phaeodactylum*
Suborder Monoraphidineae One raphe only; other valve has only a pseudoraphe or a rudimentary raphe.
FAMILY Achnanthiaceae *Achnanthes, Cocconeis*
Suborder Raphidiodinae Raphe apparent only at the terminal nodules, not extending the length of the valve.
FAMILY Eunotiaceae *Eunotia, Peronia*
PHYLUM Pyrrophyta Dinoflagellates
CLASS Dinophyceae
Subclass Adinida
FAMILY Prorocentraceae *Exuviella, Prorocentrum*
Subclass Cystoflagellata
FAMILY Noctilucaceae *Noctiluca*
Subclass Diniferida
ORDER Gymnodiniales The unarmored, or naked, dinoflagellates
FAMILY Cystodiniaceae *Cystidinium, Glenodinium*
FAMILY Gymnodiniaceae *Amphidinium, Gymnodinium*
FAMILY Polydinaceae *Polykrikos*
FAMILY Pouchetiaceae *Erythropsis, Pouchetia*
FAMILY Pronoctilucaceae *Oxyrrhis, Pronoctiluca, Protodinifer*
ORDER Peridiniales The armored, or thecate, dinoflagellates.
FAMILY Dinophysaceae *Amphisolenia, Dinophysis*
FAMILY Peridiniaceae *Ceratium, Peridinium*
FAMILY Goniaulaceae *Goniaulax*
KINGDOM PLANTAE
Subkingdom Thallophyta
PHYLUM Chlorophyta Green algae.
CLASS Chlorophyceae
ORDER Cladophorales Cylindrical multinucleate cells joined end to end, forming branched or unbranched filaments. The structure could be interpreted as coenocytic but with septa, or as a row of multinucleate cells. This structure shows some similarity to the coenocytic structure of the Codiales, which wholly lack cross walls. Sexual reproduction by conjugation of isogametes. *Cladophora*.
ORDER Codiales (Siphonales) *Codium, Caulerpa, Halimeda*.
ORDER Dasycladales: Marine, warm seas. Erect central axis with whorls of branches along the entire length or only at the top.
ORDER Siphonocladales: Multicellular thallus attached by a stem of rhizoids. Cells mononucleate and divide in a unique manner. Warm seas. *Halicystis, Valonia*
ORDER Ulvales *Ulva, Enteromorpha*.
PHYLUM Phaeophyta Brown algae.
CLASS Phaeophyceae
Subclass Cyclosporeae No gametophyte generation. The spores function as gametes instead of developing into gametophytes. The reproductive organs are in round cavities, the **conceptacles,** borne in the swollen tips of the branched thallus.
ORDER Fucales
FAMILY Durvilleaceae *Durvillea, Splachnidium*.
FAMILY Fucaceae
Ascophyllum Bladder wrack or tang. In eastern North America a dominant member of the intertidal *Balanus, Fucus, Littorina* biome (Fig. 7-5).

Fucus Rockweed. Olive-green to brown algae. Thallus tough, leathery, flattened, repeatedly forked, with numerous air vesicles. Intertidal in northern temperate zone. Tips of branches are inflated, bearing conceptacles (Fig. 7-6).

Pelvetia Occupies highest part of fucoid intertidal belt.

Pelvetiopsis

Himanthalia Sea thong. Temperate to boreal North Atlantic. Both sides.

Sargassum Gulfweed. Wide distribution in warm waters. Commonest genus of brown algae in warm waters.

Turbinaria

Subclass Phaeosporeae Alternation of gametophyte and sporophyte generation. The gametophytes bear multicellular gametangia, in which each cell contains a gamete. Some have unicellular gametangia with single gametes. The sporophyte bears unicellular sporangia, with spores in multiples of two (commonly 64 or 128). A few produce multicellular sporangia that look like gametangia.

GRADE Isogeneratae Gametophyte and sporophyte of same appearance.

ORDER Cutleriales Filaments laterally compacted, forming bladelike or disclike body. Gametophytes with groups of multicellular gametangia. Gametes markedly anisogamous. Sporophytes with unicellular sporangia. *Cutleria*

ORDER Dictyotales or **Ectocarpales** Branched, ribbon, or fanlike growth initiated by solitary or laterally adjoined apical cells. Gametophytes with patches of sex organs (sori). Female gametophyte with unicellular gametangia (oogonia). Male gametophyte with many-celled gametangia (antheridia). About 100 species, 20 genera. Predominant brown algae in tropical seas, often growing in considerable abundance.

Dictyota

Ectocarpus Slender, filamentous, branching, with abundant gametangia.

ORDER Sphacelariales Treadlike thallus, with cells arranged regularly in transverse tiers. Growth of each branch initiated by an apical cell. About 15 genera, 175 species. Mostly restricted to the Atlantic Ocean. *Sphacelaria.*

ORDER Tilopteridales Freely branching thallus. The lower part resembles *Sphacelaria;* the upper part, *Ectocarpus.*

GRADE Heterogeneratae Gametophyte always irregularly branched filaments. Sporophyte may or may not be filamentous but is not similar in appearance to the gametophyte.

Subgrade Haplostichineae Filamentous sporophytes.

ORDER Chordariales Sporophyte filamentous. Gametophyte resembles irregularly branched gametophytes of *Ectocarpus.*

Leathesia Bladderweed.

ORDER Desmarestiales Sporophyte obscurely filamentous because the branches become covered with irregularly arranged cells. Oogamous gametophytes. Both antheridia and oogonia are unicellular. *Desmarestia.*

ORDER Sporochnales Sporophytes filamentous. Oogamous gametophyte. Both antheridia and oogonia are unicellular.

Subgrade Polystichineae Nonfilamentous sporophytes.

ORDER Dictyosiphonales Sporophyte profusely branched and threadlike, with internal differentiation of tissues.

ORDER Laminariales This group includes many familiar large seaweeds known as kelps. The sporophyte may become very large, up to 30 meters long or more in some species, with differentiation of tissues internally and of the thallus into holdfast, stipe, blades, and gas floats. Sporangia are unicellular, grouped into sori on the blades.

The gametophytes are microscopic, irregularly branched filaments of rarely more than fifty cells. Male gametophytes have many antheridia, each with one antherozoid. Female gametophytes have one to six oogonia, each with one egg.

There are about 30 genera, mostly limited to the North Pacific, but *Laminaria* and a few others are found in all oceans.

Alaria (badderlocks) Olive-brown, in northern seas. Stipe bears numerous short, obovate blades basally and ends as a long, flat blade.

Cystoseira (Fig. 7-4).

Egregia (ribbon kelp) Thallus grows as a long, beltlike strap with blades and gas floats along the edges. Common along the Pacific coast of North America.

Laminaria (sea tangle) A large genus growing in all oceans. Flat, broad blade or ribbonlike thallus—single blade, with relatively short stipe and strong holdfast. Often grows just below low-tide level, where it is subject to strong wave action. Sometimes so abundant that it forms "laminaria forests" along with *Alaria*.

Macrocystis One of the giant kelps of the Pacific coast of North and South America. Numerous blades, each with a bladder at its base, borne at regular intervals along the branched stem. Blades about 80 by 40 cm. irregularly corrugated with denticulate margin. Monotypic for *M. pyrifera*. *Macrocystis* often attains extremely large size. Reports of specimens 213.5 meters long have been made. The largest reliable actual measurement was 42.5 meters. Perennial on rocky bottoms in 7 to 30 meters, especially on open coast, where there is a continuous swell. Forms kelp beds that may protect nearby shore from excessive wave action.

South of Point Conception, California, *Macrocystis* is practically the only kelp in the kelp beds. North of Point Conception, it tends to become mixed with, or largely replaced by, *Nereocystis*.

Microcladia (Fig. 7-4).

Nereocystis (bladder kelp, bullwhip) Second most abundant giant kelp of the Pacific coast of North America. The hollow stem may be many meters long, ending in a single large gas float from which a cluster of long blades floats at the surface of the water. May exceed 30.5 meters in length. *Nereocystis* is an annual. During the growing season in March to June it may grow as much as 25.5 cm. a day. The main species is *N. leutkeana*.

Odonthalia (Fig. 7-4).

Pelagophycus Another giant kelp. May attain a length of 30.5 meters or more.

Postelsia (sea palm) Prominent in lowest intertidal zone and top of subtidal zone, on exposed rocks with strong wave action. Stands erect on a thick, cylindrical, flexible stipe, with a cluster of blades arising from the apex. Looks like a small palm tree, about 0.3 to 0.6 meter high, growing on the rocks and exposed to the heaviest surf, where relatively few organisms can maintain themselves.

ORDER Punctariales Sporophyte bladelike, saccate, without internal differentiation of tissues. Medium size. *Asperococcus, Scytosiphon.*

PHYLUM Rhodophyta Red algae.

CLASS Rhodophyceae

Subclass Bangioideae

ORDER Bangiales About 15 genera and 60 species. Cell division occurs anywhere in the body. The fertilized zygote divides directly into **carpospores.**

Bangia fusco-purpurea Grows in uppermost levels of the littoral zone. Tolerant to desiccation and warming.

Porphyra (laver) Several species. Grows as thin red or purple blades, tough and leathery, one or two cells thick on rocks in the intertidal zone. Most conspicuous and abundant member of the Bangiales. Of commercial importance for food.

Subclass Florideae Several orders, about 385 genera, 2400 species. Body filamentous, but the filaments are often laterally compacted to give the macroscipic secondary forms; cell division only in terminal cells of the branches.

ORDER Nemalionales The most primitive of the Florideae. No tetrasporophyte. Gonimoblast filaments develop from the carposporangium. About 35 genera. The taxonomy of this order is in a particularly uncertain condition and in need of a thorough reworking. *Nemalion, Acrochaetium, Gloiopeltis.*

ORDER Gelidiales Gonimoblast filaments grow directly from the carpogonium. No auxiliary cell. Tetrasporophyte present in life cycle. About 6 genera.

Gelidium Much-branched cartilaginous fronds. Cystocarps immersed in swollen branchlets. Includes the agar weed *G. cartilagineum,* an important source of agar-agar.

In the remaining orders the tetrasporophyte is present in the life cycle, and the gonimoblast filaments develop from an auxiliary cell rather than from the carpogonium.

ORDER Gigartinales The auxiliary cell is an ordinary vegetative cell of the plant body, not borne on a special filament. About 70 genera.

Chondrus A rather coarsely branched cartilaginous, dark purple seaweed.

On the rocky North Atlantic coasts of Europe and North America *Chondrus crispus,* or Irish moss, forms a broad band of low dense growth in the lowest part of the intertidal zone below the band of *Mytilis* (Fig. 2-3). This growth shelters a multitude of small animals, and the stemlike portions are usually coated with encrusting bryozoans (*Membranipora, Microporella,* and others).

Gigartina A large, chiefly Pacific genus, with fleshy cartilaginous compressed fronds.

Iridophycus Forms rather extensive glistening purplish sheets in sheltered low rocks, intertidal areas. Lanceolate to ovate blades.

Hypnea Often found in subtropical waters. Thick growth just above sublittoral zone below the fringe of coralline algae and zoanthids.

Ahnfeltia

Phyllophora A major source of iodine.

Agardhiella

ORDER Ceramiales The auxiliary cell from which the gonimoblast filaments arise is a special cell at the base of a carpogonial filament. It arises after fertilization. Many species are delicate, lacelike, finely divided forms with fruiting bodies. About 160 genera.

Ceramium Large, widely distributed genus, in both Atlantic and Pacific. Delicate, finely divided.

Dasya Feathery appearance due to many slender branches from larger axial filaments.

Polysiphonia Thallus is filamentous. Much-branched, specialized antheridial branches, and complex carpogonium and cystocarp; characteristic tetraspores. *Rhodomela.*

ORDER Cryptonemiales The auxiliary cell is borne on a special filament resembling a carpogonial filament. About 85 genera.

Pikea

Aeodes Large scattered isolated sheets hanging down from point of attachment on intertidal rocks when uncovered at low tide. Found above *Champia-Vermetus* zone, and extends above, through, and below the *Lithothamnion* zone.

Schizomenia

Hildenbrandia

This order also includes the coralline algae, family **Corallinaceae,** distinguished by its calcareous habitat. The thallus becomes hardened and brittle from deposition of calcium.

Corallina Freely branching nodulose growth.

Lithothamnion Encrusting habit.

Porolithon Rocklike appearance. Commonly the dominant organism on exposed seaward upper edge of coral reefs.

ORDER Rhodymeniales Auxiliary cell is a special cell borne adjacent to the base of a carpogonial filament, developing there prior to fertilization. Ooblastoma filament short. About 25 genera.

Champia

Halosaccion

Rhodymenia

Sphaerococcus Worm moss, or wormweed.

Subkingdom Tracheophyta (vascular plants)

PHYLUM Bryophyta Liverworts, mosses, and their allies.

PHYLUM Pteridophyta Ferns and fern allies.

PHYLUM Spermatophyta Seed plants.

Subphylum Gymnospermae (Coniferophyta) Conifers and their relatives.

Subphylum Angiospermae (Anthophyta) Flowering plants.

CLASS Monocotyledonae Monocots, or seed plants, with only one seed leaf. Leaves mostly parallel veined.

CLASS Dicotyledonae Dicots, or seed plants, in which the embryo has two seed leaves, or cotyledons. Leaves mostly net veined.

Both classes of flowering plants have given rise to some marine forms. Those mentioned in this book follow.

I. Truly marine angiosperms

CLASS Monocotyledonae Includes salt marsh grasses such as *Spartina*.

ORDER Naiadales (Helobieae)

FAMILY Hydrocharitaceae *Halophila, Enhalus, Thalassia.*

FAMILY Posidoniaceae *Posidonia.*

FAMILY Ruppiaceae *Ruppia.*

FAMILY Zannichelliaceae *Zannichellia. Althenia, Amphibolis, Cymodocea, Halodule, Syringodium.*

FAMILY Zosteraceae *Zostera, Phyllospadix.*

The last four families are sometimes united as family Potamogetonaceae, of which they would then constitute subfamilies.

II. Mangrove complex

CLASS Dicotyledonae Includes salt marsh plants such a pickleweed, *Salicornia,* and marsh rosemary.

ORDER Myrtales

FAMILY Rhizophoraceae (the typical mangroves) *Rhizophora, Bruguiera, Ceriops, Kandelia.*

FAMILY Combretaceae *Lumnitzera.*

FAMILY Lythraceae *Sonneratia.*

ORDER Geraniales

FAMILY Meliaceae *Carapa.*

ORDER Polemonales

FAMILY Verbenaceae *Avicennia.*

FAMILY Acanthaceae *Acanthus.*

ORDER Primulales

FAMILY Myrsinaceae *Aegiceras.*

ORDER Rubiales

FAMILY Rubiaceae *Scyphiphora.*

CLASS Monocotyledonae

ORDER Phoenicales

FAMILY Arecaceae (Palmae) *Nipa*

KINGDOM PROTISTA Animal-like protists.

PHYLUM Protozoa Unicellular animals. Four major groups, long treated as classes of one phylum, are generally recognized. Recent students of the group tend to break it up into two or three phyla, treating the ciliates as one phylum, the other three classes as another, or in some cases setting up additional phyla for the sporozoans and for the sarcodina plus the zoomastigophora, and excluding the phytomastigophora, which are treated as phyla or classes of algae. In this outline the major groups of phytomastigophora are treated as algae (see Euglenophyta, Chrysophyta, Pyrrophyta), but for simplicity the classical arrangement of placing the remainder into four classes is retained.

CLASS Flagellata (Mastigophora) Most groups of the Protomonadina and many of the

Polymastigina are represented in marine as well as fresh waters. The free-living ones are mostly small colorless flagellates with one to three flagella, those with two or three flagella often having one of them directed posteriorly as a trailing flagellum. Species of *Bodo, Monas,* and others familiar from freshwater, often in somewhat stagnant waters with high organic content, have also been reported in marine waters or brackish estuarine waters. Likewise, some of the types with a single flagellum surrounded by a delicate cytoplasmic collar at the base, such as *Monosiga,* are found in marine as well as fresh waters. Most of the above are holozoic forms, feeding on bacteria, minute algal cells, or particles of organic detritus.

Numerous species of the family Trypanosomidae (*Leptomonas, Herpetomonas, Trypanosoma,* etc.) have been reported from marine animals, as have also the ubiquitous trichomonads (*Trichomonas,* etc.) (Polymastigina: Trichomonadidae).

One of the more famous flagellates, mentioned in many zoology textbooks, is *Pterospongia,* described as a mass of collared cells and amoebocytes occurring as a colony embedded in a common gelatinous matrix. It was thought to perhaps represent an intermediate condition between flagellates of the family Choanoflagellidae and the sponges. According to deLaubenfels, the validity of this genus and concept is very dubious. It was found only once, and it is uncertain whether it was a bona fide flagellate colony, a fragment of a sponge, or some other organism.

CLASS Sarcodina Free-living amoebas, heliozoans, and other Sarcodina are abundantly represented in marine as well as fresh waters. Two groups, the Foraminifera and the Radiolaria, are of particular importance because they are abundant in, and limited to, marine waters, and because they possess calcareous or siliceous skeletons. These skeletons contribute to and characterize various marine sediments. Since they are fossilized readily, they are an important part of the geological record.

ORDER Foraminifera With chambered test, which may be calcareous, arenaceous, or mixed. About 45 families have been distinguished. Genera mentioned in the text: *Globigerina, Globigerinoides, Globorotalia, Hastigerina* (Fig. 9-2).

Some of the principal families and genera follow.

FAMILY Saccamminidae Normally only one-chambered; rarely, a series of loosely attached similar chambers. Normally has only one opening. *Saccammina.*

FAMILY Ammodiscidae Two-chambered, a proloculum and one longer undivided tubular chamber. They are arenaceous, but with much cement material, and are usually yellowish or reddish brown. *Ammodiscus.*

Most foraminifera have several to many chambers, variously arranged. They may be arenaceous with foreign particles incorporated in a cementing material or calcareous, perforate or imperforate, smooth or spined. The aperture may be simple, toothed, radiate, or variously shaped. Chambers may be arranged in straight, curved, or spiralled series, concentrically, or biserially, and early chambers in some forms are embraced by and hidden by later ones. Septa may be single or double and in the latter case may form a complex canal system. The walls may be labyrinthine.

FAMILY Textulariidae Test usually biserial, multilocular, and arenaceous. Walls are not labyrinthine, and aperture is not radiate. *Textularia, Bolivina, Clavulina, Tritaxia, Virgulina, Bulimina,* and others.

FAMILY Lituolidae Test arenaceous and multilocular with labyrinthine wall, not conical. *Lituola, Trochammina, Haplophragmoides.*

FAMILY Valvulinidae Triserial, at least in young microspheric forms, and arenaceous. Aperture is usually toothed. *Valvulina, Verneuilina.*

FAMILY Nonionidae Mostly very small, discoidal, planispiral, with one series of chambers, which are not divided into chamberlets. Aperture is not radiate. *Nonion.*

FAMILY Camarinidae Large, planispiral, and perforate. Chambers are divided into chamberlets, and septa double with canal system developed. Aperture is not radiate. *Camarina.*

FAMILY Nummulinidae Calcareous, symmetrical, lenticular, or discoidal, and finely tuberculated. They are many chambered, the chambers spirally or concentrically arranged, and may have a supplementary shell. This family is prominent in some fossil Eocene and Oligocene strata in the Meditteranean region and southeast Asia. *Nummulina, Nummulites, Orbitoides, Archaediscus, Polystomella,* and others.

FAMILY Rotaliidae Calcareous, perforate, and typically spiral. They are coiled so that all chambers except those of last convolution show from above. Some have double septa and canal system. Numerous genera. *Rotalia, Patellina, Planorbulina, Spirillina.*

FAMILY Lagenidae Calcareous with smooth glassy- or vitreous-appearing surface. They are perforate, and the aperture is typically radiate and terminal. Test is straight or planispiral, not trochoid. They have no interseptal skelton or canal system. *Lagema, Crystellina, Nodosaria.*

FAMILY Miliolidae Test calcareous, imperforate. Chambers are partly coiled in varying planes around a longitudinal axis and may partially or wholly envelope earlier chambers. They are porcellaneous. Many genera. *Miliola, Biloculina, Quinqueloculina, Spiroloculina, Cornuspira, Triloculina.*

FAMILY Globigerinidae Pelagic, globular, and perforate. Inflated chambers are spirally arranged and appear as an irregular grapelike cluster. Some have many fine flotation bristles. Aperture is simple and may be single or multiple. Septa are single. No supplementary skeleton or canal system. Numerous genera. *Globigerina, Globorotalia, Hastigerina.*

ORDER Radiolaria Four suborders are commonly recognized. The first is sometimes separated from the Radiolaria and treated as an independent group of equivalent rank with Radiolaria and Heliozoa.

Suborder Acantharia Skeleton with long, radiating spines, mostly of strontium sulfate and often united outside animal by latticework. Spines radiate from center of central capsule, regularly arranged; each surrounded by a circle of contractile elements, the myonemes, as they leave the calymma, constituting a hydrostatic mechanism. Extremely young are uninucleate, but fully grown possess numerous small nuclei arranged in two or three layers. Inner cytoplasm often colored because of assimilation of pigments from prey. Symbiotic zooxanthellae and sometimes aberrant parasitic dinoflagellates present. Animal usually spherical and rather large.

Suborder Nassellaria (= Monopylea or Monopylaria) Oval uninucleate forms. Skeleton siliceous, characteristically in the form of a tiara or helmet. Extracapsular cytoplasm nonpigmented, but inner cytoplasm may be colored. Central capsule has a single large aperture through which axopodia protrude, grouped into bundles.

Suborder Phoeodaria (= Tripylea) Central capsule commonly has double membrane; three apertures, or pore fields—a basal one, the astropyle, and two lateral parapyles. A mass of dark pigment, the phoeodium, found near astropyle. Skeleton extracapsular and composed of hollow, siliceous elements containing a considerable amount of organic matter—an organic silicate or an amorphous silica. No zooxanthellae and no axopodia. Animal is polykaryotic and multiplies by binary fission after division of nuclei. Sometimes more than one central capsule. Formerly all radiolarians with more than one central capsule were grouped into suborder Polycyttaria but are now distributed among various other suborders, and this name is descriptive rather than taxonomic.

Suborder Spumellaria (= Peripylea or **Peripylaria)** Central capsule pierced all over with pores. Skeleton absent or composed of concentrically arranged spicules or latticed spheres; radial spines or spicules may be attached to shell in some species.

Genera mentioned: *Dictyacantha, Acanthostaurus, Thalassicolla, Carposphaera* (Figs. 9-3 and 9-4).

ORDER Heliozoa Sun animalcules. Usually rounded. They have many fine filose pseudopodia radiating in all directions from the body. Some are stalked and attached to a substrate.

CLASS Sporozoa All are endoparasites of animals; they lack special locomotor organelles, though some can move by a gliding motion. Most have complex life histories involving cycles of asexual reproduction by schizogony, followed by gametocyte formation, fertilization of the female gametocyte, and the formation of spores or sporozoites after meiosis; cycles commonly involve two or more host species. Widespread among marine fish and invertebrates as well as terrestrial and freshwater animals. *Gregarina* (Fig. 8-12, *A*).

CLASS Ciliata (Infusoria) An immense group of protozoans characterized by the possession of cilia, and by a unique nuclear dimorphism. The macronucleus is a "vegetative" nucleus which, during cell division, simply constricts into two parts. The micronucleus is responsible for transmission of the genetic information and undergoes mitosis. Ciliates also undergo a unique sexual process, conjugation, during which two individuals are joined together. The micronucleus of each undergoes meiosis, and one of the resulting haploid nuclei migrates to the other individual, uniting with one of its stationary haploid micronuclei, forming a new diploid nucleus. The two individuals, now genetic identical twins, separate and are the beginning of a new clone. Individuals from the same clone usually will not conjugate with each other. The macronucleus degenerates during or soon after conjugation and is replaced by a new one derived from a division product of the new micronucleus. All major groups of ciliates are abundantly represented in marine environments. Of special note among marine ciliates are the tintinnids, which build tube-, vase-, or bowl-like loricas, usually fastened to some stationary or floating object. The planktonic species are widespread in all oceans. Ciliates mentioned: *Condylostoma, Diophrys, Galeia, Loxophyllum, Metopus, Pleuronema, Tracheloraphis, Tracheolostyla* from the endopsammon of sediments (Fig. 13-2).

Most ciliates belong to subclass Euciliata.

Subclass Suctoria includes forms that are ciliated during the dispersal phase but that settle down as sessile, often stalked, forms without cilia, but with suctorial, commonly knobbed, tentacles with which they capture and suck in the contents of other protozoa or micrometazoans.

KINGDOM ANIMALIA Metazoa.

PHYLUM Porifera Sponges.

CLASS Demospongiae By far the great majority of sponges found in shallow shelf and intertidal waters belong to this huge class. It is characterized by having either a horny fibrous skeletal material of **spongin** or **siliceous spicules** of types other than triaxon (six-rayed), or both. Those which have spicules commonly have two types of spicules: **megascleres,** which form the actual supportive skeleton, and **microscleres,** which are small spicules in the mesenchyme not playing a directly supportive role. Flagellated chambers of Demospongiae are usually small and rounded, and the organization of the water channels is of the complex **leuconoid** type.

ORDER Keratosa Commercial sponges, or bath sponges, are large sponges of this order, in which spicules are lacking. The elastic horny fibrous skeleton, cleaned and freed from other organic remains, comprises the sponge of commerce. In recent years the demand for sponges has decreased because of the development of rubber and synthetic substitutes.

CLASS Calcarea Found in coastal marine waters and characterized by calcareous spicules with one to four rays, not differentiated into megascleres and microscleres. Most of the calcareous sponges are relatively small whitish vase-shaped or tubular sponges of the **asconoid** or **syconoid** type of organization. In the ascon type of sponge the incurrent pores lead directly into the main chamber, or spongocoel, of the vaselike or tubular sponge. This chamber is lined with choanocytes and bears the osculum at its upper end. The sycon type is somewhat more complex but can be visualized as an ascon type in which the wall has become complexly folded so that it consists of alternating incurrent canals and excurrent canals. Choanocytes line the excurrent canals, which open into the spongocoel. The small sponge widely sold under the name *Grantia* is of this type. The ostia in sycon sponges are internal, penetrating the walls separating incurrent and excurrent canals. The outer ends of the incurrent canals constitute new incurrent openings.

CLASS Hexactinellida Glass sponges, usually found on soft substrates in deeper water.

CLASS Sclerospongiae Recently proposed (Hartman and Goreau, 1970) to include the coralline sponges. These sponges bear suggestive resemblances to the ancient fossil stromatoporoids (Fig. 12-23).

Sponge genera shown: *Chalina* (Fig. 12-22), *Halichondria* (Fig. 12-21), *Astrosclera* (Fig. 12-23), *Speciospongia* (Fig. 13-22), *Pheronema* (Fig. 13-24), *Euplectella* (Fig. 13-25).

PHYLUM Cnidaria (Coelenterata) One of the major marine phyla. Also a few freshwater forms such as the familiar *Hydra* and the freshwater medusa *Craspedecusta*. There are three large, well-defined classes: Hydrozoa, the hydroids and their relatives; Scyphozoa, the true jellyfish; and Anthozoa, the sea anemones, corals, gorgonians, etc.

CLASS Hydrozoa Typically exist in two forms, the polypoid generation and the medusa. The polypoid is typically sessile, colonial with many feeding polyps, and asexual. It gives rise to free medusae, or to sessile medusoids, by budding. The medusa is typically a small jellyfish-like form but craspedote (with a velum). Gametes produced by the medusae unite to form zygotes that develop into ciliated planula larvae, which settle on suitable substrate and grow directly into the polypoid colony. Either generation may be much reduced or even absent in some species.

ORDER Hydroida Hydroids.

Suborder Gymnoblastea (Tubulariae) Naked hydroids. The polyps do not have a hydrotheca (non-cellular–secreted cuplike extension of the perisarc into which they can withdraw). Medusae are mostly deep bodied, are thimble or bell shaped, and may have ocelli at tentacle bases, but not statocysts. Usually with only four (sometimes six or eight) radial canals. Gonads on manubrium or hanging into subumbrellar space from top of this space, beside the manubrium. In the medusae-based classification—suborder Anthomedusae.

Representative genera best known in the polypoid generation are *Bougainvillia, Clava, Coryne, Coryomorpha, Eudendrium, Garveia, Hydractinia, Pennaria, Perigonimus, Podocoryne, Syncoryne, Tubularia.*

Genera best known as medusae are *Cladonema* and *Eleutheria* (with branched tentacles); *Cytaeis, Leuckartiara, Pandea, Polyorchis, Rathkea, Sarsia, Tiara, Zanclea.*

Suborder Calyptoblastea (Campanulariae) Polyps surrounded by a hydrotheca. Medusae commonly flatter, saucer or dish shaped, with the gonads on the radial canals. Statocysts present, no ocelli. Commonly four radial canals, but some have numerous radial canals. Suborder Leptomedusae.

Some genera best known in the polypoid generation: *Abietinaria, Aglaophenia, Obelia, Plumularia, Sertularia.*

Genera best known as medusae: *Aequorea, Eirene, Eutima, Gonionemus, Mitracoma, Phialidium.*

ORDER Trachylina Craspedote medusae with tentacles emerging above the bell margin. Statocysts present. Polypoid generation reduced or absent. Includes the Trachymedusae and Narcomedusae.

Representative Trachymedusae: *Aglantha, Aglaura, Geryonia, Halicreas, Liriope, Trachynema.*

Representative Narcomedusae: *Aegina, Cunantha, Cunina, Pegantha, Solmundella.*

ORDER Hydrocorallina Polypoid generation forms massive, rather smooth, corallike calcareous skeleton. Medusae very small. Strong nematocysts. *Millipora, Stylaster.*

ORDER Siphonophora Commonly divided into three groups on the basis of the possession of swimming bells and floats: the **Calycophora,** with swimming bells (nectophores), the **Physonectae,** with both floats and swimming bells, and the **Cystonectae,** with floats but no swimming bells.

Hyman (1940) divides the group into two suborders, the **Calycophora,** in which the upper end of the colony possesses one or more swimming bells but not a float, and the **Physophorida,** in which the upper end of the colony is a pneumatophore, or float. Swimming bells may or may not be present in this group.

The following examples will serve to illustrate some of the common types.

Suborder Calycophora Upper end of colony with one or more swimming bells (nectophores) but no floats. Other zooids commonly arranged along a long stem.

Abyla Upper end of colony of two unlike bells, replaceable by reserve bells. Upper bell has a rectangular prismatic apical facet and is smaller than lower bell; bears somatocyst and oil cavity. Cormidia at equal intervals along stem, and each siphon has a bract.

Diphyes Has two large acuminate swimming bells at upper end of stalk bearing cormidia. An oceanic, widely distributed genus. Numerous species.

Lensia Common cosmopolitan genus.

Monophyes A single anterior nectocalyx, representing primary nectocalyx of larva. No secondary replacement of nectocalyces.

Muggiaea Single nectocalyx which is a secondary pyramidal one, replacing primary caplike larval nectocalyx. Widely distributed genus.

Suborder Physophorida A float, or pneumatophore, is present. Nectophores may or may not be present. The advent of the float above the swimming bells has led to the evolution of forms with shorter stems, crowding the individuals together around a short, massive coenosarc. The cormidia lose their individuality. This line of evolution leads to the Cystonectae, in which nectophores and bracts have disappeared and the various zooids are crowded together at the base of the large float. The Physonectae show intermediate stages of this evolutionary trend, still having nectophores as well as a float and exhibiting varying degrees of shortening and crowding of the stem and its zooids.

SECTION **Physonectae** Both floats and swimming bells present.

Agalma Polygastric, with long tubular stem-bearing numerous siphons, palpons, and bracts. Nematocalyces are numerous and biserial. Pneumatophore has radial pouches. Colony much longer than broad, with an apical float of a simple type in which inner and outer sacs, chitinous lining, funnel, and simple gas gland are clearly evident. Swimming bells are in two alternating rows closely pressed together. Long stalk-bearing numerous cormidia composed of gasterozooids, male and female gonozooids, palpons, and large bracts.

Forskalia Polygastric, having long tubular stem with numerous siphons, palpons, and bracts. Each siphon has a branched tentacle. Nectocalyces are numerous, multiserial, and strobiliform in several spiral rows. Pneumatophore has radial pouches.

Physophora Biserial nectocalyces, followed by a tight cluster of cormidia with large gastrozooids, palpons, and clusters of small gonophores, below which trail long tentacles bearing cnidosacs on short stalks. No bracts.

Stephalia Short colony with large, broad float surrounded basally by a corona of nectocalyces, below which many siphons with tentacles are closely crowded, surrounding one larger central siphon. Smaller gastrozooids and gonozooids clustered between siphons. Corona of nectocalyces is interrupted by one larger aurophore.

SECTION **Cystonectae** A large apical pneumatophore, or float, and absence of nectocalyces and bracts characterize this group. Pneumatocyst has an apical stigma.

Cystalia Monogastric, with one large siphon and a tentacle, surrounded by a corona of smaller siphons. Pneumatophore has no radial septa or hypocystic villi.

Epibula Polygastric, with short, inflated, spirally convoluted stem. Cormidia ordinate in a spiral ring protected by a corona of palpons. Pneumatophore without pericystic radial pouches but with hypocystic villi.

Rhizophysa Polygastric, with a long stem, bearing in its median ventral line numerous monogastric cormidia with a single palpon and tentacle. Pneumatophore large, with radial pericystic pouches and hypocystic villi.

Solacia Polygastric, with a long stem, bearing in its ventral median line numerous polygastric cormidia. Pneumatophore is large, without radial pericystic pouches but with hypocystic villi.

Physalia Best-known of all Siphonophora, the Portuguese man-of-war. Polygastric, with a short, inflated stem horizontally expanded along ventral side of large horizontal pneumatophore. Cormidia are in a multiple series along ventral side of trunk and are usually dissolved, that is, so crowded together that they have lost their individuality as cormidia. Pneumatophore is large and has a chambered dorsal crest but is without radial septa or hypocystic villi.

The animals are a beautiful bright blue to azure purple, and the float measures about 305 cm. long by 15 cm. wide. Mature ones have as many as 1000 polyps (a float, gastrozooids, gonozooids, and dactylozooids). The tentacles trail from 6 to 15 or more meters in the water, and are armed with powerful nematocysts capable of killing fish. When a fish or another prey is taken, it is brought up to the mass of gastrozooids beneath the bell. They spread their mouths out over the prey, forming a sort of communal stomach into which digestive enzymes are poured, so that the victim is rapidly digested. Nematocysts may retain their ability to sting after the death of the animal. Their poison is a protein nerve poison. Fiddler crabs are particularly sensitive to seawater extracts of the tentacle. Curiously enough, there are certain small fish that can swim among the tentacles with immunity. For example, *Nomeus gronovii,* the man-of-war fish, follows *Physalia* about, finding protection from larger predators among the tentacles.

ORDER Chondrophora

Porpita Brightly colored and pelagic, and occurring in warm seas. Large central-feeding zooid surrounded by smaller gastrozooids and reproductive zooids, with a ring of slender dactylozooids near margin. Disc has a central float, or pneumatocyst.

Velella Complex float, polythalamous, composed of numerous concentric rings and bearing a diagonal vertical crest, or sail. Marginal tentacles simple, without cnidophores. Gonostyles with mouths.

Velella, like the Portuguese man-of-war, floats on the surface of warm seas, with the tentacles hanging down into the water. In life they are bright blue. In the tropical Pacific they sometimes occur in vast numbers. The gastrodermis is filled with symbiotic zooxanthellae, which perhaps supplement the diet of small animals with products of their photosynthesis. *Velella* is a favorite prey of the carnivorons pelagic gastropod *Janthina.* In the eastern North Pacific the diagonal sail is set in such a way that if the animal is placed with its long axis north and south, the sail runs from northwest to southeast. In the western North Pacific the sails run from northeast to southwest. In the South Pacific these relations are reversed.

Perhaps they are all one species, with both types occurring in the Central Pacific but sorted out by the winds into "right- and left-handed" specimens on the two sides of the ocean.

CLASS Scyphozoa True jellyfish. Large tetramerous acraspedote medusae with gastric filaments and marginal rhopalia.

ORDER Semaeostomae Most of the familiar jellyfish or temperate and boreal waters. Bell, bowl, or saucer shaped. Marginal tentacles and four oral arms present. Bell not furrowed.

Aurelia Moon jelly is probably best known and most widely studied of all jellyfish.

Abundant in coastal waters over most of the world. In many respects it is, however, an "atypical" jellyfish. Radial canals are much branched, forming a complex system of interradial, adradial, and perradial canals. Tentacles are numerous and relatively short and are not used in capturing large prey. *Aurelia* is a flagellar-mucus feeder; mucus secreted on exumbrella catches small plankton organisms and particles of detritus, is concentrated on lappets at edge of bell, and is licked off by long oral arms; currents produced by strong flagella in grooves of oral arms then transport mucus and its contained food particles to coelenteron.

Scyphozoans feeding on larger prey may use gastric filaments to help pull it into coelenteron and further subdue it with additional nematocysts.

Cyanea Widespread and readily recognized by long tentacles arising in eight groups, in V- or W-shaped areas, near outer edge of subumbrella rather than from bell margin. Oral arms long, wide, and voluminous. When ripe, four large bunches of gonads may be partly evaginated between oral lobes and bunches of tentacles. Both radial and circular muscles in the subumbrella. *Cyanea* exhibits bioluminescence but has no ocelli and does not react noticeably to changes in illumination. Eight rhopalia are located in marginal indentations. Swimmers brushing against this jellyfish often experience a burning sensation. *C. arctica* is the common large red jellyfish of the Atlantic coast. It sometimes attains a size of more than 2 meters across, weighing nearly a ton, and is the largest known coelenterate.

Chrysaora Genus contains common large jellyfish of wide distribution. Eight rhopalia, twenty-four marginal tentacles, and thirty-two or forty-eight marginal lappets. Stomach divided into sixteen radial pouches, eight rhopalar pouches being much narrower than eight tentacular ones. Exumbrella has numerous minute nematocyst warts. This genus also exhibits bioluminescence. It is exceptional in being hermaphroditic. Scyphistoma stage is well developed, often strobilating into thirteen to fifteen ephyrae. Many species have been named, but the specific characters are vague and unreliable. Some authorities regard the related genera *Dactylometra* and *Kuragea* as merely growth stages of this medusa.

Pelagia Widely distributed and prominent. Medusae large, often brightly colored, and bioluminescent. No polyp stage, and genus has both oceanic and neritic distribution. *Pelagia* has eight tentacles and eight rhopalia. Margin scalloped into sixteen marginal lobes, which are themselves partially cleft. Exumbrella covered with warts of nematocysts. Stomach has sixteen radial pouches, each with two unbranched canals leading to lappets. Large individuals of *Pelagia* are commonly a meter or more in diameter.

ORDER Cubomedusae Bell squarish. Tentacles arise from pedalia at the corners of the bell. Strong swimmers; actively predaceous.

Carybdea Only four tentacles, each arising from a pedalium at corners of medusa. Each rhopalium has two large and four or five small ocelli.

Tripedalia Three pedalia, each bearing a tentacle, at each corner of bell.

Chiropsalmus Each corner of bell has a single large thick pedalium that branches into several smaller ones, each bearing a tentacle. These jellyfish are much feared by swimmers in many tropical areas.

Tamoya Large medusae resembling *Carybdea* in having only four pedalia, each with one tentacle. In this genus, bell is taller than wide, and exumbrella is prominently sculptured in the male. It also differs from *Carybdea* in having four perradial mesenteries joining stomach to subumbrella.

ORDER Coronatae Bell with furrow dividing upper domelike portion from lower skirt-like part. Tentacles on pedalia alternating with marginal lappets. Medium-size to small jellyfish. Tend to occur well below the surface.

Atorella Six rhopalia and six tentacles alternating between each pair of lappets. Circular muscles weak. Four or six gonads.

Periphylla Sixteen pedalia, four of which bear rhopalia and twelve of which bear tentacles. Sixteen marginal lappets group into four pairs of rhopalar and four pairs of tentacular lappets. *P. hyacinthina* cosmopolitan in all oceans. Bell purple or maroon. Usually from deep water but occasionally occurs near surface.

Atolla Sixteen to thirty-two tentacles and rhopalia; rhopalia bear pedalia that form a circlet below tentacle-bearing ones. Eight gonads. Twice as many marginal lappets as rhopalia. Bell rather shallow and dark red or purple. Bathypelagic.

Linuche (Linerges) Eight rhopalia and eight tentacles, and sixteen marginal lappets. Subumbrellar surface has hernia-like protuberances. Wide radial gastric pouches. No circular canal. Central stomach opens by four perradial ostia into a ring sinus that breaks up into sixteen branching radiating pouches in the lappets.

Nausithoë Eight rhopalia and eight tentacles. No marginal lappets and no hernia-like pouches on subumbrella. Radiating stomach pouches are simple.

ORDER Rhizostomae Medium to large jellyfish, firm. No marginal tentacles. No main mouth, but many small mouth openings along inner surface of oral arms. Mostly tropical or subtropical.

Catostylus Mouth arms without clubs, filaments, or other appendages. Sixteen radial canals (eight rhopalar and eight adradial); rhopalar canals extend to bell margin, and adradial canals end in ring canal; ring canal gives off a network of anastomosing vessels on both sides, some of which connect with radial canals but not directly with stomach. Subumbrellar circular muscles only partially interrupted at eight chief radii.

Crambione Similar to *Catostylus,* but circular muscles of subumbrella not interrupted at radii, and oral arms bear numerous nematocyst-loaded filaments. A pair of ocelli and a large furrowed sense pit at each rhopalium.

Rhizostoma Mouth arms bear long clublike terminal appendages (about as long as mouth arms themselves) and also at upper end bear frilled scapulets, or shoulder ruffles, with additional mouth openings, near bell. Eight rhopalar and eight adradial canals, as well as a blind network of canals, all extend to border. A blind network of canals is present between each pair of radial canals.

Rhopilema Similar to *Rhizostoma* but with long oral arms, bearing numerous clubs and filaments in addition to the large terminal appendage, and very large scapulets. Most northerly occurring rhizostome along Atlantic coast, occasionally straying into Long Island Sound.

Stomolophus Bell deep, covering upper part of mouth-arm complex, so that well-developed scapulets all up under bell. Persistent central mouth, and mouth arms do not bear terminal appendages. General appearance of a thick toadstool on a thick fleshy short stem.

Cassiopea One of the most remarkable of jellyfish, attaining a rather large size, up to 2 or 3 feet or more in diameter. Bell flattened, edge of flattened area comprised of especially tall columnar epithelium. Animals are sluggish and spend most of their time upside down on floor of shallow tropical lagoons, often so numerous that they nearly coat floor. Bell pulsates slowly, and flattened top portion may act in suckerlike manner to anchor them. Much-branched oral arms, with many vesicular appendages, look for all the world like some rather dense vegetable growth. Remarkable color variation from one individual to another. By means of mucus and nematocysts, small animals are caught in bushy mass, swept by flagellar action into mouths, which open widely to receive them, and carried by flagella in gastric channels to stomach. There they are seized by gastric filaments. Rejected particles carried by other flagellar tracts back down brachial canals and discharged through mouths. Thus *Cassiopea* is a flagellar-mucus feeder somewhat like *Aurelia,* but details of mechanism wholly different. Exposed tissues of oral arms and subumbrella also full of zooxanthellae, which are exposed to sunlight as jellyfish lies inverted and probably contribute to nutrition of their host. This makes analogy of bushy mass of oral arms with a vegetable growth even more apropos. Numerous species have been named, but variability of individuals taken in any one lagoon is such that specific characteristics are difficult to delimit. *Cassiopea* may well ,be regarded as benthic rather than planktonic.

Thysanostoma Mouth arms elongate and bear three rows of frilled mouths from their base to lower end. Eight rhopalar canals, a distinct ring canal, and anastomosing networks of canals between radial canals.

Mastigias Small rhizostomes with globular or nearly spherical bell, only a few inches in diameter and marked with light spots. Oral arms bear long purple terminal appendages. These small jellyfish often abundant at certain times of the year in shallow tropical waters of Indo-Pacific.

ORDER Stauromedusae (Lucernariida) Sessile trumpet-shaped medusae with the aboral part of the bell drawn out into a stalk by which they are attached to surf grass or seaweeds. Found in cool to cold littoral waters. *Haliclystus* (Fig. 9-24).

CLASS Anthozoa Polypoid generation only, no medusae. Oral disk reflected into gastric cavity (coelenteron) as a gullet, connected to the body wall by muscular radical septa dividing the peripheral part of the coelenteron into radial chambers. Tentacles borne on the oral disk.

Subclass Alcyonaria Colonial, each polyp with eight pinnate tentacles and eight complete septa (reaching all the way from the body wall to the gullet). Soft corals, gorgonians, sea pens, sea fans, sea pansies, sea whips, red coral, etc.

Subclass Zoantharia. Solitary or colonial. Tentacles not pennate, varying widely in number and arrangement in different genera. Septa differ in number and arrangement from those of the Alcyonaria. A much less homogeneous group than the Alcyonaria. Sea anemones, zoanthids, stony corals, etc.

PHYLUM Ctenophora Pelagic, gelatinous, biradially symmetrical animals with eight rows of comb plates running from near the aboral pole over the surface toward the oral pole. Classified into two classes and five orders. The members of all but one of the orders are planktonic. These classes and orders will be briefly characterized, together with some of the more familiar genera.

CLASS Tentaculata Ctenophores with tentacles or oral lobes. In the adults of some groups, the tentacles may be so minute and reduced as to be difficult to find, or even absent.

ORDER Cydippida Retain their general larval characteristics into adult stage. Globular or egg-shaped and have two well-developed tentacles. Gastrovascular canals simple, blind, and do not ramify out into a peripheral canal system. Comb rows usually equal in length or nearly so.

Pleurobrachia Body globular or egg-shaped and circular in cross section. Comb rows extend most of distance from apical to oral poles. Bases of tentacle pouches diverge, rather than lying close to pharynx on each side.

Hormiphora Similar to *Pleurobrachia* in general appearance, the most easily seen difference being that bases of tentacle pouches lie closely adjacent to pharynx for most of their length.

Lampetria Comb rows short, not reaching oral third of body. Tentacle pouches small, lying at level of oral ends of comb rows. Tentacle openings in oral half of body. Stomach broad and long, and mouth can be widely opened and used to creep on, thus foreshadowing permanent condition in order Platyctenea.

Euplokamis Body taller and subcylindrical, meridional canals and costae running its whole length. Mouth not dilatable. Tentacle roots near middle level of body.

Moseria Rather similar to *Pleurobrachia,* but body may be somewhat compressed in tentacular plane. Comb rows wider, and mouth surrounded by a broad collar, forming a retractile tube. Costae continue down to level of collar.

Ganescha Of special interest. While showing, in the main, cydippid structure, genus has some features similar to those of order Lobata. Dawydoff (1936).

Euchlora This genus attracted attention because of the finding of functional nematocysts, giving rise to speculation that they represented a rather direct link with

Cnidaria. However, Komai (1951) has presented evidence that nematocysts exogenous in origin and derived from eating medusae. Similar cases in which functional nematocysts found in certain flatworms and nudibranchs had been derived from eating hydroids.

ORDER Lobata Commonly rather large, have a delicate structure, and difficult to collect and preserve in good condition. Body commonly helmet-shaped and compressed in tentacular plane and expanded in sagittal plane into large lobes, with the result that subsagittal comb rows decidedly longer than subtentacular ones. Ends of subtentacular rows produced into four heavily ciliated extensions, auricles, or lappets, a pair on each side of mouth. Ciliated grooves extending from mouth to lappets contain additional tentacles. Four interradial canals arise directly from funnel. No tentacle sheaths, and tentacles small or absent.

Bolinopsis Lobes medium-sized, and windings of meridional vessels in them are simple. Adradial canals pass directly into upper ends of meridional canals; auricles short. Common along Atlantic coast of United States.

Eucharis (Leucothea) Large oral lobes, long, slender auricles, long main tentacles, and two deep, narrow pits extending aborally from near tentacle bases. Body surface has conspicuous papillae tipped orange or brown. A warmwater genus.

Kiyohimea Large, fragile, and colorless. Strongly compressed and with a pair of earlike processes in transverse plane at aboral end; auricles rather stout and not long. Lappets of moderate size, with complex meandering canals in them. Tentacles greatly degenerated.

Mnemiopsis Large lobes, the lobes and auricles arising at almost same level as funnel; auricles long and ribbonlike. Deep grooves extend to level of statocyst. Bioluminescent. Common along Atlantic coast in summer and fall.

Ocyropsis (Ocyroë) Distinguished by large, muscular oral lobes whose movements serve as chief swimming mechanism. Abundant in Caribbean area and Gulf of Mexico; also present in Indo-Pacific.

ORDER Cestida Strongly compressed in tentacular plane and extended in sagittal plane, becoming ribbonlike, or beltlike. Upper edge bordered by long subsagittal comb rows; mouth and statocyst on opposite edges of middle of belt; subtentacular meridional canals run lengthwise through center of the ribbon; lower border has small tentacles. Some become more than a meter in length. Subtentacular comb rows rudimentary.

Cestum Venus's-girdle. Tentacles and tentacle sheaths beside mouth; tentacles reduced to mere tufts of filaments; two rows of short tentacles along whole oral edge of body.

Velamin four subtentacular canals do not arch up but proceed directly outward in equatorial plane. Does not get as large as *Cestum* and is only about 15 cm. long. Occurs off Atlantic coast.

ORDER Platyctenea Aberrant, benthic, creeping ctenophores compressed in oral-aboral plane; superficially resemble polyclad flatworms. The young have comb rows and two tentacles with tentacle sheaths; these may be lost in adult in some genera. Ventral surface homologous to ciliated pharynx of other ctenophores, opened out to provide a broad, flat, ciliated creeping surface.

Coeloplana (Kowalevsky, 1892) Flat, oval, elongated in tentacular plane. No comb rows in adult. Four rows of erectile papillae on aboral surface.

Ctenoplana (Korotneff, 1886) Comb rows present. No aboral erectile papillae.

Gastrodes (Komai, 1922) Immature stages parasitic in Thaliacea. Develops into free cydippid larva, which settles to bottom, loses comb rows, and flattens out as creeping platyctenid.

CLASS Nuda, ORDER Beroida Without tentacles at any stage of development. Mouth and pharynx wide and long, occupying most of body. Meridional canals have numerous side branches often brightly bioluminescent. Body compressed in tentacular plane.

Beroë Comb rows not extended full length of animal. *B. cucumis* cosmopolitan. Oval, with a somewhat constricted mouth. Side branches of meridional canals end in blind twigs. Beautifully bioluminescent; rose tinted. Feeds largely on other ctenophores. *B. forskali* another large, cosmopolitan species. Body large, up to 10 to 12 cm. long. More or less triangular, with mouth occupying broad end. Side branches of meridional canals form an anastomosing network. *B. gracilis* more elongate and cylindrical; about 25 mm. long, and 8 to 10 mm. wide. Costae extend about two thirds of body length.

Neis Pharyngeal canals branch at mouth and form a complete circular canal around mouth.

PHYLUM Mesozoa Enigmatic parasites of various marine invertebrates. Body comprised principally, or only, of ciliated epithelial cells and internal reproductive cells. Thought by some to represent a very primitive metazoan group. Recent authorities tend to separate the two classes into two separate phyla on the grounds that the orthonectids are more complex than originally supposed and not really related to the dicyemids.

CLASS Orthonectida Parasitic as synctial plasmodia in the tissue spaces of brittle stars, and various other marine invertebrates. Nonfeeding ciliated males and females released into the sea, where mating occurs. *Intoshia, Rhopalura, Stoechartrum.*

CLASS Dicyemida Vermiform parasites of the renal organs of benthic cephalopod molluscs. Body consists of an axial cell, containing reproductive cells and larvae in various stages of development, surrounded by a single layer of ciliated epithelial cells, and modified into a headlike calotte at the anterior end. *Dicyemennea* (Fig. 8-12, B), *Dicyema, Conocyema.*

PHYLUM Gnathostomulida Minute wormlike ciliated animals recently found to constitute an important part of the endopsammon in and below the RPD layer in fine sediments *Guzthostomolz, Gnathustomula, Nanognathia, Onychognathia, Pterognathia* (Figs. 13-1 and 13-3).

PHYLUM Platyhelminthes Flatworms.

CLASS Turbellaria Mostly free living. Includes the familiar planarians and their relatives. Development direct. Many small turbellarians are found in sediments and wet sands as a part of the endopsammon. Others are common among seaweeds, hydroids, and other marine growths. A few, such as the family Umagillidae, found in the intestines of sea urchins, are parasitic.

ORDER Acoela Small turbellarians without well-developed digestive tract. Ventral mouth opens into a mass of digestive cells. Some marine genera. *Afrontia, Convoluta, Ectocotyla, Otocelis* (ectocommensal on echinoderms), *Parotocelis, Polychoerus, Proporus.*

ORDER Rhabdocoela: Turbellarians with a simple saclike or straight digestive tract. (This group is broken up into two or more orders by various authorities.) Many marine and freshwater genera. A few representative marine genera follow.

Free living: *Acrorhynchus, Gyratrix, Macrorhynchus, Otoplana, Plagiostomum, Polycystis.*

Parasitic: *Graffilia* (on marine snails and the shipworm *Teredo*), *Paravortex* (mantle cavity of bivalves), *Umagilla* (and others of family Umagillidae, small red rhabdocoels found in echinoderms).

ORDER Tricladida: Gut trifurcated, one anterior and two posterior-lateral branches. Muscular proboscis exsertable from ventral mouth. The familiar freshwater planarians belong here. Marine forms are grouped into a subgroup termed Maricola. *Bdelloura* (ectocommensal on *Limulus*), *Procerodes, Uteriporus.*

ORDER Polycladida Comprises large active predaceous flatworms found under rocks and on reefs, etc. (Fig. 12-12). *Apidioplana, Diplosolenia, Enantia, Eurylepta, Hoploplana, Leptoplana, Notoplana, Planocera, Plehnia, Stylochus, Thysanozoon, Yungia.*

CLASS Trematoda Flukes. Parasitic flatworms in or on vertebrates. Most trematodes, both monogenetic and digenetic, are rather specific in their host ranges and for this reason are useful adjuncts to the study of the phylogeny, biogeography, and ecology of their vertebrate and invertbrate hosts. The same applies to cestodes, and most other parasites.

Subclass Monogenea Ectoparasitic flatworms on fishes, on the body surface, or in the mouth or among the gills. Some occur on other hosts such as turtles, or in the bladders of amphibians. Possess a large posterior opisthaptor, sometimes divided into six suckers or many suction areas and in some also provided with chitinous hooks. Development direct by way of a ciliated swimming larva that finds the next host individual. *Aspidogaster, Calceostoma, Chimaericola, Diplozoon, Epibdella, Gyrodactylus, Hexostoma, Microcotyle, Monocotyle, Onchocotyle, Polystomum, Tristomum, Udonella.*

Subclass Digenea Typical flukes. Internal parasites of vertebrates. Life cycle complex, including larval stages in molluscs, usually gastropods, and, in some, an encysted stage in or on a third host.

Nanophyetus (= *Troglotrema*) (Fig. 8-13).

Digenetic trematodes: Only a few representative genera of the many described from marine vertebrates will be mentioned.

Schistosoma (and other related blood flukes) are found in some marine mammals (pinnipeds) and birds. They differ from other flukes in having separate sexes as adults, and fork-tailed cercaria larvae. The cercaria sometimes invade the skin of swimmers causing swimmer's itch. Other larval stages occur in gastropods.

Derogenes Unusually wide geographic and host range: found in marine fish in both Atlantic and Pacific and in a wide range of latitudes, making it one of the most cosmopolitan widely distributed genera of flukes.

Levinseniella Unusual in that the adults are reported both from fish (toadfish) and from shore birds such as sanderlings and godwits.

Cryptocotyle The larvae cause "pigment spots" of marine fish.

CLASS Cestoda: The tapeworms. Body in most is strobilated, consisting of a scolex by which the worm attaches to the intestinal lining of its host, and a series of segments or proglottids, each with a complete set of both male and female reproductive organs. No digestive tract. *Ouwensia* (Fig. 8-12, *C*) may be a larval stage of some cestode.

Subclass Cestodaria Nonstrobilized cestodes with ten-hooked embryos (decacanth). Superficially they resemble flukes but lack digestive tract. Mostly in primitive fish, elasmobranchs, and sturgeons. *Amphilina, Archigetes, Caryophyllaeus, Gyrocotyle.*

Subclass Cestoda (Eucestoda) Tapeworms with a series of proglottids, each with a complete set of reproductive organs when mature. An attachment organ, the scolex, is at the anterior end. Embryos are six-hooked (hexacanth).

ORDER Pseudophyllidea Scolex with two (rarely one) muscular grooves or bothria and some with protrusible spiny proboscides. Genital pores of proglottids are ventral, and a uterine pore is present. Eggs are operculate, and the onchosphere has a ciliated embryophore.

Diphyllobothrium The broad fish tapeworm of man and various carnivores belongs here. Marine species occur in pinnipeds, dolphins, sea cows, and some marine birds.

ORDER Tetraphyllidea Scolex with four lappetlike outgrowths, bothridia; suckers or proboscides may be present. Genital pores are lateral and no primary uterine pore. Eggs are not operculate, and the onchosphere lacks a ciliated embryophore. Primarily in elasmobranch fish, but some occur in other fish and even amphibia and reptiles. *Acanthobothrium, Phyllobothrium.*

ORDER Tetrarhynchidea Scolex with four long, protrusible, spiny proboscides. Retractor muscles lead to pouches at the posterior end of a long necklike region. Two to four bothria are present. Genital openings are lateral, and no uterine pore is present. In spiral valves of elasmobranch fish; larvae occur in cephalopods and other invertebrates eaten by elasmobranchs. *Tetrarhynchus.*

ORDER Cyclophyllidea Includes most of the tapeworms of terrestrial vertebrates. Scolex with four suckers, and in some a single rostellum armed with one or more circles of hooks. Genital pores are lateral; no uterine pore.

PHYLUM Nemertea Rubberband worms. Similar in general grade of construction to the flatworms but differing in possession of a complete digestive tract with posterior anus, and of a unique proboscis, the rhynchodaeum, kept retracted most of the time in a coelomlike space, the rhynchocoel, dorsal to the digestive tract. Predaceous, capturing prey by means of the proboscis. Mostly intertidal and subtidal in shallow waters. Some bathypelagic, and a very few freshwater or terrestrial. Genera mentioned: *Nectonemertes, Pelagonemertes, Planctonemertes* (bathypelagic genera). Some common genera of intertidal and subtidal nemertean worms are *Amphiporus, Carinoma, Cerebratulus, Emplectonema, Lineus, Micrura, Nectonemertes, Tubulanus. Malacobdella* is unusual among nemerteans in having adopted a parasitic mode of life. It is found in the mantle cavity of bivalve molluscs and has developed a posterior sucker, giving it a leechlike appearance.

PSEUDOCOELOMATE GROUPS Most pseudocoelomate animals have been placed in a large phylum, Aschelminthes, by many zoologists. This assemblage is so diverse and so lacking in any well-defined, unifying characteristics that others prefer to use the term aschelminth in a descriptive sense without taxonomic rank and to treat the major subdivisions as phyla. The common characteristics are the presence of a pseudocoel type of body cavity; complete digestive tract; secreted noncellular cuticle, beneath which is a syncytial type of epithelium, with the nuclei usually collected into special tracts, or areas; and determinate cleavage, with early attainment of definitive somatic cell number, and further growth being by cell enlargement rather than continued cell divisions. Many show posterior adhesive glands.

PHYLUM Aschelminthes

Subphylum Gastrotricha Small group of about 400 species, part of the microbenthos of both marine and fresh waters. Some are part of the endopsammon, others crawling and swimming about aquatic vegetation. The cuticle is usually modified in the form of scales, and the locomotor cilia are restricted to the ventral surface. Marine species have been reported from *Chaetonotus, Aspidophorus, Heterolepidoderma,* and members of the order Macrodasyoidea.

Subphylum Rotifera Rotifers, or wheel animalcules. Cilia restricted to anterior trochal discs, used for swimming and to create a feeding current. Pharynx with a unique grinding organ, the mastax. Very abundant in freshwater. Several brackish water or marine species.

Forms mentioned: *Brachionus, Keratella (=Anuraea), Asplanchna, Notholoca.*

Subphylum Kinorhyncha (= Echinodera) Small benthic worms living in marine muds in coastal waters. Less than 1 mm. long. Cuticle divided into segmentlike zonites often ornamented with spines, but internal organs not segmented. Head distinct, retractile. *Echinoderes, Pyenophyes.*

Subphylum Nematoda By far the largest group of aschelminths, the roundworms. One of the most ubiquitous and widely distributed groups of animals on earth. About 10,000 described species, and many still undescribed. More than 35 genera of free-living intertidal nematodes have been described. Most animals have one or more parasitic or commensal species.

PHYLUM Acanthocephala The thorny-headed worms. Entirely parasitic group, the adults in the gut of vertebrates, the young in various invertebrates. Body with spacious body cavity, no digestive tract, a retractile proboscis armed with chitinous hooks with which they anchor to the gut mucosa. Body wall with peculiar lacunar system of fluid-filled channels. Sexes separate. The relationships of the acanthocephalans are unclear.

COELOMATE PHYLA The remaining groups of animals either possess a true coelom, a body cavity derived as a mesodermal space lined with peritoneum, or if the coelom is secondarily obliterated, are related to and derived from coelomate groups. The coelom is wholly different in origin from the pseudocoel, which is a distinct acquisition on its own, not an intermediate stage in the development of a coelom.

LOPHOPHORE-BEARING GROUPS Four phyla of protostomate coelomates possess a characteritic food-collecting structure, the lophophore. This circular or horseshoe-shaped fold of the body wall encircles the mouth and bears numerous ciliated tentacles, which are hollow outgrowths of the body wall, each containing an extension of the coelom. Ciliary tracts on the tentacles create a feeding current that brings small planktonic organisms to the mouth.

PHYLUM Phoronida Wormlike, marine, living in chitinous tube either buried in sand or attached to rocks. About 15 species. *Phoronis, Phoronopsis.*

PHYLUM Entoprocta The nodding heads. A small group of about sixty species of sessile forms somewhat resembling Bryozoa, but with coelom obliterated by growth of parenchymal tissue. Solitary or colonial, often commensal on other animals such as certain polychaetes, chitons, and others. *Barentsia, Loxosoma, Pedicellina, Urnatella.*

PHYLUM Bryozoa (Ectoprocta) Colonial, each zooid housed in a box or tubelike zooecium. Anus outside the lophophore; coelom extensive. Recent Gymnolaemata are commonly divided into three orders.

ORDER Ctenostomata Zooecium not calcified. The colony is gelatinous. The operculum has been described as composed of a fringe of setae forming a thin, membranous operculum that closes the orifice on withdrawal. Actually, there is a membranous collar that folds in definite creases. Often the creases, but not the membrane between them, are visible, giving rise to the illusion of a fringe of setae. No avicularia or vibracula are formed. Common examples of this group are *Alcyonidium, Amathia, Bowerbankia, Flustrella.*

ORDER Cyclostomata Includes forms such as *Crisia,* with a tubelike calcareous zooecium having the wide circular orifice at the end of the tube and no avicularia or operculum. *Crisia, Lichenopora, Plagioecia, Tetracycloecia, Tubulipora.*

ORDER Cheilostomata By far the largest order. Contains forms with a boxlike, calcareous skeleton with operculum. Usually the zooids are differentiated or polymorphic, such modified zooids as **avicularia, vibracula,** and **oecia** commonly being present. Avicularia are modified zooids commonly resembling a bird's head on a stalk and seem to play a role in defense of the colony and keeping it clean. Vibracula are movable, sticklike structures that probably play a defensive role. Oecia are specially modified zooids, usually with a larger zooecium than the others, containing egg cells. Two suborders are recognized—those with somewhat elastic zooecia and no compensatorium, forming the suborder **Anasca,** of which the erect colonies of *Bugula* are the most familiar; and those with rigid zooecia and compensatorium, forming the suborder **Ascophora.** The first zooid of a colony, which is sometimes distinctive in structure and position in the colony, has been termed the **ancestrula.**

The planktonic larva developed from fertilized eggs is a peculiar flattened triangular **cyphonautes** larva. A few of the more familiar genera are *Antropora, Bicellaria, Catenaria, Cellularia, Cheilopora, Chlidonia, Eschara, Eurystomella, Flustra, Lepralia, Membranipora, Micropora, Notamia, Parasmittina, Peristomella, Retepora, Scrupocellaria, Tricellaria.*

ORDERS Trepostomata and **Cryptostomata** Two extinct orders of bryozoans that seem to be limited to Paleozoic times, although some trepostomes of questionable identity have been reported from the Triassic. The Cryptostomata appeared in the Ordovician, were most abundant in the Devonian and Mississippian, and died out in the Upper Permian.

The oldest ctenostome is the peculiar threadlike *Marcusodictyon* from the Ordovician of Estonia, where the Gulf of Finland and the Baltic Sea meet and border the Soviet Union. Cyclostomata (Stenolaemata) began in the Ordovician, were the predominant bryozoans in the middle Mesozoic, and have since then been secondary to the Cheilostomata, which seem to have originated during the Jurassic in European seas and rapidly extended their range, being the dominant group of Ectoprocta throughout the world today.

PHYLUM Brachiopoda Lamp shells. Bivalved, superficially resembling bivalve molluscs but with the shells dorsal and ventral. Mantle cavity mostly occupied by large, double-spiraled lophophore.

CLASS Articulata (= Testicardines) By far the largest class. Possesses hinged valves. The peduncle passes out through the ventral valve (Fig. 12-37) or in some cases is absent. The lophophore bears an internal calcareous skeletal support, and the digestive tract is blind posteriorly, with no anus.

CLASS Inarticulata (= Euardines) No shell hinge, the shells fitting together more like two saucers. The peduncle passes between the shells, and an anus is present posteriorly between the shells. The best-known members of the Inarticulata are the Lingulidae, which are rather large brachiopods with a long contractile peduncle. They are found in shallow, muddy bottoms rather than on rocky substrate. They have relatively soft, vascular, horny, somewhat rectangular shells fringed with long, stiff setae.

Approximately 1400 genera and 25,000 species of brachiopods have been described, of which the living, or recent, forms constitute about 75 genera and 250 species. Authorities are not yet in general agreement concerning the classification within the group into orders, suborders, families, etc., especially within the huge group of articulate brachiopods. Pending better clarification of relationships within the group. Cooper's nonphylogenetic arrangement into **impunctate, pseudopunctate,** and **punctate Articulata** is a convenient and more readily determinable grouping than the various orders and suborders that have been proposed.

In the punctate brachiopods the inner layers of the shell are pierced by many tiny holes into which minute projections of the mantle extend, increasing the mantle surface—the main respiratory organ. Pseudopunctate brachiopods have minute rods of calcite, called **taleolae,** projecting into and often beyond the inner surface. Exfoliation of the shell leaves pits, or pseudopunctae, where these rods have been lost. The impunctate species have neither punctae nor taleolae.

Some of the more familiar genera of recent brachiopods follow.

Articulata: *Argiope, Atretia, Cistella, Frenulina, Hemithyris, Terebratella, Terebratulina.*

Inarticulata: *Discina, Discinisca, Glottidia, Lingula.*

PHYLUM Annelida The annelids, or segmented worms. Four principal groups are recognized, the Archiannelida, the Polychaeta, the Oligochaeta, and the Hirudinea. These groups are variously arranged into classes and orders, indicating the differences of opinion as to the degree of relationship between the groups, based on use of different characters for primary division. Thus we have:

Class Archiannelida		Class Archiannelida
Class Chaetopoda		Class Polychaeta
Order Polychaeta	or	Class Clitellata
Order Oligochaeta		Order Oligochaeta
Class Hirudinea		Order Hirudinea

In other arrangements the Archiannelida are united with the Polychaeta, and three classes—the Polychaeta, Oligochaeta, and Hirudinea—are recognized. Some writers class the Myzostomida (peculiar parasites of crinoids) as polychaetes. Others regard them as constituting a separate class, or even phylum. We shall simply list the groups mentioned as classes without implying any bias as to which arrangement best expresses the relationships.

CLASS Archiannelida Mostly small members of the interstitial fauna. Setae and parapodia reduced or absent. Larval ciliation may persist. These worms have been regarded as primitive, but several authorities regard their structural peculiarities largely as adaptations for interstitial life, and as the persistence of larval characteristics, and consider them to be an order of Polychaeta. *Polygordius, Protodrilus, Dinophilus, Nerilla.*

CLASS Polychaeta By far the greatest number of annelids belong to this predominantly marine group. Between 5000 and 6000 species have been described, and placed in some thirty to forty families. They are extremely diverse in structure and habits, ranging from active foraging predators to sessile plankton feeders, burrowing substrate ingesters and gatherers of fine organic detritus. Most have planktonic trochophore larvae. Most polychaetes bear parapodia, with setae, at the sites of each segment, although they are reduced or absent in some burrowing forms. Polychaetes with well-developed parapodia with acicula and setae, prostomium-bearing sensory structures, and pharynx with jaws or teeth are termed errant polychaetes, or Errantia. Those in which the porstomium is reduced, without sensory appendages, but the head region is provided with feeding and respiratory palps, tentacles, radials, etc., with reduced parapodia without acicula or compound setae, a pharynx without teeth or jaws, and the body often showing marked regional differentiation, are termed Sedentaria. These two terms should probably be regarded as descriptive rather than taxonomic, although they have been used by many writers as subclasses or suborders of the polychaetes. A few of the more prominent and conspicuous families and genera follow.

Subclass Errantia

FAMILY Aphroditidae Sea mice. *Aphrodita, Laetmonice.*

FAMILY Polynoidae Scale worms. *Polynoe, Harmothoe, Halosydna, Arctonoe.*

FAMILY Tomopteridae Planktonic pelagic polychaetes. *Tomopteris.*

FAMILY Nereidae Clam worms, etc. *Nereis, Neanthes.*

FAMILY Syllidae *Syllis, Odontosyllis, Autolytus.*

FAMILY Nephtyidae *Nephtys.*

FAMILY Glyceridae *Glycera.*

FAMILY Eunicidae *Eunice, Marphysa, Palola, Lysidice, Onuphis, Diopatra.*

Subclass Sedentaria

FAMILY Spionidae *Spio, Scolelepis, Polydora.*

FAMILY Chaetopteridae *Chaetopterus.*

FAMILY Cirratulidae *Cirratulus, Cirriformia, Ctenodrilus.*

FAMILY Opheliidae Blood worms, etc. *Ophelia, Euzonus.*

FAMILY Arenicolidae Lug worms. *Arenicola, Abarenicola.*

FAMILY Terebellidae *Terebella, Amphitrite.*

FAMILY Sabellidae *Sabella, Eudistylia, Schizobranchia, Pseudopotamilla.*

FAMILY Serpulidae *Serpula, Spirorbis, Spirobranchus.*

FAMILY Capitellidae *Capitella, Notomastus, Heteromastus.*

CLASS Hirudinea Leeches. *Pontobdella.*

PHYLUM Echiuroidea Spoon worms. These large stout worms are unsegmented, with anterior spatulate or troughlike proboscis or prostomium, ciliated ventrally. Gut is much longer than body, and the anus is posterior. The body cavity is a spacious schizocoel. *Arenchite, Bonellia, Echiurus, Listriolobus, Saccosoma, Thalassema, Urechis.*

PHYLUM Sipunculoidea Stout unsegmented worms. Anterior end is an eversible introvert bearing tentacles at its extremity. The gut is coiled and the anus is dorsal, at the base of the introvert. The body cavity is a spacious schizocoel. No setae. *Dendrostoma, Golfingia, Phascolosoma, Physcosoma, Sipunculus.*

PHYLUM Priapuloidea Stout unsegmented worms with large plump eversible introvert covered with rows of small spines, no tentacles. One or two large caudal respiratory appendages. The body cavity is spacious, apparently a true coelom. *Halicryptus, Priapulus.*

PHYLUM Mollusca About 80,000 living species and 35,000 fossil molluscs have been described. They are grouped into five (or eight if the subclasses of Amphineura are treated as classes) classes.

CLASS Amphineura

Subclass Polyplacophora Chitons. Mostly found on intertidal and subtidal rocks. The eight shell plates give a superficial appearance of segmentation, and allow flexibility in adapting the body to rock contours, or rolling up for protection. *Chiton, Mopalia, Ischnochiton, Tonicella, Kathrina, Chaetopleura, Cryptochiton, Placiphorella.*

Subclass Aplacophora Wormlike, without shell, mostly collected by dredging. *Neomenia, Proneomenia, Crystallophrisson.*

Subclass Monoplacophora Peculiar limpetlike forms, long known only as fossils, but rediscovered in deepwater dredging in the 1950s. The musculature, respiratory structures, nephridia, gonads, pericardium, etc. seem to show segmentation, but the indicated segmentation, as shown by the different organ systems, does not correspond with each other. *Neopilina.*

CLASS Gastropoda The snails, slugs, limpets, conchs, whelks, nudibranchs, etc. A huge class, characterized primarily by the phenomenon of torsion, during the larval stages, resulting in the rotation of the visceral hump 180 degrees, bringing the mantle cavity, gills, and heart to the front. More than 50,000 living and fossil species are known.

Subclass Prosobranchia Mantle cavity anterior; shell and usually an operculum present; mostly dioecious.

ORDER Archaeogastropoda (Aspidobranchia) Usually two bipectinate gills, or if only one, it is bipectinate, shell coiled, or limpetlike. *Acmaea, Diodora, Haliotis, Patella, Trochus, Calliostoma, Turbo, Tegula, Margarites, Nerita.*

ORDER Mesogastropoda (Pectinobranchiata) With single monopectinate gill, one auricle, one nephridium. *Littorina, Vermetus, Janthina, Crepidula, Strombus, Lambis, Cypraea, Polinices.*

ORDER Neogastropoda (Stenoglossa) Mostly carnivorous gastropods, with a siphon canal or notch, monopectinate gill, concentrated nervous system, and a proboscis with radula and three large teeth in a transverse row. *Murex, Purpurea, Thais, Conus, Voluta, Oliva, Nassarius, Olivella.*

Subclass Opisthobranchia Gastropods in which some detorsion has occurred during growth. Reduction of shell and mantle cavity common; one gill or none. Head often with a pair of cephalic tentacles, the rhinophores. Nudibranchs, tectibranchs, bubble shells, sea hares, sacoglossans, etc. *Acteon, Hydatina, Bulla, Pyramidella, Aplysia, Alderia, Limacina, Doris, Archidoris, Dialula, Chromodoris, Tritonia, Hermissenda, Dendronotus, Rostangia, Tethys,* and many others.

Subclass Pulmonata Mostly terrestrial or freshwater forms. Snails, slugs, pond snails. Hermaphroditic, no operculum, no gills, nervous system concentrated and symmetrical.

PLANKTONIC GASTROPODS

Prosobranchia

Streptoneura

Heteropoda

Principal genera: *Atlanta, Carinaria, Pterotrachea*

Others: *Janthina*

Opisthobranchia

Pteropoda

Principal genera: *Cavolina, Clio, Clione, Clionopsis, Creseis, Clumbulia, Herse (Cuvierina), Diacria, Euclio, Halopsyche, Spiratella (Limacina), Styliola*

Nudibranchiata *Phyllirhoë*

Some of the planktonic Gastropoda can be classified by the following key:

1. No shell present 2
   Shell present 5
   Pseudoshell present; cartilaginous subepithelial pseudo shell formed by connective tissue; larva with calcareous shell and operculum—Cymbuliidae
2. A series of external gills on each side of the body—*Phyllirhoë*
   No external gills along sides of body 3
3. Foot compressed laterally, forming a ventral finlike structure—*Pterotrachaea*
   Foot expanded anteriorly into a pair of fins 4

4. Body ovate, rounded posteriorly—*Halopsyche (Anopsia)*
   Body elongate—*Clione*
5. Surface-living, using a mass of bubbles as a float; foot not greatly modified; shell purplish or violet, body red—*Janthina*
   Not as above 6
6. Shell covering most of body, spiral; foot operculate 7
7. Large eyes present—*Atlanta*
   Eyes absent; foot modified to form two large fins—*Spiratella (Limacina)*
8. Shell not spiral, covers most of body 9
   Shell spiral, but small, forming a small cap on the end of the soft body—*Carinaria*
9. Shell with lateral spines or angles well back on the shell—*Cavolina*
   Shell without lateral spines or angles, or if present, they are near the anterior end of the shell 10
10. Shell narrow, conical—*Creseis*
    Shell broadened at anterior end if conical; or tubular 11
11. Shell conical; if lateral spines or angles are present, they are at anterior end of shell—*Clio*
    Shell tubular, with parallel or nearly parallel sides
    Most of length is subcylindrical—*Herse (Cuvierina)*

Suborder Pteropoda

SECTION **Gymnosomata**

Daicria quadridentata In tropical and subtropical waters. South China Sea. Java Sea. Common but not abundant; rarely more than 100 per 1000 cubic meters of water. Occasional in higher subtropical and transitional waters but absent from subarctic.

Clione antarctica Fairly important carnivorous member of antarctic macroplankton. Shows less seasonal change in abundance than most of the herbivorous zooplankton species. Fins are relatively short.

Clione limacina Sea butterfly. Also, *C. papilionacea,* occurring in North Atlantic, arctic, and cold boreal waters.

Halopsyche gaudichaudi An oddly shaped, nearly spherical form found in Indo-Pacific—South China Sea basin in densities of 100 to 400 per 1000 cubic meters.

SECTION **Thecosomata**

Styliola subula An offshore tropical-subtropal species.

Cavolina gibbosa An offshore species, common in warm waters, with a peculiar type of bitropical or amphitropical distribution—not abundant in truly equatorial waters as in warm water on both sides.

Cavolina longirostris A rather heavy-shelled tropical neritic form often found in high concentrations in areas such as Gulf of Thailand, near margins of gulf but also offshore in South China Sea.

Cavolina unicinata Both in Indo-Pacific and in eastern warm Atlantic waters, but not in Mediterranean.

Clio pyramidata Cosmopolitan in warm and temperate waters; wide distribution in North and South Atlantic, Mediterranean, around Africa, and in Indo-Pacific and South China Sea, as well as scattered occurrences in Central Pacific.

Creseis acicula and *C. vergula* Long, slender, conical, sharply pointed shells that look much like detached rostrums of crustacean nauplii. Tropical and subtropical waters. When abundant in a plankton haul, sharp, slender shells protruding through meshes in net make net painful to handle and are a considerable nuisance during removal of plankton samples from net.

Spiratella antarctica and *S. balea* Coldwater species of antarctic regions.

Spiratella helicina Coldwater subarctic species in both Atlantic and Pacific. An indicator of arctic or subarctic water masses. Said to occur in the Antarctic.

Spiratella inflata Widespread warmwater and temperate-water oceanic species.
Spiratella retroversa Oceanic cool waters.

CLASS Bivalvia (Pelecypoda) Bivalve molluscs.

Subclass Protobranchia Primitive forms with simple nonfolded gill filaments. Foot with a flattened ventral surface. Palps often developed into proboscides that can be extended beyond the shell for food gathering. *Nucula, Nuculana, Solemya, Yoldia, Malletia.*

Subclass Lamellibranchia or **Polysyringia** Gill filaments folded, and adjacent filaments attached by ciliary or tissue junctions. Most bivalves are members of this subclass.

ORDER Taxodonta Gill filaments with ciliary junctions only; no interlamellar junctions. Hinge teeth taxodont. Mantle margins not fused. *Acra, Anadara, Barbatia.*

ORDER Filibranchia or **Anisomyaria** Mussels, oysters, pectens, etc. Mostly sessile (except pectens, which can swim), foot small, anterior adductor muscle much smaller than posterior, or absent. No siphons. *Nytilus, Modiolus, Ostrea, Crassostrea, Pinctada, Anomia, Lima, Pecten, Pinna, Lithophaga.*

ORDER Eulamellibranchia Gills with interlamellar and intralamellar tissue junctions highly developed so that the interlamellar spaces are divided into vertical water tubes, and the blood is carried in vertical vessels within these junctions. Hinge teeth telodont, or reduced. The majority of bivalves are in this order. *Cardium, Mercenaria, Petricola, Tagelus, Dosinia, Donax, Tridacna, Solen, Ensis, Panope, Mya, Hiatella, Pholas, Teredo, Tresus, Tivela, Lyonsia, Pandora, Clavagella.*

Subclass Septibranchia Pistol shells. Deepwater forms with gills reduced to or replaced by perforated muscular septum between inhalant chamber and suprabranchial cavity. *Cuspidaria, Poromya.*

CLASS Scaphopoda Tooth shells. *Dentalium.*

CLASS Cephalopoda Octopuses, squids, cuttlefish, and their relatives. The most highly specialized and developed of molluscs.

Subclass Tetrabranchiata Four species of *Nautilus* are the only living remnants of this formerly great group. They are more primitive in many respects than other modern cephalopods. They have four ctenidia, four auricles, four nephridia, no branchial hearts, and no ink sac. Eye does not have a lens, and arms are numerous, arranged in groups, and without suckers. A large coiled external shell present, divided into successive chambers, all of which, except the last, are empty, gas filled, and connected to body by a centrally located membranous tube, the siphuncle.

Fossil members of order Nautiloidea number about 2000 species, dating from the Cambrian and reaching their maximum development in Silurian and Devonian. Shells vary from straight through bent, from loosely to closely coiled. Septa are concave, usually with straight or undulate margins. Siphuncle variable in position and sometimes large.

Subclass Ammonoidea This group was even larger, existing from Upper Silurian through Cretaceous or perhaps into beginning of Tertiary; they reached their greatest development during Mesozoic. Their shells differ from those of nautiloids in that margins of septa usually more or less intricately plicated and siphuncle usually marginal. About 6000 species have been described.

Subclass Coleoidea (Dibranchiata) Cuttlefish, squid, octopuses. Includes all modern cephalopods except *Nautilus*. Shell internal or absent. There are two ctendidia, two kidneys, and two branchial hearts. Siphon tubular. Mouth surrounded by either eight or ten arms, which bear suckers, hooks, or both, and is equipped with horny mandibles resembling a parrot's beak. An ink sac is present. Eye has a crystalline lens.

ORDER Decapoda Squid, cuttlefish, etc. In this large group, arms are of two kinds: eight regular tapering arms, usually connected to each other by the web, and usually bearing suckers their entire length; and two commonly longer arms, the tentacles, which are slender and of rather uniform diameter except for an expansion near end, which bears suckers, hooks, or both. Body usually torpedo-shaped, with a pair of posterior lateral fins. Most are pelagic predatory species.

Suborder Belemnoidea (Phragmophora) Extinct. Internal shell has a well-developed apical portion, the rostrum, a dorsal plate-like pro-ostracum, and a partially enclosed phragmocone, divided into chambers.

Suborder Sepiodea Shell has a reduced rostrum, rudimentary phragmocone, and modified pro-ostracum. Tentacles can be withdrawn into special pits, coelom well developed. Body usually short, broad, and saclike. *Serpia, Idiosepius, Spirula, Sepiola.*

Suborder Teuthoidea Common squid. Shell or pen is a flattened, chitinous plate buried in dorsal mantle tissue. Body usually elongate. Tentacles long. Representative genera are *Loligo, Cranchia, Architeuthis, Chiroteuthis* (Fig. 11-7), *Onycotheuthis.*

ORDER Vampyromorpha Vampire squid. Small deepwater octopus-like forms with eight arms united by a web, but also two small retractile tentacles. *Vampyroteuthis.*

ORDER Octopoda Octopuses. Eight arms connected by web, but no tentacles. Body usually globular, without shell or fins. Mostly benthic. Representative genera are *Octopus, Eledone, Eledonella, Vitreledonella, Amphitretus, Cirroteuthis, Argonauta, Benthoctopus.*

PHYLUM Arthropoda

Subphylum Trilobitomorpha Trilobites (Fig. 18-1).

Subphylum Chelicerata

CLASS Merostomata Horseshoe crabs. *Limulus* (Fig. 13-14.)

CLASS Arachnida, ORDER Acarina—Halacaridae. Marine mites.

CLASS Pycnogonida Sea spiders. *Discoarachne* (Fig. 12-57).

Subphylum Mandibulata

CLASS Crustacea

Subclass Cladocera

FAMILY Polyphemidae

Podon Carapace does not enclose legs; four pairs of legs crowded together. Abdomen slender, ending in two caudal spines. Head large, bluntly rounded, and has one eye. Head and thorax separated by a dorsal depression (Fig. 9-14).

Evadne Head and thorax set off by a dorsal depression. Body oval to piriform and pointed behind. Abdomen does not protrude.

Penilia Carapace encloses body but is broadly open ventrally. Six pairs of legs.

Subclass Ostracoda

ORDER Myodocopa Sars Shell notched in front, giving play to antennae. Basal joint of second antennae usually wide; exopodite its longest ramus, composed of many segments.

FAMILY Cypridinidae Exclusively marine. Shell hard, antennual notch at or below middle of dorsoventral extent of shell. First antennae strongly flexed at articulation between basal and second joints; they bear large olfactory hairs. Second maxillae jawlike and carry a large respiratory plate. Genera include *Cypridina* M. Edw., *Gigantocypris* Mull., *Cylindroleberis* Brady, *Sarsiella* Norman, etc.

FAMILY Halocypridae Exclusively marine. Shell calcified. Notch above middle of dorsoventral extent of shell. Numerous unicellular glands open along margins of shell. Mandibles with well-developed masticatory process. First maxillae jawlike. No eyes. Pelagic. Genera include *Conchoecia* Dana, which is circumglobal, occurring in all major oceans, *Halocypris* Dana, etc.

FAMILY Polycopidae Minute marine ostracods living at sea bottom. Shell valves round or oval, antennal notch weak or undeveloped, and first maxillae not jawlike.

ORDER Podocopa Shell flattened along the ventral border and not notched in front. Second antennae have long, leglike endopodite and short, much-reduced exopodite. Mandibles have short palp that may be biramous. First maxillae are jaws and carry a large comblike setose plate, movements of which promote respiration, though plate does not itself act as a gill.

FAMILY Bairdiidae Marine forms with shell much arched dorsally and two halves asymmetrical. Crop forms a well-developed masticatory mill.

FAMILY Cyprididae Occurs predominantly in freshwater. Some occur in brackish or neritic marine environments.

FAMILY Cytherellidae Regularly arranged, thickened bands in posterior part of body, suggesting segmentation but apparently not corresponding with limbs. Outer ramus of second antennae larger than that in other Podocopa. Divisions of caudal furca are lamellar and fringed with stout bristles. A representative genus is *Cytherella* Bosq.

FAMILY Cytheridae Marine, with strongly calcified shell. Second antennae have strongly flexed endopodite ending in strong, clawlike setae; exopodite elongate and slender, bent at tip, and perforated by duct of a large, unicellular "spinning gland," which Müller says spreads a web of fibers over surrounding objects. Divisions of caudal furca small and weakly developed. No heart. Numerous genera.

Subclass Copepoda

ORDER Calanoida Major articulation between last thoracic and first abdominal somites. Second antennae biramous; first antennae commonly extremely long, nearly equal to, or exceeding the body length, and composed of more than twenty segments. Heart usually present. Last pair of legs of male asymmetrical and modified for clasping and transfer of spermatophores; in species in which female carries eggs, only a single ovisac present, carried ventrally beneath abdomen. This suborder contains the majority of important planktonic species of copepods.

The families of calanoid copepods were grouped into two series by Giesbrecht (1892)—the **Amphascandria** and the **Heterarthrandria.** In the Amphascandria the antennules are symmetrical or nearly so in both sexes, those of the male bearing more sense organs (aesthetasks) than those of the female. The last pair of legs of the male are modified for grasping and transfer of spermatophores, whereas those of the female may be normal, reduced, or absent. The eyes are fused into a typical copepod single median eye, without a cuticular lens. Last thoracic and first abdominal segments of the male are not markedly asymmetrical.

The Heterarthrandria are characterized by having one of the antennules (usually the right) of the male geniculate and modified as a prehensile organ, and the last thoracic and first abdominal segments asymmetrical. The last legs of the male are markedly asymmetrical, the right often developed into a remarkable pincerlike structure used in transferring spermatophores. The eyes are often large and separate and in some are provided with cuticular lenses. The fifth feet in the female are of normal or reduced size but are not absent.

SERIES **Amphascandria**

Calanus Leach One of the most abundant, widespread, and important genera of copepods. Most are rather large species, some attaining a length of ¼ inch; most are about the size of a rice grain. Appendages much alike in both sexes, though left fifth leg of male larger than the right. Metasome consists of five distinct segments. Exopodites of second to fourth feet bear two spines on outer margin; second segment of basipodite of first swimming feet is smooth, without a hooklike armature.

C. finmarchicus Probably the most intensively studied of any copepod species. Body about the size of a rice grain, rather transparent, with a varying amount of red color caused by red oil drops; some, infected with parasitic *Peridinium,* are much redder than usual. The name *C. finmarchicus* formerly used in a broader sense than now, including closely related *C. glacialis,* which tends to replace *C. finmarchicus* in high arctic latitudes, and *C. helgolandicus,* which appears to be more temperate-water species.

C. hyperboreus Occurs in colder parts of range with *C. finmarchicus;* somewhat more stenothermal, tends to occur only in colder parts of range or in slightly deeper water, and does not range as far south. Readily distinguished from *C. finmarchicus* by more acute angles of posterior edge of metasome.

C. tenuicornis Widely distributed species in slightly warmer waters; transitional between boreal and temperate in character; overlaps much of range of *C. finmarchicus*.

C. simillimus Characteristic of cold waters of Southern Hemisphere, except coldest parts of Antarctic; occupies much the same ecological position in Southern Hemisphere that *C. hyperboreus* does in Northern Hemisphere; occurs in cool waters on both sides of Antarctic Convergence.

Undinula Scott, 1909 Large copepods, much like *Calanus,* but differing in segmentation of antennules (eighth and ninth segments partly fused in females) and in details of swimming feet, especially fifth pair; in fifth pair of swimming feet of female, both exopodite and endopodite of both feet have three segments; in male, endopodite of left fifth foot rudimentary (one small segment) or absent, but exopodite has three segments and ends in powerful prehensile organ; in right fifth foot both rami have three segments. *Undinula darwini* and *U. vulgaris* among most frequently taken copepods in tropical waters.

Microcalanus Small calanoids, metasome only three apparent segments. In North Atlantic waters, as along Norwegian coast, over the year as a whole it is one of the four numerically most significant zooplankters, although its small size (less than 1 mm. long) makes it less significant in total volume, or biomass. Tend to reproduce faster than larger calanoids and may have more generations per year, in the same area; maxima tend to occur from January to March. *M. pusillus*.

Nannocalanus Sars, 1925 Chief species, *N. minor,* formerly included in genus *Calanus,* is widely distributed in tropical and temperate waters of major oceans, where it is often one of the commonest copepods present; eurythermal and euryhaline, and although most frequently occurring near surface, exhibits considerable bathymetrical range. A small pointed decurved rostrum. Last segment of metasome strongly emarginate in dorsal view, but the posterior angles do not flare out, seeming rather to be curved or wrapped rather closely around urosome. Two forms, or varieties, often recognized, *N. major* and *N. minor,* exhibit slight differences in proportions and other details.

Chiridius Giesbrecht, 1893 No obvious rostrum, although in ventral view of "forehead" a small point can be discerned that may represent one. Metasome has three apparent segments, the last one produced into spines.

C. obtusifrons A high arctic species commonly found in waters of subzero temperatures.

C. poppei Taken in equatorial Pacific.

C. gracilis May be a variety of *C. poppei.*

Chirundina Giesbrecht, 1895 Rostrum moderately developed and pointed, and forehead has a distinct crest. In female last segment of metasome ends in small knob on each side; no fifth legs: male without knobs on last segment of metasome and has well-developed fifth legs. Taken in tropical waters.

Euaetideus Fourth and fifth thoracic somites fused so that metasome has only three apparent segments, the last of which is deeply emarginate and drawn out into posteriorly directed points in dorsal view. Rostrum well developed. Urosome small compared to prosoma, spines of last segment of metasome extending back about half its length. *E. giesbrechti*.

Euchirella Giesbrecht, 1888 Rather large calanoids living in deeper strata. Rostrum, if present, single-spined; metasome has four (or three) fully developed segments. Male has enlarged, elongate fifth legs, right leg forming a large grasping organ. Antennules reach to about end of urosome. Several species.

Gaetanus Giesbrecht, 1888 Distinct median spine on front of forehead; rostrum has single point or small bifurcation at tip. Mostly bathypelagic. In some species last segment of metasome is strongly excavated in dorsal view and drawn out into sharp points. Antennules often much longer than endopodite, the latter being much shorter.

Calocalanus Giesbrecht, 1888 Urosome has two or three segments. Caudal furcae extend laterally or are asymmetrically developed and bear remarkably long, plumose seta or setae. Antennules usually bear long bristles and plumose sense organs and last segment is long. Female often more conspicuously decorated than male.

Eucalanus Dana, 1852 Female without fifth feet. Frontal part of head not produced into a head or rostrum. Telson fused with, not separated from, caudal furcae. Body often elongate, cephalosome occupying about two thirds of total length and marked off from greater part of its length by a shallow emargination, and thus appears somewhat like an arrowhead on a broader shaft.

Euchaeta Philippi, 1843 Large, predaceous copepods. Female has four abdominal segments and long inner furcal setae; male abdomen has five segments. No fifth legs in female; those of male have a long styliform process on right side, and left fifth foot has exopodite developed into complex hand for grasping spermatophores; endopodite rudimentary or absent. *E. norvegica* one of the largest, most beautiful species abundant in North Atlantic, attaining a body length of about 8 mm; rather transparent; has a red color similar to that of *Calanus finmarchicus*. Female carries dark blue eggs in a single egg sac beneath abdomen. *E. norvegica* separated from this genus by Scott (1909) on the basis of differences in male fifth feet and female maxillipeds and made the type of a separate genus, *Paraeuchaeta*. *E. marina* and several other species abundant in tropical waters.

Megacalanus, Bradycalanus, Neocalanus, etc. Large, rather typical calanoids, mostly bathypelagic. Segregated into a subfamily, Megacalaninae.

Pseudocalanus Boeck *P. elongatus* the second most frequently taken species of copepod associated with swarms of *Calanus finmarchicus* in North Atlantic. *P. minutus* is a coldwater species, abundant around Cape Cod Bay all year and around other cold, neritic water, but it also extends into oceanic areas.

Scolecithrix Brady, 1883 Usually fairly large calanoids. Female without fifth feet. Hind surface of rami of third and fourth legs has spines; first legs one segment; second legs have two segments. Apex of maxillipeds bears a tuft of wormlike sensory filaments. In *S. danae* formalin-preserved samples of appendages have a delicate mauve tint, enabling them to be recognized at once in a mass of other copepods. Widely distributed. Several species.

Arietellus Giesbrecht, 1889 Head and hind corners of thorax usually pointed. Last thoracic and first abdominal segments commonly asymmetrical. Fifth feet are characteristic, and metasome has four segments. Several species.

Augaptilus Giesbrecht, 1889 Head distinct from first thoracic segment. Urosome three-segmented in female, five-segmented in male. Rami of all legs three-segmented. Grasping antenna of male may be on either right or left side.

SERIES **Heterarthrandria**

Centropages Kroyer, 1848 Neritic surface-living species. Last metasomal segment has flared, pointed projections. Antennules have twenty-four segments. All legs except left fifth of male have both endopodites and exopodites of three segments, male left fifth exopodite having two segments; male has large pincerlike modified right fifth leg for transferring spermatophores to female. Metasome has five segments. Widely distributed. *C. hamatus* occurs abundantly in North Atlantic associated with *Calanus finmarchicus, Tortanus discaudatus,* etc. in cold boreal waters.

Eurytemora Giesbrecht, 1881 Another chiefly neritic genus, also occurring in brackish and fresh waters. Abdomen of male has five segments; of female, three segments. Antennules as long as prosoma, and have twenty-four joints. Fifth feet uniramous; endopodites of legs three and four have two segments.

Temora Baird, 1850 Caudal furca long and slender, six times as long as wide, or more; furcal bristles short. Metasome has four segments, posterior corners drawn out into lateral points. Abdomen compact appearing, cylindrical, with long furcal rami. A widespread genus. *T. discaudata* and *T. longicornis* best-known species.

Heterorhabdus Giesbrecht Widespread.

Metridia Boeck, 1865 Widespread genus of essentially herbivorous, diatom-eating copepods. Urosome usually relatively long and slender; late copepodids of both sexes have a four-segmented urosome, but adult female has a three-segmented urosome because of fusion of first two, and adult male has a five-segmented urosome because of division of fourth. Fifth legs uniramous and have three or four segments in female and five in male. Basal segment of endopodite of second leg is deeply invaginate, with spines along the side of the invagination. Metasome has four segments. *M. longa* typical member of boreal community, frequenting deeper strata at about 40 to 50 meters, and sometimes ascending to surface. Essentially a coldwater high oceanic or open-water species, with *M. lucens* as a member of the *Calanus finmarchicus–Euchaeta norvegica* association.

Pleuromamma Giesbrecht, 1898 Distinguished from *Metridia* and other related copepods most readily by presence of a circular brown-pigmented knoblike luminous organ on one side of first segment of metasome; metasome has four segments. Like *Metridia,* various species tend to occupy strata below surface, although they ascend at times. Some are characteristic of warmer waters.

Anomalocera Templeton, 1837 Surface-living, warmwater or temperate-water forms. *A. patersoni* of coastal Atlantic waters blue in color. Both sides of cephalosome and corners of last segment of metasome are drawn out into lateral points, giving a characteristic appearance.

Labidocera Lubbock, 1853 Rather small calanoids. Metasome has four segments; abdomen of female has two or three segments, abdomen of male has five; antennules of female have twenty-three segments. Pair of dorsal eyes and a median ventral eye are present, dorsal eyes having cuticular lenses. Fifth legs of female are biramous, with unequal one-segmented rami; those of male are uniramous and chelate. Posterior corners of last segment of metasome drawn out into lateral points, and in some species lateral points on cephalosome.

Pontella Dana, 1846 Head with paired cuticular lenses plus additional lenses on anterior and posterior surfaces of rostrum. Median ventral element also present, forming a projection on ventral surface behind rostrum. Metasome has five segments. Fifth legs of female are biramous and reduced; in male, the left is uniramous and the right is chelate. Endopodites of first legs have three segments; those of second and third legs, two segments. Cephalosome has lateral hooks, or points.

Tortanus Giesbrecht, 1898 Metasome has five segments. One large dorsal eye; no cuticular lenses. Antennules of female have seventeen segments. Fifth legs are uniramous. Endopodites of first to fourth legs have two segments. Eggs planktonic and are extruded individually, not carried in egg sac. *T. discaudatus* a prominent member of the *Calanus finmarchicus–Tortanus–Centropages* surface-waters assemblage in North Atlantic.

Acartia Dana Slender, colorless, nearly transparent body. Metasome has four segments and rather uniform thickness. One large dorsal eye, without cuticular lens. Fifth legs uniramous; endopodites of ordinary swimming legs have two segments. Several species are among the commonest, most important neritic copepods. Wide occurrence.

ORDER Cyclopoida Major articulation between fifth and sixth thoracic somites. Second antennae uniramous, and first antennae usually clearly shorter than body and composed of twenty or fewer (usually seven to sixteen) segments. Ovisacs are paired and carried symmetrically, one at each side of abdomen. Includes numerous benthic and parasitic forms, as well as planktonic species. The cyclopoids, like the calanoids, have body divided into a broad anterior prosoma and a narrow urosome. Since major articulation occurs between fifth and sixth thoracic somites, segment that would normally bear fifth pair of swimming feet is included in urosome. In cyclopoids fifth pair of feet are usually uniramous and vestigial, without an enlarged basal segment.

Female carries eggs in two ovisacs, one on either side of urosome. Antennules are shorter than those in calanoids, usually being clearly shorter than the body and composed of only six to seventeen segments. Cyclopoids abundant in freshwater and to a lesser extent in brackish waters, and only a few species have become truly marine; of these, some are benthic or parasitic, so that number of planktonic marine cyclopoids is small compared with that of calanoids.

Oithona Baird, 1843 By far most abundant and widespread genus of marine cyclopoids; occurs in all oceans, and in many regions ranks as one of commonest and most important genera of copepod present. Several species. Body is elongate-ovate, and much like that of *Cyclops* in general appearance. First antennae in part are rather indistinctly segmented and bear long sensory bristles; second antennae have two segments. Urosome moderately long. Bristles of caudal furca not plumose. Details of mandibular palp and of maxillipeds differ from those of *Cyclops* (Fig. 9-10). Especially abundant in neritic waters.

O. nana One of the most abundant copepods all year off California coast and is also reported from many other parts of world (English Channel, Monaco, Bay of Algiers, etc.).

O. brevicornis Occurs on both Atlantic and Pacific coasts of America.

O. plumifera Appendages of female have long feathery scarlet setae. Numerous in tropical and warm waters.

O. similis Cosmopolitan and occurs off California coast and in North Sea, Irish Sea, Kara Sea, and Barents Sea.

Several marine cyclopoids have a pair of eye lenses on frontal margin, giving them appearance of possessing built-in opera glasses.

Sapphirina Second antennae have four segments. Paired eye lenses on frontal margin. Semitransparent ovate or subovate depressed body; short lamelliform furcal joints.

Copilia Similar to *Sapphirina,* but second antennae have three segments.

Oncaea Philippi, 1843 Body piriform. Antennules have six segments. Paired lenses are on frontal margin.

Corycaeus Dana, 1846 Body elongate and subcylindrical. Abdomen has one or two segments. Rather large lenses on frontal margin of cephalic segment.

ORDER Harpacticoida Metasome-urosome articulation between fourth and fifth thoracic segments, but body is often tapered rather gradually so that there is not as distinct a break between the metasome and urosome as in other copepods. Abdomen may be nearly as broad as thorax. Antennules short, rarely more than eight or nine segments, and second antennae are biramous; both second antennae of male prehensile and used to hold female. Mostly benthic, a few are parasitic and a few planktonic Those benthic in habit creep about, on, or in the substrate, among sand grains, etc. Some appear at times in plankton. A few holoplanktonic in habit. In male both second antennae prehensile, used for grasping female. No heart present. Eye is median, simple. Anterior pair of feet sometimes modified as second maxillipeds. Some planktonic forms. Eggs carried in single ovisac beneath abdomen of female.

Euterpina Norman, 1903 Widespread in temperate and warm coastal seas. Both rami of first legs have two segments. *E. acutifrons* an abundant species.

Metis Phil. Exopodites of first legs have three segments; the endopodites, two segments. One or both rami are broadened and modified.

Macrosetella Scott, 1909 First legs have three-segment exopodites and two-segment endopodites. Neither ramus broadened or modified. Body slender and cylindrical. Prominent beaklike rostrum.

Microsetella Brady and Robertson Both rami of first legs have three segments. Urosome as wide as metasome at its articulation.

Tigriopus Small red copepods found in high tide pools of the upper littoral zone.

Clytemnestra Dana, 1852 First legs have exopodites of only one or two segments; endopodites much longer and have three segments. Both rami of other legs have three segments. Body rather slender and depressed, with thoracic segments rather expanded and produced, and pointed posteriorly; triangular in outline when seen from above with blunt-pointed rostrum.

Aegisthus Giesbrecht, 1891 In first four feet both rami have three segments; fifth feet elongate, with two segments. Furcal setae long, setigerous spines, each bearing a plumose seta at jointed distal end. A well-defined, pointed rostrum is present.

PARASITIC COPOPODS A few examples are shown in Fig. 8-11.

Subclass Cirripedia The barnacles and their relatives. Order Thoracica. *Balanus, Chthamalus, Lepas, Mitella, Pollicipes,* and the parasitic orders Acrothoracica, Ascothoracica, and Rhizocephala, the most famous of which is the rhizocephalan *Sacculina. Peltogaster* is a similar form parasitizing hermit crabs.

Subclass Malacostraca

SUPERORDER Hoplocarida

ORDER Stomatopoda. Mantis shrimps. *Squilla* (Fig. 8-7).

SUPERORDER Peracarida

ORDER Mysidacea

Anchialina Wide distribution in tropical shallow waters. Characteristic in Barrier Reef lagoons, Java Sea, etc.

Hemisiriella Tropical seas; Indo-Pacific.

Eucopia Dana A typically deep-sea genus, black.

Gnathophausia W. Suhm Deep-sea form with long, spearlike rostrum.

Austromyces Czern A Mediterranean and Caspian genus.

Boreomyss Sars Arctic abyssal species.

Several brackish-water genera of Sarmantic derivation are characteristic of the Caspian Sea (*Katamysis* Sars, *Lomnomysis* Czern, *Mesomysis* Czern, *Metomysis* Sars, and others).

Mysis Latr (Fig. 9-13).

Siriella Dana Pelagic.

ORDER Cumacea *Diastylis* (Fig. 9-19).

ORDER Isopoda

Suborder Chelifera (Tanaidacea) *Tanais.*

Suborder Flabellifera *Limnoria* (gribbles, etc.), *Cirolana, Sphaeroma, Colanthrua, Cymothöe.*

Suborder Valvifera *Idothea, Astacilla, Arcturus.*

Suborder Asellota *Munna, Jaera, Jaeropsis.*

Suborder Oniscoidea (Ligiidae) *Ligia.*

PARASITIC ISOPODS *Ione* (Fig. 8-11).

ORDER Amphipoda

Suborder Gammaridea *Gammarus, Corophium, Chelura, Melita, Orchestia, Hyale, Ampithoe.*

Suborder Caprellidea *Caprella, Cyamus* (skeleton shrimps and whale lice).

Suborder Hyperiidea *Hyperia, Phronima.*

SUPERORDER Eucarida

ORDER Euphausiacea

FAMILY Euphausiidae Thoracic legs well developed, all more or less alike, and second or third pair not markedly elongated nor ending in brushes or claws. Maxillary palps simple. Eyes little constricted if at all, and not bilobate. Photophores present.

Euphausia Dana Last two pairs of thoracic limbs reduced to mere bristlelike processes hidden among bushy gills and without endopodites. Nauplius eye persists throughout life.

Pseudoeuphausia P. latifrons is a dominant neritic species occurring in tropical seas. In Indo-Pacific it is related to *Nyctiphanes.*

Nyctiphanes A reflexed leaflet on base of peduncle of antennules in front of eyes; peduncle of antennules of male noticeably stouter than that of female. Seventh and eighth thoracic limbs of female rudimentary and uniramous; seventh have two long segments, and eighth rudimentary and lacking setae. Female carries eggs in two egg sacs. Copulatory organs of male have leaflike inner lobe with finely serrated outer margin.

Meganyctiphanes Holt and Tattersol Similar to *Nyctiphanes* but usually larger. Only eighth legs vestigial; sixth and seventh biramous. Peduncle of first antennae of male not noticeably stouter than that of female.

Thysanopoda M. Edw. No reflexed leaflet on base of peduncle of antennules. Only eighth legs vestigial; sixth and seventh are alike and biramous. A widespread genus, some species of which are bathypelagic or at least spend most of their time in relatively deep water.

FAMILY Nematoscelidae Eyes usually constricted and bilobate, with dorsal and ventral lobes. Second or third legs especially elongated and end in a brush or claw. Photophores present.

Nematobrachion Calman Third legs elongated and end in a terminal group of spines.

Nematoscelis Sars, 1885 Second pair of legs greatly elongated and slender, ending in bristles on last segment or last two segments. Eyes large and bilobate. Last five gills branched. Second and third segments of antennules longer and more slender in females than in males. Seventh legs biramous in female, with two-segment endopodite; seventh legs uniramous in male. Male copulatory organ has three processes on inner lobe: spinelike process nearly straight and parallel to the others, and lateral process never hooked.

Stylocheiron Sars, 1885 Third legs greatly elongated, penultimate segment broadened and often armed with strong spines, forming with distal segment a grasping organ. Eyes large and constricted into two parts, lower part well behind upper.

Thysanoëssa Brandt Second pair of legs often much elongated and thicker than the others; last two pairs of legs reduced, last pair being uniramous. Rostrum usually well developed; flagella of antennules short; two distal segments of peduncle of antennules more slender in female. Last three pairs of gills have two branches. Numerous species, widely distributed.

FAMILY Bentheuphausiidae Deepwater forms with imperfect eyes. No legs much longer than the rest. Maxillary palps have three segments. No luminiferous organs.

Bentheuphausia Sars.

ORDER Decapoda Crabs, lobsters, crayfish, hermit crabs, shrimps, etc. Distinguished by the fact that the first three pairs of thoracic limbs are modified as maxillipeds and the remaining five pairs as legs—hence the name Decapoda. The majority of the large familiar crustaceans belong to this group.

Suborder Natantia Typical shrimps. Body usually somewhat compressed laterally; first abdominal segment not smaller than the others; rostrum well developed; large antennal scale present; pleopods well developed and adapted for swimming. *Penaeus, Sergestes, Lucifer, Alphaeus, Palaemon, Palaemonetes, Hippolyte, Stenopus.*

Suborder Reptantia First abdominal segment smaller than the rest; rostrum depressed, reduced, or absent; first pair of legs usually larger than the others and modified as chelipeds; pleopods, when present, not well adapted for swimming. Lobsters, crayfish, crabs, etc.

SECTION **Macrura** Abdomen extended, well developed, with large uropods and telson forming a strong tail fan. *Homarus, Nephrops, Panulirus, Jasus, Scyllarus, Ibacus.* The Thalassinidea, including ghost shrimps, *Callianassa, Upogebia,* and *Thalassina,* are burrowing forms often classed with the Anomura because the first and third legs differ from those of typical macrurans.

SECTION **Anomura** Abdomen variously formed, reflexed beneath thorax as in crabs, or soft and twisted asymmetrically, or secondarily symmetrically; pleura and tail fan usually reduced or absent; third legs never chelate; fifth legs commonly reduced and turned upward. *Coenobita, Birgus, Clibanarius, Pagurus, Lithodes, Galathea, Munida, Petrolisthes, Porcellana, Pachycheles, Hippa, Emerita, Blepharipoda.*

SECTION **Brachyura** True crabs. Abdomen reduced and tightly flexed beneath thorax; carapace fused with epistome; first legs in form of chelipeds; third legs never chelate; eyes usually lateral to second antennae. *Dromia, Dorippe, Calappa, Cancer, Portunus, Callinectes, Carcinus, Xantho, Menippe, Pinnotheres, Pinnixa, Oxypode, Uca, Grapsus, Pachygrapsus, Geocarcinus, Maja, Inachus, Pugettia, Loxorhynchus, Libinia.*

CLASS Insecta (Hexopoda) Insects are by far the most numerous species of terrestrial and freshwater animals. Very few have become adapted to the marine habitat, although several are found in the intertidal, and on beaches. The tropical water strider *Halobates* (Hemiptera: Gerridae) is found on open seas. Some of the insects commonly occurring in the intertidal rocks, on beaches, in estuarine waters, etc. follow.

ORDER Collembola

FAMILY Poduridae *Anurida.*

ORDER Hemiptera

FAMILY Corixidae *Trichocorixa.*

ORDER Coleoptera

FAMILY Carabidae *Thallasotrechus.*

FAMILY Eurystethidae *Eurystethes.*

FAMILY Staphylinidae *Diaulota, Liparocephalus.*

FAMILY Melyridae *Endeodes.*

FAMILY Cicindelidae Tiger beetles.

ORDER Diptera

FAMILY Tipulidae *Limonia.*

FAMILY Culicidae *Aedes.*

FAMILY Chironomidae *Camptocladius, Paraclunio, Telmatogeton, Eretmoptera, Tethymyia.*

FAMILY Dolichopodidae *Aphrosylus.*

FAMILY Anthomyidae Kelp flies. *Fucellia, Coelopa.*

FAMILY Canaceidae *Canaceoides.*

FAMILY Ephydridae *Ephydra.*

SMALLER PROTOSOME PHYLA

PHYLUM Sipuncula (Sipunculoidea) Peanut worms. *Sipunculus, Dendrostomum, Phascolosoma, Golfingia, Phascolion.*

PHYLUM Echiura (Echiuroidea) Spoon worms. *Echiurus, Urechis, Listriolobus Thalassema, Ikeda, Bonellia.*

PHYLUM Priapulida (Priapuloidea) *Priapulus, Tubiluchus.*

PHYLUM Pogonophora Long, slender tube-building worms with no digestive tract, and with one or more anterior tentacles with which prey is captured and digested. Deepwater, mostly near the outer edges of the continental shelves or on the continental slopes. *Polybrachia, Siboglinum, Lamellisabella, Spirobrachia.*

PHYLUM Tardigrada Bear animalcules. Mostly freshwater forms, in damp mosses, etc.; a few marine interstitial species.

DEUTEROSTOME PHYLA Phyla in which embryonic cleavage is radial and indeterminate, the mesoderm enterocoelous in origin, and the mouth arising as a separate opening at some distance anterior to the blastophore.

PHYLUM Chaetognatha Arrowworms. The principal genera can be distinguished as follows:

1. With two pairs of lateral fins—*Sagitta*
   With one pair of lateral fins—*Spadella* 2
2. Head cap with a pair of lateral tentacles
   Head cap without lateral tentacles 3
3. Trunk segment encased in a thick collaret—*Pterosagitta*
   Trunk segment not so encased 4
4. Lateral fin extends to the last third of the tail segment—*Krohnitta*

Lateral fin not extending posteriorly more than half the length of the tail segment 5

5. Tail segment constitutes about one third or more of the total body length. Two rows of teeth on each side of the head—*Heterokrohnia*

Tail segment constitutes less than one third of the total body length. One row of teeth on each side of the head—*Eukrohnia*

Members of *Eukrohnia* are reported to carry their eggs in a "marsupium" formed from the lateral fins.

Following are notes on some species of the principal genera.

Eukrohnia fowleri Bathypelagic

Eukrohnia hamata Möbius In Pacific, abundant in cold waters and subarctic, occasional in transitional waters, and absent from subtropical water masses. In Atlantic, occurs near surface in high latitudes, and only in deeper water in middle latitudes. In antarctic summer it is one of three dominant species of zooplankton in parts of Antarctic; here it shows seasonal rather than diurnal vertical movements, descending to 500 to 1000 meters or more in winter. A widespread coolwater and coldwater species, showing submergence in warm latitudes.

Eukrohnia subtilis Grassé A warmwater species common in subtropical Pacific waters but absent from subarctic waters.

Heterokrohnia mirabilis Ritter-Zahony.

Krohnitta pacifica Arda High oceanic, cosmopolitan in tropical waters.

Krohnitta subtilis Grassé Cosmopolitan in warm and temperate oceans.

Pterosagitta draco Krohm Cosmopolitan in warm and temperate oceanic waters.

Sagitta arctica (possibly a var. of *S. elegans*) In cold arctic waters. Two distinct age groups usually present, one group about half grown when the other spawns; spawning probably in autumn; two years to mature.

Sagitta bedoti Beraneck In tropical neritic regions, Indo-Pacific. Mainly in coastal regions under oceanic influence.

Sagitta bipunctata Quoy and Gaimard Abundant in warm and temperate oceanic waters. Cosmopolitan. Absent from cold waters.

Sagitta decipiens Fowler Mesoplanktonic, cosmopolitan. In temperate and warm oceanic waters.

Sagitta elegans Verrill Cosmopolitan, abundant. Cold and cold-temperate species. An indicator of oceanic water masses in both North Atlantic and North Pacific. In many areas it has three or four broods a year, those spawned at different seasons growing into different-sized adults. Shows diurnal vertical migration during spring and summer. In English Channel their presence indicates northern and oceanic waters and has been correlated with best commercial fishing. Apparently the only chaetognath to reproduce successfully in Gulf of Maine. Over Georges Bank an anticyclonic eddy maintains a year-round population in the region. Associated with medusa *Cosmetira pilosella* and often with *Sagitta serratodentata*. Tends to be absent at times and places where *S. setosa* is abundant, and vice versa (Russell, 1932).

Sagitta enflata Grassé Cosmopolitan in tropical and warm waters. Extends north to Plymouth in summer when warm southern Atlantic water masses move into area. Is carried in Gulf Stream as for north as Gulf of Maine but cannot live there long.

Sagitta exuina Shows seasonal changes in depth distribution, the upper limit, or occurrence, parallel to or slightly below the 11-degree isotherm. A coolwater species.

Sagitta ferox Doncaster Indo-Pacific tropical and equatorial regions. Seems to compete with *S. robusta*. In areas where either one is dominant, the other is rare or absent.

Sagitta hexaptera d'Orbigny Cosmopolitan in warm and temperate oceanic waters. Often associated with *S. serratodentata,* salps, and doliolids.

Sagitta lyra Krohn Subtropical and warm temperate waters. Less strictly tropical than *S. enflata*.

Sagitta macrocephala Bathypelagic.

Sagitta robusta Doncaster Tropical (see *S. ferox* above).

Some of the species illustrating various types of distributional patterns are tabulated in Table 9-1.

PHYLUM Echinodermata The only large phylum that is entirely marine. Starfish, sea urchins, sea cucumbers, brittle stars, sea lilies, etc.

CLASS Asteroidea Starfishes. *Asterias, Pisaster, Leptasterias. Pycnopodia, Acanthaster, Henricia, Asterina, Patiria, Astropecten, Oreaster, Goniaster, Linckia, Culcita.*

CLASS Echinoidea Sea urchins, cake urchins, sea biscuits, sand dollars, heart urchins, etc. *Arbacia, Echinus, Strongylocentrotus, Paracentrotus, Diadema, Toxopneustes, Lytechinus, Cidaris, Colobocentrotus, Echinocardium, Meoma, Moira, Spatangus, Mellita, Leodia, Dendraster, Clypeaster, Fibularia, Rotula, Encope.*

CLASS Holothuroidea Sea cucumbers. *Holothuria, Cucumaria, Thyone, Stichopus, Psolus, Scotoplanes, Caudina, Leptosynapta, Molpadia.*

CLASS Ophiuroidea Brittle stars, serpent stars, basket stars, etc. *Ophiura, Ophioneres, Amphipholis, Ophiothrix, Amphiodia, Ophioplocus, Ophiohelus, Amphiura, Ophiocomina, Ophiactis, Asteronyx, Gorgonocephalus.*

CLASS Crinoidea Sea lilies, etc. *Antedon, Neometra, Ptilocrinus, Cenocrinus.*

PHYLUM Hemichordata Acorn worms and pterobranchs.

CLASS Enteropneusta Acorn worms. *Balanoglossus, Saccoglossus, Glossobalanus, Ptychodera, Spengelia, Glandiceps.*

CLASS Pterobranchia *Rhabdopleura, Cephalodiscus, Atubaria.*

PHYLUM Chordata

Subphylum Protochordata

CLASS Urochordata Differ from other chordates in lack of segmentation and in complete lack of a coelom, even during development.The coelom has been "lost" during the evolution of Urochordata, since it is present in other deuterostomes and was almost certainly present in the ancestral chordate stock. Tunicates, salps, and appendicularians.

ORDER Ascidiacea Tunicates. *Ascidia, Molgula, Clavelina, Ciona, Styela, Perophora, Botryllus, Pyura, Metandrocarpa, Distaplia, Eudistoma, Sigillinaria, Amaroucium, Polyclinum.*

ORDER Thaliacea Pyrosomids and salps. *Pyrosoma, Salpa, Doliolum.*

ORDER Larvacea (Appendicularia)

Fritillaria "Head" longer and more slender than that in other genera, usually somewhat constricted near middle at level of attachment of tail. Anterior part contains mouth, pharynx, and esophagus; middle part, heart and stomach; and posterior part, gonads, a spherical ovary, and behind ovary a single large saccular testis. By far the largest genus.

Oikopleura "Head" stout, short. Branchial openings well back in region of rectum.

Tail tapers posteriorly, ending in a single point. Testes usually paired, one on each side of median ovary.

Appendicularia "Head" short and oval. Tail rather broad and bifid or notched at end.

Kowalevskia Tail broadest in middle, tapering both ways, ending conical or pointed. Branchial openings elongate longitudinally.

CLASS Cephalochordata Lancelets. *Branchiostoma (Amphioxus), Asymmetron.*

Subphylum Vertebrata

CLASS Pisces Fishes (used in the broad sense here).

Subclass Agnatha (Monorhina) Primitive jawless fish, possessing no arches supporting paired fins. Nasal aperture single, median, and on top of head.

ORDER Ostracodermi (Ostracophora) Extinct armored fishlike vertebrates that probably arose in Ordovician, were best represented in Silurian and Devonian periods. Head large, with a large bony shield. Finlike appendages at posterior lateral corners of head shield. Eyes and single median nostril on top of head. Posterior part of body slender, usually scaly, and ended in a heterocercal tail. Suborders included Anaspida, Antiarcha, Cephalaspidormorphi, and Osteostraci (Aspidocephali), the last named containing the best-preserved forms.

ORDER Cyclostomata (Marsipobranchii) Lampreys and hagfish. Mouth a circular jawless sucking disc armed with rasping teeth. No paired appendages, and no scales. Skeleton cartilaginous; notochord persistent throughout life; chondrocranium incomplete, without roof; vertebrae represented by neural arches (Figs. 11-10 and 11-11). Suborders include Petromyzontes (Hyperoartia), the lampreys; and Myxinoidea (Hyperotreta), the hagfish.

The cyclostomes are of considerable interest as living representatives of a primitive line of fish adapted to a semiparasitic mode of life and showing an eel type of racial senescence. The ammocoetes larvae of lampreys show striking resemblance to the Cephalochordata. Sea lampreys have been of economic importance after their invasion of the Great Lakes through the Welland Canal and the destruction by them of the important whitefish and trout fisheries in these lakes. The United States and Canada have spent about $12 million in control measures over the last twelve years. The economic loss to the fisheries has been much greater.

All remaining fish have sometimes been grouped together into a contrasting group, the **Gnathostomata,** or jawed fish. Recent authorities, however, regard this assemblage as composed of three primary groups, in this book accorded subclass rank, equivalent to that of the **Agnatha.**

Subclass Placodermi Extinct late Silurian to early Permian sharklike fish, some of which had some bone in internal skeleton as well as bony dermal plates, and scales. Biting jaws and paired fins present.

ORDER Acanthodii (Acanthodea) There were up to seven pairs of paired fins, these and dorsal and anal fins each with a strong spine along leading edge. These fish apparently evolved as river-dwelling types and only in Devonian period began going to sea. Multiple and variable number of paired fins of particular interest in connection with fin-fold theory of origin of paired fins, as are also broadly attached fins of next group, the Cladoselachii.

ORDER Cladoselachii (Pleuropterygii) Sharklike but with mouth in more terminal position than in modern sharks. First gill bars modified into jaws. Pectoral and pelvic fins essentially lateral folds, broad and unconstricted at base, and supported by nearly parallel rays. Genera include *Cladoselache, Symmorium, Cratoselache.*

ORDER Coccosteomorphi (Arthrodira) Head and forepart of body protected by bony plates, those of movable head shield articulated with body shield in such a way that to open mouth, lower jaw remained rigid while loosely hinged skull moved up. What appear to be teeth are only bony projections of skull and jaw. Homologies of jaws and fins to those of other fish uncertain.

Subclass Chondrichthyes Sharks, rays, and chimaeras. Skeleton cartilaginous. Intestine spiral-valved. Males have claspers.

ORDER Elasmobranchii (Euselachii) Modern sharks and rays. Gill clefts have separate external openings. Skin usually has placoid scales.

Suborder Selachii (Squali, Pleurotremi) The sharks. Pectoral fins not united to head. Gill openings at sides of more or less cylindrical body. The numerous families are grouped into three assemblages.

SECTION **Galeoidea** A large group comprising the majority of the most familiar sharks, including most of those known to attack people. Most of them voracious carnivores. But the largest of all, the whale shark and the basking shark, which attain a length of 18 meters or more and weigh several tons, are harmless plankton eaters. Sharks of this group have five gill slits, two dorsal fins, and an anal fin. No fin spines. Rostrum triradiate.

FAMILY Carchariidae (Odontaspidae, Galeidae, Mustelidae) Tail much shorter than body and has a notch below tip of upper lobe. Eyes have an inner eyelid. At least one of the gill slits above base of pectoral fins; gill slits do not occupy most of width of head. Spiracles small to obsolete. Several species are large, fierce sharks dangerous to man.

Carcharias lamia Cub shark of Atlantic, a large shark with a reputation for attacking man.

Galeocerdo arcticus Tiger shark, a widely distributed, dangerous shark in warm waters. Teeth large, obliquely set, and serrated on both edges.

Prionace glauca Great blue shark. Abundant in tropical seas, occasional off West Coast of United States. Grows to 6 meters long. No spiracle. May attack man.

Galeus galeus Tope, or oil shark, found in European waters. A rather small shark. Spiracle present. No pit at root of tail. Teeth large and coarsely serrated.

Galeus (Galeorhinus) zyopterus Soupfin shark, found off California coast. Fins boiled for a gelatin used by Chinese in making soup.

Mustelus californicus Gray smooth hound. Common off southern California.

Triakis semifasciatum Leopard shark. About 1 to 1.5 meters long. Gray, with black cross bands across back and round black spots along sides. Teeth have one long median and one or two smaller lateral cusps.

Reniceps (Sphyrna) zygaena Hammerhead shark. Widely distributed in warm seas. Grows to about 4.5 meters long. Subfamily Sphyrninae comprises hammerheads, in which head is flattened and extended laterally, with eyes at ends of extensions.

Odontaspis Sand shark, shovel-nosed shark. Small sharks living along sandy coasts. Sometimes separated as family Odontaspidae.

FAMILY Carcharhinidae Requiem shark.

Carcharhinus Small- to medium-sized sharks, much feared because they are aggressive and persistent. Several species.

FAMILY Lamnidae Mackerel shark, porbeagle, etc. Typical members large fierce pelagic sharks with stout body, wide mouth, large teeth, and wide gill openings. First dorsal fin large, second small; tail lunate; prominent keel on each side of caudal peduncle; small anal fin present.

Carcharodon Large, voracious, widely distributed shark with sharp, flattened, triangular, finely serrated teeth. Some species attain a length of over 9 meters, and certain fossil species even larger.

Lamna Porbeagle, mackerel shark. Somewhat smaller than *Carcharodon,* about 2.5 to 3 meters long. Pointed nose and crescentic tail. Bluish black above, becoming abruptly white below. Teeth long, fanglike, and smooth edged, with a small projection from base.

Isurus Mackerel shark. Large, voracious; teeth slender, with entire edges and no basal cusps.

Subfamily Alopiinae In thresher sharks dorsal lobe of tail greatly elongated, making tail constitute over half total length. No keel on caudal peduncle. They are said to thrash tail about violently in schools of small fish, stunning and injuring many, which can then be eaten at leisure. Genus *Alopias*.

Subfamily Cetorhininae Often treated as a separate family, this subfamily comprises basking sharks, genus *Cetorhinus* (Fig. 11-12, *A*), remarkable for their large size, often 12 meters long or more, and long gill slits that extend up whole side of head. Teeth minute and gill rakers are well developed, similar to whalebone. These huge, harmless sharks strain plankton from the water. The liver is large and about 60% oil. A single shark liver may yield several barrels of oil, used as a substitute for cod liver oil, for its high vitamin content.

FAMILY Orectolobidae Includes carpet shark, *Orectolobus,* with a somewhat flattened body and mottled skin, and whale shark, *Rhineodon* (Fig. 11-12, *B*), largest of all sharks, growing up to 18 meters long or even more and weighing several tons. Like huge basking shark, whale shark is also a harmless plankton feeder with small teeth and large whalebonelike gill rakers. *Ginglymostoma,* the nurse shark, a large, pelagic shark of tropical seas, is also sometimes included in this family.

FAMILY Scyliorhinidae (Scyllidae) Cat shark, spotted dogfish, swell shark, etc. Usually rather small sharks, found in fairly deep water. Two dorsal fins, without spines; anal fin present.

SECTION **Notidanoidea (Diplospondyli)** A relatively small group, mostly deep-water sharks with only one dorsal fin and with six or seven gill slits. Rostrum simple.

FAMILY Chlamydoselachidae Frilled shark. Elongated eellike body up to 2 meters long or more.

FAMILY Hexanchidae Six-gilled shark, cow shark. Some large, up to 6 meters long or more. Viviparous.

SECTION **Squaloidea** Five or six gill slits. Two dorsal fins, each with a stout spine at its leading edge or, if no spine present, then no anal fin present either. Rostrum simple.

FAMILY Heterodontidae (Cestraciontidae) Bullhead shark. Five gill slits on each side. Dorsal fins have strong spines; anal fin present.

Heterodontus Port Jackson shark. Small shark with posterior teeth modified into dense pavement for crushing mollusc shells.

Pristiophorus Saw shark. Small east Asian and Australian sharks with snout like that of a sawfish, but with smaller teeth. Gill openings lateral (those of true sawfish are ventral).

FAMILY Squalidae Dogfish shark, sleeper shark, spiny dogfish, etc. Dorsal fins have stout spine on leading edge of each; no anal fin. Five gill slits on each side.

Squalus Spiny dogfish. Type most widely sold by biological supply houses for comparative anatomy and zoology classes throughout United States. They are abundant in all seas, often occurring in large schools and hence easily collected in great numbers.

Centrophorus A deep-water demersal form.

FAMILY Somniosidae (Dalatiidae) Sleeper shark. No spines on dorsal fins; no anal fin present. Five gill slits on each side. These rather large sluggish demersal species live on muddy or clay bottoms, coming to surface only to eat. If hooked, they offer no resistance. Head small, with weak jaws and small teeth. Fins small and weak. Skin covered with minute tubercles. These sharks grow to about 7.5 meters in length. They are probably largely carrion eaters and eat also some live fish. Belief that they bite large chunks out of whales is probably erroneous. Genus *Somniosus*.

FAMILY Squatinidae Angel shark. This curious shark seems somewhat intermediate in character between typical sharks and rays. Body flattened, somewhat raylike, with large broad pectoral and pelvic fins. Head set off from body by a deep notch, in which five gill slits open; spiracles on top of head and in back of eyes. These demersal bottom-feeding fish have flattened teeth for crushing molluscs, crabs, sea urchins, and other animals with shells or exoskeletons. Large ones become 2 to 2.5 meters long. They are found especially in tropical and subtropical seas and in Mediterranean. Genus *Squatina*.

SECTION **Batoidea (Raiai, Hypotreme)** Rays, skates, and their relatives. Pectoral fins are produced forward, joining sides of head. Gill openings ventral. Hyomandibular cartilage purely suspensory. Body commonly much flattened dorsoventrally. Demersal fish, usually lying on, or partly buried in, the bottom much of the time. Typical rays are broad and flat, and swim by undulations of wide paired fins, a different mode of swimming from that of other fish.

FAMILY Pristidae

Pristis Sawfish. Large elongate sharklike rays with a flat elongate snout bearing lateral stout toothlike structures on each edge. Sawfish grow to as much as 6 meters in length and are usually found in warm, shallow seas, especially of tropical America and Africa, and often around river mouths. Gill slits ventral.

FAMILY Dasyatidae (Trygonidae) Whip-tailed rays, or stingrays. Typical broad ray-like flattened body; tail slender, usually bearing one or more spines. Spines, together with a poisonous, slimy secretion on skin, can inflict intensely painful wounds to the unwary swimmer who happens to step on a stingray in shallow waters. Numerous genera. Representative species include *Dasyatis dipterurus* J. & G., *Pteroplatea marmorata* Cooper, *Urobatis halleri* Cooper, and *Aetobatus californicus* Gill.

FAMILY Mobulidae Manta rays (Fig. 11-13), horned rays. Often large, with anterior part of pectoral fins produced forward into so-called "cephalic fins," or horns.

Teeth occur only in lower jaw. Found in tropical seas. Viviparous, usually giving birth to only one young at a time. *Manta birostris.*

FAMILY Rajidae Skates. Typical broad flat raylike body; tail is thicker than in the Dasyatidae and unarmed; two dorsal fins, set back on the tail well back of the hind tips of the pelvic fins; snout is longer than interorbital distance, and commonly pointed. Largest genus is *Raja*. Numerous species. Often with short patches of spines, or thornlike prickles, on dorsal surface or along top of tail.

FAMILY Rhinobatidae (Platyrhinidae) Guitarfish, fiddlefish. Body more leongated and sharklike, tail thicker. Found in warm seas. Examples include *Rhinobatus, Platyrhinoides.*

FAMILY Torpedinidae (Narcobatidae) Electric rays, or torpedo fish. Dark-colored rays with broad circular anterior disc. Skin smooth, without scales or prickles, large electric glands at base of pelvic fins on dorsal side. Giant electric ray of North Atlantic. *Torpedo nobiliana* can deliver pulses of 50 amperes at 50 to 60 volts, sufficient to electrocute a large fish. Smaller electric rays of genera *Narcine, Narcobatus,* and *Tetranarce* widespread. Electric organs derived from modified muscle, which forms masses of large, hexagonal cells on either side of head at base of pectoral fins.

ORDER Holocephali Chimaeras, or ratfish. A formerly abundant order of Chondrichthyes, especially in Mesozoic. Of the four families, only one—family Chimaeridae with three recent genera. *Chimaera, Callorhynchus,* and *Harriotta*—still exists. In contrast to sharks, gill clefts open into a chamber with only one external opening covered by a fleshy, or soft, cartilaginous operculum. Skin soft, without scales. First dorsal fin bears a long, sharp spine. Head high, compressed, with a small, narrow mouth, and teeth reduced to a few plates. Lateral line system prominent. Body tapers off into a long, narrow heterocercal or whiplike tail. Upper jaw immovably fixed to skull, and lower jaw articulates directly with skull. Male bears a peculiar stalked knobbed denticulated clasping organ in middle of forehead, in addition to having large claspers of usual elasmobranch type extending back from cloacal region.

Holocephali are demersal fish, genus *Harriotta* being found only in deep waters. In *Callorhynchus,* snout is produced into a pendent tactile organ. Species *C. callorhynchus* is the elephant fish of temperate to cold southern waters. Perhaps the most familiar species are *Chimaera (= Hydrolagus) colliei* of western coast of North America and *C. monstrosa,* king of herrings, or rabbitfish, of North Atlantic.

Subclass Osteichthyes These fish differ from elasmobranchs in having a bony skeleton, or bony dermal plates. Dermal scales without dentin, or enamel, of placoid scale. Intestine without a spiral valve (except for sturgeons). A urinary bladder present. No claspers occur on pelvic fins. Operculum composed of bony elements. Many species have a single or paired air bladder, a saclike diverticulum from pharynx, primitively an accessory respiratory as well as hydrostatic organ (in most modern fish possessing an air bladder it is purely hydrostatic in function). These fish appeared in Devonian and at first dominated freshwater and only later oceans.

GRADE **Choanichthyes (= Sarcopterygii)** Lobe-finned fish. Paired fins have a prominent scale-covered lobe of flesh and bone at base.

ORDER Crossopterygii (= Branchioganoidei) Coelacanths and their relatives. The most abundant fish during Devonian but then declined so that there were only a few species during Mesozoic. They were long regarded as extinct until 1938, when a specimen of coelacanth *Latimeria* was caught off mouth of the Cahlumna River in South Africa. Later others were taken near Comoro Islands off Madagascar in Indian Ocean. This was one of the most startling discoveries of a "living fossil." A detailed study of this primitive fish has been made by J. Millot and his students. Skeletal elements of fin lobes not homologous to limb bones of higher vertebrates. Air bladder has bony plates, which may act as a resonator. Unlike most Crossopterygii, it has no internal nares.

ORDER Dipnoi Lungfish characterized by fins with an archipterygial skeletal pattern, presence of internal nares, an air bladder that remains connected to pharynx and that may act as a type of accessory respiratory organ, and a more complex heart than that of most fish, with a pulmonary circulation. Skeleton largely cartilaginous, and there is a persistent unconstricted notochord. There are dermal bone and dermal fin rays. Body covered with imbricate, or overlapping, cycloid scales. Dipnoans were of worldwide distribution during Jurassic period but now confined to widely separated localities in Africa, Australia, and South America. Found in freshwater.

GRADE **Actinopterygii** Ray-finned fish. Projecting portion of paired fins supported by dermal rays, and usually not much of a fleshy lobe appears at base. These fish appeared in mid-Devonian. They were rare at first compared to lobe-finned fish, but today they are dominant group in both fresh and marine waters.

ORDER Polypterini

Polypterus Freshwater fish, bichir of Nile.

ORDER Chondrostei Paddlefish and sturgeon. These fish have a long rostral snout with ventral mouth like sharks. Heterocercal caudal fin. Fossil ancestral forms had an ossified skeleton, but living species have only a completely cartilaginous one, which is a secondary degenerate condition. Paddlefish live in freshwater, but sturgeon are found both in freshwater and in marine environments. Sturgeon, family Acipenseridae, are primitive and sharklike in numerous ways. Fin skeleton sharklike, and there are no functional spiracles; there is a persistent unconstricted notochord covered by a fibrous sheath; vertebrae are without centra. Mouth toothless and protrusible. Skin tough, with five rows of hard, bony dermal plates. Sturgeon usually rather large fish, becoming several feet long. They are valued as food; egg mass, or roe, is considered a great delicacy—caviar—and air bladder has been used in manufacture of isinglass.

There are sensory barbels ventrally on snout. Sturgeon are demersal bottom-feeding fish that thrust their protrusible mouth into soft substrate in search of small food animals. Some become large, one of the largest taken weighing more than 1½ tons.

ORDER Holostei Garpikes and bowfins. Freshwater fish with ossified skeleton. Dominant in Jurassic; now only few left. Approach teleosts in structure.

ORDER Teleostei This order comprises the dominant group of fish today—an extremely large and diverse assemblage constituting about 90% of all living fish. Teleosts appeared in Jurassic period, underwent a rapid adaptive radiation, and reached a climax in Cenozoic period. Teleosts represent end products of fish specialization and have not given rise to types of vertebrates beyond their own group. Scales have lost most or all of their ganion and bone and have become relatively small and thin, with a tendency to overlap; some groups have lost scales altogether. Internal skeleton is usually well ossified; vertebrae amphicoelous, and tail usually homocercal. Swim bladder used as a hydrostatic organ. Commonly there are slender bones lying between muscles. Some of the more important groups of teleosts follow.

Suborder Isospondyli (Clupeiformes) Includes large parts of Abdominales, Physostomi, and Cycloidei of various classifications. It is the most primitive and generalized group of teleosts, as well as one of the most important. In typical members pelvic fins abdominal in position, well behind pectoral. Scales, when present, cycloid. Air bladder, when present, joins esophagus by an open duct. All or most of fin rays are soft, not developed into spines. About 25 families.

SECTION **Clupeoidea** Soft-finned, mostly marine or anadromous fish, with no adipose fin and no photophores.

FAMILY Clupeidae The major family, which includes numerous well-known commercially important species such as herring, sardine, shad, and menhaden. Mostly plankton feeders with terminal mouth, narrow, compressed body, forked tail, deciduous scales, teeth small or lacking, and gill rakers long and slender. They lack a lateral line. Dorsal fin single, in middle third of body, well in front of origin of anal fin. Head without scales. Descriptions of representative genera follow.

Alosa (Pomolobus) Shad. Relatively deep-bodied, anadromous fish. Occurs along Atlantic seaboard of United States and carries out spawning runs up rivers in spring. Some also occur in Europe. The species *A. pseudoharengus* is the alewife; *A. sapidissima,* the white shad.

Brevoortia Menhaden, or mossbunker (Fig. 11-17, *A*). Along Atlantic and Gulf coasts of United States.

Clupea Herring. Among the most important of all commercial fish. Occur in extensive schools, especially at spawning time. Feed largely on copepod *Calanus* and occur chiefly in water masses heavily populated by it. Worldwide distribution. *Clupea harengus* (Fig. 11-17, *B*) is the common herring of North Atlantic. Often found in huge schools containing many millions of individuals, which greatly facilitates large-scale commercial fishing for them. Many of them are sold as "sardines." They spawn near coast in shallow water, where substrate firm or gravelly and there are stones or other objects to which eggs can adhere. A distinct race also occurs in the North Pacific, regarded by some as a separate species, *C. pallasii*. Regional populations seem to remain distinct without a great amount of mixing. Sindermann (1957) found that parasitic and fungal infections of populations from Gulf of St. Lawrence and those from along coast of Maine and southward were distinct from each other. Blood group studies on different populations also indicate this. Each female lays about 10,000 eggs in sticky masses.

Sardinella Sardine, pilchard.

Sardinops Pacific sardine.

FAMILY Chanidae

Chanos The milkfish. Extensively cultivated in tambaks or marine impoundments in warm parts of western Pacific from fry collected in ocean. Important in southeast Asia, from Indonesia to Philippines.

FAMILY Engraulidae Anchovies. Small herringlike fish distinguished by their large mouth, the mouth angle extending to back of eye, and a pointed snout that extends in front of mouth. Most species not over 5 inches long. Extensively fished for pickling or canning, like sardines, etc. Many species, found mostly in warm temperate or shallow tropical waters, often in brackish water or around river mouths. Huge populations occur in Peru Current in recently upwelled water where temperature is 14° to 18° C., shifting both laterally and vertically with movements of water having this range. Typical genus, *Engraulis*. Rich, peculiar flavor. Related families are Alepocephalidae (smooth head) and Dussumieriidae (round herring).

SECTION **Salmonoidea** Salmon, trout, etc. Usually medium-sized fish, larger than herring, with an adipose fin between main dorsal fin and tail. Lateral line present. No photophores. Many marine species anadromous, ascending rivers to spawn. Valued as game and commercial fish.

FAMILY Salmonidae Mouth large, hind edge of maxillary extending to or beyond eye in adult. Base of dorsal fin shorter than length of head; pelvic fins have scaly appendage above base; dorsal fin has less than twenty-six rays. Eyes situated laterally. Stomach has more than eleven pyloric ceca.

Salmo Atlantic salmon and most freshwater trout. North Pacific salmon differ in numerous respects, such as a greater number of anal fin rays, branchostegal rays, pyloric ceca, and gill rakers, and are commonly separated as the genus or subgenus *Oncorhynchus*. They spend most of their lives in the sea, two to three years or more. For discussion of their spawning migrations, see pp. 128 and 129.

Related families include Coregonidae (whitefish, mostly freshwater), Thymallidae (grayling), Osmeridae (smelts), Galaxiidae (southern hemisphere "trout" and New Zealand kokopu), Argentinidae (argentines, capelons), Bathylagidae (deep-sea smelts), and Opisthoproctidae (barreleyes).

SECTION **Elopoidea** Tarpon and bonefish. Large fish, the tarpon often becoming 6 feet or more in length. Good game fish, but flesh is considered rather poor.

SECTION **Stomiatoidea** Numerous deepwater fish, with rows of photophores on body. Angle mouths, hatchetfish, stomiatoids, and viperfish.

SECTION **Esocoidea** Pike, blackfish, mudminnow. Mostly freshwater fish.

SECTION **Osteoglossoidea** Bony tongues, mooneyes.

SECTION **Mormyroidea**

SECTION **Giganturoidea**

Suborder Iniomi (Myctophiformes) Resemble suborder Isospondyli in having soft fins, the pelvic fins abdominal in position in most members, scales cycloid, and an adipose fin usually present. The premaxillary bone forms the margin of the upper jaw. Many are bathypelagic species with large mouth and eyes, slender bodies, numerous photophores.

FAMILY Myctophidae (Scopelidae) Lantern fish. Perhaps the most common and widespread of the fish at intermediate depths in the oceans, sometimes coming to the surface at night or in stormy weather. On the East Coast, aggregations of myctophids were found to be responsible for an unusual, rather lumpy deep-scattering layer. Related families include Synodontidae (lizardfish), Alepisauridae (lancet fish), Sudidae (deep-sea fish), and Gonostomidae (Fig. 11-19) (angle mouth).

Suborder Lyomeri (Saccopharyngiformes) Gulper eels.

Suborder Ostariophysi (Cypriniformes) Four anterior vertebrae strongly modified, often ossified together, and support a chain of small bones connecting air bladder with ear. This group includes three of the largest families of predominantly freshwater fish, among which are many popular aquarium fish. Included are such forms as characins, tetras, suckers, minnows, carp, loaches, electric eels, and catfish. One group of catfish, the Ariinae (Tachysurinae), are marine and known as sea catfish.

Suborder Apodes (Anguilliformes) Eels. This large group of fish shows syndrome known as "eel type of degeneration or racial senescence." They have snakelike bodies, with no pelvic fins, and some of them have no pectoral fins, either. Shoulder girdle usually separate from skull. Scales rudimentary or absent, the body being smooth, and in some, slimy. In some the dorsal, caudal, and anal fins are continuous around the tail; others have a separate caudal fin.

FAMILY Anguillidae

Anguilla Circumtropical except in western Americas. Includes the common European and American eels. See pp. 129 and 130 for an account of their spawning migration.

FAMILY Muraenidae Moray eels. Large voracious, often pugnacious eels. Some species brightly colored. They have no paired fins. Gill openings small and round. Jaws narrow and strong and armed with knifelike teeth. Back of head elevated, giving eels a mean look. They occur in all warm seas, especially around coral reefs (Fig. 11-18).

Other families include Moringuidae (worm eels), Nemichthyidae (snipe eels), Congridae (conger eels), Ophichthyidae (snake eels), and Echelidae.

Suborder Heteromi (Notacanthiformes) Spiny eels.

Suborder Synentognathi (Beloniformes) Slender or elongated fish usually found in surface waters. Head pointed and one or both jaws are produced; fins soft, supported by rays only, the dorsal fin being set far back on body, similar to and above anal fin; pelvic fins abdominal in position; lateral line low on body, scales cycloid. In many, tail emarginate, lower lobe being longest, presumably giving maximum push in jumping from water. Most taste good.

FAMILY Exocoetidae Flying fish. In flying fish, pectoral fins long and firm and used in gliding. These fish swim near surface in all tropical waters, some species being cosmopolitan and found in all tropical oceans and others restricted to particular regions such as Indo-Pacific. They leap into the air, usually to escape predators or avoid path of a boat, spread their winglike pectorals and glide for considerable distances. Longest flights probably not more than a fourth of a mile. In the so-called four-winged flying fish, pelvic fins are only a little behind center of body and are also spread, giving additional lift during gliding. Largest and farthest-flying of flying fish are of this type, reaching about 18 inches in length. In two-winged flying fish pelvic fins smaller, farther back on body and not spread during flight. These fish seldom longer than 15 to 18 cm. Longest flights are made across wind in a fairly strong breeze.

FAMILY Belonidae (needlefish) Billfish, garfish, and snipefish These voracious fish have slender bodies, large eyes, and elongated jaws forming a beaklike process set with sharp teeth. They are silvery, somewhat greenish above, and swim near surface in small groups, preying on smaller fish. Large species grow 1 to 1.5 meters in length; most are smaller than that. They are abundant along shores of almost all warm seas. Many have green bones.

FAMILY Scombresocidae Sauries.

FAMILY Hemiramphidae Halfbeaks. Lower jaw is produced beyond head as a toothless pointed process. They are small, seldom exceeding a foot in length, and often swim along just below surface at an angle to surface. They are small-mouthed and mainly plankton eaters.

Suborder Anacanthini (Gadiformes) Cod and grenadiers. Tail isocercal, lacking slight structural asymmetry found in homocercal tail of most teleosts. Fins soft without spines; there are three separate dorsal and two separate anal fins. Chin bears a barbel in most species. Mostly demersal, swimming near bottom much of the time, preying on bottom invertebrates and small fish.

FAMILY Gadidae Cod. Atlantic cod, *Gadus morrhua,* is one of the best-known and most-fished commercial species. It is a cold-water boreal species abounding around offshore banks on both sides of Atlantic. Cod frequently reach a weight of 50 pounds and may sometimes reach as much as 200 pounds and a length of about 2 meters. Mouth large, upper jaw projecting slightly beyond lower. Body covered with small, smooth scales. Cod range along continental shelves and banks, usually swimming near bottom in schools, devouring any animals of appropriate size they can find—crabs, molluscs, fish, and squid, sometimes chasing the latter two up into higher water. Several closely related fish, including haddock *(Melanogrammus aeglefinus),* pollack *(Pollachius pollachius),* and tomcod, are also much valued as commercial fish.

Lingcod differ from typical cod in having a more elongate body and only two dorsal fins and one anal fin, posterior dorsal fin being much longer than anterior. Eyes large, and chin barbel well developed. European ling, *Molva,* reaches a length of about 1.5 meters and is extensively used for food, especially in Iceland. In American ling, or hake *Phycis,* pectoral fins are reduced to long two-branched filaments arising near throat. European hake and American silver hake. *Merluccius* spp. also belong in this group.

FAMILY Macruridae Grenadiers, or rattails. These deep-sea demersal fish are found on outer portions of continental shelves and in deeper water. Head is large, often angular, with a small barbel on chin. Pectoral and pelvic fins occur almost one above the other, with a short anterior dorsal fin above pectorals; posterior dorsal and anal fins are long and low, sometimes vestigial. Body is long, covered with small scales, tapering into a slender ratlike tail devoid of a caudal fin. They are numerous both in species and individuals.

Suborder Solenichthyes (Gasterosteiformes, Hemibranchii) Several families of fish with reduced branchial apparatus, and usually with a small mouth at end of an elongated tubular snout, are included here. Some are among the most bizarre-appearing fish. They suck in planktonic organisms. Most of them rather slender, elongate fish.

FAMILIES Aulostomidae and Fistulariidae Trumpet fish (Aulostomidae) and cornetfish (Fistulariidae) found in tropical shore waters. In both groups, fins soft rayed, and dorsal fin set well back, over anal fin. Cornetfish are slender but sometimes reach a length of 1 to 1.5 meters, although they are usually smaller.

Trumpet fish grow to 0.3 to 0.6 meter. In genus *Aulostomus* there is a row of short spines in midline of back, suggestive of spines of sticklebacks, which are included by some authorities in this order. It seems probable that fishes of this order descended from spiny-rayed rather than soft-rayed fish and that present lack of spines in members of group is secondary rather than primitive.

FAMILY Syngnathidae The best-known and most bizarre fish of this order belong to family Syngnathidae, which includes pipefish and sea horse. In these are rather small fish, the body is protected by variously developed bony plates, developed as rings, keels, and ridges, giving body a stiff, often angular, character. Gills in tufts on branchial arches, and snout long and tubular, with a small mouth at end. Teeth either absent or minute. Male has a ventral pouch behind vent where female deposits eggs, which are then incubated and protected by male. Female, in this case, initiates courtship and mating. Swimming is commonly a very slow, gliding motion effected by weak fins. Some can also swim somewhat faster for short distances by eellike movements of body.

Syngnathus Pipefish (Fig. 11-20) are long, slender fish, found commonly among seaweed and eelgrass along shores of all warm seas, often in protected waters of bays and estuaries. Some species extend into temperate waters. A few may even be found in essentially freshwater at head of estuaries. *S. pelagicus* extends across warm portion of North Atlantic, associated with *Sargassum*.

Hippocampus Sea horses are peculiar pipefish with much shortened, thickened, rigid, and laterally flattened bodies. Neck constricted, and head bent down at right angles to axis of body, the whole appearance being much like that of knight in a chess set, but sea horse has a long, tubular snout and a tapering, prehensile tail devoid of a caudal fin. Sea horses commonly adopt an erect position propelled by a fanlike motion of small dorsal fin, which here takes over propelling functions usually assigned to caudal fin, albeit rather inefficently.

Other families include Aulorhynchidae, tubesnouts; Centriscidae, shrimpfish; and Macrorhamphosidae, snipefish. *Centriscus,* the shrimpfish, rather translucent and has peculiar habit of assuming a vertical position in water, head downward. Small schools, when alarmed, swim off together rapidly, still maintaining this position.

Suborder Salmopercae (Percopsiformes) Trout perch, salmon perch.

Suborder Allotriognathi (Lampridiformes) Oarfish, ribbonfish, crestfish, and opahs. Soft-rayed fish with protrusible jaws differing structurally from those of other fish. These fish tend to be laterally flattened and often reach a considerable size.

FAMILY Lamprididae

Lampris Moonfish. Sometimes up to 2 meters long and weighing over 500 pounds.

FAMILY Trachypteridae Oarfish and ribbonfish. Bandlike body; largest (genus *Regalecus*) may attain a length of 9 meters.

FAMILY Veliferidae Crestfish. Dorsal fin in these fish, and in some Trachypteridae, forms a flaglike crest.

Suborder Berycomorphi (Berciformes) Squirrelfish, soldierfish, beardfish, and berycids. This ancient group is known from Upper Cretaceous and probably close to ancestral types for Acanthopterygii.

Squirrelfish, or soldierfish (Holocentridae), are abundant in shallow tropical waters around world. They are moderate-sized fish with two incompletely separated dorsal fins, anterior one longer and bearing strong spines. Peduncle narrow, and tail firm and well forked. Anal fin bears four spines, of which first is small and third long, strong, and sharp. Pelvic fins under or only slightly behind pectorals and consist of a spine and seven flexible soft rays. Lower hind edge of opercle also bears a backwardly directed spine. Scales are rather large and ctenoid. Squirrelfish are usually mostly red, with longitudinal white or yellow stripes. Red color consistent with hypothesis that squirrelfish are shallow-water invaders from fish commonly inhabiting greater depths. Many fish, as well as other animals, inhabiting somewhat deep-

er, dimly lit waters, are red in color compared with their shallow-water relatives.
Related families Polymixiidae (beardfish) and Berycidae do inhabit somewhat deeper water.

Suborder Zeomorphi (Zeiformes) Dories and boarfish. Spiny-rayed, thin-bodied, laterally compressed fish living at intermediate depths. The John Dory is relished as a gourmet item. They grow to be about 3 feet long.

Suborder Microcyprini (Cyprinodontiformes) Includes many small freshwater fish, such as cavefish, killifish, topminnows and livebearers, and four-eyed fish *(Anableps)*. Numerous members of Cyprinodontidae also abundant in coastal salt and brackish waters as well as freshwater, especially in warm regions. Although soft rayed, they are regarded as related to spiny-rayed groups and to have lost spines secondarily. Most of them only a few inches long, differing from true minnows in having rounded rather than forked tails. Males of many species more brightly colored than females, especially during breeding season.

Suborder Acanthopterygii (Perciformes or **Percomorphi)** A large and diverse group that includes several rather distinct assemblages, given below.

SECTION **Perciformi** Perchlike fish. Symmetrical fish with spinous anterior and soft-rayed posterior dorsal fins, often more or less confluent and of about equal length; pelvic fins almost directly under pectorals, and consist of one spine and five branched rays. Scales moderate to small and often rough. Tail rounded, truncate or emarginate, or strongly forked. Several families and numerous important commercial and game fish. Some of the more important families are as follows.

FAMILY Percidae Perch (Fig. 11-16), darters, walleyes. Mostly freshwater fish.

FAMILY Centrarchidae Freshwater sunfish and American bass.

FAMILY Kyphosidae Chiefly herbivorous. Found in warm seas. Basslike.

FAMILY Serranidae Sea bass family. A large, diverse family of mostly marine, carnivorous fish commonly living near bottom in shallow to moderately deep waters. Some attain considerable size, and most of them excellent food fish. Included are striped bass, black sea bass, true groupers, jewfish, hamlet, sergeant fish or robalo, and many others. A few are anadromous or even landlocked. Scales usually ctenoid. Anal fin never has more than three spines. They are especially abundant in warm seas. Largest ones, certain species of groupers or jewfish, sometimes attain a length of 1.5 to 2 meters and weigh up to 800 pounds. Many serranids are hermaphroditic, having an ovotestis. Most such species function as males when young, then later as females as they grow larger, but in at least one genus functional sperm and eggs produced at same time.

FAMILY Lutianidae Snappers. Active carnivorous fish with large mouths, strong teeth, and usually deeper, more compressed bodies than serranids. Tail emarginate. They are fast swimmers and travel in small schools, preying on smaller fish. Abundant in warm waters. Species occurring in somewhat deeper waters tend to be bright red or rose colored. Shallow-water species often greenish or variously marked. Commonly about 0.6 meter long. In related grunts the inside of mouth bright red. Grunts make a peculiar grunting noise when taken from water. They tend to be somewhat smaller and to have weaker teeth than snappers. Many species occur in West Indies, in Caribbean.

FAMILY Sparidae Porgies, sea breams, scup, sheepshead, etc. Related to snappers and grunts, but some of the teeth along sides of jaws flattened and constitute large, blunt molars. Front teeth often incisor-like.

FAMILY Leiognathidae Tend to be more active swimmers, with well-forked tail, weakly developed teeth, and unusually protractile mouth that can be extended or retracted.

FAMILY Sciaenidae Croakers, drumfish, squeteague, weakfish or sea trout, kingfish, maigres, etc. A large family, best represented along tropical shores of eastern Pacific but generally distributed along sandy shores of all temperate warm seas. Most feed near bottom, but weakfish *(Cynoscion)* swims actively higher up. Members of this family usually have a large and complicated air bladder, and some, especially croakers and drumfish, produce characteristic peculiar sounds with it. Body shape and size and size of mouth are variable in this family. Croakers *(Micropogon)* and kingfish *(Mentricirrhus)* tend to have small mouths, lower jaw somewhat shorter than upper and often with small barbels on chin. Croakers have unusually large, smooth otoliths or free ear bones.

FAMILY Mullidae Goatfish. Common in tropical seas. Dorsal fins approximately equal size; caudal fin forked. Snout elongate and curved, usually projecting slightly beyond lower jaw; a pair of long fleshy barbels on chin. European red "mullet" one of better-known species.

FAMILY Apogonidae Cardinal fish. Short-bodied, usually less than a foot long. Dorsal fins separate, of about same size. Large, oblique mouth. Unusually large eyes.

FAMILY Carangidae Pompano, jacks, crevalles, scad, amberjack, moonfish, cavalla, pilot fish, etc. Spinous dorsal fin much shorter than soft-rayed dorsal behind it, and spines weakly developed. Premaxillaries protractile, upper jaw not fused with snout. Lateral line has a strong arch anteriorly. Caudal peduncle keeled, often with enlarged scutes over keel. More or less compressed narrow body, tapering to slender peduncle and widely forked tail. Most carangids fast swimmers, often near surface. Abundant in warm seas, some migrating north in summer.

FAMILY Coryphaenidae Dolphins. Among the most truly pelagic of fish. They occur in all warm seas, subsisting largely on flying fish when far from land. Body streamlined, blue above and bright yellow below, with an even-bordered, undifferentiated dorsal fin running full length of back. In older males a high, narrow crest gives forehead a vertical profile. (These should not be confused with small porpoises, which have also often been called dolphins.) Dolphins grow to be about 6 feet long and usually swim in small groups, at or near surface.

FAMILY Pomatomidae Bluefish and relatives. Bluefish of Atlantic and Indian oceans is a highly predaceous, schooling fish somewhat resembling an amberjack but with no keel on caudal peduncle and with ctenoid scales that are slightly rough. It is a widely distributed species, valued for food along Atlantic seaboard.

FAMILY Embiotocidae Surf perch. Pacific distribution, especially off coasts of North America and Japan. Viviparous, giving birth to a relatively small number of young rather than laying eggs.

FAMILY Notothenidae Found in Southern Hemisphere.

The following three families, among them, contain a majority of the well-known brightly colored coral-reef fish, or coral fish.

FAMILY Chaetodontidae Butterfly fish. Deep, laterally compressed body. Small mouth at end of short, snoutlike extension of head, for picking off small organisms from crevices in coral. Dorsal fins continuous, and their bases and sometimes also base of anal fin covered with fine scales. A large number of brightly colored, conspicuous small fish are found around coral reefs in tropical seas.

FAMILY Labridae Wrasses. Most wrasses also colorful coral fish. They have a small mouth and thickish lips. Jaws armed with strong, usually pointed teeth, and throat armed with blunt grinding teeth. Most of them have rather elongate, compressed bodies, large scales, continuous dorsal fins, and a deep, compressed tail, deeply forked in adults of some species. Many wrasses swim much of the time with a rowing motion of pectoral fins, using tail when extra speed needed. A few exceptional species are found outside tropics and become larger than most species. Blackfish, or tautog *(Tautoga onitis),* of United States Atlantic coast, and burgall, or cunner *(Tautogolabrus adspersus),* which extends as far north as Labrador, are examples. Some wrasses are cleaning symbionts.

FAMILY Scaridae Parrot fish. Rather similar to wrasses but usually thicker bodied, and with jaw teeth fused to form a hard, parrotlike beak, capable of biting off bits of coral. Usually average a little larger than wrasses. Parrot fish may be ecologically important in preventing overgrowth of algae and other seaweed on tropical reefs, since they tend to graze off any beginning growth that occurs, often leaving characteristic marks of their teeth on calcareous surface. If submerged coral rocks, etc. are protected from such grazing, they usually develop an algal cover much richer than any found on exposed areas nearby.

SECTION **Bathyclupeoidea**

FAMILY Bathyclupeidae Deep-sea fish.

SECTION **Acanthuridea**

FAMILY Acanthuridae Surgeonfish, Moorish idol, etc. Usually deep bodied, compressed, with narrow peduncle, emarginate to forked tail. Several have a sharp-edged spine on caudal peduncle that can be erected forward and outward or depressed into a groove. Largely herbivorous. Fine, inconspicuous scales.

SECTION **Scombroidea** Mackerel, tuna, billfish or swordfish, albacore, sailfish, bonitos, etc. Swift-swimming pelagic fish with elongate fusiform streamlined bodies. Two dorsal fins followed, in mackerel and tuna, by a series of small finlets. Pelvic fins thoracic in position. Tail is emarginate, on a slender peduncle. Many large and highly prized both as game fish and for food.

FAMILY Scombridae Mackerel. Pelagic fish, lacking a swim bladder, which gives them greater facility in going rapidly from one depth to another but means that they must swim to maintain their level in water.

Scomber Spinous dorsal fin short and set considerably ahead of soft second dorsal fin. Following second dorsal fin and anal fin are a series of small finlets. A pair of small keels present at bases of tail lobes, but none in center of peduncle. Mackerel are medium-sized or small fish usually not exceeding 2 feet in length and weighing 3 or 4 pounds. They swim in close-ranked schools, sometimes of great size, and near surface. North Atlantic mackerel *(S. scombrus)* found throughout summer, when they go north to spawn in somewhat cooler waters than those in which most of family are found. In winter, schools disappear.

Scomberomorus (= Cybium) Contains somewhat larger fish, with a central keel on peduncle and spinous dorsal fin extending along back to near front of second dorsal fin. Spanish mackerel *(S. maculatus)* attains about 10 pounds and is regarded as a choice food fish. King mackerel *(S. cavalla)* may reach 1.5 meters in length and weigh up to 100 pounds.

FAMILY Thunnidae Tuna (Fig. 11-14), bonitos, skipjacks, albacore, etc. Closely related to mackerel, sometimes being classed as a subfamily of Scombridae. Most are somewhat larger, often deeper-bodied fish. Like mackerel, some, at least, also lack a swim bladder. They are pelagic, voracious, carnivorous fish, usually found in large schools, and form basis for some of the world's most important fisheries.

Thunnus Contains large typical tuna such as bluefin tuna, also called horse mackerel or great albacore *(T. thynnus)*. They have a world-wide distribution in all warm seas, often attaining a weight of 1000 pounds or more. Flesh coarse and oily but used for food, and oil also valued. In this genus pectoral fins clearly shorter than head, and body does not bear vertical or oblique stripes.

Sarda Bonitos. Robust, active fish in tropical and subtropical seas. Body elongate and has minute scales. Keel on peduncle is thick, naked. Teeth large, compressed, and strongly curved inward; tongue and vomer toothless. Numerous dark, more or less oblique stripes occur on upper part of body. Bonitos take bait readily, are good sport fish, and make good eating. They congregate in large schools and are important in economy of some areas.

Germo Albacore. In Pacific, associated with transitional waters between central and subarctic water masses, with surface temperatures of 12° to 18° C.; in summer near polar front. Pectoral fins of albacore considerably longer than those of most tuna, reaching back to level of anal fin.

Other genera include *Gymnosarda,* the little tunny and ocean bonito; *Neothunnus* (Fig. 11-14), yellow tuna, or albacore; *Parathynnus,* bigeye tuna, or ahi; and *Katsuwonus,* skipjack tuna.

FAMILY Xiphiidae (including **Istiophoridae**) Swordfish and billfish. Large pelagic fish, with rostrum drawn out into a long, sharp spear above mouth.

Istiophorus (sailfish), *Makaira* (marlins), *Tetrapturus* (spearfish), etc. Dorsal fin long, body bears scales, teeth present, and pelvic fins present though reduced to one or a few rays.

Xiphias (swordfish) Short high first dorsal fin set well forward on body and a small second dorsal set far back. There are no pelvic fins, scales, or teeth. Swordfish have small pelagic eggs, and young pass through three unlike phases while growing up. In one of these phases a long high dorsal fin is present and in another, scales, showing its close relation to sailfish-marlin group.

A series of deeper-water, mackerel-like fish leads by small steps to certain slender deepwater fish in which a small, although mackerel-like, tail is present, apparently as a vestigial structure, and finally to fish such as the Trichiuridae—slender, bright silver cutlass fish that have a long dorsal fin but no caudal fin at all, tail end tapering to a slender filament. Like swordfish, they have no pelvic fins.

SECTION **Stromateoidea**

FAMILY Stromateidae Butterfish. Small deep-bodied laterally compressed shore fish superficially resembling Carangidae but not closely related to them. Young have small pelvic fins, but these fins much reduced or absent in adults. Some pelagic species usually regarded as related to, if not to be included in, this group. One of the best known is *Nomeus gronovii,* which follows Portuguese man-of-war *(Physalia)* and finds protection from predators by swimming close up under it when alarmed.

SECTION **Gobioidea** Largest and best-known family of this group is family Gobiidae, the gobies and their close relatives. They are common in shallow waters of warm and temperate oceans everywhere. Skin seems smooth, and no lateral line is in evidence. Scales, when present, moderate to small in size and smooth; numerous species lack scales. Eyes commonly placed high on head, especially in mudskippers *(Periophthalmus)* of tropical mangrove swamps and estuaries. Mudskipper has a froglike appearance with bulging eyes mounted on top of its head. At other extreme is the blind goby *(Typhlogobius californiensis),* in which eyes rudimentary and color pale pinkish, and which lives under stones and in dark crevices in rocky areas along California coast. Gobies have two dorsal fins, anterior one with weak spines. Pectoral fins usually rounded; the pelvic fins thoracic in position, at same level as pectorals or even somewhat in advance of them. In clingfish (Gobiesociformes) pelvic fins modified to form a sucker with which fish can cling to smooth surfaces. Related families are the Eleotridae (sleepers), Microdesmidae (wormfish), and Callionymidae (dragonets).

SECTION **Trachinoidea** Weavers, stargazers, and related families. Elongate fish with large heads, mouth directed upward and eyes usually rather small and placed well forward on top of head almost over mouth. Weavers (Trachinidae) are small to medium-sized, up to about 18 inches in length, with grooved spines in dorsal fin associated with poison glands. There is also a spine on opercle. Stargazers (Uranoscopidae) sluggish benthic fish about a foot long, with an almost vertical mouth and small eyes on top of head directed upward.

SECTION **Blennioidea** Blennies and related families. A large and varied group, in which pelvic fins reduced to a small number of rays or absent. When present, they are forward in position, arising near back of mouth. Blennies mostly small fish, often found in shallow waters along shore, together with gobies. Typical blennies (Blenniidae), sometimes called combtooth blennies, lack scales and have a rather large, froglike head with a relatively small, transverse mouth and a row of comblike teeth in each jaw. Dorsal fin rather high and continuous, with weak spines in anterior half; caudal fin rounded. Klipfish (Clinidae) have scales, a somewhat less froglike head, a larger mouth, and sharper spines in dorsal fin extending along a greater part of its length. Family Pholidae includes gunnels, most of which live in colder northern waters and have a more eellike body, usually with small, inconspicuous scales. Dorsal fin is long, low, even, and spinous along its full length. Other fish related to blennies include prickle-backs (Stichaeidae) and eelpouts (Zoarcidae).

SECTION **Ophidioidea** Largest family, Brotulidae, comprises a varied assemblage of about 50 genera mostly found in rather deep water, although type genus, *Brotula,* includes several species about a foot long, found along tropical shores. Eyes usually large. Bodies elongate, and covered with fine scales. Long even soft-rayed dorsal and anal fins meet in a point around tail. Pelvic fins reduced to a pair of filaments arising beneath gill cleft. Family Carapidae includes numerous curious small fishes often commensal with molluscs or echinoderms. They are slender little fish in which tail tapers to a threadlike ending with no caudal fin. Some of them enter cloacal openings of sea cucumbers. Silvery little sand lances (Ammodytidae) may be related to this group, although their affinities not certain. They abound in loose, wet sand of many northern beaches.

Suborder Haplodoci (Batrachoidiformes)

FAMILY Batrachoididae Toadfish. Toadfish mostly shallow-water, rather sluggish, dull-colored fish about a foot long, without scales. Head large, slightly flattened, with large mouth bearing strong, blunt teeth. Fins rounded. Toadfish hide by day under rocks or in crevices, shells, etc. Eggs deposited in masses attached to surfaces in such hiding places and guarded by one or both parents until hatched.

Suborder Heterosomata (Pleuronectiformes) Flatfish, halibut, flounder, sole, turbot, etc. Demersal fish, with remarkable adaptations to this habit. Flatfish spends much time lying on, or partially buried in, bottom, on one side. Body laterally flattened and deep. Upper side darkened and variously patterned, whereas lower side pale. Most flatfish can vary tint and even pattern of upper side to blend with substrate, often remarkably well. This color change under nervous control, and eyes are necessary for it to take place properly. Although larva bilaterally symmetrical, adult has both eyes on same side of head, the upper side. Mouth also curiously and asymmetrically twisted in many of them. Most species are either right- or left-sided, but some are variable in this respect, some individuals lying on right side, others on left. When they swim, by vertical (lateral) undulations of body, they remain in same general position as when resting on substrate. Flatfish are a very successful group and have diversified into several families and many genera. Almost all of them are good eating, several being among the most highly prized table fish. Some of the better-known groups of commercial importance follow:

Halibut: species with symmetrical pelvic fins and large symmetrical mouth. Atlantic halibut *(Hippoglossus hippoglossus)* largest of flatfish, sometimes attaining a length of about 3 meters and weighing as much as 700 pounds.

Sole: flatfish with small eyes, small, crooked mouth, and often lacking pectoral fins.

Flounder: flatfish with a rather small, crooked mouth and nearly or symmetrical pelvic fins.

Halibut, sole, and flounder lie with right side upward.

Turbot: left-sided flatfish with symmetrical mouth and asymmetrical pelvic fins.

Suborder Discocephali (Echeneiformes) Remoras. Peculiar fish in which top of head somewhat flattened and occupied by a large sucking disc that also extends partly onto back. Sucking disc oval, with a central longitudinal ridge and a series of transverse ridges. This disc presumably represents a highly modified spinous dorsal fin. Remoras invariably associated with larger fish, such as sharks and marlins, to which they attach themselves by means of sucking disc, obtaining a free ride to distant waters plus protection by host and a share in host's meals. Most species are rather host-specific with regard to kinds of fish to which they habitually attach. Remoras are usually grayish in color and lack countershading found in many free-swimming pelagic fish. Soft dorsal fin and anal fin set well back, tail emarginate, and pelvic fins arise beneath pectorals. Skin covered with easily overlooked, minute scales. Relationships of remoras uncertain: they do not seem closely related to any of major families of teleosts.

Suborder Loricati (Cottiformes) Mail-cheeked fish. Huge group containing many familiar fish, medium-sized to fairly large, such as sculpins, cottids, rockfish, greenlings, scorpionfish, poachers, lumpfish, sea robins, and gurnards. Typically they are demersal fish with a large head and mouth, often bearing spines at sides of head. Spinous dorsal fin of many species bears formidable spines, which in stonefish *(Synanceja),* lionfish *(Pterois),* and a few others have poison glands associated with anterior spines, enabling them to inflict painful or even fatal wounds if they are stepped on. A bony process, or stay, of third suborbital plate extends back to or toward preopercle, whence the name mail-cheeked fish. In many species this bony process is not obvious without dissection. Although distributed over most of oceans, this group appears to have its center of origin in, and greatest number and variety of species endemic to, North Pacific. Some of the principal families follow.

FAMILY Cottidae Includes sculpins and cottids, commonly rather small fish with a large mouth and with eyes placed near the top of head. Cottids commonly rest on bottom much of the time, move forward with a sudden movement, and then come to rest again. They are voracious eaters, nearly omnivorous. Small, variously mottled and patterned cottids are among the most characteristic fish seen in tide pools along Pacific coast. *Oligocottus.*

FAMILY Scorpaenidae Common and varied in North Pacific and also largely replace cottids in tropics. Mostly larger fish, with a large, spinous head, and with strong spines in anterior part of spinous dorsal fin. Many are red in color and variously mottled. Some of larger species of *Sebastodes,* the rockfish or "rock cod," and Atlantic rosefish *(Sebastes marinus)* are of considerable market value.

FAMILY Hexagrammidae Greenlings. Common around kelp beds in North Pacific coastal waters. Valued food fish, noteworthy for presence of several well-developed lateral lines. They are usually elongate, with confluent dorsal fins extending length of back, and with spines of anterior part rather weak. Lingcod *(Ophiodon)* a related fish that attains a length of over 1 meter and a weight of up to 40 pounds. It is one of the most important food fish from cold waters along Pacific coast, in spite of bluish or greenish tint of its flesh. It has only one lateral line.

Suborder Plectognathi (Tetraodontiformes) Porcupine fish, filefish, puffers, triggerfish, mola, etc. This large, mostly tropical or subtropical group is characterized by lack of pelvic fins, which are either wholly absent or represented by a spine at end of long pelvic bone. Mouth rather small and pincerlike, and scales lacking or abnormal. They are weak swimmers; pelagic species tend to have high dorsal and anal fins and swim by moving fins from side to side. The principal families follow.

FAMILY Balistidae Triggerfish and filefish. Leathery skin and small, nonimbricate scales. In triggerfish spinous dorsal fin well forward in position and consists of one strong and two weaker spines, depressible into a groove. When erected, strong spine is locked into position by first weak spine, suggesting trigger mechanism of a gun. Filefish have more elongate, compressed bodies and a single well-developed anterior spine.

FAMILY Diodontidae Porcupine fish and puffers. Relatively short, stocky fish, body surface armed with numerous strong spines. They inflate themselves when alarmed, so that spines project in all directions.

FAMILY Tetraodontidae Puffers, or swellfish. Related to porcupine fish but without strong spines, skin being smooth or with fine prickles. They can inflate themselves with water or air until nearly spherical.

FAMILY Ostraciontidae (Ostraciidae) Trunkfish, or boxfish. Head and body are encased in bony plates, leaving only tail and fin bases free. Armor often angled or squarish, somewhat triangular in cross section. Some of them, the cowfish, have hornlike spines projecting forward from above eyes.

FAMILY Molidae Mola. Largest of plectognath fishes, attaining a length of up to 10 feet and a weight of over a ton. They have short, deep, compressed bodies with rounded outline, high, stiff dorsal and anal fins, a thick, wrinkled skin, and small mouth and eyes. Body abruptly truncated behind, bearing a small fringelike caudal fin. They are weak swimmers and are usually seen drifting on one side in upper strata of water, often far out at sea. Body for the most part has a fatty, cartilaginous character, different from flesh of most fish or other large animals. Brain unusually small and undeveloped for an animal of its size. Young fry only slightly more than a millimeter long and have a peculiar spinous appearance.

Suborder Pediculati (Lophiiformes) Anglers, frogfish, etc. Among the most specialized and peculiar marine fish. They are mostly demersal, some of them living in abyssal depths. They have a large, broad head occupying almost half body length and a large, transverse mouth, directed upward and armed with numerous long, pointed teeth. Eyes usually on or near top of head, and body tapers rather rapidly behind head. Pectoral fins broad and, in anglers, supported from short, fleshy arms. Anglers have anterior dorsal fin ray located well forward on head, modified into a luminescent lure, or bait, which is dangled over mouth, luring smaller fish to within easy reach. Anglers and frogfish are sluggish, given to hiding, and either luring or slowly stalking their prey until it is close enough to take with a quick lunge. Some of the deep-sea species have a remarkable adaptation for reproduction. Males much smaller than females, and one (or more) of them attaches itself to body of female by its mouth. It remains permanently attached for the rest of its life, organically connected to female with united vascular systems. Male's digestive organs, etc. are degenerate, only reproductive organs remaining well developed.

The sargassum fish *(Histrio)* is an exception to usual demersal habits of suborder, but there is a similarity in that it tends to hide in floating masses of sargassum weed in Sargasso Sea. It exhibits a wonderful case of mimicry, body being drawn out into irregular projections so closely resembling growth habit of sargassum weed in form and color that it is almost indistinguishable in its natural habitat.

Another relatively small, specialized group that deserves mention are the Xenopterygii (Gobiesociformes), or clingfish. These fish are small, usually not more than a few inches long, and characterized by their flattened ventral surface, which bears a large sucking disc between and behind pelvic fins involved in its structure. This is the most efficient sucking disc formed by any fish except remoras, in which sucker is of altogether different location and origin. Clingfish have a smooth, scaleless skin, large head, with eyes on top, tapering body, and soft fins, dorsal and anal fins placed well back on body. Many are found in shallow coastal or even intertidal waters, clinging to undersurfaces of stones, which they may match in color. Clingfish from deeper water are often reddish or pinkish.

CLASS Reptilia Marine reptiles were far more numerous and diverse in the past than they are today. Whole orders are now extinct. Among the more prominent and extraordinary were the plesiosaurs, thalattosaurs, ichthyosaurs, mosasaurs, thalattosuchians, and the flying reptiles, or pterosaurs. There were also a number of large turtles such as *Archelon* and *Protostega*.

Contemporary marine reptiles

ORDER Chelonia The turtles and tortoises. *Chelonia, Eretmochelys, Caretta, Dermochelys.*

ORDER Squamata Lizards and snakes.

Suborder Lacertilia Lizards. *Amblyrhynchos,* the Galápagos sea lizard.

Suborder Ophidia The snakes.

FAMILY Hydrophidae About 50 species in 15 genera, poisonous sea snakes.

CLASS Aves Birds.

ORDER Procellariiformes

FAMILY Diomedeaceae Albatross. *Diomedea.*

FAMILY Procellariaceae Petrels, fulmars, shearwaters. *Fulmarus, Ossifraga, Priocella, Puffinus, Macronectes, Oceanites, Prion, Pelecanoides.*

ORDER Sphenisciformes Penguins. *Aptenodytes, Eudyptula.*

ORDER Steganopodes Pelicans, cormorants, boobies, frigate birds, tropic birds.
FAMILY Pelicanidae Pelicans.
FAMILY Phalacrocoracidae Cormorants.
FAMILY Sulidae Boobies and gannets.
FAMILY Fregatidae Frigate birds.
ORDER Charadriiformes Shorebirds (gulls, auks, terns, puffins, sandpipers, godwits, curlews, etc.).
Suborder Charadriini
FAMILY Charadriidae Sandpipers, godwits, curlews, sanderlings, etc.
Suborder Lari
FAMILY Laraceae Gulls and terns. *Larus, Bruchigavia, Pagophila, Rhodostethia, Rissa, Xema, Creagrus, Sterna, Anous, Gygis.*
FAMILY Stercorariaceae Jaegers and skuas. *Stercorarius, Megalestris, Catharacta.*
FAMILY Rhynchopaceae Skimmers. *Rhynchops.*
SUBORDER Alcae Auks, puffins, guillemots, auklets, murres.
FAMILY Alcaceae *Alca, Fratercula, Pinguinus, Lunda, Uria, Cepphus, Alle, Synthliboramphus.*
CLASS Mammalia
ORDER Carnivora
Suborder Pinnipedia
FAMILY Otariidae Eared seals. *Callorhinus, Arctocephalus, Eumetopias, Zalophus.*
FAMILY Phocidae Hair seals. *Phoca, Pusa, Hydrurga, Leptonychotes, Mirounga, Lobodon.*
FAMILY Odobenidae Walrus. *Odobenus.*
Suborder Fissipedia Sea otter.
FAMILY Mustelidae *Enhydra.*
ORDER Sirenia Dugongs, manatees, sea cows. *Halitherium, Halianassa* (large fossil genera), *Dugong (=Halicore), Trichechus (=Manatus), Hydrodamalis (=Rhytina).*
ORDER Cetacea Whales, porpoises.
Suborder Archaeoceti (Zeuglodontia) Early fossil group.
Suborder Odontoceti Toothed whales.
FAMILY Ziphiidae Beaked whales. *Berardius, Hyperoodon.*
FAMILY Delphinidae Dolphins or porpoises. *Delphinus, Lagenorhynchus, Tursiops.*
FAMILY Phocaenidae *Phocaena, Neomeris.*
FAMILY Stenidae *Steno, Sousa, Sotalia.*
FAMILY Orcinidae (or as subfamily of Delphinidae) Killer whale. *Orcinus.*
FAMILY Platanistidae River dolphins.
FAMILY Monodontidae Beluga and narwhal. *Delphinapterus, Monodon.*
FAMILY Physeteridae Sperm whale or cachalot. *Physeter.*
Suborder Mysticeti Whalebone, or baleen, whales.
FAMILY Balaenopteridae *Balaenoptera, Magaptera.*
FAMILY Balaenidae *Balaena, Caperea.*
FAMILY Eschrichtidae *Rhachianectes.*

INDEX

Boldface type indicates page on which an organism is illustrated; t, table; n, note.

E

P

T